PHYSICO-CHEMICAL BEHAVIOUR OF ATMOSPHERIC POLLUTANTS

Commission of the European Communities

PHYSICO-CHEMICAL BEHAVIOUR OF ATMOSPHERIC POLLUTANTS

Proceedings of the Second European Symposium
held in Varese, Italy,
29 September - 1 October 1981

Organized within the framework
of the Concerted Action COST 61 A bis

Edited by

B. VERSINO and H. OTT
Commission of the European Communities

D. REIDEL PUBLISHING COMPANY
Dordrecht, Holland / Boston, U.S.A. / London, England

Library of Congress Cataloging in Publication Data

Main entry under title:

Physico-chemical behaviour of atmospheric pollutants.

    Includes index.
    1. Air–Pollution–Congresses. 2. Atmospheric chemistry–Congresses.
3. Aerosols–Congresses. I. Ott, H. II. Versino, B., 1935–
III. Commission of the European Communities.
TD881.P48   628.5'3   81–19969
ISBN 90-277-1349-9   AACR2

The Symposium was organized by the
Commission of the European Communities
Joint Research Centre Ispra
Directorate-General for Research, Science and Development, Brussels

Publication arrangements by
Commission of the European Communities,
Directorate-General Information Market and Innovation, Luxembourg

EUR 7624
Copyright © 1982 ECSC, EEC, EAEC, Brussels and Luxembourg

LEGAL NOTICE

Neither the Commission of the European Communities nor any person acting on behalf of the
Commission is responsible for the use which might be made of the following information.

Published by D. Reidel Publishing Company
P.O. Box 17, 3300 AA Dordrecht, Holland

Sold and distributed in the U.S.A. and Canada
by Kluwer Boston Inc.,
190 Old Derby Street, Hingham, MA 02043, U.S.A.

In all other countries, sold and distributed
by Kluwer Academic Publishers Group,
P.O. Box 322, 3300 AH Dordrecht, Holland

D. Reidel Publishing Company is a member of the Kluwer Group

Printed in The Netherlands

PREFACE

In this volume, the Commission of the European Communities presents the proceedings of the second European Symposium on the physico-chemical behaviour of atmospheric pollutants.

These Symposia are organized in intervals of about two years within the framework of a Concerted Action in this area, which is part of the Communities' research programme in the environmental field. The European Communities co-operate in this area with European Non-Member States under an agreement within the framework of COST (Coopération Scientifique et Technique). This Agreement (COST Project 61a bis) has been signed by the European Communities, Austria, Switzerland, Sweden and Yugoslavia.

The scope of the Concerted Action is to co-ordinate all research in the area executed in the participating countries and to collect and dissemi-nate the results. The research inventory established comprises 165 individual projects; regular meetings of 5 Working Parties permit close contacts amongst the scientists involved. The European Symposia should permit from time to time an overall review of the progress.

The first Symposium held in October 1979 [*] permitted to review the state of progress at the beginning of the Concerted Action. The second Symposium gives now an overview of the important achievements during the past two years. These are evident already from the number of contri-butions which increased from 45 in 1979 to 74 in 1981. The results collected during this period permitted the presentation of a number of important review papers.

To our mind, the present volume represents an almost complete overview of the work presently done in this field in Europe, corresponding entirely to the given tasks of co-ordinating the European research effort. It illustrates the high scientific quality of this work.

Brussels/Ispra, November 1981

B. Versino   H. Ott

[*] The proceedings of the first European Symposium on physico-chemical behaviour of atmospheric pollutants have been published by the Com-mission of the European Communities as report EUR 6621 and may be obtained on request from the Commission of the European Communities, Directorate General for Science, Research and Development, 200, rue de la Loi, B - 1049 Brussels.

CONTENTS

I.   IDENTIFICATION AND ANALYSIS OF POLLUTANTS

CHAIRMEN' S SUMMARIES

IDENTIFICATION AND ANALYSIS OF POLLUTANTS

Chairman: A. LIBERTI

DETERMINATION OF GROUND LEVEL OH CONCENTRATIONS BY A LONG PATH LASER ABSORPTION TECHNIQUE

G. Hübler, D.H. Ehhalt, H.W. Pätz, D. Perner
U. Platt, J.Schröder and A. Tönnißen

Kernforschungsanlage Jülich GmbH
Institut für Chemie 3: Atmosphärische Chemie

Introduction

In 1971 Levy / 1 / showed that even in the clean troposphere the photolysis of ozone should lead to the formation of hydroxyl radicals. This observation was of extreme importance given the fact that the hydroxyl radical, OH, reacts with a large number of trace gases. For many of them the reaction with OH is their main destruction path. Since then considerable effort has been made to determine the concentration of OH in ambient air.

Various ways have been pursued by different groups. Either direct methods / 2-4 / (laser induced fluorescence LIF, long path absorption LPA) or indirect ones / 5,6 / (^{14}CO radio-active tracer) have been used to determine the OH concentration. Additionally, there are estimates from atmospheric models / 7,8 / and from calculations / 9,10 / which try to balance the cycles of specific trace gases.

Up to now the measured and predicted hydroxyl concentrations vary widely. We feel that the most convincing way to measure OH is by a direct method.

Method and experimental

We measure atmospheric optical absorptions with such a sensitivity and resolution that OH absorptions can be extracted from a composite spectrum of all atmospheric constituents absorbing in this spectral region / 11 /. This method has given us the possibility of determining the OH concentration at ground level directly and unambiguously.

To measure trace gases spectroscopically in ambient air at
mixing ratios below 1 ppt, as expected for OH, the atom or
molecule must show specific lines with large absorption coefficients. The OH radical has two vibrational bands suitable
for spectroscopic observation in the near UV.
At room temperature all hydroxyl radicals are in the electronic ground state with no vibrational excitation. Most of the
molecules occupy the lowest four rotational levels. OH may be
excited with UV light near 280 and 308 nm, into the two lowest
vibrational levels of the first excited electronic state with
sufficiently high transition probabilities.
The transitions around 308 nm (A $^2\Sigma$ v'=0 <-- X ^{2¶} v"=0)
show the larger cross section and the atmospheric attentuation
at this longer wavelength is smaller. Since a strong light
source at 280 nm may cause an interference due to the ozone
photolysis, the three strongest rotational absorption lines
around 308 nm are preferable for observation. Because of the
resolution and the spectral bandwidth of our apparatus
(.003 nm, .028 nm) we chose the Q_1(2) line (307.9951 nm,
linewidth less than .002 nm). It shows a satellite : the
Q_{21}(2) line which is shifted .0055 nm to longer wavelength.
The simultaneous measurement of both lines allows a reliable
and definite identification of OH in the atmosphere.
For our long path differential absorption technique a laser
system is used as light source (Figure 1). An argon ion
laser is optoacoustically modulated and pumps a synchronised
jet stream dye laser. As dye we use Rhodamin 6G dissolved
in ethyleneglycol. With a two element birefringent filter the
dye laser (bandwidth about .1 nm) is tuned to 616 nm. The
second harmonic of this light coincides with the two absorption lines of the OH molecule. For frequency doubling the
light from the dye laser is focussed into an ADA crystal thermostated to 120° Celsius under 90 degree phase matching conditions. This setup generates a sequence of UV light pulses
(82 Mhz repetition rate, duration less than 50 psec) at 308 nm
with an average power of 2 mW. Our detection system with a
time resolution of 1 µsec cannot resolve the pulses at 82 Mhz

repetition frequency. Thus the light is treated as a continuous wave. To reduce the divergence and the mean flux density of the UV beam the output from the second harmonic generator is expanded in a Cassegrain type telescope to a diameter of .3 m. The UV beam is sent through the free troposphere and is reflected by a plane mirror at a distance between 1.5 and 5 km back to the laboratory. There the light is focussed on the entrance of a Spex double monochromator (.85 m focal length, dispersion .1 nm/mm, resolution .002 nm at 308 nm). The exit of the monochromator is scanned by a slotted disc at a high repetition rate. A preslit ensures that only one slit at a time scans the spectrum and the wavelength dependent intensity distribution is converted into a timevariable signal which is recorded by a photomultiplier. The electronic signal is amplified and fed into a LSI 11 minicomputer. Here the data are synchronised and added up for signal averaging. For wavelength calibration of the laser and the spectrometer, a laminar acetylene / air flame is put in the unexpanded beam to generate hydroxyl radicals. The OH is easily observed by the absorption in the two rotational lines $Q_1(2)$, $Q_{21}(2)$. To measure the concentration of OH in the free air, the burner is turned off. For 5 minute intervals the data are summed up. After each interval the spectral position is controlled with the flame.

Data reduction and results

As the laser intensity varies with wavelength and as the expected differential absorptions are in the order of 10^{-3} and 10^{-4} a proper baseline determination is required. This is made by fitting a fourth order polynomial to the spectrum with a least squares method. After dividing by the fitted profile absorption lines of the order of 10^{-4} begin to emerge. These absorptions can be identified as rotational lines of sulphur dioxide which at the observed tropospheric condition obscure the hydroxyl spectrum. Before the OH concentration can be determined, the interfering lines have to be eliminated from the measured spectrum. A reference spectrum is obtained by measuring the absorption of SO_2 in a quartz-cell. Fitted to the data points a spectrum corresponding to the actual SO_2 concentration is subtracted. Calibrated against a known amount the SO_2 concentration was verified as well by comparison measurements. After subtracting the SO_2 from the original data a new polynomial fit is run. The newly divided spectrum now shows the remaining absorptions. Their position and shape agree with the ones of the hydroxyl radical (Fig. 2).

As the OH concentration is expected to be correlated with UV-intensity, measurements were made at Jülich on almost every sunny day in 1979. Although only days with good visibility were chosen the intensity of the received light was rather poor and thus only insufficient photon statistics were achieved. Therefore we have added up the data taken between 11:00 and 16:00 CET on these days to get "monthly averages" (Table 1). However the statistics still varied and for some measurements we can only give upper limits rather than absolute values for the hydroxyl mixing ratio. These cases are indicated by a "<" sign in the tables. By 1980 our system was improved to a stage where data for single days could be obtained (Table 2). These measurements were made at Jülich, a moderately polluted area 50 km west of Köln, and at Deuselbach in the Hunsrück mountains about 150 km southwest of Köln where a rural air station of the Umweltbundesamt is located. All of our measurements made so far show rather low OH concentrations with one exception. On September 24th, 1980 we found 4×10^6 OH/cm^3. Our values are considerably lower than the other published ones. This discrepancy between our and the measurements of other groups and some calculation still have to be resolved.

Table 1: OH concentrations measured during allmost all sunny
episodes in 1979. Because of poor light intensity
only "monthly averages" for those days can be given.

Month	Path	$[OH] \ cm^{-3}$ $\times \ 10^{6}$
April	3 km	< 2
May	"	1.8
June	"	< 1.5
July	"	1.7
August	"	2
September	"	1.7

All measurements made at Jülich. Only data taken between
11:00 and 16:00 CET were considered.

Table 2: OH concentrations measured in 1980

Date	Time [CET]	Location	Path in km	$[OH] \ cm^{-3}$ $\times \ 10^{6}$
June 6	11:00-15:30	Jülich	3	< 3
July 22	11:30-15:30	"	3	2.9
July 23	13:00-15:30	"	3	< 4
Aug. 7	11:50-14:00	"	3	1.8
Aug. 15	11:30-15:10	"	3	< 1.2
Sept. 1	15:00-16:00	"	8	1.2
Sept. 3	11:00-14:30	"	3	.8
Sept. 23	12:00-14:40	Deuselbach	9.7	< 2
Sept. 24	11:00-12:30	"	9.7	4
Oct. 6	12:45-14:20	"	6	2.4
Oct. 15	12:00-15:20	"	6	.8

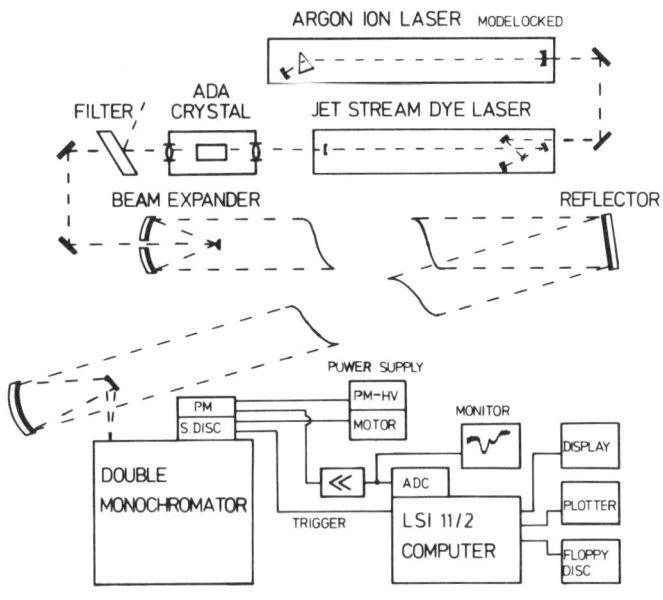

FIG. 1   EXPERIMENTAL SET UP

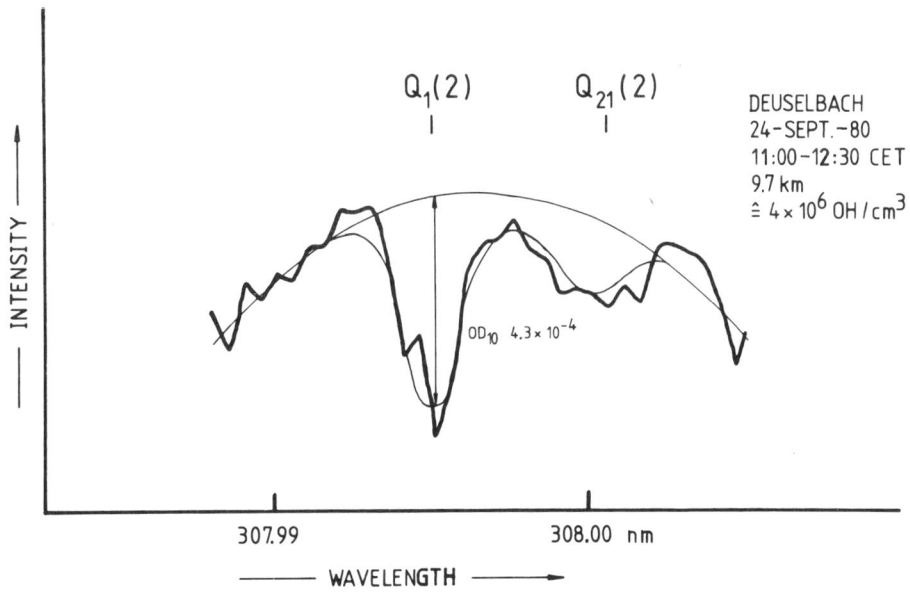

FIG. 2   REDUCED ABSORPTION SPECTRUM OF OH

References

/ 1 /   Levy II, H.
        Normal atmosphere: large radical and formaldehyde
        concentrations predicted
        Science, 173, 141-143 (1971)

/ 2 /   Davis, D.D., W. Heaps, and T. McGee
        Direct measurements of natural tropospheric levels
        of OH via an aircraft borne tunable dye laser
        Geophysical Research Letters, 3, 331-333 (1979)

/ 3 /   Wang, C.C., L.I. Davis jr., Ch. Wu, S. Japar, H. Niki,
        and B. Weinstock
        Hydroxyl radical concentrations measured in ambient
        air
        Science, 189, 797-800 (1975)

/ 4 /   Perner, D., D.H. Ehhalt, H.W. Pätz, U. Platt,
        E.P. Röth, and A. Volz
        OH radicals in the lower troposphere
        Geophysical Research Letters, 3, 466-468 (1976)

/ 5 /   Campbell, M.J., J.C. Sheppard and B.F. Au
        Measurements of hydroxyl concentration in boundary
        layer air by monitoring CO oxidation
        Geophysical Research Letters, 6, 175-178 (1979)

/ 6 /   Volz, A., D.H. Ehhalt and R.G. Derwent
        Seasonal and latitudinal variation of ^{14}CO and the
        tropospheric concentration of OH radicals
        Journal of Geophysical Research, 86, 5163-5171 (1981)

/ 7 /   Logan, J.A., M.J. Prather, S.C. Wofsy and
        M.B. McElroy
        Tropospheric chemistry: A global perspective
        Journal of Geophysical Research, 86, 7210-7254 (1981)

/ 8 /   Crutzen, P.J. and J. Fishman
        Average concentrations of OH in the troposphere, and
        the budgets of CH_4, CO, H_2 and CH_3CCl_3
        Geophysical Research Letters, 4, 321-324 (1979)

/ 9 /   Singh, H.B.
        Preliminary estimations of average tropospheric OH
        concentrations in the northern and southern
        hemispheres
        Geophysical Research Letters, 4, 453-456 (1977)
/ 10 / Neely, W. and J.H. Plonka
        Estimation of the time-averaged hydroxyl radical
        concentration in the troposphere
        Environmental Science and Technology, 12,
        317-321 (1978)
/ 11 / Platt, U., D. Perner and H.W. Pätz
        Simultaneous measurement of atmospheric CH_2O, O_3
        and NO_2 by differential optical absorption
        Journal of Geophysical Research, 84, 6329-6335 (1979)

CHARACTERIZATION OF AIRBORNE PARTICLES BY LASER MICRORAMAN SPECTROSCOPY

I. ALLEGRINI, A. INNOCENZI and A.FEBO
CNR - Istituto Inquinamento Atmosferico
Area della Ricerca di Roma - CP 10
00016 MONTEROTONDO STAZIONE (ITALY)

Summary

The possibility of using Laser Raman spectroscopy in micro analysis represents a great opportunity to open new research areas for the characterization of airborne particulate matter. In this technique, a modified optical microscope is used to select the particle of interest. The microscope objective focusses a laser beam on the particle and collects the backscattered radiations which are then analyzed in a conventional Laser Raman spectrometer. The instrument output is the Raman spectrum of the particle which can be conveniently used for its identification. Particles are directly collected on microscope slides and analyzed as such. The technique does not require special sample preparation steps and the particles can be kept in controlled atmospheres. A dedicated reaction chamber has been set up to study chemical reactions between gases and particles. Experimental results show that the Laser Raman microprobe is sensitive to most compounds of environmental interest, including organics and sulphur containing molecules. Preliminary results on simple reactions show the capability of this technique in following chemical reaction of gases on individual particles.

1.INTRODUCTION

Raman spectroscopy has not been so far widely used in analytical chemistry, even though the information carried out by a Raman spectrum can be valuable for molecule and crystal identification. This is due to the low cross-section for the Raman emission which seldom exceed $10^{-29} cm^2 molecule^{-1} sr^{-1}$. The use of Lasers as exciting sources partly compensate the small value of the molecular cross-section, but the applications of the Raman effect in chemical analysis still remain limited and the potential use in environmental analysis is restricted to the remote sensing of air pollutants (1).

Recently, coupling an optical microscope to a Raman spectrometer, high levels of sensitivity have been achieved (2,3). To evaluate the quantitative performance of the technique, it will be convenient to recall the S/N equation for this kind of spectrometry. The minimum detectable amount of a substance in particulate form through Raman spectrometry is such that:

$$N_r t \geqslant 2 \left[(N_b + N_s + N_d) t \right]^{1/2} \qquad (1)$$

where Nr,Nb,Ns and Nd are the arrival rate for Raman, background, spectral interferences and dark photons respectively. N_r can be written as:

$$N_r = N \, V \, \sigma \, K \, I_o \qquad (2)$$

where N is the number of active molecules per unit volume, V the irradiated volume, σ the Raman cross-section, K the collection efficiency (optics+electronics) and I_o the Laser irradiance (W cm^{-2}). The terms on the right side of eq.1 can be made negligible with a proper choose of the equipment, while N_r considerably approaches low values due to the small number of active molecules (NV) in the particles. The only way to compensate the loss of photons rate is an increase in I_o, i.e., by focussing the exciting Laser by using a microscope objective, which is effective in focussing the Laser beam to diffraction limit.

On introducing in eq.1 and 2 typical values of the relevant parameters, it can be shown that a 1 μm particle, irradiated with a power of 10 mW, gives a S/N of about 100 for a counting time of 1 s. Such a particle has a mass of 5.10^{-12} g, thus we are really facing with one of the most sensitive and selective analytical methods. By using this technique, is possible to study the morphology of the particle through microscopic observation, and its chemical structure through the Raman spectrum. This makes the Laser microRaman a challenging alternative to other microprobes, such as XRF, which essentially sense atoms, with the exclusion of light elements and which are insensitive to organic compounds.

The technique shares the advantages of conventional Raman spectrophotometry, with special regard to sampling which does not require complex and time consuming steps. In addition, the sample is usually analyzed in air and can be kept in controlled atmospheres. This feature is extremely useful to follow chemical transformations of the particles.

Unfortunately the method cannot be adapted to any kind of particle. Recalling eq.2, the Laser irradiance I_o must be quite large in order to have a reasonable counting rate of Raman photons. Typical irradiance values are in the order of several KW cm^{-2}. Such high irradiance levels might cause photodecomposition or thermodegradation of the sample. In addition, irradiation causes a rise in temperature which is a complex

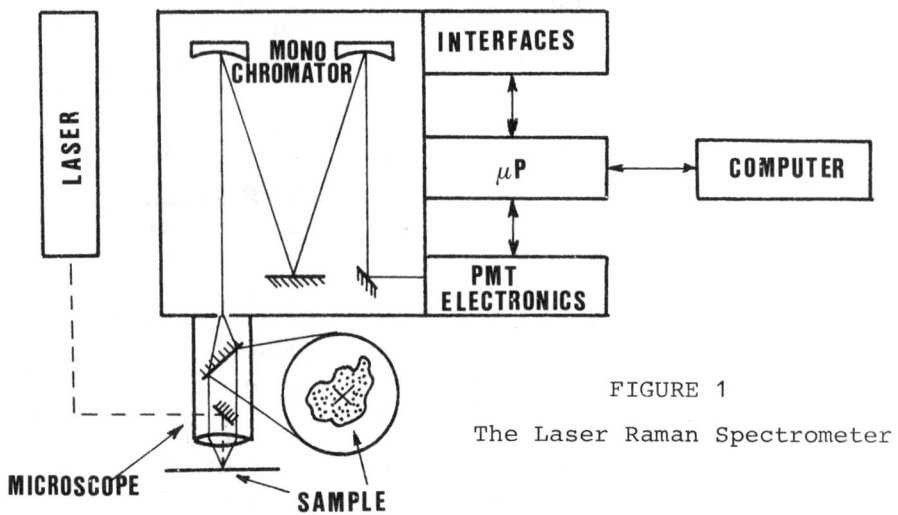

FIGURE 1

The Laser Raman Spectrometer

function of the particle size, refractive index, irradiance level and nature of the substrate (4). This effect must be taken into account if a chemical process is monitored.

Another effect, which completely mask the Raman spectrum is the sample fluorescence arising from compounds either present or adsorbed on the particles. Heating and other instrumental limitations are not however insuperable problems and the technique can be adapted to a large variety of environmental problems.

2. INSTRUMENTATION

a) The Laser Raman microprobe

The instrumentation for Raman microscopy is similar to that of a conventional spectrophotometer whose sample compartment has been replaced by an optical microscope (Fig.1).

The equipment consists of an Argon Ion Laser (Spectra-Physics-Mod 165-03) operating at powers between 10 and 300 mW. Filtering of plasma lines is achieved by a prism premonochromator. The beam enters a microscope (Nachet-Mod S400) modified in order to accept the laser beam in the objective and to be optically coupled to the Raman monochromator. A screen provides a safe and convenient way for the direct morphological observation of the particles in light transmission or reflection. A tilting mirror is used to send the light backscattered by the sample on the screen or into the Raman monochromator for spectral analysis. Particles are supported on standard slides.

The final resolution of the microprobe is about 1 μm.

The Raman monochromator is a double monochromator (Jobin Yvon-Mod HG2S) equipped with holographic gratings for superior stray light rejection which is better than 10^{-14} at 20 cm^{-1} from the exciting line. The use of a double monochromator with holographic gratings is mandatory due to the large amuont of light elastically backscattered by a single particle.

Light detection and measurement is performed with a cooled photomultiplier and a standard photon counting equipment. The dark noise is better than 10 s^{-1}. The electronics for instrumentation controls and data acquisition has been completely assembeld in laboratory and it is built around a single board microprocessor (Rockwell-AIM 65). With the use of proper interfaces a full control of the scanning and slits operations are possible. Data are collected on the "on-board" memories and are output to a minicomputer (CBM - 8032) equipped with floppy disks for storing and retrival.

b) <u>Sample treatment</u>

As said before, samples for MicroRaman spectrometry do not need preparation. Pure compounds are analyzed as such without any problem, while airborne particles, collected on membrane filters, need to be separated from the filter material which might confuse the Raman spectrum or mask it with flourescence emission. Portions of the membrane filters are supported on standard microscope slides and incinerated by using a low temperature oxigen plasma, RF excited. This procedure completely eliminates the interferences from filter materials, but it

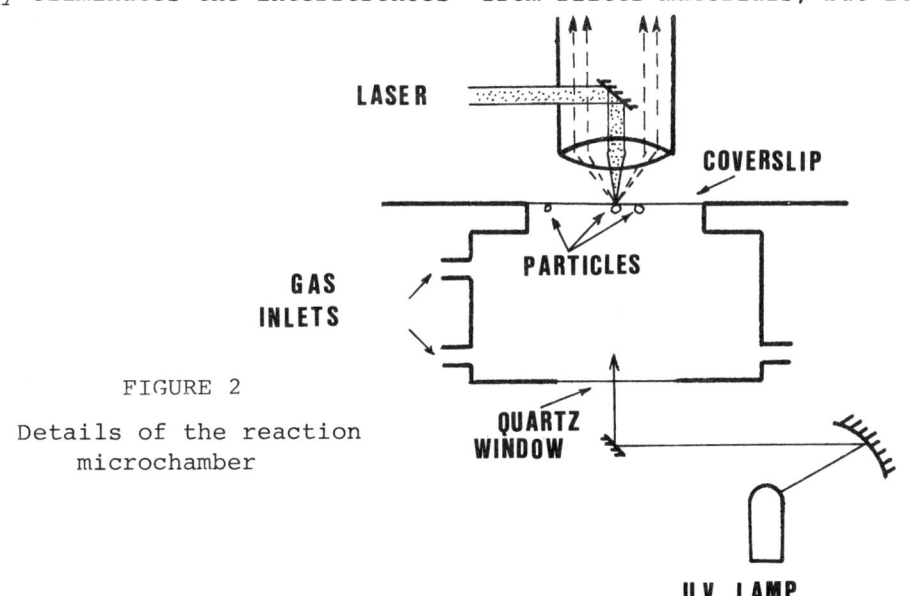

FIGURE 2

Details of the reaction
microchamber

also removes organics. In some instance that would be convenient because organics absorbed on the particles might give strong fluorescence interferences. According to our experience plasma incineration is the most convenient procedure to analyze airborne fibers and any other inorganic component.

c) Reaction microchamber

The microRaman spectrometer has been adapted to a small reaction chamber to directly follow the chemical evolution of a single particle exposed to a variety of gases (Fig.2). The chamber is completely built in Teflon and it is supported on the translation stage of the microscope. Particles are supported face-down on a standard glass cover-slip which is sealed to the chamber with silicone. This unusual optical arrangement is such that a loss of about 20% of the signal occours. The chamber incorporates facilities for UV irradiation through a quartz window and connection for inlet and outlet of the reacting gases.

3. RESULTS AND DISCUSSION

a) Sensitivity and resolution

Tests on the complete spectrophotometer have been performed with the use of pure compounds which have been finely ground and deposed on microscope slides. In order to prepare a relatively large fraction of small particles, several substances have been dissolved in proper solvents and sprayed. Aerosols particles are directly collected on slides by means of impaction.

The sensitivity of the apparatus as a function of the particle size is highly dependent upon the molecular cross-section. For instance, it is possible to obtain the entire Raman spectrum by single particles of PbO or ThO_2 of size down to 1 μm, while other compounds, such as most silicates, need samples of about 5 μm in order to have a well resolved spectrum in a reasonable time. When emission is very faint, multiple scans, microprocessor controlled, are required.

Fig.3 shows some spectra obtained by single particles in the range 2-6 μm diameter obtained from pure compounds or directly collected from the atmosphere. In the latter instance positive identification (Calcite, gypsum and quartz) has been possible through comparison with reference spectra.

Sulphates and carbonates are easily identified through the simmetric stretching vibrations of the molecular ions CO_3^{2-} and SO_4^{2-} which are located at about 1090 and 1000 cm^{-1} respectively from the exciting line. The study of the Raman lines due to lattice vibrations makes possible a differentiation among different crystal structures of the same compound. As an

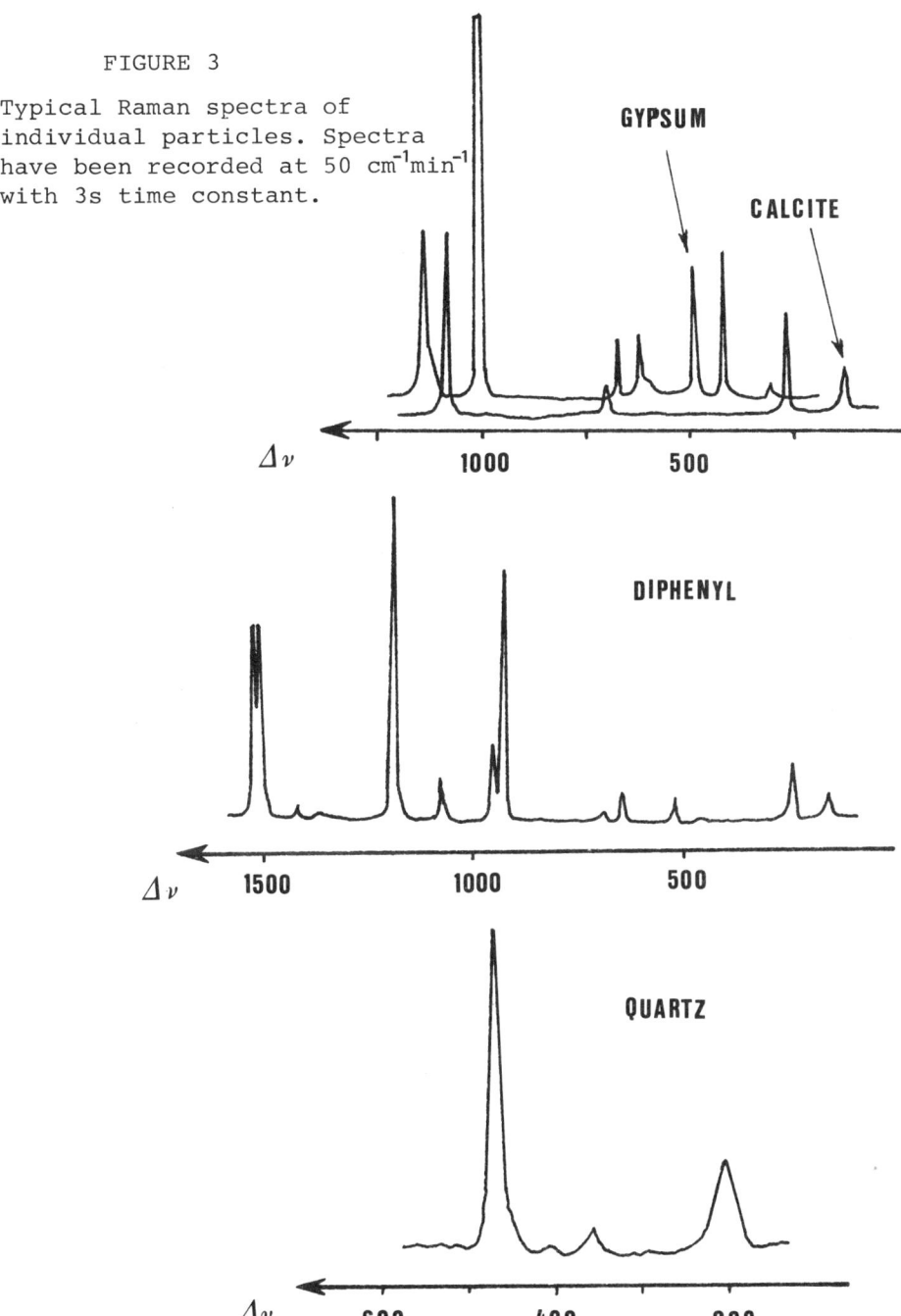

FIGURE 3

Typical Raman spectra of individual particles. Spectra have been recorded at 50 cm^{-1}min^{-1} with 3s time constant.

GYPSUM

CALCITE

$\Delta\nu$

1000   500

DIPHENYL

$\Delta\nu$

1500   1000   500

QUARTZ

$\Delta\nu$

600   400   200

example gypsum and anhydride ($CaSO_4$ monoclinic and rhombic) and calcite and aragonite ($CaCO_3$ trigonal and rhombic) can be easily differentiated.

It should be stressed that the identification of a single particle require that several lines of the spectrum are present. For an inorganic crystal there is usually a very intense line and several weaker bands whose Raman cross-sections migth be so small that they cannot be easily evidenced. In such a case, a very long measurement time is required for very small particles (1 to 5 μm diameter).

From an environmental point of view, as is well known, it is very important to have sensitive and selective methods for the analysis of sulphur containing compounds. The microRaman spectrometer fulfill this requirement being able to analyze samples in solution. Sulphites and sulphates, as well as sulphates and bisulphates are easily differentiated, while the Raman spectrum of H_2SO_4 show signals due to SO_4^{-2} at 980 cm^{-1} and HSO_4^- at about 1040 cm^{-1}. The concentration of sulphate and bisulphate ions might be extimated by using the intensity of the H_2O bending mode line located at about 650 cm^{-1} as an internal standard[6]. This procedure has been adapted to samples in a conventional Raman spectrometer and in Raman microprobes for the study of fluid inclusions in minerals (6,7). The detection limit for inclusions in minerals has been found to be about 10^3 ppm of sulphate in water, which is a figue low enough to study the chemical evolution of individual micronic droplets. Unfortunately the direct application of this procedure to environmental samples has not yet been reported.

The spectrum of diphenyl demonstrate the usefulness of the technique in the analysis of organic microparticles. Our preliminary tests did not show organics on airborne particles because of the combustion step. Several runs on particles directly collected from the atmosphere, show a strong fluorescence background which arises from organics adsorbed on the sample. In this case a very high irradiance level results in either photo and thermal decomposition and two bands, located at 1350 and 1600 cm^{-1}, appear on the spectrum. These signals are due to elemental carbon or "soot" which results from the thermodegradation of organics. In addition, it should be stressed that Raman spectra of organics are characterized by a large number of signals, thus the positive identification of etherogeneous organic particles is a highly difficult task.

b) Application to gas-particle interactions

The ability of the microRaman spectrophotometer in following the chemical evolution of individual particles when they are reacting with gases, has been tested with preliminary runs in the reaction microchamber.

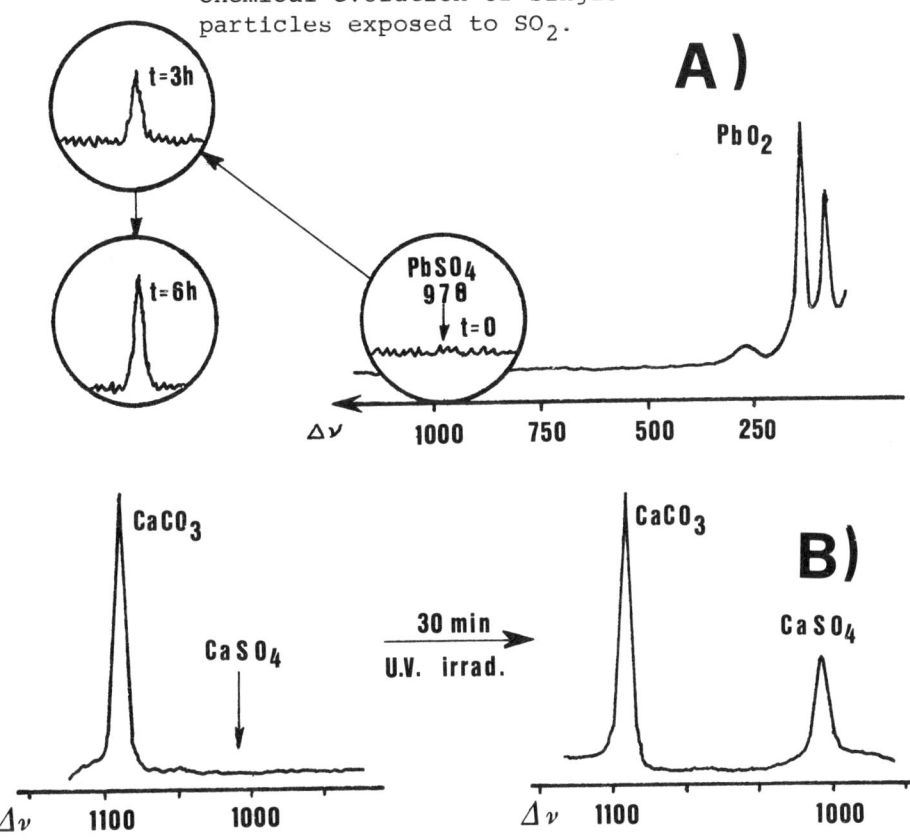

FIGURE 4

Chemical evolution of single
particles exposed to SO_2.

At the present time the reaction chamber does not yet
incorporate facilities to exactly measure the gas concentration or the environmental parameters. All tests reported do
not take into account for wall effects, while reacting gases
have been introduced in the chamber with static methods. In
both experiments the concentration of SO_2 is about 10^3 ppm.

The first reaction has been extensively used since several
years ago for the evaluation of the concentration of SO_2 by
simply measuring the gas uptake by a candle prepared with PbO_2
which is converted to $PbSO_4$. Pulverized PbO_2 has been deposed
on a cover slip and the excess shaked off, then SO_2 has been
added. The mean particle size was about 3 μm, while observations have been performed on an individual particle approximately 2x3 μm .

Fig.4a shows the evolution of the particle after a few
hours of exposure. Pure PbO_2 shows spectral features which are
characteristic of its molecular and crystal structure and does

not show any appreciable signal in the region 950-1050 cm^{-1} where there is the emission of the molecular ion SO_4^{-2} . After three hours exposure a weak signal appears at 978 which exactly matches the most intense line of $PbSO_4$. After six hours the intensity of the signals double.By using previously published relative Raman cross-sections (8), a rough extimation of the amount of $PbSO_4$ formed during the exposure can be made. Within an extimated error of 30%, the intensity ratios are such that a sulphatation rate of about 1% $hour^{-1}$ can be inferred.

Another experiment has been carried out on the reaction between $CaCO_3$ with SO_2 with the final formation of $CaSO_4$. The procedure is similar to that used for the test with PbO_2 and the results are shown in Fig.4b. A 5μm individual particle has been selected and its spectrum recorded every 30 min in the region 950-1150 cm^{-1} in order to scan both regions of interest for the emission of molecular ions CO_3^{-2} (1050-1150) and SO_4^{-2} (950-1050). After three hours exposure the particle gives the same spectrum of pure calcium carbonate, indicating that no reaction took place. After 30 min irradiation a signal at 1008 cm^{-1} appears which exactly matches the most intense line of $CaSO_4$. Even in this case a conversion rate of about 10% hour can be extimated using the same procedure adapted to the reaction between SO_2 and PbO_2.

4. CONCLUSIONS

A certain number of interesting features makes the micro-Raman spectrometry a successful approach for a wide variety of environmental problems. The direct chemical analysis of indivi-dual micrometer-size particles is now possible in several instance. Microprobing is now accessible even for organic materials and for liquid droplets, while the direct observa-tion of the chemical transformations of single particles reacting with gases is now feasible.

However, much work remain to be done either in instrumen-tation and in applications. Fluorescence effect is probably the most severe limitations for this kind of technique, but it can be reduced by using pulsed laser as exciting sources and gated detection or using exciting lasers at low frequency. Applications would include experiments on acid droplets and the further development of a microreaction chamber incorpora-ting facilities for gas and environmental parameters analysis. This would certainly open very interesting fiels of investi-gation on the physico-chemical behaviour of individual particles.

6. ACKNOWLEDGMENTS

The authors are indebited to G.Calogero for technical assistance and A.DiChiara Maggioli who assisted in the development and set-up of the electronics.

7. REFERENCES

1) Hinkley E.D., Editor "Laser monitoring of the atmosphere Topics in applied physics, Vol.14.Springer-Verlag, Hidelberg (1978).

2) Rosasco G.J., Etz E.S., Cassat W.A. Appl.Spectroscopy 29,396 (1975)

3) Delhaye M., Dhamelincourt P. J.Raman Spect 3,33 (1975)

4) Bennet H.S.,Rosasco G.J. J.Appl.Phys. 49,640 (1978)

5) Stafford R.G., Chang R.K. "absolute Raman scattering cross-sections of sulphates and bisulphates" International conference on environmental sensing and assessment. Vol.2 p.23-5 IEEE, New York (1976)

6) Cunningham K.M., Goldberg M.C., Weiner E.R. Anal.Chem. 49(1),70(1977)

7) Rosasco G.J., Roedder E. Geochim.Cosmochim.Acta 43,1907(1979)

8) Wright M.L. "Feasibility Study on in-situ monitoring of particulate composition by Raman and fluorescence scatter". EPA Contract 68-02-0594, Washington DC (1973)

AUTOMATIC BETA GAUGE INSTRUMENT FOR ATMOSPHERIC DUST MONITORING

I. ALLEGRINI, A. LIBERTI, A. FEBO
CNR- Istituto Inquinamento Atmosferico
Area della Ricerca di Roma - CP 10
00016 Monterotondo Stazione (ITALY)
M. SALMI - Facoltà Ingegneria, Università
dell'Aquila
M. BERETTA - Div. Industriale, GELMAN
Instrument S.p.A. Opera (Milano)

SUMMARY

An automatic atmospheric dust monitor is described which makes use of a beta attenuation gauge to measure the mass of particulate matter collected on membrane filters. It incorporates an electronic flow transducer whose output is also used to keep the sampling flow rate at a fixed value in order to obtain a constant linear velocity over the entire period of sampling. The instrument is fully controlled by a microprocessor which can be programmed in order to operate in several sampling modes. Performance of the beta ray gauge and the electronic flow meter are discussed on the light of the relevant parameters affecting precision, accuracy and reliability. Tests on the instrument show that the response is highly correlated with the conventional gravimetric method.

1. INTRODUCTION

Air quality management needs sensitive and accurate measurement and sampling methods. In addition, monitoring networks require simple, reliable and relatively inexpensive instrumentations. Such needs might be fulfilled by using modern electronic devices for both the measure of the relevant parameters and the signal processing. The recent advancements in electronic instrumentation and devices are such that data treatment and transmission are not a severe problem, while the active measurement of the parameters it is still a primary objective.

For instance the measurement of total suspended particulate matter (TSP) is usually performed by weighing a membrane before and after sampling a known volume of air. This procedure, simple in principle, has severe limitations. It is entirely manual and time consuming, manual handling of membranes might cause also contamination, loss of particles and even rupture of the filters. In addition the volume of air sample must be accurately known and corrected for variations in temperature and pressure. As an alternative to gravimetric method the β gauge for mass measurement can be used, which has several advantages

(1-3). Risks of contamination are highly reduced and the automatic analysis of many samples is possible.

In order to fully utilise the method for the automatic monitoring of TSP, highly accurate β gauges and electronic flow meters have developed and included in a new instrument which is completely automatic and measures TSP according to most national and international regulations. The instrument is valuable on the light of its features and the large use of computer controlled devices is such that precision, accuracy and reliability, makes it suitable for any environmental problem caused by suspended particles.

2. BETA RAYS GAUGE

The intensity of β rays transmitted through a membrane filter can be expressed by:

$$I = Io \ exp \ (-\mu x) \tag{1}$$

were x is the surface mass density (g cm^{-2}) and μ the absorption coefficient ($cm^2 g^{-1}$) which is a function of the maximum energy of β radiations. According to eq.1, by measuring I and Io and by knowing the filter surface, it is possible to obtain the total mass of particles (m) which is given by:

$$m = 3d^2 \ log \ (Io/I) \tag{2}$$

with d the filter diameter and if the radioisotopic source C^{14} (Emax = 156 Kev) is used. The measurement of the mass of TSP is therefore reduced to a measurement of β transmission in a simple gauge which includes a radioactive source and a proper detector. However in order to achieve the desired performance, it is very important to control any parameter which can affect sensitivity, precision and reliability of the technique.

For such a purpose a gauge has been assembled (Fig. 1). It includes a C^{14} source of $100 \mu Ci \ cm^{-2}$, a filter holder which is used to position the filter between the source and the detector, a detector housing capable to accept several types of detectors, and an electronic part which include an amplifier discriminator, a counting circuit with display and an electronic timer.

The amount of dust collected on a filter usually ranges between 10 and $100 \mu g \ cm^{-2}$ so that a very high sensitivity is required. The minimum detectable amount of mass deposed on membrane filters is expressed by:

$$\Delta m = 6d^2 / (t.f)^{1/2} \tag{3}$$

where d is the filter diameter, t the counting time and f the counting frequency. Being Δm proportional to the second power of diameter, most β gauges monitor particulate mass over very small surfaces.

Although this is a practical way to increase the sensitivity to μg levels, it conflicts whith most national and internacional regulations which require sampling over membrane filters of several cm^2 of selected materials and in well defined experimental sampling conditions. Alteration of the linear sampling velocity is objectable. Several devices bypass the problem by mea

suring a portion of the filter using a small radioactive source
coupled whith a small detector and scaling the results over the
entire membrane surface. According to our experience this proce
dure is not correct because usually particles are not evenly de
posed on the membranes, this effect being a function of the
physical properties of the particles. Thus, for large values of
d (Several cm) and for a reasonable counting time, the only way
to decrease Δm is an increase of the counting frequency f. That
could be done by using intense β sources and a fast detector such
as surface barrier silicon detectors which are basically p-n
junction widely used in nuclear spectroscopy. They are able to
measure events up to 10^5 Hz, but are severely limited by their
electronic noise which is of the same order of magnitude of low
energy rays used in TSP measurement. Typical distribution curves for noise and signal amplitudes as obtained by a silicon de
tector of 35 mm diameter biased at 80 v and analyzed in a MCA,
show that the pulse height distributions are such that most of
the useful signals confuse with the noise and must be discrimi
nated by using an electronic window which reduces the signal
to a magnitude comparable to that of a good Geiger Muller (GM)
tube. Noise in a silicon detector dramatically increases with
the useful surface, thus their use in instrumentation for air
pollution monitoring is limited to the analysis of small surface. In addition silicon detectors are highly sensitive to tempe
rature, thus they must be operated in a well thermostated environment. Finally the cost of this device and its associated
electronics are extremely high and this prevents a wide use in
commercial apparatuses.

    In order to keep full advatages of the low cost, simple and
reliable GM detections, several prototypes based on β attenuation have been assembled (4). They all include a C^{14} source
and thin window GM detector which assure satisfactory sensitivi
ty, very high reliability and relatively low cost. These prototypes are now serving as microbalance for membrane filters (5).

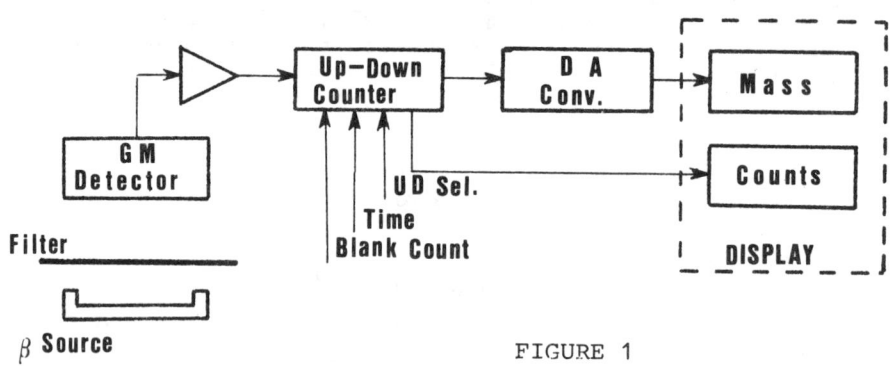

FIGURE 1

The β gauge for the direct measurement
of aerosol mass accumulated on membrane
filters.

The blank membrane filters are measured before sampling over a programmed time and the observed number of counts Io is recorded. After sampling, the membrane is again measured with the counting train in "down" mode starting from the figure obtained with blank membrane. After the counting time, the counter memory retains the difference between initial and final data. This difference is converted in analog signal and displayed on a DVM directly as mass of particles on the filter.

The measurement is rapid and sensitive because the instrument is capable to measure accumulated masses of $25\,\mu g$ in less than two minutes on 47 mm diameter membrane filters. For 25mm filters the absolute sensitivity is better than a few μg, i. e., comparable to the most sensitive analytical balances, thus applications on personal samplers, dicothomous and elutriators are practically endless. The instruments can be used with any kind of membranes and are very simple to operate, so they do not require trained technicians. The microbalances have been used to test the errors introduced in the mass measurement by variations in the particle size distribution and in the chemical nature of the particles.

Particles of different size and chemical nature have been prepared by dissolving pure chemicals in proper solvents and spraying the solutions with compressed air. By changing the solute concentration, particles of different size have been obtained. The β gauge response does not depend upon particle size in the range 0.3-7 μm, while it can be affected by the chemical nature. Indeed, β attenutation is a function of the electronic density of the absorbing material which is, in turn, proportional to the ratio of atomic number to mass (Z/M), thus low Z elements will exibit higher absorption coefficients. Experimental results (TAB I) show that deviations of about 10% should be taken into account in the presence of massive quantities of organic substances, otherwise deviations are below the statistical errors.

The pratical independence of β response to the chemical nature and size distribution of the particles, is very important from an experimental point of view, because the method does not require calibration procedures.

3. ELECTRONIC FLOWMETER

Instruments intended for TSP monitoring, should incorporate sensitive and accurate flow and volume records, which should be corrected for variations in the temperature and pressure of the air being sampled. In addition, they should operate at a constant flow rate in order to keep constant the size distribution of the particle being collected. Such features have been achieved with a flow transducer which senses flow rates by measuring the pressure drop $\Delta P = P_1 - P_2$ accross an orifice meter, by a differential pressure transducer. If P_2 is set equal to the atmospheric pressure, it can be shown that the flow rate can be expressed by:

$$Q = K \sqrt{(\Delta P / T)} \qquad (4)$$

where K is a constant and T the temperature of the air. Thus measuring ΔP by using a differential pressure transducer and T

with a temperature transducer, squaring the ratio of the electronic signals, a signal proportional to the flow rate is obtai
ned, fully corrected for variations in temperature and practically corrected for the usual changes in atmospheric pressure,
so that no further corrections are required. The resulting elec
tronic signal can be conveniently used for the measurement of
the total volume and for feedback control of the sampling train
in order to keep constant the flow rate at a given value.

Fig. 2 shows a block scheme of the electronic air sampler.
The ratio and squaring of the pressure (ΔP) and temperature (T)
signals is performed on an integrated circuit (National LH0094).
The signal is displayed on a digital voltmeter and serves as in
put to a V/F converter with a conventional counter which can be
programmed in order to advance when a given volume of air has
been sampled. The flow signal is also compared with a programmed flow rate in a differential amplifier. According to the value of the difference, two microswitches, controlling two win-
dings of a reversible small motor, are activated. The motor
open or close a needle valve, which by-passes the air flow
through the pump. In this way a complete stabilization of the
flow rate is achieved, regardless of the pressure drop on the
filter. Tests on the flow transducer show that the error is
within $\pm$ 1% while flow stabilization is possible within $\pm$ 3% at
a flow rate of 15 L min! (6).

4. AUTOMATIC BETA MONITOR

The excellent features of the βgauge and of the electronic
flowmeter, have been included in a new prototype of TSP monitor
shown in Fig. 3.

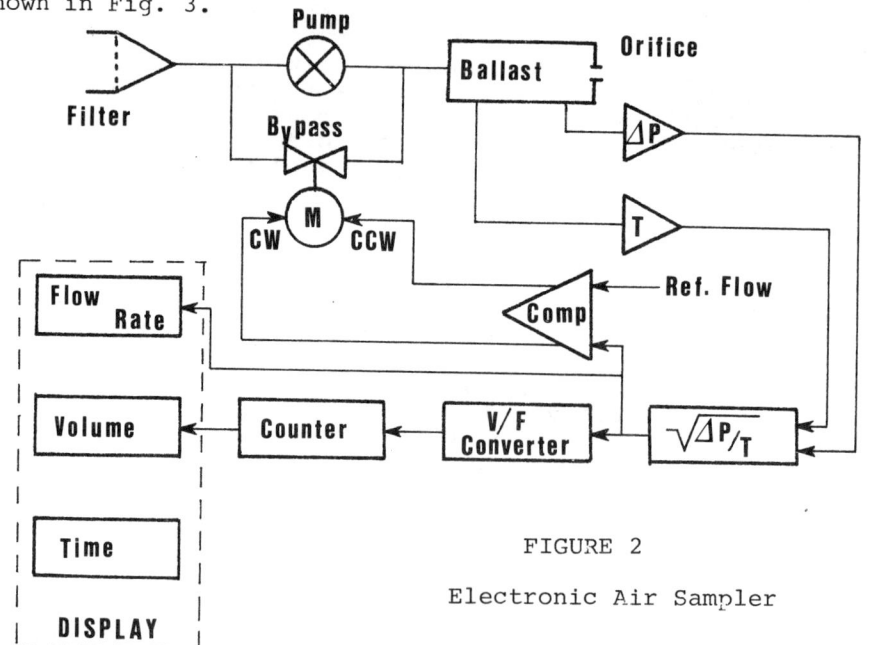

FIGURE 2

Electronic Air Sampler

Membranes are supported on plastic frames and contained in a tray which is free to move back and forth. An elevator pick-up the selected membrane up to the β source-detector assembly, where the blank membrane is measured, and up to the sampler assembly where the membrane is set in a special airtight holder. After sampling, the elevator carries the membrane on the β assembly and, after measurement, in the slide tray which advances one step and a new membrane is ready to go.

The instrument is entirely controlled by a microprocessor which enables the mechanical operations, keeps trak of the counting and sampling times, and provides the control of the electronic flowmeter. Therefore, at the end of the measurement, a small line printer records the sample number, hour and day of start and end of sampling, the volume of air which has been sampled, the accumulated mass, the TSP concentration and the statistical error. In addition, by taking advantage of the microprocessor capability, it is possible to program four operation modes, namely: A) Time program, where the membranes are collected according to sampling and waiting times. B) Volume program, where the volume of air sampled is programmed. C) Step program, where the filter is sampled and checked at programmed periods of time in order to have cumulative mass measurements over short periods of time (Peak measurement). D) Mass program where the filter is sampled and checked until a programmed minimum mass is accumulated. After any operation the membrane filters are available for conventional chemical analysis.

The use of the microprocessor dramatically simplifies the electronic set up, being able to perform all operation needed for scaling the tranducers outputs to the relevant parameters. For instance the complex electronic hardware used for flow measurement is no longer needed because the microprocessor, through an A/D converter and an I/0 port, is directly interfaced with the bare pressure transducer. A multiplex circuit provides the same operation with the temperature transducer and the "Math" routines in the microprocessor operating system will directly provide the flow rate in L min^{-1} while working as V/F converter for volume measurement.

The minimum use of the electronic hardware results in a superior reliability and ease of service which are important features of instrumentation intended for continuous and automatic monitoring of suspended particles.

Preliminar results on the use of instrument are given in Fig. 4 where the concentrations of TSP as obtained with the automatic β gauge and the conventional gravimetric method are compared. It is worth stressing that the gravimetric method makes use of a prototype of an air sampler built around a pressure transducer. Tests have been performed on program A (Time program) by selecting a sampling and waiting time of 8 and 16 hours respectively, while counting of blank and loaded membranes lasts for 5 min. Although data points are not enough for a good statistical analysis and a residual variance, typical of such comparisons, is to be taken into account, it appears that a good agreement between the two methods can be inferred.

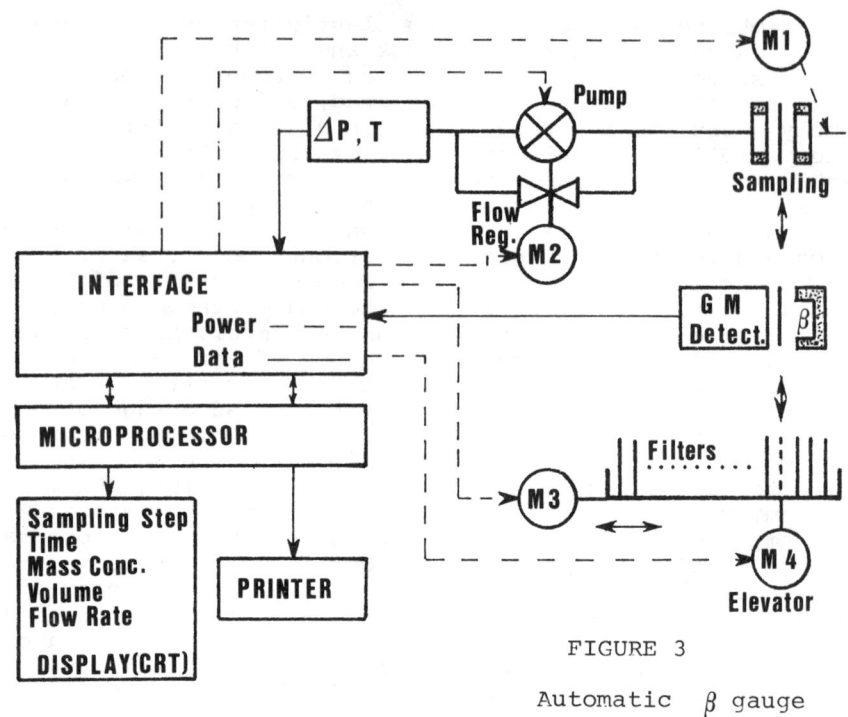

FIGURE 3

Automatic β gauge

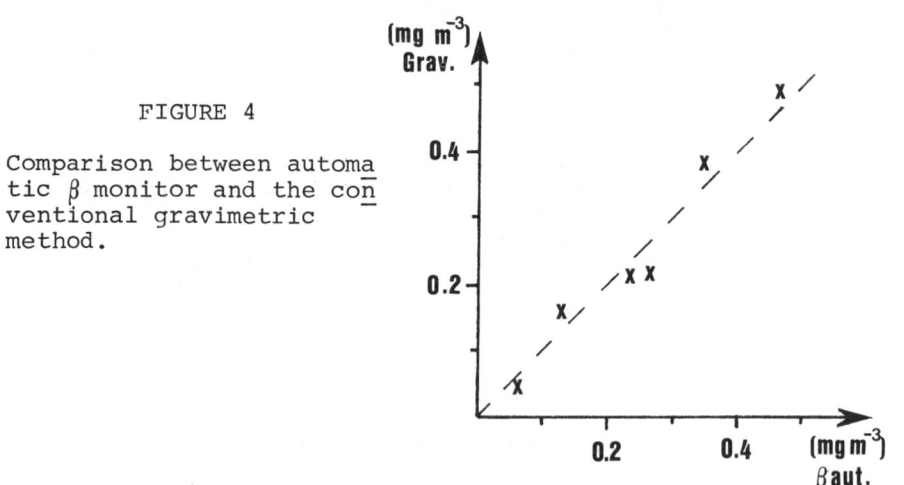

FIGURE 4

Comparison between automatic β monitor and the conventional gravimetric method.

TAB. I - Mass absorption coefficient for several
aerosolic materials as measured by means
of β gauge.

Nature of particulate	μ (cm^2 mg^{-1})
cement dust	0.277
rodhamine	0.267
ammonium fluoride	0.270
potassium iodide	0.277
lubricating oil	0.304
diacetildoxime	0.302
ambient aerosol	0.263
value expected from theory	0.267

5. CONCLUSIONS

The automatic β gauge monitoring of TSP has been proved to
be equivalent to gravimetric method in precision and accuracy,
which, trough the use of electronic transducers, are at a very
high standard. The wide use of electronic controls and data pro
cessing makes the instrument quite versatile in any medium or
large scale monitoring program. It is worth stressing that the
instrument makes use of standard membrane filters, thus particu
late matter can be analyzed in terms of chemical composition by
using conventional methods and the relative concentration can
be easily inferred scaling the absolute amount to the volume of
air which is exactly known and automatically corrected for variations in temperature and pressure.

The microprocessor can also be used for data transmission
or to store relevant parameters in a cassette in order to perform statistical analysis over a long period of time or to be
interpreted by a more powerful computer. Work on these aspects
are now on the way.

6. REFERENCES

1) J.S. Nader, D.R. Allen M. Ind. Hyg. Assoc. J.21,300 (1960)
2) P. Lilienfeld, Am. Ind. Hyg. Assoc. J.31,722 (1970)
3) Macias E.S., Husar R.B., in K.T. Whitby Ed."Fine Particles:
   Aerosol Generation, Measurement, Sampling and Analysis" p.535
   Academic Press, New York, N.Y. (1976).
4) M. Salmi, I. Allegrini, Isotopenpraxis 314, 15 (1979).
5) I. Allegrini, A. Liberti, M. Salmi, AES 5,82 (1980).
6) I. Allegrini, A. Febo, A. Di Chiara Maggioli, Acqua & Aria
   2,197 (1981).

QUALITATIVE AND QUANTITATIVE ANALYSIS OF EMISSIONS OF A
MUNICIPAL INCINERATION INSTALLATION

J. JANSSENS, L. VAN VAECK , P. SCHEPENS and F. ADAMS
Department of Pharmaceutical Sciences (Laboratory of Toxicology) and
Department of Chemistry
University of Antwerp (U.I.A.), B 2610 Wilrijk

Summary

Polychlorinated dibenzo-p-dioxins (PCDDs), dibenzofurans (PCDFs) and
their assumed precursors, the polychlorinated benzenes (PCBzs) and
phenols (PCPs), were determined in the flue gases of a municipal incinerator. The fly ash was sampled isokinetically with separation of the
coarse and fine particles by using a cyclone. Simultaneously, the pollutants in the gas phase were trapped on a porous polymer adsorber.
Chemical analysis of the extracts was performed by GC-ECD and GC-MS.
The determination of these compounds in the gas phase was essential
since their gas-particle distribution in the flue gases has shifted
extremely towards the gas phase. The concentrations varied from the
ng m^{-3} range for the PCDDs and PCDFs up to the μg m^{-3} range for the precursors. The total yearly emission of PCDDs and PCDFs by this incinerator represents the equivalent toxicological risk of ca 3 g 2,3,7,8-
tetrachlorodibenzo-p-dioxin.
Experimental evidence was obtained for the role of the PCPs and PCBzs
as precursors in the actual formation process of the PCDDs and PCDFs.
A postulated pyrosynthesis mechanism from the literature could be confirmed by the identification of several intermediates as well as by the
correlation of the concentration levels of the precursors with those of
the PCDDs and PCDFs.

1. INTRODUCTION

    Many noxious organic compounds in ambient particulate matter originate
from anthropogenic sources. Until now, the attention in environmental research has centered on the polycyclic aromatic hydrocarbons (PAHs), of
which several have shown to be carcinogenic (1,2,3,4). Recently, several
polychlorinated aromatic compounds have been identified in the fly ash from
municipal incinerators (5,6,7). The presence of the following compounds has
been reported: polychlorobenzenes (PCBzs),polychlorophenols (PCPs),polychlo-
robiphenyls (PCBs), polychloronaphtalenes (PCNs),polychlorodibenzo-p-dioxins
(PCDDs) and polychlorodibenzofurans (PCDFs) (8). Some representants of the
classes of PCDDs and PCDFs are extremely hazardous (9,10,11).
    It is also known from model experiments that PCDDs and PCDFs are formed
by pyrolysis of PCBzs, PCPs, PCB and polychlorodiphenylethers (PCDPEs)
(12,13,14,15). From these findings a tentative formation mechanism has been
set up (16). Although the generation of PCDDs and PCDFs by refuse incineration is generally accepted to involve similar reaction steps, this has
never been verified experimentally until now.
    Therefore, this research has focussed on a systematical monitoring of
the PCDDs and PCDFs and their possible precursors, the PCBzs and PCPs,
emitted by a municipal incinerator. The purpose of the experiments was not

only the evaluation of the potential health hazard from the PCDDs and PCDFs released by this type of pollution source, but also the definition of the role of the PCBzs and PCPs as precursors for the PCDDs and PCDFs in the actual conditions of the waste burning process. Table I summarizes the selected compounds, their structure formulas, the number of isomers and the abbrevations used throughout this work.

general structure formula	number of chlorine atoms	general name	selected compound (number of isomers)	abbrevation
$\bigcirc\!\!-Cl_x$	$x = 1 - 6$	polychlorobenzene	x = 4 (3) 5 (1) 6 (1)	tetra-CBz penta-CBz hexa-CBz
$\bigcirc\!\!-Cl_x$ OH	$x = 1 - 5$	polychlorophenol	x = 3 (6) 4 (3) 5 (1)	tri-CP tetra-CP penta-CP
$Cl_x\bigcirc O \bigcirc Cl_y$	$x + y = 1 - 8$	polychlorodibenzo-p-dioxin	x + y = 4 (22) 5 (14) 6 (10) 7 (2) 8 (1)	TCDD P_5CDD H_6CDD H_7CDD OCDD
$Cl_x\bigcirc O \bigcirc Cl_y$	$x + y = 1 - 8$	polychlorodibenzofuran	x + y = 4 (38) 5 (28) 6 (16) 7 (4) 8 (1)	TCDF P_5CDF H_6CDF H_7CDF OCDF

TABLE I : Structure formula, number of isomers and abbrevation of the selected compounds.

2. EXPERIMENTAL

2.1. Sampling

The samples were collected in a modern municipal waste incinerator, located in Beveren, Belgium. The installation is of the 'Alberti-Fonsar' type and is designed for the combustion of household refuse as well as light industrial waste with significantly higher calorific value. The plant has two simultaneously operating furnaces with a total capacity of 4.3 ton h^{-1}. Data about the amount and composition of the treated waste were available only for 1979. The installation worked continuously five days a week at about 65 % of its maximum capacity, thus combusting about 16000 ton of household refuse and 2100 ton of light industrial waste.

The sampling was performed during February and March 1980. The feed was mainly household refuse, which was burned without prior recuperation of any material, such as metal, glass, etc. . The sampling system is shown in figure 1, and was operated under isokinetic conditions. It includes a BCURA cyclone probe (Airlow Developments), which separates the particles according to their equivalent aerodynamic diameter(cut-off at 50 % collection efficiency = 0.5 μm (17)). The coarses are collected into a hopper (1a), the finer particles are retained by a glass wool plug (1c). The gas phase compounds are collected out of a small fraction of the total volume of flue gases sampled, by an adsorption column (c) , containing about 1.5 g of Tenax G.C. adsorbent. The sampling rate through the gas trap is about 10

ml min $^{-1}$, whereas the total sampling rate is about 150 ml min $^{-1}$.Additio-
nal condensation systems are provided to protect the pumps and the gas
volume meters. The volumes, as they had been measured, were converted to
standard conditions according to the directions of the Belgian Institute
for Normalisation (18). Typical sampling periods were 4 h , during which
the sampling system was consecutively placed at four points (for one hour
each) in the duct between the electrostatic precipitator and the chimney.

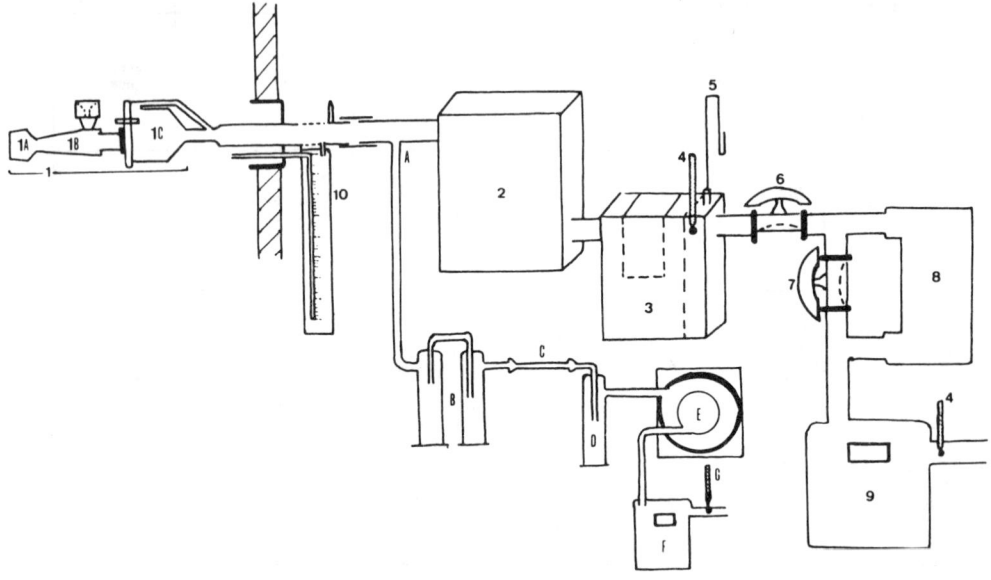

FIGURE 1 : Scheme of the sampling train : 1.dust collector (1a hopper, 1b
cyclone, 1c filter holder); 2.water cooled condensation box;
3.ice cooled condensation box; 4,G.thermometers; 5,10.manometers
6,7.ball valves; 8,E.pumps; 9,F.gas volume meter; B,D.condensa-
tion flasks; C.Tenax gas adsorption trap.

2.2. Chemical analysis

The samples were extracted with t-butylmethylether and acetone consecutively. A clean-up procedure was elaborated, based on the one described
by Buser et al.(6). Figure 2 summarizes the sample pretreatment.

The PCDDs and PCDFs were determined, using a Hewlett-Packard 5992 A
quadrupole GC-MS with electron impact source, coupled to a dual drive floppy disc data system (Hewlett-Packard, 9885). The samples were eluted on a
glass column of 1.8 m length and 1.5 mm internal diameter, packed with 3 %
OV 101 on Gaschrom Q (100-120 mesh). The temperature was programmed from
240 to 260 °C at 10 °C min $^{-1}$. Quantitative determinations were carried out
using multiple ion detection. Each compound was measured on two selective
ions, usually the $(M + 2)^{+\cdot}$ and the $(M + 4)^{+\cdot}$. Calibration curves were
constructed daily using 3 reference solutions containing TCDD and H_6CDD
(19) . As internal standard Co-ral $^{®}$ (0-3-chloro-4-methylcoumarin-7-yl-0,0-
diethylphosphorothioaat) was added. For quantitation of the compounds not
available as a reference, the following assumptions were made:
1.the same relative response factor was assumed for equally chlorinated
PCDDs and PCDFs;
2.the amount of penta-, hepta- and octa-chlorinated isomers was calculated

by inter- or extrapolation , after correction for the isotope effect.
Qualitative analyses were carried out using the mass chromatography method (scan from 50 to 500 m/z). The oven temperature was programmed from 140 to 260 °C at 6 °C min^{-1} .

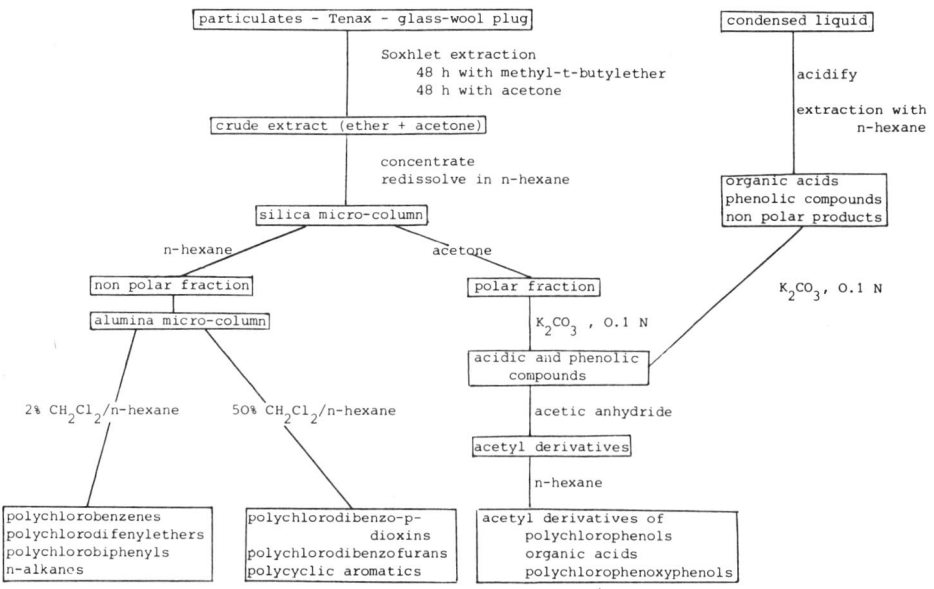

FIGURE 2 : Sample preparation scheme.

3. RESULTS AND DISCUSSION

3.1. Total emission and toxicological assessment

Table II summarizes the whole of the emission data (sum of the gas phase and the particulates) for the polychlorinated compounds. The concentration range is derived from the results of twelve samples, most of them collected under conditions, typical of normal operations. Typical values are in the range of ng m^{-3} for the PCDDs and PCDFs, and of μg m^{-3} for the PCBzs and the PCPs. These findings are in agreement with the data from other authors (20)

It was also found that the concentration ratios of the different individual PCBzs were nearly constant, irrespective of the total concentration (cfr. table III). No influence was observed from the composition of the treated refuse. In contrast, the behaviour of the PCPs differs entirely from that of the PCBzs : the concentration

Compound	Typical concentration range (ng m^{-3})
1,2,3,5- and 1,2,4,5-tetra-CBz	650 – 850
1,2,3,4-tetra-CBz	650 – 950
penta-CBz	2500 – 3600
hexa-CBz	1400 – 2400
2,4,5-tri-CP	D.L.– 100
tri-CP (all other isomers)	1000 – 2000
2,3,4,6- and 2,3,5,6-tetra-CP	2000 – 3000
2,3,4,5-tetra-CP	150 – 350
penta-CP	1900 – 3300
TCDD	0.1– 7
P$_5$CDD	1 – 12
H$_6$CDD	6 – 64
H$_7$CDD	25 – 150
OCDD	50 – 200
TCDF	2 – 30
P$_5$CDF	5 – 61
H$_6$CDF	10 – 60
H$_7$CDF	10 – 85
OCDF	10 – 70

TABLE II : Typical concentration range of the selected pollutants.

Date	Code	Fraction (%) found for				Total PCBz-concentration ($\mu g\ m^{-3}$)
		1,2,3,5- and 1,2,4,5-tetra-CBz	1,2,3,4-tetra-CBz	penta-CBz	hexa-CBz	
21/2/80	1	17	14	46	23	5.4
26/2/80	2	13	13	47	28	5.0
14/3/80	3	10	9	45	36	2.8
18/3/80	4	3	13	42	43	0.5
20/3/80	5	12	13	48	26	6.8
21/3/80	6	12	11	47	29	6.3
25/3/80	7	13	13	46	28	2.4
26/3/80	8	8	12	46	33	8.7
27/3/80	9	11	12	46	31	7.8
28/3/80	10	14	16	44	25	5.9
Average		11 ± 4	13 ± 2	46 ± 2	30 ± 6	5.2 ± 2.6

TABLE III : Concentration ratios of the PCBzs.

ratios of the different PCPs strongly changed and seemed to be related to the moisture content of the flue gases. Table IV shows some average data. Low moisture levels always correspond to an increased ratio of the emitted amount of tri-CP as compared to that of penta-CP. This might be due to a selective removal of the tri-CP by the water, injected to cool down the flue gases: the water solubility of tri-CP is 1.5 g l^{-1}, but only 80 mg l^{-1} for penta-CP. However, this tentative explanation needs to be confirmed by additional experiments.

Number of samples	moisture content (ml m^{-3})	Average fraction (%) found for		
		all tri-CPs	all tetra-CPs	penta-CP
1	< 45	70	19	12
2	45-60	46	27	27
4	60-75	23	40	37
1	> 75	10	29	60

TABLE IV : Concentration ratios for the PCPs. Average data according to the moisture content of the flue gases.

Table V illustrates the distribution of the different compounds between the gas phase, the coarse (> 0.5 μm) and fine particulates (< 0.5 μm). In the chimney, only a minor fraction of the total concentration is found on the particulates, as could be expected from the relatively high temperature at the sampling point (270°C) and the vapour pressure of the compounds under study. The implications of these observations for the sampling technique are obvious: the analysis of fly ash samples is in no way relevant for the total emission. Moreover, attention must be drawn to a sampling artefact: considering the high temperature, the elution of volatile compounds out of the collected particles, especially those < 0.5 μm, during the sampling is very likely to occur. This may be the cause of important losses (21,22).

The gas-particle distribution is also important with respect to the toxicological implications. Indeed since these compounds are found predominantly in the gas phase at the elevated stack temperature, it can be expected that during the cooling of the flue gases in the atmosphere, the polychlorinated compounds will

Compound	Fraction of the total concentration (%)		
	vapour phase	particulate phase > 0.5 μm	< 0.5 μm
1,2,3,5- and 1,2,4,5-tetra-CBz	97.7	2.2	0.1
1,2,3,4-tetra-CBz	97.2	2.5	0.3
penta-CBz	95.9	3.8	0.3
hexa-CBz	93.8	5.4	0.8
2,4,5-tri-CP	99.0	1.0	D.L.
tri-CP (all other isomers)	99.9	0.1	D.L.
2,3,5,6- and 2,3,4,6-tetra-CP	99.7	0.2	0.1
2,3,4,5-tetra-CP	98.9	0.6	0.5
penta-CP	99.3	0.5	0.2
H$_4$CDD	82.5	15.1	2.4
H$_2$CDD	88.3	8.0	3.7
OCDD	91.3	1.9	6.8
TCDF	99.6	0.4	D.L.
P$_5$CDF	98.2	1.2	0.7
H$_6$CDF	83.2	16.8	D.L.
H$_7$CDF	89.4	5.2	5.4
OCDF	93.0	D.L.	7.0

D.L. = concentration beneath the detection limit.

TABLE V : Gas-particle distribution of the chloroaromatic compounds.

undergo gas-particle conversion processes.This results in the enrichment of these products, especially in the respirable particles. Although it is known that for several organics a relatively high stable gas phase concentration can build up in the atmosphere (23), one should expect a quantitative gas-particle conversion, especially for the less volatile PCDDs and PCDFs. The gas-particle conversion was found to occur very fast : at a distance of 1 km from a municipal incinerator, penta-CP was no longer detectable in the vapour phase, but could be quantified in the particulate matter (23). However, further investigations of ambient aerosols are necessary.

From the total emission data, it was also attempted to estimate the toxicological risk, associated with the release into the environment of PCDDs and PCDFs by the incineration of municipal waste. The total amount of chlorinated organic pollutants, emitted during one year by this incinerator was calculated from the measured emission and the quantity of combusted waste. The results are summarized in table VI. Using literature data, the emission of PCDDs and PCDFs could be converted into an equivalent amount of the highly toxic analog 2,3,7,8-TCDD, which would have the same biological activity (20,24). In this way, one can estimate the health risk from the accumulated yearly emission by the incinerator under study. This corresponds to an emission of about 3 g 2,3,7,8-TCDD. In order to evaluate these data: at Seveso about 2.5 kg 2,3,7,8-TCDD was released in one moment. However, it has to be emphasised that such a comparison needs an interpretation with due reserve. Indeed, at Seveso the whole amount was released within a very short time span and contaminated a rather limited area. It clearly concerned an acute intoxication. The municipal incinerator emits during a period of one year only 0.12 % of the total amount released in Seveso. Furthermore, there is only one incinerator every 500 km^2 in Belgium at the moment.

In conclusion, it can be stated that the main problem is certainly not the possibility of an acute intoxication but one might question seriously the health risks caused by a long term exposure to low concentration levels of PCDDs and PCDFs

Compound	Average emission		Yearly emission
	mg h^{-1}	mg ton^{-1}	g
1,2,3,5- and 1,2,4,5-tetra-CBz	37	13	240
1,2,3,4-tetra-CBz	41	14	250
penta-CBz	150	51	920
hexa-CBz	93	32	580
2,4,5-tri-CP	9	3	50
tri-CP (all other isomers)	150	52	940
2,3,4,6- and 2,3,5,6-tetra-CP	140	48	870
2,3,4,5-tetra-CP	14	5	90
penta-CP	160	54	980
TCDD	0.14	0.05	1
P_5CDD	0.44	0.15	3
H_6CDD	2.3	0.8	14
H_7CDD	6.7	2.3	42
O_8DD	5.6	1.9	34
TCDF	1.4	0.5	9
P_5CDF	3.0	1.0	18
H_6CDF	2.6	0.9	16
H_7CDF	3.0	1.0	18
O_8DF	2.8	1.0	18

TABLE VI : Average and yearly emission of the selected compounds in the incinerator of Beveren.

3.2. Formation processes of PCDDs and PCDFs

A lot of research has been invested in the understanding of the generation of PCDDs and PCDFs during pyrolysis in the presence of oxygen of different materials such as PVC, PCBzs, PCPs, PCBs and PCDPEs (25,13,12,14, 15). The results have been combined to the consistent reaction mechanism, summarized in figure 3 (16). It is generally assumed that the presence of PCDDs and PCDFs in the emission of waste incinerators is the result of very similar reactions. However, the validity of the reaction scheme has never

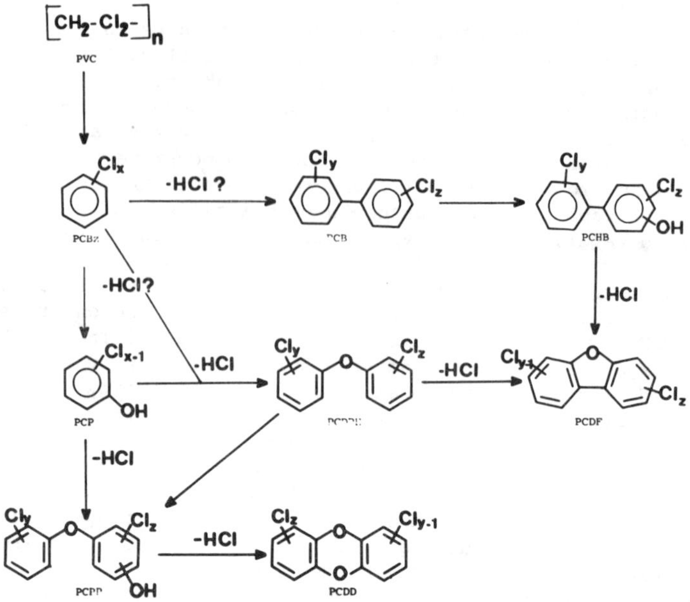

FIGURE 3 : The formation mechanism for the PCDDs and PCDFs
from PVC, PCBzs and PCPs, as derived from the
pyrolysis experiments (16).

been verified experimentally in the actual conditions of refuse burning
until now.

Essential to the formation mechanism is its strong emphasis on the
role of chlorinated organic compounds in the waste treated. Considering the
wide applications of PVC, household refuse should certainly contain a considerable amount of this product. In contrast, PCPs and PCBzs cannot be
expected to occur invariably in the refuse, as these products have less
applications (as bactericide or fungicide for example). Moreover, the contribution of individual PCBzs to the total PCBz emission was observed to be
nearly constant: it was not influenced by the concentration levels in the
flue gases or by the composition of the treated refuse. These findings
strongly suggest the occurrence of an alternative formation process for
these products, similar to that of the PAHs. The generation of PAHs in
incomplete combustion processes has been studied intensively: it involves
the recombination of small initial radical fragments (containing two carbon
atoms). The different analogs are formed in characteristic concentration
ratios, depending on the combustion temperature and - to a less extent -
on the composition of the material burned.

In view of these findings, it was attempted to relate the PCBz formation to the content of organic chlorinated compounds of the burned material.
Therefore, several analyses were performed on fly ash samples, obtained
from the combustion of wood, coal and heating oil. Preliminary results
indicated that the chemical nature of the chlorine containing compounds is
only of minor importance for the formation of PCBzs, confirming again the
possibility of an alternative formation mechanism.

Although the formation of PCBzs is likely to occur according to a
similar mechanis as the one of the PAHs, strong evidence was found in favour
of the reaction scheme in figure 3. Several combined extracts were analysed

qualitatively by GC-MS: several well-known, non chlorinated compounds were identified, such as aliphatic and polyaromatic hydrocarbons, which are characteristic of but not specific for combustion processes. In addition, a number of mono-chloro-PAHs was found, but no indications were given for polychlorosubstitution. Moreover, several PCDPEs and PCPPs as well as some PCBs could be characterised; these are the intermediates mentioned in figure 3. An estimation of their concentrations showed that most of them were only minor constituents of the flue gases in comparison to the PCDDs and PCDFs. However, this should be interpreted with due reserve, because of the complete lack of reference compounds (undoubtly their mass spectrometric response will not be the same as for the PCDDs and PCDFs),whereas the sample pretreatment was not optimised for quantitative recovery of these products.

The relevance of these results is clear: the mechanism, shown in figure 3, and hence the essential role of PCPs and PCBzs as precursors, is now supported by strong experimental evidence. Furthermore, there is no longer any doubt about the presence of intermediates such as the PCDPEs, which can interfere with the GC-MS determination of the PCDFs (26). In order to avoid significant positive errors in the quantitation of these hazardous pollutants, the sample pretreatment procedure has to include a suitable separation step, irrespective of the fact that this is time consuming and therefore complicates routine analyses.

In addition to the qualitative evidences for the formation of PCDDs and PCDFs in incinerators,it was also attempted to confirm this relationship by correlating the quantitative data. However, at the moment only a rather tentative interpretation can be given: a strictly mathematical treatment and a quantification of the various relationships could not be performed because of the rather limited number of data now available and because of the uncertainty about the selective wash-out of the lower chlorinated PCPs in the waterspray cooling tower. Moreover, the formation of the PCPs out of the PCBzs is not understood well. Nevertheless, in spite of these limitations, one can still discuss several aspects of the reaction mechanism. The following questions can be asked:
1.do the PCDFs originate preferably from the intermolecular reactions of two PCBzs, thus implying that the PCBs are the intermediates, or, in contrast, are the PCPs also involved in the initial reaction step, giving the PCDPEs as intermediates?
2.what is the actual role of the PCPs? And do the measured concentrations provide a valid basis for the total PCP precursor concentration? Indeed, it is unknown whether or not the PCPs are generated in a kind of first order mechanism from the PCBzs. Furthermore, the assumed removal of the tri-CPs in the waterspray tower will only affect the tri-CP/PCDD or PCDF relationships if the following reactions occur before the cooling tower.

Figure 4 summarizes the concentrations of some PCDFs, PCDDs, PCPs and PCBzs in seven samples. A comparison of the bar-graphs of the PCDFs with those of the corresponding PCBzs indicates no correlation at all, which suggests an essential role for the PCPs. This is clearly illustrated by the concentration profiles of H_6CDF and H_7CDF, showing a striking similarity with the one of tetra-CP. However, the concentration of the appropriate PCBz analog has also to be taken into account. The increased tetra-CBz concentrations in the samples 8 to 10 explain why H_6CDF does not follow the tetra-CP profile strictly. The same holds for H_7CDF: the increased concentration of penta-CBz as compared to hexa-CBz, will favour tetra-CP as a precursor. The contribution of the reactions between tetra-CP and tetra-CBz (formation of H_6CDF) and penta-CBz (formation of H_7CDF) respectively, as opposed to the other possible reactions, is dominating. This is a logical consequence from the low concentrations of one of the precursors in the other reactions

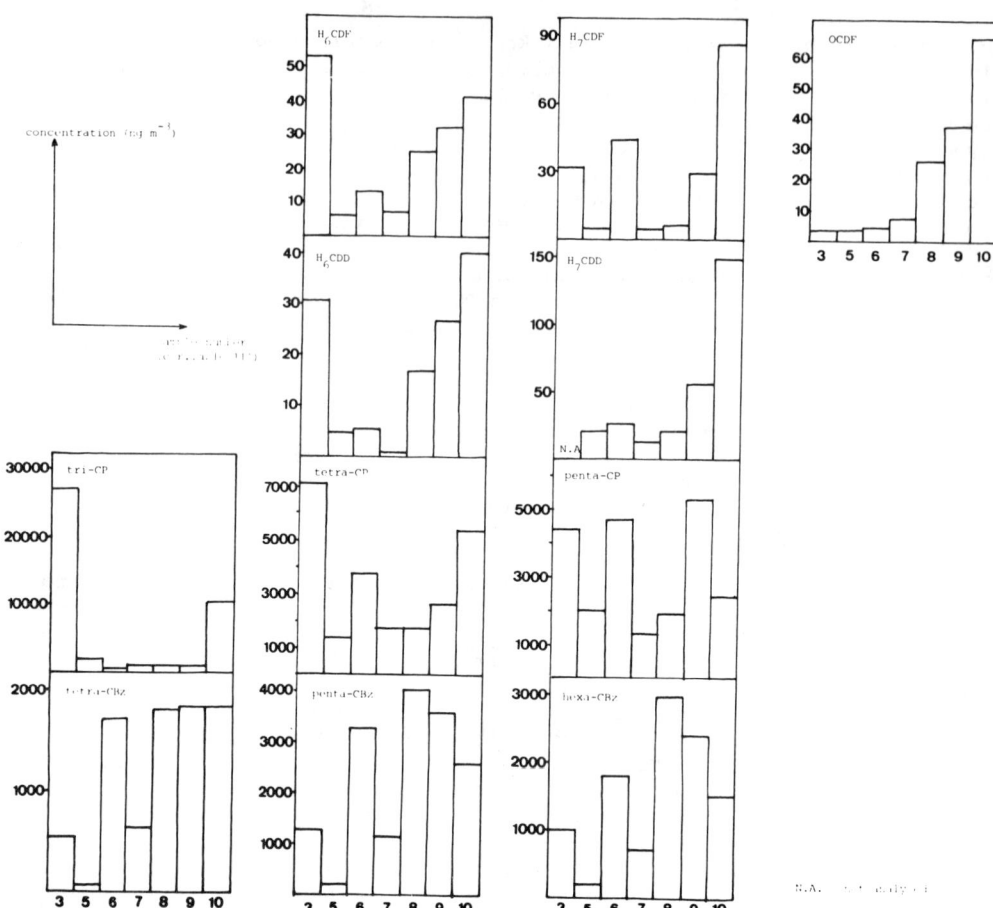

FIGURE 4 : Bar-graphs of some PCDFs, PCDDs, PCPs and PCBzs.

(tri-CBz, di- and tri-CP). The high tri-CP concentration in sample 3 increases the amount of H_6CDF but hardly changes that of H_7CDF. For OCDF, it can be expected on the basis of purely statistical reasons, that the reaction between tetra-CP and hexa-CBz will be more important than the one between penta-CP and penta-CBz. Indeed, the strong decrease of the penta-CP concentration in sample 10 does not prevent the OCDF concentration from increasing further. TCDF and P_5CDF were not presented in the figure, because of the low concentrations (usually beneath detection levels) of their precursors.

For TCDD and P_5CDD too, the precursor concentrations are beneath the detection limit, which prevents the setting up of a balanced explanation, whereas for OCDD, an insufficient number of measurements is available at the moment. The concentration profiles of H_6CDD and H_7CDD resemble those of the corresponding PCDFs, which could point to the fact that the PCDPEs act as a common intermediate. However, no decisive answers can be given by a merely qualitative investigation of the correlations, since the PCDF concentrations strongly depend on the PCP levels as well.

Although several ideas from this discussion have to be confirmed by further experiments, these preliminary results strongly indicate the key position of the PCPs in the formation reactions of both the PCDFs and the

PCDDs in a waste incinerator. The PCP concentrations measured after the waterspray cooling system provide a good basis for correlation with the PCDD and PCDF levels in the flue gases. The intermolecular reactions of the parent compounds PCPs and PCBzs hence take place after the cooling process, but only if the assumption of a selective wash out of tri-CP by water is correct. It is clear that in further investigations first priority must be given to this hypothesis : it would allow a direct control on the formation of the lower chlorinated and highly toxic PCDDs and PCDFs.

4.CONCLUSION

The incineration of municipal waste is a source of chlorinated organic compounds : the PCPs and PCBzs are released in the $\mu g \, m^{-3}$ concentration range, the PCDDs and PCDFs in the $ng \, m^{-3}$ range. In stack, these compounds are mainly found in the gas phase (> 80 %). Therefore, the analysis of the fly ash provides in no way a meaningful basis for the assessment of the emissions. One may expect the almost quantitative incorporation of the highly toxic PCDDs and PCDFs into the respirable particles when they are released into the atmosphere.

The amount of PCDDs and PCDFs emitted could be converted into a toxi-cologically significant figure : the biological activity of the total amount of PCDDs and PCDFs yearly released by this waste incinerator is estimated to be equivalent to that of about 3 g 2,3,7,8-TCDD. However, this result should be interpreted with due reserve, because of the assumptions and extrapolations involved.

The individual PCBz analogs are found in nearly constant ratios. Considering the PCPs, the tri-CP concentration increased strongly with the decreasing moisture content of the flue gases. This was tentatively con-tributed to a wash-out effect in the waterspray tower, resulting in a selective removal of the more water soluble lower chlorinated PCPs.

For the first time, clear experimental evidence could be obtained for the assumption of the PCBzs and PCPs being the actual precursors of the PCDDs and PCDFs when waste is burned : several intermediates could be de-tected, such as the PCDPEs and the PCPPs, but also a logical and consistent relationship could be established between the concentrations of PCDDs and PCDFs and those of their corresponding precursors.

Finally, further research will be needed to understand the exact in-fluence of the moisture content on the tri-CP concentration. Indeed, tri-CP is the main precursor of the most toxic PCDDs and PCDFs. The waterspray tower could offer a unique possibility to reduce or even prevent the re-lease of these extremely hazardous compounds.

Acknowledgments

This research was part of the R & D program "Environment-Air" of the Belgian Government (Ministerie van Wetenschapsbeleid,Unit 5.5). We thank the National Foundation of Scientific Research(N.F.W.O.) for granting the degree of 'aspirant-navorser' to dr. Van Vaeck and for their financial support. Dr. Delvaux of the University of Louvain-la-Neuve (U.C.L.) is acknowledged for supplying us with the TCDD and H_6CDD standards.

References

1.R.P.Hangebrauck, D.J.Von Lehmden and J.E. Meeker, J.A.P.C.A.14, 267 (1964)
2.S.E.Hrudey,R.Perry and R.A.Wellings, Environ. Res. 7, 294 (1974)
3.I.W.Davies,R.M.Harrison,R.Perry,D.Ratnayaka and R.A.Wellings,Env.Sci. Technol. 10, 451 (1976)
4.G.Broddin,L.Van Vaeck and K.Van Cauwenberghe, Atmos. Environ.11,1061 (1977)
5.K.Olie,P.L.Vermeulen and O.Hutzinger, Chemosphere 6, 455 (1977)
6.H.R.Buser,H.-P.Bosshardt and C.Rappe, Chemosphere 7, 165 (1978)
7.Dow Chemical Company,'Trace Chemistries of Fire',Rapport M.L.-A.M. 78-63/3 (1978)
8.G.A.Eiceman,R.E.Clement and F.W.Karasek, Anal.Chem. 51, 2343 (1979)
9.G.L.Sparschu, F.L.Dunn and V.K.Rowe, Food Cosmet. Toxicol.9, 405 (1971)
10.A.S.Kende,J.J.Wade,D.Ridge and A.Poland, J. Org. Chem.39, 931 (1974)
11.W.F.Greenlee and A. Poland in 'Dioxin:Toxicological and Chemical Aspects' eds.F.Cattabeni,A.Cavallaro and G.Galli,S.P. Medical and Scientific Books,Halsted Press/ J.Wiley and Sons Inc.,New York (1978)
12.H.G.Langer,T.P.Brady,L.A.Dalton,T.W.Shannon and P.R.Briggs in 'Chlorodioxins : Origin and Fate' ,ed. H.Blair,Advances in Chemistry Series 120 (1972)
13.H.R.Buser, Chemosphere 8, 415 (1979)
14.H.R.Buser and C.Rappe, Chemosphere 8, 157 (1979)
15.R.Lindahl,C.Rappe and H.R.Buser, Chemosphere 9, 351 (1980)
16.J.W.A.Lustenhouwer,K.Olie and O.Hutzinger, Chemosphere 9, 501 (1980)
17.P.G.W.Hawksley,S.Badzioch and J.H.Blackett in 'Measurements of Solids in Flue Gases' , The Institute of Fuel , London, 1977
18.Belgisch Instituut voor Normalisatie, NBN rapport X44-002 , 1977
19.the standards were obtained from dr. E.L. Delvaux, U.C.L., Belgium.
20.A.Cavallaro,G.Bandi,G.Invernizzi,L.Luciani,E.Mongini and A.Gorni, Chemosphere 9, 611 (1980)
21.L.Van Vaeck and K.Van Cauwenberghe in 'Advances in Mass Spectrometry' Volume III , ed. A.Quayle, Heyden & Son Ltd. London , 1980
22.J.Janssens, thesis, Universitaire Instelling Antwerpen (U.I.A.) ,1978
23.W.Cautreels and K.Van Cauwenberghe, Atmos.Environ. 12,1133 (1978)
24.O.Hutzinger,K.Olie,J.W.A.Lustenhouwer,A.B.Okey,S.Bandiera and S.Safe, Chemosphere 10, 19 (1981)
25.B.Ahling,A.Bjørseth and G.Lunde, Chemosphere 7, 799 (1978)
26.H.R.Buser, J.Chromatogr. 107, 295 (1975)

SOME MEASUREMENTS OF THE CONCENTRATION OF CHLORINE AND SULPHUR CONTAINING MOLECULES

A. Ionescu, W.D. McGrath [*]) and K.H. Becker

Physikalische Chemie /FB 9
Universität Wuppertal
Gaußstr. 20, 56 Wuppertal 1,

[*])The Queen's University of Belfast
Dept. of Chemistry
Belfast BT9 5AG Northern Ireland

Summary

Field studies were conducted in Wuppertal/ W.-Germany,
St.Moritz/ Switzerland, Helgoland isle/ W.-Germany and Tielan Bay/
Ireland in order to determine the concentrations of different
chlorine and sulphur containing molecules in the lower troposphere. A
comparison between a typical polluted atmosphere and areas with
"natural background levels" was performed. A gradient of pollution from
Wuppertal to Helgoland, St. Moritz and Tielan Bay was observed for all
measured constituents exept COS. The higher concentration of COS with
an average of 491 pptv is very likely due to local natural sources.
High concentrations of SF_6 were also observed in the St. Moritz area
with an average of 0.54 pptv.

1. INTRODUCTION

The current problem of the exact balancing of halogen and sulphur
containing substances in the atmosphere has led to intensive research
on their atmospheric concentrations and photochemistry. Although a considerable number of research groups have been actively pursuing this
subject over the past several years the extent of the problem as regards
stratospheric ozone depletions is far from clear. In this respect the role
of freons commonly used as spray can propellants, is central, since they
are essentially totally inert in the lower troposphere and would, therefore, appear to have no natural sinks.
Previously reported measurements in the Düsseldorf/Wuppertal area (9) have
shown that freons and chlorinated hydrocarbons such as CCl_4 can be detected
in air samples in the pptv concentration range.

The simultaneous concentration measurement of chemically stable and
less stable compounds in the troposphere is of importance for the determination of their atmospheric lifetimes and possible sinks. Of equal
importance is to determine whether the tropospheric concentrations of
individual compounds is increasing with time and to correlate this information with the estimated global release rates. In order to do this,
some geographical sites have to be chosen which are not to close to any
densely populated urban areas.

It is only recently that sulphur containing substances such as COS and
CS_2 which could be of importance in the atmospheric sulphur cycle have been
identified in the atmosphere (1-6). It is known that COS is relatively
stable in the troposphere and it has been assumed that COS must diffuse
into the stratosphere and play a part in the formation of sulphates.
According to recent measurements CS_2 is less stable due to photolysis under
tropospheric conditions. It has been suggested that one of the major
natural sources for CS_2 on the earth's surface is Spartina alterniflora
(8). So far no natural source for COS has been identified. Since both
halogen containing and sulphur containing gases are also produced anthropo-
genically in urban areas, mostly from aerosol propellant sprays and industrial plants, respectively, it is important to establish what one would

consider to be "natural background levels" for these trace gases and to compare these with concentration values in a typically polluted urban or industrial area. Ideally one would select several rural sites which would seem to be free from local emissions or from long range transport of pollutants. Long range transport effects would show up in a variation of the measured concentrations as a function of meteorological conditions, in particular wind speed and direction. Such effects would influence the different sites at different times. Further, comparison of the ratios of one trace gas to another should help in the identification of possible transport effects. Hence it is of interest to obtain ratios of trace gas with respect to SF_6, since this gas is evenly distributed in the tropo-sphere with the exception of sites situated close to electrical power stations and similar related installations, and it is ideally suited to gas chromatographic measurements using an electron capture detector (ECD).

In an initial attempt to identify possible "background level" loca-tions and compare concentration values for the previously mentioned atmospheric trace gases at these sites with those in an typically urban area, we selected Tielan Bay (Donegal, Ireland $54^{\circ}41'N$; $8^{\circ}47'W$), Helgoland isle (north coast of Germany $54^{\circ}11'N$; $7^{\circ}57'E$), St. Moritz (Switzerland $46^{\circ} 30^{\circ}'N$; $9^{\circ}51'E$) and Wuppertal (Rhein/Ruhr, Germany $51^{\circ} 12'N$; $7^{\circ}00'W$). Since it was not possible to carry out continuous monitoring, grab samples were taken and analysed in the laboratory.

2. EXPERIMENTAL

Concentration measurements of the atmospheric trace gases were carried out by means of GC-ECD. Details of the experimental set-up have already been described (9) and it suffices to give here an outline of the salient features of the apparatus. The column used was Chromosil 330 Supelco 1/8" O.D., of 6 ft. length and the carrier gas flow rate 40 cm^3 min^{-1} of nitrogen. This column permitted separation and concentration measurements to be made in a single run of both halogen and sulphur containing substances, thus leading to a considerable simplification of the experimental routine. The T/Sc detector (Intersmat 120 GC) was operated at temperature of $310^{\circ}C$. The air samples, contained in stainless steel vessels of 5 l volume were connected to the inlet of the gas chromato-graph by means of a gasdosing valve. An air sample of 10 - 20 ml volume was admitted into the equipment and, in order to prevent back-diffusion of the air into the detector, a stainless steel tube of approximately 4 m length fitted to the end with a needle valve was connected to the exit of the detector. Measurements were carried out under the following con-ditions: The column was kept isothermal at $25^{\circ}C$ for 1 minute and then allowed to increase in temperature from $25^{\circ}C$ to $75^{\circ}C$ at a rate of $8^{\circ}C$ min^{-1} at which point it was again held at this temperature. Peak areas and heights were measured by means of an integrator (CSI-Supergrator 3) from 3 injections of the same samples. The standard deviation determined from repeated measurements on the same sample lay between 1-4% for 10 injections. Retention times are listed in table I. Individual peaks were assigned to particular compounds from considerations of their boiling points and enthalpies of vaporization together with mass spectra data obtained by coupling of the gas chromatograph outlet with a mass spectrometer (Varian MAT 44). Utilization of the mass spectrometer for this purpose could be carried out using the same column but required preconcentration of the sample by removing nitrogen and oxygen directly in a cooled column held at $-100^{\circ}C$. Air samples were taken over a period of half a hour. A sample of representative outdoor air compressed at 120 bar in a stainless steel vessel of 10 l volume was used as a secondary standard. For the concentrations of F11, F12, SF_6, CCl_4, COS and CS_2 in this vessel no change was observed for at least a period of three months. In general, all air samples were analysed within less than three weeks. (see Table I)

3. RESULTS AND DISCUSSION

The grab samples from Tielan Bay cover the period April to October 1979, those from Helgoland May 1980, and those from St. Moritz June 1980. Each sample taken was analyzed at least three times and an average value calculated. The results obtained from the grab samples for CCl_2F_2, CCl_3F, SF_6, CCl_4, COS and CS_2 are listed in table 2. The table shows the average values for all measurements together with their corresponding standard deviations. For the location Wuppertal only the concentration values are given. More detailed information about the concentration profiles of CCl_2F_2 and CCl_3F in the Wuppertal area were published previously (9). The Wuppertal area exhibits the highest concentrations in comparison to the other locations, except for COS. The fairly high COS concentration found in the Tielan Bay area very likely indicates natural sources. (see Table II)

4. HALOGEN CONTAINING COMPOUNDS

All locations show quite similar concentration values for SF_6 as well as a high stability (low σ values). Somewhat higher concentrations of SF_6 are observed in St. Moritz with a low scatter of the data (V = 4). This can be explained by a larger number of electrical power stations in comparison with other areas. The lowest concentrations of all measured halogen containing compounds are obtained in Tielan Bay. The mean values of 161 pptv for CCl_3F and 186 pptv for CCl_2F_2 are comparable with values of background concentrations obtained in remote areas such as Alaska (10).
Helgoland isle also exhibits low concentrations of CCl_4, CCl_3F or CCl_2F_2, but the possibility of contaminations by long range transport from Northern Europe is higher as in Tielan Bay.
Because one of our goals with the present measurements was to find an appropriate place for the continuous monitoring of halogen and sulphur compounds in the lower atmosphere, we estimated the concordance between different sets of measurements in Tielan Bay, Helgoland and St.Moritz and also the correlations in this respect. A good estimation of concordance is the Kendall's coefficient of concordance W which permits the determination of the relationship between various sets of values. The values obtained in Tielan Bay and St. Moritz show a good concordance (see table III), but for the sets of SF_6, CCl_4, CCl_3F and CCl_2F_2 concentrations obtained in Helgoland a lack of concordance is observed. This lower concordance can be explained by the prevailing weather situation during the measurements. In the case of the measurements in Helgoland the ridge of a pressure maximum had moved from the North Sea to the Helgoland area. (see Table III) The general weather situation was characterized by a certain instability which was not the case for the measurements in St. Moritz and Tielan Bay. The partial correlation coefficients together with their standard deviations are also given in table III. They were calculated using the Fischer Z method, which ensures that a symmetrical skewness of the distribution is obtained above and below the measured value.
A very good correlation was observed for measurements in Tielan Bay and

either poor or no correlation in Helgoland isle. The values obtained in
St. Moritz lie inbetween.

The very low correlation of the results from Helgoland can again be
explained by the weather instability which occurred in this area during the
measurements. Under such conditions poor or negative correlations can be
expected (12). The results in Tielan Bay can be explained by long range
transport of contaminated air into this area. However, high ozone concentrations were not found at the same time (W.D. McGrath, unpublished data).-

Considering the generally good correlation coefficients which emerge
from the present work, except for Helgoland, it is of some interest to
consider the ratios of values of various halogen containing compounds.
The average values of these ratios are given in table IV. The values of
these ratios, except Wuppertal, correspond to those obtained by other
measurements (13,14). The results from Wuppertal are an exception,
especially for the $F12/SF_6$ and $F11/SF_6$ ratios which indicate a higher
contamination with CFMs in this area due to local sources. The ratios
between F11, F12 and CCl_4 are very close to those calculated from
the world production figures for these compounds (15) and to those obtained
from monitoring atmospheric concentrations over a longer period of time (16)
(see Table IV)

The concentration of ozone which was monitored in Tielan Bay from
April - Sept. 1979 followed the normal pattern of falling off during
the night and rising steadily from sunrise to its maximum around midday,
on no occasions were high ozone values obtained. The fact that one
series of measurements on halogen containing compounds (April 22nd
and 23rd 1979 showed abnormally high concentrations of these substances
(F12 = 986 pptv, F11 = 844 pptv) whereas the ozone level was very
low (on average for this geographical area) is difficult to explain.
The findings of COX et al. (17) indicated that during periods of high
concentrations of halogenated hydrocarbons the ozone levels were also abnormally high seems to depend on the general weather conditions. There
are meteorological conditions which favour long range transport of
CFMs without the production of high ozone levels in the polluted air
masses.

5. SULPHUR CONTAINING COMPOUNDS

The results of the concentration measurements for COS and CS_2 are
shown in table II. The average values for COS and CS_2 obtained in the
Tielan Bay are 491 and 201 pptv, respectively, which are in good agreement with those obtained by SANDALLS and PENKETT (3). No data are
given from St. Moritz because they lie under the limit of quantitative
determination. Recent measurements show that CS_2 is less stable due to
photolysis under tropospheric conditions. According to other recent
measurements (3,4) and the high stability (18,19), COS should be evenly
distributed with a concentration of about 500 pptv. This may indicate
that the values of 100 pptv reported for Helgoland and St. Moritz in
table II are influenced by artifacts during sampling and analysis. A
possible change of the COS concentration in the sampling vessel cannot
be excluded even by controlling the secondary standard where at 120 bar
no change of COS could be observed during a period of three months.
However, the value of 368 pptv for Wuppertal (table II) was obtained by
direct sampling of outdoor air. This measurement, therefore, should be
free from such artifacts. The COS values for Wuppertal are lower than
the previously published values from rural areas and the present value

Table I :

Retention times of halid and sulphur compounds.

Substance	Retention time (min)
SF_6	0.45
COS	0.68
CCl_2F_2 (F12)	1.00
CS_2	1.95
$CHCl_2F$ (F21)	2.43
CCl_3F (F11)	3.06
CCl_4	6.35

from Tielan Bay, even though there are possible larger anthropogenic sources in this area. From this it can be concluded that the COS background concentration is mainly determined by natural rather than by anthropogenic sources. Apart from direct anthropogenic emission the source of COS in the atmosphere is not clear, atmospheric formation of COS by atmospheric reactions of CS_2 are considered as possible sources (19,20).

The stability of CS_2 seems to be limited by the photolysis and not by OH reaction (7). A lifetime of a few hours was calculated. The CS_2 concentrations measured in the lower troposphere show a large variation from 70-370 pptv[3], another paper[5] gives values of 38 pptv with an extrapolation to $\leqslant$ 30 pptv as a background level for the northern hemisphere. Under the influence of anthropogenic sources, concentrations between 62-339 pptv were reported in this paper. Recent measurements at different heights give 30 pptv as a typical value in the boundary layer and 3 pptv in the free troposphere. The Wuppertal value of 329 obtained by direct sampling of outdoor air is certainly influenced by anthropogenic emissions and falls within the range of data published for measurements at other urban and rural areas. The values for the remote areas Tielan Bay as well as Helgoland do not seem to reflect background levels. Several reasons can be put forward to explain these high values. Either they are influenced by unidentified natural sources or by long range transport of pollutant air masses to these areas. Error arising from the grab sampling proceedure also cannot be completely eliminated especially in the case of COS. The source of carbonyl sulphide in the atmosphere is not fully understood at present. Volcanoes and fumaroles have been suggested as natural sources and various industrial processes, such as blast furnace gas and coke oven gas, as anthropogenic sources (2). Atmospheric oxidation of certain sulphur - carbon containing species such as CS_2 and $S(CH_3)_2$ might be possible sources for atmospheric COS.

Some data about industrial production of COS and CS_2 are available for the Ruhrgebiet area and Cologne (W. Germany). A production of 4.5×10^6 kg/year CS_2 is known in the Rhine-Main area and 7.1×10^6 kg/year

Table II:

The concentrations of SF_6, F12, F11, CCl_4, COS and CS_2 in air samples from different European locations (all values in pptv = 10^{-12} v/v).

location		SF_6	F12	F11	CCl_4	COS	CS_2
Wupper-tal	$\bar{X}$	0.60	744	455	449	368	329
Tielan Bay N=18	$\bar{X}$	0,47	254[*]	161[*]	186[*]	491	201
	σ	0,10	59	78	59	155	73
	V	21	23	48	32	32	36
Helgoland N=7	$\bar{X}$	0,46	299	176	108	100[**]	299
	σ	0,06	15	11	17	-	89
	V	12	5	6	16	-	30
St.Moritz N=8	$\bar{X}$	0,54	345	244	143	100[**]	100[**]
	σ	0,02	56	22	45	-	-
	V	4	16	9	31	-	-

N = Number of samples
σ = Standard deviation, $\bar{X}$ = mean value, V = coefficient of variation
-): The values are under the limit of quantitative determination
+): Only a representative sample from Wuppertal areas.
    For mean values of F12 and F11 over a long period in Wuppertal
    see ref.(9)
*): Mean values from 12 samples
**):due to the uncertainties of the integration only the mean values
    for these measurements are given

Table III:

Coefficients of concordance and correlation for
SF_6, F12, F11 and CCl_4

Location	W	r-F12/F11	r-F12/CCl_4	r-F12/SF_6
Tielan Bay	0.56	0.98±0,01	0.90±0.06	0.45±0.25
Helgoland	0.28	-0.04	0.29±0.41	0.33±0.44
St. Moritz	0.78	0.67±0.22	0.75±0.18	0.59±0.27

W = Kendalls's coefficient of concordance between values of SF_6,
    F12, F11 and CCl_4
r = coefficient of correlation.

Table IV:
The mean values of the ratios between F 12, F 11, SF_6, CCl_4, COS and CS_2

Location	F12/F11	F11/SF_6	F12/SF_6	CCl_4/SF_6	COS/SF_6	CS_2/SF_6
Wuppertal	1.64	758	1240	748	613	548
Tielar Bay	1.66	429	610	392	1063	430
Helgoland	1.71	392	665	240	217	635
St. Moritz	1.41	450	636	260	185	185

CS_2 and 3.1×10^3 kg/year COS in Ruhrgebiet and Cologne area (21). Unfortunately data from other areas of W. Germany are not available. The data of the presently reported work show that it is possible to measure simultaneously halogen and sulphur compounds as trace substances in the atmosphere. Additional data are necessary in order to meaningfully interprete the background concentrations of these substances in the troposphere.

Acknowledgements:
The financial support by the MWF (NRW), UBA, and BMFT is gratefully acknowledged.

References

1) Lovelock,J.E.(1974) in Nature, 247, 625-626

2) Hanst,P.L.,Spiller,L.L.,Watts,D.M.,Spence,J.W.&Miller,M.F.(1975) in J.Air Poll.Control Assoc., 25, 1220-1226

3) Sandalls,F.J.&Penkett,S.A.(1977) in Atm.Environment, 11, 197-199

4) Torres,A.L.,Maroulis,P.J.,Goldberg,A.B.&Bandy,A.R.(1980) in J.Geophys. Res.,85,7357-7360

5) Maroulis,P.J.&Bandy,A.R.(1980) in Geophys.Res.Lett.,7,681-684

6) Bandy,A.R.,Maroulis,P.J.,Shalaby,L.&Wilner,L.A.(1981) to be published

7) Wine,P.H.,Chameides,W.L.&Ravishankara,A.R.(1981) in Geopyhs.Res.Lett., 8,543-546

8) Aneja,V.P.,Overton,J.H.&Cu pitt,L.T.(1979) in Nature,282,493-496

9) Becker,K.H.,Ionescu,A.&Ionescu,M.(1978) in Pageoph,116,567-574

10) Robinson,E.,Rasmussen,R.A.,Krasnec,J.,Pierotti,D.&Jakubovic,M.(1977) in Atm.Environment 11, 215-223

11) Downie,N.M.&Heath,R.W.(1974):"Basic statistical methods", 4th edition, Harper International Edition, chapter 16

12) Pack,D.H.,Lovelock,J.E.,Cotton,G.&Curthoys,C.(1977) in Atm.Environment 11,329-344

13) de Bortoli,M.&Pecchio,E.(1976) in Atm.Environment 10,921-923

14) Krey,P.W.,Lagomarsino,R.J.,Toonkel,L.E.&Schonberg,M.(1976) in Health and Safety Laboratory, HASL 302, I50-I57

15) McCarthy,R.L.,Bower,F.A.&Jesson,J.P.(1977) in Atm.Environment, 11, 491-497

16) Penkett,S.A.,Brice,K.A.,Derwent,R.G.&Eggleton,A.E.J.(1979) in Atm.Environment 13, 1011-1019

17) Cox,R.A.,Eggleton,A.E.J.,Derwent,R.G.,Lovelock,J.E.& Pack,D.H.(1975) in Nature,255,118 -121

18) Cox,R.A.& Sheppard,D.(1980) in Nature 284, 330-331

19) Wine,P.H.,Shah,R.C.& Ravishankara,A.R.(1980) in J.Phys.Chem.84,2499-2503

20) Kurylo,M.J.(1978) in Chem.Phys.Lett. 58, 238-242

21) Emissionskataster Rhein-Main (1978); Luftreinhalteplan Ruhrgebiet Mitte 1980-1984, Düsseldorf 1980

ELECTROSTATIC ACCUMULATION FURNACE FOR ELECTROTHERMAL ATOMIC SPECTROMETRY:

A NEW APPARATUS FOR THE DETERMINATION OF METALS IN THE ATMOSPHERE.

G.Torsi and E.Desimoni
Istituto di Chimica Analitica dell'Università di Bari, Via Amendola I73
70I26 Bari, Italy

Summary

An apparatus is described by which the elemental analysis of atmospheric particulate matter can be easily performed by combining in situ electrostatic accumulation with electrothermal atomic absorption spectrometry. In optimum experimental conditions a collection efficiency as high as 100% was inferred. By considering the urban level of metals in air and the normal detection limits of EAAS it can be deduced that an analysis can be performed by sampling air volumes of the order of I00 ml with sampling times less than 60 s. The potentialities of the new apparatus were tested for Lead and Mercury in the air particulate matter.

1. INTRODUCTION

The determination of metals in air particulate matter is of paramount importance since highly pollutant and poisonous metals, such as for example Hg, Pb, As, Be, Cd etc., can be present in the airborne particles. Usual techniques are generally performed in two steps, sampling and analysis, with procedures which take quite a long time.

Here a new apparatus, named Electrostatic Accumulation Furnace for Electrothermal Atomic Spectrometry (EAFEAS)[1], is described by which elemental analysis of the particulate matter can be performed by combining in situ electrostatic accumulation with EAAS.The novelty of the technique stands in using the electrothermal furnace (a simple graphite tube) both as a particulate matter precipitator and as atomization device: such a combination is characterized by high sensitivity, great selectivity, rapidity and absence of contaminations.

2. THE ELECTROSTATIC FILTER

An electrostatic precipitator is a device which can capture small solid or liquid particles by charging them and making them impinge, by a suitable electric field, against a conducting surface to which they transfer their charge remaining at the same time captured. As shown in figure 1, the charge is produced by applying a high voltage between an axial "active" electrode and an outer "passive" electrode. A negative corona discharge is induced at the active electrode and the ejected electrons are transferred, via ion bombardment or ion diffusion, to the particles of the air flowing through the filter.[2]

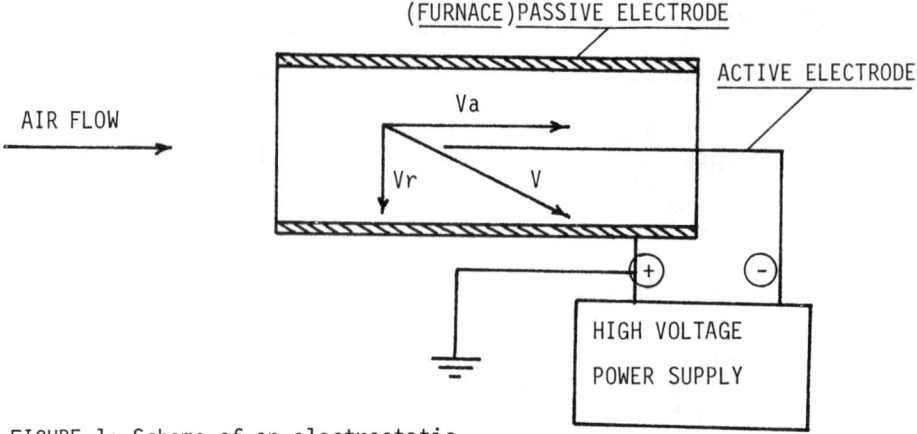

FIGURE 1: Scheme of an electrostatic precipitator.

The particles are subjected to two driving forces: the first is produced by the gas flow and gives origin to the axial velocity, Va, the second is due to the electric field and produces the radial velocity, Vr. The resulting velocity, V, makes the airborne particles to impinge on the collector electrode.

3. THE ELECTROSTATIC ACCUMULATION FURNACE

The same principle is applied to the EAF apparatus. It is mainly composed by three elements: the furnace, A, the furnace container, B, and the spectrometer positioner, C (see figure 2). The active electrode is a tapered tungsten wire, a, which slides along the axis of the atomizer, b, when the furnace is inserted in the container (i.e. during the sampling step). The high voltage is applied between the wire and the furnace. Air to be sampled is sucked from c, crosses the graphite tube and goes out from d. In the course of the atomization step the furnace is transferred onto the spectrometer positioner where is tightly maintained by two brass plates, e. The two screws f serve as electrical leads. A plexiglass cover, h, permits to maintain an inert atmosphere during the atomization.

4. CALIBRATION AND ACCUMULATION EFFICIENCY

The apparatus was tested for Lead and Mercury in air particulate matter since the two metals can be captured by electrostatic precipitators [1,3].

For a proper evaluation of the accumulation efficiency a calibration of the EAFEAS was necessary and to this aim a gas of known, constant load of metal is necessary. The calibration procedure for mercury can be schematized as follows:

- Production of a continuous stream of N_2 with constant and known content of Hg vapours : Hg^{2+} (0.I ppm in HNO_3) + $NaBH_4$ (T=I50°C) under controlled flow of N_2.

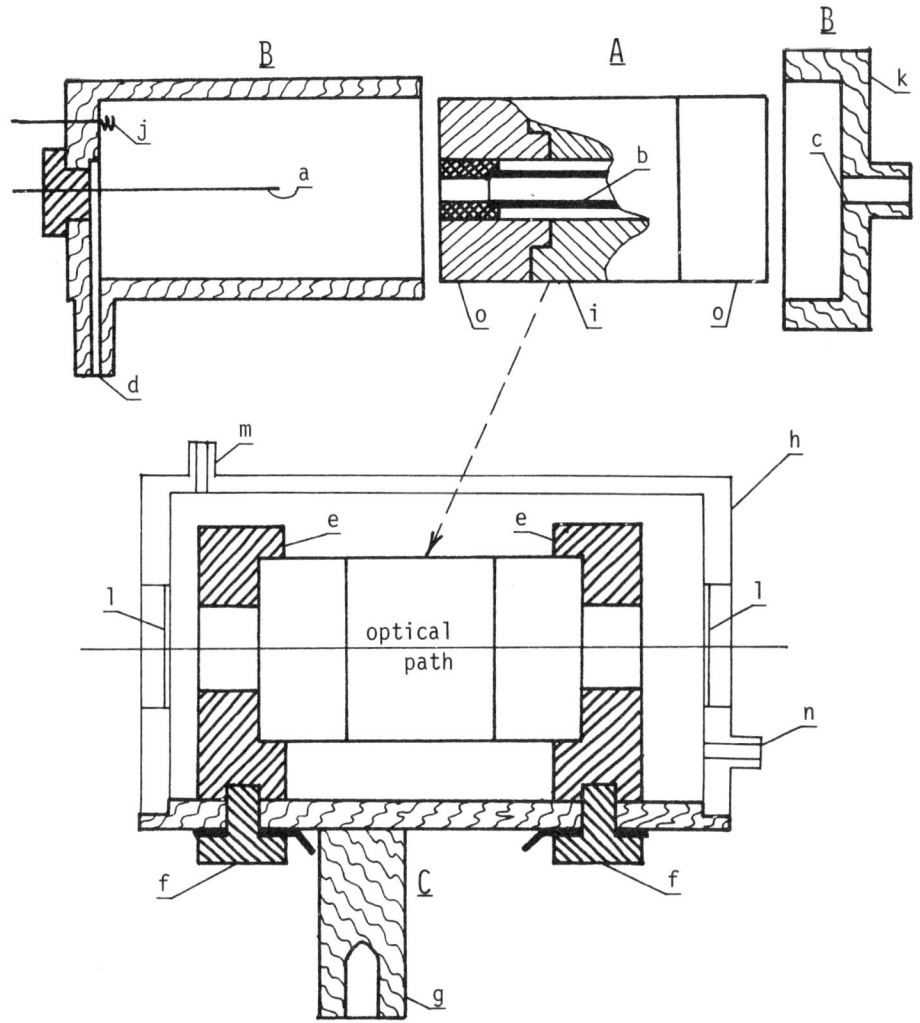

FIGURE 2: A: furnace; B: furnace container; C: spectrometer positioner.
a: tungsten wire; b:graphite tube; c:air inlet; d: air outlet;
e:brass plates; f:electrical lead screws; g: adapter for the
spectrometer flame mount; h:plexiglass cover; i: glass reinfor-
ced teflon ring; j: spring-powered contact; k:screwed cover; l:
quartz windows; m: gas inlet; n:gas outlet; o:brass blocks.

- Calibration of the absorption cell (B in figure 3) with the N_2 stream produced in the first step.
- Calibration of the furnace (D in figure 3) by means of the absorption cell calibrated in the second step.

The third step was performed as follows: a mercury vapour-containing stream (see figure 3) is produced by passing N_2 above a small Hg drop contained in a thermostated tube. The Hg-enriched flow is passed through the calibrated cell B and, after dilution 1:25 with purified air, is used in the accumulation step.

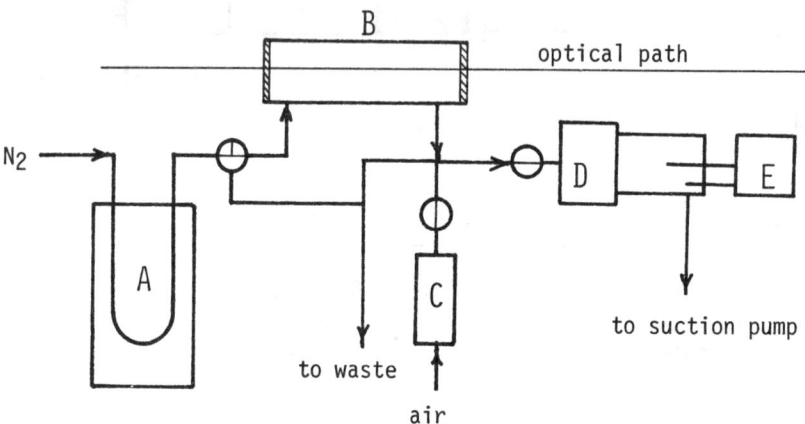

FIGURE 3: A: thermostated reaction tube; B: absorption cell; C: trap with Zn pellets; D: EAFEAS; E: high voltage power supply.

A gas with known, constant load of lead cannot be produced. Then it was decided to use the air stored in a platic bag connected to the air inlet of the EAFEAS. As known the particles load in the bag decays[2] with time: by monitoring the peak height for lead at regular intervals of time the relative particulate content can be calculated at any time and measures in different experimental conditions can be normalized to it.

In theory the efficiency of an electrostatic precipitator can be increased by increasing the applied potential or by decreasing the flow rate (that is by increasing the radial velocity and decreasing the axial velocity). In figure 4 the absorbance, normalized to the maximum, is reported as a function of the applied potential, at constant flow rate, for lead and mercury. The current flowing in the circuit is also shown. As can be seen curve B is shifted towards higher potentials than curve A. The coincidence between the onset of a measurable current and an appreciable variation in the capture efficiency (i.e. in the peak height) is a clear indication that the charging of the particles is the main factor governing the capture efficiency.

In figure 5 the absorbance, normalized to a maximum, is reported as a function of the air flow, at constant applied potential.

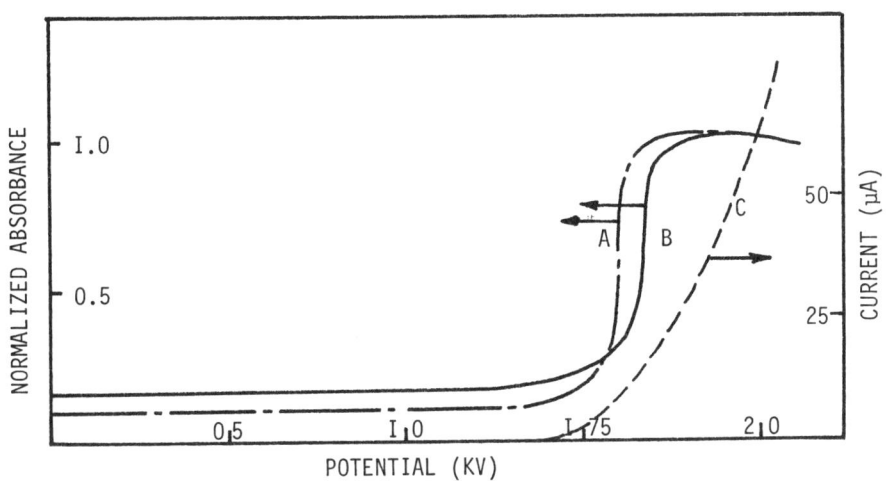

FIGURE 4: Influence of the applied potential on the absorption signal of
lead (A) and mercury (B) (at constant flow rate) and on the current (C). Curve A: flow rate=20cm/s; Curve B: 6cm/s.
The arrows indicate the ordinate to which the curves must be
referred.

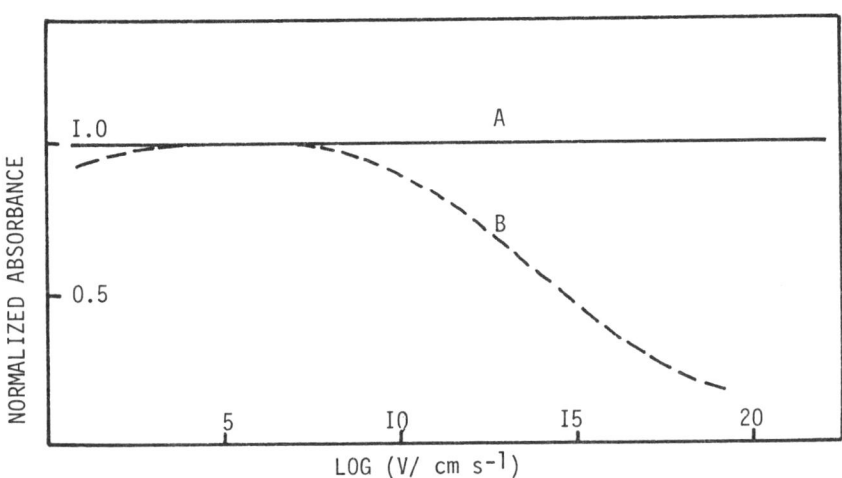

FIGURE 5: Influence of flow rate, at constant applied potential, on the absorption signal. A: Pb particles; applied potential I.8 KV
B: Hg vapours; applied potential I.95 KV.

The difference between the two curves can be explained, according to theory, on assuming that the ion diffusion mechanism prevails for the collection of mercury atoms and taking into account that the capture efficiency is an increasing function of the particle dimensions[2].

These results show that the EAFEAS can be fully tested with mercury vapours since an efficiency decrease is observable in the air flow range tested.

On using the optimum experimental conditions (see figures 4 and 5) a direct evaluation of the capture efficiency of Hg can be made: to this aim air is passed over the mercury drop as in figure 3 but the EAFEAS is placed before the absorption cell. The results are shown in figure 6. Even in the presence of very high Hg vapour concentrations (about 1 mg/m^3) the capture efficiency is more than 90%. On the other hand it is known that electrostatic filters can easily operate with efficiencies as high as 99% for collecting particles ranging from 0.I to I00 microns[4]. In particular the efficiency for mercury atoms capture likely represents the lower limit for particles capture[2] and then the above result furnishes an indirect evaluation also for the lead capture efficiency. This suggests the feasibility to use aqueous standard solutions to calibrate the EAFEAS.

From the above results it can be concluded that the EAFEAS permits a rather fast and precise measure of very low levels of metals in the atmosphere. The detection limits for Hg and Pb in air are the same which characterize the EAAS technique, i.e. 10^{-12}g. The potentialities of the technique can be easily evaluated by monitoring the fluctuations of the metal content in the authors laboratory over a day.

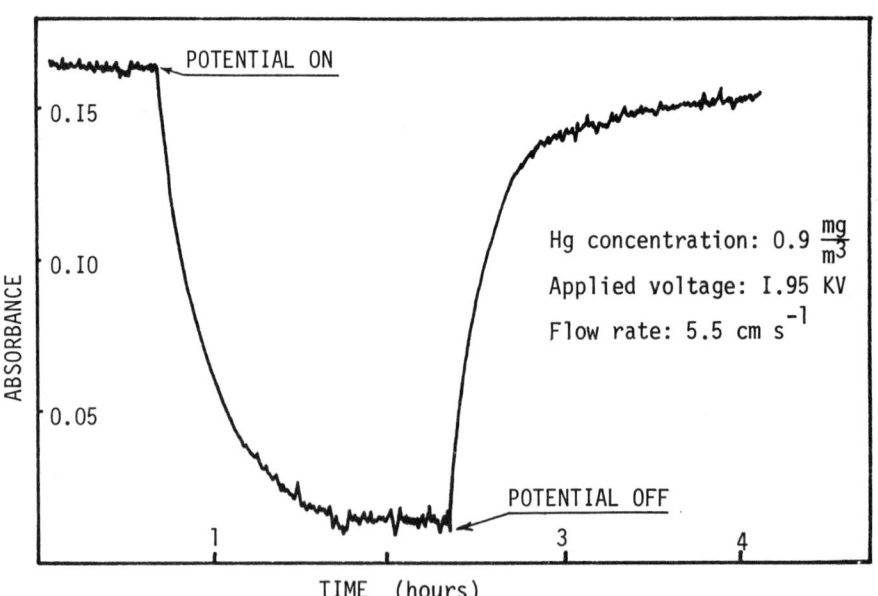

FIGURE 6: Plot Abs vs. time in the absorption cell when the potential in the EAFEAS is swithched on and off.

REFERENCES

I) G.Torsi,E.Desimoni,F.Palmisano,L.Sabbatini, Anal.Chem.,(I98I) in press.
2) H.E.Rose,A.J.Wood,"An introduction to electrostatic precipitators in theory and practice", Constable and Company LtD, London (I956).
3) O.M.G.Newman,D.J.Palmer, Nature, 526, 275(I978)
4) W.Licht,"Air Pollution Control and Engineering", Marcel Dekker Inc., New York (I980) Chapter 7.

APPLICATION OF THE TCM DENUDER FOR SO$_2$ COLLECTION

Ella E. Lewin[*] and D. Klockow
Universität Dortmund
Abteilung Chemie
D-4600 Dortmund 50
West Germany

Summary

Sulphur dioxide can be collected effectively and easily in an absorption tube covered internally with a mixture of tetrachloromercurate and malein buffer. The method was developed for cases where the gas must be removed before collecting aerosol.

Collected samples can be analyzed by a number of methods: For example, those of isotope dilution, Thorin, and West & Gaeke have been tried and found suitable. The performance of the absorption tube under different working conditions is reported.

1. INTRODUCTION

Recently, more and more attention has been paid to possible changes that may occur in the composition both of a gas and particulate phase during collection. Some gaseous compounds can be absorbed on a collected dust while volatile compounds can be released from the particles. These phenomena lead to changes in the character of the sample. Thus, Stevens and Dzubay (1) showed that when the gaseous ammonia has been removed prior to the dust, the collected aerosol showed a much lower pH. Ferm (2) constructed an absorption tube for the effective collection of ammonia and with its help showed that some gaseous ammonia can be released from the dust as well. Calculations of Brosset (3) indicate that the same can be true of other volatile compounds.

Some evidence exists (e.g. 4,5) that sulphur dioxide can react either with the filter material itself or the dust collected on it, or that some artifact sulphate is produced on a back-up filter. The presence of SO$_2$ also leads to incorrect estimates of nitrate content of aerosol (6).

In connection with the intended investigations of the possible interaction between gas and dust during collection, we felt the need of a method for removing SO$_2$ selectively. For measurements of sulphate aerosol in the presence of large amounts of SO$_2$, Durham et al. (7) used a tube coated internally with lead dioxide to remove the gas. No description of the tube performance at different sampling conditions is given (except that the sampling was apparently effective under the conditions of the experiment). Moreover, if the collected SO$_2$ samples were to be analyzed, a special uncommonly used method of analysis should be applied. Therefore, the performance of an absorption tube coated internally with tetrachloromercurate (TCM) solution was investigated for different sampling conditions. The applicability of commonly used methods of analysis was also established. The present paper describes this method of sampling and determining sulphur dioxide.

---

[*] On leave from: National Agency of Environmental Protection, Air Pollution Laboratory, Risø National Laboratory, DK-4000 Roskilde, Denmark.

2. DESCRIPTION OF THE METHOD

2.1. Separation and collection

The diffusion coefficient is strongly dependent on the molecular diameter. This fact is used for separating gas and aerosol particles in so-called absorption tubes (denuders). During laminar passage through a cylindrical tube both gas and particles diffuse to walls where they can be absorbed. Because of the much higher diffusion coefficient of a gas, depletion in its concentration is orders of magnitude higher than in the case of particles. If the walls are a perfect sink for the gas, its concentration is given by the Gormley-Kennedy equation (8):

$$\bar{c}/c_0 = 0.819 \exp(-14.6272\Delta) + 0.0976 \exp(-89.22\Delta) +$$
$$+ 0.01896 \exp(-212\Delta) \tag{1}$$

$$\Delta = (\pi D \ell)/4F$$

where c_0 is the concentration entering the tube, $\bar{c}$ = the average concentration leaving, ℓ = tube length, D = diffusion coefficient of the gas, and F = volume flow of the gas.

The diffusion coefficient for SO_2 at temperatures 0°, 20°, and 37° C was calculated by Fish and Durham (9). For the sampling flow of 40 ℓ/h and internal tube diameter of 0.6 cm (values chosen for the present investigation), the length ensuring the effective collection ($\bar{c}/c_0 = 0.01$) is 32 cm.

As mentioned above, the equation is valid for laminar flow conditions. Thus, the Reynold's number must be below 2000 (in our case it is 157). Also, to remove all the whirls caused by impaction, an uncovered part made of Teflon is mounted on the entrance of the tube. The length of this part can be determined from:

$$L = 0.05 \times d \times Re \tag{2}$$

where L is the distance necessary for the gas to be transported to create laminar flow, d = tube diameter, and Re = Reynold's number. In our case L = 4.7 cm.

The sampling arrangement, consisting of a Teflon rod, absorption tube, and filter holder with particulate filter, is shown in Fig. 1.

2.2. Preparation of tubes and analysis of samples

The tube is covered internally with 0.4 ml of 0.1 M TCM, 0.1 M malein buffer in 1:1 methanol/water solution. The mixture is injected into the tube and the solute evaporated with a constant flow of dry, clean nitrogen blown through the tube, while it is held horizontally and slowly rotated about its axis. Afterwards, the tube is closed at both ends with parafilm.

The presence of the malein buffer (10) in the absorbing mixture is necessary to neutralize hydrochloric acid produced during the reaction between TCM and SO_2. Fig. 2 shows the capacity of the 'pure' TCM and the TCM/buffer solution. The theoretical capacity of a tube prepared in the manner described is 1200 μg SO_2. After collection, the contents of the tube are removed by dissolving the cover in a known amount of water. The samples were routinely analyzed by isotope dilution analysis (IDA) (11), in which a small and known amount of radioactive labelled $BaSO_4$ is added

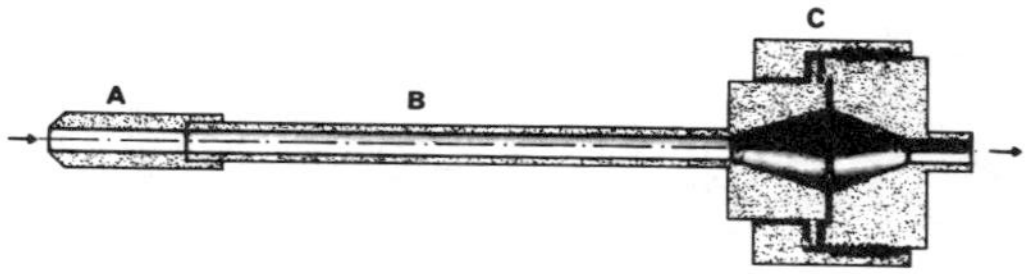

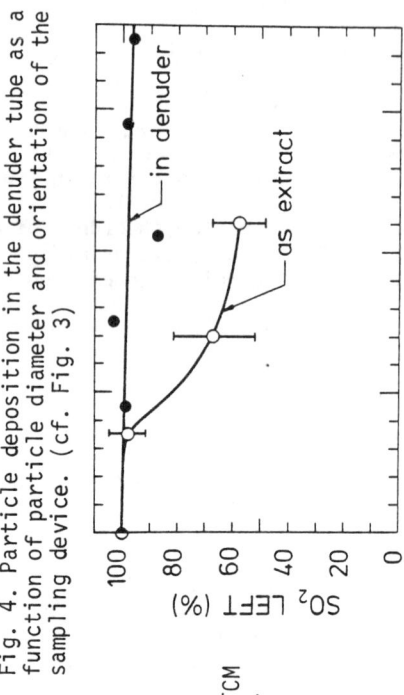

Fig. 1. Sampling device consisting of : A: Teflon rod; B: Denuder tube; C: Filter holder with filter for particle collection.

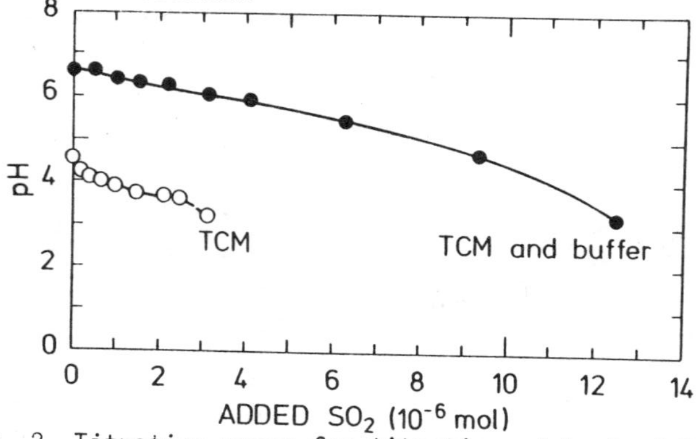

Fig. 2. Titration curve for titration of 1 ml of 0.1 M TCM or 0.1 M TCM/0.1 M malein buffer with the standard solution of SO_2

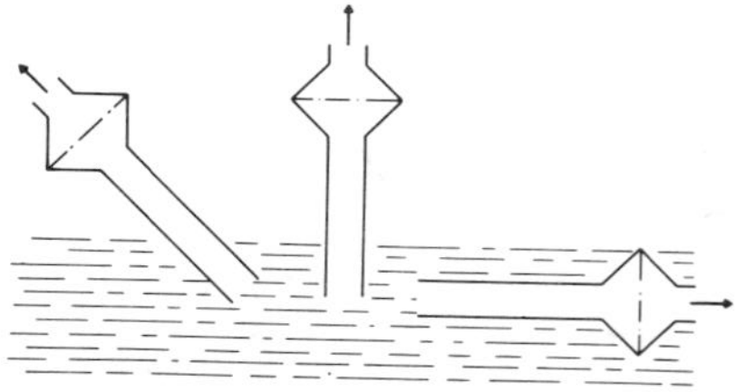

Fig. 3. Three orientations of the sampling device for investigation of the particle deposition in the denuder tube

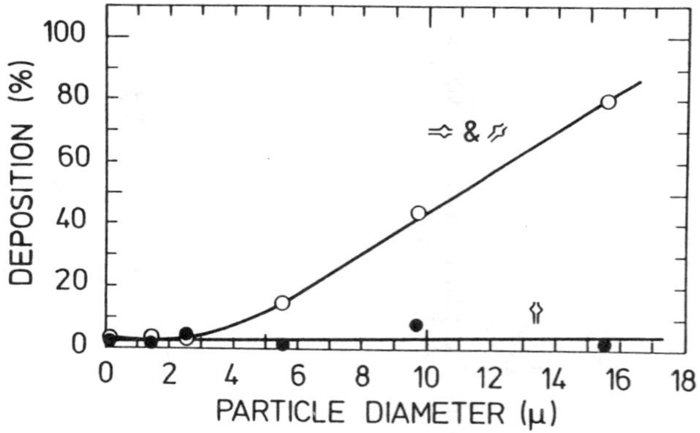

Fig. 4. Particle deposition in the denuder tube as a function of particle diameter and orientation of the sampling device. (cf. Fig. 3)

Fig. 5. Stability of the collected SO_2 stored in a sealed denuder tube or as an extract

to the sample. After acidification, barium sulphate precipitates, and the amount of radioactivity remaining in the supernatant liquid is proportional to the amount of sulphate from the sample. A series of calibration solutions is run together with each batch of samples.

To compensate for the salt effect from the TCM present in the tube extract, the same amount of absorbing agent should be added to the calibration solutions.

Also, other methods of analysis, e.g. Thorin (12) and West & Gaeke (13, 14) were applied and compared with IDA. As these methods are also based on the comparison between the sample and calibration solution, all samples should contain the same amount of TCM. Results obtained by all three methods were in very good agreement.

3. INTERFERENCES FROM PARTICLE DEPOSITION

3.1. Theoretical considerations

Deposition of particles from moving air to the surface originates from sedimentation, molecular and turbulent diffusion, impingement, and electrostatic forces. Deposition caused by sedimentation and molecular diffusion will be treated below in more detail. Other effects noted above are considered to be of minor importance: Choice of the proper material can minimize the effect of electrical forces, and turbulent deposition is avoided when the Reynold's number is below 2000.

The rate of deposition to the horizontal surface due to gravitational forces (sedimentation) is expressed as (15):

$$n/n_0 = (2/\pi) \left(2 \mu(1-\mu^{2/3})^{1/2} + \arcsin \mu^{1/3} - \mu^{1/3}(1-\mu^{2/3})^{1/2} \right) \quad (3)$$

$$\mu = 3 v_s \ell/4d \, \bar{U}$$

where n_0 is the initial number of particles, n = the number of particles deposited, ℓ = the length of the tube, d = its diameter, $\bar{U}$ = the mean velocity of flow, and v_s = the velocity of fall. Values of v_s for spheres of unit density in air at one atmosphere and $20°C$ are given in (16).

The degree of molecular diffusion depends on the diffusion coefficient D and is given by:

$$n/n_0 = 2.56 \, \Delta^{2/3} - 1.2 \, \Delta - 1.77 \, \Delta^{4/3} \quad (4)$$

(all symbols as in (1)).

Calculations made for particles with diameter 0.1 - 15.0 μ showed that their loss due to diffusion is negligible (about 0.1% for the smallest particles). For particles smaller than 1 μ, deposition due to sedimentation is unimportant, but increases strongly with particle diameter.

3.2. Experimental determination of particle deposition

The deposition of particles with diameters in the range 0.1 - 15 μ was investigated as function of particle diameter and the position of the tube in relation to the air stream. The sampling system was as in Fig. 1, the back-up filter was of the Nuclepore polycarbonate type with pore size 0.2 μ. All three parts of the sampling arrangement were analyzed for the amount of material deposited. Three orientations of the absorbing tube were investigated: horizontal, vertical, and approximately 45° (Fig. 3).

Two systems were used for generating a monodispersive aerosol: Particles with diameter smaller than 1 μ were produced in an atomizer (17) and larger particles were obtained from the Berglund & Liu generator (18).

The atomizer was filled with the solution of Na_2SO_4; the size of the particles was determined from the scanning electron microscope measurements. The amounts of sulphate deposited were found by IDA. With the procedure adopted the minimum detectable amount of sulphur in the denuder is 0.2 μg (amounts found during analysis were between 0 and 1.2 μg S/tube). For the generating aerosol in the Berglund & Liu generator dioctylphtalate and uranine solution in propanol-2 was applied (19). The particle size was determined by the concentration of the solution and the operating conditions of the instrument. The amount of aerosol deposited was determined from the fluorescence measurements. The results of the deposition tests are shown in Fig. 4. Every value is a mean of three determinations. The values found are in very good agreement with theoretical predictions. Submicron particles produced in the atomizer were not electrically neutralized before entering the tube and a part of the measured deposition could be due to electrical interactions. However, the amount of total mass deposited is negligible. The dependence of the deposition upon tube orientation for particles larger than 2 μ shows the dominant role of sedimentation in this case. Although airborne sulphate is connected mostly with submicron particles, deposition of even small amounts of large particles can be important as they will contribute strongly to the total mass collected. In any case it is recommended that the denuder be mounted vertically during the collection.

4. COLLECTION EFFICIENCY OF SO_2

The collection efficiency was investigated during laboratory and field experiments using a set-up similar to the one shown in Fig. 1. After the particulate filter, a KOH-impregnated filter was mounted to collect any unabsorbed SO_2. The collection efficiency was calculated as:

$$\varepsilon = 1 - m_2/m_1 \qquad (5)$$

where m_1 is the amount collected in the denuder and m_2 the amount found on the KOH-filter.

Table 1 shows the results of some experiments conducted under field conditions. Sampling time varied between 3 and 24 hours, the relative humidity was never lower than 50%, and temperature was between -5^0 and $+10^0C$. The sampling took place in a polluted industrial atmosphere. SO_2 concentrations calculated from the amount collected were between 5 and 80 ppb. It is seen that in most cases collection efficiency is above 90%; it can be improved further by applying longer tubes. In the two cases where the efficiency dropped below 90%, the very humid weather caused the cover to become deliquescent and droplets were observed in the tube. The surface available for absorption is then, of course, much smaller.

The reaction between TCM and SO_2 involves water; therefore, it is necessary to investigate the influence of relative humidity on the absorption efficiency. This was done in a series of laboratory experiments in which a permeation tube was used as an SO_2-source. As a check for the uncollected gas, a flame photometric detector, placed after the tube, was used. To obtain the desired humidity, dry- and water-vapour-saturated air were mixed in a proper ratio. Full absorption in the tube was achieved for relative humidities higher than 20%.

μg S collected			Efficiency %
Denuder	Back-up filter	Sum	
4.23	0.25	4.48	94
12.30	1.02	13.32	92
21.46	2.48	23.94	88
5.18	0.36	5.54	93
13.84	0.56	14.40	96
27.44	2.76	30.20	90
28.62	1.60	30.22	94
21.74	4.72	26.46	81[*]
26.62	4.04	30.66	85[*]
26.46	1.47	27.93	94

Table I. Collection efficiency of the TCM-denuder under field conditions. Back-up filter: KOH-impregnated filter.

[*] Humidity of air close to 100%, droplets were observed in the tube.

Sulphur dioxide μg S/m^3		Sulphate μg S/m^3	
Denuder	Two-filter	Denuder	Two-filter
30.50	30.40	10.77	9.31
31.52	29.26	11.77	12.80
50.76	52.05	9.84	10.04
19.66	16.66	0.99	1.36
14.58	14.86	2.34	2.15
6.25	9.50	2.62	1.47
6.11	10.83	1.00	1.62

Table II. Comparison between collection of SO_2 in the TCM-denuder and on a KOH-impregnated filter ("two-filter"). Sulphate collected on Nuclepore polycarbonate filter.

5. STABILITY OF THE COLLECTED SAMPLES

The stability of the collected SO_2 was investigated for samples obtained from the ambient air. The stability of the samples stored in closed tubes as well as that of tube extracts was determined. In the case of extract samples, the content of the exposed tube was dissolved in water shortly after the collection ended; the extracts were then stored in closed plactic bottles at room temperature. Tube samples were stored under identical conditions. The samples were analyzed by the West & Gaeke method and by IDA to determine the amount of unconverted SO_2 and total sulphur content, respectively. Fig 5 shows the results of the investigation in terms of unoxidized SO_2 found in the sample after the indicated time interval. The total amount of sulphur in the sample was unchanged in all cases and is therefore omitted from the figure.

It is evident that oxidation proceeds faster in the solution than in the closed tubes, where only a very small fraction (less than 5%) is oxidized after one month. If the method of analysis is applied where the sulphur is determined as sulphate, the velocity of oxidation plays no role.

6. COMPARISON BETWEEN COLLECTION IN DENUDERS AND ON KOH-FILTERS

Several experiments were conducted in which ambient air was collected parallel in denuders followed by a particulate filter (Nuclepore polycarbonate, 0.4 μ) and on particulate- (Nuclepore) and KOH-impregnated cellulose filters (Whatman 40) placed in series. Collection took place in an industrial atmosphere and during winter. Table 2 gives the results of the comparison, recalculated for the amount of S/m^3. The experiment should be repeated under different weather conditions and in different environments, but the preliminary results show that the methods are in good agreement, at least for these sampling conditions.

Acknowledgment: The research has been conducted in the Department of Inorganic Chemistry at the University of Dortmund. It was financed by the Heinrich Hertz Foundation and the Danish National Council for Scientific and Technical Research.

Literature

1. STEVENS, R.K., DZUBAY, T.G., RUSSWURM, G., and RICKEL, D. (1978). Sampling and analysis of atmospheric sulfates and related species. Atm. Environ. 12, 55-68 pp.
2. FERM, M. (1979). Method for determination of atmospheric ammonia. Atm. Environ. 13, 1385-1393 pp.
3. BROSSET, C. (1979). Possible changes in aerosol composition due to departure from equilibrium conditions during sampling. Proceedings of ACS/CSJ Chemical Congress, Honolulu, Hawaii, April 1-6.
4. PIERSON, W.R., HAMMERLE, R.H., and BRACHACZEK, W.E. (1976). Sulfate formed by interaction of sulfur dioxide with filters and aerosol deposits. Anal. Chem. 48, 1808-1811 pp.
5. PIERSON, W.R., BRACHACZEK, W.E., KORNISKI, T.J., TRUEX, T.J., and BUTLER, J.W. (1980). Artifact formation of sulfate, nitrate, and hydrogen ion on back-up filters: Allegheny mountain experiment. J. Air Pollut. Control Association 30, 30-34 pp.
6. HARKER, A.B., RICHARDS, L.W., and CLARK, W.E. (1977). The effect of atmospheric SO_2 photochemistry upon observed nitrate concentrations in aerosols. Atm. Environ. 11, 87-91 pp.

7. DURHAM, J.L., WILSON, W.E., and BAILEY, E.B. (1978). Application of an SO_2-denuder for continuous measurement of sulfur in submicrometric aerosols. Atm. Environ. <u>12</u>, 883-886 pp.
8. GORMLEY, P.G. and KENNEDY, M. (1949). Diffusion from a stream flowing through a cylindrical tube. Proceedings R. Irish Ac. <u>52</u>, Sec. A. 163-169 pp.
9. FISH, B.R. and DURHAM, J.L. (1971). Diffusion coefficient of sulfur dioxide in air. Environ. Lett. <u>2</u>, 13-21 pp.
10. Temple, J.W. (1929). Sodium maleat - a buffer for the pH region of 5.2 to 6.8. J. Am. Chem. Soc. <u>51</u>, 1754-1755 pp.
11. KLOCKOW, D., DENZINGER, H., and RONICKE, G. (1974). Anwendung der sub-stöchiometrischen Isotopenverdünnungsanalyse auf die Bestimmung von atmosfärischen Sulfat und Chlorid in "Background" Luft. Chem. Ing. Techn. <u>46</u>, 831 p.
12. PERSSON, G.A. (1966). Automatic colorimetric determination of low con-centration of sulphate for measuring sulphur dioxide in ambient air. J. Air Wat. Pollut. <u>10</u>, 845-852 pp.
13. WEST, P.W. and GAEKE, G.C. (1956). Fixation of sulfur dioxide as disul-fitomercurate (11) and subsequent colorimetric estimation. Anal. Chem. <u>28</u>, 1816-1819 pp.
14. SCARINGELLI, B.P., SALTZMAN, B.E., and FREY, S.A. (1967). Spectrophoto-metric determination of atmospheric sulfur dioxide. Anal. Chem. <u>39</u>, 1709-1719 pp.
15. FUCHS, N.A. (1964). The mechanics of aerosols, Pergamon Press, Oxford, 181-250 pp.
16. DAVIES, C.N. (1966). Aerosol science, Academic Press, N.Y., 393-445 pp.
17. NIESSNER, R. (1981). Ph.D. Thesis, University of Dortmund, Department of Inorganic Chemistry.
18. BERGLUND, R.N. and LIU, B.Y.H. (1973). Generation of monodispersive aerosol standards. Environ. Sci. and Technol. <u>7</u>, 147-153 pp.
19. RAO, A.K. (1975). An experimental study of inertial impactors. PhD. Thesis, University of Minnesota, Particle Technology Laboratory, publ. No. 269.

ANALYTICAL AND SAMPLING TECHNIQUES IN THE DETERMINATION OF NITRATE IN THE ATMOSPHERIC AEROSOL

D. WANGE, G. HELAS and P. WARNECK, Institut für Anorganische Chemie und Analytische Chemie, Universität Mainz and Max-Planck-Institut für Chemie (Otto-Hahn-Institut), Mainz, FRG

Summary

A procedure for the determination of the nitrate content of the atmospheric aerosol in the presence of gas phase (artifact) nitrate has been tested. Two high volume samplers were used in parallel, each unit containing three quartz fibre filters in series, in order to differentiate between aerosol nitrate on the first filter and artifact nitrate adsorbed on the first and subsequent filters. Nitrate was eluted from the filters with water and was analysed by both the brucine and the Cd/Hg reduction methods. Both are adequate, but the latter is more sensitive and thus preferable. The average concentration of aerosol nitrate found in Deuselbach, a rural background region of Germany, was 431 ± 37 ng NO_3-N/m^3. Virtues and drawbacks of the procedures are discussed.

1.   INTRODUCTION

The determination of the nitrate content of the atmospheric aerosol requires the collection of sufficient material, which usually is accomplished by deposition of filter mats with the aid of high volume samplers. In recent years it has been recognized (1-6) that gas phase nitrogen compounds, for example HNO_3, PAN, and to some extent also NO_2, become partially adsorbed on the filters and give rise to a non-aerosol contribution to the total nitrate found. The amount of excess or artifact nitrate seems to be most pronounced for glas fibres, and much less for quartz and teflon fibres.

In this laboratory, a similar effect was previously encountered with formaldehyde (7, 8). In that case, adsorption from the gase phase quickly reached equilibrium, and it was possible to differentiate between the formaldehyde of the aerosol and that due to the gas phase in two ways. One method was the use of back-up filters which provided a measure of the amount of material deposited by adsorption from the gas phase after the aerosol component had been removed by the first filter. The other method consisted of a parallel collection of aerosol by simultaneously sampling sufficiently different volumes of air with different flow rates. A plot of the amount of material deposited on the first filter of each sampler versus the air sampling volume yielded a straight line with a positive intercept on the ordinate which is equivalent to the amount of gaseous formaldehyde adsorbed on each filter. The slope of the line provided the concentration of aerosol formaldehyde.

Under certain conditions, these techniques may also be applicable

to the differentiation of aerosol nitrate and artifact nitrate. We have performed a number of experiments in this direction and describe here the procedures used and some of the results obtained.

2. EXPERIMENTAL PROCEDURES

(a) Sampling

The majority of samples were taken at Deuselbach, a regional background station of the German Environmental Agency (UBA). A few samples
were obtained on the grounds of the institute at Mainz. Aerosol was collected by means of high volume samplers using tissue quartz filters of 100 mm
diameter. Pallflex-2500 QAS and -2500 QAST filters were used, the latter
differing from the former by having been heat-treated. Filter holders were
similar to those used previously (7, 8). Each holder contained three filters in series with a spacing of 30 mm. Two samplers were operated in
parallel.

(b) Extraction of nitrate from the filters

Within a few hours after sampling, a filter was placed onto a fritted
disc, covered with 25 ml of distilled water, and soaked for a few minutes.
The resulting solution was sucked into a collection flask by means of an
aspirator. The procedure was repeated three times. Between the second and
third extraction, the filter was ultrasonically rinsed for five minutes.
The ultrasonic treatment released up to 5 µg NO_3-N which would not have
been extracted otherwise. The combined extracts were made up to 100.0 ml
of solution. A drop of chloroform was added to prevent decomposition of
nitrate by bacteria while the solution was stored for later analysis.

(c) Analysis

Two wet-chemical analytical techniques were employed: the brucine
method (9-11), and the reduction to nitrite with amalgamated cadmium
followed by a Gries reaction (12-14). In both cases, the reaction products
were determined spectrophotometrically. Each sample was analysed once with
the brucine method, using 5 ml of the sample solution, and twice with the
reduction method, using 46 ml each time. The brucine method measures the
extinction at 420 nm of the product resulting from the oxidation of brucine by nitrate in a 10 molar sulfuric acid medium. A 40 mm cuvette was
used. The procedure employed was essentially the same as procedure "B"
described by Fädrus and Malý (11). In order to obtain reproducible results
it is absolutely essential to adhere rigorously to the quantitative details
chosen because the method is sensitive to changes in the temperature, the
H_2SO_4 concentration and the time of color development. Thirty analyses can
be performed within 8 hours.

The reduction of nitrate to nitrite was achieved by sucking the
sample solution, buffered at pH = 8.0 - 8.5 with NH_4Cl/HCl, through a
reductor column filled with amalgamated cadmium granulate. A peristaltic
pump was used to obtain reproducible flow rates (1.2 ml/min). The resulting nitrite was reacted with sulfanilamide and N-(1-naphtylethylene)di-
ammonium chloride to an azo-dye the extinction of which was measured at
540 nm. 20 or 5 mm cuvettes were used depending on the concentration. Six
reductors were employed so that six samples could be treated per hour.
Each reductor had to be calibrated separately. The details of both procedures are described by Wange (15).

3. RESULTS

(a) Analytical procedures

Linear calibration curves were obtained for concentrations up to about

1 mg NO_3-N/filter, with correlation coefficients of 0.9996 and 0.997 for the reduction and the brucine methods, respectively. The deviation among the results for the three analyses of each sample with both methods was far greater, however, than the probable error calculated from the calibration lines. If the filters contained between 5 and 100 µg NO_3-N, the deviation from the mean was about 4% for both methods; if they contained less than 5 µg NO_3-N, the results of the reduction method were in agreement within 10-15%, whereas the brucine method was unreliable.

For the procedures described, sensitivities can be expressed in extinction per concentration (ext/mg NO_3-N liter^{-1}). The brucine method has a sensitivity of 0.017, the reduction method of 0.285. With the brucine method the sample volume is limited by the low pH required and fixed by the sulphuric acid concentration. The advantage of a small sample volume is lost due to the lower sensitivity. Using the same small sample volume of 5 ml with the reduction method would require a dilution by almost a factor of ten. Even then the sensitivity would still be somewhat better.

Blank value for filters not exposed to an air flow but otherwise handled in the same manner as filters on which aerosol was deposited were 1.0 ± 0.2 µg NO_3-N for QAST filters, and 3.2 ± 0.1 µg NO_3-N for QAS filters. We take the blank value of QAST filters to define the detection limit (10 µg NO_3-N/liter) of the reduction method.

(b) Field measurements

To evaluate the reproducibility of the entire technique we made a number of simultaneous measurements with two filter heads operated in parallel with the same flow rate so as to sample equal volumes of the same air mass. Results are shown in Table I for the first and the second filters. The third filters gave almost the same values as the second filters, provided adsorption equilibrium was reached. This required air sampling volumes of at least 120 m³ as will be shown below. The condition was met for the first two runs in Table I, but not for the last two. When one takes this into account, Table I shows that an adequate reproducibility is obtained for the combination of sampling and analysis. For the first filters, the precision would have been better had we corrected for the slight differences in the sampling volumes. The greater deviation of the last run is probably due to greater air pollution in Mainz. As can be seen from Table I the two filter types gave the same results. They differ only in their blank values.

Table I: Pairs of values for the amounts of NO_3-N obtained from parallel runs under identical conditions. Each filter was analysed once with the brucine method and twice with the reduction method and the results were averaged:

Location (FRG)	Volume/m³		Mass of NO_3-N/µg 1st filter		2nd filter		Deviation from the mean / % 1st	2nd
Deuselbach	150	142	62.4	58.9	54.1	57.2	2.8	2.8
	259	241	77.2[b]	72.4	43.1[b]	48.1	3.2	5.5
	96	95	50.2[b]	47.8	2.5[b]	1.5	2.4	a
Mainz	109	106	25.5	29.4	18.2	7.9	7.3	a

a   sampling volume was insufficient to achieve adsorption equilibrium;
b   QAS filters, the others were QAST filters.

Fig.I shows results of measurements performed in Deuselbach on July 31, 1980, a day with stable meteorological conditions and stationary air masses.

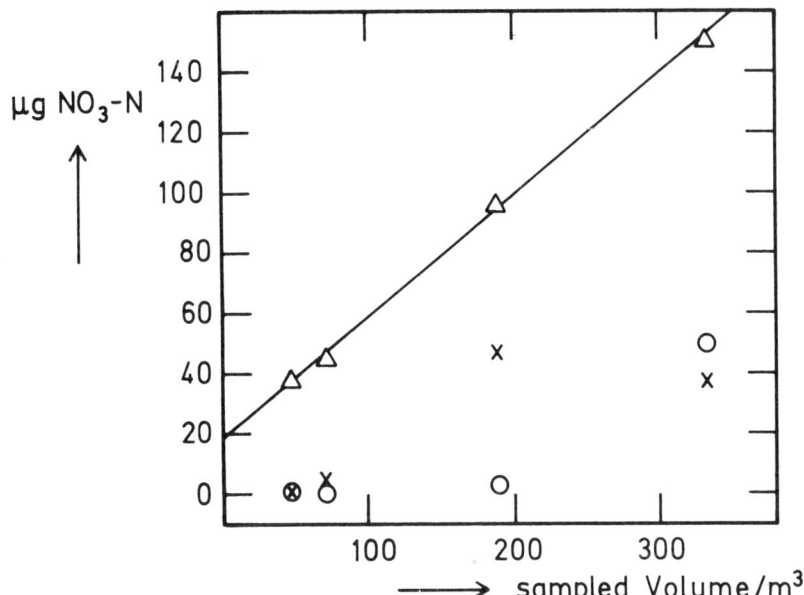

Fig. I : Mass of NO₃-N versus the air sampling volume
Δ: first filter   X : second filter   O: third filter
Data were obtained in Deuselbach on a single day
(July 31, 1980) sampling with equal flow rates

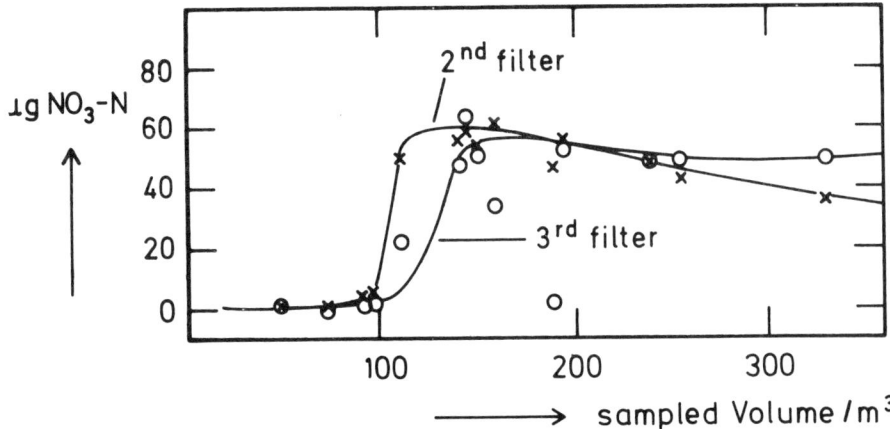

Fig.Ⅱ : Mass of NO₃-N on back-up filters versus sampling volume
X : second filter   O : third filter
Data were obtained in Deuselbach during a 2½ week
period in the summer 1980 with approximately equal
flow rates.

Flow rates were kept constant at about 22.2 m³/h. Four samples were obtained and the mass of nitrate found on each filter is plotted versus the air
sampling volume. Tha data points for the first filters fall on a straight
line as expected. A positive intercept on the ordinate indicate the extent
of artifact nitrate. The slope of the straight line, 402 $^\pm$ 12 ng NO_3-N/m³,
should correspond to the aerosol nitrate concentration on that day. Since
the aerosol is deposited almost entirely on the first filter, the nitrate
found on the subsequent filters must arise from adsorption from the gas
phase. It can be seen from the data in Figure I that a minimum volume of
air must pass through the three filters before nitrate will be registered
on the back-up filters. Figure II shows nitrate obtained on the two back-
up filters in 14 runs with equal flow rates performed on different days
during a two and a half week period in the summer of 1980. The data are
again plotted versus the air sampling volume. Saturation corresponding to
an adsorption equilibrium is reached on the second filter after 110 m³ of
air have passed through, and on the third filter after 140 m³. Initially,
the gaseous nitrate is retained almost completely on the first filter.
Figure I and other data show that the amount of nitrate on the two back-up
filters in the state of equilibrium is more than twice as high as the
intercept with the ordinate obtained by extrapolation of the straight line
resulting from the first filter measurements. We have not yet investigated
this behaviour in detail.

From a linear regression analysis of all first filter data obtained
in Deuselbach as a function of sampling volume we determined the concentration of aerosol nitrate to be 431 $^\pm$ 37 ng NO_3-N/m³ during the observation period. The value has the same order of magnitude as the results
of other authors for air in regions not directly influenced by urban
plumes (6, 16-18).

4. DISCUSSION

It has been shown that the procedures tested have the sensitivity and
the precision required for the determination of aerosol nitrate in the
atmosphere. Although the field data are somewhat limited they are in agreement with the expectation that the mass of nitrate on the first filter increases linearly with the air sampling volume whereas the amount of nitrate
resulting from the adsorption from gaseous compounds come to an equilibrium.

If the adsorption from the gas phase and the deposition of aerosol
nitrate were entirely independent processes, one would expect the amount
of nitrate residing on the back-up filters to equal that given by the intercept on the ordinate in the plot of first filter values versus the air
sampling volume. Figure I and our other results show, however, that in
most cases the values for the back-up filters are higher. It is possible
that the higher values are due to an increase of the relative humidity
caused by the pressure drop behind the first filter (2) or by particulate
matter deposited on the first filter, but this remains to be investigated.
Furthermore, it is not clear why the first filter is saturated with artifact nitrate after less than 50 m³ of air has been sampled, while the
second and third filters saturate shortly one after another between 100
and 140 m³ of sampled air.

5. ACKNOWLEDGEMENTS: We acknowledge with gratitude the hospitality extended to us by Mr. K.J. Rumpel at the station in Deuselbach. We are also indebted to Prof. R. Neeb for illuminating discussions. The work was performed as part of the program of the Sonderforschungsbereich 73, "Atmospheric

Trace Components" and has received in part financial support from the Deutsche Forschungsgemeinschaft.

6. REFERENCES

(1) C.W.SPICER et al., Sampling and analytical methodology for atmospheric particulate nitrates; Batelle-Columbus Final Report to U.S. Environm.Protection Agency (EPA-600/2-78-067) (1978).

(2) C.W.SPICER and P.M.SCHUMACHER, Interferences in sampling atmospheric particulate nitrate; Atmosph. Environ. 11, 873-876 (1977).

(3) C.W. SPICER and P.M.SCHUMACHER, Particulate nitrate: Laboratory and field studies of major sampling interferences; Atmosph. Environ. 13, 543-552 (1979).

(4) J.FORREST et al., Determination of total inorganic nitrate utilizing collection of nitric acid on NaCl-impregnated filters; Atmosph. Environ. 14, 137-144 (1980).

(5) B.R.APPEL et al., Interference effects in sampling particular nitrate in ambient air; Atmosph. Environ. 13, 319-325 (1978).

(6) L.A.ROHLACK et al., Nitrogen oxide interferences in the measurement of atmospheric particulate nitrates; Vol.1 and 2, Final report of the Austin Corporation, Austin, Texas, USA (1979).

(7) W.KLIPPEL and P.WARNECK, The formaldehyde content of the atmospheric aerosol; Atmosph. Environ. 14, 809-818 (1980).

(8) W.KLIPPEL, "Bestimmung der Formaldehyd-Konzentration im Regenwasser und am Aerosol in kontinentaler und maritimer Reinluft"; Dissertation Universität Mainz und Max-Planck-Institut f.Chemie, Mainz (1978).

(9) D.F.BOLTZ, Colorimetric determination of nonmetals; N.Y., Interscience Publications (1958).

(10) D.JENKINS and L.MEDSKER, Brucine method for determination of nitrate in ocean, estaurine, and fresh waters; Analyt. Chem. 36, 610-613 (1964).

(11) H. FADRUS und J.MALÝ, Nitratbestimmung in Wässern; Z.Anal.Chem. 246, 239-241 (1969).

(12) A.W.MORRIS and J.P.RILEY, The determination of nitrate in sea water; Anal. Chim. Acta 29, 272-279 (1963).

(13) K.GRASSHOFT, Determination of nitrate; methods of sea water analysis; 137-145 (1976).

(14) R.BENESCH, H.HARBST, Nährstoffbestimmung im Meerwasser; Technischer Bericht Nr.10, Technicon International Division S.A., 12-14 Chemin Rieu, CH-1208 Geneva (Switzerland).

(15) D.WANGE, "Zur Bestimmung und Differenzierung von an Aerosolen gebundenem Nitrat"; Diploma Thesis, Universität Mainz (1981).

(16) T.OKITA et al., Measurement of gaseous and particulate nitrates in the atmosphere; Atmosph. Environ. 10, 1085-1089 (1976).

(17) B.J.HUEBERT and A.L.LAZRUS, Tropospheric gas-phase and particulate nitrate measurements; J. Geophys. Res., accepted for publication.

(18) B.J.HUEBERT AND A.L.LAZRUS, Global tropospheric measurements of nitric acid vapor and particulate nitrate; Geophys. Res. Lett. 5, 577-580 (1978).

GASEOUS REFERENCE MATERIALS - CERTIFICATION - TRACEABILITY

H.J. PEPERSTRAETE
S.C.K./C.E.N. Nuclear Study Centre
MOL-Belgium

Abstract

The availability of primary gas standards is of the highest importance to the environmental scientist in order to perform accurate
analysis and estimate the working precision.
A number of national and international laboratory exercises approved
the lack of standardisation in this particular field and reaffirmed
the need for unambiguous normalisation of the measuring methods and
traceability of their results.

Seen the concentration of most pollutants and their lower toxicity
level in air is of the ppb or ppt level, calibration of the detectors must be performed within this low concentration range of air
contaminants. Last year considerable efforts have been paid by gas
manufacturers to provide ambient level pressurized mixtures of
pollutants in nitrogen or air. Within well determined limits of
operation they succeeded for SO_2, NO, NO_2, CO, ...

1. INTRODUCTION

Traceability to national or international standards in the field of
gas analysis is gaining more and more importance for current analytical
process controll and environmental safeguarding.

The development of gas standardisation procedures is a current oc-
cupation of some central authorities like NBS, ASTM, EPA and NIOSH in the
USA, BCR in the European Community Member Countries and ISO.

It is essential, indeed, that gas manifacturers can rely on these
authorities for assessment in relation to the involved preparation techniques and determination of there accuracy error estimates.

The results of a recent interlaboratory exercise, organized by the
Commission of the European Communities, based on the analysis of pressurized gasmixtures of NO in nitrogen at a level of about 300 ppb clearly
pictures the present state of art with regard to the availibility of and
the need for certified reference materials.

2. EXPERIMENTAL

The alignment of a measuring method consists of a number of consecutive operational steps giving rise to a final numeral result in conjunction with an estimate of the uncertainty range; thus quantifying the
sample ingredients or composition.

Actually the analytical chemist disposes of a set of highly perfec-
tionated, sophisticated and automated instrumental techniques, characterised by a remarkable high repeatability and reproducibility of the
data.

Since these instruments apply the comparison between the output respons to the sample under investigation and the equivalent respons to a well known standard sample in order to quantify its components, the analyst needs a number of reference materials allowing him to reduce the ample quantification procedure and in the meanwhile to correct for systematic errors.

The alignment of these reference gases will under these conditions permit the analytical chemist to improve the accuracy of produced data.

The technical committee 158 "gas analysis" of the International Standardisation Organisation (ISO-Geneva) published a number of standards describing the preparation and identification procedure for gasmixtures and is almost ready to produce manuals for intercomparison and verification of primary and secundary reference gasmixtures. These ISO regulations are based on the assumption that for an accuracy based measuring system it is of vital importance that the preparation methods for the reference materials are characterized in terms of accuracy, since this allows a real error estimation on the gained results.

The assumed diagram of a measuring system is given in figure 1.

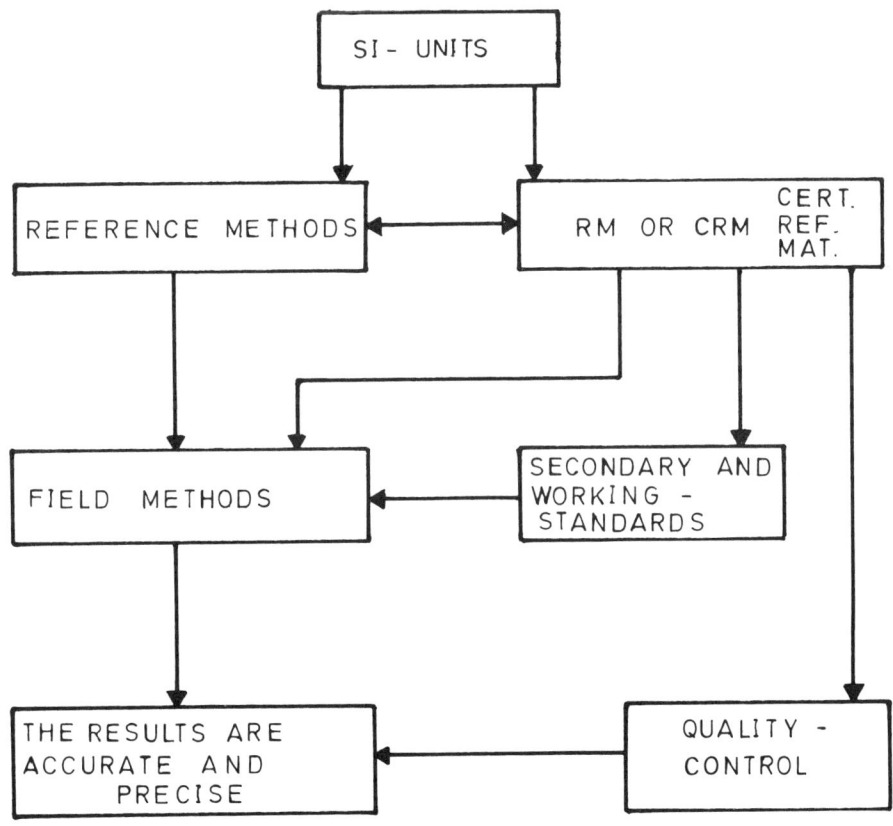

FIGURE 1

DIAGRAM OF A MEASURING SYSTEM

Actually most national laboratories only accept the gravimetric method (ISO 6142) in conjunction with an extensive range of verifications as a valid method for the preparation of their working standard gases.

The standardisation of methods for the preparation of gas mixtures on its own is not sufficient to ensure internationally comparable results; strong and intensive cooperation between international laboratories has to be established.

In this field an important role can be played by the Community Bureau of references (BCR) of the Commission of the European Communities, which already started an analytical intercalibration exercise with the following set of gasmixtures : 100 ppm CO in N_2
$$50 \text{ ppm NO in } N_2$$
$$1 \% \quad CO_2 \text{ in } N_2$$
$$50 \text{ ppm } C_3H_8 \text{ in air}$$

At present this bureau of references is working out a project dealing with the preparation of a wide range of certified reference gasmixtures covering a broad field of applications like e.g. automotive exhaust and ambient air survey. A preliminary list of the mixtures proposed to be certified is given in table 1.

TABLE I

PROPOSAL FOR BCR GASSTANDARDS

CO/N_2	CO_2/N_2	NO/N_2	C_3H_8/AIR	NO_2/AIR	SO_2/N_2	
		0.5		0.5	0.5	
9.5			9.5			ppm
50						
		95	95			
1	0.95					
4,5	4,5					%
7	15					

Reference gasmixtures, prepared under well controlled severe conditions, by selected laboratories in the European member countries will be certified under BCR control along the procedure given in figure 2.

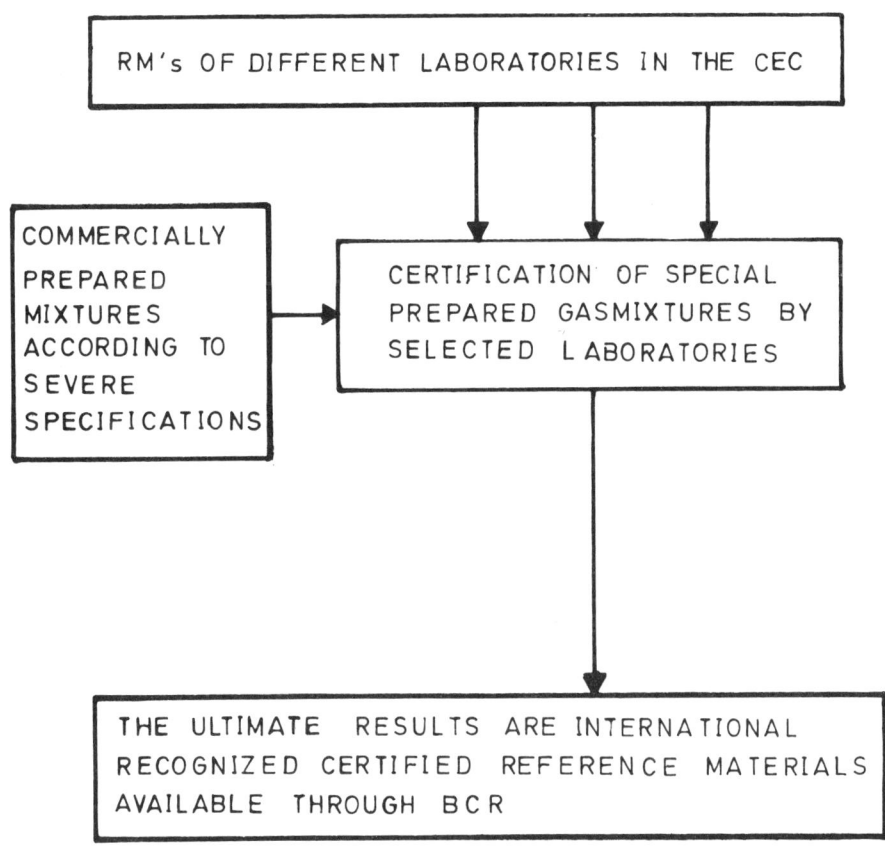

FIGURE 2

REALISATION SCHEME OF BCR - CRM's

The former description of available standard procedures for the preparation and evaluation of RM's may be misleading since it only covers the main explored area of stable gasmixtures.

The final preparation method and working procedure will be selected taking into account the practical temperature and pressure limits, the chemical composition of the mixture and eventual interactions with the inside walls.

One should further notice that a chemically reactive gasmixture will not become stable while it is pressurized in a cylinder.

In order to deal with this particular application one need to rely on dynamic preparation methods, capable of generating a gasmixture just before use. The desired degree of accuracy may help to select the best suited method of preparation.

The availability and validity of the produced reference gasmixtures and the present state of art in this field may be illustrated by the results of an intercalibration exercise, organised by the Environmental and Consumer Protection Service of the CEC In 1980.

In this exercise four cylinders with nitrogen monoxide (range 0.2-0.4ppm) were circulated to ten laboratories in seven European countries for analysis.
The participating laboratories were :

CCR-ISPRA	Commission of the European Communities Joint Research Centre, Ispra, Italy
IHE	Instituut voor Hygiëne en Epidemiologie, Brussels, Belgium
INSERM	Institut National de la Santé et de la Recherche Médicale, Vigoulet-Auzil, France
ISS	Istituto Superiore de Sanita, Rome, Italy
RISO	Riso National Laboratory, Roskilde, Denmark
RIV	Rijksinstituut voor Volksgezondheid, Bilthoven, Netherlands
S.C.K./C.E.N.	Studiecentrum voor Kernenergie, Mol, Belgium
TNO	Instituut voor Milieuhygiëne en Gezondheidstechniek, Delft, Netherlands
UBA	Umweltbundesambt Pilotstation, Frankfurt, West-Germany
WSL	Warren Spring Laboratory, Stevenage, United Kingdom

The mixtures were blended in pretreated cylinders, two by BOC and two by L'Air Liquide Belgium.
BOC used aluminium and L'Air Liquide steel cylinders.
The analytical results are presented in figures 3,4,5 and 6.

The objectives of the cylinder circulation programme were to investigate the suitability of NO in N_2-premixtures contained in pressure cylinders for quickly and inexpensively comparing the accuracy and precision of NO_x measurements at a number of laboratories in different European Community Countries. These objectives were covered successfully since the used NO standards in treated aluminium and steel cylinders remained sufficiently stable to provide a meaningful intercomparison of NO_x measuring methods and secondly the relative accuracy and precision of the measurements have been assessed. There were indications that after further harmonisation an accuracy error of $\pm$ 10 % and a precision error of $\pm$ 5 ppb for NO are achieveable.

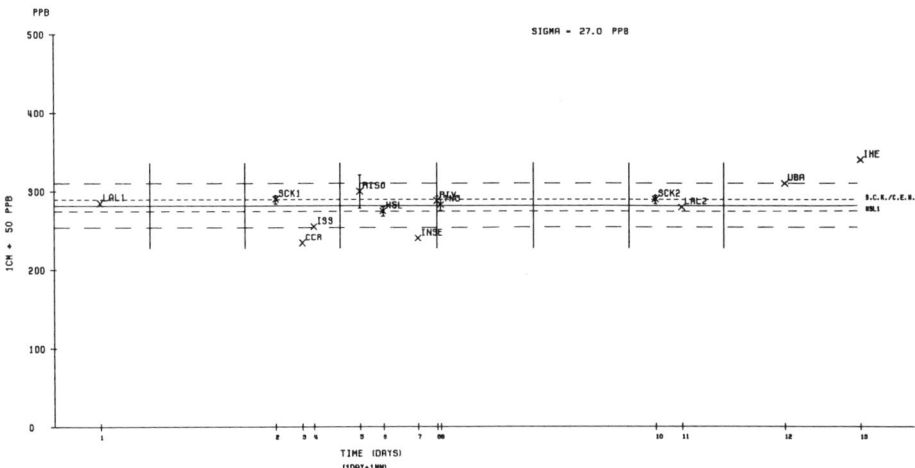

FIG. 3 - Results for NO - L'air liquide - Bottle 1

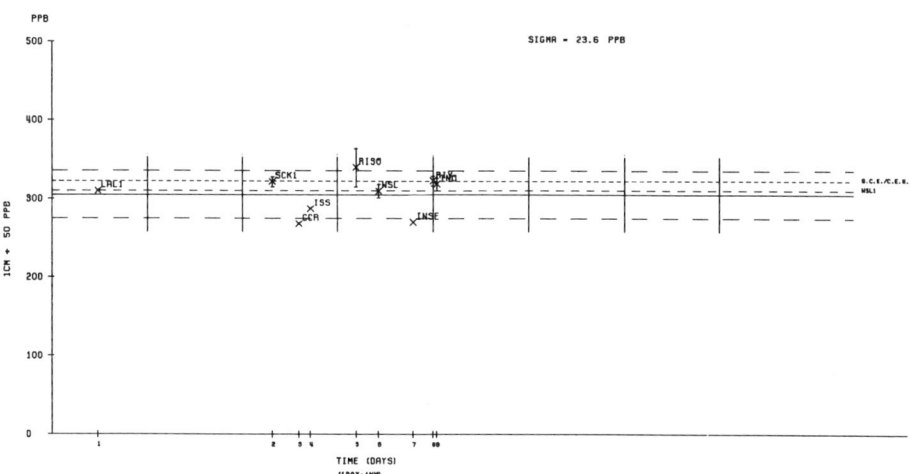

FIG. 4 - Results for NO - L'air liquide - Bottle 2

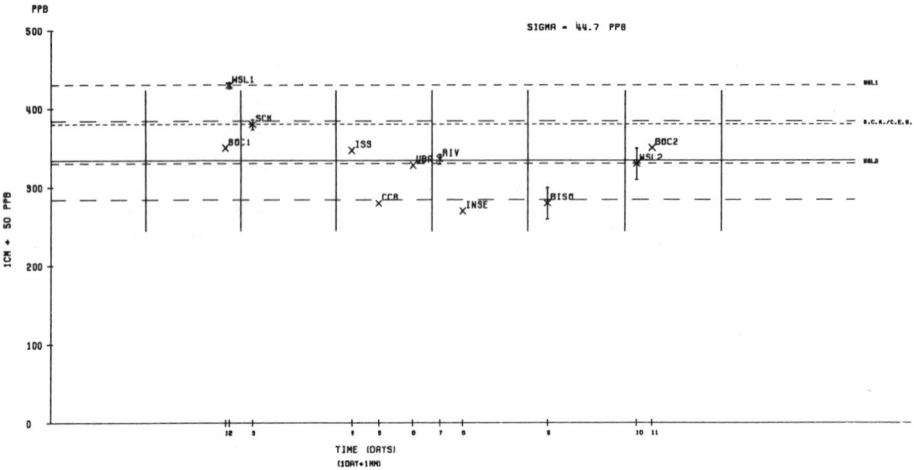

FIG. 5 - Results for NO - BOC - Bottle 29

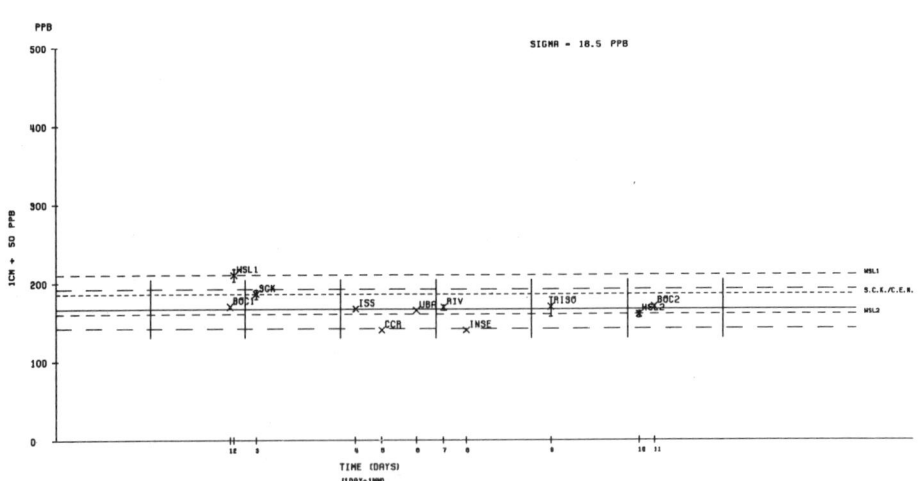

FIG. 6 - Results for NO - BOC - Bottle 45

3. CONCLUSIONS

Remarkable efforts have been paid by ISO and the CEC to assist gasmanu-
facturers and users with preparation and evaluation manuals and standards. The availability of certified standards is a need for the analytical chemist in order to assess and improve the accuracy of the data. The
environmental scientists, used to determinations of reactive and non
reactive pollutants in the ppb and ppt concentration range, should be
very careful in selecting and preparing their primary standards for gas-
analysis. An exact evaluation of the real involved accuracy error estimate is primordial.

4. REFERENCES

(1) W. Frankvoort
    Proceedings of the International Symposium on the Production and Use
    of Reference materials Berlin 1979 462-471

(2) H. Peperstraete - A. Apling - F. Magdonelle
    Proceedings of th 5TH International Clean Air Congress
    Buenos Aires 1980 (to be published)

GAS CHROMATOGRAPHIC DETECTION OF HYDROCARBONS IN THE ATMOSPHERE USING SPECIFIC GC DETECTORS AND MASS SPECTROMETRY IN SELECTED ION MONITORING MODE

M.POSSANZINI, P.CICCIOLI, E.BRANCALEONI, R.TAPPA, A.BRACHETTI

CNR - Istituto Inquinamento Atmosferico
Area della Ricerca di Roma - CP 10
00016 Monterotondo Stazione (ITALY)

SUMMARY

An analytical procedure for the simultaneous determination of hydrocarbons containing different functional groups is described. Two gas chromatographs are employed: one is equipped with a multidetection unit comprised of three gc-detectors (ECD, FID, FPD), the other is combined with a mass spectrometer. The determination of the various compounds is obtained by injecting into the GC columns equal amounts of the same sample collected on two different traps. The results obtained with this method during a monitoring campaign are also presented.

1. INTRODUCTION

Monitoring of hydrocarbons in ambient air is particularly important because of the foundamental role played by these compounds in the formation of photochemical oxidants. However, due to the complexity of the organic components present in the atmosphere, their determination is difficult and requires very selective and sensitive analytical methods. Gas chromatography with selective detectors has been found the most suitable method for the determination of organic compounds in the atmosphere and many papers dealing with the specific determination of some classes of hydrocarbons (halocarbons, PAN, sulfur containing compounds etc.) have been reported (1-3). Although in many instances the determination of a single class of hydrocarbons is sufficient for relating the ambient air quality to specific emission sources, a more real picture of the organic spectrum in the atmosphere can be obtained only by monitoring several classes of hydrocarbons simultaneously. This paper presents an analytical procedure for the simultaneous detection of hydrocarbons containing different functional groups. Hydrocarbons containing sulfur and halogen atoms can be identified in a single chromatographic run by using a multidetection unit comprised of three different detectors.
    The identification of more reactive species (such as tetramethylbenzenes) is carried out by using another chromatographic unit combined with a mass spectrometer equipped with a Selected Ion Detection system.

The determination of various compounds is accomplished by injecting into the two g.c. columns equal amounts of the same sample enriched on the trapping system.

Preliminary results obtained by using this technique will be presented.

2. EXPERIMENTAL

A Carlo Erba (Milan ,Italy) gas chromatograph model GI was modified according to the scheme of Fig. 1. A Dani (Monza, Italy) ECD-FID unit (4) was placed in parallel with the FPD originally supplied with the gaschromatograph. A three way manifold was connected to the column outlet in order to divide in two equal portions the sample eluted.

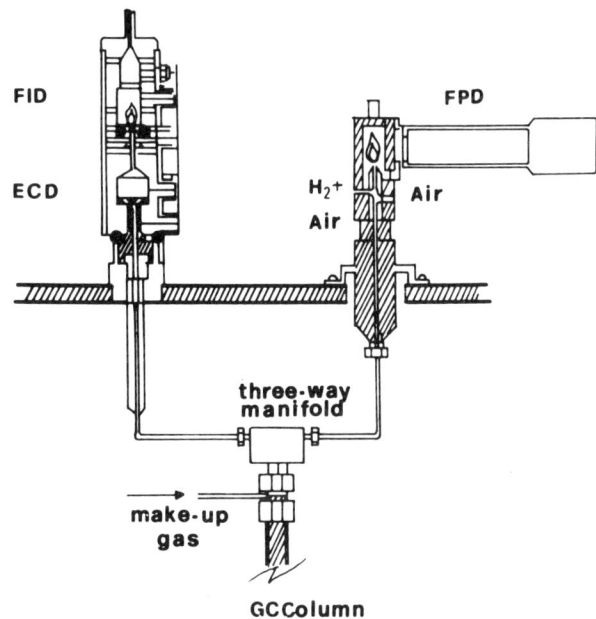

FIGURE 1   Schematic diagram of the
g.c. multidetection unit.

To avoid peaks broadening occurring within the ECD and the FPD a make-up gas flow rate of 100 ml/min was added at the column end.

The injection trap system made by using a 4-way valve was placed in series with the conventional injector and this apparatus was attached to the left side of the gc. This arrangement allows the gas chromatograph to be also used in the usual way. This is very important when quantitative determinations have to be carried out.

A cilindrical furnace equipped with a thermostatic control

TABLE I

HYDROCARBONS DETECTED IN AIR

Peak No.	Compounds	Amount (ppb CH_4)
1	C_5 unsaturated hydrocarbon	-
2	i-C_5 hydrocarbon	30
3	Benzene + i-C_6 hydrocarbon	42
4	i-C_7 hydrocarbon	4
5	i-C_7 hydrocarbon	4
6	3-methyl-hexane	3
7	Toluene	32
8	Dichloropropane	3
9	Ethylbenzene	8
10	i-C_8 hydrocarbon	
11	m-Xylene	12
12	p-Xylene	5
13	o-Xylene	5
14	n-Propylbenzene	3
15	1-methyl-3-ethylbenzene	8
16	1-methyl-ethlbenzene	4
17	i-propylbenzene	4
18	1.3.5-trimehylbenzene	6
19	Methyl-ethylbenzene	3
20	1.2.4-trimethylbenzene	20
21	Diethylbenzene	6
22	Methyl-propylbenzene+ 1.2.3-trimethylbenzene	8
23	1.2-dimethyl-ethylbenzene	6
24	1.2.3.4.Tetramethylbenzene + 1.2.4.5-Tetramethylbenzene	10
25	1.2.3.5-Tetramethylbenzene	3

was employed for thermal desorption of the traps. The furnace temperature was adjusted to 250°C.

A Dani gas chromatograph model 3900 was employed in connection with a VG 70-70F (VG Analytical, UK) double focussing mass spectrometer.

A single stage jet separator was used as gc-ms interface.

The two gc columns used (3 m x 2 mm i.d.) were packed with Carbonpack C (60-80 mesh) (Supelco, USA) coated with 0.5% SP 1000.

Two different adsorbents were employed as trapping material and they were Tenax (Enka, The Netherlands) and Carbopack C (Supelco, USA). These two materials possess not only similar specific surface areas (15 and 20 m^2/g) but also similar adsorption properties and in both cases a complete recovery of compounds having molecular weight up to C_{18} was observed with the stripping temperature employed. The traps were made with glass tubing (15 cm x 4 mm i.d.) and Vespal ferrules were employed to ensure tight connections with both the gc as well the trapping apparatus.

A novel field apparatus developed in our laboratory (5) was used for sample collection. It ensures not only the constancy of the air flow rate passing through the trap ($\pm$ 1%), but also it allows digital setting of the volume to be sampled.

Because the traps were kept at room temperature, the usual sampling volume was ranging between 3-5 l in order to be sure that compounds with molecular weight above C_5 were almost completely collected (> 80%). For this reason, sampling sites relatively close to hydrocarbons emission sources were selected.

3. RESULTS AND DISCUSSION

After two months of continuous operation carried out in different meteorological conditions, some differences were observed between Carbopack C and Tenax. In spite of the fact that Tenax traps were carefully cleaned, protected against the solar radiation and the ozone level measured during the sampling was below 50 ppb, the appearence of small but well detectable decomposition peaks with retention times very close to the compounds reported by Knoeppel et al. (6) was observed. In the same conditions Carbopack did not give rise to interfering compounds, although the trapping efficiency for the more volatile compounds was found smaller.

Fig. 2 shows the gaschromatographic profiles obtained with multidetection unit described above. The black shadowed area reported in the FID trace corresponds to the presence of decomposition peaks observed with the Tenax traps.

The chromatographic profile at the bottom of Fig. 2 corresponds to the Selected Ion Detection trace obtained by focussing the mass spectrometer on the ion with m/e 134. The most intense peaks recorded on the SID trace are relative to C_4-benzene isomers which are known to have very high photochemical reactivity.

Table I lists the compounds that have been identified and their average concentrations measured during a two months monitoring campaign carried out at Monterotondo, a very small town

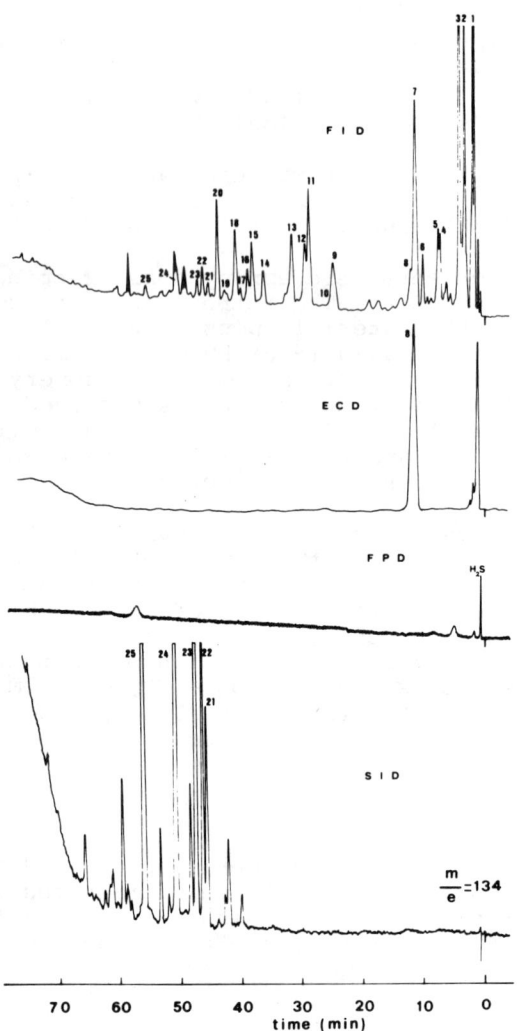

FIGURE 2   Gas chromatographic and mass spectrometric
(SID) traces obtained with the multidetection units described above during the analysis of 5 L of air sample.

close to our laboratory.

The identification of the single compounds reported in
Fig. 2 has been mainly based on the spectra recorded during se
veral gc-ms runs. In the case of isomers exhibiting very close
fragmentation patterns, the identification was made by injecting pure standards. The most important advantage of the multidetection unit described above is that of giving at once the

full spectrum of hydrocarbons present in the air sample. This unit , can be used not only for the identification of the various classes of hydrocarbons but mainly for obtaining the "finger preints" of a given air sample. These data might particularly important for establishing the relative contributions of natural and anthropogenic substances present in the atmosphere. However some inconsistencies have been observed between the expected and observed compounds.

Although the measurements of NMHC are in good agreement with the data obtained using the traps (once we take into account that hydrocarbons ranging from C_2 to C_5 are not completely recovered), few peaks were found on the FPD and ECD traces. Actually only H_2S and dichloropropane gave noticeable peaks. Because these last two detectors are from one to two order of magnitude more sensitive to halogen and sulfur containing hydrocarbons than the FID and the samples were collected on a relatively polluted area, it is surprising that no other compounds were recorded. As these compounds have also not been detected with mass spectrometry, we believe that some losses may occur within the traps packed with both materials. For these reasons our future efforts will be focussed to the recovery of polar compounds (such as mercaptans and chlorinated species) when different trapping methods are used.

LITERATURE CITED

1. F. Bruner, G. Crescentini, F. Mangani, E. Brancaleoni, A. Cappiello, P. Ciccioli; Anal. Chem. 53, 798, 1981.

2. F.J. Sandalls, S.A. Penkett, B.M.R. Jones:Preparation of PAN and its determination in the atmosphere. AERE Report R-7807 - UKAEA Harwell England - november 1974.

3. P. Ciccioli, G. Bertoni, E. Brancaleoni, R. Fratarcangeli, F. Bruner; J. Chromatography 126, 757, 1976.

4. F. Poy, J. High Resolution Chromatography 2, 244, 1979.

5. I. Allegrini, A. Febo, A. Di Chiara Maggioli; Acqua-Aria n. 2, 197, 1981.

6. H. Knoeppel, B. Versino, H. Schlitt, A. Peil, H. Schauenburg H. Vissers. Organic in air. Sampling and identification. Proc 1st Symposium "Physico-chemical behaviour of atmospheric pollutants", B. Versino and H. Ott eds. pp. 25-40, 1980.

POLYCHLORODIBENZODIOXINS AND - DIBENZOFURANS IN THE ENVIRONMENT

D. Brocco, A. Cecinato and A. Liberti
Istituto Inquinamento Atmosferico C.N.R.
Area della Ricerca di Roma
Via Salaria km. 29,300 - Monterotondo Stazione (RM)

Summary

An analytical procedure has been developed for the analysis of polychlorodioxins (PCDD) and polychlorodibenzofurans and of their precursors in environmental samples and namely in urban incinerators emissions. The variation of composition of PCDD and PCDF have been discussed in terms of the concentration of precursors and incinerators operating condition.

1. INTRODUCTION

The presence of toxic species such as chlorodibenzodioxins PCDD and chlorodibenzofurans PCDF, which can be found in the environment has been the cause of a great concern. It has been also pointed out (1) in a wide investigation on environmental samples these compounds are ubiquitous as any combustion process might contribute to their formation.

In a series of papers (2-5) it has been shown that fly ash and fumes of municipal and industrial incinerators may contain both PCDDs and PCDFs at the ppm level.

This paper reports the development of an analytical procedure to determine PCDDs and PCDFs, the results of analysis carried out upon the emissions from incinerators, located in different urban centers, and in soil samples from areas of an incinerator plant.

2. EXPERIMENTAL

The determination of PCDDs and PCDFs in environmental samples requires three steps, each of which is of a great importance: the sampling, the extraction and its clean-up, the determination through gas chromatography-mass spectrometry.

As far as the sampling concerns there are no problems for solid and liquid materials but when the determination has to be carried on fumes, a great attention has to be paid for the isokinetic sampling of particulates and for an efficient trapping of the gaseous phase. This coperation can be carried out by cold condensation of the water vapour and by trapping the effluent gases in ethylene glicol, kept at low temperature.

3. SAMPLE EXTRACTION AND PURIFICATION

The extraction from solid samples is one of the analytical steps which has been widely discussed and various procedures have been evaluated in terms of the highest recovery of PCDD they yield. Toluene extraction in a soxhlet apparatus for a long time (24 - 72 hours) of acid treated material is usually recommended. As in several cases misleading results have been reported (6), a critical evaluation has been carried out on various extraction procedures and upon the effect of temperature.

As it will be later discussed as the temperature has a determining effect upon reactions which might occur. among precursors of PCDD and PCDF, present in the sample such as phenols, polychlorinated phenols, and other a leaching procedure has been adopted as extraction process.

The fly ash or soil sample (20 - 30 g) is placed in a chromatographic column (i.d. = 2,5 cm.) equipped with a fritted glass septum. The column in connected to a reservoir containing 500 ml of benzene or a solution of benzene-methylene chloride (1 : 1 v/v) and eluted.

The extract is concentrated to 1 ml by evaporation under a nitrogen stream at room temperature and transfered in a chromatographic column (10 cm long; 0,8 cm i.d.) containing 2 g of silica gel (Merck Kieselgel 60; 70 - 230 mesh). The elution is carried aut with n - pentane (30 ml.). The eluated was carefully brought to 1 ml. by evaporation, transfered in a tailed tube and dried at room temperature. Finally 200 μl of toluene are added to make the solution ready for G.C. analysis. The PCDD and PCDF recovery is always higher than 90%.

In some cases, when the samples contained large quantities of organic compounds interfering with the PCDD and PCDF determination, the extract is subjected to a second elution clean-up.

Occasionally a further elution upon an alumina column (Woelm - basic activity gradel) with methylene chloride - n hexane in the ratio (2 : 98) and further in the ratio 50:50 was used.

In order to have information upon the role of precursors and namely of chlorophenols their determination has been carried by elution of the silica gel with acetone.

By means of TLC using benzene n-hexane (60:40 v/v) solution as eluant the chlorophenols were isolated (R_F = 0.3-0.4) and furtherly analyzed by GC.

A full scheme of the analytical procedure is shown in the next page.

4. GC AND GC - MS ANALYSIS

The purified extracts were analyzed either on packed column or on high resolution glass capillary column which were set on a chromatograph (Dani, mod. 5600, Monza, Italia) equipped with a Ni[63] electron capture detector.

The packed column (1,5 mt in lenght, 1,5 mm i.d.) filled with Supelcoport 100-120 mesh coated with 1,5% OV 17 + 1,95% QF-1 was operating at 230ºC with nitrogen carrier gas flow of 16 ml/min. Under these conditions the column had 5.500 theoreti-

Sampling

Extraction - Leaching with benzene

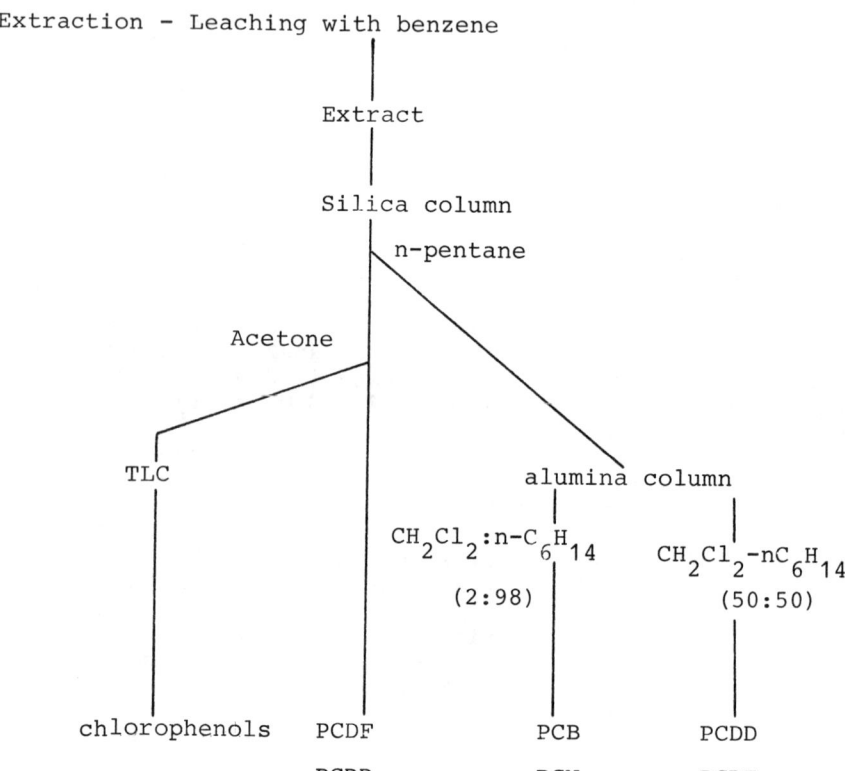

Extract

Silica column

n-pentane

Acetone

TLC

alumina column

$CH_2Cl_2:n-C_6H_{14}$
(2:98)

$CH_2Cl_2-nC_6H_{14}$
(50:50)

chlorophenols   PCDF   PCB   PCDD
                 PCDD   PCN   PCDF

Analysis by GC (packed-capillary column)
     HR mass spectrometry

cal plates for hexa-CDD.

The capillary column (25 m long; 0,26 i.d.) was coated with OV-17 as stationary phase.

The column was run at 245°C with a nitrogen carrier gas pressure of 0,4 kg/cm^2 with average linear velocity of 7 cm/second. The stream splitting of 80:1 ÷ 90:1 of the sample injected was used. For quantitative analyses the splittless injection technique or on - column injection are required. Under these conditions the column has a efficiency corresponding to 80.000 theoretical plates for hexa - CDD.

Both columns can be coupled to a mass spectrometer for fur ther peaks identification.

Severale samples have been examined also by mean framentography according to the procedure developed by Liberti (7). All results heve been reported in terms of PCDD and PCDF with the same number of chlorine atoms.

5. RESULTS AND DISCUSSION

Fig. 1 shows a typical gas chromatogram of a fly ash extract on a packed column. The PCDDs and PCDFs are eluted according to their chlorine content, but the various families are not separated. On account of the efficiency of the capillary column a higher degree of separation is obtained by using these column. In both cases and namely for quantitative work is required to carry on the analysis by selective ion monitoring and this aim is achieved by monitoring three ions (M^+, M^{+2}, and M^{+4}) for each family of PCDD and PCDF. A typical SID diagram of a fly ash sample of a urban incinerator is shown in fig. 3. The relative concentrations of various PCDD and PCDF families expressed in ppb, are the following:

TerCDF 10.9; PCDF 29.4; ECDF 37; HpCDF 58; OCDF 15

TerCDD 2.7; PCDD 7.4; ECDD 21; HpCDD 47; OCDD 92

The concentration of these species vary noticeably in the emissions of various incinerators according to the input waste material and the incinerators working conditions.

Though members of the various families of PCDD and PCDF are always present, the relative concentration may be quite different.

Some data summarized from fly ash analysis of urban incinerators which burn wastes with a different treatment are collected in table I.

It seems that one main factor might be the hydrochloric acid concentration in the fumes the higher its value the larger being the yield of the more chlorinated species.

Since HCl originates primarly from PVC plastics and other chlorinated hydrocarbons, the high content of HCl in the incineritors emission is due to the high halogen plastic content.

When these plastics are removed, as it is carried out in re cycling and compost production plants, the concentration of higher chlorinated species sharply decreases.

The emissions temperature plays also an important role on the physical state of the variuos species. It has been found that the higher the fumes temperature, the more abundant the con

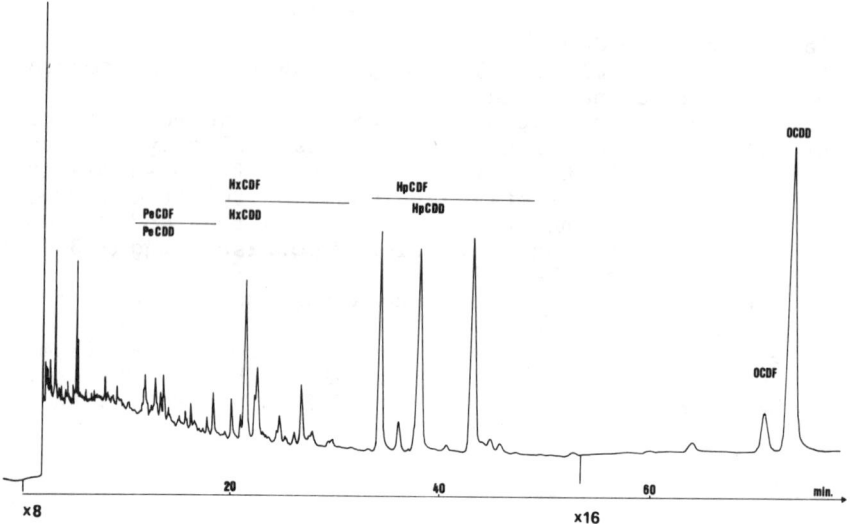

FIG. 1

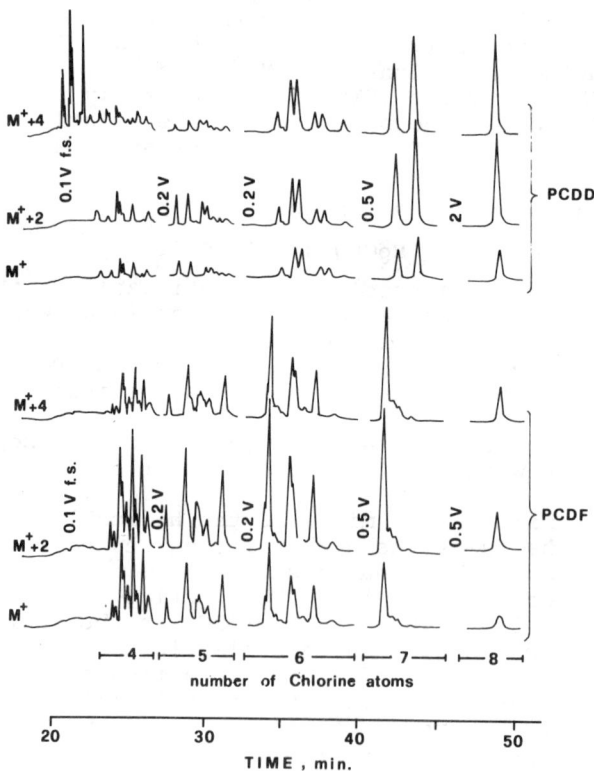

← FIG. 2

centration of chlorinated species as vapours. When water scrubbing is used for fumes abatement and the fumes temperature is low, a higher concentration of chlorinated species is found in the abatement sludge.

TABLE I

PCDD, PCDF anal HCl concentration in emissions of incenerators.

Incinerators	He-CDD -CDF (ppb)	Hp-CDF (ppb)	Hp-CDD (ppb)	OCDF (ppb)	OCDD (ppb)	HCl (ppm)
A_1	90	68	145	65	152	10-200
A_2	39	30	42	31	118	10-200
A_3	135	50	85	35	90	10-200
B	—	—	6	—	12	0-10
C	—	—	5	—	5	0-10

A = inc. withot any treatment
B = inc. after recycling
C = inc. after compost production.

To obtain information upon the impact the chlorinated species might have upon the environment soil samples of various areas around one incinerator have been collected and analyzed. The concentration fo members of various species of samples taken at various distances from the stack are summarized in Table II.

TABLE II

PCDD and PCDF concentration (ppt) in soil sampled at various di stances from the stack.

Sample	distance m.	$HpCDF_1$	$HpCDF_2$	$HpCDD_1$	$HpCDD_2$	OCDF	OCDD
1	500	50	—	50	250	600	1000
2	1000	100	50	60	60	100	120
3	1000	10	10	10	10	10	20
4	2000	—	—	tracce	tr.	tr.	tr.

This investigation shows that PCDF are in several cases more abundant than PCDD. This finding in consistent with what has been previously stated (8) that phenols and polyphenols present in most vegetables residues are definitely precursors of PCDD and PCDF. These precursors in the presence of chlorine donor species yield these species. It has been shown that by burning vegetable residues into a glass flask under an air flow to which chlorine was added large amounts of PCDF are found. This reaction which definitely occurs in the incinerator might as well take place in the extraction process if a specificic care is not taken to prevent it. The conversion of chlorophenols into PCDD and OCDF, which might occur in the extraction process, can be observed in the following way: the concentration of chlorophenols in an incinerator fly ash is in the range of 0.15 µg/g. If the extract obtained with various solvents (benzene, toluene, xilene) is made to reflux in a soxhlet apparatus their concentration sharply decreases and increases in correspondance the concentration of PCDD and PCDF.

REFERENCES

1) The chlorinated Dioxin Task Force,Michigan Division Dow Chem. Midland, Mich., "The Trace Chemistries of Fire" 1978.

2) K. Olie, P.L. Vermeulen and O. Hutzinger, Chemosphere,7,455 (1977).

3) A.Cavallaro et al. Chemosphere, 9, 611 (1980).

4) A.Liberti, D.Brocco, A.Cecinato and M. Possanzini, Mikrochimica Acta (Wien), 18, 271 (1981).

5) G.A.Eiceman, R.E. Clement and F.W. Karasek, Anal. Chem., 51, 2343 (1979).

6) R.M. Kooke, J.W.Lustenhouwer, K. Olie and O. Hutzinger, Anal Chem. 53, 461, (1981).

7) A.Liberti, E. Brancaleoni, P. Ciccioli, A. Cecinato, Euroana lysis IV - Helsinki,1981.

8) A. Liberti, D. Brocco, Rome Dioxin Symposium 1980,Pergamon in press.

QUANTITATIVE DETERMINATION OF TERPENES EMITTED BY CONIFERS

H. KNÖPPEL, B. VERSINO, A. PEIL, H. SCHAUENBURG, and H. VISSERS
Joint Research Center Ispra
Commission of the European Communities

Summary

In the framework of research aimed at an assessment of the relative importance of antropogenic and natural organic compounds in photo-chemical particle formation, the quantity of terpenes emitted by different conifer species under the climatic conditions of the Po Valley has been determined.

The sampling and analysis procedure using TEFLON bag enclosure of live biomass portions, adsorption of emitted compounds on TENAX sampling columns and GC-MS analysis is described.

Results are discussed and compared to similar measurements performed in the U.S.A. Taking in consideration the large variations of individual emission values and the temperature dependence of the emission rate, values reported here and measured in the U.S.A. are in good agreement.

1. INTRODUCTION

In the framework of research aimed at an assessment of the relative importance of antropogenic and natural organic compounds in photochemical particle formation in the Po Valley (1, 2) organic compounds emitted by conifers have been qualitatively and quantitatively evaluated in order to

- identify the more important emitted compounds as a prerequisite for a proper choice of model compounds for photochemical bag experiments;

- be able to follow the concentration of individual emission constituents during photochemical experiments;

- verify, whether the quantitative emission data reported by Americal workers (3 - 6) are valid also under the geographical and climatic conditions of the Po Valley.

Conifers emit essentially terpenes (3 - 7), though isoprene and a few aromatic compounds such as estragol are also emitted by some species in appreciable quantities (7).

We report here on measurements aimed at a quantitative assessment of terpenic compounds emitted by different conifer species under the climatic conditions of the Po Valley for comparison with values measured by different authors in the U.S.A. (3 - 6).

2. EXPERIMENTAL

Fig. 1 shows schematically the sampling device for organic compounds emitted from trees as used in our experiments.

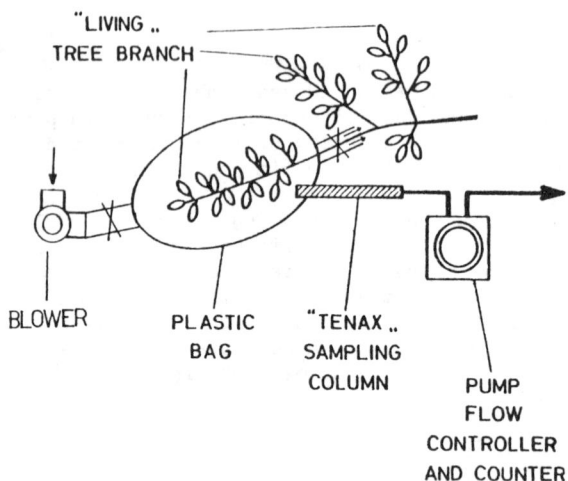

Fig. 1. - Scheme of the experimental arrangement for the rate
determination of terpene emission by conifers

Branch portions of about 200 g fresh weight were enclosed in a
TEFLON bag of 120 - 130 liters volume. When no samples were taken,
the bag was partially opened and a blower passed outside air at a rate
of approximately three bag volumes per minute through the bag. The bag
was closed during sampling for 10 - 45 minutes and the blower was separated from the bag. During the last 2 - 10 minutes of the closure
time 5 - 10 liter portions of bag air were sampled at a rate of 1 -
2 l/min.

Temperature was measured inside the bag near to the conifer needle
surface. When bags were in a sunny position, temperature sometimes
increased by 1 - 3 °C after bag closure. In these cases a mean temperature was established taking into consideration the ratio of the time
interval during which the temperature increase was observed and the
total bag closure time.

From preliminary experiments resulted that even using clean
reconstituted air for the bag sampling procedure typical ambient air
pollutants are detected in the sample. They are presumably adsorbed
on or absorbed in the conifer needles surface. Using ambient air for
the bag sampling procedure therefore does not introduce additional
complications, but guarantees a realistic air constitution.

Analysis by GC-MS prevented from possible errors in determining the
conifer emission products. Fig. 2 shows a glass capillary chromatogram
of the particularly rich emission of Pinus Ponderosa. A persilylated
50 m column (0.3 mm int. dia., OV1 coated, 1,1 μm film thickness) was
used for separation. The temperature was programmed from ambient temperature to 250 °C at a rate of 4 °C/min.

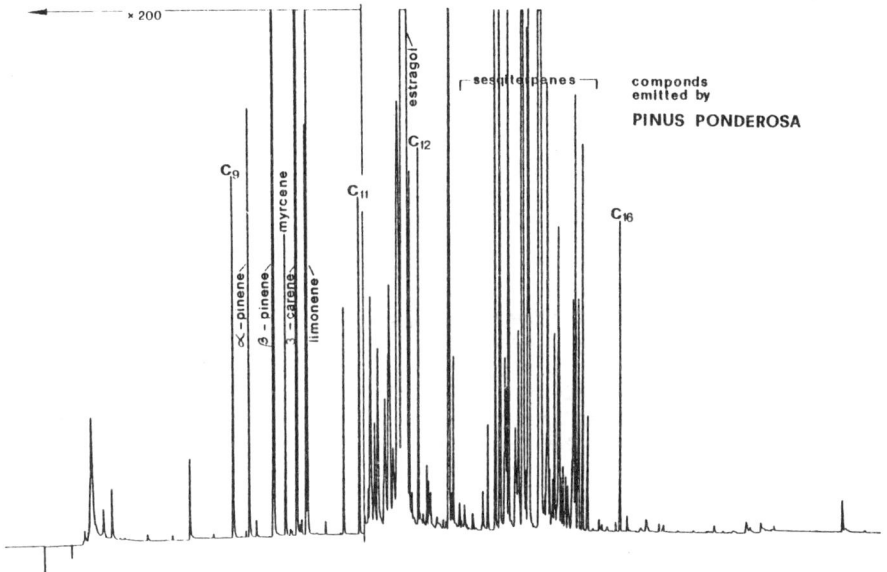

Fig. 2 - Glass capillary gas chromatogram of the organic compounds
         emitted by Pinus Ponderosa. For the chromatographic con-
         ditions see text.

The combination of retention indices - even if of limited
precision - and of mass spectra turned out to be very efficient
for a unique characterisation of individual terpenes. As shown in
Fig. 2, four n-alkanes (C_9, C_{11}, C_{12} and C_{16}) were added to each
sample as references for the retention index calculation following
a procedure described in another paper of this volume (8) and as
internal standards for the quantitative determination of the sampled
terpene amounts. The reference alkanes were added by injection of
1 μl of a methanol solution on the TENAX adsorption column before
sampling. Methanol was used as a solvent due to its low retention
volume on the TENAX column. During sampling it was nearly quantitatively eluted. The four n-alkanes were chosen in order to cover the
elution range of the terpenes without producing interferences.

The retention indices of the terpenes showed maximum deviations from the mean values of approximately half an index unit as shown in Table I. Positive deviations could become larger if the quantity of a compound introduced into the gas chromatograph exceeded 200 - 500 ng (split ratio 1 : 4), depending on the individual compound.

Table I. Programmed Temperature GC retention indices of some terpenes commonly emitted by conifers*

COMPOUND	NUMBER OF MEASUREMENTS	MEAN VALUE	STANDARD DEVIATION	MAX. DEVIATION	
α–PINENE	11	924.97	0.28	− 0.33	+ 0.61
CAMPHENE	20	936.12	0.23	− 0.27	+ 0.56
β–PINENE	10	963.10	0.28	− 0.45	+ 0.58
MYRCENE	17	982.14	0.23	− 0.40	+ 0.62
Δ3–CARENE	10	999.33	0.37	− 0.48	+ 0.69
LIMONENE	17	1016.29	0.26	− 0.38	+ 0.52

* determined with a persilylated OV–1 coated glas capillary column (length 50 m, int. dia. 0,3 mm, film thickness 1,1. μm)

Quantitation of the sampled emission product amounts was based on the flame ionization detector (FID) signal which was recorded simultaneously with the mass spectra by means of a dedicated minicomputer based data system. The system is also used to scan the quadrupol type mass spectrometer (RIBER R10–10 type), to calculate GC peak areas and retention indices, to select mass spectra of eluted compounds and to compare them together with the respective retention indices with a small library containing our own data obtained from previous experiments.

As a result of the analysis procedure the data system prints a report giving on one page all relevant experimental parameters and on a second page the analysis result. An example of this second page is shown in Fig. 3. The report gives for each detected compound the GC retention index and time, the GC peak area in counts and in percent, the number of the corresponding mass spectrum and the time at which it has been registrated. This time differs by 0.06 ± 0.02 min from the GC time due to a difference in transfer time between the GC column outlet and the FID respectively the mass spectrometer and since the spectra usually are scanned at 1 sec intervals. In case the library comparison gives a positive result the identification is also printed.

Based on the analysis report, the weight W_E of the sampled emission compounds was determined by summing all terpene GC area counts giving ΣT and all reference alkane area counts giving ΣA using the relation

$$(3) \quad W_E = 1{,}18 \cdot \frac{\Sigma T}{\Sigma A} \cdot W_A$$

where W_A is the weight of the injected reference n-alkanes and 1,18 is an experimentally determined calibration factor.

GC REPORT

===================================
DATE: 23. 7. 80   RUN NR: 793

PEAKNR	RET. IND	RET. TIME	AREA	PERCENT	MS PEAK	MS TIME	
13	515.46	3.09	319714	0.695			
14	523.93	3.16	167571	0.364	132	3.18	
15	532.59	3.24	209499	0.456	137	3.28	
16	551.72	3.41	256181	0.557	147	3.48	
	650.73				205	4.63	BENZENE
32	754.26	7.01	240716	0.523	326	7.05	TOLUENE
56	851.48	10.74	119282	0.259	514	10.82	443-586 M-P-XYLENE
68	899.74*	12.80	1541342		616	12.85	N-NONANE
73	926.91	13.98	25526230	55.505	674	14.02	ALPHA-PINENE
77	935.90	14.37	1050910	2.285	694	14.42	412/1103 CAMPHENE
85	964.00	15.61	10100360	21.963	756	15.65	BETA-PINENE
90	981.97	16.40	2389686	5.196	796	16.45	MYRCENE
95	994.93	17.15	1378781	2.998	833	17.20	DELTA-3-CARENE
99	1013.28	17.78	830012	1.805	865	17.83	763/893 TERPENOL
101	1015.91	17.90	838909	1.824	871	17.95	LIMONENE
112	1073.55	20.40	138295	0.301	996	20.45	763/1032
113	1080.45	20.69	145405	0.316	1011	20.75	
117	1099.81*	21.51	1696225		1052	21.58	N-UNDECANE
119	1109.35	21.91	134094	0.292	1072	21.98	
124	1139.88	23.18	318669	0.693	1135	23.23	
128	1160.62	24.03	79273	0.172	1178	24.10	
131	1169.14	24.37	141955	0.309	1195	24.43	ESTRAGOL
138	1199.92*	25.61	1814617		1257	25.68	N-DODECANE
194	1443.43	34.67	655101	1.424	1710	34.73	
203	1492.38	36.37	821725	1.787	1795	36.42	753/1840
210	1526.04	37.50	126658	0.275	1851	37.55	
225	1598.78*	39.86	1621363		1970	39.93	N-HEXADECANE

END GC: 41.05MIN   TOTAL:100.000PERCENT

TOTAL PEAKNR: 26

Fig. 3 – Data system report for a GC-MS analysis of organic compounds
emitted by a conifer tree

The plant emission rate E is calculated using the equation

$$(4) \qquad E = \frac{W_E}{B \cdot F}$$

where B is the dry leaf biomass weight of the branch (determined by
drying the needles of the analysed branches at 70 °C to constant weight)
and F is given by the equation

$$(5) \quad F = \tau + R \cdot (t - \tau) + \frac{\tau}{R} \cdot (1 - R) \cdot \ln(1 - R)$$

with t = total bag closure time, τ = time used for sampling,
and R = ratio of the sample volume and the initial bag volume.

Formula (5) results from integration of the equation

$$(6) \quad W_E = -\int_{t-\tau}^{t} C(x) \frac{dV}{dx} \, dx = -\frac{R \cdot V_o}{\tau} \int_{t-\tau}^{t} C(x) \, dx$$

with (6a)
$$C(x) = C(t-\tau) + \int_{0}^{x} \frac{dC}{dy} \, dy \quad \text{and}$$

$$(6b) \quad C(t-\tau) = \frac{E \cdot B}{V_o} \cdot (t - \tau) \, , \quad (6c) \quad \frac{dC}{dy} = \frac{E \cdot B}{V_o \cdot (1 - R \cdot y/\tau)}$$

with C (x), C (y) = terpene concentration at time x or y,
V = bag volume at time x, V_o = initial bag volume.

3. RESULTS AND DISCUSSION

Table II summarizes the obtained results. They scatter widely between
3 and nearly 300 ng per gram of dry needle biomass and per minute.

Table II. Summary of the experimental data of conifer emission rate
measurements

Date	Plant name	Emission ng / g · min	Start time	Bag closure time	Bag temperature °C	Bag position (1)	Weather condition (2)
5.6.	Picea Abies 1	14	14.30	45'	25	2	1
12.6.	Picea Abies 1	3	9.50	34'35"	20	2	1
12.6.	Picea Abies 1	10	14.10	34'24"	25	2	2 − 3
24.6.	Picea Abies 2	9	10.51	25'46"	22	1	1 − 2
24.6.	Picea Abies 2	33	14.30	26'41"	25	1	2
30.6.	Picea Abies 2	200	14.32	16'30"	30,5	1	1
17.7.	Abies Douglas 1	84	8.51	20'55"	19	2	1
17.7.	Abies Douglas 2	280	11.48	17'30"	32	1	1
17.7.	Abies Douglas 2	260	14.04	12'45"	30	1	1 − 2
2.7.	Pinus Sylvestris 1	7	12.03	27'20"	22	1 − 2	1
2.7.	Pinus Sylvestris 1	32	14.15	20'35"	28	1	1
2.7.	Pinus Sylvestris 1	16	16.43	20'	25	2	1
14.7.	Pinus Sylvestris 2	(85)	10.18	20'20"	22	2	2 − 3
14.7.	Pinus Sylvestris 2	42	11.57	21'30"	25	1 − 2	2 − 3
14.7.	Pinus Sylvestris 2	38	17.03	21'14"	24	2	3
3.7.	Pinus Ponderosa	(45)	9.22	21'	23	1	1
3.7.	Pinus Ponderosa	47	11.40	20'40"	27	1	1
3.7.	Pinus Ponderosa	56	14.32	21'30"	27,5	1	1
3.7.	Pinus Ponderosa	18	17.01	20'40"	28	1	1
22.7.	Pinus Taeda 1	14	12.09	10'32"	32	1	1
22.7	Pinus Taeda 1	13	14.06	10'32"	33	1	1
22.7.	Pinus Taeda 1	4	16.47	15'18"	27	2	1
23.7.	Pinus Taeda 2	41	14.30	15'23"	29,5	(1−) 2	1
23.7.	Pinus Taeda 3	28	16.00	10'38"	32,5	1	1

(1) in sun = 1; in shadow = 2
(2) sunny = 1; partially cloudy = 2; overcast = 3

Many factors apparently influence the emission rate. Emission rate varies with the plant type, but may also be different for different trees of the same type (e.g. Pinus Sylvestris 1 and 2) and even for different branches of the same tree (e.g. Pinus Taeda 1, 2 and 3). The influence of other factors results more clearly from mean values and will be discussed below.

There is one important error source which may produce exceedingly high emission values. Emission of damaged plant surfaces (e.g. cuts, broken needles) is dramatically increased for a period of 20 - 40 minutes after plant lesion. Though during the enclosing procedure care was taken not to damage the enclosed branch, it cannot be excluded that particularly some of the long and delicate pine needles have been broken. This type of artifact probably has affected the values obtained for the first samples of Pinus Sylvestris and Pinus Ponderosa which are high with respect to the values obtained subsequently at higher temperatures in contrast to the generally observed temperature dependence (see below). For values measured later in a day an influence of this artifact is much less probable. Between different samples the bag was not removed but only opened towards the blower and around the branch in order to allow ambient air to flow through the bag (see above).

The emission of Abies Douglas is considerably higher than the emission of all other measured emissions. In fact, the very soft needles of Abies Douglas exhaled a much more pronounced fragrance than the other conifers under consideration.

Table III reports a number of mean values which have been calculated omitting the two above mentioned pine emission data.

The emission rate values have been classified according to the bag temperature during sampling and the mean values of different temperature intervals have been reported in Table III.1. The values obtained for Douglas fir have not been included, since only values of the lowest and highest temperature class are available and, hence, would have distorted the overall picture.

In spite of the wide scattering of individual values in the different groups, a distinct temperature dependence results for the terpene emission in agreement with analog findings by American authors (3 - 6). Using the emission values of Table II for those branches for which values at a minimum of two different temperatures have been measured, a least squares calculation gives the following exponential emission - temperature relationship:

(7) $$E_{exp}(t) = \text{constant} \cdot \exp(b \cdot t).$$

If E_{exp} is expressed in $ng \cdot g^{-1} \cdot min^{-1}$ and t in °C, a value of $b = 0.25 \pm 0.04$ results for all plants with the exception of Abies Douglas, for which a value of $b \approx 0.1$ appears more appropriate.

Table III.1 Mean emission rate values as a function of bag temperature

Number of values	temperature range $[°C]$	mean temperature $[°C]$	Emission rate range $[ng \cdot g^{-1} \cdot min^{-1}]$	mean emission rate $[ng \cdot g^{-1} \cdot min^{-1}]$
3	20 – 22	21.3	3 – 9	6
6	24 – 25	24.8	10 – 42	25
5	27 – 28	27.5	4 – 56	31
5	29.5 – 33	31.5	13 – 200	59

Table III.2. Mean emission rate values corrected to 30 °C

values measured by	emission rate $[ng \cdot g^{-1} \cdot min^{-1}]$	temperature $[°C]$
JRC ISPRA		
mean of all values including Douglas fir	92	30
mean of all values excluding Douglas fir	67	30
Douglas fir	247	30
Picea Abies	80	30
Pinus Silvestris	96	30
Ponderosa pine	78	30
Pinus Taeda	17	30
R.R. ARNTS et al (3)		
Loblolly pine	70	30
R. RASMUSSEN (4)		
Ponderosa pine	61	30 – 32
P.R. ZIMMERMANN (5)		
different conifers	79	30
D.T. TINGEY et al (6)		
Slash pine	110	30
Cryptomeria	50	30

Equation (7) has been used to correct the measured emission rate values reported in Table II to a temperature of 30°C in analogy to the procedure used by Tingey (6) and by Zimmerman (5)

(8)
$$E(30°C) = E_{measured} \frac{E_{exp}(30°C)}{E_{exp}(t°C)}$$

where $E_{measured}$ is the emission value measured at the temperature t (°C).

Different means of these corrected emission rate values are reported in Table III.2 and compared to values obtained by American workers (3 - 6).

In view of the considerable emission rate differences between different plant types, different plants of the same type and even different branches of the same plant, the agreement between the values obtained by this study and the values reported by the American authors is very satisfactory. The difference between the value reported here and by RASMUSSEN for Ponderosa Pine is well within the differences observed between different trees of the same species. Interesting for any comparison of the relative impact of naturally emitted terpenes on photochemical processes in America and Europe is that the emission rates of species more characteristic for the U.S.A. like Slash, Ponderosa and Loblolly pine and species more typical for Central Europe like Picea Abies and Pinus Sylvestris do not show significant differences.

Table III.3 compares the emission rate values (corrected for a temperature of 30°C) measured in shadowy (code 2 in Table II) and sunny (code 1 in Table II) positions excluding values obtained for Douglas fir, which would too much bias the mean values. Essentially no difference results. This is in agreement with findings of Tingey et al.(6) who, during experiments in an air exchange chamber and using artificial light sources, could not observe a marked dependency of emission on light intensity.

Table III.3 Mean of temperature corrected (30°C) emission rate values

measured in shadow	(6 values) :	59 ng g^{-1} min^{-1}
measured in sun	(10 values) :	61 ng g^{-1} min^{-1}

In conclusion, an assessment of the amount of naturally emitted terpenes will depend rather critically on a proper evaluation of the temperature distribution of the emitting biomass. Of minor importance will be whether it is based on emission rate values determined in Europe or in the U.S.A.

REFERENCES

1. C. LOHSE, H. STANGL, B. VERSINO, B. NICOLLIN, G. OTTOBRINI and H.RAU
   Proceedings of the First European Symposium 'Physico-Chemical
   Behaviour of Atmospheric Pollutants', B. VERSINO and H. OTT ed.,
   Commission of the European Communities, doc. EUR 6621, Brussels-
   Luxembourg 1980, pp. 150 - 156

2. H. STANGL, C. LOHSE, M. PAYRISSAT, B. VERSINO, B. NICOLLIN,
   G. OTTOBRINI and H. RAU, ibid. pp. 472 - 478

3. R.R. ARNTS, R.L. SEILA, R.L. KUNTZ, F.L. MOWRY, K.R. KNOERR,
   A.C. DUDGEON: Proceedings of the Fourth Joint Conference on Sensing
   of Environmental Pollutants, March 1978, pp. 829 - 833

4. R.A. RASMUSSEN: J. of the Air Pollution Control Association, 22
   (1972) pp. 537 - 543

5. P.A. ZIMMERMAN: E.P.A. Contract No. DU-77-1063, Final Report,
   May 1978

6. a) D.T. TINGEY, M. MANNING, H.C. RATSCH, W.F. BURNS, L.C. GROTHANS,
   and R.W. FIELD: U.S.E.P.A. Corvallis, Oregon, EPA Report CERL-45,
   August 1978
   b) E.W. PETERSON, D.T. TINGEY: Atmospheric Environment 14
   (1980) 79-81

7. H. KNÖPPEL, B. VERSINO, H. SCHLITT, A. PEIL, H. SCHAUENBURG,
   and H. VISSERS: Proceedings of the First European Symposium
   'Physico-Chemical Behaviour of Atmospheric Pollutants'
   B. VERSINO and H. OTT ed., Commission of the European Communities,
   doc. EUR 6621, Brussels-Luxembourg 1980, pp. 25 - 40

8. H. KNÖPPEL, M. DE BORTOLI, A. PEIL, H. SCHAUENBURG, and H. VISSERS,
   this volume, pp. 99-109.

THE DETERMINATION OF LINEAR PTGC RETENTION INDICES FOR USE IN ENVIRONMENTAL ORGANICS ANALYSIS

H. KNÖPPEL, M. DE BORTOLI, A. PEIL, H. SCHAUENBURG and H. VISSERS
Joint Research Center Ispra
Commission of the European Communities

Summary

The identification of many environmental organic pollutants, particularly of isomers, during GC-MS analysis is difficult or even impossible using only their mass spectra for identification. The simultaneous use of GC retention data in many cases allows to overcome this problem.

A method is described which allows to obtain relative retention data from linear temperature programmed GC (PTGC) runs. The method is based on polynomial interpolation, does not require the presence of a complete series of homolog straight chain alkyl compounds in the sample and may use also non-alkane sample compounds as references for the calculation of relative retention data.

The reproducibility of indices obtained under constant experimental conditions using apolar columns has been assessed as well as the influence of different experimental parameters.

1. INTRODUCTION

Linear programmed temperature gas chromatography (PTGC) using glass or fused silica capillary columns combined with mass spectrometry has become a standard method in environmental organics analysis. Identification of individual compounds in general is based on mass spectra. Often, however, mass spectra alone do not afford unambiguous identification. This is particularly true for a number of isomeric compounds with identical or very similar mass spectra such as alkylbenzenes (contained in car exhaust gases) and terpenes (emitted by plants). Moreover, in complex pollutant mixture analysis the quality of mass spectra of minor constituents is often poor and doubts remain. In these situations the agreement of gas chromatographic (GC) retention data of an unknown and a reference compound is a valuable confirmation (1). The direct use of retention times, however, is of only limited value since

- retention times change as soon as GC parameters such as initial temperature, temperature program, flow, stationary phase film thickness (which tends to diminish slightly with time in many cases) or the GC column itself are changed;

- particularly with older equipment insufficient control of some of the GC parameters may yield a scarce reproducibility of the retention times even maintaining the experimental parameters nominally unchanged;

- for the above reasons retention times are not appropriate for reference use over longer periods of time or for storage in a reference library.

Some of these limitations can be by-passed using relative retention data, like the linear PTGC retention index (2). The calculation of this index, however, is generally based on the linear interpolation between retention times of neighbouring straight chain homologs (in general

n-alkanes) and therefore requires the presence of a complete homolog series
of compounds in the sample. In view of the complexity of many environmental
samples it is often impossible to add the complete nor-alkane (or other
homolog) series to the sample without creating interferences with the
sample itself.

On the other hand the relationship between the homologs' carbon numbers (which by definition are proportional to the index) and their retention times is not strictly linear as supposed by linear interpolation.
This is demonstrated by fig. 1. The figure shows a linear programmed
temperature glass capillary gas chromatogram of a reference mixture containing the nor-alkanes from C_6 to C_{24}. The dashed line is a plot of the
alkane retention indices (by definition 100 x carbon number) versus
retention times and is distinctly curved. Therefore indices of the same
compound calculated by linear interpolation between different pairs of
homologs differ in general.

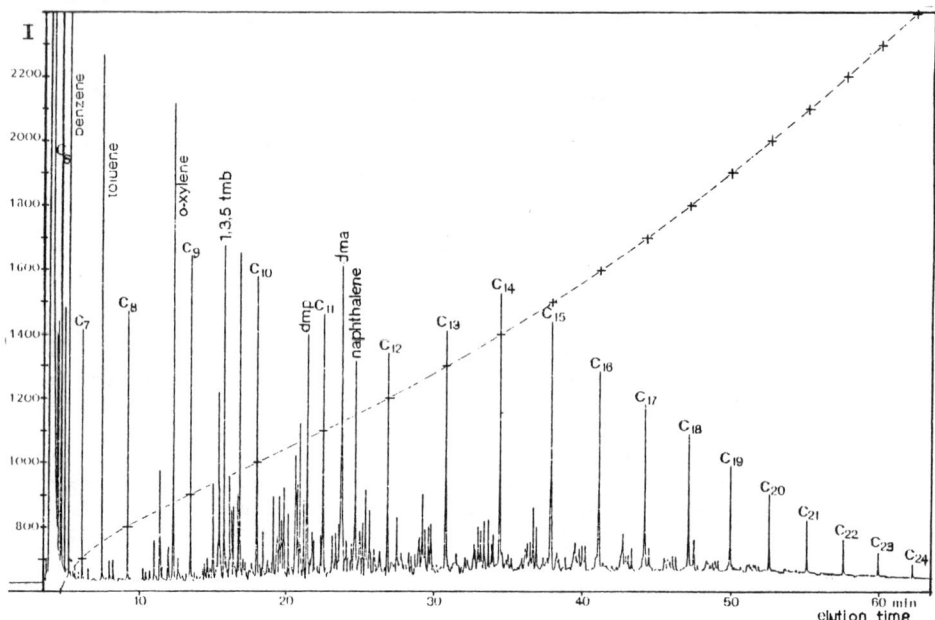

Fig. 1. Glass capillary chromatogram of a reference mixture containing
nor-alkanes (C_6 – C_{24}) and further compounds like indicated
(tmb = 1,3,5-trimethylbenzene, dmp = 2,6,-dimethylphenol, dma =
2,6,-dimethylanilin)

Table 1 reports as an example the indices of some nor-alkanes as obtained
by linear interpolation between neighboured and more distant homologs.
They deviate clearly from the defined values.

In isothermal gas chromatography a much better linearity is found
between the nor-alkane (or other straight chain homologs) carbon number
and the logarithm of the respective net retention times. However, even
in this case slight deviations from linearity have been observed and
corrected by a second and third order polynomial approach (3,4).

Table 1 – Comparison of PTGC retention indices calculated by [1] linear and by polynomial interpolation

alkane carbon n°	retention time $\underline{/}$min$\underline{/}$	by definition	retention indices calculated by linear interpolation		calculated by polynomial interpolation [2]
6	3.849	600	<u>600.0</u>	<u>600.0</u>	<u>600.0</u>
7	5.496	700	667.3	653.0	698.9
8	8.743	800	<u>800.0</u>	757.6	799.3
9	13.168	900	897.6	<u>900.0</u>	<u>900.1</u>
10	17.808	1 000	<u>1 000.0</u>	1 008.0	1 000.7
11	22.253	1 100	1 103.2	1 111.5	1 100.7
12	26.424	1 200	<u>1 200.0</u>	1 208.6	1 200.3
13	30.351	1 300	1 302.8	<u>1 300.0</u>	<u>1 300.0</u>
14	34.063	1 400	<u>1 400.0</u>	1 417.7	1 400.0
15	37.543	1 500	1 502.9	1 528.1	1 499.9
16	40.824	1 600	<u>1 600.0</u>	1 632.2	1 599.7
17	43.939	1 700	1 702.5	1 731.0	1 699.6
18	46.903	1 800	<u>1 800.0</u>	1 825.0	1 799.7
19	49.728	1 900	1 902.4	1 914.0	1 899.8
20	52.421	2 000	<u>2 000.0</u>	<u>2 000.0</u>	<u>2 000.0</u>

1) underlined values have been used as references and are those given by definition in the case of linear interpolation. In the case of polynomial interpolation the corresponding retention time values have been used for adaptation of the master polynomial.

2) see section 'description of the method'.

This paper describes a computer-based method for the determination of PTGC retention indices using higher order polynomial for representation of the relation between the carbon atom number and elution time of straight chain homologs. The method has been designed in order to allow for interpolation between larger elution time differences than those of neighbouring non-alkanes and for the simultaneous use of compounds of different classes as references. This way a sample has not to contain a complete homolog straight chain compound series for the index calculation and the addition of references to a sample can be limited to relatively few compounds chosen in order not to interfere with the sample.

2. DESCRIPTION OF THE METHOD

The calculation of the retention indices for the constituents of an environmental (or other) sample is performed in two steps.

1. A reference mixture containing a complete series of non-alkanes covering the elution time range of interest is separated under the same chromatographic conditions as the sample (s) for which indices have to be determined. Using the non-alkane retention times t_n (n = number of carbon atoms), a "master" polynomial of degree m (m $\leq$ 13) is calculated which gives the retention indices I as a function of retention time t and approximates the points $(t_n / 100 \cdot n)$ using the least squares method :

$$(1) \qquad I = \sum_{i=0}^{m} a_i \cdot t^i \quad, \text{ where I meets the condition}$$

$$(2) \qquad \sum_{n=1}^{l+k-1} (I_n - 100 \cdot n)^2 = \text{MIN !}$$

k = number of reference alkanes $\geq$ m + 2 ,
l = number of carbon atoms of the first of the reference alkanes.

The minimum condition (2) yields m + 1 linear equations for the m + 1 coefficients a_i of equation (1) :

$$(3) \qquad \sum_{i=0}^{m} a_i \left(\sum_{n=1}^{l+k-1} t_n^{i+j} \right) = 100 \cdot \sum_{n=1}^{l+k-1} n \cdot t_n^j \quad ; \quad j = 0, 1, \ldots, m$$

Using standard mathematical methods, the coefficients a_i are calculated from equations (3) by means of a computer program. Equation (1) allows then to calculate an index I for any compound contained in the reference mixture. A master polynomial needs to be determined only once for a given set of chromatographic conditions.

2. To each sample subsequently analysed under the same chromatographic conditions a number of k' = k reference alkanes (or other compounds contained in the reference mixture) are added, chosen in a way not to interfere with the sample compounds, and their retention times t'_n and

indices I'_n $(n = 1,...,k';$ n has here no longer the meaning of a carbon atom number) are used to adjust the $(k' - 1)$ first coefficients of the master polynomial in the following way: The indices of the sample compounds are calculated using the equation

$$(1') \quad I' = \sum_{i=0}^{k'-2} a'_i \cdot t'^i \quad + \quad \sum_{i=k'-1}^{m} a_i \cdot t'^i$$

a'_i = coefficients of the polynomial which are adapted

t', I' = retention times respectively indices of sample compounds

In analogy to the construction of the master polynomial the minimum condition

$$(2') \quad \sum_{n=1}^{k'} \left[\sum_{i=0}^{k'-2} a'_i \cdot t'^i_n \quad + \quad \sum_{i=k'-1}^{m} a_i \cdot t'^i_n - I'_n \right]^2 = MIN!$$

yields $k'- 1$ linear equations

$$(3') \quad \sum_{i=0}^{k'-2} a'_i \cdot \left[\sum_{n=1}^{k'} t'^{i+j}_n \right] = \sum_{n=1}^{k'} \left[I'_n \cdot t'^j_n - \sum_{i=k'-1}^{m} a_i \cdot t'^{i+j}_n \right]$$

for the $k'- 1$ coefficients a'_j $(j = 0,..., k'-2)$.

The rationale of this adaptation procedure is that the shape of the curve $I(t)$ will not change drastically as long as the chromatographic parameters remain nominally constant.

Computer programs have been written which calculate the master polynomials and perform the adaptation. They have been implemented using the Finnigan INCOS GC-MS data system. It is noteworthy, that retention times are linearly transformed befor index calculation in a way to shift the time '0' to the center of the elution time interval of the reference compounds. This way a loss of precision by round off errors during evaluation of equations (3) and (3') can be avoided and polynoms up to the 13th order can be elaborated.

3. RESULTS AND DISCUSSION

A series of measurements have been performed with a test mixture similar to that shown in fig. 1. in order to check the reproducibility of the indices obtained by the described method on apolar capillary columns and to assess the influence of the covered elution range and a change of the chromatographic column on the index performance.

A Hewlett Packard 5880 series gas chromatograph equipped with a 5880A series GC terminal (level 4) and a flame ionization detector has been used for the determination of retention times. Retention times have been transferred to the Finnigan INCOS data system for index calculations.

All measurements have been performed using helium as carrier gas and the following temperature program: 5 min isothermal at 35°C, subsequently progrmmed to 250°C at a rate of 4°C/min.

Index reproducibility for repetitive measurements. The reproducibility of indices calculated using master polynomials and adapted polynomials has been determined injecting 1 μl amounts of the reference mixture 10 times on each of four OV-1 coated columns (for specifications see table 5). The results for each of these columns were comparable.

Table 2 summarizes the results obtained for column 2 (table 5). The table reports for the listed compounds mean values of the ten retention times and indices. Three index values have been calculated: one using master polynomials (fitted through the complete n-alkane series) and two values using polynomials adapted to one of the ten master polynomials with different sets of references (underlined in table 2), including non-alkane compounds.

Together with the retention times and indices the maximum deviations from the means are reported, expressed in percent of the retention time resp. index difference of the two n-alkanes adjacent to each compound (in case this is a n-alkane of course the maximum deviation has to be referred to half this difference). This way the maximum deviations of retention times and of indices can be directly compared to one another. In the case of indices the percent values are identical with the absolute maximum deviations, since the index difference between neighboured n-alkanes is 100 by definition. Maximum rather than standard deviations have been reported in view of their greater practical importance.

With the exception of benzene and toluene, index values calculated by means of master polynomials are 8 -> 10 times better reproducible than retention times. The case of benzene and toluene is discussed below. Using the first set of references for the polynomial adaptation, indices in the range nonane - heptadecane are reproduced within 0,1 index units or better. The same is true for indices in the range benzene - dodecane using the second set of reference compounds. For comparison, SOJAK and RIJKS (6) obtained for the Kováts (7) retention index of o-xylene on a Squalane coated stainless steel capillary at 70°C an index value of 877,48 with a standard (not maximum) deviation of 0,01 index units.

Considering the presented results a combined use of two sets of references appears to be most appropriate in order to obtain best possible results for a wide elution time resp. temperature range. Benzene, toluene and o-xylene in general need not to be added to environmental samples as references since they are ubiquitous pollutants which are easily identified.

Influence of the elution time range covered by the reference compounds on index performance. The relatively bad reproducibility of the indices obtained for benzene and toluene using the set of master polynomials can be explained by the following effect: both compounds elute near to the isothermal portion of the chromatogram where the I(t) curve follows a law which is different from that ruling the linear programmed range. As can be observed in figure 1 the I(t) curve has an inversion point near n-nonane. Obviously compounds eluting before the corresponding retention time move already markedly on the column during isothermal time. (Even starting programmation immediately would not change essentially this situation unless lower starting temperature could be used.)

The described behaviour of the I(t) curve introduces a 'stress' on the fitting polynomials which tend to oscillate between the first reference n-alkanes, as suggested by table 3.

Table 2. COMPARISON OF THE REPRODUCIBILITY OF RETENTION TIMES AND INDICES a)

compound name	mean retent. time (min)	DMAX b) (%)	mean retention indices I and maximum deviations DMAX b) obtained by adaptation to the underlined values c)					
			I	DMAX (%)	I	DMAX (%)	I	DMAX (%)
n-hexane	2,564	0,18	600,00	0,02	600,15	0,81	600,37	0,12
BENZENE	3,611	0,25	646,16	0,93	646,06	0,57	646,15	0,00
n-heptane	5,367	0,32	700,00	0,01	700,01	0,43	699,99	0,04
TOLUENE	7,641	0,45	751,83	0,37	751,89	0,35	751,82	0,00
n-octane	9,994	0,52	799,99	0,00	799,99	0,23	799,94	0,02
o-XYLENE	13,691	0,57	874,94	0,04	874,93	0,03	874,93	0,02
n-nonane	14,905	0,56	900,00	0,00	899,99	0,00	900,00	0,02
1,3,5-TRIMETHYLBENZENE	17,425	0,58	953,23	0,07	953,23	0,03	953,24	0,05
n-decane	19,559	0,51	999,99	0,00	1000,00	0,04	1000,00	0,05
N,N-DIMETHYLANILIN	22,301	0,60	1062,62	0,05	1062,63	0,05	1062,62	0,06
2,6-DIMETHYLPHENOL	22,958	0,58	1078,07	0,04	1078,09	0,05	1078,07	0,06
n-undecane	23,879	0,58	1100,00	0,00	1100,01	0,03	1099,99	0,00
NAPHTHALENE	26,211	0,67	1157,10	0,06	1157,11	0,06	1157,10	0,07
n-dodecane	27,902	0,62	1199,98	0,00	1199,97	0,02	1199,99	0,00
n-tridecane	31,654	0,74	1300,01	0,00	1300,00	0,06	1300,13	0,10
n-tetradecane	35,159	0,82	1399,97	0,01	1399,98	0,08	1400,24	0,22
n-pentadecane	38,460	0,78	1500,02	0,00	1500,05	0,04	1500,32	0,25
n-hexadecane	41,586	0,95	1599,98	0,00	1600,00	0,00	1599,99	0,00
n-heptadecane	44,548	1,04	1700,00	0,00	1700,02	0,06	1699,24	0,71
ANTHRACENE	46,208	1,21	1758,43	0,14	1758,44	0,10	1756,88	1,45
n-octadecane	47,360	1,09	1799,99	0,00	1799,99	0,16	1797,69	2,13

a) The reported retention time and index values are averages of ten GC runs.

b) DMAX is given in % of the elution time resp. retention index difference of adjacent n-alkanes. For n-alkanes half this difference is taken. In the case of indices this difference is 100 by definition, i.e. DMAX is also the absolute maximum deviation.

c) Indices of the first column have been obtained by calculating 'master' polynom all ten runs. Indices of the second and third column have been calculated by adapting one of the master polynomials to the other nine runs.

Table 3. RETENTION INDICES AS A FUNCTION OF POLYNOMIAL ORDER
AND COVERED NOR-ALKANE RANGE

compound name	indices calculated with polynomial[a] of order				
	5	7	9	11	13
n-hexane	–	–	–	–	599,99
n-octane	–	–	–	799,99	799,99
o-xylene	–	–	–	874,20	874,93
1,3,5-trimethyl benzene	–	–	–	953,44	953,23
n-decane	–	–	999,99	999,99	999,99
naphthalene	–	–	1157,20	1157,14	1157,12
n-dodecane	–	1199,99	1199,99	1199,99	1199,99
n-tetradecane	1400,00	1400,00	1399,99	1399,98	1399,98
n-hexadecane	1600,00	1600,00	1599,99	1599,98	1599,98
anthracene	1758,47	1758,47	1758,48	1758,49	1758,49
n-nonadecane	1899,99	1899,99	1900,00	1900,00	1900,00
n-eicosane	2000,00	1999,99	1999,99	1999,99	1999,99

[a] also intermediate n-alkanes not listed in the table have been used for the calculation

The table reports indices calculated by means of master polynomials of increasing order, covering increasing elution time ranges with eicosane as common upper limit. Polynomials for ranges from decane upwards, though of relatively low order, fit exactly the n-alkane indices and yield a constant index value for anthracene. Polynomials fitting also to lower alkane references still fit well the n-alkane indices, however yield varying indices for o-xylene, 1,3,5-trimethylbenzene and naphthalene, variations being the larger the earlier a compound elutes. Index variations for o-xylene and 1,3,5-trimethalbenzene go in opposite directions confirming the above mentioned oscillation of the polynomials. Obviously polynomial interpolation works best for retention index calculation, if it covers only an elution temperature range where an isothermal influence is excluded.

Influence of column change on index performance. The influence of column change on isothermal retention index performance has been recognized (7,8) as one of the major reasons for variations exhibited by relative retention data from different sources and the resulting reduced suitability of retention indices as reference data for identification. SOJAK et al. (7) assume, that the low reproducibility of isothermal retention data is mainly due to adsorption interactions of the chromatographed compounds at gas – stationary phase or stationary phase – column wall interfaces and to an influence of the stationary phase film thickness on retention data. This second influence has been found more important for more polar liquid phases (8).

Table 4. COMPARISON OF RETENTION INDICES OBTAINED WITH DIFFERENT OV-1 COATED CAPILLARY COLUMNS [a]

compound name	column			
	1	2	3	4
BENZENE	646,45	646,16	646,25	649,57
TOLUENE	753,09	751,83	751,69	750,91
o-XYLENE	874,57	874,94	874,90	876,06
1,3,5-TRIMETHYL BENZENE	952,94	953,23	953,18	954,13
N,N-DIMETHYL ANILIN	1061,44	1062,62	1062,55	1064,24
2,6-DIMETHYL PHENOL	1078,59	1078,07	1078,22	1078,75
NAPHTHALENE	1154,08	1157,10	1156,91	1160,29
ANTHRACENE	1749,69	1758,43	1757,79	1765,99

[a] for column parameters see table 5

Table 5. EXPERIMENTAL PARAMETERS OF THE COLUMNS MENTIONED IN TABLE 4 [a]

column nr.	column material	column length (m)	int. dia. (mm)	film thickness (μm)	desacti- vation	flow at 35°C (ml/min)	split ratio
1	fused silica	50	0,2	0,2	b)	1,0	1/9,3
2	soft glass	24	0,3	1,1	b)	3,7	1/5,4
3	"	25	0,3	1,1	b)	3,8	1/5,7
4	"	30	0,3	1,1	b)	3,7	1/6,5

[a] all columns were run isothermally for 5 min at 35°C and subsequently temperature programmed to 250°C at 4°C/min

[b] col. 1 is a commercial persilylated column; col. 2-4 have been persilylated according to the procedure described by GROB (5)

Unpolar capillary columns, as mainly used in environmental analysis, should therefore be most promising with respect to retention index reproducibility. Moreover the development of well disactivated unpolar columns has made considerable progress troughout the last few years (9).

Under these aspects it is interesting to compare the PTGC retention indices obtained by the described method on different unpolar columns. Table 4 reports the mean values of the indices of non-alkane compounds contained in the reference mixture obtained from ten repetitive injections (for maximum deviations from the means see table 2). Column and further experimental parameters are summarized in table 5. Columns 2, 3 and 4 were all produced following the same procedure (5). With the exception of anthracene, indices measured on columns 2 and 3 agree within 0,2 index units, a very satisfactory result. Column 4 however, yields considerably larger deviations, in particular for naphthalene and anthracene. In view of the similarity of all other parameters a different activity and/or a inhomogeneity of the stationary phase of this column may be the reason.

The indices obtained for column 1 differ also most from those of column 2 and 3 for naphthalene and anthracene, however in the opposite direction compared to that observed for column 4. In view of the many differences between this column and the other tree ones it is not possible to attribute this deviation to any particular parameter.

An interesting observation is, that the indices of 2,6-dimethylphenol agree within half an index unit for all columns. Since this compound is the most acidic one of those compared in table 4 all columns might be slightly acidic, a fact usually observed for fused silica columns.

The agreement of the indices observed for benzene, toluene and o-xylene on the different columns compares well with that of Kovàts indices determined on squalane for the same compounds. The following values have been collected from literature and corrected to a temperature of 70°C by SOJAK and RIJKS (6): benzene: 641,6 - 645,3; toluene: 748,9 - 751,9 and o-xylene: 874,1 - 877,8 . These values have however been obtained using different carrier gases and column materials.

4. CONCLUSION

Using the described method for the determination of linear PTGC retention indices, for repetitive injections on a OV-1 coated column and constant experimental conditions an index reproducibility of 0,1 index units or better has been obtained.

The method works best for compounds which start to move into the column only at temperatures above the initial temperature, i.e. which elute at at least 40°- 50° higher temperatures.

Deviations of indices measured on different apolar columns are comparable to those obtained isothermally. In view of the rapid progress of the development of apolar, temperature stable capillary columns with reproducible design parameters and of the progress in GC instrument performance an inter-column reproducibility of PTGC retention indices of about 0,5 index units appears to be obtainable in the next future.

REFERENCES

1. H. KNÖPPEL, B. VERSINO, A. PEIL, H. SCHAUENBURG, H. SCHLITT
   and H. VISSERS: Proceedings of the First European Symposium
   "Physico-Chemical Behaviour of Atmospheric Pollutants",
   B. VERSINO and H. OTT ed., Commission of the European Communities, doc. EUR 6621, Brussels-Luxembourg 1980 (pp. 25-40)

2. W.H. HARRIS and H.W. HABGOOD, "Programmed Temperature Gas
   Chromatography", John Wiley & Son, Inc., New York / London /
   Sydney 1966, p, 153 ff.

3. F.J. HEEG, R. ZINBURG, H.J. NEU and K. BALLSCHMITER,
   Chromatographia $\underline{12}$ (1979) 451 - 58.

4. F.J. HEEG, R.ZINBURG, H.J. NEU and K. BALLSCHMITER,
   Chromatographia $\underline{12}$ (1979 790-798.

5. a. K. GROB, G. GROB and K. GROB Jr.; Journal of HRC & CC,
      2,(1979) 31 - 35
   b. K. GROB, G.GROB and K. GROB Jr.; Journal of HRC & CC,
      2 (1979) 677 - 678
   c. K. GROB and G. GROB; Journal of JRC & CC, 3 (1980) 197
   d. K. GROB, Journal of HRC & CC, 3 (1980) 493 - 496

6. L. SOJAK and J.A. RIJKS, J. Chromatogr., 119 (1976) 505 - 521

7. E. KOVATS, Advan. Chromatogr., 1 (1966) 229

8. L. SOJAK, V.G. BEREZKIN, and J. JANAK, J. Chromatogr. 209 (1981)
   15 - 20

9. T. SHIBAMOTO, K. HARADA, K. YAMAGUCHI and A. AITOKU,
   J. Chromatogr., 194 (1980) 277 - 284

10. K. GROB, Journal of HRC & CC, 2 (1979) 599 - 604

CHEMICAL AND PHOTOCHEMICAL REACTIONS

Chairman: R. A. COX

ATMOSPHERIC REACTIVITY OF OXYGENATED MOTOR FUEL ADDITIVES

RICHARD A. COX AND ANNMARIE GOLDSTONE

Environmental and Medical Sciences Division, A.E.R.E.,
HARWELL, OX11 ORA, U.K.

Summary

The OH-radical initiated photo-oxidation of the oxygenated compounds
methyl-tert-butyl ether (MTBE), tert-butyl alcohol (TBA) and ethanol,
has been investigated. These compounds are increasingly being used as
additives for octane improvement in motor fuels, and can enter the
atmosphere by evaporative emission. The rate coefficients for OH attack
on these compounds has been determined using a relative rate technique
involving photolysis of HONO as a source of OH. The following values
were obtained at 295K:

$$k(OH + MTBE) = (2.5 \pm 0.5) \times 10^{-12} \ cm^3 \ molecule^{-1} \ s^{-1}$$
$$k(OH + TBA) = (1.0 \pm 0.2) \times 10^{-12} \ cm^3 \ molecule^{-1} \ s^{-1}$$
$$k(OH + C_2H_5OH) = (3.1 \pm 0.5) \times 10^{-12} \ cm^3 \ molecule^{-1} \ s^{-1}$$

The major products of the oxidation of MTBE and TBA were tentatively
identified at t-butyl formate and acetone, and mechanisms for the
formation of these products have been suggested. Implications of the
results for the atmospheric degradation of these oxygenated additives
are briefly discussed

INTRODUCTION

The reduction of lead containing additives in petrol in response to
air quality regulations and the requirements of installed emission control
equipment (catalytic convertors) has led to other methods of octane
number improvement. The addition of oxygenated compounds, alcohols and
ethers, to petrol, in amounts up to 20vol% is becoming a widespread practice for octane improvement of motor fuels for sale in Europe and the USA[1]
This implies widespread dissemination of such materials and their high volatility can lead to evaporative emission to the atmosphere.

In common with all volatile organic compounds, the major fate of oxygenates in the atmosphere is expected to be homogeneous photo-chemical oxidation. They can consequently contribute to photo-oxidant (ozone) formation under appropriate weather conditions. Since alcohols and ethers do not absorb visible or near U.V. sunlight,[2] the primary criterion selected to assess their atmospheric reactivity with respect to photo-oxidant formation is their rate of reaction with hydroxyl radicals, OH.[3] Secondary factors which need to be considered are the stoichiometry for NO to NO_2 oxidation during oxidative degradation and the nature of the oxidation products[4]

In the present paper, a study of the reaction of OH radicals with three oxygenated compounds, Methyl t-butyl ether (MTBE), t-Butanol (TBA) and Ethanol under simulated atmospheric conditions is reported. The rate coefficients for OH attack were determined by a relative rate method using HONO photolysis as a source of OH radicals in synthetic air[4,5] Preliminary investigation of the products of the reactions was carried out and a mechanism for the oxidative degradation of these compounds under atmospheric conditions has been proposed. The implications of the results for the behaviour of oxygenated additives in the atmosphere is discussed.

EXPERIMENTAL

The OH-initiated photo-oxidations were conducted in a 'smog chamber' comprising basically a 200 L Tedlar bag (Dupont) irradiated by 20 x 20 watt fluorescent "Blacklamps' (Philips TL 20/08).

OH radicals were produced by the photolysis of part-per-million concentrations of nitrous acid vapour (1 ppm = 2.45 x 10^{13} molecules cm^{-3} at 760 torr and 298 K) prepared in an $N_2 + O_2$ diluent at atmospheric pressure and room temperature (295 $\pm$ 2 K) as described before.[4-6] The steady state [OH] is controlled initially by the reactions

$$HONO + h\nu \rightarrow OH + NO \qquad\qquad I_{abs}$$
$$OH + HONO \rightarrow H_2O + NO_2 \qquad\qquad (1)$$

As NO and NO_2 build up, OH is also removed by:

$$OH + NO \ (+M) \rightarrow HONO \ (+M) \qquad\qquad (2)$$
$$OH + NO_2 (+M) \rightarrow HONO_2(+M) \qquad\qquad (3)$$

If an organic substrate is present it may react with OH in competition with reactions (1)-(3).

The dilute mixtures of HONO together with organic substrate were irradiated for up to 30 min. Samples were withdrawn periodically for gas-chromatographic analysis of the organic reactants and products using both FID and ECD detection. Separation was achieved on columns packed with PEG 400 on Chromosorb W either 1, 3 or 5 ft. in length, depending on the analysis. The response of the FID detector for the major components was calibrated with authentic samples, and the overall accuracy of the measurements was $\pm$ 20% or better. HONO, NO and NO_2 were measured using a commercial analyser (TECO Model 12A) employing $NO + O_3$ chemiluminescence.[5]

The rate coefficients were determined from measurements of the decay of the compound of interest together with a reference compound of known reactivity with OH. The concentrations at any given time are related by the following expression, obtained from integration of the rate equation for the two compounds:

$$\log \frac{S_0}{S} = \frac{K_s}{K_r} \log \frac{R_0}{R} \qquad (i)$$

where S and K_s are the concentration and OH reaction rate coefficient for the substrate, and R and K_r those for the reference compound. A plot of equation (i) should be linear if OH is the only species causing decay of the two compounds, and the slope gives the rate constant ratio K_s/K_r directly.

Products were identified as far as possible by comparison of retention times of pure samples on the G.C.

RESULTS

1. The reaction of OH with MTBE

Mixtures of HONO + MTBE together with either ethene or n-Hexane as reference compound were irradiated. Six experiments were performed with initial MTBE concentration between 0.34 and 3.7ppm. The decay of MTBE was slower than the reference compounds but quite readily measurable. The results are plotted according to eq(i) in Fig.1. where it will be seen that a good linear relationship was obtained for both ethane and n hexane as reference compound. Table 1 shows the values of the rate coefficient for reaction (4) obtained from the slopes and using

$$OH + CH_3OC(CH_3)_3 \rightarrow products \qquad (4)$$

the consensus values for $K(OH + C_2H_4) = 8.0 \times 10^{-12}$ cm^3 molecule^{-1} s^{-1}[7]
and $K(OH + n-C_6H_{14}) = 5.9 \times 10^{-12}$ cm^3 molecule^{-1} s^{-1}[8].

Two products were detected in the photolysis of MTBE-HONO-Air
mixtures. A minor product had a retention time corresponding to acetone.
The major product, eluting after acetone, was tentatively identified as
t-butyl formate, $HCOOC(CH_3)_3$, by comparison of its retention times with
that of $HCOOCH_3$, CH_3COOCH_3 and $CH_3COOC(CH_3)_3$.

2. The reaction of OH with TBA

The rate coefficient for reaction (5) between OH and TBA was

$$OH + (CH_3)_3COH \rightarrow products \qquad (5)$$

determined relative to ethene. Three different mixtures of TBA (1.7-3.5ppm)
together with ethene ($\sim$ 3ppm) + HONO were irradiated. The decay of TBA
was relatively slow and since analysis of TBA was less reproducible than
the other substrates used in this study, accuracy of measurement of the
concentration-time data was less. Nevertheless the data plotted in
Fig.2 show reasonable concordance with equation (1) and allow the value
of K_1 to be determined from the slope, see Table 1.

The only detectable product in the photolysis of TBA-HONO-Air
mixtures was acetone. Calibration of the detector for acetone allowed the
absolute yields of this product to be measured. This was compared to the
amount of TBA removed, by plotting the ratio log $[TBA]_o/([TBA]_o-[CH_3COCH_3]_t)$
according to eq.(i), see filled points Fig.2. If acetone is formed in
1:1 stoichiometry in reaction (2), these data should coincide with the
plot based on TBA concentration-time data. The data indicate that at
least 80% of reaction (2) forms acetone.

3. The reaction of OH with Ethanol

The rate coefficient for reaction (6) between OH and ethanol was
determined relative to propene.

$$OH + C_2H_5OH \rightarrow products \qquad (6)$$

TABLE 1 – Rate Cofficients for Reaction of Oxygenates
with OH at 295 ± 2K

Compound	Reference	K_s/K_r	K_s cm^3 molecule^{-1} s^{-1} X 10^{12}
MTBE	C_2H_4	0.33 ± 0.06	2.6 ± 0.5
	n-Hexane	0.44 ± 0.07	2.4 ± 0.4
TBA	C_2H_4	0.125±0.015	1.0 ± 0.2
Ethanol	C_3H_6	0.13 ± 0.02	3.1 ± 0.5

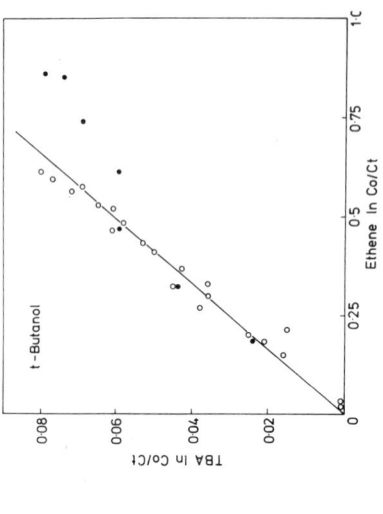

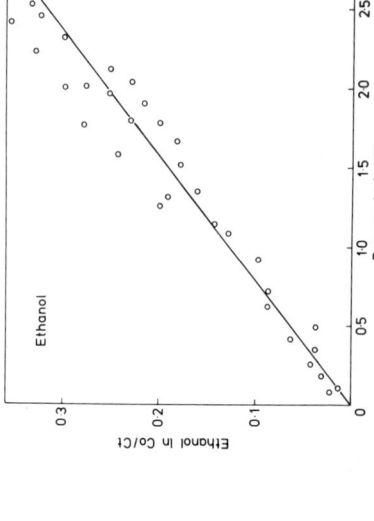

Fig. 2 – Plot of concentration-time data for TBA+C_2H_4
according to eq(i). Filled points show plot with TBA
concentrations calculated from acetone product yields
i.e. $[TBA]_o - [CH_3COCH_3]_t$

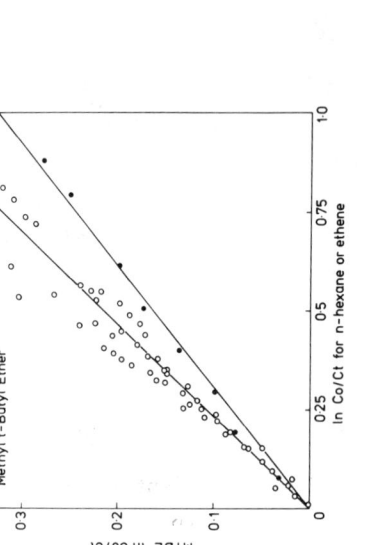

Fig. 3 – Plot of concentration-time data for Ethanol+C_3H_6
according to eq.(i)

Fig. 1 – Plot of concentration-time data for MTBE+n-
hexane (open points) and MTBE+C_2H_4 (filled points)
according to eq. (i)

Fig.3 shows a plot of the concentration-time data from 3 experiments with initial C_2H_5OH concentrations of 2.8-11.2ppm according to eq.(i). From the slope of this plot and using a value for $K(OH + C_3H_6) = 2 \times 10^{-11}$ cm^3 molecule^{-1} s^{-1}[8] we obtain $K_3 = (3.1 \pm 0.5) \times 10^{-12}$ cm^3 molecule^{-1} s^{-1}.

The major product of the reaction of OH with C_2H_5OH under simulated atmospheric conditions has been previously shown to be acetaldehyde[9]. Since this is also a product from the OH + C_3H_6 reaction, formation of CH_3CHO from reaction (6) would not be measured in these experiments.

DISCUSSION

The rate coefficient determined for the reaction of OH with ethanol is an excellent agreement with previous measurements of this rate coefficient at $\sim 300K$ having both direct and indirect techniques, which give a mean value of 3.2×10^{-12} cm^3 molecule^{-1} s^{-1}.[8] This lends support to the validity of the present technique for OH rate measurements. Rate coefficients for reaction of OH with MTBE and TBA have not been reported before but the values are reasonable in terms of the expected reactivity of the C-H bonds in these compounds. The rates are low relative to OH attack on propene, a typical reactive hydrocarbon component of motor exhaust and n-hexane, a typical component of evaporative emissions. Thus these oxygenates, are of relatively low photochemical reactivity for oxidant formation with relatively long atmospheric residence times ($3\frac{1}{2}$ days for MTBE, $14\frac{1}{2}$ days for TBA, based on a mean atmospheric OH concentration of 10^6 molecules cm^{-3}).

The product formation data provides information relevant to the oxidative degradation mechanism of the simple ethers and alcohols. In ethers it has been found that C-H bonds adjacent to the -O atom react more readily than the equivalent C-H bonds in hydrocarbons[8]. Initial attack on MTBE is therefore more likely on the $-O-CH_3$ group.

$$MTBE \xrightarrow{OH} t\text{-}Bu\text{-}O\text{-}\overset{\cdot}{C}H_2 \xrightarrow{O_2, NO} (CH_3)_3C\text{-}O\text{-}CH_2\text{-}\overset{\cdot}{O} + NO_2$$

The alkoxy radical so produced can either decompose or react with O_2:

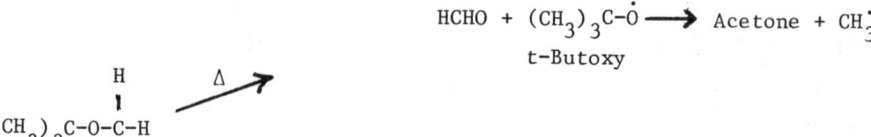

$$HCHO + (CH_3)_3C\text{-}\overset{\bullet}{O} \longrightarrow \text{Acetone} + CH_3^{\bullet}$$
$$\text{t-Butoxy}$$

$$(CH_3)_3C\text{-}O\text{-}\overset{\overset{\displaystyle H}{|}}{\underset{\underset{\displaystyle O}{|}}{C}}\text{-}H \qquad \overset{\Delta}{\nearrow} \qquad \overset{+O_2}{\searrow}$$

$$HCOOC(CH_3)_3 + HO_2 \xrightarrow{NO} OH + NO_2$$
$$\underline{\text{t-Butyl formate}}$$

Since t-Butyl formate is the major product observed, it is clear that reaction with O_2 predominates under atmospheric conditions. The small amounts of acetone formed may reflect a minor decomposition pathway but this product could also be formed by fragmentation of the alkoxy radical formed following attack of OH on the $(CH_3)_3C$- group. The dominant pathway, forming t-Butyl formate, gives a stoichiometry factor of 2 for NO to NO_2 conversion.

The major product from TBA oxidation was acetone and this could be formed following initial abstraction from an O-H or a C-H bond in TBA. However, from the small amount of information available, the O-H bonds in alcohols are apparently less readily attacked than primary and secondary C-H bonds.[9,10] This would indicate that attack occurs on the $(CH_3)_3C$- group in TBA, which is consistent with the relatively low value found for K_3. A possible mechanism for acetone formation is as follows:-

$$TBA \longrightarrow \overset{\overset{\displaystyle CH_3}{|}}{\underset{\underset{\displaystyle CH_3}{|}}{\overset{\bullet}{C}H_2\text{-}C\text{-}OH}} \longrightarrow \overset{\overset{\displaystyle CH_3}{|}}{\underset{\underset{\displaystyle \overset{\bullet}{O}\ \ CH_3}{|\ \ \ |}}{CH_2\text{-}C\text{-}OH}} \longrightarrow HCHO + \overset{\overset{\displaystyle CH_3}{|}}{\underset{\underset{\displaystyle CH_3}{|}}{{}^{\bullet}C\text{-}OH}}$$

$$\Big\downarrow O_2$$

$$CH_3COCH_3 + HO_2$$

The final step involving the reaction of an α-hydroxy radical with O_2 to give HO_2 directly has been shown to be rapid compared to addition of O_2 to give the corresponding peroxy radical.[9] This mechanism gives a stoichiometry factor of 2 for NO to NO_2 oxidation.

In considering the overall environmental impact of volatile organics, the fate of the oxidation products needs to be considered. Acetone which is formed in oxidation of both TBA and MTBE is relatively unreactive toward OH and is photolysed only slowly[5]. The reactivity of tert-butyl formate toward OH is not known. However the esters of the higher aliphatic acids are relatively unreactive[8] as is also formic acid[11]. Therefore it seems likely that tert-butyl formate is also unreactive. The formate esters are on the other hand rather corrosive and their presence in the atmosphere as secondary pollutants at significant concentrations may add to the undesirable effects of photochemically generated pollution.

Acknwoledgement.

This work was carried out as part of a programme of atmospheric pollution research sponsored by the U.K. Department of the Environment.

REFERENCES

1. Oxygenates as Fuels for Spark Ignition Engines' Brit.Technical Council of the Motor and Petroleum Industries Report BTC/F1/80 May (1980).

2. Calvert, J.G. and Pitts, J.N., Jr., 'Photochemistry' J.Wiley, New York (1966).

3. Darnall, K.R., Lloyd, A.C., Winer, A.M., and Pitts, J.N. Jr, Environ. Sci.Technol. 10 692 (1976).

4. Cox, R.A. Derwent, R.G., and Williams, M.R., Environ.Sci.Technol. 14 57 (1980).

5. Cox, R.A. Patrick, K.F., and Chant, S.A., Environ.Sci.Technol., 15 587 (1981).

6. Cox, R.A., J.Photochem 3 175 (1974).

7. Baulch, D.L., Cox. R.A., Hampson, R.F., Kerr, J.A., Troe, J., and Watson, R.T., J.Phys.Chem.Ref.Data.

8. Atkinson, R., Darnall, K.R., Lloyd, A.C., Winer, A.M., Pitts, J.N. Jr. Advances in Photochem. 11 375 (1979).

9. Carter, W.P.L, Darnall, K.R., Graham, R.A., Winer, A.M., and Pitts, J.N. Jr., J.Phys.Che,. 83 2305 (1979).

10. Radford, H.E. Chem. Phys. Lett., 71(1980).

11. Zetsch, C. presented at COST 61a Working Party 2 Discussion Meeting, Leuven, Belgium, Feb. 1981.

THE RATE CONSTANT FOR THE REACTION OF THE HYDROXYL RADICAL WITH HO_2NO_2

I. Barnes, V. Bastian, K.H. Becker, E.H. Fink, and F. Zabel

Physikalische Chemie/ FB 9, Universität-Gesamthochschule Wuppertal,
56 Wuppertal 1, Germany

Abstract

A rate constant ratio of 0.268 has been determined for the reactions
of OH radicals with peroxynitric acid and propene at 295 K and a total
pressure of 1 torr ($M = N_2$) by application of FTIR spectroscopy and
a relative rate method in a 420 l reaction chamber. Using the literature
value $2.6 \times 10^{-12} cm^3$ molecule^{-1}s^{-1} for the rate constant of the reaction
of OH with n-butane in combination with results from additional measure
ments on the relative reaction rates of OH with propene and n-butane at
1 torr, a value of $K_4 = 4.1 \times 10^{-12} cm^3$ molecule^{-1}s^{-1} is obtained for
the title reaction at 295 K.

1. Introduction

Attention has recently been focused on the role played by peroxynitric acid HO_2NO_2 (PNA) in stratospheric chemistry and in particular on
its reaction with OH as a potentially important sink for odd hydrogen
(2). Niki and co-workers (1) were the first to identifiy PNA in the gas
phase using Fourier transform infrared (FTIR) spectroscopy which may
be formed in the atmosphere by recombination of HO_2 and NO_2:

(1) $HO_2 + NO_2 (+M) \rightarrow HO_2NO_2 (+M)$

The most important sink processes controlling its lifetime are reactions
(2)-(4):

(2) $HO_2NO_2 (+M) \rightarrow HO_2 + NO_2 (+M)$

(3) $HO_2NO_2 + h\nu \rightarrow products$

(4) $HO_2NO_2 + OH \rightarrow H_2O + NO_2 + O_2$

Recent results on K_1 (3,4), K_2 as a function of temperature and pressure
(5) and K_3 ($\lambda \geqslant 290$ nm) (6,7) are available and preliminary results
on K_4 (5,8,9) have been reported. Based on one dimensional model calculations using absorption cross sections for PNA obtained by Molina
and Molina (7) and $K_4 = 3 \times 10^{-12} cm^3$molecule^{-1}s^{-1}, Sze and Ko (10)
predict a peak concentration of PNA of 1.6 ppb at a height of 28 km.

Their study also showed that the impact of PNA on tropospheric ozone chemistry was dependent on the reaction products of reaction (3) and the values of K_3 and K_4.

In the present work we have determined K_4 at 294 K and 1 torr using a relative rate method. The study was carried out in a reaction chamber coupled to an FTIR spectrometer which was used to monitor the concentration-time behaviour of reactants and products.

2. Experimental

The reaction chamber is a 420 l glass cylinder (150 cm long, 80 cm in diameter) with aluminium end plates. It is equipped with a teflon coated fan and can be evacuated to less than 10^{-4} torr (1 torr = 133 Pa). Connected to the chamber was a glass gas handling system for the production of PNA. NO and hydrocarbons were admitted by syringe injection and the N_2 dilutant gas was directly taken from cylinders. The reaction mixture was analyzed by in-situ infrared absorption spectroscopy using a built-in multireflection White mirror system of 50.4 m pathlength coupled to an FTIR spectrometer (Nicolet 7199, Globar light source, HgCdTe detector). The experimental device has been described in detail elsewhere (11).

Once PNA has been admitted into the reaction chamber, it is subject to homogeneous and heterogeneous thermal decomposition. At 298 K and atmospheric pressure k_2 is $0,1\ s^{-1}$ (5). However, due to the rapid reverse reaction (1), the actual life time of PNA is approximately 1 hour in our reaction chamber under these conditions. The equilibrium reactions (1) and (2) are disturbed by the bimolecular reaction of HO_2 with itself and by wall reactions of PNA and HO_2 resulting in the observed slow PNA decay. This behaviour has been discussed in detail by Graham et al. (5,12). When NO is added to this mixture, HO_2 is rapidly converted to OH via reaction (5),

(5) $HO_2 + NO_2 \rightarrow NO_2 + OH$, $\quad k_5 = 1\times10^{-11} cm^3 molecule^{-1} s^{-1}$ at 298 K (13,14)

and PNA disappears within several seconds. This technique has been used as a convenient "dark" source of OH radicals in relative rate measurements (15). However, the high pressure relative rate method as applied in the previous work cannot be used for the determination of k_4 since PNA is not consumed solely by the reaction with OH but also by thermal decomposition. For this reason, a low pressure of 1 torr where k_2 is small, and a two-stage process have been applied in the present work: PNA is mixed with a hydrocarbon, then diluted with nitrogen and finally

an excess of NO is added to initiate the reaction. OH is formed via reactions (2) and (5) and both PNA and the hydrocarbon compete for OH radicals. When approximately half of the total reaction time is over, a larger
amount of the same hydrocarbon is added. The OH radicals are now consumed
mainly by the hydrocarbon (RH),

(6) RH + OH → products;   RH = C_3H_6(6a), n-C_4H_{10} (6b),

and the PNA decay is essentially controlled by its thermal decomposition
alone. The concentrations of both PNA and the hydrocarbon are measured
as a function of time by FTIR spectroscopy. From the experimental concentration-time profiles, the ratio k_4/k_6 can be determined (see discussion
below). For several reasons which will be outlined propene has been used
as the hydrocarbon in most experiments. A few measurements have been
performed with n-butane in order to confirm the consistency of the
data. From the ratio k_4/k_6, k_4 can be obtained when k_6 is known. k_{6a}
seems to be well known at atmospheric pressure, but inconsistent results
are reported for the torr region (16). For this reason, two relative
rate measurements have been performed with propene and n-butane being
simultaneously exposed to PNA/NO mixture at 1 torr total pressure.
From these experiments, the ratio k_{6a}/k_{6b}) was obtained. The rate constant k_{6b} is well established and pressure independent. The preparation
of PNA and other experimental technicalities have been previously described (27).

2. Results and discussion

Fig. 1 shows a typical plot of the PNA and C_3H_6 concentrations
as a function of time selected from a series of six experiments performed
on PNA/C_3H_6/NO mixtures. Within both of the individual reaction stages
I and II logarithmic concentration-time profiles may be approximated
by straight lines for PNA and propene.

The difference in the experimentally determined first-order PNA
decomposition rates for stages I and II can be represented by equation I:

(I) $$\frac{d(\ln[PNA])}{dt}\bigg|_I - \frac{d(\ln[PNA])}{dt}\bigg|_{II} = k_4 \left([OH]_I - [OH]_{II} \right)$$

where OH_I and OH_{II} are the steady-state hydroxyl radical concentrations for the respective experimental steps and are given by

$$\left(\frac{d(\ln [C_3H_6])}{dt}\middle/ k_{6a}\right)_I \quad \text{and} \quad \left(\frac{d(\ln [C_3H_6])}{dt}\middle/ k_{6a}\right)_{II}$$

This leads to equation II:

$$(II) \quad \frac{k_4}{k_{6a}} = \frac{\left(\dfrac{d \ln [PNA]}{dt}\right)_I - \left(\dfrac{d \ln [PNA]}{dt}\right)_{II}}{\left(\dfrac{d \ln [C_3H_6]}{dt}\right)_I - \left(\dfrac{d \ln [C_3H_6]}{dt}\right)_{II}}$$

from which the results summarized in table I have been evaluated. Typically 15 ppm PNA was reacted with 30 ppm NO in the presence of 4 ppm (stage I) and 15 ppm (stage II) propene respectively. At a total pressure of 1 torr ($M = N_2$) and $(295\pm1)K$ an average of $k_4/k_{6b}=$ (0.268 ± 0.026) has been obtained.

With the inclusion of the wall loss reaction of PNA,

$$(7) \quad HO_2NO_2 \xrightarrow{\text{wall}} \text{products},$$

in the reaction mechanism the validity of equation II is dependent upon the following assumptions:
1) only a minor fraction of the equilibrated HO_2NO_2 is dissociated at any time throughout the reaction,
2) decay in PNA is first order,
3) the main hydrocarbon decomposition pathway is OH attack,
4) within each reaction stage the OH concentration is constant,
5) no reactant reformation during the course of the reaction,
6) the wall loss reaction rate coefficient k_7 is constant for the duration of a single experiment.

Fig. 2 shows $\ln[C_3H_6]$ plotted as a function of $\ln[n-C_4H_{10}]$ for two measurements performed with $PNA/NO/C_3H_6/n-C_4H_{10}$ mixtures at a total pressure of 1 torr. The slope gives the rate constant ratio $K_{6a}(1 \text{ torr})/$ K_{6b}. A value of (5.9 ± 0.6) was obtained for this ratio from the two experiments using least squares analysis. This method for obtaining rela-

Table I. Summary of experimental results

(T = 295 ± 1 K, P_{total} = 1 torr, total reaction time 30 min, concentrations in units of 10^{14} molecules cm^{-3}, k_4 in cm^3 molecule^{-1} s^{-1}, error limits correspond to 1 standard deviation)

Run	RH_1	RH_2	$\dfrac{d \ln[PNA]}{dt}$ (I)	$\dfrac{d \ln[PNA]}{dt}$ (II)	$\dfrac{d \ln[RH]}{dt}$ (I)	$\dfrac{d \ln[RH]}{dt}$ (II)	$\dfrac{k_4}{k_{6a}}$ (1 torr)	$\dfrac{k_{6a}(1\ torr)}{k_{6b}}$	$\dfrac{k_4}{k_{6b}}$	k_4
1	C_3H_6	-	9.55	7.42	11.67	4.40	0.293	-	-	-
2	C_3H_6	-	10.97	8.52	15.47	6.03	0.260	-	-	-
3	C_3H_6	-	8.95	7.68	5.90	1.14	0.267	-	-	-
4	C_3H_6	-	8.82	7.45	6.18	1.34	0.233	-	-	-
5	C_3H_6	-	10.00	8.29	11.32	3.58	0.221	-	-	-
6	C_3H_6	-	8.30	6.52	7.48	1.16	0.282	-	-	-
							mean value: 0.268 (+0.026)			4.1×10^{-12}
7	C_3H_6	$n\text{-}C_4H_{10}$	-	-	-	-	-	5.82(+0.48)	-	-
8	C_3H_6	$n\text{-}C_4H_{10}$	-	-	-	-	-	6.02(+0.66)	-	-
								mean value: 5.9		
9	-	$n\text{-}C_4H_{10}$	12.15	9.44	2.42	0.27	-	-	1.26	-
10	-	$n\text{-}C_4H_{10}$	10.25	8.48	1.51	0.67	-	-	2.11	-
11	-	$n\text{-}C_4H_{10}$	10.52	8.52	1.68	0.52	-	-	1.71	-
								mean value: 1.70(+0.43)		4.4×10^{-12}

tive OH-rate constants is exactly similar to that recently employed
by us at 760 torr (27). Although the measurements at 760 torr were not
time-resolved a comparison of some of the the relative OH-rate constants,
as shown in table II, against literature values obtained using diverse
experimental methods is highly favourable adequately demonstrating the
correctness of the chemical principles behind the method.

Table II

Reactivities of some organic compounds towards OH radicals

$(K_{OH+Ethene} = 8.0x10^{-12} cm^3 molecules^{-1}s^{-1})$

Substance	PNA Method (15)	Literature data*
Propane	0.25	0.24 - 0.31
n-Butane	0.32	0.35
n-Pentane	0.48	0.38 - 0.63
Ethene	1	1
Propene	3.1	3.0 - 3.4
1-Butene	3,8	3.9
Iso-Butene	7.3	7.0
Ethanol	0.46	0.46
Propanol	0.72	0.69
Isopropanol	0.91	0.86

*The given literature data includes values from all experimental methods
taken from a review by Atkinson et al. (28).

With $k_{6b} = (2.6 \pm 0.3) \times 10^{-12} cm^3 molecule^{-1}s^{-1}$ as taken from the
literature (19,20,21) and with the ratios k_4/k_{6a} and k_{6a}/k_{6b} at a total
pressure of 1 torr from this work, k_4 can be calculated using eq. (III)

$$(III) \quad k_4 = \left(\frac{k_4}{k_{6a} \text{ (1 torr)}} \right) \times \left(\frac{k_{6a} \text{ (1 torr)}}{k_{6b}} \right) \times k_{6b}$$

From (III), $k_4 = (4.1\pm1.0) \times 10^{-12} cm^3 molecule^{-1}s^{-1}$ is obtained for the
rate constant of the reaction of OH with peroxynitric acid at (295±1)K

and a total pressure of 1 torr ($M=N_2$). The stated error limits are obtained according to the theory of propagation of errors with a 2σ error for k_4/k_{6a} (1 torr) and estimated total errors for k_{6a}(1 torr)/k_{6b} and k_{6b} and correspond to the estimated total error for k_4.

Naturally k_4 could be obtained more directly if a non-pressure dependent OH-rate coefficient such as for OH + n-butane is used in place of the pressure dependent OH + propene rate constant. Several relative rate measurements using PNA/NO/n-butane mixtures gave an average of $k_4 = 4.4 \times 10^{-12} cm^3 molecule^{-1} s^{-1}$. Although there was a large scatter in the results due to the slow reaction of OH with n-butane which necessitated small Δ(n-butane) measurements the k_4 value is fully consistant with that obtained from the PNA/NO/propene work.

To date there have been three studies of the OH+PNA rate constant. From studies on the thermal decomposition of PNA using n-butane to scavange the OH radicals Graham et al. (5) have deduced an upper limit of $k_4 \leqslant 3 \times 10^{-12} cm^3 molecule^{-1} s^{-1}$. Using a molecular modulation technique Littlejohn and Johnston (8) have reported a preliminary value of $(2\pm1) \times 10^{-12} cm^3 molecule^{-1} s^{-1}$ and finally, Trevor et al. (26) have obtained a value of $(4.0\pm1.0) \times 10^{-12} cm^3 molecule^{-1} s^{-1}$ from a laser-flash photolysis resonance-fluorescence study at 298 K and 3-15 torr. Our present result compliments the value obtained by Trevor et al. and is also consistent, within the combined error limits, with the two former results.

4. Conclusions

Combination of the value for k_4 obtained in this study with values from other recent work on the OH+PNA reaction would suggest a room temperature value of $k_4 = 4 \times 10^{-12} cm^3 molecule^{-1} s^{-1}$. The magnitude of k_4 seems to cofirm recent model calculations which predicted that the reaction of PNA with OH is both a very important removal process for PNA and an additional sink for OH in the lower stratosphere.

Acknowledgements

The authors wish to express their thanks to Dr. H. Niki (Ford Motor Co., USA) for discussions on the atmospheric relevance of PNA reactions and to Dr. I. Barker (SRI International, USA) for providing a preprint of his paper (26) on reaction (4). Financial support by the "Bundesminister für Forschung und Technologie" (UC/FKW 23) is gratefully acknowledged.

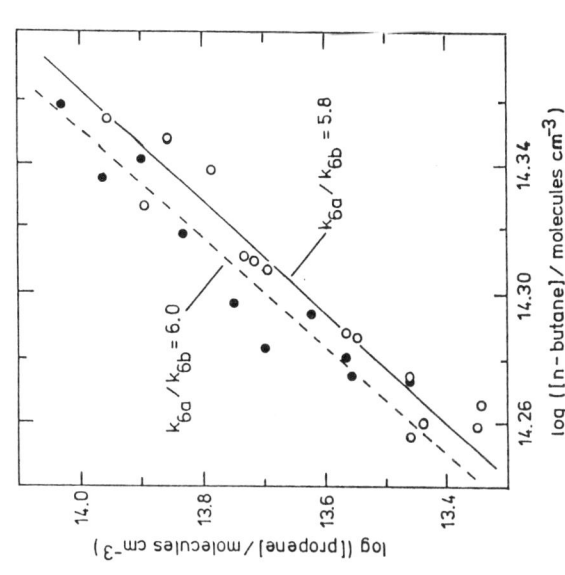

Fig. 2 – Determination of k_{6a}/k_{6b} at 1 torr and 296 K
from run 7 (○,−) and run 8 (●,−−), see table I.

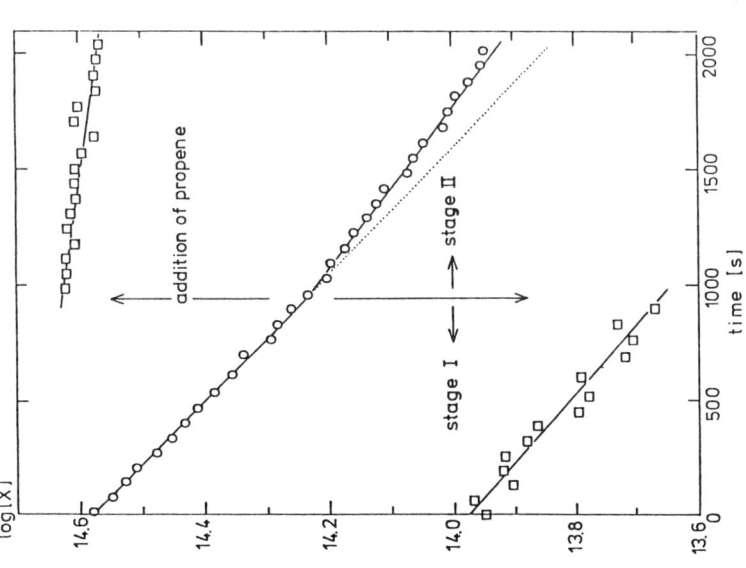

Fig. 1 – Concentration (molecules cm^{-3})-time profiles
for X = PNA (○) and X = propene (□); (run 6, see table I)

References

1) H. Niki, P.D. Maker, C.M. Savage, and L.P. Breitenbach, Chem. Phys. Lett. 45 (1977) 564
2) J.P. Jesson, L.C. Glasgow, D.L. Filkin, and C. Miller, Geophys. Res. Lett. 4 (1977) 513
3) C.J. Howard, J. Chem. Phys. 67 (1977) 5258
4) R.A. Cox and K. Patrick, Int. J. Chem. Kinet. 11 (1979) 635
5) R.A. Graham, A.M. Winer, and J.N. Pitts, Jr., J. Chem. Phys. 68 (1978) 4505
6) R.A. Graham, A.M. Winer, J.N. Pitts, Jr., Geophys. Res. Lett. 5 (1978) 909
7) L.T. Molina, and M.J. Molina, J. Photochem. 15 (1981) 97
8) D. Littlejohn and H.S. Johnston, abstract of a paper, published in Eos 61 (1980) 966
9) P. Trevor, J.S. Chang, and J.R. Barker, paper presented at the 14th Informal Conference on Photochemistry, March 30-April 3, 1980
10) N.D. Sze and M.K.W. Ko, Atmospheric Environment, accepted for publication 1981
11) I. Barnes, K.H. Becker, E.H. Fink, V. Kriesche, J. Wildt, and F. Zabel, Proceedings of the "First European Symp. on the Physico-chem. Behaviour of Atmospheric Pollutants", Ispra, Italy, Oct. 16-18, 1979, B. Versino and H. Ott, eds.
12) R.A. Graham, A.M. Winer, R. Atkinson, and J.N. Pitts, Jr., J. Phys. Chem. 83 (1979) 1563
13) I. Glaschick-Schimpf, A. Leiss, P.B. Monkhouse, U. Schurath, K.H. Becker, and E.H. Fink, Chem. Phys. Lett. 67 (1979) 318
14) D.L. Baulch, R.A. Cox, R.F. Hampson, Jr. J.A. Kerr, J. Troe, and R.T. Watson, J.Phys.Chem. Ref.Data 9 (1980) 295
15) I. Barnes, V. Bastian, K.H. Becker, E.H. Fink, and F. Zabel, accepted for publication in Atmospheric Environment 1981
16) R.F. Hampson, Jr. and D. Garvin, "Reaction Rate and Photochemical Data for Atmospheric Chemistry - 1977 "National Bureau of Standards Special Publication, 513, 1978
17) Y.S. Chang, J.H. Shaw, E. Niple, J.G. Calvert, W.H. Chan, S.Z. Levine, and W.M. Uselman, EPA Interim Report, Ohio State University, 1977
18) N.R. Greiner, J. Chem. Phys. 53 (1970) 1070
19) F. Stuhl, Z. Naturforsch., 28a (1973) 1383
20) R.A. Perry, R. Atkinson, and J.N. Pitts, Jr., J. Chem. Phys. 64 (1976) 5314
21) G. Paraskevopoulos and W.S. Nip, Can.J.Chem. 58 (1980) 2146
22) R. Atkinson and J.N. Pitts, Jr., J.Chem.Phys. 63(1975) 3591
23) A.C. Lloyd, K.R. Darnall, A.M. Winer, J.N. Pitts, Jr., J.Phys.Chem. 80 (1976) 789
24) A.R. Ravishankara, S. Wagner, S. Fischer, G. Smith, R. Schiff, R.T. Watson, G. Tesi, and D.D. Davis, Int.J.Chem.Kinet. 10 (1978) 783
25) K. Hoyermann and R. Sievert, Ber.Bunsenges.Physik.Chem. 83(1979) 933
26) P.L. Trevor, G. Black, and J.R. Barker, submitted to J. Phys. Chem. 1981
27) I. Barnes, V. Bastian, K.H. Becker, E.H. Fink and F. Zabel, submitted to Chem. Phys. Lett. 1981.
28) R. Atkinson, K.R. Darnall, A.C. Lloyd, A.M. Winer and J.N. Pitts Jnr., Adv. Photochem. 11(1979)375.

RATE CONSTANTS FOR REACTIONS OF OH WITH CARBONIC ACIDS

C. Zetzsch and F. Stuhl
Physikalische Chemie I, Ruhr-Universität
D-4630 Bochum 1; W.-Germany

Summary

The reactions of OH with carbonic acids (formic, acetic, propionic,
and butyric acid) were investigated at room temperature using pulsed
vacuum uv photolysis of H_2O for the production and resonance fluorescence for the detection of OH. Static and dynamic methods were used to
control the reactant concentrations. Using the dynamic method the di-
merisation of the carbonic acids in the gas phase above the liquid was
taken into account. The following rate constants, $(k \pm 3\sigma)$, were obtained for formic, acetic, propionic, and butyric acid (in units of
10^{-13} cm^3s^{-1}) : (3.2 ± 0.2), (6.0 ± 0.8), (16 ± 1.2), and (18 ± 1.6).
Static measurements performed with the first three members of the
homologous series are in good agreement with the dynamic measurements.
The rate constants were found to be not dependend on total pressure
(30-300 Torr, He or Ar), H_2O pressure (0.03-0.3) Torr), and on flash
energy (3-25 J).

1. INTRODUCTION

Carbonic acids are minor constituents in the exhaust of technical
combustion processes. They are found for example in the exhaust of automobiles. The lower carbonic acids (formic, acetic, propionic, and butyric
acid) are generated in the combustion of plastics and refuse. These acids
have been also identified in tobacco smoke and in the plume of forest
fires. Natural sources are plant volatiles and odors of animal waste (1).
The industrial production of formic acid exceeded 10^5t in 1972 world wide
(2); the US production of acetic acid exceeded $1.5 \cdot 10^6$t in 1979 (3). In the
Ruhr-area in the region of Dortmund more than 38 t of formic acid and more
than 500 t of acetic acid are emitted per annum (4), in the region of
Duisburg, Oberhausen, Mülheim more than 8 t/a of butyric acid (5). Carbonic
acid vapors irritate mucous membranes, e.g. the eyes. With increasing
chain length the smell becomes disgusting. The olfactometric threshold for
butyric acid is of the order of 50 ppt for human beings, for dogs it is
found to be 6 orders of magnitude lower (6).
    In the atmosphere carbonic acids are assumed to be intermediates in
the photooxidation of hydrocarbons. Tropospheric photooxidation cycles leading from methane, ethylene, methanol, and formaldehyde to formic acid are
shown in figure 1. In all cases shown the first step is the attack of OH by
abstraction or by addition followed by the addition of O_2 to the radical
formed in the first step. The resulting peroxi radical either undergoes
unimolecular decay (leading to formaldehyde in the case of methanol) or
attack by NO to form an alkoxy radical (which decomposes splitting the C-C
bond in the case of ethylene (7). Further steps lead to formaldehyde, and
partly to formic acid (8).
    The radical $\overset{\bullet}{C}H_2OH$ appears to be an important intermediate in the for-
mation of formic acid. It should be noted here, that other loss processes

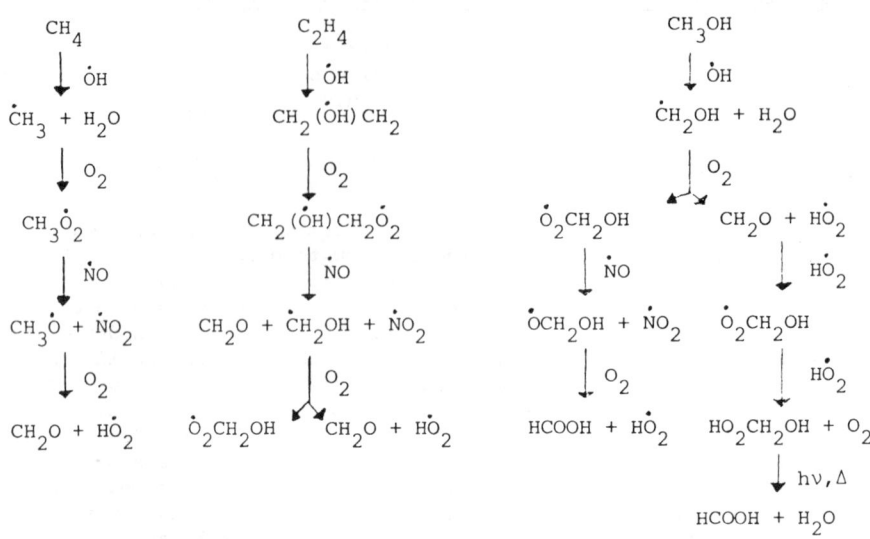

Fig. 1: Examples of chemical reaction paths in the tropospheric photooxida-
of alkanes, alkenes, alcohols, and aldehydes demonstrated by methane, ethylene, methanol, and formaldehyde. Formaldehyde is included in the reaction
scheme for methanol (right hand column). For formaldehyde, besides attack by
HO_2, other major loss processes (attack by OH or photolysis) are known which
are not included in this figure.

are known for ethylene and formaldehyde e.g. ozonolysis of ethylene or
attack by OH and photolysis of formaldehyde. Also the Criegee intermediate
of the ozonolysis of propene, CH_2O_2, has been postulated to rearrange to
formic acid (9), and a still unknown fraction of all organic molecules will
be converted to carbonic acids. Formic acid has been detected in the Los
Angeles smog (10) and in smog chambers (11) at concentrations of up to
72 ppb. The involvement of formic acid in the formation of aerosols has
been discussed previously (12).

The chemical lifetime is of importance for the understanding of the
fate of the carbonic acids in the atmosphere. Since OH radicals have been
recognized to be among the most reactive species in the atmosphere (1), we
have studied their reactions with carbonic acids.

2. EXPERIMENTAL

The apparatus has been described elsewhere (13), and only the main
features will be mentioned here. OH radicals are generated by pulsed vacuum
uv photolysis ($\lambda > 105nm$) of mixtures of H_2O, inert gas, and reactants. Time
resolved decays of the radicals are monitored by resonance fluorescence of
OH (14). To enhance the detection sensitivity photon counting and multi-
channel scaling was used. Usually the fluorescence signals from 64 single
decays were accumulated. Since the acids are strongly adsorbed at the reactor walls, care was taken to determine the pressure of these reactants.
Therefore, the reactants were introduced into the reactor either in the
order reactant, water, inert gas (static method) or by flowing the gas mixtures slowly through the reactor vessel (dynamic method) (15).

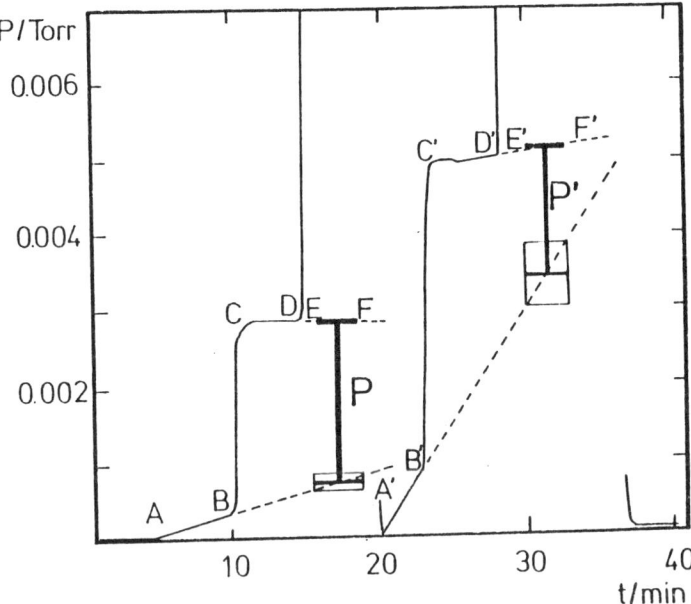

Fig. 2: Change of pressure during the course of two successive measurements with propionic acid. At A pumping is finished at B propionic acid is introduced to the vessel until the desired pressure is reached at C. At D, H_2O and inert gas are introduced. The pulsed photolysis begins at E and is finished at F. The other experiment starts at A'. The estimated uncertainty of the determination of reactant pressure P (due to adsorption, desorption, and leak rate) is indicated by the height of the rectangles.

The static method appeared to be not very useful for liquids with low vapor pressure. The limitations of the method are demonstrated in figure 2. The curve shows the variation of pressure monitored by a strip chart recorder during the filling procedure of two successive experiments. At A the pumping cycle was finished by closing the main valve of the diffusion pump. The following linear increase of pressure is due to the leak rate (including desorption from the walls). At B propionic acid is introduced into the vessel through a needle valve which is closed at C. The resulting change of pressure is monitored for several minutes in order to improve the subsequent extrapolation. At D water vapor and inert gas are introduced. A short time is allowed for mixing by diffusion and the pulsed photolysis is performed during the time interval between E and F. After the measurement the vessel is evacuated to 10^{-5} Torr by the diffusion pump using a baffle, cooled by liquid N_2. The partial pressure P (heavy bars in figure 2) of propionic acid during the photolysis, is determined by extrapolation of the pressures as shown in figure 2, (dashed lines). The uncertainty of P is estimated by the variation of the extrapolated pressures during the photolysis. This is indicated in figure 2 by the height of the rectangle at the lower end of the bar denoting P. At the beginning of the next measurement cycle (at A') the apparent leak rate has increased by almost a factor of three. This can be explained by desorption of H_2O (and of formic acid) from the previous experiment. Although this increased apparent leak rate can be diminished by prolonged pumping cycles this is time consuming and unpractical for getting a sufficient number of measurements per day. This procedure becomes increasingly difficult with increasing reactant pressures for the rate of adsorption of the acid increases with partial pressure.

To overcome these difficulties a dynamic method of introducing the reactants was developed. Using this method, the reactant mixture is introduced under steady slow flow conditions, the cell volume being exchanged every 50 s. Small fractions of the inert gas are saturated at known temperatures and at atmospheric pressure. At the same temperature the main flow of inert gas dilutes the small saturated flow in order to prevent condensation in case saturation is performed at elevated temperatures. Reactant concentrations are calculated from the appropriate equations of vapor pressures of the acids, taking into account the dimerization of the molecules in the vapor phase in equilibrium with the liquid phase (16).

3. RESULTS

Initial concentrations of OH were of the order of $10^{11} cm^{-3}$. Reactants were added in large excess over the concentration of OH radicals. In all cases the radicals follow pseudo first order kinetics i.e. the intensity of the resonance fluorescence, I, which is proportional to the concentration of OH decays exponentially.

Static Measurements

Examples for pseudo first order decays of OH in the presence of formic acid are shown in figure 3. The initial concentration of OH in this figure corresponds to approximately $5 \cdot 10^{10} cm^{-3}$. The decays of OH can be fol-

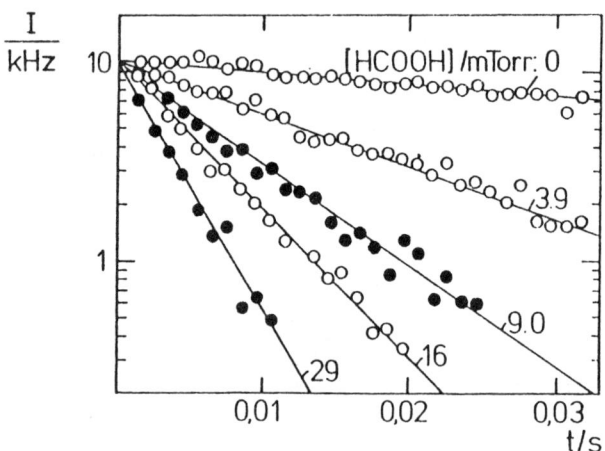

Fig. 3: First order decays of the resonance fluorescence intensity in the presence of various pressures of formic acid. Experimental conditions: 200 Torr He, 0.05 Torr H_2O, 3 J flash energy.

lowed by more than a factor of fifty in concentration. The observed decays are exponential (straight lines in figure 3) thus defining a decay rate, τ^{-1}.

Rate constants, k, are determined from plots of the decay rate vs. the reactant concentration $[R]$, according to the equation:

$$\tau^{-1} = \tau_o^{-1} + k \, [R].\qquad(I)$$

In this equation, τ_o^{-1} is the decay rate in the absence of reactant. It contains losses of OII by diffusion from the observation zone and by recombination and disproportionation of OH radicals (17). Plots of decay rates, τ^{-1}, vs. the concentration of reactants are shown for the three lowest members of the carbonic acids (formic, acetic, and propionic acid) in figure 4 for static measurements.

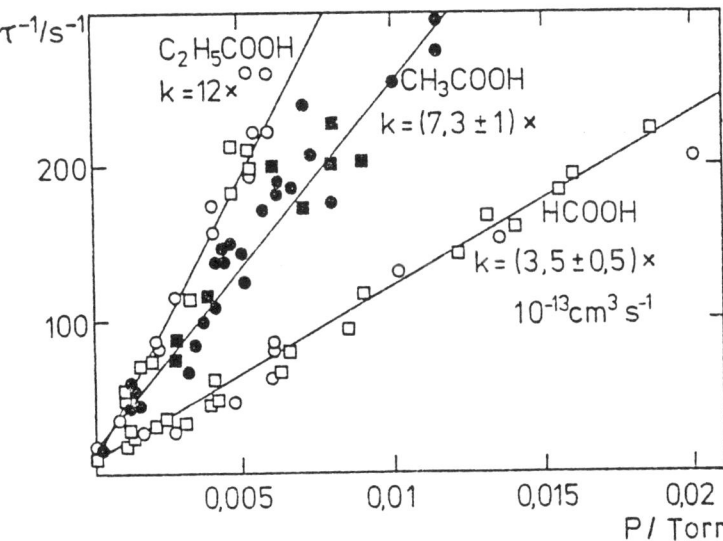

Fig. 4: Increases of the decay rate upon addition of carbonic acids for static measurements: formic, acetic, and propionic acid. Experimental conditions: 0.05 Torr H_2O, 25 J flash energy. O, ● : 100 Torr Ar, □, ■ : 100 Torr He. The quoted error limits refer to three standard deviations.

The decay rate is seen to increase linearly with the partial pressure of the carbonic acids. A linear regression of the data in figure 4 yields the values for the rate constants, $k \pm 3\,\sigma$, displayed in the figure. Figure 4 shows measurements with 100 Torr Ar (circles) and 100 Torr He (squares) as inert gas. In the case of formic acid the flash energy was varied from 3 to 25 J. Different grades of purity (technical or p.a.) of formic acid were used, and additional measurements (not shown in the figure) were performed at 200 Torr He pressure. Within the error limits these measurements are in agreement with those displayed in figure 4. With acetic acid the independence of total pressure and initial concentration of OH was checked at 10, 50, and 300 Torr Ar and 0.01 and 0.2 Torr H_2O pressure. With propionic acid the independence of total pressure and initial OH concentration was checked at 10, 50, and 250 Torr Ar using 0.01 Torr H_2O to generate OH. No significant changes were observed within the scatter of the experimental data. In the case of propionic acid no error limits are given in the figure, since the errors caused by the extrapolation of the reactant pressure could be hardly estimated.

Dynamic Measurements

When saturating a carrier gas with vapor of carbonic acid the association of the molecules in the gas phase has to be taken into account. The
equilibrium between monomeric and dimeric molecules is calculated to favor
dimeric molecules at the pressure of the saturated vapor in a wide temperature range (18). However, at the partial pressures used in the reaction
vessel, monomeric molecules are by far the major species present. Higher
degrees of association can be neglected considering available data for
acetic acid (19). Figure 5 shows the results of the dynamic measurements.

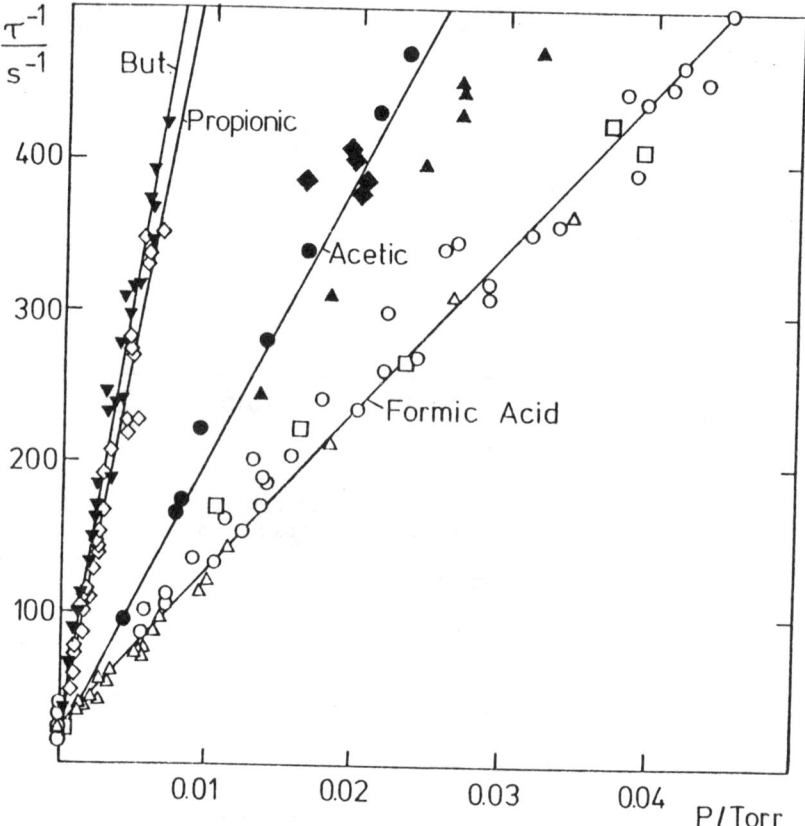

Fig. 5: Dependence of the decay rate on the concentration of the carbonic
acid for the dynamic filling procedure. Experimental conditions:
Δ : 25 Torr Ar, 0.03 Torr H_2O; O : 100 Torr Ar, 0.12 Torr H_2O; $\square$: 100 Torr
Ar, 0.2 Torr H_2O; flash energy 16 J, (formic acid); $\blacktriangle$, $\bullet$, $\blacklozenge$: 20, 100, and
500 Torr Ar, 0.1 Torr H_2O; flash energy 3 J, (acetic acid); $\lozenge$: 100 Torr Ar,
0.02 - 0.22 Torr H_2O; flash energy 16 J, (propionic acid); $\blacktriangledown$: 30, 100, and
300 Torr Ar; 0.04 - 0.1 Torr H_2O; flash energy 16 J, (butyric acid).

The decay rate of OH can be seen to vary linearly with the partial pressures of the carbonic acids. The partial pressures in figure 5 were estimated using twice the saturated vapor pressure. The corresponding constants of the Antoine equations and the respective references are given in table I. To verify this procedure the inert gas flow saturated with carbonic acid was fed through two successive cold traps at dry ice temperature. The amount of carbonic acid was determined by the weight of the condensed liquid and compared to the amount expected by assuming total saturation by dimeric molecules. The fractions observed/expected, α, determined from more than five runs are given in table I.

Rate constants for the reactions of OH with carbonic acids are calculated from the slope of the plots in figure 5 according to equation I and given at the bottom of table I. Error limits of three standard deviations are given in table I to characterize the reproducibility of the measurements.

Table I: Vapor pressure data log $P/Torr = A - B/(T/^{\circ}C + C)$, experimental degree of saturation including association, α, and rate constants, $k \pm 3\sigma$, in units of $10^{-13}cm^3s^{-1}$.

Acid	Formic	Acetic	Propionic	Butyric
A	4.97536	8.3953	6.4033	8.71017
B	541.738	2148.97	950.175	2433.014
C	137.051	273.15	130.343	255.189
References	20,21	22	20,23	20,24
$\alpha \pm 3\sigma$	0.98+0.15	0.997+0.03	0.995+0.05	–
$k \pm 3\sigma$	3.2+0.2	6 + 1.2	16 + 1.2	18 + 1.6

4. DISCUSSION

The static measurements displayed in figure 4 show a reasonable reproducibility although the handling of carbonic acids becomes increasingly difficult with increasing boiling point. Because of the difficulties encountered with propionic acid no static experiments were performed with butyric acid. The dynamic measurements of figure 5 are found to be in good agreement with the static measurements. Also for the dynamic measurements no significant dependence of the rate constants on the initial concentration of OH (to a first approximation proportional to the H_2O concentration) or on the total pressure was observed.

The overall precision of the rate data in the dynamic measurements is estimated to be of the order of $\pm$ 30%, including the calibration of the flowmeters, the pressure measurement, variations of the room temperature, and of the vapor pressure of the carbonic acids (due to the saturator temperature). Additional uncertainties are introduced by the equilibrium between monomer and dimer acid molecules. In the dynamic measurements this equilibrium is assumed to be shifted quickly to the monomer side in the reaction vessel. If dimeric molecules are more reactive than monomeric molecules they may contribute to the decay of OH. For that case the observed rate constants might be regarded as upper limits for the monomeric molecules. On the other hand the reactivity of the hydrogen bonded dimers can be expected to differ not significantly from the reactivity of monomers. Such influences of dimers might be investigated in temperature dependent studies, which were not accessible during these experiments.

The photolysis of the carbonic acids can form radicals which might react with OH. However the variation of the flash intensity did not affect

the OH decay rate. Unfortunately, vuv absorption coefficients of the
parent molecules seem to be unknown altough absorption bands have been
identified (25). Assuming a similar absorption coefficient for the acids
and for water the concentration of the radicals formed from the parent
molecules is estimated bo be at least ten times smaller than $\lfloor OH \rfloor$. This
small concentration of radicals is believed to barely influence the OH
decays.

Rate constants for the reactions of carbonic acids with OH have not
been reported in the literature. The reactivity of the lowest member of
the carbonic acids, formic acid, is found in the present study to be comparable to ethane. With increasing chain length the reactivity of the
carbonic acids increases due to the number of C-C- bonds and due to the
decreasing bond energies according to the inductive effect. From the rate
constants of the present study estimates of limits of atmospheric residence
times can be calculated for these acids. Using a global average tropospheric concentration (1) of OH of $4 \cdot 10^5 cm^{-3}$ the residence time of formic
acid is estimated to be of the order of 3 months.

However, besides reactions with OH radicals, other important loss
processes of these intermediates will be rainout and aerosol formation (12).
Analysis of aerosols with respect to carbonic acids (26) might elucidate
tropospheric oxidation cycles of hydrocarbons.

Acknowledgement

This work was supported by the Umweltbundesamt and by the Fonds der
Chemischen Industrie.

References

1) T.E. Graedel, Chemical Compounds in the Atmosphere, Academic Press,
   New York 1978
2) Ullmanns Encyklopädie der technischen Chemie, Bd. 7 u. 9, Verlag
   Chemie, Weinheim 1975
3) W. Storck, Chem. Eng. News 58, (18), 34 (1980)
4) Luftreinhalteplan Ruhrgebiet Ost 1979-1983, Ministerium für Arbeit,
   Gesundheit und Soziales des Landes NRW, Düsseldorf 1979
5) Luftreinhalteplan Ruhrgebiet West 1978-1982, Ministerium für Arbeit,
   Gesundheit und Soziales des Landes NRW, Düsseldorf 1978
6) W. Leithe, Umweltschutz aus der Sicht der Chemie, Wissenschaftliche
   Verlagsgesellschaft, Stuttgart 1975
7) H. Niki, P.D. Maker, C.M. Savage, and L.P. Breitenbach, J. Phys. Chem.
   82, 135 (1978)
8) F. Su, J.G. Calvert, and J.H. Shaw, J. Phys. Chem. 83, 3185 (1979)
   F. Su, J.G. Calvert, H.H. Shaw, H. Niki, P.D. Maker, C.M. Savage, and
   L.P. Breitenbach, Chem. Phys. Letters 65, 221 (1979)
9) H. Akimoto, H. Bandow, F. Sakamaki, G. Inoue, H. Hoshino, and M. Okuda,
   Environ. Sci. Technol. 14, 172 (1980)
10) P.L. Hanst, W.E. Wilson, R.K. Patterson, B.W. Gray, Jr., L.W. Chaney,
    and C.S. Burton, Environ. Prot. Agency, Washington, D.C. 1975
11) J.N. Pitts, Jr., B.J. Finlayson-Pitts, and A.M. Winer, Environ. Sci.
    Technol. 11, 568 (1977)
12) G.E. Likens, Chem. Eng. News 54 (48), 29 (1976)
13) C. Zetzsch, Habilitationsschrift, Universität Bochum 1977
    C. Zetzsch and F. Stuhl, Ber. Bunsenges. Phys. Chem. 80, 1354 (1976)
14) F. Stuhl and H. Niki, J. Chem. Phys. 57, 3617 (1972)
15) A. Wahner and C. Zetzsch, this volume, p. 138.

16) J.M. Prausnitz, Molecular Thermodynamics of Fluid-Phase Equilibria, Prentice-Hall, Englewood Cliffs 1969
17) C. Zetzsch, Proceedings of the International Workshop on the Test Methods and Assessment Procedures for the Determination of the Photochemical Degradation Behaviour of Chemical Substances, Berlin, 2. - 4.12.1980 (in press)
18) A.D.H. Clague and H.J. Bernstein, Spectrochim. Acta 25A, 593 (1969), G.C. Pimentel and A.L. Mc Clelland, The Hydrogen Bond, Freemann, San Francisco 1960
19) H.L. Ritter and J.H. Simons, J. Amer. Chem. Soc. 67, 757 (1945)
20) T. Boublik, V. Fried and H. Hala, The Vapour Pressure of Pure Substances, Elsevier, Amsterdam 1973
21) G.W.A. Kahlbaum, Z. Physik. Chem. 13, 14 (1894)
22) C.R.C. Handbook of Chemistry and Physics, 52nd Ed., The Chemical Rubber Company, Cleveland, Ohio, 1972
23) G.W.A. Kahlbaum, Ber. 16, 2476 (1883)
24) J.J. Jasper and G.B. Miller, J. Phys. Chem. 59, 441 (1955)
25) S. Bell, T.L. Ng, and A.D. Walsh, J. Chem. Soc. Faraday Trans. II, 71, 393 (1975)
26) M. Termonia, X. Monseur, G. Alaerts, A. DeMeyer, P. Dourte, and J. Walravens, Proc. First European Sympos. Physichochemical Behaviour of Atmospheric Pollutants, Ispra 1979, p. 52, B. Versino, H. Ott, Eds., CEC, Brussels 1980

REACTIONS OF DISUBSTITUTED BENZENES WITH OH IN THE GAS PHASE :
BENZENE, O-, M-, P-DICHLOROBENZENE, AND P-CHLOROANILINE

A. Wahner and C. Zetzsch
Physikalische Chemie I, Ruhr-Universität
D-4630 Bochum 1; W.-Germany

SUMMARY

To estimate tropospheric lifetimes of chlorinated aromatics, absolute
rate constants for the reactions of these compounds with OH radicals
were determined. In the experiments OH radicals were generated at low
initial concentrations of about $5 \cdot 10^{10} cm^{-3}$ using pulsed vacuum uv
photolysis of H_2O ($\lambda > 125$ nm) at a flash energy of 2 J. Using resonance fluorescence for the detection, decays of OH were observed for
more than two orders of magnitude. Reliable control of the concentration of reactants with low volatility (vapor pressures of 0.01 mbar)
was obtained by saturating slow flows of inert gas at regulated temperatures. The following values of rate constants, k, were obtained
at room temperature in the presence of Ar above 100 Torr (in units of
$10^{-13} cm^3 s^{-1}$): benzene, (8.8 ± 0.4); o-, m-, p-dichlorobenzene, $(4.2
\pm 0.2)$, (6.9 ± 0.2), (3.3 ± 0.3); p-chloroaniline, (810 ± 70). The
influence of the substituents and of the orientation on the reactivities is discussed.

1. INTRODUCTION

     Chlorinated aromatics are produced in large quantities world-wide
(chlorinated benzenes 123000 t/a FRG 1969 (1), chlorobenzene 150000 t/a
USA 1978 (2), o-dichlorobenzene 40000 t/a USA 1976 (2)). They are widely
spread both in industry and household because of their frequent application
as solvents, heat conductors, dielectric materials, deodorizers and pesticides. At present o- and p-dichlorobenzene are used at an estimated amount
of 32000 t/a (3) world-wide to mask ordours of waste water in industry and
household (sanitary facilities) and to control moth. In the course of production processes about 407 t o-dichlorobenzene and about 550 t p-dichlorobenzene enter the environment per annum in the USA (2). Chlorinated
benzenes are also released to the environment as intermediate products
during the production of diazonium salts and various herbicides and they
are also present in the end products as impurities. Chloroaniline is a
degradation product of phenylurea herbicides.
     Thus it is not surprising that p-dichlorobenzene has been identified
in human blood and adipose tissue (3). Dichlorobenzenes could be detected
even in drinking water (New Orleans (3)) and in urban air (Berlin approx.
1 ppb (4)). After having taken up by skin, lungs and gut the halogenated
benzenes accumulate in adipose tissue, since they appear to be highly
soluble in lipids. Acute poisoning by dichlorobenzene affects the central
nervous system (depressions, paralysis) (5) whereas the chronic absorption
of these chemicals causes liver and kidney damages (5,6). In some cases
leukemia has also been observed (3). In contrast to dichlorinated benzenes
there are several studies which prove the cancerogenic effect of p-chloroaniline and benzene (3,5,6). Dichlorobenzene volatilizes from waste water
in a short time (7) and is distributed in the troposphere. It is important

to know the rate of degradation of these chemicals in the atmosphere; for this knowledge enables us to assess the extent of the danger caused by these chemicals and possibly to restrict them. Up to now studies of gas-phase reactions of dichlorobenzenes and p-chloroaniline have not been published. Brown et al. (7) show that o- and p-dichlorobenzene are resistant to O_3 in air but that they react with OH radicals.

In the present work rate constants for the reactions of OH radicals with o-, m-, p-dichlorobenzene, benzene and p-chloroaniline are determined using pulsed vacuum uv-flash photolysis of H_2O and resonance fluorescence detection of OH.

2. EXPERIMENTAL

The resonance fluorescence technique for the detection of OH, originally developed by Stuhl and Niki (8), is used to monitor OH radicals generated by pulsed vacuum uv photolysis of H_2O. Figure 1 shows a cross section of the apparatus, which is described elsewhere (9) in detail. The reaction chamber (RC) is made of black anodized aluminium. The flash lamp (FL), the microwave lamp (MWL), and the photomultiplier tube (PM) are attached to the reaction chamber at right angles. The wavelength of the flash light generated by discharging the condenser (C) is limited to $\lambda >$ 125 nm by a CaF_2 window (W). The flash lamp is flushed by N_2 at 1 atm. The light of the transition OH $(A^2\Sigma^+ \rightarrow X^2\Pi)$ is produced by a microwave lamp (MWL) and is focussed into the centre of the reaction chamber (RC) by means of a quartz coated concave aluminium mirror (MI) and two plano-convex quartz lenses (L) (Suprasil, Heraeus, f = 50 nm). The resonance fluores-

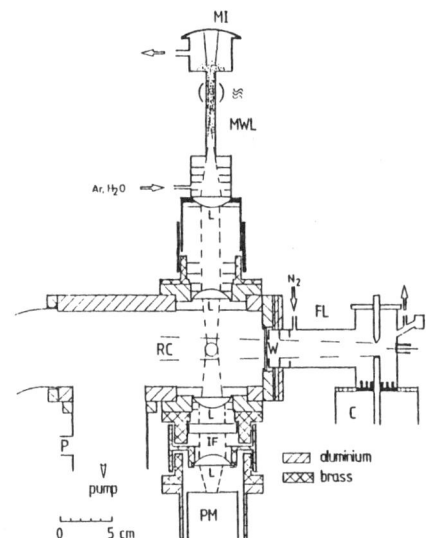

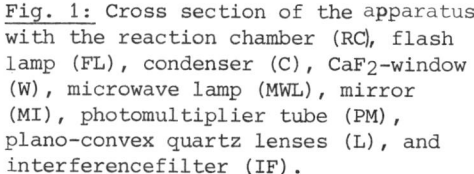

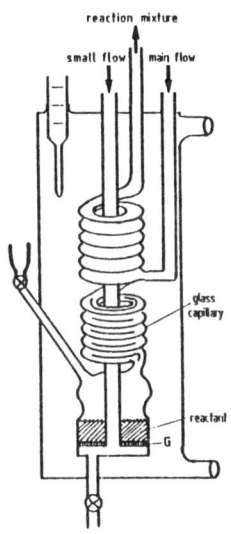

Fig. 1: Cross section of the apparatus with the reaction chamber (RC), flash lamp (FL), condenser (C), CaF_2-window (W), microwave lamp (MWL), mirror (MI), photomultiplier tube (PM), plano-convex quartz lenses (L), and interferencefilter (IF).

Fig. 2: Cross section of the saturator. A small flow of the inert gas is saturated by passing it through a frit (G) and bubbling it through the reactant above the frit. A glass capillary minimizes the diffusion of the reactant into the main gas flow.

cence light of the excited OH radicals generated in the reaction chamber is focussed on the cathode of the photomultiplier tube (PM) (EMI 9789 QB) by means of two plano-convex quartz lenses (L) (Suprasil, f = 50 nm) through an interference filter (IF) (Omega Optical Inc., λ_{max} = 309.5nm, FWHM = 4 nm, $T_{max} \doteq$ 55%). The arrangement of light baffles shown in figure 1 minimizes the scattered light and limits the light paths (dashed lines). Wood's horns made of glass (painted black outside) are attached as light traps (opposite to the microwave lamp, to the photomultiplier, and to the flash lamp). The diameter of the reaction volume in the centre of the cell observed by the photomultiplier is estimated to 1.7 cm.

The time dependence of the concentration of the OH radicals is monitored by the technique of photon counting of the resonance fluorescence signal. The limits of this technique for higher inert gas pressures arise from the facts that the excited OH radicals are quenched by collisions of inert gas and that the Rayleigh scattering of the light of the microwave lamp increases (10).

In order to ensure the exact adjustment of the reactant concentration in the reaction chamber (11) and to avoid reactions with generated products the measurements are carried out under slow flow conditions . The reactant mixtures are produced by a dosing system similar to that used by Biermann et al. (12). According to this system the flows of inert gas are determined by rotameters (13). The reactant and water vapor are dosed by passing a slow flow of inert gas through a saturator at atmospheric pressure. The saturator is shown in figure 2. Its temperature can be adjusted from 233 K to 373 K by a thermostat or a cryostat with an accuracy of $\pm$ 0.1 K. In the saturator a part of the inert gas is saturated by passing it through a frit (G) and bubbling through the reactant above the frit. The gas flow and the bubble size are small enough to establish a thermodynamic equilibrium between the gas phase and the liquid (or solid) chemicals.

A glass capillary (length: 400 mm, inside diameter: 1.5 mm) connecting the gas space of the saturator with the main gas line suppresses the diffusion of the reactant into the main gas flow. The concentration of the reactant caused by diffusive transport is 6 orders of magnitude lower than the vapor pressure of the reactant. Condensation of the reactant is avoided by diluting the saturated inert gas flow inside the thermostated saturator at the same temperature. This arrangement allows to saturate at a temperature above room temperature. Liquids and solids with low volatility (e.g. p-chloroaniline) can be dosed exactly.

To remove O_2 from the inert gas, Ar (Messer Griesheim, 99.999%), the inert gas was fed through Oxisorb (Messer Griesheim). The organic impurities in Ar, mainly methane (gaschromatographic analysis (14)), produce a negligible effect on the lifetime of OH radicals. H_2O was destillated twice and degassed prior to use, benzene (Oekanal, Riedel de Haën) was used without further purification. The dichlorinated aromatics and p-chloroaniline were mainly devolatilized from their components of higher volatility by pumping off 50% of the substance. The gaschromatographic analysis of these chemicals showed impurities produced by portions < 1 % of lower volatility. The portion of impurities (other isomers) in m-dichlorobenzene was 1.3 %.

The accuracy of the dosing depends on the accuracy of the used vapor pressure equation. The literature data for the chemicals investigated are shown in figure 3. The Antoine constants and references are given in table I. Figure 3 also shows the vapor pressures (symbol O) determined at the temperatures used for the saturation by freezing out and weighing the chemicals. For the process of freezing out the gas flow coming from the dosing system was not fed into the reaction chamber but fed through two successive

Tab. I : Constants of the Antoine equation
lg P/Torr = A - B / (C + t/°C)

chemical	A	B	C	reference
benzene	9.1064	1885.9	244.2	(15)
o-dichlorobenzene	7.07123	1650.129	213.367	(19)
m-dichlorobenzene	7.22086	1690.7	218.4	(18)
p-dichlorobenzene	7.23948	1703.2	218	(18)
p-chloroaniline	14.223	4706.	273.15	(20)
water	8.184254	1791.3	238.1	(15)

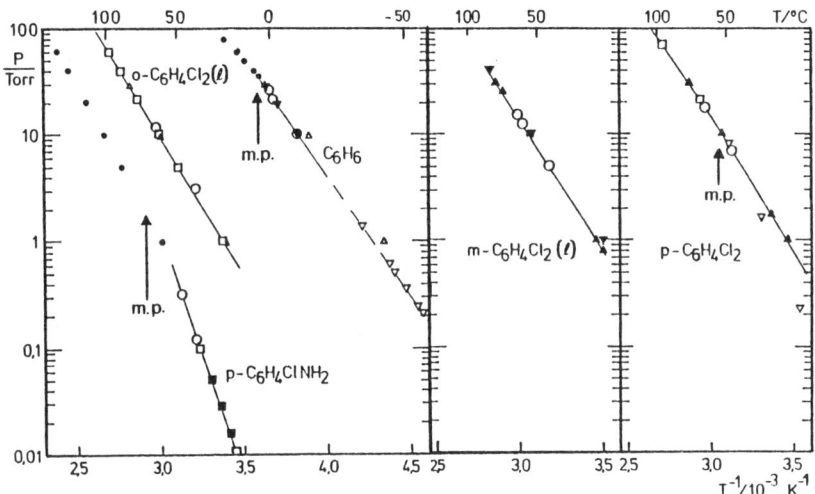

Fig. 3: Literature data on vapor pressures for the chemicals investigated.
Benzene: ● Young (17), ▼ Wilhoit et al. (15), ∇ Mündel (16), ▲ Dreisbach
(18), Δ Dreisbach (18) calculated; o-dichlorobenzene: □ Ohe (19), Δ Dreisbach (18) calculated; p-chloroaniline: □ Landolt Börnstein (20), ■ Klöppffer (21), ● Stull (22); m-dichlorobenzene: ▲ Dreisbach (18), ▼ Weast (23);
p-dichlorobenzene: ▲ Dreisbach (18), □ Ohe (19), ∇ Darkis (24). The vapor
pressures determined in this work are shown by the symbol O.

traps cooled down to -78 °C. Up to five runs were performed at two
different temperatures of the saturator (equal to those used during the
determination of the rate constants). Agreement between the results of
weighing and the vapor pressure equations was found within ± 2%. When
changing the reactant concentration in the photolysis vessel a newly fixed
reactant concentration was obtained in the vessel within 15 to 20 minutes
at an accuracy of 5%. This time response is in accordance with the exchange
in the photolysis vessel which takes approximately one minute.

3. RESULTS

In all measurements a large excess (> 100 fold) of the reactant over
the OH radicals is applied. The resonance fluorescence intensity (which is
proportional to the concentration of OH) decays according to a first order
rate law. The temporal behaviour can be described by the equation

$$I = I_o \cdot \exp \; (-t/\tau) \qquad (I).$$

-141-

The intensity I has to be corrected for the constant background signal resulting from reflexions and from Rayleigh scattering by Ar (28 kHz at 100 Torr Ar pressure). Figure 4 shows exponential decays of OH in the presence of various concentrations of p-chloroaniline in a semilogarithmic diagram. The decays are averaged from 100 flashes at 2 J flash energy using a dwell time of 5,2 or 1 ms. The gas mixture photolyzed consists of 100 Torr Ar, 0.05 Torr H_2O and O Torr, $1.1 \cdot 10^{-5}$, $1.9 \cdot 10^{-5}$ or $2.4 \cdot 10^{-5}$ Torr p-chloroaniline. The decays of OH in figure 4 can be monitored for more than 2 orders of magnitude in concentration. In the reactions of benzene and dichlorobenzenes with OH similar exponential decays of OH were observed at least for one order of magnitude in concentration. At long reaction times the decay slows down. This result concerning benzene agrees with the observations by Perry et al. (25).

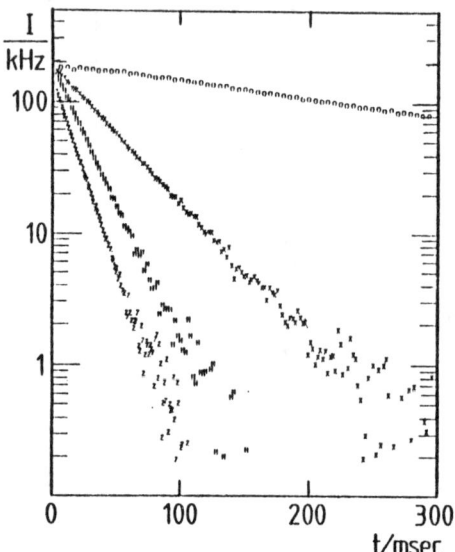

Fig. 4: Semilogarithmic plot of the resonance fluorescence intensity of OH vs. time in the presence of p-chloroaniline (symbol O,X,H,Z: O Torr, $1.1 \cdot 10^{-5}$, $1.9 \cdot 10^{-5}$, $2.4 \cdot 10^{-5}$ Torr), 100 Torr Ar and 0.05 Torr H_2O. The decays are averaged from 100 flashes at 2 J flash energy using a dwell time of 5 ms (symbol O), 2 ms (symbol X,H) or 1 ms (symbol Z).

The decay rates, τ^{-1}, obtained from the slopes of semilogarithmic plots in all cases increase linearly with the concentration of the reactant, $[R]$, according to the equation

$$\tau^{-1} = \tau_0^{-1} + k\,[R] \qquad \text{(II)}.$$

Rate constants, k, are determined from the slope of the linear plot of the decay rate vs. the concentration of the reactant. The slope which is determined by means of least squares fit yields the values of the rate constants, k, for the reactions

$$OH + R \xrightarrow{k} \text{products}$$

The intercept of the ordinate τ_0^{-1} is caused by diffusion losses of OH radicals from the observed volume and by losses caused by reactions of OH radicals with each other. In all series of measurements this intercept is within the expected limits (τ_o^{-1} = 2-12 s^{-1}).

Figures 5 and 6 show plots of the experimental data plotted according to equation II. The values of rate constants obtained at room temperature (294 ± 1 K) and error limits (three standard deviations) are given in table II).

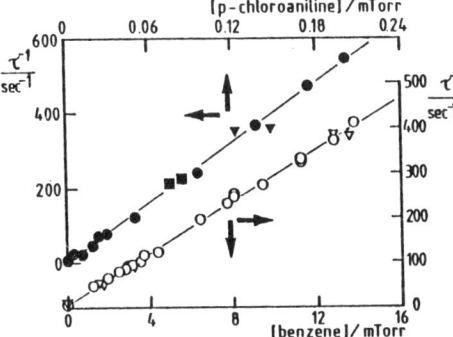
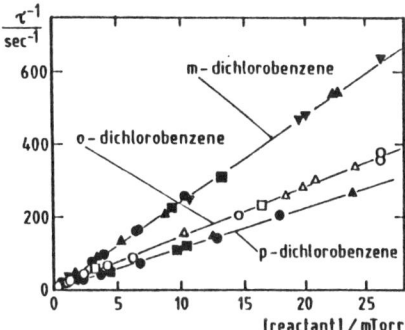

Fig. 5: Dependence of the decay rate on the concentration of p-chloroaniline (upper part) and benzene (lower part). Experimental conditions: flash energy 2 J, p-chloroaniline: ●,■ 100, 40 Torr Ar, 0.05 Torr H₂O, ▼ 300 Torr Ar, 0.2 Torr H₂O; benzene: O 100 Torr Ar, 0.05 Torr H₂O, ∇ 130 Torr Ar, 0.1 Torr H₂O.

Fig. 6: Dependence of the decay rate on the concentration of the dichlorobenzenes. Experimental conditions: meta-: ●,▼,▲ 100,100, 50 Torr Ar, 0.05 Torr H₂O,■ 30 Torr Ar, 0.03 Torr H₂O, flash energy 2 J except ▼ (11 J); ortho-: O,△ 100, 200 Torr Ar, 0.05 Torr H₂O, □ 50 Torr Ar, 0.09-0.02 Torr H₂O, flash energy 2 J; para-: ●,▲,■ 100,250,50 Torr Ar, 0.03-0.05 Torr H₂O, flash energy 2 J.

Tab. II: Rate constants obtained in the present study and experimental conditions:

reactant	$\dfrac{[Ar]}{Torr}$	$\dfrac{[H_2O]}{Torr}$	$\dfrac{k \pm 3\sigma}{10^{-13}\ cm^3s^{-1}}$
p-chloroaniline	40-300	0.01-0.2	(810 ± 70)
benzene	100-130	0.05-0.1	(8.8 ± 0.4)
o-dichlorobenzene	50-200	0.02-0.09	(4.2 ± 0.2)
m-dichlorobenzene	30-200	0.03-0.05	(6.9 ± 0.2)
p-dichlorobenzene	50-250	0.02-0.05	(3.3 ± 0.3)

The data for p-chloronaniline shown in figure 5 (upper part) were obtained at different pressures of Ar (40, 100, and 300 Torr) and different H₂O pressures (0.01, 0.05, and 0.2 Torr). No dependence of the rate constant on the concentrations of Ar and H₂O was observed. The data for benzene are also included in figure 5 (lower part). All measurements (performed at 100 and 130 Torr Ar pressure and 0.05 and 0.1 Torr H₂O pressure) are found to fit to the same correlation.

Figure 6 shows the decay rates of the OH resonance fluorescence intensity as a function of the partial pressures of the three different dichlorobenzenes. The observed decay rates did not depend on the Ar pressure or on the H₂O concentration (see table II). Additional measurements were carried out with m-dichlorobenzene using a flash energy of 11 J instead of 2 J (used normally). Also these measurements fit to the linear regression in figure 6. Cutting off the spectrum of flash lamps by a reflection filter (200 nm $>\lambda>$ 160 nm) the rate constants of p-dichlorobenzene did not change the rate constants either.

4. DISCUSSION

The precise dosing of the reactants at known concentrations is a main condition to obtain reliable values of rate constants. The check of the dosing system (by freezing the reactants in cold traps and determining the weight of the condensed material) yielded good agreement with vapor pressure data from the literature (as shown in figure 3). This verifies the total saturation of the small flows of inert gas in the saturator at known temperatures and gives support to the vapor pressure data used.

The rate constants were found to be independent of the concentration of water vapor. This indicates a constant stoichiometry of the reactions equal to unity (one OH radical is consumed per reactant molecule), since the photolytically formed OH concentration is to a first approximation proportional to the water concentration. In addition the rate constants were found to be independent of the flash energy, which was varied by a factor of five in the case of m-dichlorobenzene. Hence photolysis products of the aromatics may not be involved significantly in the kinetics of the observed decays of OH radicals and a contribution of radical radical reactions can be ruled out. In the case of p-dichlorobenzene a small contribution of radical radical reactions could be observed at short reaction times when photolysing with wavelengths $\lambda >125$ nm. The slight curvature of the decay curves (in the semilogarithmic diagram) disappeared when photolysing at 200 $>\lambda>$ 160 nm (reflection filter). The rate constant obtained with the reflection filter did not differ significantly from the value obtained with the CaF$_2$ window.

Data for the rate constants of the reactions of OH with the dichlorobenzenes and p-chloroaniline could not be found in literature. The rate constant obtained for benzene in the present study ($0.88 \pm 0.04) \cdot 10^{-12}$cm^3 s^{-1} is found to be significantly lower than previous literature data ($1.24 \pm 0.12) \cdot 10^{-12}$cm^{3}s^{-1} (26). But recent determinations by Lorenz and Zellner $(0.8 \pm 0.2) \cdot 10^{-12}$cm^{3}s^{-1} (27) and Rinke and Zetzsch ($1.0 \pm 0.08) \cdot 10^{-12}$cm^3 s^{-1} (28) using He as inert gas agree within the error limits of the present study.

The reactivities of the aromatics given in the present study differ by more than two orders of magnitude. This may be interpreted assuming electrophilic addition of OH to the aromatic ring. In a previous publication (10) a correlation between the ionization potential and the reactivity against OH was proposed in accordance with similar correlations for the reactions of O atoms with aromatics (29) and of OH radicals with olefins (30). Rate constants for the reactions of substituted benzenes with OH radicals at room temperature are given in table III arranged in the order of increasing ionization potential. These data are displayed in figure 7 in a semilogarithmic diagram of the rate constants vs. the ionization potential. Figure 7 shows a reasonable correlation for most aromatics. This straight line is a linear regression plot of the displayed data with the exception of nos. 13, 16, 18, 17, 19, 21, 23 (the cresols, the dichlorobenzenes, and benzaldehyde); benzene (no. 22) and p-chloroaniline (no. 2) fit well to the straight line in the diagram. P-chloroaniline reacts slower than aniline due to the negative inductive effect of the chlorine atom. A considerable increase in reactivity is observed for both substances in

Tab. III: Ionizationspotentials (39) and rate constants of differently
substituded aromatics

No.	chemical	IP/eV	$k/10^{-12}cm^3s^{-1}$	reference
1	aniline	7.69	119	28
2	p-chloroaniline	7.77	81	this work
3	methoxybenzene	8.21	19.6	25
4	1.2.4 trimethylbenzene	8.27	30.8-40.0	25,26,31
5	1.3.5 trimethylbenzene	8.40	47.2-62.4	25,26,31
6	p-xylene	8.45	10.5-15.3	25,26,31,32
7	1.2.3 trimethylbenzene	8.48	21.3-33.3	25,26,31
8	phenol	8.50	28.2	28
9	m-xylene	8.56	20.6-24.0	25,26,31-33
10	o-xylene	8.56	12.0-15.3	25,26,31,32
11	n-propyl benzene	8.72	5.86-6.1	32,33
12	i-propyl benzene	8.73	6.1-7.79	32,33
13	m-cresol	8.75	67	34
14	ethylbenzene	8.76	7.65-8.0	32,33
15	toluene	8.82	3.9-6.4	25,26,31,35
16	o-cresol	8.93	34.1-47	25,34
17	p-dichlorobenzene	8.94	0.33	this work
18	p-cresol	8.97	52	34
19	o-dichlorobenzene	9.07	0.42	this work
20	1.2.4 trichlorobenzene	9.1	0.52	28
21	m-dichlorobenzene	9.12	0.69	this work
22	benzene	9.24	0.88-1.59	25,26,31,35 this work
23	benzaldehyde	9.53	16	36
24	benzonitrile	9.705	0.33	37
25	hexafluorobenzene	9.90	0.22	32

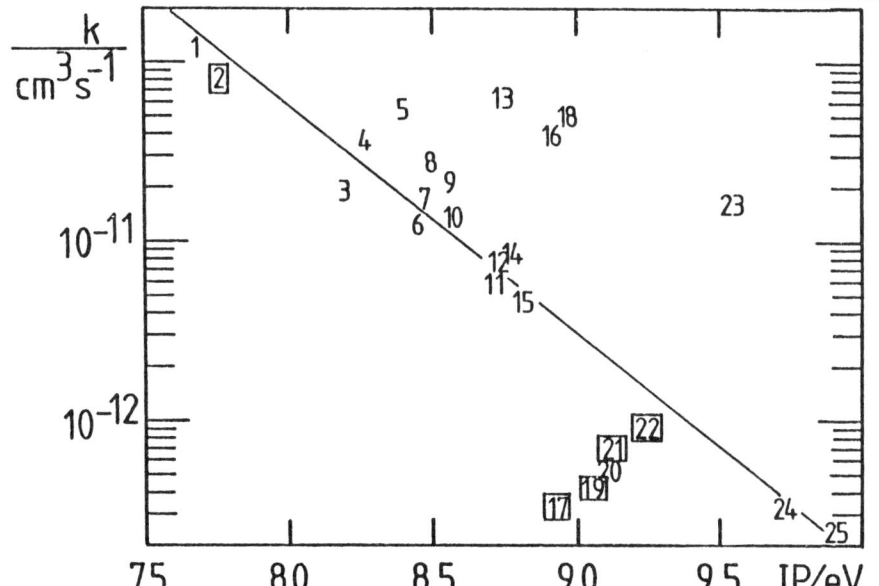

Fig. 7: Semilogarithmic plot of the rate constants of differently substituted aromats vs. the ionization potential. Numbers see tab. III.

comparison with benzene due to the mesomeric effect from the lone pair
electrons of the amino group. The deviations of the cresols (nos. 13,16,
and 18) and benzaldehyde (no. 23) from the correlation might be explained
by contributions of H-abstraction to the reaction mechanism. This argument
is especially valid for benzaldehyde because of the weak aldehydic C-H
bond.

The ranking of the three isomeric dichlorobenzenes does not fit to
the correlation in figure 7. This different behaviour in the electrophilic
addition of OH to the aromatic ring can be explained by both the negative
inductive and electron delivering positive mesomeric effect of the chlorine
atoms. All dichlorobenzenes are less reactive than benzene, due to the
negative inductive effect. The ranking in reactivity (m->o->p-) is plausible
regarding the possible number of mesomeric structures (assuming σ-bonded
addition complexes).

m- dichlorobenzene

o- dichlorobenzene

p- dichlorobenzene

The most reactive m-dichlorobenzene is able to form the highest number of
mesomeric structures, i.e. the addition complex is the most stable. The
other isomers, o- and p-dichlorobenzene have the same number of mesomeric
structures. But in the case of p-dichlorobenzenes all free positions at
the aromatic ring are found to be meta positions with respect to one of the
Cl atoms. Steric hindrance is more effective with p-dichlorobenzene than
with o-dichlorobenzene. In the latter case two non-meta positions are
available increasing the reactivity of o-dichlorobenzene. These considerations are confirmed by the experimental data.

The correlation in figure 7 enables us to predict rate constants (10).
The prediction should be used with caution if deactivating and ortho, para
directing substituents are involved (like the halogens). From the rate
constants determined in the present study upper limits of the persistence
of these chlorinated aromatics can be estimated assuming reaction with OH
as the major tropospheric loss process and neglecting possible temperature
dependences of the rate constants. Using a global average tropospheric concentration of $[OH] = 4 \cdot 10^5 cm^{-3}$ (38), the tropospheric residence times of

the dichlorinated benzenes are found to vary from approximately 1 month
(o-dichlorobenzene to 3 months (p-dichlorobenzene). The persistence of
p-chloroaniline is found to be less than one day.

Acknowledgement
    Financial support of this work by the Bundesminister für Forschung
und Technologie is gratefully acknowledged.

References

1) F. Korte, "Ökologische Chemie", G. Thieme Verlag, Stuttgart 1980
2) H.B. Singh, L.J. Salas, A.J. Smith and H. Shigeishi, Atmospheric
   Environment 15, 601 (1981)
3) S.A. Ware and W.L. West, "Investigation of Selected Potential
   Environmental Contaminants: Halogeneted Benzenes", Final Report,
   Office of Toxic Substances, U.S. Environmental Protection Agency,
   Washington D.C. 1977
4) E. Lahmann, B. Seiffert and W. Dulson, Bundesgesundheitsblatt 21, 75
   (1978)
5) L. Roth and W. Daunderer, "Giftliste: Gifte, krebserzeugende, gesundheitsschädliche und reizende Stoffe", Verlag Moderne Industrie,
   München 1978
6) D. Henschler, "Gesundheitsschädliche Arbeitsstoffe-Toxikologisch-
   arbeitsmedizinische Begründung von MAK-Werten", Verlag Chemie,
   Weinheim 1979
7) S.L. Brown, F.Y. Chan , J.L. Jones, D.M. Lin, K.E. McCaleb, F. Mill,
   R.N. Supios and D.F. Schendel, "Research Program on Hazard Priority
   Ranking of Manufactured Chemicals : Phase II", Final Report, Stanford
   Research Institute, Menlo Park, California 1975
8) F. Stuhl and H. Niki, J. Chem. Phys. 57, 3671 (1972)
9) A. Wahner, Diplomarbeit, Ruhr Universität Bochum 1981
10) C. Zetzsch, Proceedings of the International Workshop on the Test
    Methods and Assessment Procedures for the Determination of the Photo-
    chemical Degradation Behaviour of Chemical Substances, Berlin, 1980
    (in press)
11) C. Zetzsch and F. Stuhl, Second European Symposium on "Physico-
    Chemical Behaviour of Atmospheric Pollutants", Varese 1981
12) H.W. Biermann, C. Zetzsch and F. Stuhl, Ber. Bunsenges. Phys. Chem. 82,
    633 (1978)
13) H. Merkel, Reglungstechn. Prax. 20, 317 (1978)
14) we thank Dr. P. Bruckmann (Landesanstalt für Immissionsschutz, Essen)
    for performing the analysis
15) R.C. Wilhoit and B.J. Zwolinsky, "Handbook of Vapor Pressure and Heat
    of Vaporization of Hydrocarbons and Related Compounds", API 44-TRC,
    Publications in Science and Engineering, New York 1971
16) C.F. Mündel, Z. Phys. Chem. 85, 435 (1913)
17) S. Young, Sci. Proc. Soc. Dublin, N.S. XII, 374 (1909)
18) R. Dreisbach, "Physical Properties of Chemical Compounds", Adv.
    Chem. Ser. 15, New York 1955
19) S. Ohe, "Comparts aided Data Book of Vapor-Pressures", New York 1965
20) Landolt-Börnstein, II. Band, Teil 2 a, Springer Verlag, Berlin 1969
21) W. Klöppffer, "Merkblätter über Referenzchemikalien", Battelle In-
    stitute, Frankfurt 1979
22) D.R. Stull, Ind. Eng. Chem. 39, 517 (1947)
23) R.C. Weast, "Handbook of Chemistry and Physics", CRC-Press, Cleveland
    1976

24) F.R. Darkis, H.E. Vermillion and P.M. Gross, Ind. Eng. Chem. <u>32</u>, 946 (1940)

25) R.A. Perry, R. Atkinson and J.N. Pitts, J. Phys. Chem. <u>81</u>, 296 (1977)

26) D.A. Hansen, R. Atkinson and J.N. Pitts, J. Phys. Chem. <u>79</u>, 1763 (1975)

27) K. Lorenz and R. Zellner, Proceedings of the International Workshop on the Test Methods and Assessment Procedures for the Determination of the Photochemical Degradation Behaviour of Chemical Substances, Berlin, 1980 (in press)

28) M. Rinke and C. Zetzsch, Regionales Photochemiker Treffen, Schriftenreihe d. Inst. f. Strahlenchem. <u>1</u>, 15 (1980)

29) R.E. Huie and J.T. Herron, Prog. Reac. Kin. <u>8</u>, 1 (1975)

30) J.S. Gaffney and S.Z. Levine, Int. J. Chem. Kin. <u>10</u>, 1197 (1979)

31) G.J. Doyle, A.C. Lloyd, K.R. Darnall, A.M. Winer and J.N. Pitts, Environ. Sci. Technol. <u>9</u>, 273 (1975)

32) A.R. Ravishankara, S. Wagner, S. Fischer, G. Smith, R. Schiff, R.T. Watson, G. Tesi and D.D. Davis, Int. J. Chem. Kin. <u>10</u>, 783 (1978)

33) A.C. Lloyd, K.R. Darnall, A.M. Winer and J.N. Pitts, J. Phys. Chem. <u>80</u>, 789 (1976)

34) R. Atkinson, K.R. Darnall and J.N. Pitts, J. Phys. Chem. <u>82</u>, 2759 (1978)

35) D.D. Davis, W. Bollinger and S. Fischer, J. Phys. Chem. <u>79</u>, 293 (1975)

36) H. Niki, P.D. Maker, C.M. Savage and L.P. Breitenbach, J. Phys. Chem. <u>82</u>, 132 (1978)

37) C. Zetzsch, to be presented at: Nato Advanced Study Institute, "Chemistry of the unpolluted and polluted troposphere", Corfu 1981

38) T.E. Graedel, Chemical Compounds in the Atmosphere, Academic Press, New York 1978

39) H.M. Rosenstock, K.Draxl, B.W. Steiner and J.T. Herron, J. Phys. Chem. Ref. Data <u>6</u>, Supp. 1 (1977)

PRODUCT ANALYSIS OF CHEMICAL REACTIONS RELEVANT TO

ATMOSPHERIC CHEMISTRY

P.N. GHOSH, F. RAKOCZI, A. BAUDER and Hs.H. GUENTHARD
Laboratory for Physical Chemistry
Swiss Federal Institute of Technology
8092 Zürich, Switzerland

Summary

Reactions of olefines with ozone and partly with added
nitrogen dioxide have been run in linear reactors or mi-
croreactors with residence times between 30 sec and
1 msec. The reaction mixtures have been analyzed by freezing the products into an argon matrix at 4 K and by taking infrared and electron spin resonance spectra. Alternatively the reaction mixture was pumped through the absorption cell of a microwave spectrometer and the products identified by their rotational spectra. Unstable
radical species as shortlived intermediates could be detected for the first time with electron spin resonance.
Among the relatively stable reaction products many new
and unexpected species have been found from infrared and
microwave spectra. Addition of nitrogen dioxide to the
reaction produces a number of NO_2-containing compounds
including peroxyacetylnitrate (PAN) even in the dark.

1. INTRODUCTION

In the initial phase of photochemical smog formation, reactions of NO_x and oxygen build up a reservoir of NO_2 and O_3
under sunlight irradiation. The two reactive species then
start chains of chemical reactions with hydrocarbons, olefins,
aldehydes, ketones and other compounds present in polluted
air. A large number of products are formed during these reactions, some of them being highly toxic to the biosphere and
man. So far only a limited number of products could be identified, some of them appearing only in very small concentrations.
The reaction mechanisms are poorly understood at present. Many
of the reactions were included more on the basis of simple
guesses than on thorough knowledge. For a few of the reaction
chains the kinetics of the initial reactions are known from
laboratory measurements. Kinetic constants are used for modelling the time evolution of photochemical smog formation.

In this paper we report the results of a series of laboratory measurements with main emphasis on identification of
the products. Since the reaction system in the photochemical
smog is extremely complicated, we investigated simplified
systems composed of an olefin which was reacted with ozone

with or without nitrogen dioxide added simultaneously. Matrix-isolation infrared spectroscopy and microwave spectroscopy were used for the identification of products by comparison of the spectra of reaction mixtures with reference spectra of pure compounds. Isotopically substituted species of the reactants were occasionally employed in order to elucidate certain steps of the reaction mechanisms. Electron spin resonance was a unique tool for a sensitive detection and identification of a number of radical species for the first time (1).

2. INFRARED AND MICROWAVE SPECTROSCOPY OF LINEAR REACTOR EXPERIMENTS

The reaction mixtures of olefin and ozone with or without nitrogen dioxide were passed through a long tube forming a linear reactor if radial diffusion is neglected. At the end of the linear reactor the reaction mixture was subjected to spectroscopic measurements. In the experiments with infrared (IR) detection, argon was added as an inert diluent gas which served as matrix when the reaction mixture was frozen onto a window cooled with liquid helium. No diluent gas could be used for observing the microwave (MW) spectra due to the broadening of absorption lines by foreign gases. Table I shows the conditions for the linear reactor experiments. The reaction system has been investigated for different olefins. Those

Table I. Typical Conditions of Linear Reactor Experiments

Type of experiment	Spectroscopic detection	Initial composition (Torr)				Residence time (s)	Temperature (K)
		$p(Olefin)$	$p(O_3)$	$p(NO_2)$	$p(Ar)$		
1	IR	20	20	0	700	60	298
2	IR	10	20	10	700	60	298
3	MW	2	2	0	0	30	298
4	MW	1	3	1	0	30	298

most thoroughly studied include ethylene and cis-2-butene. The products identified for these reaction systems are collected in Tables II and III. Positive identification of products from infrared spectra of reaction mixture containing more than a dozen different compounds is only possible for the narrow absorption lines of matrix isolated molecules. There are some differences between the results from infrared and microwave detection. Occasionally such differences might

Table II. Products of the Reaction of Etylene and Ozone with or without Addition of Nitrogen Dioxide.[a]

Compound	Without NO_2 detection by[b]		With NO_2 detection by
	IR	MW	MW
CO	+ +	-	-
CO_2	+ +	-	-
H_2O	+ +	+ +	+ +
HCHO	+ +	+ +	+ +
HCOOH	+ +	+ +	+ +
$\overline{CH_2OCH_2}$	-	+	+
CH_3OH	-	+	+
CH_3CHO	+	+	+
$(CHO)_2O$	+	+	+
CH_2OHCHO	+	+	-
HNO_3	-	-	+ +
HONO, cis and trans	-	-	+
CH_3ONO, cis and trans	-	-	+
CH_3ONO_2	-	-	+

[a] Estimated abundance in reaction mixture: + + > 1 mol%, + < 1 mol%,
  - not detected.

[b] See (2).

be due to changes in the product spectrum for highly exothermic reactions if no diluent gas is present for efficiently thermalizing excited molecules by collisions. Most of them, however, can be traced back to different sensitivities for products in the two spectroscopic detection methods. Therefore the two methods complement each other.

Comparison of the results for the reactions with and without NO_2 addition shows that the main products of the ozonolysis of olefins were still found in the reactions with NO_2 added. This was rather unexpected from model calculations of the kinetic behaviour, since reaction velocities for the two reactions at 298 K

Table III. Products of the Reaction of cis-2-Butene and Ozone with or without Addition of Nitrogen Dioxide

Compound	Without NO_2 detection by[a]		With NO_2 detection by	
	IR	MW	IR	MW
CO	+	−	+	−
CO_2	+	−	+	−
H_2O	+	+	+	+
HCHO	+	+	+	+
CH_4	+	−	+	−
HCOOH	+	+	+	+
CH_3OH	+	+	+	+
$(HCO)_2O$	+	−	+	−
CH_3CHO	+	+	+	+
CH_3COOH	+	+	+	+
CH_2OHCHO	−	+	−	−
CH_3COOOH	+	+	−	−
$CH_3COOCHO$	+	+	+	−
$(CH_3)_2$ C=C=O	−	+	−	−
$CH_3CH\overset{\diagdown}{\underset{O}{}}CH\,CH_3$, cis and trans	−	+	−	−
$CH_3COCH_2CH_3$	−	+	−	−
$(CH_3CO)_2O$	+	−	−	−
$CH_3\,CH\overset{\diagup O\diagdown}{\underset{O-O}{\,\,}}CH\,CH_3$	+	+	+	−
HNO_3	−	−	+	+
N_2O_5	−	−	+	−
CH_3ONO, cis and trans	−	−	+	−
CH_3ONO_2	−	−	+	+
CH_3COONO_2	−	−	+	−
$CH_3COOONO_2$ (PAN)	−	−	+	−

[a] See (3).

$$C_2H_4 + O_3 \rightarrow \text{products} \quad (1); \quad k_1 = 1.7 \times 10^{-18} \text{ cm}^3 \text{ s}^{-1} \text{ molecule}^{-1} \quad (4);$$

and

$$NO_2 + O_3 \rightarrow NO_3 + O_2 \quad (2); \quad k_2 = 3.2 \times 10^{-17} \text{ cm}^3 \text{ s}^{-1} \text{ molecule}^{-1} \quad (5);$$

are widely different. Furthermore it should be pointed out, that the reaction mixtures were not irradiated by light. Nevertheless, products like peroxyacetylnitrate were formed in the dark.

3. ELECTRON SPIN RESONANCE DETECTION OF RADICALS

Radical intermediates in the reaction of olefins with ozone have often been postulated. But only hydroxyl radicals were observed through their chemiluminescence emission. Direct evidence for the presence of further radical species was recently obtained in matrix electron spin resonance (ESR) of reaction mixtures. The reaction between ethylene and ozone diluted by argon was carried out at 500 Torr in a special microreactor which allowed residence times between 1 ms and 1 s. The reaction mixture was then trapped in an argon matrix at 18 K. ^{17}O labelled ozone was used in some runs in order to facilitate the assignment of the radical species with the help of hyperfine splittings. The results of the analysis of the ESR spectra are given in Table IV.

Table IV. Radicals Detected by Matrix ESR in the Reaction of Ethylene with Ozone.

Radical species	$HO_2^\cdot$ and / or $RO_2^\cdot$	H·	$CH_3^\cdot$	OH·
Identification	+ +	+	+	-

The main radical species are found to be hydroperoxyl and / or alkylperoxyl. The ESR spectra do not allow a discrimination between these two species. Hydrogen and methyl radicals were observed only if molecular oxygen is completely eliminated from the ozone used in the reaction. Hydroxyl radicals could not be detected under the conditions for the reactions. The hydroxyl chemiluminescence was reported for reactions carried out with the total pressure not exceeding a few Torr.

ACKNOWLEDGMENT

We acknowledge gratefully financial support by the Swiss National Science Foundation (Project No. 2.219-0.79 and 2.612-0.80), by the Bundesamt für Wissenschaft und Forschung (COST 61a bis) and by the authorities of the Swiss Federal Institute of Technology.

REFERENCES

(1) F. Rakoczi and Hs.H. Günthard, Chem. Phys. Lett. 67, (1979) 173.

(2) H. Kühne, S. Vaccani, A. Bauder and Hs.H. Günthard, Chem. Phys. 28 (1978) 11.

(3) H. Kühne, M. Forster, J. Hulliger, H. Ruprecht, A.Bauder and Hs.H. Günthard, Helv. Chim. Acta 63 (1980) 1971.

(4) J.T. Herron and R.E. Huie, J. Phys. Chem. 78 (1974) 2085.

(5) D.L. Baulch, R.A. Cox, R.F. Hampson, J.A. Kerr, J. Troe and R.T. Watson, J. Phys. Chem. Ref. Data 9 (1980) 295.

RATE CONSTANT OF THE REACTION $CH_3 + O_2 \rightarrow CH_2O + OH$

V.Fonderie and J.Peeters
Department of Chemistry, University of Leuven,Belgium

Summary

The bimolecular reaction

$$CH_3 + O_2 \rightarrow CH_2O + OH \qquad (r.1)$$

was investigated in a fast-flow reactor at a temperature of 400 K and a pressure of 2 torr using molecular beam sampling and mass spectrometric analysis.

A known amount of CH_3 radicals, produced by thermal dissociation of CH_3Br on a hot tungsten filament, was mixed with O_2 and the resulting [OH] was measured at a reaction time of $4 \cdot 10^{-3}$ s. Helium was used as carrier gas.

The rate constant was deduced from the OH production rate, derived from the observed [OH] and from the known OH wall loss rate constant; termination of OH on the wall of the pyrex reactor was the dominant OH removal process in our experimental conditions.

A statistical analysis of the OH signals observed over a period of about an hour, both with and without CH_3 radicals added to the O_2 stream, has led us to conclude that the rate constant k_1 is smaller than $7.5 \cdot 10^8$ cm^3 $mole^{-1}$ s^{-1} with a statistical confidence of 97.5%.

1. INTRODUCTION

The reaction of methyl radicals with molecular oxygen is of importance in the oxidation of hydrocarbons and in the chemistry of the upper and lower atmosphere.

The reaction

$$CH_3 + O_2 \rightarrow CH_2O + OH \qquad (r.1)$$

has been proposed to explain the formation of formaldehyde in the combustion of methane (1,2); other authors have questioned the direct formation of formaldehyde in this reaction (3-5).

In the ambient atmosphere CH_3 is mainly produced by the reaction of methane with hydroxyl radicals while in the stratosphere CH_3 is also created in the reaction of methane with $O(^1D)$ atoms (6,7). The resulting methyl radicals react with molecular oxygen.

In atmospheric photooxidation systems the three body reaction

$$CH_3 + O_2 + M \rightarrow CH_3O_2 + M \qquad (r.2)$$

is generally accepted as a CH_3 sink; the rate constant for the methylperoxyl formation is fairly well established. On the other hand, the reported values for the rate constant of the bimolecular reaction (r.1) diverge by several orders of magnitude (5,8-11). The low values imply a negligible contribution of the bimolecular process to the total CH_3 destruction in the atmosphere while the higher values would entail a significant contribution of that reaction, even at atmospheric pressure.

In this study an upper limit of the rate constant of the bimolecular reaction was derived directly from the measurement of the rate of OH production by the $CH_3 + O_2$ reaction.

2. EXPERIMENTAL

The experiments were carried out in a conventional fast-flow reactor made of pyrex (inner diameter 1.6 cm). Methyl radicals were generated by thermal decomposition of methylbromide, diluted in He as carrier gas, on a current-heated tungsten filament array placed in the reactor. The array consists of coiled tungsten wires (0.1 mm diameter) spot welded on tungsten rods that serve as support as well as current supply leads. The rods are mounted coaxially around a central injector tube as shown in fig.1.

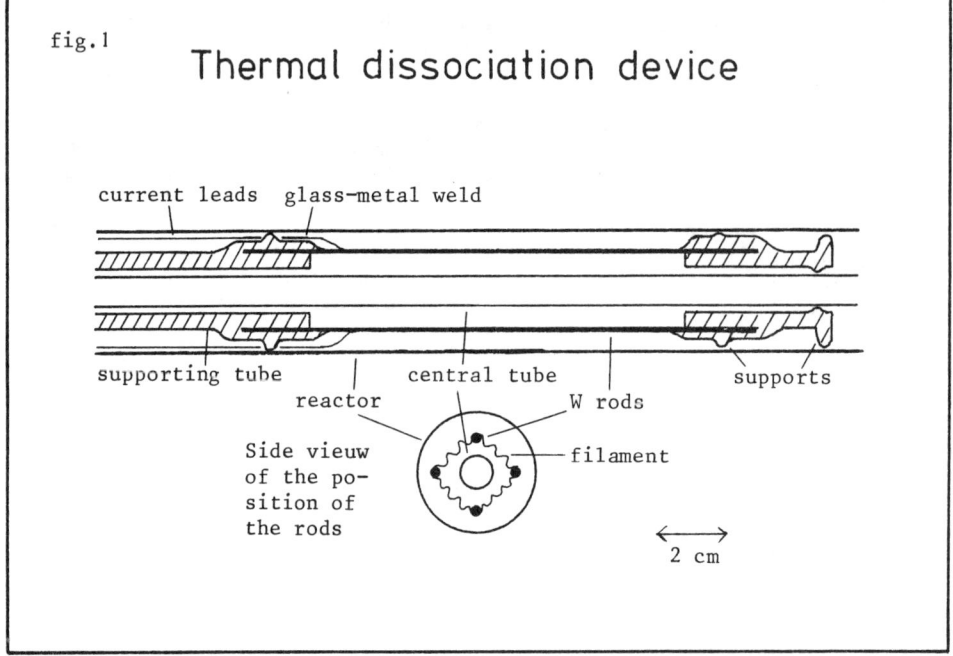

fig.1

Thermal dissociation device

Molecular oxygen was added to the reaction mixture through the movable central injector. The pressure was measured downstream of the reactor with a Datametrics capacitance manometer; it was held constant at 2.00 ± 0.01 torr. The temperature in the reactor was monitored with a thermocouple mounted in the coaxial inlet tube.

The thermal radical source is located 20 cm upstream of the sampling point; oxygen molecules are added to the main gasstream 8 cm downstream of the thermal source such that the kinetic zone has a length of 12 cm (fig.2). The linear flow velocity, at the average temperature of 400 K in the kinetic zone, is 3120 cm s^{-1} leading to a reaction time of $3.8 \, 10^{-3}$ s.

The concentrations of radicals and stable species were measured by molecular beam sampling and mass spectrometric analysis using an Extranuclear Laboratories quadrupole filter. Beam modulation and phase sensitive detection was applied. The sampling and detection techniques were described previously (4,12).

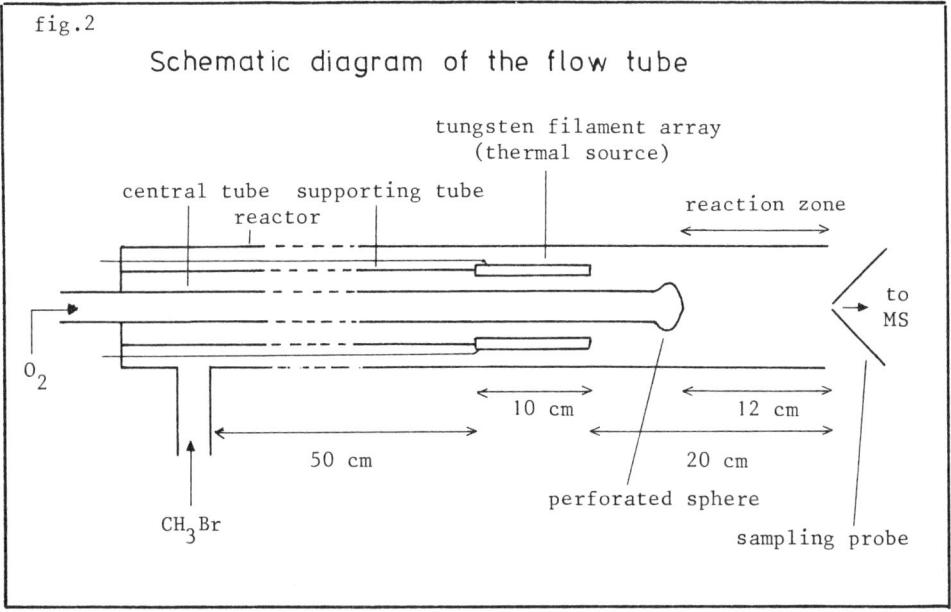

fig.2

Schematic diagram of the flow tube

tungsten filament array
(thermal source)

central tube   supporting tube
reactor

reaction zone

to
MS

O_2

10 cm   12 cm

50 cm   20 cm

CH_3Br

perforated sphere

sampling probe

Radicals and atoms were monitored at ionising electron energies of only 2 or 3 eV above the corresponding ionisation potentials in order to a-
void interference by fragment ions. However, because of the low appearance
potential of CH_3^+ ions formed from CH_3Br, this fragmentation process could
not be completely suppressed even at the low energy of 11.5 eV at which
CH_3 was monitored. The contribution of the CH_3^+ fragment ions was derived
from the CH_3^+/CH_3Br^+ fragmentation ratio measured in the absence of CH_3
radicals.

For stable components the sensitivity factors S_x were obtained by direct measurement of the mass spectrometric output signal i_x for a known
concentration of the species in the reactor:

$$S_x = i_x / [X]$$

The sensitivity factor for CH_3 at 11.5 eV was taken equal to that for CH_4
at the same excess electron energy above the corresponding ionisation
potential, multiplied by the ratio of the ionisation cross sections at
70 eV. The validity of this estimation procedure was verified by the measurement of $[CH_3]$ in a fuel rich $C_2H_6/H_2/O_2$ flame based on the reaction

$$CH_3 + H_2 \rightleftharpoons CH_4 + H \qquad (r.3)$$

balanced at a given point in that flame (13). The CH_3 sensitivity factor
derived from the balanced reaction (r.3) and that from the estimation method outlined above agreed to within 30 %.

The calibration of OH is based on measurements on the $H + O_2$ system,
investigated in the same reactor with the H atoms generated by a microwave
discharge in a H_2/He mixture. Details will be given in the next section.
H atoms finally were calibrated by the discharge on/off method.

The electron energy, the ionisation potential and the sensitivity factors for the species of interest in this study are given in table I.

Table I

component	electron energy (eV)	IP (eV)	sensitivity factor (μV/mole cm^{-3})
H	16.2	13.6	$3.20 \ 10^{12}$
OH	15.2	13.2	$1.1 \ 10^{14}$
CH_3	13.0	9.8	$4.46 \ 10^{14}$
	11.5		$1.12 \ 10^{14}$

The experiments were carried out with an initial CH_3Br concentration of $3 \ 10^{-10}$ mole cm^{-3} and $[O_2] = 1.6 \ 10^{-8}$ mole cm^{-3} at the mean reaction zone temperature of 400 K. The thermal source produces CH_3 up to a concentration of $1.12 \ 10^{-12}$ mole cm^{-3} at the beginning of the kinetic zone. Since the reactor was not cooled the thermal source results in an inhomogeneous temperature profile with the temperature ranging from 500 K at the start of the kinetic zone to 300 K at the sampling probe.

3. RESULTS AND DISCUSSION

The rate constant for the reaction

$$CH_3 + O_2 \rightarrow CH_2O + OH \qquad (r.1)$$

was determined by the measurement of the amount of hydroxyl radicals formed after a fixed reaction time. Measurement of the rate of production of OH was preferred to that of CH_2O because of the interference at m/e = 30 due to C_2H_6 produced in the fast recombination of CH_3:

$$CH_3 + CH_3 \rightarrow C_2H_6 \qquad (r.4)$$

and because of the possibility of CH_2O formation in side reactions.

A. OH production

In order to be sure that any detected OH signal is to be ascribed to the $CH_3 + O_2$ reaction, it must first be ascertained that possible OH production in side reactions is negligible.

In principle OH could also be formed via the following reaction scheme

		k ($cm^3 mole^{-1} s^{-1}$) at T = 400 K	ref.
$CH_3 + O_2 + M \rightarrow CH_3O_2 + M$	(r.2)	$4 \ 10^{16}$ ‡	
$CH_3O_2 + CH_3 \rightarrow 2 \ CH_3O$	(r.5)	$3.6 \ 10^{13}$	(14)
$2 \ CH_3O_2 \rightarrow 2 \ CH_3O + O_2$	(r.6)	$9.4 \ 10^{10}$	(15,16)
$CH_3O_2 + Br \rightarrow CH_3O + BrO$	(r.7)	$4.5 \ 10^{12}$	(15,17)
$CH_3O + Br \rightarrow CH_2O + HBr$	(r.8)	$4 \ 10^{12}$	(18)
$CH_3O + O_2 \rightarrow CH_2O + HO_2$	(r.9)	$4 \ 10^{8}$	(19)
$CH_3 + HO_2 \rightarrow CH_3O + OH$	(r.10)	$2 \ 10^{13}$	(20)

‡ units: $cm^6 mole^{-2} s^{-1}$

The rate constant for reaction (r.2) is the result of a combination of our experimental results at room temperature with the temperature dependence reported by Washida and Bayes (9). The rate constants for reactions (r.7) and (r.8) are assumed equal to the k values for the corresponding NO reactions. For reaction (r.5) no information about the temperature dependence of the rate constant is available; the indicated value was measured at room temperature. The k_{10} value is estimated (20).

The ratio of OH production via (r.10) to the OH production via (r.1) is equal to

$$k_{10}[HO_2] \; / \; k_1[O_2]$$

Due to the low initial CH_3 concentration and to the rather short reaction time the intermediates CH_3O_2, CH_3O and HO_2 will react only to a small extent so that their concentrations can be deduced from their rate of formation:

$$[CH_3O_2] = k_2\int[CH_3][O_2][M]dt$$
$$[CH_3O] \simeq 2 \, k_5\int[CH_3O_2][CH_3]dt$$
$$[HO_2] = k_9\int[CH_3O][O_2]dt$$

A simple numerical calculation shows that for this system the $[HO_2]/[O_2]$ ratio is at most 3×10^{-7} after a reaction time of 3.8×10^{-3} s implying a negligible OH formation via reaction (r.10).

B. OH destruction

The hydroxyl radical can be destroyed by the following reactions:

		k $(cm^3mole^{-1}s^{-1})$ at T = 400 K	ref.
$CH_2O + OH \rightarrow CHO + H_2O$	(r.11)	6×10^{12}	(21)
$CH_3 + OH \rightarrow$ products	(r.12)	$<5.6 \times 10^{13}$	(22)
$C_2H_6 + OH \rightarrow C_2H_5 + H_2O$	(r.13)	2.8×10^{11}	(23)
$CH_3Br + OH \rightarrow CH_2Br + H_2O$	(r.14)	5×10^{10}	(24)
$OH \rightarrow$ wall	(r.15)	$500 \, s^{-1}$	

As shown in section C the rate constant for OH removal on the wall is $500 \, s^{-1}$ at 400 K. Reactions (r.11) and (r.13) are negligibly slow, even when all CH_3 would be transformed to formaldehyde or to ethane. With a mean CH_3 concentration of 5×10^{-10} mole cm^{-3} in the kinetic zone reaction (r.12) proceeds at a rate of $30 \, s^{-1}$ while reaction (r.14), with $[CH_3Br] = 3 \times 10^{-10}$ mole cm^{-3}, has a pseudo first order rate constant of $15s^{-1}$. Consequently the dominant destruction process for hydroxyl radicals is termination on the wall of the reactor.

C. Determination of k_{15} and of S_{OH}

To connect a value for the rate constant k_1 to the observed OH signals the sensitivity factor for OH has to be known as must be also the rate constant for the major OH removal mechanism i.e. wall termination. Both were determined by measurements on the $H + O_2$ system in the same reactor and at the same O_2 concentration as used for the thermal $CH_3 + O_2$ experiments.

The $H + O_2$ system is described by the following reactions:

$H + O_2 + M \rightarrow HO_2 + M$	(r.16)	$9.9 \ 10^{15}$ ‡	
$HO_2 + H \rightarrow OH + OH$	(r.17a)	$2.5 \ 10^{13}$	(25)
$\rightarrow H_2 + O_2$	(r.17b)	$1 \ 10^{13}$	(25)
$\rightarrow H_2O + O$	(r.17c)	$7 \ 10^{11}$	(25)
$HO_2 + HO_2 \rightarrow H_2O_2 + O_2$	(r.18)	$2 \ 10^{12}$	(15,26-28)
$HO_2 + OH \rightarrow H_2O + O_2$	(r.19)	$1.8 \ 10^{13}$	(15)
$OH + OH \rightarrow H_2O + O$	(r.20)	$1.2 \ 10^{13}$	(15)
$OH + O \rightarrow H + O_2$	(r.21)	$2.3 \ 10^{13}$	(15)
$HO_2 + O \rightarrow OH + O_2$	(r.22)	$1.9 \ 10^{13}$	(15,29)
$H_2O_2 + OH \rightarrow H_2O + O_2$	(r.23)	$4.8 \ 10^{11}$	(15,30)
$H \rightarrow$ wall	(r.24)	$22 \ s^{-1}$	
$OH \rightarrow$ wall	(r.15)	to be determined	

‡ units: $cm^6 mole^{-2} s^{-1}$

The k_{16} value was obtained in a seperate study and corresponds with the recommended value of Baulch (15) for a $23\%O_2/77\%He$ mixture. The rate of H removal by wall termination was also measured in seperate experiments by changing the volume/wall ratio of the reactor. Details of these experiments will be given elsewhere.

Signals for OH and H were measured as a function of the distance between mixing and sampling points for initial concentrations of $5.2 \ 10^{-11}$ mole cm^{-3} for H and $2.5 \ 10^{-8}$ mole cm^{-3} for O_2 at a total pressure of 2 torr He and at room temperature. The measurements were indeed carried out at room temperature since in the thermal experiments the sampling point is at that temperature.

A computer simulation was made for the reaction mechanism given above with initial concentrations of H and O_2 identical to those of the experiments. Calculations were carried out for different values of k_{15}, the rate constant for OH wall loss. That value of k_{15} was selected that resulted in the best fit of the computed with the measured OH profile shape and also in a good match of the absolute H profiles. The goodness of fit for OH is shown in fig.3 and fig.4. It can be seen from the figures that the fit is least good at short reaction times; the probable reason is back diffusion of O_2, resulting in earlier onset of the reaction than expected from the position of the O_2-inlet tube end. The "best" k_{15} value obtained in this way is $350 \ s^{-1}$ at 300 K. Taking into account the approximate dependence of k_{15} on temperature (31) k_{15} can be evaluated to be about $500 \ s^{-1}$ at 400 K.

Comparison of the calculated OH concentrations and the measured OH signals at different reaction times yields the sensitivity factor for OH: $S_{OH} = 1.1 \pm 0.4 \ 10^{14}$ $\mu V/mole \ cm^{-3}$, a result that corresponds fairly well (to within 30 %) to the sensitivity factor obtained in other experiments on the basis of the titration reaction

$$H + NO_2 \rightarrow OH + NO \qquad (r.25)$$

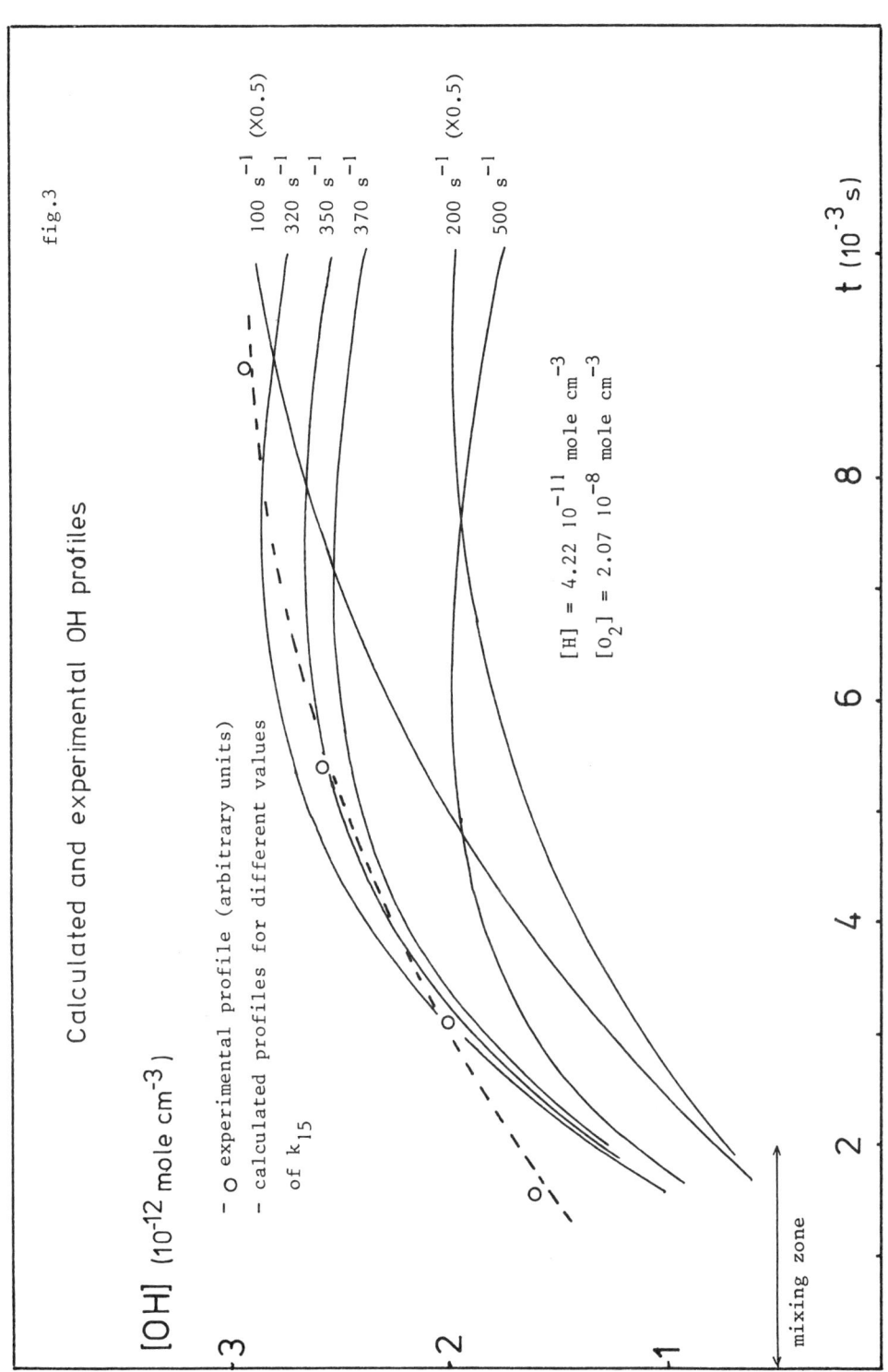

Calculated and experimental OH profiles

fig.3

[OH] $(10^{-12}$ mole cm^{-3})

- o experimental profile (arbitrary units)
- calculated profiles for different values of k_{15}

$[H] = 4.22 \ 10^{-11}$ mole cm^{-3}
$[O_2] = 2.07 \ 10^{-8}$ mole cm^{-3}

100 s^{-1} (X0.5)
320 s^{-1}
350 s^{-1}
370 s^{-1}
200 s^{-1} (X0.5)
500 s^{-1}

mixing zone

$t \ (10^{-3}$ s)

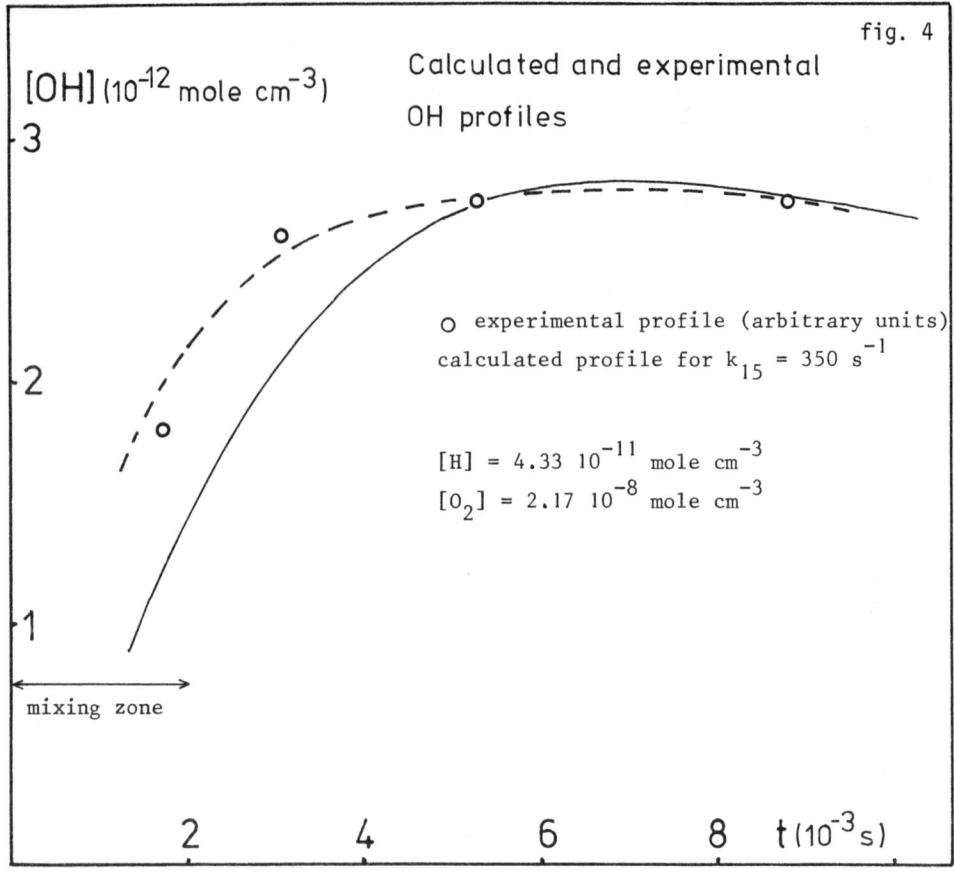

fig. 4

[OH] $(10^{-12}$ mole cm$^{-3})$

Calculated and experimental
OH profiles

○ experimental profile (arbitrary units)

calculated profile for k_{15} = 350 s^{-1}

$[H]$ = 4.33 10^{-11} mole cm^{-3}

$[O_2]$ = 2.17 10^{-8} mole cm^{-3}

mixing zone

2   4   6   8 $t (10^{-3}$s)

D. The rate constant of reaction (r.1)

The thermal experiments, with $[CH_3]$ = 1.12 10^{-12} mole cm^{-3} and $[O_2]$ = 1.56 10^{-8} mole cm^{-3}, showed an OH signal of 1.10 ± 0.40 μV at a reaction time of 3.8 10^{-3} s. The blank experiment (without CH_3) gave a signal of 1.04 ± 0.61 μV in the same experimental conditions.

The stated standard deviations of the signals were evaluated from statistical considerations regarding the response of a phase sensitive detector to a random pulse train input. The output current of the mass filter corresponds indeed to a Poisson distribution. The output of the phase sensitive detector is a sequence of a train of positive pulses of ions from both the molecular beam and the background gas during the first half of the cycle, followed by a train of negative pulses of solely background gas ions during the second half period. The mean output is proportional to the molecular beam ion intensity, whereas the standard deviation from the mean after a total time t depends on the total ion count. The standard deviation σ for an observation time t can also be expressed in terms of the measured RMS noise signal V_{RMS} (32):

$$\sigma = V_{RMS} \frac{\sqrt{4RC}}{\sqrt{t}}$$

where 4RC is the reciprocal equivalent noise bandwith of the low pass filter. The latter was equal to 125 s and the total observation time t was 96 minutes.

The concentration of hydroxyl radicals at the sampling point follows from the difference of the signals with and without methyl radicals:

$$i_{OH} = \frac{\overline{i_{OH}}(\text{with } CH_3) - \overline{i_{OH}} (\text{without } CH_3)}{\pm T\sqrt{\sigma^2_{\text{with } CH_3} + \sigma^2_{\text{without } CH_3}}}$$

the last term indicating the confidence limits as given by the Student's T distribution (33). For T = 1 this term indicates the standard deviation. This gives for the OH signal 0.06 ± 0.72 µV in terms of standard deviation and 0.06 ± 1.60 µV with a 95% confidence interval. Upper limits of respectively 0.79 and 1.16 µV can therefore be connected to the observed signals, corresponding to OH concentrations of respectively $7.5 \; 10^{-15}$ and $1.5 \; 10^{-14}$ mole cm^{-3}. It is clear that the indicated values are essentially the result of the combined standard deviations or 95% confidence limits on the signals; therefore they can be regarded as the detection limits for OH for the given registration time and confidence interval.

In a computer simulation of the reaction sequence given below, with the experimental initial CH_3 and O_2 concentrations, the k_1 input value was adjusted so that after a reaction time of $3.8 \; 10^{-3}$ s the calculated OH concentration corresponded to the measured one.

$$CH_3 + O_2 \rightarrow CH_2O + OH \qquad (r.1) \qquad k_1 \text{ to be determined}$$
$$CH_3 + O_2 + M \rightarrow CH_3O_2 + M \qquad (r.2)$$
$$CH_3 + CH_3 \rightarrow C_2H_6 \qquad (r.4) \qquad k_4 = 2.2 \; 10^{13} \; cm^3 mole^{-1} s^{-1}$$
$$CH_3 + OH \rightarrow \text{products} \qquad (r.12)$$
$$OH \rightarrow \text{wall} \qquad (r.15)$$

The rate constants for most of the reactions were already indicated. The k_4 value was taken from a room temperature investigation in our laboratory; Parkes (34) found a negligible dependence of k_4 on temperature up to 450 K.

This procedure leads, with $[OH] = 7.5 \; 10^{-15}$ mole cm^{-3} - based on the standard deviation on the OH signal - to a "maximal" value for k_1 of $4 \; 10^8$ cm^3 mole^{-1} s^{-1}. In terms of the 95% confidence limits the "maximal" OH concentration of $1.5 \; 10^{-14}$ mole cm^{-3} gives an upper limit for k_1 of $7.5 \; 10^8$ cm^3 mole^{-1} s^{-1}.

4. CONCLUSION

The upper limit for the rate constant of the reaction of CH_3 with O_2 to form CH_2O and OH was found to be $7.5 \; 10^8$ cm^3 mole^{-1} s^{-1} at 400K. This value was determined with a statistical confidence of 95% which means that in a new identical experiment the statistical probability to find a higher upper limit is only 2.5%. The low rate constant indicates that the bimolecular reaction is negligible in comparison to the CH_3O_2 formation at pressures above 0.2 torr.

Our result confirms the values reported by Kurylo et al. (10) and by Golden et al. (5) and is in disagreement with the results of Bayes (9) and of Washida (11) who favour a much higher value.

5. ACKNOWLEDGMENTS

The authors wish to express their gratitude to the Belgian "Fonds voor Kollektief Fundamenteel Onderzoek" for its finantial support. J.P. is indebted to the "Nationaal Fonds voor Wetenschappelijk Onderzoek" of which he is a Research Associate; V.F. thanks the "Instituut voor Aanmoediging van het Wetenschappelijk Onderzoek in Nijverheid en Landbouw" for granting a doctoral fellowship.

6. REFERENCES

(1) A.A.Westenberg and R.M.Fristrom; J.Phys.Chem., 65, 591 (1961)
(2) I.Glassman; "Combustion", Academic Press, New York (1977)
(3) C.P.Fennimore; "Chemistry in Premixed Flames", The International Encyclopedia of Physical Chemistry and Chemical Physics, Vol.5, Pergamon Press, New York (1964)
(4) J.Peeters and G.Mahnen; 14th Symposium (International) on Combustion, p.133, The Combustion Institute, Pittsburgh (1973)
(5) A.C.Baldwin and D.M.Golden; Chem.Phys.Lett., 55, 350 (1978)
(6) P.Crutzen; Can.J.Chem., 52, 1569 (1974)
(7) J.Heicklen; "Atmospheric Chemistry", Academic Press, New York (1976)
(8) N.Basco, D.G.L.James and F.C.James; Int.J.Chem.Kinet., 4, 129 (1972)
(9) N.Washida and K.D.Bayes; Int.J.Chem.Kinet., 8, 777 (1976)
(10) O.Klais, P.C.Anderson, A.H.Laufer and M.J.Kurylo; Chem.Phys.Lett., 66, 598 (1979)
(11) N.Washida; J.Chem.Phys., 73, 1665 (1980)
(12) C.Vinckier and W.Debruyn; 17th Symposium (International) on Combustion, p.623, The Combustion Institute, Pittsburgh (1979)
(13) J.Peeters and C.Vinckier; 15th Symposium (International) on Combustion, p.969, The Combustion Institute, Pittsburgh (1975)
(14) D.A.Parkes; Int.J.Chem.Kinet., 9, 451 (1977)
(15) D.L.Baulch, R.A.Cox, R.F.Hampson Jr., J.A.Kerr, J.Troe and R.T.Watson; J.Phys.Chem.Ref.Data, 9, 225 (1980)
(16) C.S.Kan, J.G.Calvert and J.H.Shaw; J.Phys.Chem., 84, 3411 (1980)
(17) A.R.Ravishankara, F.L.Eisele, N.M.Kreutter and P.H.Wine; J.Chem.Phys., 74, 2267 (1981)
(18) L.Batt,R.D.McCulloch and R.T.Milne; Int.J.Chem.Kinet., Symp.1, 441 (1975)
(19) J.R.Barker,S.W.Benson and D.M.Golden; Int.J.Chem.Kinet., 9, 31 (1977)
(20) M.B.Colket, D.W.Naegeli and D.M.Golden; 16th Symposium (International) on Combustion, p.1023, The Combustion Institute, Pittsburgh (1977)
(21) R.Atkinson and J.N.Pitts Jr.; J.Chem.Phys., 68, 3581 (1978)
(22) T.J.Sworski, C.J.Hochanadel and P.J.Ogren; J.Phys.Chem., 84, 129 (1980)
(23) N.R.Greiner; J.Chem.Phys., 53, 1070 (1970)
(24) D.D.Davis, G.Machado, B.Conaway, Y.Oh and R.T.Watson; J.Chem.Phys., 65, 1268 (1976)
(25) W.Hack, H.Gg.Wagner and K.Hoyermann; Ber.Bunsenges Phys.Chem., 82, 713 (1978)
(26) R.A.Cox and J.P.Burrows; J.Phys.Chem., 83, 2560 (1979)
(27) R.A.Graham, A.M.Winer, R.Atkinson and J.N.Pitts Jr.; J.Phys.Chem., 83, 1563 (1979)
(28) B.A.Trush and J.P.T.Wilkinson; Chem.Phys.Lett., 66, 441 (1979)
(29) W.Hack, A.W.Preuss, F.Temps and H.Gg.Wagner; Ber.Bunsenges Phys.Chem., 83, 1275 (1979)
(30) U.C.Sridharam, B.Reimann and F.Kaufman; J.Chem.Phys., 73, 1286 (1980)
(31) A.A.Westenberg and N.de Haas; J.Chem.Phys., 58, 4061 (1973)
(32) T.Coor; J.Chem.Ed., 45, A533 (1968)
(33) J.R.Green and D.Margerison; "Statistical Treatment of Experimental Data", Physical Sciences data 2, Elsevier, Amsterdam (1978)
(34) D.A.Parkes, D.M.Paul and C.P.Quin; J.C.S.Faraday Trans.I, 72, 1935 (1976)

REACTIONS OF THE TRIPLET STATE OF KETONES WITH MOLECULAR OXYGEN

M.B. FOLEY and H.W. SIDEBOTTOM
Chemistry Department, University College, Dublin, Ireland.

Summary

The quantum yield of product formation in the photooxidation of acet-one and biacetyl has been studied at atmospheric pressure and as a function of excitation wavelength. The results are compared with emission and lifetime data from previous investigations on these photochemical systems. The data indicate that the reaction of triplet carbonyl compounds with oxygen occurs predominately via an energy transfer pathway and that the contribution to triplet removal by a direct oxidation channel is small. It is suggested that the photo-induced atmospheric removal of ketones occurs predominately via radical formation in a dissociative primary process.

1. INTRODUCTION

Ketones are emitted into the atmosphere both as a consequence of their use as solvents and as partial oxidation products of hydrocarbon combustion processes. Ketones are also produced in secondary reactions by the oxidation of hydrocarbons in the presence of NO_x under ambient condit-ions (1). α - Dicarbonyls arise mainly from hydroxyl radical attack on aromatic hydrocarbons present in polluted atmospheres (2). The available evidence suggests that carbonyl compounds are removed from the atmosphere by either reaction with hydroxyl radicals or by photo-induced oxidation (1,3). However, major uncertainty remains concerning the photolysis of ketones under atmospheric conditions particularly with respect to the quantum efficiency of radical production. Since photolysis is likely to be an important sink process for these species in the atmosphere and in addition provides a source of free radicals, which can subsequently prom-ote photochemical oxidation, accurate quantum yield data for carbonyl photooxidation is desirable. Available literature data indicates that the photodissociative mechanism for carbonyl compounds is a complex function of pressure, temperature and excitation wavelength (4). Hence accurate quantum yield data for carbonyl photooxidation is required over the comp-lete solar spectrum and at atmospheric pressure.

The present work is concerned with the photooxidation of acetone and biacetyl at atmospheric pressure and over a range of excitation wave-lengths.

2. EXPERIMENTAL

A conventional mercury-free greaseless vacuum system was used for all the experiments and pressure measurements were made with an MKS Baratron capacitance manometer (MKS 220A). Acetone (Hopkins and Williams, Analar grade) and Biacetyl (Fluka, Puriss) were bulb-to-bulb distilled on the vacuum system prior to use. The resulting reagents showed no impurities on gas-chromatographic analysis. The oxygen and nitrogen were both Matheson ultra high purity products and were used without further purification.

The reaction mixtures were photolysed in a cylindrical quartz reaction vessel, (length 10 cm and volume 220 ml), using a Hanovia 500-W medium pressure mercury source. A Bausch and Lomb high-intensity monochromator was used to isolate the selected wavelength regions. Transmitted intensities of the exciting radiation were measured on an R.C.A. 935 phototube. Products were identified by comparison of retention times with those of authentic samples, and confirmed by means of gas-liquid chromatographic coupled mass spectrometry. All quantitative analyses were carried out using a Gow-Mac gas density balance chromatograph. Quantum yields for acetone photooxidation were calculated against the carbon monoxide yield from acetone photolysis at 130° C taking $\phi_{CO} = 1$ in the absence of oxygen at both 280 and 313 nm. Oxygen-free biacetyl was used as an actinometer for determination of the quantum yields of biacetyl photooxidation using the data from the work of Sheats and Noyes (5) and Bell and Blacet (6).

A number of photolysis experiments were performed using in situ infrared analysis. The T-shaped reaction vessel was located in the sample compartment of a Perkin Elmer Model-137 spectrometer, with the sodium chloride windows in the infrared beam (path length 10 cm). The mercury arc and monochromator were in the perpendicular axis (path length 10 cm). Calibration curves of optical density against pressure were obtained for pure samples of the reactants and oxidation products.

3. RESULTS AND DISCUSSION

Biacetyl

The major products detected in the gas phase photooxidation of biacetyl in the range 280 to 436 nm were carbon dioxide, formaldehyde, methanol and carbon monoxide. As the photolysis times increased the HCHO yields decreased and the CO yields increased. Thus it appears that CO is a secondary product resulting from formaldehyde decomposition. Table I lists the initial quantum yields for $(CH_3CO)_2$ consumption together with those for CO_2 and CH_3OH formation. The yields show some scatter but essentially were independent of the $(CH_3CO)_2$ or O_2 pressure. Although the data at 436 nm and 360 nm were found to be equal within experimental error further increases in excitation energy produced higher values for the quantum yields. The results from the present study are somewhat higher than those reported previously by Padnos and Noyes (7) ($\phi_{CO_2} = 0.18$) for excitation at 436 nm but at significantly lower pressures (<30 torr).

There are two separate absorption regions in the UV-visible spectrum of biacetyl (8,9). Both bands have been assigned to n → π* electronic transitions (10). The first absorption is discrete but extremely complex and occurs in the 340 to 470 nm region. The second absorption lies between 220 and 320 nm and is continuous in nature. Strong evidence exists that biacetyl excited into the first excited singlet state undergoes unimolecular intersystem crossing to excited vibrational levels of the corresponding triplet state with near unit efficiency at a rate $\sim 10^8$ sec^{-1}(11). Thus except for a small amount of fluorescence, ($\phi_f \sim 2.5 \times 10^{-3}$) any further decay must originate from the triplet state (12). For total pressures greater than approximately 100 torr vibrational relaxation is complete and the quantum yield of phosphorescence, ($\phi_p = 0.149$) is independent of excitation wavelength (11,13). Dissociation is negligible at temperatures less than 30° C, ($\phi_D < 0.001$) and intersystem crossing to the ground state is the major removal pathway for vibrationally relaxed triplet biacetyl molecules (5).

TABLE I : Quantum Yields of Product Formation in the Photooxidation
of Biacetyl at 28 ± 3° C.

Pressure[a] (torr)		Quantum Yields[b]		
$(CH_3CO)_2$	O_2	CO_2	CH_3OH	$(CH_3CO)_2$
λ_{ex}=436nm				
10.2	2.04	0.11	0.03	–
10.1	56.9	0.08	0.02	–
10.1	200	0.10	0.02	–0.06
20.7	23.9	0.10	–	–
1.20	20.8	0.09	–	–0.05
λ_{ex}=360nm				
1.04	24.6	0.12	–	–0.06
10.4	23.0	0.11	–	–
24.3	22.9	0.09	0.02	–
λ_{ex}=313nm				
1.31	194	0.19	–	–0.10
24.3	20.6	0.21	–	–
λ_{ex}=280nm				
1.56	27.4	0.62	0.09	–0.39
19.5	26.6	0.54	0.08	–

(a) Total pressure was made up to 760 torr with N_2
(b) $(CH_3CO)_2$ was monitored in situ by infrared spectroscopy;
CO_2 and CH_3OH were analysed by gas chromatography after
irradiation was terminated.

Groh (14) has shown that oxygen does not effect the fluorescence
emission of biacetyl and hence reactions of the singlet atate with oxygen
are unimportant. Quenching of the triplet state by oxygen has been shown
to be extremely efficient having a bimolecular rate constant of 5.66×10^8 1
$mole^{-1}$ sec^{-1} at 25° C (15). For the oxygen pressures used in the present
work triplet molecules will be totally removed by reaction with oxygen.
The following mechanism adequately accounts for the reactions of biacetyl
following excitation within the first absorption band (340 – 470 nm).

$$(CH_3CO)_2 \; + \; h\nu \; (340 - 470 \text{ nm}) \rightarrow {}^1B_n \rightarrow {}^3B_m \overset{M}{\rightarrow} {}^3B_o$$

$$^3B_o \; + \; O_2 \; \rightarrow \; \text{Products}$$

The mechanism for quenching of the first excited triplet state of biacetyl
may be either physical or chemical in nature. Electronic energy transfer,
leading to excitation of oxygen from its ground state to one of its low-
lying singlet states, is expected to be a significant removal pathway.

$$^3B_o \; + \; {}^3O_2({}^3\Sigma_g{}^-) \; \rightarrow \; (CH_3CO)_2 + O_2({}^1\Sigma_g{}^+); \; \Delta H = -18 \text{ kcal/mole}$$

$$\rightarrow \; (CH_3CO)_2 + O_2({}^1\Delta_g) \; ; \; \Delta H = -34 \text{ kcal/mole}$$

It is also possible that interaction of the triplet state with oxygen can
lead directly to chemical changes in the system. The photooxidation
products observed in this work are consistent with the reaction scheme outlined below.

$$^3B_o \; + \; O_2 \qquad \rightarrow CH_3\overset{O_2{\cdot}}{\underset{}{\overset{\overset{O{\cdot}}{|}}{C}}}{-}\overset{O}{\overset{\|}{C}}CH_3 \; \rightarrow \; CH_3\overset{O}{\overset{\|}{C}}O_2{\cdot} \; + \; CH_3\overset{O}{\overset{\|}{C}}{\cdot}$$

$$CH_3\overset{O}{\overset{\|}{C}}\cdot \quad + \quad O_2 \quad \rightarrow \quad CH_3\overset{O}{\overset{\|}{C}}O_2\cdot$$

$$2CH_3\overset{O}{\overset{\|}{C}}O_2\cdot \quad \rightarrow \quad 2CH_3\overset{O}{\overset{\|}{C}}O\cdot \quad + \quad O_2$$

$$CH_3\overset{O}{\overset{\|}{C}}O\cdot \quad \rightarrow \quad CH_3\cdot \quad + \quad CO_2$$

$$CH_3\cdot \quad + \quad O_2 \quad \rightarrow \quad CH_3O_2\cdot$$

$$2CH_3O_2\cdot \quad \rightarrow \quad CH_3OH \quad + \quad HCHO \quad + \quad O_2$$

$$\rightarrow \quad 2CH_3O\cdot \quad + \quad O_2$$

$$CH_3O\cdot \quad + \quad O_2 \quad \rightarrow \quad HCHO \quad + \quad HO_2\cdot$$

$$CH_3\overset{O}{\overset{\|}{C}}O_2\cdot \quad + \quad CH_3O_2\cdot \rightarrow \quad CH_3\overset{O}{\overset{\|}{C}}O\cdot \quad + \quad CH_3O\cdot \quad + \quad O_2$$

The low values found for the quantum yields of photooxidation at 360 and 436 nm, Table I, indicate that energy transfer is the dominant reaction channel for reaction of thermalised triplets with molecular oxygen.

Dissociation of biacetyl following excitation into the second absorption region (220 - 320 nm) has been investigated in detail by Bell and Blacet (6) and Sheats and Noyes (5). There appears to be little triplet state involvement under these conditions and the emission yields are negligible. The quantum yields of decomposition are less than unity even at the shortest wavelengths studied. The authors suggest that rapid dissociation occurs from the initially formed second excited singlet state and that internal conversion to the ground state accounts for the energy not reflected in the primary dissociation yield. Quantum yields of decomposition in this region are both wavelength and pressure dependent. Also at shorter wavelengths the acetyl radicals generated contain sufficient vibrational excitation to spontaneously decompose.

$$(CH_3CO)_2 \quad + \quad h\nu(220\text{-}320nm) \rightarrow \quad 2(1 - \alpha)(CH_3\overset{O}{\overset{\|}{C}}\cdot) \quad + \quad 2\alpha(CH_3\cdot) \quad + \quad 2\alpha(CO)$$
where α is a value from 0 to 1

For pressures in the region of 50 torr Bell and Blacet (6) report values

of $\quad \phi_D \quad = \quad 0.27 \quad$ and $\quad \alpha \sim 0.4 \quad$ at 280 nm

and $\quad \phi_D \quad = \quad 0.074 \quad$ and $\quad \alpha \sim 0.1 \quad$ at 313 nm

Photooxidation of biacetyl in this wavelength region presumably arises from reactions of the radicals formed in the photodissociation process,

$$(CH_3CO)_2 \quad + \quad h\nu(220\text{-}320nm) \rightarrow \quad CH_3\overset{O}{\overset{\|}{C}}\cdot \quad + \quad CH_3\overset{O}{\overset{\|}{C}}\cdot$$

$$\text{or} \quad \rightarrow \quad CH_3\overset{O}{\overset{\|}{C}}\cdot \quad + \quad CH_3\cdot \quad + \quad CO$$

The present data for the quantum yields of photooxidation of biacetyl agree reasonably well with those observed for the photodecomposition of biacetyl alone. However strict comparison is difficult due to pressure effects on the decomposition yields. Decomposition of vibrationally excited acetyl radicals prior to reaction with oxygen would be expected to reduce the yield of carbon dioxide in the photooxidation process. The data suggest this may be the case at 280 nm.

The results from the present investigation indicate that removal of biacetyl via a photooxidation pathway in the atmosphere will mainly arise from excitation into the second absorption band where photodecomposition of biacetyl is important.

Acetone

The products detected from the photooxidation of acetone at 280 and 313 nm were carbon dioxide, formaldehyde, methanol and carbon monoxide. As in the biacetyl photooxidation system the majority of the CO formed appears to be the result of secondary reactions involving HCHO. The quantum yield values given in Table II were found to be independent of acetone or oxygen pressure, but showed a significant rise with increasing excitation energy. The data at 313 nm are in good agreement with those previously reported by Kirk and Porter (16) (ϕ_{CO_2} = 0.19) and Pearson (17) (ϕ_{CO_2} = 0.14).At this wavelength these authors found that the photooxidation yields were independent of the total pressure (16,17). However a strong pressure dependence for the quantum yields has been reported at 280 nm (16).The results showed a progressive decrease in ϕ_{CO_2} from 1.12 to 0.71 as the total pressure was increased from 24 to 77 torr. The lower averaged value of ϕ_{CO_2} = 0.47 determined in this work at a total pressure of 760 torr is consistent with this observation.

TABLE II :   Quantum Yields of Product Formation in the Photooxidation
of Acetone at 28 ± 3° C.

Pressure[a] (torr)		Quantum Yields[b]		
CH_3COCH_3	O_2	CO_2	CH_3OH	CH_3COCH_3
λ_{ex} =313nm				
1.26	23.4	0.14	0.04	−0.13
20.3	29.5	0.13	0.03	−
67.9	19.8	0.12	−	−
100	26.8	0.15	0.03	−
59.8	204	0.11	−	−
λ_{ex} =280nm				
1.09	20.4	0.46	0.11	−0.59
20.0	5.74	0.44	−	−
103	20.7	0.52	0.15	−

(a) Total pressure was made up to 760 torr with N_2
(b) CH_3COCH_3 was monitored in situ by infrared spectroscopy; CO_2 and CH_3OH were analysed by gas chromatography after irradiation was terminated.

The first absorption band of acetone extends from 230 to 350 nm and is associated with a singlet-singlet n → π* transition. Emission from acetone excited within this region extends from 380 to 470 nm (18). The emission consists of fluorescence from the singlet state having a lifetime of approximately 3 x 10⁻⁹ sec (19) and phosphorescence from the corresponding triplet state with a lifetime of about 2 x 10⁻⁴ sec (20,21). The quantum yield of triplet formation at 313 nm has been found to be close to unity (22,23) and thus acetone excited into the first excited singlet

state at 313 nm undergoes unimolecular intersystem crossing to excited
vibrational levels of the corresponding triplet state with near unit
efficiency. The singlet state lifetime (19) and quantum yield of fluorescence, $\phi_f \sim 2 \times 10^{-3}$, (24) are reported to be essentially independent of
excitation wavelength and pressure. Thus acetone excited into high vibrational levels of the singlet state must also undergo intersystem crossing
to the triplet state with almost unit efficiency. The relative efficiencies of phosphorescence, intersystem crossing to ground state and dissociation are all strongly dependent on the vibrational energy of the excited
triplet molecule. As a consequence they vary with the absorbed wavelength
and pressure. For excitation at 313 nm vibrational relaxation of the
triplet state is complete for pressures greater than a few torr. The
quantum yield of dissociation is 0.25 (25) and the phosphorescence yield
approximately 0.019 (24). At higher excitation energies there is competition between decomposition and vibrational deactivation of the excited
triplet molecules. Phosphorescence from vibrationally excited triplets is
unimportant since only the relaxed triplet would survive long enough for
emission to occur. Thus the phosphorescence efficiency drops dramatically
in going from excitation at 313 to 280 nm (24).

It is well recognised that the phosphorescence of acetone is efficiently quenched by oxygen whereas the singlet emission is unaffected (24).
Thus any interaction between oxygen and excited acetone molecules must
involve only the triplet state. From the above considerations photolysis
of acetone-oxygen mixtures at 313 nm must involve the reaction of vibrationally relaxed triplets with molecular oxygen. The relatively low quantum
yields observed suggest that the energy transfer process

$$^3A_0 \;+\; ^3O_2(^3\Sigma_g^-) \;\rightarrow\; CH_3COCH_3 + O_2(^1\Sigma_g^+); \;\; \Delta H = -36 \text{ kcal/mole}$$

$$\rightarrow CH_3COCH_3 + O_2(^1\Delta_g) \;;\;\; \Delta H = -52 \text{ kcal/mole}$$

is considerably more efficient than the direct reaction channel

$$^3A_0 \;+\; O_2 \qquad \rightarrow CH_3\underset{\overset{|}{O_2\cdot}}{\overset{\overset{O\cdot}{|}}{C}}CH_3 \;\rightarrow\; CH_3\overset{O}{\overset{\|}{C}}O_2\cdot \;+\; CH_3\cdot$$

The products observed suggest a mechanism for the reactions of the peroxyacetyl and methyl radicals which is entirely analogous to that outlined
for the direct reaction of biacetyl triplets.

Excitation of acetone at 280 nm in the presence of oxygen results in
significantly higher photooxidation yields. This can be rationalised in
terms of the decomposition of highly vibrationally excited triplet
molecules,

$$^3A_m \qquad \rightarrow CH_3\overset{O}{\overset{\|}{C}}\cdot \;+\; CH_3\cdot$$

$$^3A_m \;+\; M \qquad \rightarrow {}^3A_0 \;+\; M$$

Even at atmospheric pressure this reaction is presumably important
relative to vibrational relaxation within the triplet manifold. Acetyl
radicals formed at short wavelengths may contain sufficient vibrational
energy to decompose in an early vibration. This would lead to the reduction in the quantum yield of CO_2 formation relative to that for acetone
loss.

The above results indicate that the majority of the photo-induced
removal of acetone under solar irradiation arises from absorption in the

shorter wavelength region of the spectrum where direct decomposition into radicals is important.

4. REFERENCES

1. K.Demerjian, J.A.Kerr and J.G.Calvert, 'Chemistry of Photochemical Smog', Adv.Environ.Sci.Technol., 4, 1 (1974).
2. K.R.Darnall, R.Atkinson and J.N.Pitts Jr., J.Phys.Chem., 83, 1943 (1979).
3. W.P.L.Carter, A.C.Lloyd, J.L.Sprung and J.N.Pitts Jr., Int.J.Chem. Kinet., 11, 45 (1979).
4. A.C.Lloyd, 'Tropospheric Chemistry of Aldehydes', N.B.S. special publication 557 (1979).
5. G.F.Sheats and W.A.Noyes Jr., J.Am.Chem.Soc., 77, 1421 (1955).
6. W.E.Bell and F.E.Blacet, J.Am.Chem.Soc., 76, 5332 (1954).
7. N.Padnos and W.A.Noyes Jr., J.Phys.Chem., 68, 464 (1964).
8. J.W.Sidman and D.S.McClure, J.Am.Chem.Soc., 77, 6461, 6471 (1955).
9. J.C.D.Brand and A.W.H.Mau, J.Am.Chem.Soc., 96, 4380 (1970).
10. E.Drent and J.Kommandeur, Chem.Phys.Letters, 14, 321 (1972).
11. G.M.McClelland and J.T.Yardley, J.Chem.Phys., 58, 4368 (1973).
12. H.W.Sidebottom, C.C.Badcock, J.G.Calvert, B.R.Rabe and E.K.Damon, J.Am. Chem.Soc., 94, 13 (1972).
13. S.S.Collier, D.H.Slater and J.G.Calvert, Photochem.Photobiol., 7, 737 (1968).
14. H.J.Groh Jr., J.Chem.Phys., 21, 674 (1953).
15. C.O.Concheanainn, M.B.Foley and H.W.Sidebottom, J.Photochem., 15, 185 (1981).
16. A.D.Kirk and G.B.Porter, J.Phys.Chem., 66, 556 (1962).
17. G.S.Pearson, J.Phys.Chem., 67, 1686 (1963).
18. G.W.Luckey, A.B.F.Duncan and W.A.Noyes Jr., J.Chem.Phys., 16, 407 (1948)
19. G.M.Breuer and E.K.C.Lee, J.Phys.Chem., 75, 989 (1971).
20. W.E.Kaskan and A.B.F.Duncan, J.Chem.Phys., 18, 427 (1950).
21. M.B.Foley and H.W.Sidebottom, Cost 61A bis Working Party 2 Discussion Meeting, Leuven, Belgium (1981).
22. R.B.Cundall and A.S.Davies, Proc.Roy.Soc.Lond., A290, 563 (1966).
23. R.E.Rebbert and P.Ausloos, J.Am.Chem.Soc., 89, 1573 (1967).
24. J.Heicklen, J.Am.Chem.Soc., 81, 3863 (1959).
25. H.E.O'Neal and C.W.Larson, J.Phys.Chem., 73, 1011 (1969).

DECOMPOSITION OF THE t-BUTOXY RADICAL

L. BATT and G.N. ROBINSON
University of Aberdeen, Scotland.

Summary

By allowing the t-butoxy radical to decompose in the presence of nitric oxide, it has been possible to determine a rate constant for decomposition by the measurement of the relative rates (2) and (3):

$$t-BuO + NO \longrightarrow t-BuONO \qquad (2)$$

$$t-BuO + M \longrightarrow Me + Me_2CO + M \qquad (3)$$

Process (3) is clearly pressure dependent. The value of $k_3(\infty)$ has been determined in the presence of the inert gases CF_4, SF_6 N_2 and A and a value of k_3 interpolated for atmospheric conditions. The results may be compared with those for other relevant alkoxy radicals at room temperature.

1. INTRODUCTION

In this study it has been our intention to examine the decomposition of the t-butoxy radical under as wide a range of experimental conditions as possible, in order to obtain detailed information on the reaction and hence, reliable and accurate Arrhenius parameters. Di-t-butyl peroxide was used as a thermal source of t-butoxy radicals. Suitable concentrations of nitric oxide were added to allow the combination reaction (2) to compete at a rate comparable with the rate of decomposition of the t-butoxy radical (3):

$$(t-BuO)_2 \longrightarrow 2t-BuO \qquad (1)$$

$$t-BuO + NO \longrightarrow t-BuONO \qquad (2)$$

$$t-BuO + M \longrightarrow Me_2CO + Me \qquad (3)$$

$$Me + NO \longrightarrow Products \qquad (4)$$

The addition of inert gases (25 - 1500 Torr) allowed us to examine the fall-off behaviour of reaction (3) and to determine experimentally the high pressure limiting rate constant for this reaction over the temperature range 402 - 443K. Rice - Ramsperger - Kassel - Marcus (RRKM) calculations have also been carried out for the reaction and we report here the Arrhenius parameters obtained in this way.

2. EXPERIMENTAL

Di-t-butyl peroxide (Fluka, pract. grade) was purified by bulb-to-bulb distillation at 273K. The resulting material contained less than 0.3%

total impurity. Nitric oxide (Matheson, C.P. grade) was purified by
freezing on to molecular sieve 13X, pumping and thawing, until the material
appeared white (around six cycles). Carbon tetrafluoride (Matheson) was
purified as before [1] and dried by condensing on to phosphorus pentoxide.
Sulphur hexafluoride (Matheson) was treated in the same say as carbon
tetrafluoride, while nitrogen (British Oxygen Co., white spot grade) and
Argon (British Oxygen Co.) were passed through the deoxygenating column [1].
and stored over phosphorus pentoxide. t-Butyl nitrite, used for
identification and calibration purposes, was prepared as described
previously [2]. Other compounds used for this purpose, were obtained from
standard sources.

The apparatus used was a static system [2]. Pressures of di-t-butyl
peroxide and nitric oxide were measured out separately in a calibrated
volume and transferred to a one litre mixing vessel. They were allowed to
warm up to room temperature and the required pressure of inert gas added.
After mixing for at least $1\frac{1}{2}$ hours (up to 48 hours for pressures greater
than one atmosphere) the reactants were expended into the reaction vessel.
At the end of each run, the reaction vessel contents were transferred into
two efficient traps at 76K. The inert gas was then removed by pumping at
an appropriate temperature (76K for nitrogen and argon, 90K for carbon
tetrafluoride and 143K for sulphur hexafluoride) and the contents of the
traps remaining at that temperature were distilled into a small tube and
allowed to warm and mix before analysis. A Perkin-Elmer Sigma 2 gas
chromatograph fitted with a flame ionisation detector and using helium
carrier gas was used for all analyses. A 2m column containing 30%
dinonyl phthalate on chromosorb W (acid washed, 80 - 100 mesh) at 353K was
used. Tests showed that no reaction between components occurred during
storage, collection or analysis. The concentrations of both peroxide
and nitric oxide were in the range $1.2 - 2.0 \times 10^{-5}$ M and the extent of
decomposition was $\sim 8\%$ except for runs at 443K, which were as short as
practicable (60 s), giving an extent of decomposition of $\sim 12\%$.

3. RESULTS

Gas liquid chromatographic analysis revealed that the major products
were acetone and t-butyl nitrite. Small amounts of isobutene were also
detected ($\sim 1\%$ of the nitrite). U-V spectrophotometric analysis of a
white solid which collected in the sample tube revealed that this product
was the cis dimer of nitrosomethane [3]. An additional product,
t-butanol, was detected when water vapour was present either as a result of
a leak or when the appropriate inert gas was not adequately dried. It was
apparent that the alcohol formed as a result of the hydrolysis of the
nitrite formed in the reaction, since a high alcohol yield was associated
with a low nitrite yield. Runs which gave a measurable alcohol peak were
discounted.

It was known from our previous work [2] that the t-butyl nitrite
formed was stable under the reaction conditions used. The rate constant
for reaction (3) was calculated via the relationship:

$$R_3/R_2 = [Me_2CO]/[t\text{-}BuONO] = k_3/k_2[NO]_{av}$$

The value of k_2 was known from our previous work on the decomposition of
t-butyl nitrite to be $10^{10.4 \pm 0.2}$ s^{-1} [2]. $[NO]_{av}$ was calculated on the
basis that every t-butoxy radical led to the removal of a molecule of
nitric oxide, either by reaction (2) or (4) so that:

$$[NO]_{av} = [NO]_{init} - \tfrac{1}{2}([Me_2CO] + [t\text{-}BuONO])$$

The value of k_3 at a given temperature was found to be strongly dependent upon the pressure of added inert gas. This can be attributed to the pressure dependence of reaction (3) since it is known that both reactions (1) and (2) are pressure independent in the pressure range covered in the present work [2], [4], [5]. Fall-off curves are shown in figures 1 and 2. The points shown are mean results of at least two determinations. Results obtained using nitrogen and argon as inert gases were not as reproducible as those with carbon tetrafluoride and sulphur hexafluoride but it was clear that nitrogen and argon are less efficient as energy transfer agents than the other two gases.

A packed reaction vessel, instead of a spherical reaction vessel, was used at 414K for a number of runs to determine whether any surface reaction was occurring in the system. The values of k_3 obtained at various pressures of carbon tetrafluoride are shown in figure 2. As may be seen, the values obtained lie within the scatter of the points obtained using the spherical reaction vessel. It is concluded that no significant heterogeneous reaction occurs in this system. In order to obtain rate constants for reaction (3) at infinite pressure at each temperature examined, the method recommended by Oref and Rabinovitch [6] was used, since the Lindemann extrapolation procedure leads to too low values of $k_3(\infty)$. The method used involves plotting $1/k_3$ versus $p^{-\alpha}$ where $0 < \alpha < 1$, and finding the value of α which gives the best straight line for the experimental values of $1/k_3$. A simple computer program was used to determine the best value of α at each temperature. In carrying out this procedure, values of k_3 which were less than $0.1\ k_3(\infty)$ were omitted from the calculation [6]. Figure 3 shows the best straight lines obtained by this method, and gives the corresponding values of α. The values of $k_3(\infty)$ obtained from the intercepts of these lines are plotted in figure 4. A least squares analysis of the results yield $\log (A_3^\infty/s^{-1}) = 14.6 \pm 0.6$ and $E_3(\infty) = 15.9 \pm 1.2$ Kcal mol^{-1}.

A computer program was used to carry out RRKM calculations for the reaction. This program is a modification of programs kindly supplied by Professors B.S. Rabinovitch, H.M. Frey and Dr. R. Walsh. The frequencies and other parameters used are listed in table 1. Frequencies for the t-butoxy radical are based upon those for t-butanol [7] and di-t-butyl peroxide [8]. The frequencies for the activated complex were adjusted to obtain the best fit with experimental fall-off curve obtained with carbon tetrafluoride as the added inert gas at 403K. The computed fall-off curve is given with the experimental curve in figure 1. The computed high pressure Arrhenius parameters which gave the best fit to the experimental curve were $\log (A_3(\infty)/s^{-1}) = 14.9$, $E_3(\infty) = 16.6$ Kcal mol^{-1}, in agreement with the experimentally determined parameters, within the error limits. The computed and observed curves for sulphur hexafluoride as the additive gas, also shown in figure 1, were also found to be in good agreement with each other using the parameters listed in table 1. The predicted fall-off curve for nitrogen is also shown in figure 1 for comparison. It was not possible to match all the fall-off curves satisfactorily with one set of input parameters. When the same input parameters which gave this good fit at 403K were tried at the higher temperatures, the computed curves were not steep enough.

The results obtained in the present work may be used to derive values for k_1, the rate constant for the decomposition of di-t-butyl peroxide, since the total yield of acetone plus t-butyl nitrite gives a direct measure of the number of t-butoxy radicals produced. In table 2 the

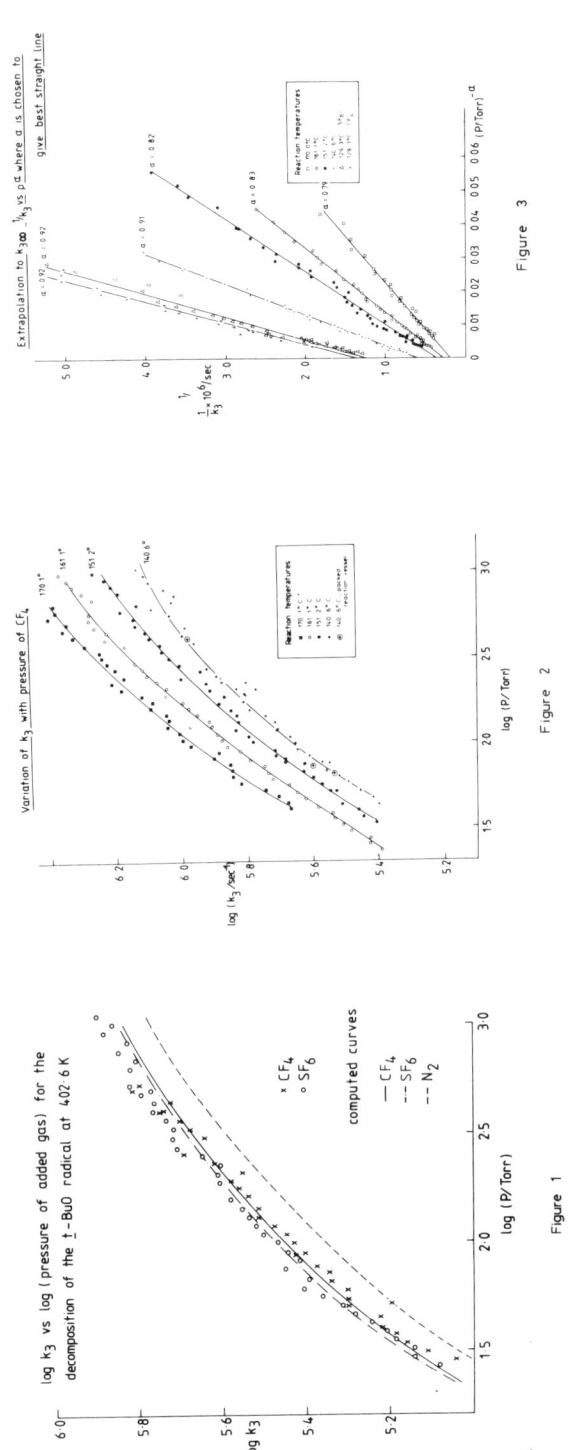

Figure 1

log k_3 vs log (pressure of added gas) for the decomposition of the $\underline{t}$-BuO radical at 402·6 K

x CF₄
o SF₆

computed curves
— CF₄
-- SF₆
-- N₂

Figure 2

Variation of k_3 with pressure of CF₄

Figure 3

Extrapolation to k_{300}, $\frac{1}{k_3}$ vs p^{-a} where a is chosen to give best straight line

Table 1   Vibration Frequencies of t-Butoxy Radical and Activated Complex used in RRKM Program.

Molecule		Complex	
frequency	degeneracy	frequency	degeneracy
2980	6	2980	6
2900	3	2900	3
1460	6	1460	6
1350	4	1350	4
1220	2	1220	2
1110	2	1110	1
1010	3	1010	2
920	1	920	1
750	1 (reaction coordinate)	530	2
450	3	450	1
350	2	250	2
250	3	220	2
		120	1
		110	1

Input parameters for RRKM program

$E(O) = 15.18$ Kcal mol^{-1} $\log(A/s^{-1}) = 14.89$ (403K)

E(Arrhenius) $- E(O) = 1.44$ Kcal mol^{-1} (403K)

values of k_1 obtained in this way are listed. An Arrhenius plot of these results is shown in figure 5. A least square treatment of these results gives $k_1 = 10^{15.5 \pm 0.53}$: $10^{37.0 \pm 1.0}/\theta$ s^{-1}. This is in excellent agreement with the result of Batt and Benson [9], who found $k_1 = 10^{15.6}$ $10^{-37.4 \pm 0.5}/\theta$ s^{-1}.

4. DISCUSSION

Because of the lack of any significant heterogeneous reaction, and since our values of k_1 agree so well with the values that would be predicted [9], we feel confident that the values of k_3 obtained by this method will be reliable, bearing in mind the precision of k_2. The question arises as to what is the best method of obtaining the high pressure limiting Arrhenius parameters from a series of fall-off curves. The determination of $k_3(\infty)$ by the method of Oref and Rabinovitch [6] seems to offer a simple solution to the problem of curvature encountered in the traditional extrapolation methods which tended to over - or underestimate high pressure limiting rate constants. One problem with the method of Oref and Rabinovitch arises when the points are scattered: at the low pressure end of the range this scatter is accentuated. Also the cutoff point (only values of $k_3 > 0.1 \, k_3(\infty)$ are included) is not clear-cut unless

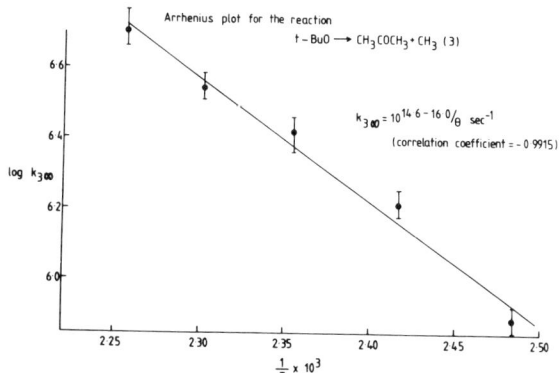

Figure 4

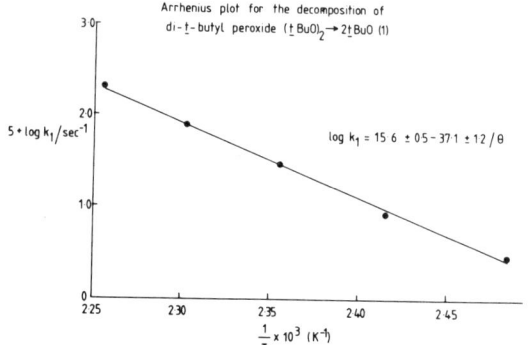

Figure 5

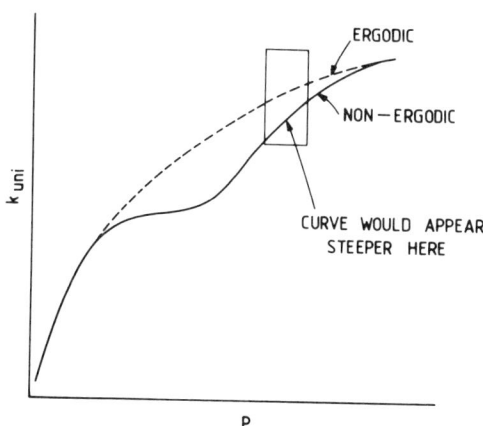

SECTION OF A UNIMOLECULAR REACTION FALL-OFF
CURVE

Figure 6

Table 2   Values of k_1 obtained from the total yield [acetone] +
[t-butyl nitrite] at various temperatures.

Temperature °C	Run time sec	% decomposition	k_1 sec^{-1}	number of runs
129.3	2500	7.18 ± 2.61	$(2.98 ± 1.14) \times 10^{-5}$	98
140.6	1000	8.26 ± 1.43	$(8.62 ± 1.57) \times 10^{-5}$	42
151.2	300	8.44 ± 1.72	$(2.94 ± 0.63) \times 10^{-4}$	40
161.1	100	7.46 ± 1.84	$(7.75 ± 2.01) \times 10^{-4}$	88
170.1	60	11.63 ± 5.02	$(2.06 ± 0.98) \times 10^{-3}$	66

a large number of determinations are carried out at the low pressure end of
the range. One way to circumvent these problems is to draw a smooth
curve through the experimental points and read off a number of points from
this curve. In this way it is possible to get a perfect straight line for
a certain value of α when $1/k_3$ is plotted against $p^{-\alpha}$. In fact it was
found that the values of $k_3(\infty)$ obtained when this method was used were
very close to the values obtained when the raw values of k_3 and p were
treated as described earlier. RRKM theory provides a very good model
for unimolecular reactions, and when an experimental fall-off curve can be
fitted by a computed curve, the high pressure limiting Arrhenius parameters
are produced. In the present work it was found that an excellent fit
could be obtained at the lowest temperature of the reaction (the most
intensively studied temperature) giving $\log [A_3(\infty)/\text{s}^{-1}] = 14.9$ and
$E_3(\infty) = 16.6$ Kcal mol^{-1}. It is not clear why the higher temperature curves
appear too steep to be consistent with these parameters. In fact it was
not found possible to fit the higher temperature curves since they are so
close together and steep. A possible suggestion might be that these
curves are sections of non-ergodic fall-off curves (figure 6) [10].
However to some extent we are in a dilemma as to which $A_3(\infty)$ and $E_3(\infty)$
values should be quoted although both are in agreement within the quoted
experimental error.

    We may now consider the likely nature of the transition state for this
reaction on the basis of this result. In terms of a thermochemical
formulation of the transition state theory, the pre-exponential factor (A)
for a unimolecular process is given by:

$$A = e k\bar{T}/h \; e^{\Delta S^\ddagger/R}$$

where k is the Boltzmann constant, $\bar{T}$ is the mean temperature of the
experiments, h is Planck's constant, $\Delta S^\ddagger$ is the entropy of activation and
R is the gas constant = 1.987 cal deg^{-1} mol^{-1}. If $A = 10^{14.6}$ s^{-1}, $\Delta S^\ddagger$
must be of the order of 6 cal deg^{-1} mol^{-1}. $\Delta S^\ddagger$ may be evaluated as
follows: the contributions to $\Delta S^\ddagger$ will be –

1. A symmetry contribution of $R \ln 3 = 2.2$ cal deg^{-1} mol^{-1}.

2. Contributions due to the changes in bond vibration frequencies.
   The frequencies which would be expected to be changed upon going
   to the transition state will be:

   (a) one C – C stretch (~ 750 cm^{-1}) becomes the reaction coordinate
   $\Delta S^\ddagger_{400} = -0.5$ cal deg^{-1} mol^{-1}.

(b) the C – O stretch$_{\perp}$ (~ 1040 cm^{-1}) becomes a C —— O stretch (~ 1400 cm^{-1}) $\Delta S^{\ddagger}_{400} = -0.2$ cal deg^{-1} mol^{-1}.

(c) two CH$_3$ rocks (~ 1000 cm^{-1}) will have their frequencies reduced to around 700 cm^{-1} $\Delta S^{\ddagger}_{400} = 0.6$ cal deg^{-1} mol^{-1}.

(d) the hindered rotation of the departing methyl group becomes a free rotation. If the barrier to rotation of the methyl group is ~ 4 Kcal mol^{-1} (similar to the barrier of 3.6 Kcal mol^{-1} to the rotation of a methyl group relative to an isopropyl group [11]) a contribution to $\Delta S^{\ddagger}_{400}$ of $+ 1.4$ cal deg^{-1} mol^{-1} results.

3. Contribution due to the increase of the moment of inertia when the C – C bond stretches. If $\Delta S^{\ddagger}$ is to be ~ 6 e.u., the contribution due to the increasing moment of inertia must be of the order of 2.5 e.u. implying an increase in the C – C bond length of a factor of ~ 2, from 1.52 Å [12] to ~ 3 Å.

The A factor of $10^{14.6}$ s^{-1} represents a relatively loose transition state therefore.

5. CONCLUSIONS

Despite not being able to precisely fit the experimental fall-off curves at every temperature, it is concluded that the computed high pressure Arrhenius parameters best represent the rate constants for the decomposition of the t-butoxy radical. As mentioned previously, the Arrhenius parameters obtained from the least squares analysis of the points plotted in figure 4, agree with the RRKM model values within the error limits. However the plot, or perhaps better the least squares analysis, will be subject to the errors involved when obtaining parameters over a relatively small temperature range, particularly when the activation energy is low. It is planned to extend the temperature range down to room temperature using t-butyl nitrite as a photochemical source of t-butoxy radicals. Any excess energy carried by the t-butoxy radical would have to be quenched by added inert gas. We also plan to examine in detail, the fall-off behaviour of the isopropoxy radical. Both from preliminary experimental results and on theoretical grounds, we expect to be able to follow the decomposition further into the fall-off region than the t-butoxy radical under the same experimental conditions and look for any possible ergodic behaviour.

Our results may be used to calculate the rate constant for the decomposition of the t-butoxy radical at room temperature. They predict that at 298K $k_3(\infty) = 8.2 \times 10^2$ s^{-1}. At one atmosphere (N$_2$) RRKM theory predicts a value of 6.5×10^2 s^{-1}, a factor of 1.27 below its high pressure limit. The results obtained in the present work underline the necessity of considering the pressure dependence of a unimolecular reaction before drawing conclusions about the kinetic parameters and the relative importance of the reaction paths for the alkoxy radical. This result also applies to the alkoxy radicals produced from butadiene [13], n-butane and simple ketones [14] and ethene and 2-butene [15].

6. ACKNOWLEDGEMENT

G.N.R. is indebted to the S.E.R.C. (U.K.) for a research award. We are grateful to B.S. Rabinovitch, H.M. Frey and R. Walsh for providing RRKM computing programs and to R. Walsh and W.A.J. Bryce for providing assistance with the computing.

7. REFERENCES

1. L. Batt and G.N. Rattray, Int. J. Chem. Kinet., 11, 1183-1196 (1979).

2. L. Batt and R.T. Milne, Int. J. Chem. Kinet., 8, 59 (1976).

3. B.G. Gowenlock and J. Trotman, J. Chem. Soc., 4190 (1955).

4. L. Batt and F.R. Cruickshank, J. Phys. Chem., 70, 723 (1966).

5. L. Batt and R.D. McCulloch, Int. J. Chem. Kinet., 8, 491-500 (1976).

6. I. Oref and B.S. Rabinovitch, J. Phys. Chem., 72, 4488 (1968).

7. See D.M. Golden, R.K. Solly and S.W. Benson, J. Phys. Chem., 75, 1333 (1971).

8. D.C. McKean, J.L. Duncan and R.K.M. Hay, Spectrochim. Acta, 23A, 605 (1966).

9. L. Batt and S.W. Benson, J. Chem. Phys., 36, 895 (1962).

10. I. Oref and B.S. Rabinovitch, Accts. Chem. Res., 12, 166 (1979).

11. S.W. Benson, Thermochemical Kinetics, 2nd edition, John Wiley and Sons, New York, 1976.

12. J.D. Swalen and C.C. Costain, J. Chem. Phys., 31, 1562 (1959).

13. V. Carassiti, C.A. Bignozzi, C. Chiorboli and A. Maldotti, Proceedings of First European Symp. on Atmos. Pollutants, Eds. B. Versino and H. Ott, Ispra, 1979, p.221-225.

14. R.A. Cox, K.G. Patrick and S.V. Chant, Cost 61A bis Discussion, Leuven, February 1981.

15. H. Niki, P.D. Maker, C.M. Savage and L.P. Breitenbach, J. Phys. Chem., 82, 135 (1978).

CHLORINE PHOTOSENSITIZED OXIDATION OF CHLOROMETHANES, -ETHANES and -ETHYLENES.

J. OLBREGTS and G.R.DE MARE.

Chimie Physique Moléculaire, Faculté des Sciences,
Université Libre de Bruxelles, 50, ave F.D.Roosevelt,
B.1050 Brussels (Belgium).

Summary

From the study of the chlorine photosensitized oxidation of chloromethanes, -ethanes and -ethylenes it was possible to develop a general reaction mechanism characterized by the following features :

-the occurence of a long chain oxidation can be explained by the thermochemistry of the $RO^{\cdot}$ radicals

-the oxidation involves a biradicalar chain propagation

$$2\ RO_2^{\cdot} \rightarrow 2\ RO^{\cdot} + O_2$$

-depending on the particular experimental conditions, one observes a competition between mono- and biradicalar chain terminations.

Numerous rate constants have been estimated for reactions such as

$$R^{\cdot} + O_2 \quad \rightarrow \quad RO_2^{\cdot}$$

$$RO_2^{\cdot} \quad \rightarrow \quad R^{\cdot} + O_2$$

$$RO_2^{\cdot} + R^{\cdot} \rightarrow \quad RO_2R$$

$$RO_2^{\cdot} + RO_2^{\cdot} \rightarrow \quad RO_2R + O_2$$

$$RO_2^{\cdot} + RO_2^{\cdot} \rightarrow \quad 2RO^{\cdot} + O_2$$

$$RO_2^{\cdot} + Cl^{\cdot} \rightarrow \quad products$$

$$RO_2R \quad \rightarrow \quad 2RO^{\cdot}$$

where $R = CCl_3,\ C_2H_5,\ C_2H_3Cl_2,\ C_2HCl_4,\ C_2HCl_5$

1. INTRODUCTION

At the beginning of the sixties, the chemical kinetics group headed by Prof. Goldfinger in Brussels started a research programme concerning the chlorine photosensitized oxidation of halocarbons. The purpose was clearly stated in the

first paper(1) of this series:"chlorine is a convenient photo-
sensitizer for photo-induced oxidations permitting the study
of these reactions at much lower temperatures than usual". A
new important justification of the interest of these reaction
systems appeared about 10 years later, when in 1974 Molina and
Rowland(2) suggested that the build up of chlorofluorocarbons
in the atmosphere could present a threat to stratospheric ozone.
Quite different systems have been studied in our group; a mechanism has been developped that enables us to understand the
different observed behaviours and a number of rate constants
for elementary reaction steps have been estimated. The purpose
of the present paper is to give a synthesis of the papers we
have published on this subject during these last 15 years.

2. GENERAL MECHANISM FOR THE CHLORINE PHOTOSENSITIZED OXIDA-
TION AND THE OXYGEN INHIBITED PHOTOCHLORINATION OF HALO-
CARBONS.

When reaction mixtures of chlorine, oxygen and chloroethanes or -ethylenes are irradiated at 435.8 nm one observes
either only the inhibited photochlorination (3-5), or both the
inhibited photochlorination and the sensitized oxidation occuring simultaneously (4,5-8). In the latter case, the high values observed for the quantum yields clearly indicate that
both reactions occur via a long chain mechanism involving the
same radicals. Furthermore, these simultaneously occuring reactions present different values for the order against absorbed
light intensity, i.e. 1/2 for the chlorination and 1 for the
oxidation. From this observation it is clear that i) the chain
termination steps must involve two chain carriers, ii) the chain
propagation of the chlorination, one chain carrier and iii) the
chain propagation of the oxidation two chain carriers. This
latter conclusion can in no way be avoided even if it appears
surprising at first sight. It led P.Goldfinger and G.Huybrechts
(9) to propose a reaction of 2 $RO_2^{\cdot}$ radicals as the chain propagating step of the oxidation. The mechanism has been completed
taking into account the different observed behaviours. The
first stages of this mechanism consist in the formation of R·
radicals either through hydrogen atom abstraction by chlorine
atoms from a saturated halocarbon, or by the addition of a chlorine atom on the double bond of an olefinic one. This follows
the well known mechanism (10) of the photochlorinations which
may be written in a simplified way as

$$Cl_2 \xrightarrow{h\nu} 2\ Cl^{\cdot}$$

$$Cl^{\cdot} + A \xrightarrow{(2)} R^{\cdot} \qquad\qquad Cl^{\cdot} + RH \xrightarrow{(2')} R^{\cdot} + HCl$$

$$R^{\cdot} + Cl_2 \xrightarrow{(3)} RCl + Cl^{\cdot}$$

$$R^{\cdot} + Cl \xrightarrow{(7)} RCl \qquad\qquad\qquad\qquad (i)$$

$$R^{\cdot} + R^{\cdot} \xrightarrow{(8)} product(s)$$

Here A and RH stand respectively for an olefinic or a saturated halocarbon.

In the presence of oxygen, this mechanism must be completed by the following steps

$$R^{\cdot} + O_2 \xrightarrow{(9)} RO_2^{\cdot}$$

$$RO_2^{\cdot} \xrightarrow{(-9)} R^{\cdot} + O_2$$

$$RO_2^{\cdot} \xrightarrow{(wall\ 1)} \ldots\ R^{\cdot}$$

$$RO_2^{\cdot} \xrightarrow{(wall\ 2)} \text{terminating step} \qquad (ii)$$

$$RO_2^{\cdot} + R^{\cdot} \xrightarrow{(10)} RO_2R$$

$$RO_2^{\cdot} + RO_2^{\cdot} \xrightarrow{(11)} RO_2R + O_2$$

$$RO_2^{\cdot} + RO_2^{\cdot} \xrightarrow{(12)} 2\ RO^{\cdot} + O_2$$

$$n\ RO^{\cdot} \xrightarrow{(13)} \text{terminating step}$$

$$RO^{\cdot} \xrightarrow{(14)} \text{oxidation product + chain carrying radical or atom.}$$

$$RO_2^{\cdot} + Cl^{\cdot} \xrightarrow{(16)} \text{terminating step}$$

3. CONDITION TO OBSERVE A LONG CHAIN CHLORINE PHOTOSENSITIZED OXIDATION.

As stated above a long chain chlorine photosensitized oxidation is not observed in each reaction system. For example, the quantum yield of oxidation for ethane or 1,2-dichloroethane does not exceed unity while it goes up to a few hundred for tetra- or pentachloroethane under favorable conditions. These different behaviours can be explained by the reaction mechanism (i, ii) if one considers the competition between reaction steps 13 and 14. When the RO$^{\cdot}$ radicals, produced in reaction 12, react more rapidly in a terminating step (such as 13) than in a chain propagating step yielding the oxidation products (such as 14), the long chain oxidation is not observed and vice versa. The fact that one of these two steps is faster than the other one can be correlated (4) to the thermochemistry of the decomposition of the RO$^{\cdot}$ radicals. Three kinds of bonds, i.e. C-C, C-Cl, C-H, could be dissociated in these radicals. Nevertheless, the C-H bond cleavage has never been observed under our experimental conditions for any chlorinated ethane or ethylene. The C-C or C-Cl bond dissociation occurs when the corresponding reaction is exothermic by at least 15 kcal.mol^{-1}. For example, these reactions are not observed

$$\Delta H, \text{kcal.mol}^{-1}$$

		ΔH, kcal.mol^{-1}
$CH_3CH_2O^{\cdot}$	$\rightarrow CH_3CHO + H^{\cdot}$	17
	$\rightarrow CH_2O + CH_3^{\cdot}$	11
$CH_2ClCHClO^{\cdot}$	$\rightarrow CH_2ClCHO + Cl^{\cdot}$	-4
	$\rightarrow CH_2ClCClO + H^{\cdot}$	2
	$\rightarrow CH_2Cl^{\cdot} + CHClO$	-8

$$\begin{array}{lr} & \Delta H, kcal.mol^{-1} \\ CCl_3CHClO^{\cdot} \rightarrow CCl_3CHO + Cl^{\cdot} & -4 \\ \phantom{CCl_3CHClO^{\cdot}} \rightarrow CCl_3CClO + H^{\cdot} & 6 \end{array}$$

while the following ones definitely occur

$$\begin{array}{lr} CCl_3CHClO^{\cdot} \rightarrow CCl_3^{\cdot} + CHClO & -16 \\ CCl_3CCl_2O^{\cdot} \rightarrow CCl_3CClO + Cl^{\cdot} & -17 \\ \phantom{CCl_3CCl_2O^{\cdot}} \rightarrow CCl_3^{\cdot} + COCl_2 & -20 \end{array}$$

From the known or estimated (using group property additivity rules (11)) value (4) of the heat of formation of the $RO^{\cdot}$ radicals and of their dissociation products it is possible to predict correctly the occurence of long chain oxidation as well as the nature of the oxidation products whenever only one type of $R^{\cdot}$ radical is formed in the reactions of chloroethanes and -ethylenes. This is summarized in Table I.

Table I : Chlorine photosensitized oxidation of chloroethanes and chloroethylenes.

Chloroethanes	Long chain oxidation products
C_2H_6	not observed
$1,2-C_2H_4Cl_2$	not observed
$1,1,1-C_2H_3Cl_3$	not observed
$1,1,1,2-C_2H_2Cl_4$	$COCl_2$, CO, HCl
$1,1,2,2-C_2H_2Cl_4$	$CHCl_2COCl$, HCl
C_2HCl_5	$CCl_3COCl, COCl_2$, HCl

Chloroethylenes	
C_2H_4	not observed
$1,2-C_2H_2Cl_2$	not observed
C_2Cl_4	CCl_3COCl, $COCl_2$

4. APPLICATION OF THE MECHANISM TO DIFFERENT SYSTEMS.

a. Ethane (3)

As we have just discussed, the only reaction observed when irradiating reaction mixtures of chlorine, ethane and oxygen is an inhibited photochlorination. The experimental technique (1) we use to study the kinetics of these reactions is suitable for reaction rates ranging between about 0.1 and 10 torr.min^{-1}. Therefore, the experimental conditions must be chosen to satisfy this condition. According to the above mechanism the rate of chlorination is given by

$$v_{Cl} = k_3 (R^{\cdot})(Cl_2) \qquad \text{(iii)}$$

The value of the rate constant k_3 is very large for ethane (3):

thus, to meet the needs of the experimental technique, this reaction must be studied at low radical concentration. One way to do this, is to use a low light intensity : under such conditions, the concentration of all kinds of radicals in the system remains very low, favouring monoradicalar reactions. The $C_2H_5O_2$ radicals formed in reaction 9 are then rapidly destroyed in a reverse reaction which has been assumed to occur at the wall because it involves no activation energy (see reaction "wall 1" in scheme (ii)). The slower biradicalar reactions 10, 11 and even 12 act as chain termination steps, which associated to the monoradicalar propagations 2 and 3 lead to an inhibited photochlorination of 1/2 order against absorbed light intensity.

A second way to produce low concentrations of chain carrying radicals, is to work at normal light intensities but with high enough amounts of added oxygen so as to transform most of the $C_2H_5 \cdot$ radicals into $C_2H_5O_2 \cdot$ radicals. Under these conditions the reversibility of reaction step 9 becomes negligible and this reaction appears to be the monoradicalar chain termination : the reaction rate is then first order against absorbed light intensity. Up to 427K, the gas phase decomposition of $C_2H_5O_2 \cdot$ (reaction -9) with an activation energy of 28 kcal. mol^{-1} (12), does not play a significant role in this reaction.

b. 1,2-dichloroethane (5)

1,2-dichloroethane is another example of a chlorocarbon for which the oxidation quantum yield is lower or equal to unity. The value of the rate constant k_3 is still high (13), although it is smaller than in the case of ethane. The inhibited photochlorination must thus be studied at rather low $R \cdot$ concentrations. This has been achieved using not too high absorbed light intensities associated with relatively high oxygen pressure to decrease the ratio of $(R \cdot)/(RO_2^\cdot)$ radicals. Monoradicalar reactions still play an important role. The reaction -9 has an activation energy of 24 kcal.mol^{-1} (5) : at temperatures lower than 430 K, this reaction is too slow to give a significant reversibility of reaction 9, which then acts as a chain termination. At higher temperatures, reactions 9 and -9 are at thermodynamic equilibrium and the chain termination occurs through the wall destruction of the $RO_2^\cdot$ radicals (reaction "wall 2" in scheme (ii)). In any case the chain terminations are first order in radicals, yielding a first order rate against absorbed light intensity.

c. 1,1,1,2-tetrachloroethane, pentachloroethane, trichloroethylene and tetrachloroethylene (1,3,6-8).

These four systems have very similar behaviours. Increasing concentrations of oxygen lead successively to the inhibited photochlorination, the competition between chlorination and oxidation and finally the chlorine photosensitised oxidation. Under our experimental conditions, the more important reaction of $RO_2^\cdot$ radicals remains the monoradicalar reaction -9. Nevertheless the biradicalar reaction 12 (followed by

reaction 14) is an important chain propagation for the oxidation. The chain terminations 7 and 8 corresponding to the non
inhibited chlorination, are observed in conjunction with reaction 10 at the lower oxygen pressures. At the higher ones,reac-
tion 11 becomes progressively predominant. All these chain terminations are biradicalar ones and are common to both the chlorination and the oxidation. The chain propagating steps are
mono- and biradicalar respectively for the chlorination (reaction 3) and for the oxidation (reaction 12): this explains the
orders 1/2 and 1 observed respectively for these simultaneous
reactions. It results from the mechanism, that the inhibited
photochlorination and the photosensitized oxidation rates should
not depend on the way used to produce the $R^\cdot$ radicals. Therefore
these reaction rates should be the same under identical experimental conditions, for C_2HCl_5 and C_2Cl_4, both yielding the same $C_2Cl_5^\cdot$ radicals. This has been effectively verified to a good
approximation (7). Nevertheless, a small but systematic difference is observed between the rates of both systems, C_2HCl_5
reacting slower than C_2Cl_4. It has been assumed that because
of the higher value of k_2 (relative to C_2Cl_4) compared to k_2'
(relative to C_2HCl_5), the ratio $(Cl^\cdot)/(R^\cdot)$ would be greater in
the case of C_2HCl_5 and that chain terminations involving Cl
atoms could play a role in this latter system.

d. Chloroform (14)

If the assumption of chain terminations involving Cl atoms (see just above) is correct, such reactions would be more
important in the case of chloroform where the $(Cl^\cdot)/(R^\cdot)$ ratio
is still greater. The study of the oxygen inhibited photochlorination and of the chlorine photosensitized oxidation of chloroform (14) enabled us to give evidence for a chain termination
by recombination of $RO_2^\cdot$ with $Cl^\cdot$ (reaction 16 in scheme (ii)).
This reaction system appears to be quite complicated because
of the occurence of five different chain terminating steps
(reactions 7,8,10,11 and 16): under these conditions it is not
possible to write a simple rate equation neither for the inhibited chlorination nor for the induced oxidation. Nevertheless,
under particular experimental conditions, it is possible to
check different relationships between the rates of the simultaneous chlorination and oxidation, and to obtain reliable values
for the rate constants. The observations can be summarized as
follows: at temperatures below 350K, because of the very low
value of k_{-9}/k_9 the ratio $(R^\cdot)/(RO_2^\cdot)$ is very small and therefore neither chlorination nor chain termination 10 is observed;
the chain termination 11 is dominant and the importance of chain
termination 16 increases with decreasing chloroform pressure.
At about 400 K, k_{-9}/k_9 becomes high enough to have non negligible concentrations of $R^\cdot$ radicals, which are responsible for
the chlorination and the observation of chain termination 10.
At 426 K, k_{-9}/k_9 is so high that the $RO_2^\cdot$ radicals convert rapidly to $R^\cdot$ and that the oxidation is relatively unimportant
compared with the inhibited chlorination. This means that the
chain termination 7 and 8 characteristic of the non inhibited
photochlorination, can no longer be neglected.

5. SENSITIVITY OF THE PHOTOCHLORINATION TO THE INHIBITING EFFECT OF OXYGEN.

In the text above, we talked about small or high oxygen pressures without precising what this means quantitatively. In most of the cases discussed here, an equilibrium exists between the concentration of $R^\cdot$ and $RO_2^\cdot$ radicals which is regulated by reactions 9 and -9 (or eventually "wall 1"); this means that

$$(RO_2^\cdot)/(R^\cdot) = (k_9/k_{-9})(O_2) \qquad (iv)$$

The higher the value of k_9/k_{-9}, the lower the O_2 concentration needed to attain a certain ratio $(RO_2^\cdot)/(R)$ and the higher the sensitivity of the photochlorination to the inhibiting effect of oxygen. The parameter which essentially controls the value of the ratio k_9/k_{-9}, is the activation energy associated with reaction -9.[9] This activation energy seems to depend only on the number of chlorine atoms on the carbon bearing the oxygen, i.e.

$$E_{-9} = 28 \text{ kcal.mol}^{-1} \qquad \text{for } CH_3CH_2O_2^\cdot$$

$$E_{-9} = 24 \text{ kcal.mol}^{-1} \qquad \text{for } CH_2ClCHClO_2^\cdot \text{ or } CCl_3CHClO_2^\cdot$$

$$E_{-9} = 18\text{-}20\text{kcal.mol}^{-1} \qquad \text{for } CHCl_2CCl_2O_2^\cdot \text{ or } CCl_3CCl_2O_2^\cdot$$

The sensitivity of the photochlorination of ethane to the inhibiting effect of oxygen will thus be the highest. Oxygen pressures of about 1 part in 10^6 of the reactants may cause a decrease in the rate of photochlorination by about a factor of 10. To obtain the same inhibiting effect in the photochlorinations of C_2HCl_3, C_2Cl_4, $C_2H_2Cl_4$ or C_2HCl_5 one needs oxygen up to a few percent of the reactants. The cases of 1,2-dichloroethane and 1,1,1,2-tetrachloroethane are intermediate ones. For the one carbon radical $CCl_3O_2^\cdot$, E_{-9} was measured to be 22 kcal.mol^{-1}. This system was studied over a large temperature range (303-426 K). In that case, the variation of k_{-9} is rather important and explains the observations that the photochlorination of chloroform is almost unaffected by the presence of a few percent oxygen at 426 K while its rate is reduced by several orders of magnitude under the same conditions at 303 K.

6. AFTER-EFFECT.

Neither the oxygen-inhibited photochlorination nor the chlorine photosensitized oxidation do occur before irradiating the reaction mixtures. In most of the systems studied (except for C_2H_6 and $C_2H_4Cl_2$), as soon as the reactants have been irradiated, the reaction continues after the actinic light is cut off (6,7,8,14). This has been called an "after-effect" and has been explained (6) by the thermal reactions of the peroxide RO_2R (formed in reactions 10 and 11) :

$$\begin{array}{c} (15) \\ RO_2R \rightarrow \quad 2\ RO \\ (wall) \\ RO_2R \rightarrow \quad \text{stable products} \end{array} \qquad (v)$$

where reaction 15 initiates the dark reaction.
These additional reactions permit one to represent the reaction rates during the course of the reaction up to at least 30% consumption of the reactants.

7. APPENDIX : MEASURED RATE CONSTANTS

The following list contains all the rate constants that have been measured in this laboratory. The activation energies (E) are expressed in $kcal.mol^{-1}$, the preexponential factors (A) in the units : liter, mole, second.

A. Photochlorination of chloro-methanes, -ethanes and ethylenes.

Reactions			E	$log_{10}A$	Temp.(K)	ref.
$Cl + CH_4$	$\rightarrow$	$CH_3 + HCl$	3.9	10.7	360–475	15
$Cl + CD_4$	$\rightarrow$	$CD_3 + HCl$	5.75	10.72	304–461	a
$Cl + CH_3Cl$	$\rightarrow$	$CH_2Cl + HCl$	3.1	10.5	360–475	15
$Cl + CH_2Cl_2$	$\rightarrow$	$CHCl_2 + HCl$	3.1	10.4	360–475	15
$Cl + CHCl_3$	$\rightarrow$	$CCl_3 + HCl$	3.35	10.2	360–475	15
$Cl + CDCl_3$	$\rightarrow$	$CCl_3 + DCl$	4.85	10.34	360–475	b
$Cl + C_2D_6$	$\rightarrow$	$C_2D_5 + DCl$	1.28	10.71	303–433	c
$Cl + C_2H_5Cl$	$\rightarrow$	$CH_2ClCH_2 + HCl$	1.5	10.05	303–453	19
$Cl + C_2H_5Cl$	$\rightarrow$	$CH_3CHCl + HCl$	1.5	10.55	303–453	19
$Cl + 1,1-C_2H_4Cl_2$	$\rightarrow$	$CHCl_2CH_2 + HCl$	3.4	10.00	303–453	19
$Cl + 1,1-C_2H_4Cl_2$	$\rightarrow$	$CH_3CCl_2 + HCl$	1.9	9.95	303–453	19
$Cl + 1,2-C_2H_4Cl_2$	$\rightarrow$	$CH_2ClCHCl + HCl$	3.1	10.80	323–423	19
$Cl + 1,1,1-C_2H_3Cl_3$	$\rightarrow$	$CCl_3CH_2 + HCl$	3.6	9.40	323–423	19
$Cl + 1,1,2-C_2H_3Cl_3$	$\rightarrow$	$CHCl_2CHCl + HCl$	3.7	10.15	323–423	19
$Cl + 1,1,2-C_2H_3Cl_3$	$\rightarrow$	$CH_2ClCCl_2 + HCl$	3.1	9.95	323–423	19
$Cl + 1,1,1,2-C_2H_2Cl_4$	$\rightarrow$	$CCl_3CHCl + HCl$	3.55	9.80	323–423	19
$Cl + 1,1,2,2-C_2H_2Cl_4$	$\rightarrow$	$CHCl_2CCl_2 + HCl$	3.4	10.1	323–438	19
$Cl + C_2HCl_5$	$\rightarrow$	$C_2Cl_5 + HCl$	3.55	9.65	336–421	19
$Cl + C_2H_4$	$\rightarrow$	C_2H_4Cl	–	10.6	310	20
$Cl + C_2H_3Cl$	$\rightarrow$	$C_2H_3Cl_2$	–	10.8	322	21
$Cl + C_2HCl_3$	$\rightarrow$	$CHCl_2CCl_2$	0	9.75	433–497	22
$Cl + C_2Cl_4$	$\rightarrow$	C_2Cl_5	0	10.1	310–348	23
$C_2H_4Cl^x$	$\rightarrow$	$C_2H_4 + Cl$	–	9.3	310	20

Reaction					
$CH_2ClCHCl_2$	$\rightarrow$	$CH_2CHCl + Cl$	20.71	14.33	433–510	24
$CHCl_2CCl_2$	$\rightarrow$	$C_2HCl_3 + Cl$	20.4	13.7	433–497	22
CCl_3CHCl	$\rightarrow$	$CCl_2CHCl + Cl$	16.4	11.6	350–405	25
C_2Cl_5	$\rightarrow$	$C_2Cl_4 + Cl$	16.08	13.51	310–421	23
$CCl_3 + Cl_2$	$\rightarrow$	$CCl_4 + Cl$	5.	8.74	304–425	26
C $HCl_2CCl_2 + Cl_2$	$\rightarrow$	$C_2HCl_5 + Cl$	5.18	8.77	363–497	22
$CCl_3CHCl + Cl_2$	$\rightarrow$	$CCl_3CHCl_2 + Cl$	4.6	8.35	360–420	8,25
$C_2Cl_5 + Cl_2$	$\rightarrow$	$C_2Cl_6 + Cl$	5.43	8.31	360–520	27
$CCl_3 + HCl$	$\rightarrow$	$CHCl_3 + Cl$	11.3	8.65	303–425	26
$C_2Cl_5 + HCl$	$\rightarrow$	$C_2HCl_5 + Cl$	10.8	8.1	385–490	15
$Cl + CCl_4$	$\rightarrow$	$CCl_3 + Cl_2$	20.0	10.93	303–425	26
$Cl + C_2HCl_5$	$\rightarrow$	$CHCl_2CCl_2 + Cl_2$	17.9	10.8	433–497	22
$Cl + C_2Cl_6$	$\rightarrow$	$C_2Cl_5 + Cl_2$	19.45	11.3	385–490	15
$Cl + CCl_3$	$\rightarrow$	CCl_4	0	10.8	303–425	26
$Cl + CHCl_2CCl_2$	$\rightarrow$	product(s)	–	10.85	497	22
$Cl + C_2Cl_5$	$\rightarrow$	product(s)	0.06	11.03	385–490	15
$CCl_3 + CCl_3$	$\rightarrow$	C_2Cl_6	0	9.66	303–425	26
$CHCl_2CCl_2 + CHCl_2CCl_2$	$\rightarrow$	product(s)	–	9.3	371	26
$C_2Cl_5 + C_2Cl_5$	$\rightarrow$	product(s)	0.08	8.66	360–520	27
$2(cyclo-C_6F_{10}Cl)$	$\rightarrow$	product(s)	–	8.4	314	28
$Cl + Cl + Cl_2$	$\rightarrow$	$Cl_2 + Cl_2$	0	9.85	552	29

For those rate constants concerning steps of the chlorination
of methane, ethane, ethylene and their chlorinated derivatives,
that are not listed in the preceeding table, estimates have
been published in ref.10.
a) Calculated from the ratio of the rate constants obtained
   from the study of the competitive chlorination of $CHCl_3$
   and CD_4 (16) and the value of the rate constant for
   $CH_4 + Cl$ (17)
b) Calculated from the ratio of rate constants obtained from
   the study of the competitive chlorination of $CHCl_3$ and
   $CDCl_3$ (16) and the value of the rate constant for
   $CHCl_3 + Cl$ (17)
c) Calculated from the ratio of rate constants obtained from
   the study of the competitive chlorination of C_2H_6 and
   C_2D_6 (16) and the value of the rate constant for
   $C_2H_6 + Cl$ (18)

B. Estimates of the rate constants of the reaction steps of
   the oxygen inhibited photochlorination and the chlorine
   photosensitized oxidation of chloro-methanes, -ethanes and
   -ethylenes

Reaction					
$C_2H_5 + O_2$	$\rightarrow$	$C_2H_5O_2$	0	8.8	253-427	3
$CH_2ClCHCl + O_2$	$\rightarrow$	$CH_2ClCHClO_2$	0	7.8	321-489	5
$CHCl_2CCl_2 + O_2$	$\rightarrow$	$CHCl_2CCl_2O_2$	0	8.0	363-403	1,6
$C_2Cl_5 + O_2$	$\rightarrow$	$C_2Cl_5O_2$	0	8.0	353-373	7
$CCl_3\,O_2$	$\rightarrow$	$CCl_3 + O_2$	22		400-426	14
$CH_2ClCHClO_2$	$\rightarrow$	$CH_2ClCHCl + O_2$	24	13.0	321-489	5
$CHCl_2CCl_2O_2$	$\rightarrow$	$CHCl_2CCl_2 + O_2$	20	14.5	363-403	1,6
CCl_3CHClO_2	$\rightarrow$	$CCl_3CHCl + O_2$	23.9		360-420	8
$C_2Cl_5O_2$	$\rightarrow$	$C_2Cl_5 + O_2$	18	14.5	353-373	7
$CCl_3+CCl_3O_2$	$\rightarrow$	$CCl_3O_2CCl_3$	0	9.2	400-426	14
$CHCl_2CCl_2+CHCl_2CCl_2O_2$	$\rightarrow$	$CHCl_2CCl_2O_2CCl_2CHCl_2$	0	10	363-403	1,6
$C_2Cl_5+C_2Cl_5O_2$	$\rightarrow$	$C_2Cl_5O_2C_2Cl_5$	0	9.9	353-373	7
$2\ CCl_3O_2$	$\rightarrow$	$CCl_3O_2CCl_3+O_2$	0	8.0	303-426	14
$2\ CHCl_2CCl_2O_2$	$\rightarrow$	$CHCl_2CCl_2O_2CCl_2CHCl_2+O_2$	0	8.0	363-403	1,6
$2\ C_2Cl_5O_2$	$\rightarrow$	$C_2Cl_5O_2C_2Cl_5+O_2$	0	7.7	353-373	7
$2\ CCl_3O_2$	$\rightarrow$	$2\ CCl_3O + O_2$	0	10.7	303-426	14
$2\ CHCl_2CCl_2O_2$	$\rightarrow$	$2\ CHCl_2CCl_2O + O_2$	0	10.0	363-403	1,6
$2\ C_2Cl_5O_2$	$\rightarrow$	$2\ C_2Cl_5O + O_2$	0	9.9	353-373	7
$CCl_3O_2 + Cl$	$\rightarrow$	products	0	10.1	303-400	14
$CHCl_2CCl_2O_2CCl_2CHCl_2$	$\rightarrow$	$2\ CHCl_2CCl_2O$	27	12.5	363-403	1,6
$CCl_3CHClO_2CHClCCl_3$	$\rightarrow$	$2\ CCl_3CHClO$	35.5	15.5	360-420	8
$C_2Cl_5O_2C_2Cl_5$	$\rightarrow$	$2\ C_2Cl_5O$	26		353-373	7

References

1. G.Huybrechts, G.Martens, L.Meyers, J.Olbregts and K.Thomas, Trans.Faraday Soc., **61**, 1921 (1965).

2. M.J.Molina and F.S.Rowland, Nature, **249**, 810 (1974).

3. P.Goldfinger, G.Huybrechts, G.Martens, L.Meyers and J.Olbregts, Trans.Faraday Soc., **61**, 1933 (1965).

4. L.Bertrand, L.Exsteen-Meyers, J.A.Franklin, G.Huybrechts and J.Olbregts, Int.J.Chem.Kinet., **3**, 89 (1971).

5. L.Bertrand, J.Bizongwako, G.Huybrechts and J.Olbregts, Bull.Soc.Chim.Belges, **81**, 73 (1972).

6. G.Huybrechts and L.Meyers, Trans.Faraday Soc.,**62**, 2191(1966).

7. G.Huybrechts, J.Olbregts and K.Thomas, Trans.Faraday Soc., **63**, 1647 (1967).

8. D.Gillotay and J.Olbregts, Int.J.Chem.Kinet.,**8**, 11 (1976).

9. P.Goldfinger and G.Huybrechts, Chemical Kinetics and Chain Reactions (N.N.Semenov's 70th Anniversary Volume) Nauka,

Moscow, p.323 (1966).

10. G.Chiltz, P.Goldfinger, G.Huybrechts, G.Martens and G.Verbeke, Chem.Rev., 63, 355 (1963).

11. S.W.Benson, Thermochemical Kinetics, Wiley, NY (1968).

12. S.W.Benson, J.Amer.Chem.Soc., 87, 972 (1965).

13. F.S.Dainton, D.A.Lomax and M.Weston, Trans.Faraday Soc., 58, 308 (1962).

14. J.Olbregts, J.Photochem., 14, 19 (1980).

15. P.Goldfinger, G.Huybrechts and G.Martens, Trans.Faraday Soc., 57, 2210 (1961).

16. G.Chiltz, R.Eckling, P.Goldfinger, G.Huybrechts, H.S.Johnston, L.Meyers and G.Verbeke, J.Chem.Phys.,38, 1053(1963).

17. J.H.Knox, Trans.Faraday Soc., 58, 275 (1962).

18. J.H.Knox and R.L.Nelson, Trans.Faraday Soc., 55, 937(1959).

19. C.Cillien, P.Goldfinger, G.Huybrechts and G.Martens, Trans. Faraday Soc., 63, 1631 (1967).

20. J.A.Franklin, P.Goldfinger and G.Huybrechts, Ber.Bunsenges. Physik.Chem., 72, 173 (1968).

21. J.Olbregts, Bull.Soc.Chim.Belges, 83, 73 (1974).

22. G.Huybrechts, L.Meyers and G.Verbeke, Trans.Faraday Soc., 158, 1128 (1962) and L.Bertrand, J.A.Franklin, P.Goldfinger and G.Huybrechts. J.Phys.Chem., 72, 3926 (1968).

23. J.A.Franklin, G.Huybrechts and C.Cillien, Trans.Faraday Soc., 65, 2094 (1969).

24. G.Huybrechts, J.Katihabwa, G.Martens, M.Nejszaten and J.Olbregts, Bull.Soc.Chim.Belges, 81, 65 (1972).

25. J.Olbregts, Int.J.Chem.Kinet., 11, 117 (1979).

26. G.R.De Maré and G.Huybrechts, Trans.Faraday Soc., 64, 1311 (1968), and G.R.De Maré and G.Huybrechts, Chem.Phys.Letters, 1, 64 (1967).

27. S.Dusoleil, P.Goldfinger, A.M.Mahieu-Vander Auwera, G.Martens and D.Vander Auwera, Trans.Faraday Soc.,57,2197(1961).

28. L.Bertrand, G.R.De Maré, G.Huybrechts, J.Olbregts and M.Toth, Chem.Phys.Letters, 5, 183 (1970).

29. G.Chiltz, R.Eckling, P.Goldfinger, G.Huybrechts, G.Martens and G.Simoens, Bull.Soc.Chim.Belges, 71, 747 (1962).

LABORATORY KINETIC INVESTIGATIONS OF THE TROPOSPHERIC OXIDATION OF SELECTED INDUSTRIAL EMISSIONS

B. FRITZ, K. LORENZ, W. STEINERT and R. ZELLNER
Institut für Physikalische Chemie, Universität Göttingen,
3400 Göttingen, FRG

Summary

Emissions from industrial sources are characterized by a large variety of pollutants including inorganic compounds of high toxicity. However, very little is yet known regarding the lifetime and mechanism of the photo-oxidation of such species in the troposphere. As part of an investigation of the reactive removal of such emissions by homogeneous gas reactions, we report in this study direct rate measurements for the reactions of OH radicals with C_2H_4O (ethene oxide), PH_3 and HCN using an excimer laser photolysis/resonance fluorescence (LPRF) system. The properties of the rate constants $k(p,T)$ are being used to propose probable oxidation mechanism of these species as applicable to tropospheric conditions.

1. INTRODUCTION

Emissions from chemical industries constitute one of the important sources of anthropogenic pollutants. Although the total amount of emissions from industrial sources is small compared to other sources (energy production, transport), industrial emissions are responsible for a large variety of pollutants including compounds of high toxicity.

Research in the photo-oxidation of pollutants has so far largely been concentrated on organic species, i.e. hydrocarbons and their partially oxidized compounds. This emphasis has been guided by the importance of these species in photochemical smog formation and the generation of secondary pollutants. Very little, however, is yet known about the corresponding behaviour of inorganic species. The main questions that would have to be asked are: i) what is the rate of degradation and ii) what are the respective degradation pathways; i.e. how does toxicity degrade in an oxidation route under tropospheric conditions. Obviously, not all these questions can be answered in a simple way and without making recourse to a number of elaborate laboratory investigations. In the present work we have therefore concentrated on rate measurements only. Since there is conclusive evidence from both direct kinetic studies and smog-chamber type of experiments that the OH radical is the primary reagent in homogeneous tropospheric oxidation, all the rate measurements reported here involve the OH radical. Reaction product identification and an analysis of the oxidation pathway will be the aim of future work in our laboratory. It will be shown here though, what tentative information in this direction can be deduced from measurements of the temperature and pressure dependence of the individual rate coefficients, i.e. from the type of the primary reaction, whether H abstraction or OH addition.

The emissions we have concentrated on in this work are: C_2H_4O (ethene oxide), PH_3, and HCN. They are generally toxic compounds and their industrial use is as insecticides (C_2H_4O, PH_3, HCN), and as intermediate products in organic synthesis (C_2H_4O, HCN). Their amounts emitted range from 30 t/a (PH_3) to several hundred t/a (C_2H_4O, HCN) in highly industrialized areas (i.e. Cologne - Düsseldorf, FRG) [1].

2. EXPERIMENTAL

The experiments reported in this work are part of an investigation of direct kinetic measurements of reactions of OH radicals with anthropogenic pollutants. A schematic representation of the laser photolysis/resonance fluorescence (LPRF) experiment is provided in Fig. 1.

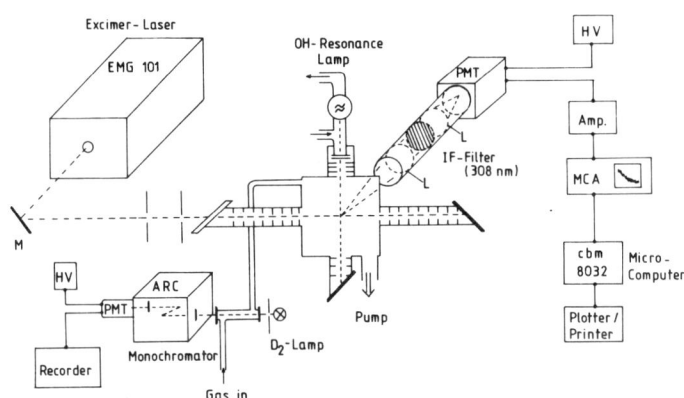

Fig. 1   Laser flash photolysis/resonance fluorescence (LPRF) experiment

An excimer laser is used as a short pulse, monochromatic photolysis light source. Due to the availibility of various UV laser lines, the photolysis precursor of OH can be chosen as to avoid as much as possible unwanted photolysis or photoexcitation of the stable reagent. Potential OH sources in laser photolysis are summarized in Fig. 2. For most of the experiments described below HNO_3 has been used as OH precursor at wavelengths 193 nm (ArF) and 248 nm (KrF). In the latter case an absorption cross section of 2×10^{-20} cm^2 is sufficient for the generation of OH with a typical laser energy of 50 - 100 mJ/puls and without submitting undue high concentrations of the precursor (p(HNO_3) ≤ 100 mTorr). From independent calibration experiments using

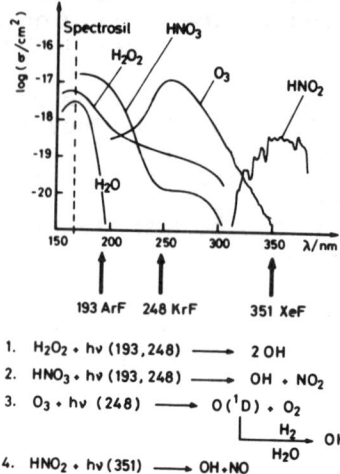

Fig. 2 Photolytic precursors of OH radicals in excimer laser
flash experiments

NO_2 photolysis and measurements of the attenuation of the laser
beam across the reaction cell, we estimate that our initial OH
concentrations are around 1×10^{12} cm^{-3}. The detection of OH
is by time resolved fluorescence in the $A^2\Sigma^+ - X^2\Pi - (0,0)$ band,
using a conventional microwave discharge source and photoelectric detection. Including signal averaging (of at least 50 laser pulses), an OH detection limit (corresponding to S/N = 1)
of 1×10^{10} cm^{-3} is estimated.

The present set-up is limited to measurements at total
pressures up to 200 Torr of Ar due to fluorescence quenching.
For this reason we are routinely backing this experiment by a
conventional flash photolysis/resonance absorption set-up,
which we have used in previous experiments on OH reactions [2,3]
and which allows measurements up to 1 atm, even with N_2 as di-
lutent gas. The majority of the results reported here for OH +
HCN have been obtained in this system.

3. RESULTS AND DISCUSSION

In all LPRF experiments we have used HNO_3 as OH precursor.
Since the reaction between OH and HNO_3

(1) $OH + HNO_3 \longrightarrow H_2O + NO_3$

is not negligible even at 100 mTorr of HNO_3, this gives rise to
a (small) additional term in the overall first order rate constant for the decay of OH in the presence of any other reactant:

$k_{1st} = \Delta \ln[OH] / \Delta t = k_{diff} + k_1[HNO_3] + k_R[Reactant]$

Moreover, since there is some disagreement in the literature
value of k_1 [4,5], we have firstly aimed at a re-determination

of this rate constant. From measurements in the range T =
250 - 435 K and over a ten-fold variation of HNO_3 concentration (which was determined by in-situ absorption measurements
in the gas mixture prior to entering the reaction cell) we obtain

$$k_1 = (1.5 \pm 0.4) 10^{-14} \exp(622/T) \ cm^3/molecule \ s$$

This is somewhat larger than previous measurements by Margitan
et al. and Smith and Zellner [4], but in extremely good agreement with the result of Wine et al. [5]. In particular, the
negative temperature dependence reported in [5], is being confirmed. In an attempt to further identify the nature of the
reaction in terms of its temperature dependence, we have also
carried out measurements at higher pressures (up to 1 atm of
N_2). It is found that over the pressure region 10 Torr - 1 atm
k_1 increases by about a factor of 2, indicating that (1) is not
a direct H atom abstraction reaction, but presumably occurs via
intermediate complex formation. Detailed results of this work
will be reported elsewhere [6].
    For the purpose of studying reactions of OH with other
reactants in the presence of HNO_3, the term $k_1 \cdot [HNO_3]$ has to be
known for any particular run. In the present study we have
dealt with this by measuring k_{1st} over a wide range of reactant concentration but constant $[HNO_3]$. In this case the contribution of reaction (1) to the overall OH decay occurs together with k_{diff} - the first order rate coefficient for the
loss of OH out of the excitation volume due to transport and
flow - in a relatively small and constant intercept in k_{1st} vs.
[Reactant] plots.

OH + C_2H_4O (ethene oxide) $\longrightarrow$ products

This reaction has been studied at total pressures around
10 Torr (Ar) and over a temperature region T = 297 -435 K.
Fig. 3 is a summary of first order rate constants vs. C_2H_4O
concentration. The slope of these lines corresponds to the
second order rate coefficient, for which we obtain at room
temperature:

$$k_2 (297 \ K) = (8.0 \pm 1,6) 10^{-14} \ cm^3/molecule \ s,$$

where the error limits pertain to the $\geq$ 90 % confidence limit
(3σ). The only other measurement of this rate constant has
been reported by Zetzsch [7] in acceptable agreement with our
result.
    Fig. 4 is an Arrhenius representation of k_2. The line
shown in this figure corresponds to the expression

$$k_2 (T) = (1,1 \pm 0.4) 10^{-11} \exp(- 1460/T) cm^3/molecule \ s$$

It is of course of interest to elucidate what the primary reaction step and the subsequent oxidation mechanism in the troposphere may be. Unfortunately, there appears to be no value
available in the literature for the heat of formation of the
(cyclic) C_2H_3O radical and hence the C-H bond strength in C_2H_4O.

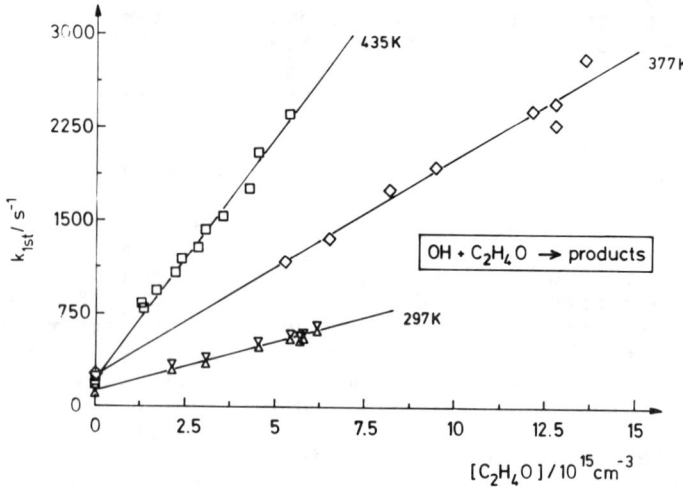

Fig. 3   Dependence of first order rate coefficients on C_2H_4O concentration

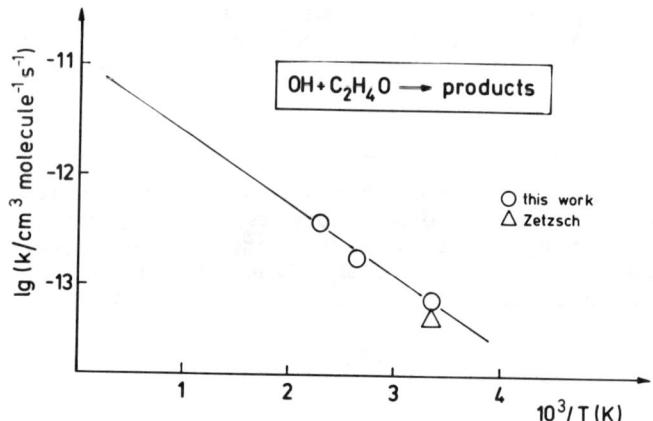

Fig. 4   Arrhenius representation of k_2

Although H atom abstraction is certainly an exothermic reaction path, it is questionable, whether the C-H bond strength in C_2H_4O is reduced suffciently compared to C_2H_4 (due to a transition form sp^2 hybridization to a distorted sp^3) to be in accordance with a 2.8 kcal/mol activation energy for H abstraction by OH. The general order of the activation energy found here, however, appears very reasonable in the light of a 5.8 kcal/mol activation energy found for the corresponding O atom reaction [8]. We are therefore tempted to write the primary reaction as

$$(2) \quad OH + C_2H_4O \longrightarrow H_2O + C_2H_3O$$

The subsequent oxidation of C_2H_3O under tropospheric conditions may then be expected to proceed in either of the two following ways (Fig. 5)

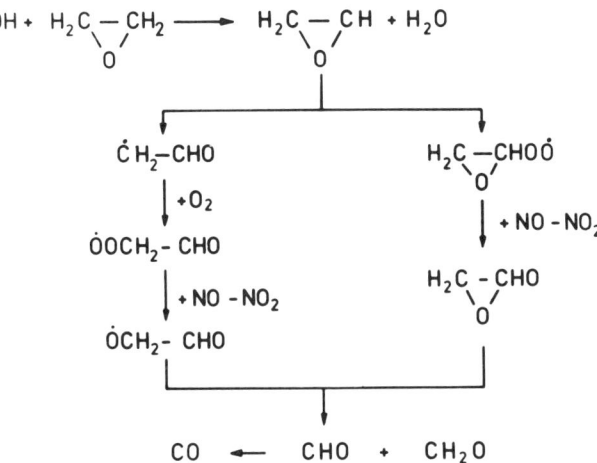

Fig. 5   Probable tropospheric oxidations pathways of ethene oxide

In either case we may expect CH_2O and CO as stable end products. Both mechanisms differ in at what stage along the oxidation chain ring opening occurs. The formation of the linear vinyloxi ($\dot{C}H_2CHO$) radical from cyclic C_2H_3O may be prompted by the high ring strain energy in C_2H_4O (26.9 kcal/mol [9]). This is supported by a recent study of the $F + C_2H_4O$ reaction in which CH_2CHO in high yield is detected by laser induced fluorescence [10]. If we assigne to CH_2CHO a heat of formation of 6 kcal/mol [10], we calculate for the overall reaction

$$(2) \quad OH + C_2H_4O \longrightarrow H_2O + CH_2CHO$$

an exothermicity of 48.6 kcal/mol. Hence, the low activation energy observed here may at least in part be due to the additional energy gain in the isomerization of cyclic C_2H_3O.

$$\underline{OH + PH_3 \longrightarrow H_2O + PH_2}$$

We have studied this reaction at 10 Torr total pressure of Ar in the temperature range T = 249 - 438 K. First order rate constants are again strictly proportional to $[PH_3]$. (Fig. 6) The temperature dependence, however, is extremely weak. From the Arrhenius representation (Fig. 7) we deduce

$$k_3(T) = (2.7 \pm 0.6)10^{-11} \exp(-155/T) \text{ cm}^3/\text{molecule s}$$

which corresponds to a room temperature rate constant of

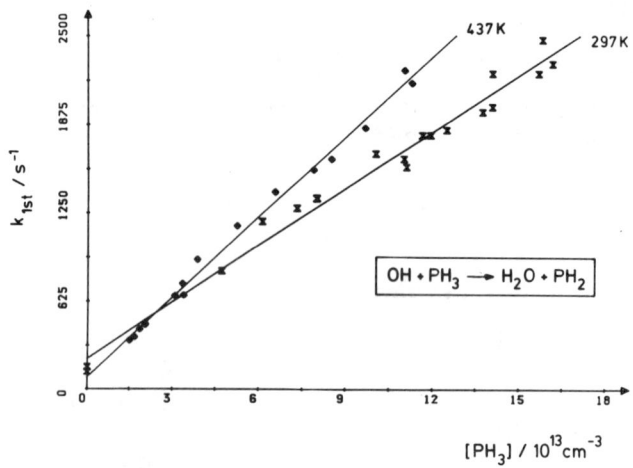

Fig. 6 Dependence of first order rate coefficients on
PH$_3$ concentration

$$k_3(297\ K) = (1.6 \pm 0.3)10^{-11}\ cm^3/molecule\ s$$

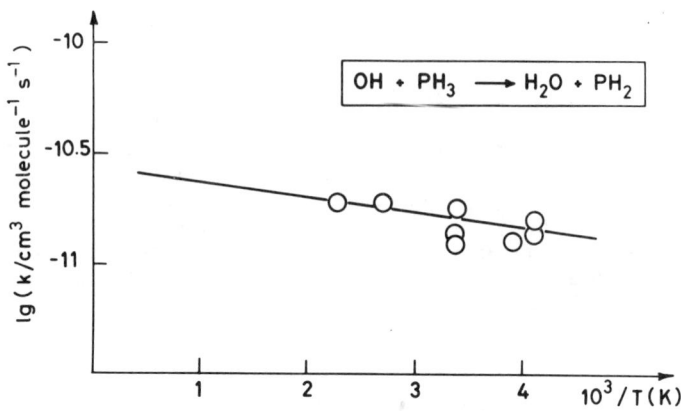

Fig. 7 Arrhenius representation of k$_3$

Hence, PH$_3$ will be degraded very quickly in the atmosphere,
having a lifetime of less than 1 day comparable to the most
reactive organic compounds (i.e. C$_3$H$_6$, CH$_3$CHO, Xylene).
    The magnitude of the rate constant as well as thermoche-
mistry ($\Delta H_f(PH_2) = (33 \pm 2)$ kcal/mol [11]) suggest that the
primary reaction proceeds by H atom abstraction:

(3)   OH + PH$_3$ $\longrightarrow$ H$_2$O + PH$_2$; $\Delta H_R = (- 35 \pm 2)$ kcal/mol.

Concerning the fate of PH_2 in the troposphere, one may only speculate by making recourse to a chemically similar species, NH_2. It is well known that NH_2, unlike alkyl radicals, shows little tendency to combine with O_2 [12,13] ($k(NH_2 + O_2)$ < 2×10^{-18} cm^3/molecule s [12]). Its dominant oxidation reaction therefore is with O_3 ($k(NH_2 + O_3)$ = 1.8×10^{-13} cm^3/molecule s [14]). Although the products of this reaction are largely undefined, HNOH (or H_2NO) are suggested as primary products [12,14]. Hence, intuitively we may expect the following reaction for PH_2:

$$PH_2 + O_3 \longrightarrow HPOH + O_2$$

Addition of NO_2 and subsequent hydrolysis could then yield hypophosphorous acid, H_3PO_2.

OH + HCN $\longrightarrow$ products

Measurements of this reaction have shown that the rate constant increases with total pressure. In order to determine $k(p,T)$ we have therefore performed the majority of our experiments in a conventional flash photolysis/resonance absorption system, which can be operated at pressures up to 1 atm. The result of this investigation is shown in Fig. 8, where the effective second order rate coefficient is plotted against the N_2 pressure

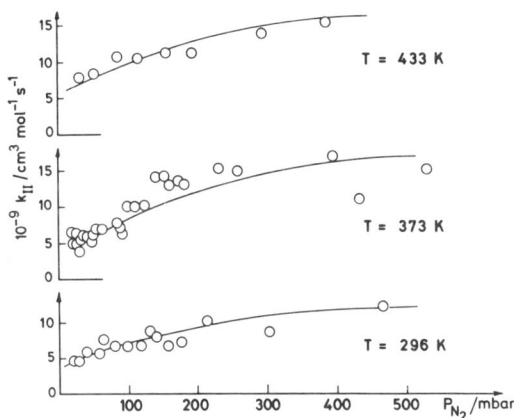

Fig. 8 Pressure dependence of the second order rate coefficient for OH + HCN $\longrightarrow$ products

The overall temperature dependence is extremely weak. In the high pressure limit, appropriate to the conditions in the lower troposphere, we obtain

$$k_4(T) = (1.2 \pm 0.5)10^{-13} \exp(-400/T) \ cm^3/\text{molecule s}$$

At higher temperatures, however, we observe a strong increase

of temperature dependence (Fig. 9), which we attribute to a change in reaction mechanism. A discussion of this will be given elsewhere [15].

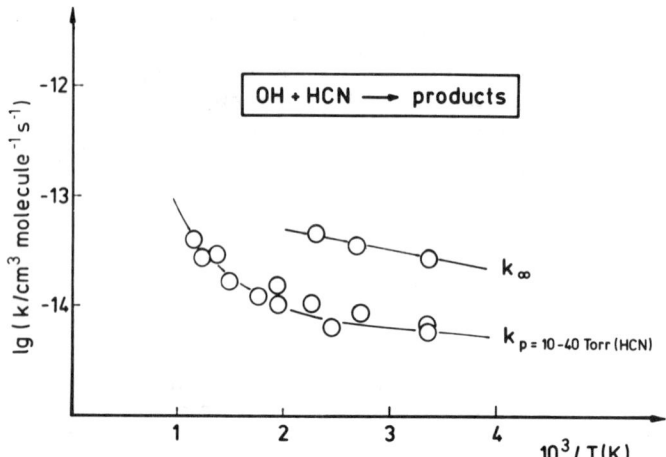

Fig. 9   Arrhenius representation of k_4

The form of k(p,T) as observed at lower temperatures leads us to conclude that the reaction between OH and HCN proceeds via addition of OH to the CN bond.

(4) $OH + HCN \longrightarrow HCNOH$

From the pressure dependence observed on the time scale of our experiments we estimate an exothermicity of this step of $\geq$ 15 kcal/mol. The rate constant for reaction (4) at room temperature and 1 atm pressure is

$$k_4(298 \text{ K}) = (3 \pm 1)10^{-14} \text{ cm}^3/\text{molecule s}$$

yielding a tropospheric lifetime of HCN in the order of 1 year.
     For the degradation of HCN in the troposphere the following mechanism may be proposed (Fig. 10). This is delineated in close proximity to the corresponding oxidation mechanism of hydrocarbons. The important feature of it is that CO + NO are probable final oxidation products and hence the CN bond is permanently split.
     It may be asked whether HCN can be considered representative for the oxidation of nitriles. We have hence also measured the rate constant for the reaction between OH and CH_3CN. At T = 295 K and a total pressure of 7 Torr we obtain

$$k (OH + CH_3CN) = (2.4 \pm 0.3)10^{-14} \text{ cm}^3/\text{molecule s}$$

This compares with the k_4^∞ value for OH + HCN. However, the corresponding value for k_4 at 7 Torr is only ~ $4 \times 10^{-15} \text{ cm}^3/$molecule s. Hence, the larger rate constant observed for CH_3CN may hint   towards a different reaction mechanism, probably

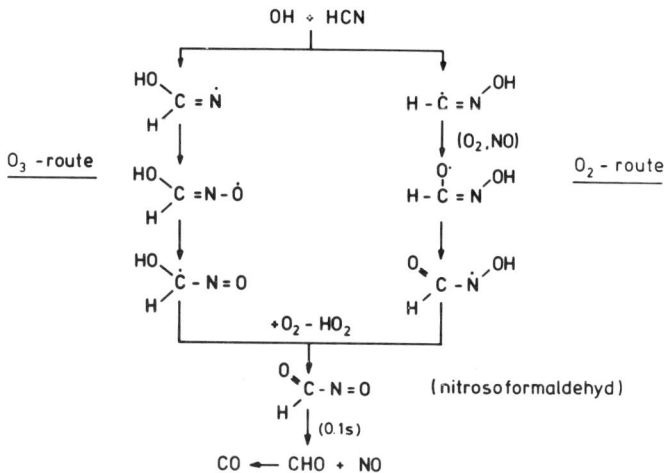

Fig. 10 Probable mechanism of the trophospheric oxidation
of HCN

resulting in initial H abstraction. The CN bond can then only
degrade later on in the oxidation chain.

ACKNOWLEDGEMENT

The authors are grateful to Prof. H. Gg. Wagner for continuous interest and valuable discussion. Support of this work
by the German Federal Environmental Agency (Umweltbundesamt),
by Bundesministerium für Forschung und Technologie (BMFT) and
by 'Heerdt-Lingler-Stiftung' is gratefully acknowledged.

REFERENCES

1. H. Gg. Wagner, R. Zellner, Ang. Chem. Int. Ed. Eng. 18,
   663 (1979)
2. R. Zellner, W. Steinert, Int. J. Chem. Kin. 8, 397 (1976);
   V. Handwerk, R. Zellner, Ber. Bunsenges. phys. Chem. 82,
   1161 (1978); G. Wagner, R. Zellner, Ber. Bunsenges. phys.
   Chem. (1981) in press.
3. B. Fritz, R. Zellner, paper presented at Chemiedozententagung (1980), Erlangen, FRG; B. Fritz, Diplomarbeit,
   Göttingen University (1980)
4. J. J. Margitan, F. Kaufman, J. G. Anderson, Int. J. Chem.
   Kin. Symp. No. 1, 281 (1975); I. W. M. Smith, R. Zellner,
   ibid. 341 (1975)
5. P. H. Wine, A. R. Ravishankara, N. M. Kreutter, R. C.
   Shah, J. M. Nicovich, R L. Thompson, D. J. Wuebbles,
   J. Geophys. Res. 86, 1105 (1981)
6. B. Fritz, K. Lorenz, R. Zellner, in preparation
7. C. Zetzsch, paper presented at Chemiedozententagung
   (1980), Erlangen
8. U. Welzbacher, Diplomarbeit, Göttingen University, 1971

9. S. W. Benson, 'Thermochemical Kinetics', J. Wiley, New York 1976
10. G. Inoue, H. Akimoto, J. Chem. Phys. $\underline{74}$, 425 (1981)
11. cited in 'Handbook of Chemistry and Physics', 60th Ed. 1979, CRC-Press
12. R. Lesclaux, M Demissy, Nouv. J. Chim. $\underline{1}$, 443 (1977); H. Kurasawa, R. Lesclaux, Chem. Phys. Lett. $\underline{72}$, 437 (1980)
13. S. W. Benson, 18th Symp. (Int.) Combustion, p. 882 (1981)
14. W. Hack, O. Horie, H. Gg. Wagner, Ber. Bunsenges. phys. Chem. $\underline{85}$, 72 (1981)
15. B. Fritz, W. Steinert, R. Zellner, to be published

OXIDATION RATES OF SO₂ IN DELIQUESCENT SALT AEROSOLS

A.G. CLARKE, P.T. WILLIAMS and M. RADOJEVIC
Department of Fuel and Energy, Leeds University, Leeds, U.K.

Summary

New experimental results have confirmed previous work showing that
the rate of oxidation of SO_2 in salt aerosols above their deliquescence points are several orders of magnitude higher than in pure
water with traces of catalysts. The salts include $MgCl_2$, NaCl and
$(NH_4)_2SO_4$ in addition to the well known catalyst $MnSO_4$. The
possible reasons for these high rates are surveyed. The influence
of the high ionic strength on the sulphite equilibria are considered
and improved methods of calculation indicated. Metal complex
formation and, in chloride solutions, the elevation of the pH by HCl
desorption are included in the model calculations. Other factors in
addition to the well established cation catalysis include salt
effects on the ion-ion reactions and anion catalysis by Cl^-. This
is believed to be due to oxygen atom transfer reactions involving
ClO^-.

1. INTRODUCTION

Studies of the aqueous oxidation of sulphur dioxide may generally be
put into two categories - measurements in relatively pure water with traces
of catalysts at levels of the order of $10^{-4}M$ or below, and measurements in
solutions of salts, probably in aerosol form, within which the concentrations can be up to 5M. The former studies are directly relevant to the
situations in cloud, fog and rain droplets while the latter studies are
directly relevant to absorption and oxidation of SO_2 by deliquescent
aerosols such as $(NH_4)_2SO_4$, and NaCl and sea salt at high humidities. The
main physical difference between the two situations is that the mass
concentration of the liquid phase in the atmosphere is several orders of
magnitude lower in the case of the deliquescent aerosols, except near 100%
RH.   This fact has led to the oxidation of SO_2 within such aerosols being
generally discounted in comparison with other mechanisms such as homogeneous
photochemical oxidation and oxidation in clouds.   Comparison between the
two sets of experimental data however shows that the reaction rates in
concentrated salt solutions are several orders of magnitude higher than in
pure water even with traces of catalysts such as Mn^{2+}.   Application of
pure water rate constants to situations such as marine aerosols is
therefore not valid and the possibility of SO_2 oxidation within such
aerosols needs closer examination.   The aim of this paper is to focus
attention on the experimental results for deliquescent aerosols, to discuss
the qualitative reasons for the high rates of reaction observed and to
indicate improved methods of calculation which must be employed if the
solution equilibria and kinetics within concentrated salt solutions are to
be properly interpreted.

2. COMPARISON OF EXPERIMENTAL RESULTS

Within our own laboratory we have studied the oxidation of SO_2 in aerosols of $MgCl_2$, NaCl and $(NH_4)_2SO_4$ at humidities up to 85% RH and gas phase concentrations of 1-20 ppm SO_2. The aerosols were prepared by spray atomisation into a 1 m^3 chamber and the reaction with radioactively labelled $^{35}SO_2$ studied by removal of filtered particulate samples for analysis using liquid scintillation counting. The details are presented elsewhere (1). The aerosols absorbed about 12±2 mg SO_2 per gram of water associated with the particles over a reaction time of 140 minutes. This was independent of the SO_2 concentration used.

Table I shows a comparison of our findings with those of other workers and with some results for trace metal catalysis and pure water oxidation. In some cases the figures have had to be estimated from data given in the original papers.

Table I.   Comparison of reaction rates in deliquescent aerosols

Reference	Electrolyte	SO_2 Conc. (ppm)	SO_4^{2-} formed mol $(kgH_2O)^{-1}$	Time (min)	Average Rate m.mol. $(kgH_2O)^{-1}$ min^{-1}
Johnstone & Moll(2)	NaCl	250	0.71	2	355
Cheng et al.(3)	NaCl	18	0.34	60	5.7
Matteson et al.(4)	$MnSO_4$	67	0.69	10	69
Haury et al.(5)	$MnSO_4$	1-2	0.71	480	1.5
Kaplan et al.(6)	$(NH_4)_2SO_4$ + $MnSO_4$	0.7	0.0025	0.75	3.3
Clarke & Williams(1)	$MgCl_2$, NaCl, $(NH_4)_2SO_4$	1-20	0.16	140	1.1
Barrie & Georgii(7)	10^{-4}M $MnCl_2$	0.9	7.5×10^{-4}	15	0.05
Beilke et al.(8)	Water + Buffer pH 3.1	(1)			6×10^{-8}

The 6 studies using concentrated salt solutions consistently give average rates of SO_4^{2-} formation greater than 1 m.mol.$(kgH_2O)^{-1}min^{-1}$ and the apparent final pH values are below 1. These rates are about 2 orders of magnitude faster than Barrie and Georgii's rates for 10^{-4} $MnCl_2$ with 0.9 ppm SO_2 which dropped below 10^{-5} mol.l^{-1}.min^{-1} once the pH reached 2. The pure water rates at low pH (Beilke et al. (8)) are six orders of magnitude lower still at a corresponding SO_2 concentration.

3. FACTORS AFFECTING THE OXIDATION RATE

The reasons for the high reaction rates must be sought in the wide range of factors which influence aqueous SO_2 oxidation. For clarity these are listed in Table II.

The discussion will not cover all these factors in detail but only those which have been most relevant in the consideration of our own work.

Ionic Strengths and Activity Coefficients

Calculation of the solution equilibria requires a knowledge of the activity coefficients. The salts listed in Table I, when in equilibrium

Table II.   Factors affecting SO_2 oxidation in aqueous solutions

---

Oxidant species	O_2, O_3, H_2O_2, NO_2 or HNO_2
Neutralizing agents	NH_3, alkaline metal oxides HCl desorption from chloride solutions
Ionic strength	Effects on S(IV) solution equilibria Salt-effects on ion-ion reaction kinetics
Catalysts	Complex formation with S(IV) species - most metals M^{2+} Promotion of electron transfer to/from SO_x^{n-} species Mn^{2+}, Fe^{3+} Promotion of O-atom transfer to/from SO_x^{n-} species Cl^-, Br^-, solid carbon Synergistic effects between several catalysts or between catalyst and different oxidants.
Inhibitors	Mechanisms as for catalysts but non-reversible thereby terminating the chain reaction.

---

with water vapour at humidities above their deliquescence point, but below
95% RH will have concentrations of about 2-5 molar.   To avoid problems of
actual volumes and densities of solutions the discussion is best carried
out in terms of molalities or $mol.(kg H_2O)^{-1}$.   Molal activity coefficients
for most of the pure salts of interest are available (9).   However what
are required are the activity coefficients of species such as H^+, HSO_3^-,
SO_3^{2-} when present as trace components in the highly concentrated solutions.
This problem is discussed by Robinson and Stokes (9) and we have adopted
one of the approaches discussed there, namely that of Guggenheim (10).
The mean activity coefficient of an electrolyte having a formula $M_{\nu_+} X_{\nu_-}$
is taken to be

$$\log_{10} \gamma_{M,X} = - A|Z_+ Z_-| I^{\frac{1}{2}}/(1 + I^{\frac{1}{2}}) + F$$

where A is a numerical constant (0.5085 at 25°C), Z_+ and Z_- are the ionic
charges and I is the ionic strength.   F is a sum of functions representing
the interactions between all possible pairs of oppositely charged ions in
the mixture:-

$$F = (\nu_+/(\nu_+ + \nu_-)) \sum_{X'} \lambda_{M,X'} m_{X'} + (\nu_-/(\nu_+ + \nu_-)) \sum_{M'} \lambda_{M',X} m_{M'}$$

where X' represents all possible anions, M' all possible cations with
molalities $m_{X'}$ and $m_{M'}$.   For solutions containing only a single electrolyte at molality m

$$F = (\nu_+ \nu_-/(\nu_+ + \nu_-)) \lambda_{M,X} m$$

so that the term in brackets is unity for 1:1 and 2:2 electrolytes or 4/3
for 2:1 or 1:2 electrolytes.   For an electrolyte $M_{\nu_+} X_{\nu_-}$ as a trace in an
electrolyte M'X' at molality m'

$$F = (a\lambda_{M,X'} + b\lambda_{M',X}) m'$$

where the constants a and b take the following values:-   1:1 or 2:2

electrolytes - a = n/2, b = 1/2; 2:1 electrolytes - a = 2n/3, b = n/3;
1:2 electrolytes - a = n/3, b = 2n/3.

To apply these formulae so as to obtain realistic estimates of the
activity coefficients at high ionic strength we have adopted the following
approach. The coefficients $\lambda_{M,X}$ etc. have been chosen to yield the
correct value of the experimental activity coefficients at I = 6. In the
absence of data on HSO_3^- or SO_3^{2-} these have been assumed to behave as Cl^-
and SO_4^{2-} respectively. So, for example, to calculate γ for H^+, HSO_3^- as a
trace component in $MgCl_2$ we have

$$F = (\lambda_{H^+,Cl^-} + \lambda_{Mg^{2+},HSO_3^-}/2)m_{MgCl_2}$$

We find

$$\lambda_{H^+,Cl^-} = 0.145, \lambda_{Mg^{2+},HSO_3^-} = \lambda_{Mg^{2+},Cl^-} = 0.279 \text{ and } F = 0.285\, m_{MgCl_2}$$

Similar calculations can be carried out for other components. For some
sulphate salts slightly negative λ values are required to fit the experimental data but we have preferred to adopt $\lambda = 0$ in such cases. The
results applied to $MgCl_2$ and NaCl solutions are shown in Figs 1A and 1B
respectively.

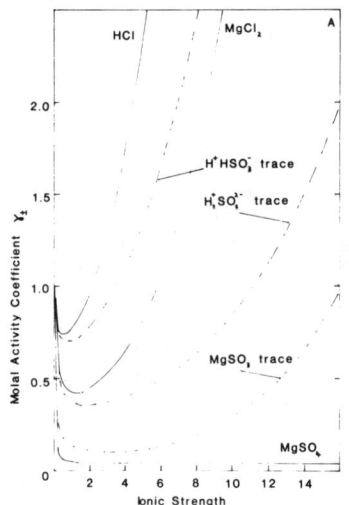 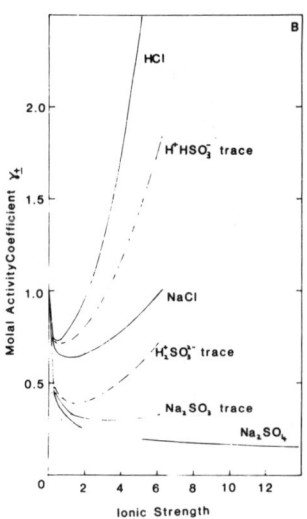

Fig. 1.   Activity coefficients for sulphite species in $MgCl_2$(A) and NaCl
(B) solutions

Complex Formation
Table III shows the relative concentrations of complexes $M^{2+}SO_3^{2-}$ or
$M^+SO_3^{2-}$ in 5 molal solutions of various salts using appropriate activity
coefficients together with the indicated equilibrium constants, SO_3^{2-} again
being assumed to behave as SO_4^{2-}.

Table III.   Complex formation in 5 molal salt solutions

Solution	NaCl	$(NH_4)_2SO_4$	$MgCl_2$	$MnSO_4$
Complex	$NaSO_3^-$ trace	$NH_4SO_3^-$	$MgSO_3$ trace	$MnSO_3$ trace
Eq. Const. K	5.0	12.7	143	191
C(complex)/C(SO_3^{2-})	0.97	2.0	469	0.8

It is clear that for Mg^{2+} the complex predominates over free SO_3^{2-} at high salt concentrations whilst with Na^+ and NH_4^+ the complex and the free SO_3^{2-} have comparable concentrations. Low activity coefficients reduce the value for Mn^{2+} despite the high K_A. Since the complexes have lower ionic charge we can anticipate lower activation energies for reactions with other negatively charged ions and faster rates than for free SO_3^{2-}.

Solution Equilibria

The equilibrium between gaseous SO_2 and the liquid phase concentrations of the various species can be calculated, as has been done by many authors, by setting up the hydrogen ion balance, substituting the appropriate expressions for the species concentrations in terms of $[H^+]$ and the equilibrium constants, and solving the cubic equation that results. The equilibria and the constants we have used are given in Table IV. They include the HSO_4^-/SO_4^{2-} and $M^{n+}/MSO_4^{-(2-n)}/SO_4^{2-}$ reactions of sulphate formed in the oxidation reaction. HCl desorption was also permitted in chloride containing solutions and this is discussed in the next section. Atmospheric CO_2 absorption was permitted but was generally found to be unimportant in most of the cases considered where the pH was below 5.

Table IV. Equilibria and equilibrium constants at 25°C

1.	$SO_2(g)$	$= H_2SO_3$	H	$=$	1.24 mol. l^{-1}. atmos^{-1}
2.	H_2SO_3	$= H^+ + HSO_3^-$	K_2	$=$	1.74×10^{-2} mol. l^{-1}
3.	HSO_3^-	$= H^+ + SO_3^{2-}$	K_3	$=$	6.24×10^{-8} mol. l^{-1}
4.	$M^{n+} + SO_3^{2-}$	$= MSO_3^{-(2-n)}$	K_4	$=$	(see Table III) l. mol^{-1}
5.	HSO_4^-	$= H^+ + SO_4^{2-}$	K_5	$=$	1.04×10^{-2} mol. l^{-1}
6.	$M^{n+} + SO_4^{2-}$	$= MSO_4^{-(2-n)}$	K_6	$=$	(see Table III) l. mol^{-1}
7.	$HCl(g)$	$= HCl(l)$	H_{HCl}	$=$	19 mol. l^{-1}. atmos^{-1}
8.	$HCl(l)$	$= H^+ + Cl^-$	K_8	$=$	1.3×10^6 mol. l^{-1}
9.	$CO_2(g)$	$= CO_2(l)$	H_{CO_2}	$=$	3.4×10^{-2} mol. l^{-1}. atmos^{-1}
10.	$CO_2(l) + H_2O$	$= H^+ + HCO_3^-$	K_{10}	$=$	4.45×10^{-7} mol. l^{-1}

Fig. 2A shows the influence of salt concentration on the SO_3^{2-} and HSO_3^- concentrations at equilibrium with 1 ppm SO_2. Fig. 2B shows the corresponding data after 0.1 molal H_2SO_4 has been formed in the oxidation reaction. As can be seen there are changes of 1 - 2 orders of magnitude in the concentrations. At low concentrations and at high concentrations with NaCl and $(NH_4)_2SO_4$ the concentrations of SO_3^{2-} are increased due to the low activity coefficients. With $MgCl_2$ the activity coefficients at high concentrations become > 1 and depress $[SO_3^{2-}]$. However if HCl desorption is included as indicated by the dashed lines in Fig. 2 the concentrations of both HSO_3^- and SO_3^{2-} are markedly increased. This is especially significant for $MgCl_2$.

Complex formation results in a net addition of S(IV) to the solution. The concentrations of the complexes in 5 molal salt solution can be obtained from the end points of the SO_3^{2-} curves in Fig. 2 and the ratios of complex to free SO_3^{2-} given in Table III.

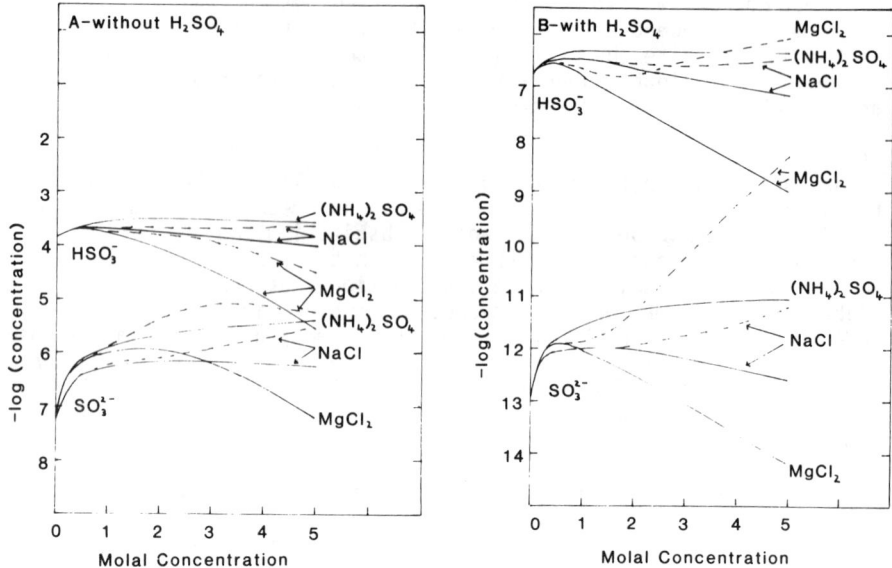

Fig. 2. The variation of $[SO_3^2]$ and $[HSO_3^-]$ with salt concentration for 1 ppm SO_2. A. Initial Conditions. B. After 0.1 molal H_2SO_4 formed. The dashed curves indicate the results if HCl desorption from 1 mg salt m^{-3} is permitted.

HCl Desorption

In calculating the overall hydrogen ion balance for solutions containing chloride ions allowance can be made for HCl loss as follows. Assuming the mass concentration of dry salt aerosol is z ($g.m^{-3}$), the molality of the droplets formed m and the molecular weight of the salt w, then the mass of water associated with the droplets is z/wm ($kg.m^{-3}$). The equilibrium partial pressure of HCl gas is given by

$$P_{HCl} = [H^+][Cl^-]\gamma_{H^+,Cl^-}^2 / H_{HCl} K_{HCl}$$

where H_{HCl} is the Henry's Law constant for HCl absorption to HCl(liq) and K_{HCl} is the equilibrium constant for HCl(liq) dissociation into H^+ and Cl^-. Assuming P_{HCl} is zero before desorption begins and that only a small fraction of Cl^- in the solution is lost then the net reduction in $[H^+]$ can be expressed as

$$\Delta[H^+] = 40.90 \frac{m \ w \ [H^+][Cl^-]}{z \ H_{HCl} \ K_{HCl}} \gamma_{H^+,Cl^-}^2 \qquad \text{at } 25°C$$

For a pure salt MCl_n, $[Cl^-] = nm$ and $\Delta[H^+] \propto m^2$. Table V gives some values for NaCl solutions typical of our experimental situation, 1 mg salt m^{-3}, and within the atmosphere, 10 μg salt m^{-3}.

Table V. The effect of HCl desorption on the pH of 5 molal
Sodium Chloride aerosols.

H$_2$SO$_4$ present molal	1 ppm SO$_2$, 1 mg NaCl m^{-3}			0.01 ppm SO$_2$, 10 µg NaCl m^{-3}		
	No desorption	With desorption		No desorption	With desorption	
	pH$_a$	pH$_a$	P$_{HCl}$(ppb)	pH$_a$	pH$_a$	P$_{HCl}$(ppb)
0	3.83	4.21	0.018	4.79	5.90	3.9x10^{-4}
10^{-3}	2.54	3.31	0.14	2.54	5.23	1.7x10^{-3}
10^{-2}	1.56	2.32	1.4	1.56	4.24	1.7x10^{-2}
10^{-1}	0.65	1.35	13	0.65	3.24	0.16

The pH$_a$ values are calculated from the hydrogen ion activity rather
than concentration. In our experimental situation P$_{HCl}$ could have
reached 10 - 20 ppb and the increase in pH$_a$ would have been 0.7. In the
atmospheric situation the effect is greater due to the higher gas to
liquid ratio for the aerosol system. pH$_a$ changes of over 2 units occur
with P$_{HCl}$ values of only 10^{-3} ppb once 10^{-3} molal H$_2$SO$_4$ has been formed.

Kinetic Salt Effects
    The oxidation mechanism involves several steps in which two negati-
vely charged ions react. According to the activated complex theory of
reaction rates the number of collision complexes formed depends on their
free energy which is affected by surrounding ions. A 3- or 4- complex
will have a very low activity coefficient at high ionic strength corresponding to a decreased free energy. An acceleration of the reaction rate
may therefore be expected. Some numerical calculations were given in an
earlier note (Clarke (11)).

Catalysis
    The catalysis of SO$_2$ oxidation by Mn^{2+} has received most attention.
It is generally taken to involve complex formation with SO$_3^{2-}$ and also
electron transfer to and from the metal ion. In concentrated salt
solutions where the reaction rate is already fast small amounts of Mn^{2+}
have little effect. This was observed by Kaplan et al. (6) for an
aerosol of (NH$_4$)$_2$SO$_4$ with a few percent by weight of MnSO$_4$ at 90% and
also by Clarke and Williams (1) for 360 ppm Mn^{2+} in the MgCl$_2$ droplets at
50% RH. In both cases the rates were only accelerated by factors of
about 2 - 3. The high oxidation rates in MnSO$_4$ aerosols are no doubt
partly due to the catalytic effect of the cation but this does not help
to interpret the high rates in other aerosols. Cations such as Na$^+$, NH$_4^+$
and Mg^{2+} are unlikely to be involved in electron transfer reactions.
    To clarify this situation Clarke and Radojevic (12) studied the
oxidation of sodium sulphite in bulk solutions containing added electrolytes. The results clearly demonstrated that the anions Cl$^-$ and Br$^-$
have a catalytic effect and the oxidation rate was found to be proportional to S(IV)^2 [Cl$^-$]$^{1.3}$ [H$^+$]$^{-0.5}$. Since neither of these ions can
undergo reaction with O$_2$ their influence must come later in the oxidation
mechanism. It is suggested that they could be involved in oxygen atom
transfer reactions from and to SO$_x^{n-}$ species which occur in the Backstrom
mechanism or its variations proposed by several authors:-

$$SO_3^{2-} + M^{n+} \rightarrow SO_3^{-\cdot} + M^{(n-1)+}$$

$$SO_3^{-\cdot} + O_2 \rightarrow SO_5^{-\cdot}$$

$$SO_5^{-\cdot} + SO_3^{2-} \rightarrow SO_4^{-\cdot} + SO_4^{2-}$$

or

$$SO_5^{-\cdot} + Cl^- \rightarrow SO_4^{-\cdot} + ClO^-$$

$$ClO^- + SO_3^{2-} \rightarrow Cl^- + SO_4^{2-}$$

$$SO_4^{-\cdot} + SO_3^{2-} \rightarrow SO_4^{2-} + SO_3^{-\cdot}$$

Such an involvement of ClO^- and BrO^- ions is consistent with the finding that neither F^- or I^- have a catalytic effect. Neither of these ions form stable oxyhalide ions. The discovery that carbon particulates within aqueous droplets can catalyse sulphite oxidation (Chang et al. (13)) may be cited as another example of the promotion of O-tranfer. In that case SO_3^{2-} may react with a surface oxygen complex.

In addition to such cation and anion catalysis there may be synergistic effects either between two catalysts (e.g. Mn^{2+} and Fe^{2+}, see (7)) or between a catalyst and a specific oxidant species. For example Clarke and Williams (1) found that Cu^{2+} acted as an inhibitor for the oxidation of $MgCl_2$ droplets. NO_2 acted as an effective additional oxidising agent. But Cu^{2+} and NO_2 together gave an extremely high rate of reaction.

4. CONCLUSIONS

A number of factors have been considered which, taken together, can probably account for the very high oxidation rates of SO_2 in deliquescent salt aerosols. Extrapolation to environmental situations is not easy but it is clear that the reaction rates in marine aerosols, for example, will be higher than generally believed. Peterson and Seinfeld (14) used pure water rate constants for O_2 and O_3 oxidation in a simulation of marine aerosols. Their conclusion that the oxidation is too slow to be of much significance should therefore be reassessed.

ACKNOWLEDGEMENT

The financial support of the Science Research Council in the form of research grants and a studentship for P.T.W. is gratefully acknowledged.

REFERENCES

1. A.G. Clarke and P.T. Williams. Submitted to Atmospheric Environment.
2. H.F. Johnstone and A.J. Moll. Ind. Eng. Chem. 52, 861 (1960).
3. R.T. Cheng, M. Corn and J.O. Frohliger. Atmos. Env. 5, 987 (1971).
4. M.J. Matteson, W. Stöber and H. Luther. Ind. Eng. Chem. Fund. 8, 677 (1969).
5. C. Haury, S. Jordan and C. Hofman. Atmos. Env. 12, 281 (1978).
6. D.J. Kaplan, D.M. Himmelblau and C. Kanaoka. Atmos. Env. 15, 763 (1981).
7. L.A. Barrie and H.W. Georgii. Atmos. Env. 10, 743 (1976).
8. S. Beilke, D. Lamb and J. Müller. Atmos.Env. 9, 1083 (1975).
9. R.A. Robinson and R.H. Stokes. Electrolyte Solutions, 2nd Edition revised, Butterworths, London (1965).

10. E.A. Guggenheim. Phil. Mag.
11. A.G. Clarke.   Atmos. Env. (1981). In press.
12. A.G. Clarke and M. Radojevic.   Submitted to Atmospheric Environment.
13. S.G. Chang, R. Toossi and T. Novakov.   Atmos. Env. 15, 1287 (1981).
14. T.W. Peterson and J.H. Seinfeld.   Adv. in Env. Sci. and Tech. 10, 125 (1980).

PHOTOCHEMICAL EXPERIMENTS UNDER SIMULATED ATMOSPHERIC CONDITIONS

C. LOHSE, H. STANGL, M. PAYRISSAT, H. RAU, G. OTTOBRINI, B. NICOLLIN
Commission of the European Communities
Joint Research Center, Ispra Establishment

Summary

Irradiation of ambient air in 1 m^3 Teflon bags under different meteorological conditions with the sun as light source demonstrates the photochemical
origin of ambient ozone at Ispra. No evidence could be found in favour of
ozone transport from polluted areas or from the unpolluted troposphere.
Preliminary experiments indicate that the system CO - NO_x could very well
be an important ozone precursor in ambient ispra air.

1. INTRODUCTION

The tropospheric ozone budget has since the classical paper by Junge
(1) been the subject of numerous investigation. Junge described ozone as
a rather inert stratospheric tracer. The main source of tropospheric ozone
resulted from transport of ozone rich air from the stratosphere into the
troposphere, with the major loss process being defined as reaction on contact with land surface areas or oceans. This view is favoured by Junge
and Czeplak (2); Pruchniewiez (3); Fabian et al. (4); Chatfield et al. (5);
Newell (6); Reiter (7); Danielsen et al. (8) and Singh et al. (9).
Another position now held is that gas phase photochemical production
and destruction of ozone in the troposphere may be more important than
injection of ozone from the stratosphere. This view is held by Chamaides
et al. (10); Crutzen (11); Gidel et al. (12); Fishman et al. (13), (14),
(15), (16). The proposed ozone formation reactions are the following:

$$CO + OH \longrightarrow CO_2 + H \quad (1)$$
$$H + O_2 + M \longrightarrow HO_2 + M \quad (2)$$
$$HO_2 + NO \longrightarrow OH + NO_2 \quad (3)$$
$$NO_2 + h\nu \longrightarrow NO + O \quad (4)$$
$$O + O_2 + M \longrightarrow O_3 + M \quad (5)$$
$$Net: \quad CO + 2O_2 + h\nu \longrightarrow CO_2 + O_3 \quad (6)$$

The above investigations have been concentrated on the unpolluted troposphere above the boundary layer, e.g. from 2 km to 10 km above sea level, leaving the observed ozone values in the lower 2 km of the unpolluted troposphere as a somewhat open question. Thus we do not have a very accurate
picture of the stratospheric contribution to the ground level ozone values,
and how this contribution could be taken into consideration for the various
air quality standards for oxidants. This question is particularly important for rural or semi-rural areas such as Ispra, where the influence of
anthropogenic activity on ambient oxidant levels is difficult to demonstrate or even quantify, but where nevertheless high ozone values are
observed.
Reported here are ozone data collected during the first 8 months of 1981,
using as experimental technique the irradiation of ambient air samples
in Teflon bags.

2. EXPERIMENTAL

1,2 m^3 Teflon bags were used in all experiments. The bags were prepared by heat sealing of the Teflon film material followed by an outside mechanical support of the seal. This gives quite durable bags, which can be used outside for several weeks even under windy conditions. The support consists of two strips of 2 mm Teflon, one on each side of the seal, held together by stapler for each 2 cm. This prevents the seal from breaking, and no synthetic organic compound is touching the surface of the bag. The method makes it possible to obtain bags filled with high purity air. We do not find it possible to use Teflon bags outside without reinforcement of the seal, it is much too fragile.

The purity level of the air inside the Teflon bag has been investiga-ted by GC/MS using $C_6 F_6$ as internal standard ($C_6 F_6$ has been found photochemically very stable). Approximately ten compounds could be detected, most of them typical of gasoline-diesel fractions. The major compound was identified as toluene, 0,36 ppb, and the level of all impurities together was less than 2 ppb.

It is important that the bags do not touch any synthetic organic material during the experimental procedure. For instance leaving an inflated bag on a PVC coated floor during the weekend gave a 1.5 ppb level of vinylchlorid inside the bag; similar, a 5 cm long polystyrene tubing in the air supply system gave 2.0 ppb styrene inside the bag.

Irradiation of NO/NO_2 mixtures and zero air in Teflon bags (sunlight irradiations) did not give a net production of ozone, even if only NO_2 was used.

Irradiation of zero air alone resulted in all cases in a small production of ozone, typical 10 - 15 ppb in 5 hours with intense sunshine. We are at present not able to relate the amount of ozone to any other measured parameter.

Ambient air samples were fanned into the bags and, whenever possible, bags were filled in parallel to secure identical air samples in the experiments. The bags were filled at 08.30 - 09.00 hours local time followed by sunlight irradiation until 16.30 hours. The bags were cleaned by washing with air after the experiments.

3. RESULTS AND DISCUSSION

The results from the irradiations are presented in Table I and Table II. Experiment N° 1, Bag 1 and Bag 2 (E 1 - B 1 - B 2), E 2 - B 1 - B 2, E 3 - B 1 - B 2, E 4 - B 1 - B 2, E 7 - B 2, E 8 - B 2, E 13 - B 2 and E 15 - B 2 are confrontations of ozone formation between a morning air parcel inside a 1 m^3 Teflon bag and the ozone formation in the free outside air on a typical hazy and sunny day at Ispra. The results show that the ozone formation inside and outside the bag is very much the same. As the outside air in the evening cannot possibly be the same as taken for the morning bag sample, these findings demonstrate a very homogeneous ozone formation in the air masses surrounding Ispra. We suggest a photochemical origin of the ozone with the sunlight acting on a ubiquitous precursor, whereas transport of ozone seems to be excluded.

TABLE I
Ozone formation from irradiation of ambient Ispra air in 1 m^3 Teflon bags

Experiment n°		Ozone values, p.p.b. morning = before, evening = after irradiation			Comments
		Bag 1	Bag 2	outside air	
1	morning	21	21	24	Bags filled in parallel
	evening	32	37	37	
2	morning	7	7	9	Bags filled in parallel
	evening	21	21	25	
3	morning	34	34	30	Bags filled in parallel
	evening	60	60	53	
4	morning	18	18	17	Bags filled in parallel
	evening	36	36	38	
5	morning	0*	0*	15	*Bags from experiment n° 4
	evening	23	23	55	left in lab. over weekend
6	morning	–	–	–	
	evening	71	71	75	
7	morning	35*	36**	35	*Bag from exp.6 left overnight
	evening	46	80	71	**Fresh ambient air sample
8	morning	30*	15	16	*Ambient air sample,
	evening	39	60	60	irradiated the day before
9	morning	17	17	17	Föhn. Excellent visibility.
	evening	108	109	43	Bags filled in parallel
10	morning	3	3	3	Föhn. Excellent visibility.
	evening	86	90	41	Bags filled in parallel
11	morning	14	13	17	Föhn
	evening	128	130	47	Bags filled in parallel
12	morning	30*	38**	38	Föhn.*Ambient air from prior
	evening	42	105	52	evening, not irradiated. **Fresh air
13	morning	30*	40	45	*Air sample from a pine
	evening	55	125	125	forest
14	morning	45*	53	45	*Air sample from a pine
	evening	75	130	100	forest
15	morning	14*	23	20	*Air sample from a pine
	evening	45	93	88	forest

TABLE II

Ozone formation from irradiations of ambient Ispra air
in 1 m^3 Teflon bags with the addition of CO and NO_x

Experiment N°	Bag	Bag content	O_3 max. ppb
16	–	Outside air. (free atmosphere)	105
	A	Outside air + 2 ppm CO	129
17	–	Outside air. (free atmosphere)	56
	A	Outside air + 15 ppb NO_x	60
	B	Outside air + 15 ppb NO_x + 1 ppm CO	70
	C	Outside air + 15 ppb NO_x + 5 ppm CO	88
18	–	Outside air. (free atmosphere)	95
	A	Zero air	23
	B	Outside air + 15 ppb NO_x	120
	C	Outside air + 15 ppb NO_x + 1 ppm CO	128
	D	Outside air + 15 ppb NO_x + 5 ppm CO	152
19	–	Outside air. (free atmosphere)	85
	A	Outside air + 1 ppm CO	135
	B	Outside air + 1 ppm CO + 8 ppb NO_x	162
20	–	Outside air. (free atmosphere)	60
	A	Outside air + 1 ppm CO	95
	B	Outside air + 1 ppm CO + 8 ppb NO_x	87
21	–	Outside air. (free atmosphere)	100
	A	Zero air + 1 ppm CO	15
22		Outside air. (free atmosphere)	45 – 50 (wind)
	A	Zero air + 15 ppb NO_2	25
	B	Zero air + 15 ppb NO_2 + 5 ppm CO	50

The photochemical origin of the ozone is supported by another finding. The ozone formation was on the few very clear days with a Föhn event found to be much more efficient inside the bag than in outside air (E 9 - B1 -B2, E 10 - B 1 - B 2, E 11 - B 1 - B 2, E 12 - B 2), e.g. the difference was most marked during episodes of strong mixing of the outside air. Under these experimental conditions, one would expect dilution of the precursor in the outside air, with a subsequently reduced formation of ozone in the outside air, what in fact is found. The generally higher rate of reaction is due to the increase in light intensity.

The above evidence leaves in our opinion only a very small possibility for the transport of ozone: The outside air contained in all experiments less ozone or the same amount of ozone as the bag sample. If transport was important, one would at least on a few occasions expect an ozone rich air parcel to surround the bag. This has to our surprise not been observed until now.

The homogeneity of the ozone forming processes and the lack of evidence of transport is very difficult to explain if primary anthropogenic emission was the precursor for the ozone, whereas an ubiquitous and slow reacting compoing would give the observed ozone formation pattern.

The compound(s) cannot be very reactive as we have demonstrated (17) that the integrated sun intensity is a limiting factor for ozone formation, and shown that our ozone maximum arrives in the late afternoon as compared to the ozone maximum observed around noon in areas with heavy road traffic and high concentration of reactive compounds.

The limiting effect of the sun intensity shows in addition that the ozone precursors are not used up during one day of irradiation. On the other hand they cannot be present in higher concentrations, as the reactivity of an air sample is much less on the second day of irradiation (E 5 - B 1 - B 2, E 7 - B 1, E 8 - B 1).

The results of E 13, E 14 and E 15 are interesting as they demonstrate the low level of ozone precursors inside an air sample taken 1 m above ground level in a pine forest compared to an air sample taken 50 m above ground level on the roof of a building.

We want to suggest CO as a possible precursor for the ozone formation reacting via the reactions (1) - (5). We do know that the currently accepted rate constants for these reactions (16) are apparently too slow, however the results presented in Table II demonstrate nevertheless that CO seems to have a reactivity more or less as the compound(s) responsible for ozone formation in Ispra air. It is in this connection interesting to note that the air above acidic soil (pine forest) known as a sink for CO (18) shows reduced reactivity.

We are at present not able to measure very low concentrations of CO to demonstrate a correlation between CO consumption and ozone production of an air sample inside a Teflon bag. The method will however soon be available in our laboratory. Until then and even though the results presented here are not completely consistent, we want to suggest that CO should be taken seriously into consideration as a possible precursor for rural ozone.

References

1. C.E. JUNGE
   Tellus, 14, 363-377 (1962)

2. C.E. JUNGE and G. CZEPLAK
   Tellus, 20, 422 (1968)

3. P.G. PRUCHNIEWICZ
   Pure Appl. Geophys., 106-108, 1058-73 (1973)

4. P. FABIAN and P.G. PRUCHNIEWICZ
   J. Geophys. Res., 82, 2063-73 (1977)

5. R. CHATFIELD and H. HARRISON
   J. Geophys. Res., 82, 5969-76 (1977)

6. R.E. NEWELL
   Quart. J. Roy. Meteorol. Soc., 89, 372 (1963)

7. E.R. REITER
   Proceedings of the Joint Symposium on Atmospheric Ozone,
   Environm. Prot. Agency, Dresden, 1976

8. E.F. DANIELSEN and V.A. MOHNEN
   J. Geophys. Res., 82, 5867-77 (1977)

9. H.B. SINGH, F. L. LUDWIG and W.B. JOHNSON
   Atmospheric Environm., 12, 2185-96 (1978)

10. W.L. CHAMEIDES and J.C.G. WALTER
    J. Geophys. Res., 78, 8751-60 (1973)

11. P.J. CRUTZEN
    Tellus, 26, 47-57 (1974)

12. L.T. GIDEL and M. SHAPIRO
    J. Geophys. Res., 85, 4049-58 (1980)

13. J. FISHMAN and P.J. CRUTZEN
    J. Geophys. Res., 82, 5897-5906 (1977)

14. J. FISHMAN and P.J. CRUTZEN
    Nature, 274, 855-58 (1978)

15. J. FISHMAN, W. SEILER and P. HAAGENSON
    Tellus, 32, 456-63 (1980)

16. J. FISHMAN, S. SOLOMON and P.J. CRUTZEN
    Tellus, 31, 432-46 (1979)

17. H. STANGL, C. LOHSE, M. PAYRISSAT, B. VERSINO, B. NICOLLIN,
    G. OTTOBRINI and H. RAU, Proceedings of the First European Symposium of
    Physico-Chemical Behaviour of Atmospheric Pollutants
    B. VERSINO and H. OTT ed., Ispra, 16-18 October 1979

18. R.E. INMAN, R.B.INGERSOLL & E.A. LEVY, Science, 172, 1229-31 (1971)

OZONE FORMATION IN A SMOG CHAMBER AT CONSTANT AND VARIABLE LIGHT INTENSITY

K. HENRICH, H. LIPPMANN, U. SCHURATH AND W. WENDLER
Institut für Physikalische Chemie der Universität Bonn, D-53 Bonn,
Wegelerstr. 12, Federal Republic of Germany

Summary

Smog chamber experiments were carried out in authentic urban air,
to establish a data base for the validation of a lumped mechanism of
sufficient simplicity to be incorporated into a detailed atmospheric
transport model. The smog chamber was connected to a large Tedlar bag
which served as storage volume for urban air of constant composition,
either for consecutive fillings of the evacuable chamber, or simply
for dilution (dilution rate of the smog chamber: 3.7 %/h). The 124
experiments comprise simulations at constant light intensity, and
summer day simulations. The latter were run for at least 24 h, using
a switching program for the 16 fluorescent lamps around the glass
chamber, to approximate the time pattern of the NO2 photolysis frequency on a typical summer day. Multiday irradiations were also
carried out. The results are discussed and compared with the preliminary version of a lumped mechanism.

1. INTRODUCTION

    Computer models which handle both transport and chemical reactions of
trace gases in the atmosphere are essential for assessing the effect of
control strategies on oxidant formation. However, the chemistry of polluted air is extremely complex, the main cause being the structural diver-
sity of the non methane hydrocarbons (NMHC's) which occur in variable
mixing ratios. Their reactions with OH radicals trigger an avalanche of
radical reactions with largely unknown branching ratios and rate constants.
The quality of a model depends on the proper representation of this complex scheme of reactions which produce peroxi radicals for NO oxidation,
and thus for oxidant buildup. These considerations have led to the formulation of two distinct classes of chemical models:
    1) Detailed models incorporate as many organic trace gases as pos-
sible, using individual degradation schemes for each (1). Reasonable
estimates must be made of unknown branching ratios and rate constants, and
of the radical pairings which may or may not be omitted from the scheme.
    2) Lumped models incorporate a fairly complete set of inorganic
reactions, but the organic chemistry is reduced to a few symbolic hydrocarbons and aldehydes, and their degradation schemes. This keeps the
number of reactants and reactions reasonably low. Different lumping
schemes have been devised (2,3,4).
    Detailed models are clearly more attractive to the reaction kineti-
cist, because they can be constantly updated with improved rate constants
and degradation schemes. However, the necessarily very large number of
reactions make these models too voluminous on a computer to be suitable
for integration into a more detailed transport model. The requirement of
few non-steady-state species and not too many reactions can be met by
lumped mechanisms, but there is no way of assessing their "correctness" on
an a priori basis. Therfore, comparison with smog chamber data has become

the most widely used (though disputed) method of model validation.

This paper summarizes results of 124 smog chamber simulations in authentic urban air, and discusses their suitability for model validation.

2. EXPERIMENTAL

The smog chamber is a Duran glass cylinder of 0.425 m^3 volume (5), shown schematically in <u>fig. 1</u>. The interior is illuminated by 16 fluorescent lamps which emit strongly in the NO2 dissociation region 300 - 400 nm and provide an NO2 photolysis frequency of 0.52/min. The spectrum is intensity deficient in the region of aldehyde and ozone photodissociation, the 313 nm mercury lines being just weakly transmitted by the Duran glass. A water cooler and blower keep the chamber temperature typically below 30°C, but higher temperatures can be reached on hot days. The chamber can be evacuated to 10 mtorr, then filled from a Tedlar bag which contains 10 m^3 urban air when freshly inflated. The bag was kept in a dark room. The ana-

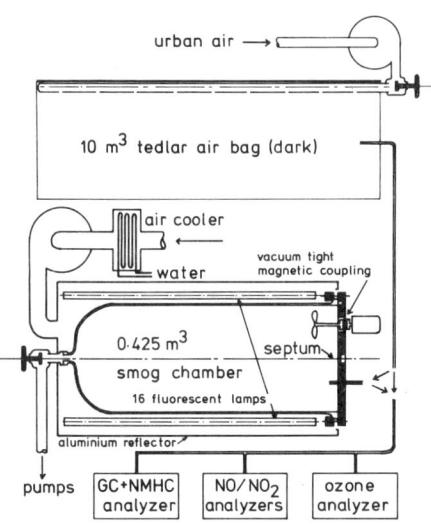

Figure 1: Schematic of the smog chamber and Tedlar bag.

lytical instruments consisted of chemiluminescent analyzers for ozone, NO, and NO2, of a Siemens U 180 automatic GC for hydrocarbons, and temporarily of an Enraf Nonius Total NMHC Analyzer. PAN was measured in 2 ml air samples by GC/ECD. Two NO/NO2 analyzers were run in parallel: one was equipped with a photochemical converter which is specific for NO2 (6); the other contained a hot carbon converter which partially reduces the organic nitrogen compounds and nitric acid also, thus giving higher NOx readings in the presence of these compounds. Because of the high sampling rates of the pre-concentrating NMHC analyzers, these could be only used prior to and after the simulations. The chemiluminescent analyzers were run continuously, causing a chamber dilution of 3.7 %/h which was made up by air from the Tedlar bag. The ozone instrument was calibrated by the UV absorption technique (7), while NO and NO2 diluted in Argon were injected into the reactor with gas tight syringes to calibrate the NOx analyzers. The first order decay constant of 1 ppm ozone in synthetic air was less than 0.1/h, but a value of 0.2/h was found for the decay of 25 ppb ozone in urban air after 15 h of irradiation. This provides an upper limit for the heterogeneous loss rate of ozone during the simulations. NO, NO2 and/or a standard mix of hydrocarbons could be injected into the chamber prior to the simulations to increase the trace gas load above that of the urban air in the Tedlar bag. The 15 hydrocarbons in the mix were typical of urban air, consisting of alkanes, alkenes and aromatics in a carbon atom mixing ratio 100:43:74. Aldehydes were not added.

3. IRRADIATIONS AT CONSTANT LIGHT INTENSITY

Smog chamber experiments are often carried out at a constant light intenstiy. The observed ozone maxima as function of the initial NOx and

NMHC concentrations are compared with computer generated isopleths. We used 5 Tedlar bag fillings of urban air for 56 irradiations with all 16 fluorescent lamps. The results are shown in figure 2. The hatched circles represent irradiations with authentic urban air (chamber air = dilution air). In the other runs, NO+NO2 and/ or the NMHC mix were added to the chamber, but not to the dilution air in the bag. In series of experiments with the same matrix air, the effect of increasing the NOx and NMHC concentrations separately was tested. The results were horizontal and vertical cuts through the isopleths. Fig. 3 shows the dependence of ozone on NOx at constant NMHC $\cong$ 200 ppbC; it differs from current lumped models (e.g. the EKMA mechanism (2), or the carbon bond mechanism (4)) which predict a de-

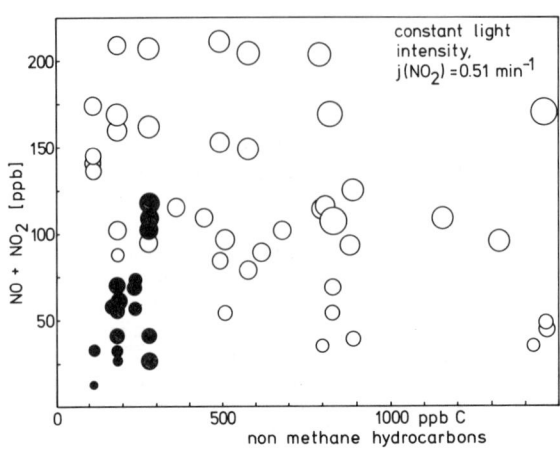

Figure 2: Ozone maxima at constant light in-tensity.The diameters are proportional to the ozone maxima (cf. fig. 8 for the scale).

crease of the ozone maximum at higher NOx concentrations. In contrast, the dependence of ozone on NMHC at (nearly) constant NOx, fig. 4, is much weaker than predicted by these models (chamber effects such as the true NO2 photolysis frequency, the heterogeneous ozone loss rate, and the dilu-tion rate, were of course taken into account in the calculations).

It should be mentioned that the NO2 concentration in the Tedlar bag decreased slowly over storage times of 10 days, while the change in NO was slow and equal to the theoretical oxidation rate by molecular oxygen. At the same time, a change in reactivity occured: In a series of 6 irradia-tions with consecutive chamber fillings from the same matrix air in the

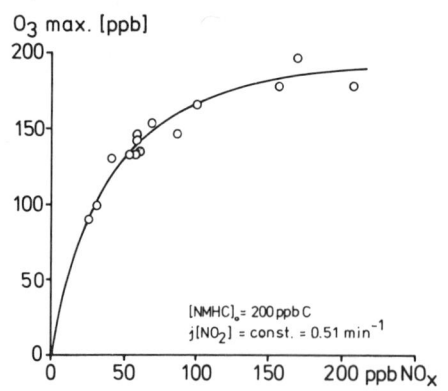

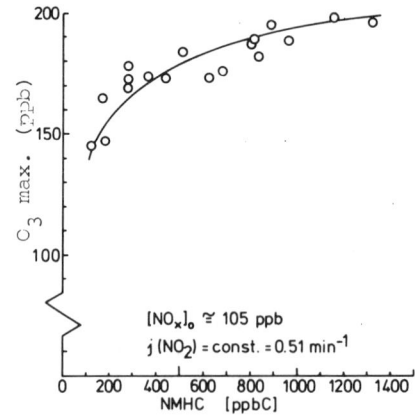

Figure 3: Ozone maixma as function of NOx at constant light intensity

Figure 4: Ozone maxima as function of NMHC at constant light intensity and (nearly) constant NOx.

Tedlar bag, the time of the NO2 maximum decreased from initially 65 min. to 8 min. after 3 days of storage. The ozone maximum was also affected, but in the opposite sense: it decreased from 155 ppb initially to 132 ppb after 3 days, probably because of the NO2 loss in the Tedlar bag during storage (cf. fig. 3). The enhancement of the NO oxidation rate at short irradiation times could be modelled surprisingly well on the assumption that about 50 % of the NO2 lost in the bag was transformed into HONO, which photolyzes at a rate of 0.08/min., generating OH radicals for the initiation of the chain reactions.

4. INTENSITY DEPENDENCE OF THE OZONE MAXIMUM

The photochemical reactor offers the possibility of irradiating urban air of essentially constant trace gas contents with different light intensities, by switching on different numbers of the fluorescent lamps. The ozone maxima of 8 simulations which were carried out within 5 days are plotted in fig. 5 as function of the NO2 photolysis frequency. Consecutive smog chamber fillings from the same urban air mass in the bag were irradiated. The NOx concentration given in fig. 5 was read off the NOx detector with the hot carbon converter; the sum of NO and NO2 was 50 ppb, decreasing to 40 ppb during the 5 day period. The results indicate that the ozone maximum in the chamber depends on the light intensity less than linearly. Qualitatively the same result was obtained by a model calculation (the lumped mechanism of this calculation was not identical with the updated version referred to below, but the intensity dependence is quite insensitive to the details of the model).

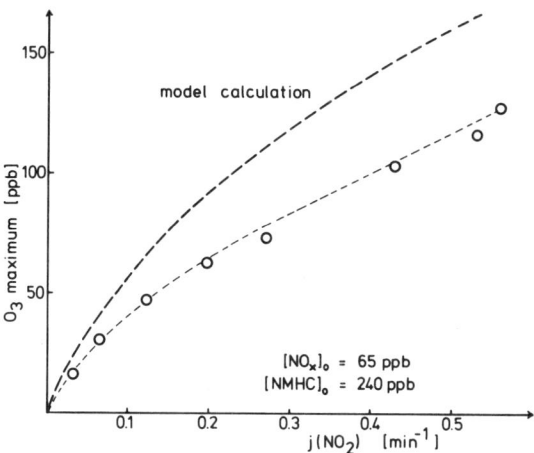

Figure 5 : Dependence of the ozone maximum on the NO2 photolysis frequency.

5. SUMMER DAY SIMULATIONS

Irradiations at constant light intensity are difficult to compare with the real atmosphere: The reactive trace gases (NOx, olefins) are so rapidly consumed that the steady state between chemical conversion and replacement by dilution is established after 5 to 10 hours, depending on the amount of NOx and NMHC injected into the chamber in excess of their concentrations in the Tedlar bag, and thus in the dilution air. Fig. 6 illustrates the simulation of a typical summer day in the smog chamber: The lamp switching program models the NO2 photolysis frequency as function of time after sunrise for 7 July, 50°N, cloudless sky.

Fig. 7 summarizes the ozone maxima for summer day simulations with genuine urban air (chamber air = dilution air). From these experiments onward, all simulations were carried out with freshly sampled urban air, to avoid interference from HONO and possibly other reactive species which might build up in the Tedlar bag. Only the dilution air was supplied by

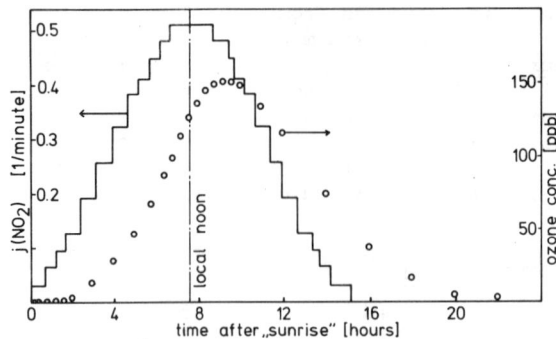

Figure 6: Simulation of the NO2 photolysis frequency as function of time after "sunrise" in the smog chamber. A typical ozone profile obtained with urban air containing 98 ppb NO, 17 ppb NO2 and 470 ppb NMHC is also shown.

the air bag. The NMHC:NOx ratio of the randomly collected air samples ranged from 18:1 to 3.2:1, the average being about 6:1. In order to cover a wider range of precursor concentrations, NOx and/or NMHC were injected into the chamber in other experiments. This complicates the situation, because the trace gas contents of the dilution air differs at random from the initial concentrations in the smog chamber. Figure 8 summarizes the results of 57 simulations, including those with enhanced trace gas concentrations in the chamber. The ozone maxima in unaltered urban air, which are summarized separately in figure 7, are set off by hatching.

It is instructive to compare the dependence of the ozone maxima on NMHC at (approximately) constant NOx, figure 9, with the corresponding data for constant light intensity, figure 4. It is obvious that figure 9 is in much better agreement with the strong dependence of the ozone maxima on NMHC predicted by current models. This is attributed to the accelerating effect of the reactive hydrocarbons on ozone buildup, which is important when both the total dose and the duration of the irradiation are limited on a natural day, but plays a minor role in irradiations at high constant light intensity, and long duration.

The dependence of the ozone maxima on NOx for summer day simulations, which corresponds with figure 3, is not shown here. For a range of NMHC concentrations between 310 and 354 ppbC, the ozone maxima as function of NOx are of the order of 140 ppb around 90 ppb NOx, decreasing again for still higher NOx concentrations, in clear contrast to figure 3 where such a decrease is absent, but in agreement with current chemical models.

6. MULTIDAY IRRADIATIONS

The more we learned about the peculiarities of lumped models, the more we realized that the development of a useful model ("useful" being defined as a model which permits exptrapolations to be made beyond the validation range) is in fact a parameter fitting exercise. Although the inorganic reactions and some other important radical reactions

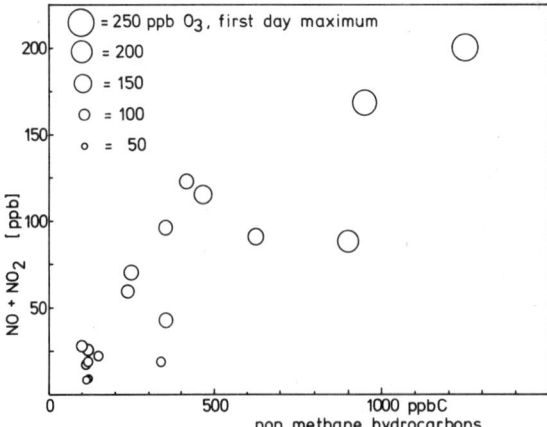

Figure 7: Ozone maxima for summer day simulations in genuine urban air.

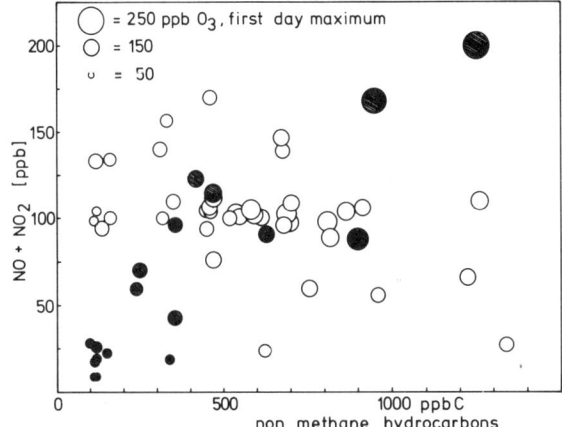

Figure 8: Same as figure 7, but including simulations with enhanced NOx and NMHC concentrations in the smog chamber. The ozone maxima in unaltered urban air are set off by hatching.

Figure 9: Ozone maxima for summer day simulations as function of NMHC at (nearly) constant NOx.

are given their true rate constants, several other parameters in the mechanism can be adjusted arbitrarily within certain limits. Among these are the photolysis frequencies of the aldehydes which are not known for most smog chambers, their initial concentrations, the lumping of the atmospheric hydrocarbon mix into "representative" species, the radical yields of their reactions with ozone, and some heterogeneous reactions in smog chambers which cannot be measured.

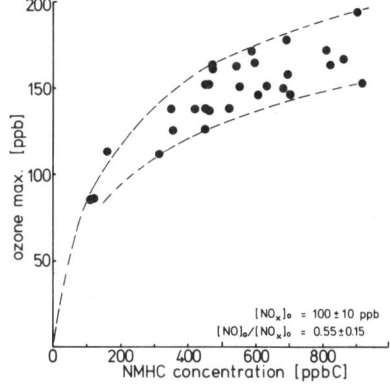

It is not too difficult to devise and adjust a model which can reproduce the ozone maxima as function of NMHC and NOx for summer day simulations within experimental scatter. It is much more difficult to also reproduce the concentration profiles of ozone, NO, and NO2 correctly. Multiday irradiations put a model to a still more rigorous test, since the depletion of the more reactive trace gases on the first day is expected to develop its effects on the second and following days. It depends primarily on the hydrocarbon and aldehyde lumping schemes whether a model is capable of reproducing these effects.

We have carried out 4 multiday simulations (between 3 and 4 days long) with freshly sampled urban air, using the Tedlar bag for dilution. Figure 10 shows the ozone profiles of the first and second days of such an experiment, for initial concentrations of 98 ppb NO, 17 ppb NO2, 250 ppbC alkanes + alkenes, and 220 ppbC aromatic hydrocarbons in the smog chamber as well as in the Tedlar bag. The dotted lines are computed ozone profiles for a particular lumped mechanism which assumed initial concentrations of 4 ppb HONO, 38 ppb "formaldehyde", and 28 ppb "acetaldehyde". The concentrations and photolysis frequencies of these compounds were varied within reasonable limits to match the rapid oxidation of NO after sunrise which is characteristic of the summer day simulations (cf. figure 11a). The calculated ozone maxima are too high, but the particular model does a reasonably good job in simulating the characteristic shapes of the ozone profiles on the first, second, and third day (not shown). The NO and

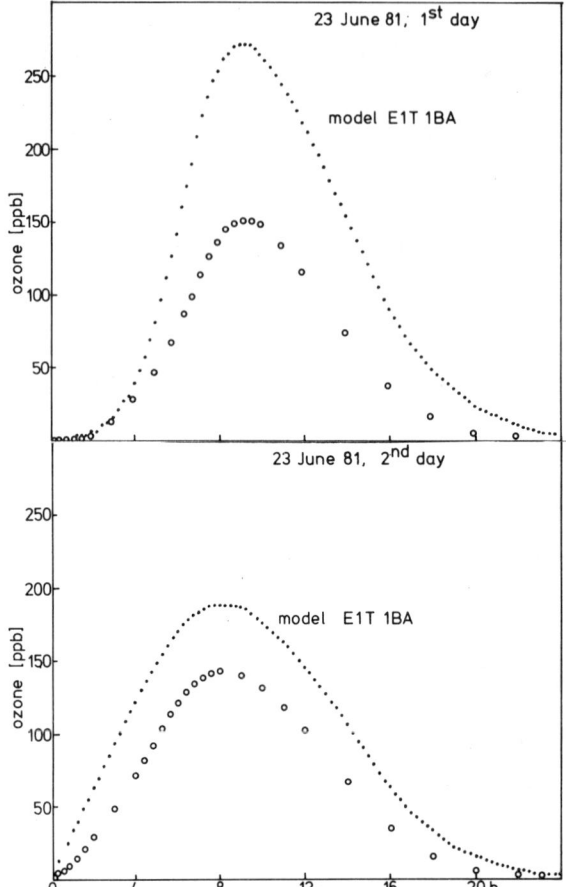

Ozone profiles
of a multiday irradiation
of urban air (see text
for details). The dotted
lines are modelling
results.

NO2 profiles of the first
day are plotted in figure
11a. The calculation is
in qualitative agreement
with experiment, but the
NO2 loss later in the day
is overestimated. This is
brought out more clearly
still in figure 12 which
shows various NOx profiles
of the second day. The
NOx deficit of the model
calculation is probably
responsible for the in-
correct ratio of the ozone
maxima on the first and
second days (dotted lines
in figure 10). Also, an
unrealistically large
amount of NOx is tied up
as PAN in the particular
model, although the hot
carbon converter did
indeed suggest the pre-
sence pf NOx compounds
other than NO and NO2.
The corresponding NOx
profiles of the first
day, figure 11b, show that the "excess NOx" remains negligible until about
7 hours of irradiation, but does not correspond with the PAN concentration
which is plotted in arbitrary units in figure 11b. We estimate from an
earlier calibration of the GC/ECD that the PAN concentration did not exceed
1 - 2 ppb in this particular experiment.

7. TEMPERATURE EFFECTS

We mentioned earlier that the "noon" temperature in the smog chamber did
not exceed 30° C under normal conditions. However, during the multiday
irradiation discussed in the last section, the ambient temperature in
the laboratory happened to exceed 32° C, causing the smog chamber temperature to increase even further. The temperature profile of the second day
is plotted in figure 13.

It is well known now that PAN is thermally unstable, or in other
words, that the equilibrium constant of the reversible reaction

$$(1,-1) \qquad PAN \;\; \underset{k-1}{\overset{k1}{\rightleftarrows}} \;\; CH_3CO_3 + NO_2$$

has a strong negative temperature coefficient (8), as illustrated by the
solid line in figure 13. This explains why the concentration porfile of
PAN (dotted line in figure 13) exhibits a double maximum. At temperatures

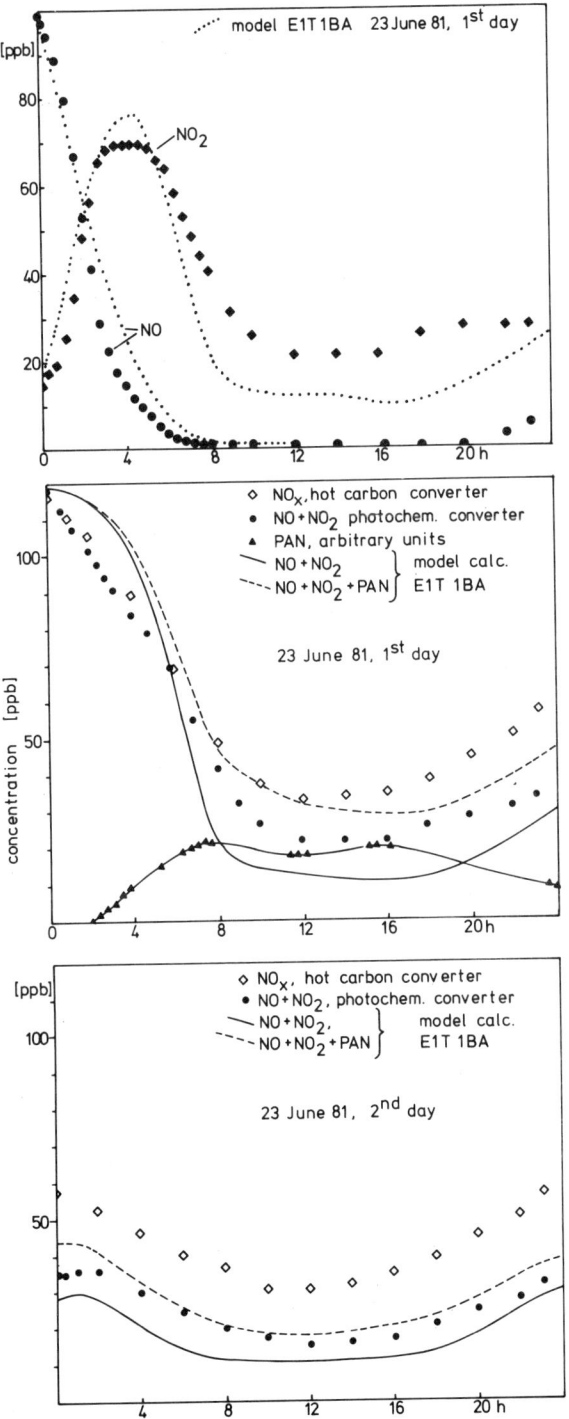

Figure 11a:

NO and NO2 profiles for the first day of a multi-day simulation. The dotted lines are modelling results.

Figure 11b:

NOx and PAN profiles for the same day, including some modelling results.

Figure 12:

Same as figure 11b, but for the second day. The PAN profile is shown in figure 13.

in excess of about 300 K, the loss rate of the peroxy acetyl radicals by reactions other than (-1) becomes comparable to, and finally exceeds their creation rate, causing the PAN concentration to pass through a minimum at the time of the highest temperature in the smog chamber, which occured about 9 hours after "sunrise".

Temperatures as high as in this simulation do not normally occur in the atmosphere of Middle Europe. However, the smog chamber runs with large temperature variations during "daylight hours" allow a further test to be made of the lumped mechanism. For the purpose, the current experimental version of our mechanism (alluded to in the previous sections) was modified to accommodate the temperature dependence of reaction (1) (and of the thermal decomposition of N2O5), as well as the observed temperature profile of figure 13. This

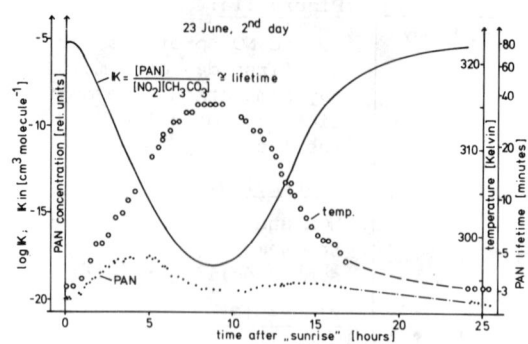

Figure 13:

PAN and temperature profiles
for the second day. The solid
line is the calculated equilibrium constant of PAN,
which is also proportional
to the lifetime of PAN for
irreversible thermal decomposition, e.g. in the presence of a large excess of
NO.

resulted in an equal enhancement of the daily ozone maxima by 15 %, but
did not shift the positions of the maxima, nor did it change the general
appearance of the profiles, except for a slight improvement of the curvature in the ozone profiles during the "early morning" hours of the second
and third days. The NO and NO2 profiles did not change significantly.

PAN concentration profiles can also be generated by our model. However, the organic reaction mechanism of the current version is based on
a mixture of two symbolic "propenes" of equal structure (termed PROP1 and
PROP2, the latter being a factor 5 slower in its OH reactions, and completely unreactive towards ozone), simulating the complex reactivity
spectrum of the atmospheric hydrocarbon mix. It is difficult to compare
PAN measured in the smog chamber with the symbolic "PAN" of the model,
which stands for various organic nitrogen compounds. In fact, the calculated PAN profiles did not agree with the observed ones, except for the
double maximum which did show up in the temperature dependent calculation.

In the model calculations, "PAN" continues to be produced during the
early night, in disagreement with experiment, figure 13. The reason is
found in the reaction of PROP1 with ozone, which is assumed to produce OH
radicals, among other stable and unstable products. Their reactions with
acetaldehyde and propene constitute a peroxiacetyl radical source. In
reality, ozone reactions seem to be rather poor OH sources. However,
complete elimination of this doubtful OH source from the mechanism resulted in a failour of the model to reproduce the rapid buildup of ozone
before noon. We suspect that the same experience has been made by other
designers of lumped models who always include this OH source.

8. CONCLUSIONS

Smog chamber simulations have been carried out in real urban air
down to low levels of pollution, in order to furnish a data base for the
validation of lumped chemical models. In addition to changing the NOx and
NMHC concentrations, the light intensity and its time pattern have also
been varied systematically. The validation can be further refined by
including multiday irradiations and temperature dependences.

It seems extremely difficult, if not impossible, to devise a lumped
model which is capable of modelling this wide range of conditions equally
well. However, even if such a model were found, the question arises
whether it could model the real atmosphere as well. Some peculiarities
of smog chambers, such as the heterogeneous ozone loss and dilution
effects, can be measured and made allowance for in calculations. Others,
however, are treated as fitting parameters in the model, and cannot be
easily adjusted to atmospheric conditions. E.g., the proper formulation

of the nature and strengths of the various radical sources poses a problem. The photolysis frequencies of the aldehydes which fit the chamber dara are higher than their true values, as far as these are known, and cannot be simply replaced by atmospheric data.

We conclude that these shortcomings limit the predictive capabilities of lumped models significantly. They are useful in conjunction with detailed transport models, to obtain qualitative information about the buildup and distribution of oxidants in an area, but quantitative predictions should be regarded with caution.

Acknowledgement:

This project is supported by the Umweltbundesamt. A grant of the Minister für Wissenschaft und Forschung des Landes Nordrhein-Westfalen for the development of the photochemical NO2 converter and detector is gratefully acknowledged.

REFERENCES

(1) R.G. Derwent and Ø. Hov, Env. Sci. Technol. 14, 1360 (1980)
(2) G.Z. Whitten, H. Hogo and M.C. Dodge, EPA-600/8-78-014a (1978)
(3) A.H. Falls, G.J. McRae and H.J. Seinfeld, Int. J. Chem. Kinet. XI, 1137 (1979)
(4) G.Z. Whitten, H. Hogo and J.P. Killus, Env. Sci. Technol. 44, 690 (1980)
(5) U. Schurath, Kapitel VII in "Forschungsbericht 79-104 02 502/03/04", im Auftrag des Umweltbundesamtes, August 1979
(6) B. Jesser, Diplomarbeit, Bonn 1979; H.-J. Goede, Diplomarbeit, Bonn 1980
(7) K.H. Becker, A. Heindrichs und U. Schurath, Staub-Reinhalt. Luft 35, 326 (1975)
(8) U. Schurath and V. Wipprecht, 1st European Symposium Phys. Chem. Behaviour of Atmospheric Pollutants, Proceedings edited by B. Versino and H. Ott, p. 157 (1980)

ATMOSPHERIC PHOTOCHEMISTRY: KINETICS AND MECHANISM OF REACTIONS BETWEEN AROMATIC OLEFINS AND HYDROXYL RADICAL.

C. CHIORBOLI, A. MALDOTTI, C.A. BIGNOZZI and V. CARASSITI
Centro di Fotochimica del CNR, Istituto Chimico dell'Università - Via Borsari 46, 44100 FERRARA, ITALY.

SUMMARY

Rate constant values of the reaction of OH radical with styrene, alfa- and beta-methyl styrene and beta-dimethyl styrene, have been obtained by a relative method using isooctane as reference hydrocarbon; the k_{OH} values are (units: cm^3 molecule^{-1} sec^{-1}): $(5.3\pm0.5)\times10^{-11}$, $(5.3\pm0.6)\times10^{-11}$, $(6.0\pm0.6)\times10^{-11}$, $(3.3\pm0.5)\times10^{-11}$ respectively. The principal oxidation products of these aromatic olefins are: Benzaldehyde - Formaldehyde for styrene, Benzaldehyde - Acetophenone for alfa-methylstyrene, Benzaldehyde - Acetaldehyde for beta-methylstyrene, and Benzaldehyde - Acetone for beta-dimethylstyrene; a kinetic treatment of the experimental data shows that the aromatic olefins are stoichiometrically oxidized to the above products, suggesting that the OH attack occurs only on the aliphatic moiety of the molecule; in the reaction mechanism, this rate determining OH attack leads to the formation of 1-hydroxy-2-phenyl-2-ethenyl radical.

INTRODUCTION

Aromatic hydrocarbons constitute a significant fraction of the total amount of organic compounds in ambient air. However, at present, the status of the knowledge of the kinetics and mechanism of the elementary reactions involved in the atmospheric photooxidation of aromatic hydrocarbons is insufficient to incorporate this class of compounds into a simulation model of air pollution. In this field, the determination of the kinetics and mechanism of the reactions between hydroxyl radical and aromatic olefins is of particular relevance.

In this paper, we report, some results about the mechanism of OH attack on aromatic olefins and the rate constants of reaction of phenyl-ethene (styrene), 2-phenyl-1-propene (amethylstyrene), trans 1-phenyl-1-propene (βmethylstyrene) and 2-methyl-1-propene (βdimethylstyrene) with hydroxyl radical.

EXPERIMENTAL

The apparatus and techniques employed have been previously described in detail (1), and will only be briefly summarized here. Irradiation of the hydrocarbon $-NO_x-$ air system was carried out in a polimethylmetaacrilate smog-chamber of 760 1 volume. The light intensity, measured as the rate of NO_2 photolysis in air (2), was 0.10 ± 0.02 min^{-1}; during the irradiation, the chamber was kept at 298 ± 2 K by external air conditioning. Relative humidity was fixed at about 65%.

All gaseous reactants were sampled with a vacuum system from BAKER cylinders and injected into the chamber using a PTFE line.

Liquid reactants were injected using microsyringes.

Aromatic olefins, ketones, benzaldehyde and isooctane (reference hydrocarbon) were measured with a PERKIN-ELMER F. 17 gas-chromatograph with flame ionization detector, using an all glass two meter 1/4 inch column packed with 8% fluid silicone on chromosorb W 80-100 mesh.

The NO_x and O_3 concentrations were measured with a Monitor-Labs mod. 8440 chemiluminescence NO_x analyzer and a Monitor-Labs mod. 8410 chemiluminescence O_3 analyzer. The two chemiluminescence analyzer were calibrated with a Monitor-Labs 8500 dynamic calibrator.

The initial reactant concentrations used were: aromatic olefins 1.1 ppm, isooctane 1.1 ppm, NO 1.7 ppm and NO_2 0.1 ppm.

RESULTS AND DISCUSSION

The utilization of the relative method (3) in the determination of the rate constants of reactions 1), 2), 3) and 4):

$$C_6H_5CH=CH_2 \quad + \quad OH \xrightarrow{k_1} \text{products} \qquad 1)$$

$$C_6H_5C(CH_3)=CH_2 \quad + \quad OH \xrightarrow{k_2} \text{products} \qquad 2)$$

$$C_6H_5CH=CHCH_3 \quad + \quad OH \xrightarrow{k_3} \text{products} \qquad 3)$$

$$C_6H_5CH=C(CH_3)_2 \quad + \quad OH \xrightarrow{\quad k_4 \quad} \text{ products} \qquad \text{4)}$$

requires a negligible O_3 concentration during all the irradiation time, so that losses of aromatic olefins via reaction with O_3 are negligible. Since NO inhibits the building up of O_3 concentration through the fast reaction 5)

$$NO \quad + \quad O_3 \xrightarrow{\hspace{2cm}} NO_2 \quad + \quad O_2 \qquad \text{5)}$$

the choice of a suitable NO to hydrocarbon initial concentration ratio is sufficient to avoid the competition of O_3 with OH for reaction with aromatic olefins.

The absolute rate constant of reactions 1) to 4) may be derived from the rate of disappearance of the investigated hydro carbon relative to that of isooctane (2,2,4-trimethylpentane), using a reported value of $3.73 \times 10^{-12} cm^3 molecule^{-1} sec^{-1}$ (4) for the reaction

$$isoC_8H_{18} \quad + \quad OH \xrightarrow{\quad k_6 \quad} \text{products} \qquad \text{6)}$$

The so obtained absolute rate constant values are $(5.3 \pm 0.5) \times 10^{-11} cm^3 molecule^{-1} sec^{-1}$ styrene, $(5.3 \pm 0.6) \times 10^{-11} cm^3 molecule^{-1} sec^{-1}$ αmethylstyrene, $(6.0 \pm 0.6) \times 10^{-11} cm^3 molecule^{-1} sec^{-1}$ βmethylstyrene, $(3.3 \pm 0.5) \times 10^{-11} cm^3 molecule^{-1} sec^{-1}$ βdimethylstyrene.

When the aromatic olefin - NO mixtures in air were irradiated the principal oxidation products of the hydrocarbons were found to be: benzaldehyde and formaldehyde from styrene; acetophenone and formaldehyde from αmethylstyrene; benzaldehyde and acetaldehyde from βmethylstyrene; benzaldehyde and acetone from βdimethylstyrene. These findings are agreement with some partial previous results (5).

Complications met in the calibration procedure for acetaldehyde and formaldehyde did not allow the quantitative determination of their concentrations during the photochemical runs so that quantitative results were obtained only for benzaldehyde, acetophenone and acetone.

Formation of oxidation products other than those reported

above has not been observed, in particular no oxidation products of the aromatic ring were detected.

These results, together with the high values of k_{OH} for all the four aromatic olefins (similar to those of conjugated aliphatic olefins), suggested that, in reactions 1)-4), OH attack occurs only on the aliphatic moiety of the organic molecule.

Since to oxidation mechanism of these compounds seems to be similar to that of other olefins, the overall oxidation process can be well summarized as follows:

$$\underset{C_6H_5}{\overset{R}{>}}C{=}C\underset{R''}{\overset{R'}{<}} \; + \; OH \xrightarrow{\;k_i\;} \; \underset{C_6H_5}{\overset{R}{>}}C{=}O \; + \; \underset{R''}{\overset{R'}{>}}C{=}O \qquad 7)$$

where R, R' and R'' stand for H or CH_3; i = 1, 2, 3, 4.

When one of the carbonilic compounds formed is benzaldehyde, as in the cases of styrene, βmethylstyrene or βdimethylstyrene, the concentration of the produced aromatic aldehyde increases up to a maximum value and then decreases, indicating that the aldehyde also undergoes oxidation.

We have observed that, in the absence of nitrogen oxides, benzaldehyde is not decomposed during the irradiation; therefore its disappearance should be attributed to a reaction with hydroxyl radical:

$$C_6H_5CHO \; + \; OH \; \xrightarrow{\;k_8\;} \; products \qquad 8)$$

Reactions 7) and 8) can be schematized as follows:

$$\underset{C_6H_5}{\overset{H}{>}}C{=}C\underset{R''}{\overset{R'}{<}} \; + \; OH \xrightarrow{\;k_i\;} \alpha\left(\underset{C_6H_5}{\overset{H}{>}}C{=}O\right) \; + \; OH \xrightarrow{\;k_8\;} products \; 9)$$

where k_i = 1, 3 or 4, $\alpha \leq 1$.

Since the OH radical concentration can be considered to be constant for the first 4 - 5 hours of irradiation, (where the aromatic olefin decay is practically exponential) a kinetic treatment of eq.9) allows the calculation of aldehyde concentration as a function of time through eq.10):

$$\left[C_6H_5CHO \right] = \frac{\alpha\, k_i \left[C_6H_5CHCR'R'' \right]_o}{\left(k_8 \left[OH \right] - k_i \left[OH \right] \right)} \left(e^{-k_i \left[OH \right] t} - e^{-k_8 \left[OH \right] t} \right) \quad 10)$$

A sample calculation for styrene is reported in fig. 1, where the concentration time profile (continuous line) calculated from eq. 10) with $\alpha = 1$ is compared with that experimentally obtained (circles).

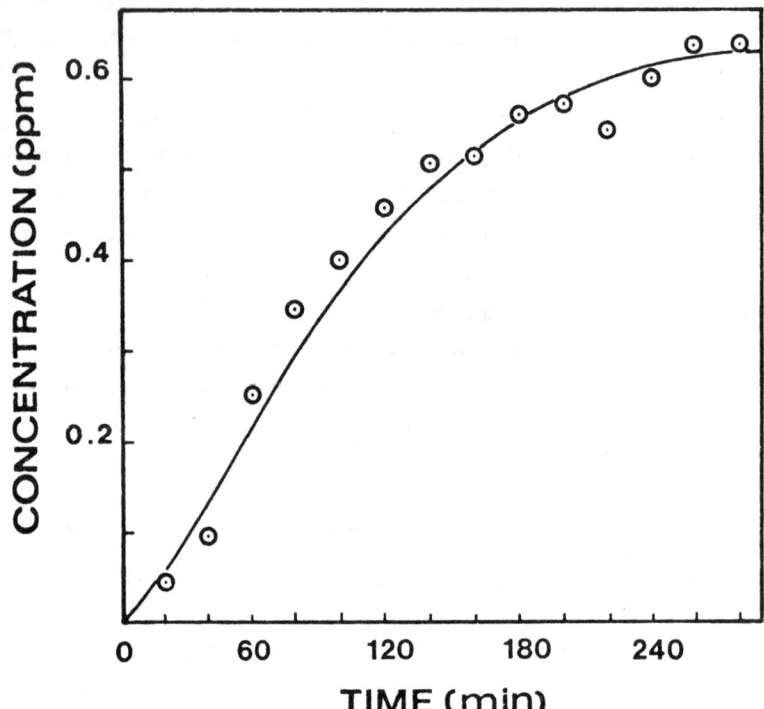

FIG. 1 - Tipical fit of the experimental benzaldehyde
concentration - time profiles (circles) using
eq. 10) (continuous line); with $\alpha = 1.0$,
$k_i = 5.3 \times 10^{-11}$ cm^{3}molecule^{-1}sec^{-1},
$k_c = 12.8 \times 10^{-12}$ cm^{3}molecule^{-1}sec^{-1} (see text).

As to the other carbonilic compounds (formaldehyde, acetaldehyde, acetophenone and acetone) formed, complications met in the calibration procedure for formaldehyde and acetaldehyde (see above), as well as the lack in the knowledge of k_{OH} for acetophenone and acetone, did not allow similar comparisons between calculated and experimental concentration - time profiles. However, qualitative results seem to indicate that the behavior of all the carbonilic compounds is quite similar to that of benzaldehyde.

The best fit to the experimental data for all the concentration-time profiles of benzaldehyde calculated from eq. 10) is $\alpha = 1.03 \pm 0.15$ for styrene, $\alpha = 0.98 \pm 0.14$ for βmethylstyrene and $\alpha = 1.05 \pm 0.15$ for βdimethylstyrene (the indicate error limits are the overall error limit calculated over five runs for each aromatic olefins).

REFERENCES

(1) - A. Maldotti, C. Chiorboli, C.A. Bignozzi, C. Bartocci and V. Carassiti: Int.J. Chem. Kinet., 12, 905 (1980).

(2) - C.H. Wu, H. Niki: Environ. Sci. Technol., 9, 46 (1975).

(3) - G.J. Doyle, A.C. Lloyd, K.R. Darnall, A.M. Winer and J.N. Pitts: Environ. Sci. Technol., 9, 238 (1975).

(4) - N.R. Greiner: J. Chem. Phys., 53, 1070 (1970).

(5) - J.M. Heuss, W.A. Glasson: Environ. Sci. Technol., 2, 1109 (1970).

AEROSOLS

Chairman: G. MADELAINE

RESULTS OF 11-YEARS' MEASUREMENTS OF AEROSOL

PARTICLE SPECTRA AT 0.7, 1.8, AND 3.0 KM ALTITUDE

R. REITER (Director), R. SLADKOVIC
and W. CARNUTH
Fraunhofer-Institut für Atmosphärische Umweltforschung
Garmisch-Partenkirchen, FRG

Summary

For more than 11 years, the aerosol particle spectrum is recorded per
working day at our 3 neighboring mountain stations Garmisch (0.7 km),
Wank peak (1.8 km), and Zugspitze peak (3.0 km). Used are therefor
5-fold double stage impactors after Junge and Jaenicke. Evaluation of
particle numbers deposited on the five slides happens through the automatic, computer-controlled Zeiss Viodeomat TV-Microscope.
Represented and interpreted is the dependence of the particle spectra
at the 3 stations on altitude, season, total air pollution, vertical
exchange, visibility, and relative humidity.

1. BRIEF DESCRIPTION OF THE RESEARCH WORK

The study was designed to statistical processing and meteorological
parameterization of the aerosol particle spectrum obtained from simultaneous
measurements at three neighboring mountain stations of about 1 km height
difference each (highest-situated station 3 km a.s.l.). The measured values
and their parameterization shall provide the means to infer later on for a
larger geographical area around our stations on the basis of known meteorological conditions the particle size spectrum and to derive therefrom predictions about visibility conditions. The large-scale validity of the measurement results has been verified through comparing the fine structures
of radiosonde ascents at our Institute with those from the German Weather
Service at Munich. The linear distance between the two radiosondes is
nearly 90 km. From the clearly positive results of these comparisons it can
be concluded that under certain meteorological conditions (no front systems
close by) the applicability of our data obtained at 1.8 km (about 830 mb)
and 3.0 km a.s.l. (about 710 mb) is valid in practice for a range of 200 km
distance at least from the west across the north to the east and thus specifically at and above the boundary layer, i.e., between 2 - 3 km altitude
above sea level where the influence of near-ground sources of air pollution
is negligible. Yet, it is but natural that <u>within</u> the boundary layer, mainly between ground and 500 - 800 m above, also horizontal inhomogeneities
are to be expected.

For some of the many important parameters used for classifying the data material aerosol spectra have been plotted graphically in order to increase in this manner the clearness of the results. Fig. 1 shows the location
of our 3 stations on a map.

For the measurement of the particle size spectra a five-fold so-called
double-stage impactor is used which has been developed by Junge and Jaenicke
(1). In contrary to the conventional cascade principle, using jets with

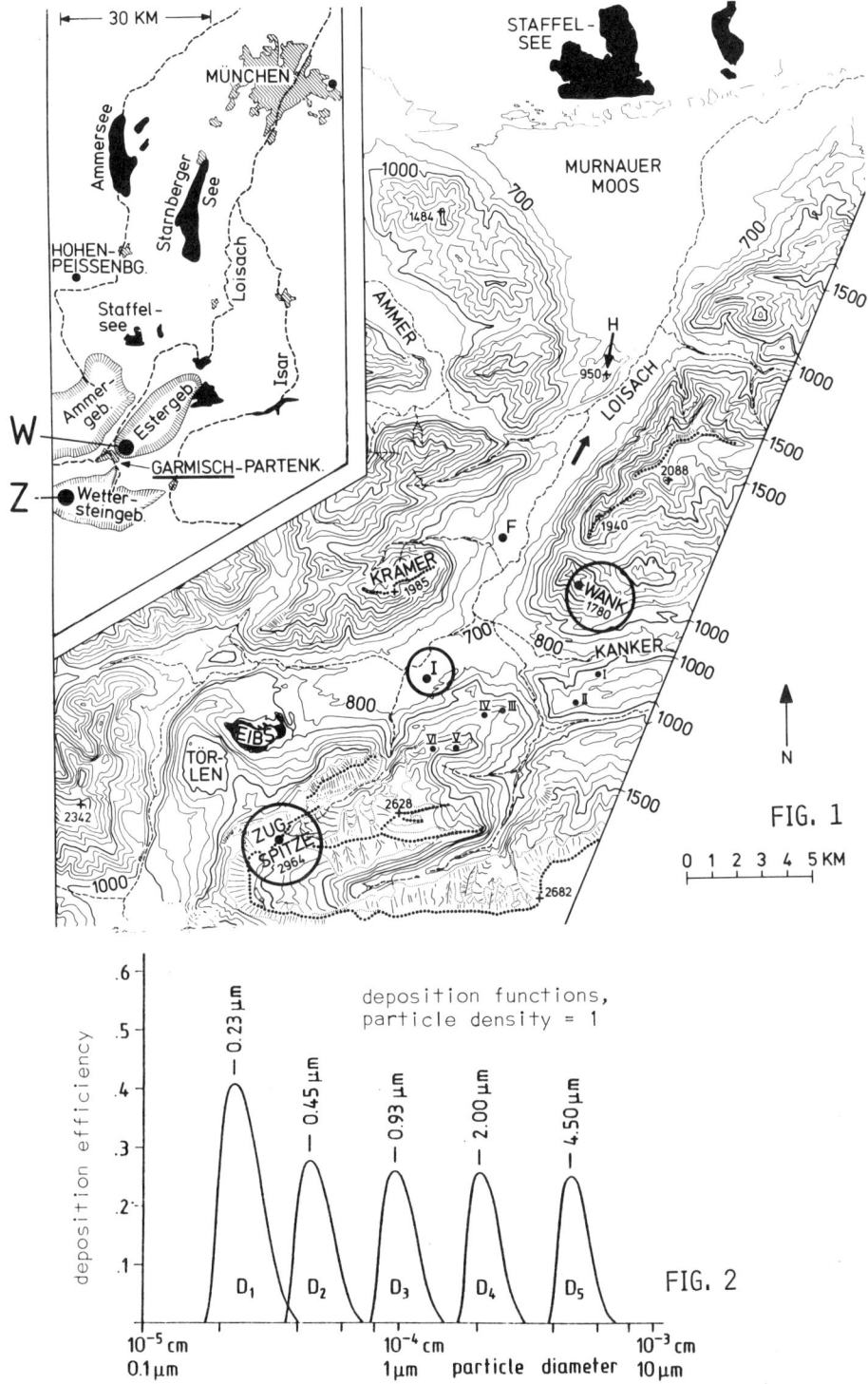

FIG. 1

FIG. 2

deposition functions,
particle density = 1

successively smaller slit widths in series,one separate impactor consisting of two identical nozzles in series, with separate pump, flow and air volume meter is used for each size range. Only the particle deposit behind each second nozzle is evaluated. This impactor type is more complicated than the cascade impactor but offers some essential advantages. As the air stream through each single unit is independent of each other, the flow rate can be chosen individually for optimum performance and for achieving the desired deposition size range. Furthermore, the exposition time can be set independently for each partial impactor in order to avoid too thin (bad statistics) and too thick (particle superposition) deposits. Finally, this impactor type allowed the development of an electro-optical device for the in situ measurement of the light extinction of the deposited particles and automatic turning off the pumps when a certain extinction level is reached (Carnuth and Dorn, (2)). This extinction measurement device is used at the valley station (later at the mountain stations, too) for obtaining optimal air volumes for microscopic counting automatically. We still use the microscope for the evaluation of the particle deposits. The deposition functions $f_i(D)$, i = 1...5, i.e. the probabilities for a particle with diameter D to be deposited on the slide have been calculated using mathematical procedures developed by Jaenicke. With the deposition functions thus depending on the atmospheric pressure, different results would be obtained for the three station levels with the other parameters kept constant. In order to get identical functions for all altitudes, slightly different flow rates, which have been found by calculations, are used at the three stations.

The deposition functions depend additionally on the effective particle density if geometric instead of aerodynamic diameters are wanted. According to Jaenicke's extensive calibration measurements, we assumed unit particle density in our calculations.

Figure 2 shows the resulting deposition functions for the five partial impactors. The deposition maxima are located at diameters of 0.23, 0.45, 0.93, 2.0, and 4.5 microns, respectively. Using densities other than one in the calculations would result in a shift of the curves parallel to the diameter axis by an amount approximately equal to the square root of the density ratio, toward smaller diameters if the density is increased, and vice versa.

According to Jaenicke (1), we chose a distribution function consisting of four Junge-type power law functions between five fixed diameter values D_i (i = 1...5), which would result in maximum number of deposited particles for an average size distribution and are thus slightly smaller than the afore mentioned maximum diameters of the deposition functions: .23, .45, .93, 2.0, and 4.5 microns. The values of the distribution function dn/dlogD at these diameters D_i are denoted N_i. The partial power law functions between these points are equivalent to straight lines in the usual log-log graph. The total size distribution is definitely and completely described by the five N_i values and we use them for the presentation of our particle size spectra in this report.

The N_i's are calculated by means of an iterative mathematical procedure. We start with approximative values $N_{i,o}$, which are defined below. With the distribution function defined by these starting values and using the deposition functions $f_i(D)$ we calculate theoretical numbers of deposited particles per unit air volume, $m_{i,o}$, for each impactor unit.Then we calculate a first set of corrected N_i values, $N_{i,1} = N_{i,o} \cdot n_i/m_{i,o}$. With these new numbers we repeat the calculations, getting values $N_{i,2}$, and so on, until the deviation between the measured (n_i) and calculated numbers of particles ($m_{i,k}$) is less than 1%. For the initial $N_{i,o}$ numbers any value could be used in principle, for example $N_{i,o} = 1$, but at least one iter-

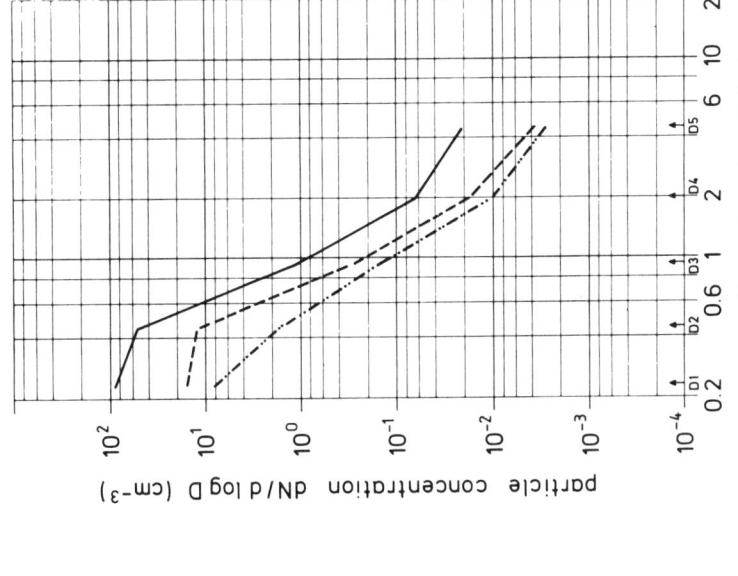

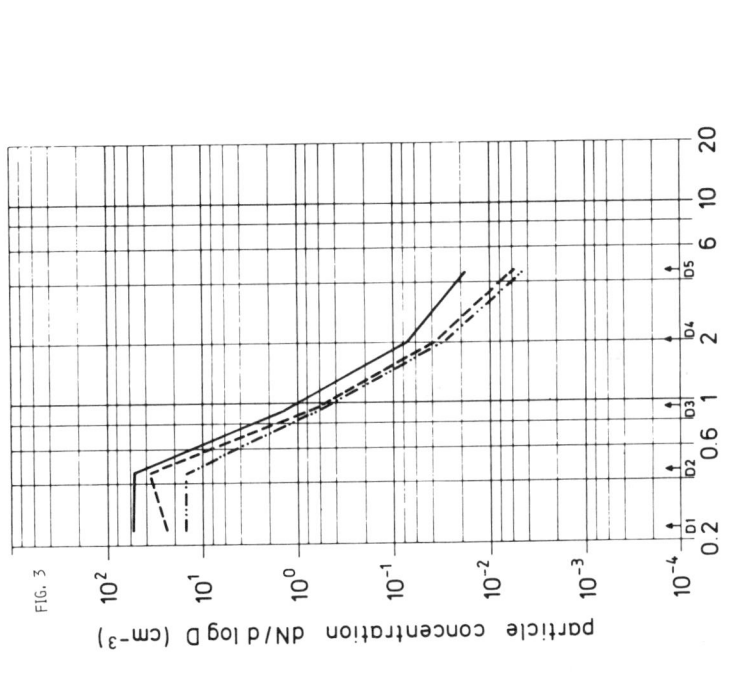

Fig. 3 : Particle spectrum in Summer at the 3 stations. Fig. 4 : same in Winter. Difference Fig. 3/4 : N° of particles D<0,3 in Winter higher than in Summer within the boundary layer (——) ; above the mixing layer all particle sies much lower in concentration than inside. This is true for all ranges, stronglw for the smaller particles.

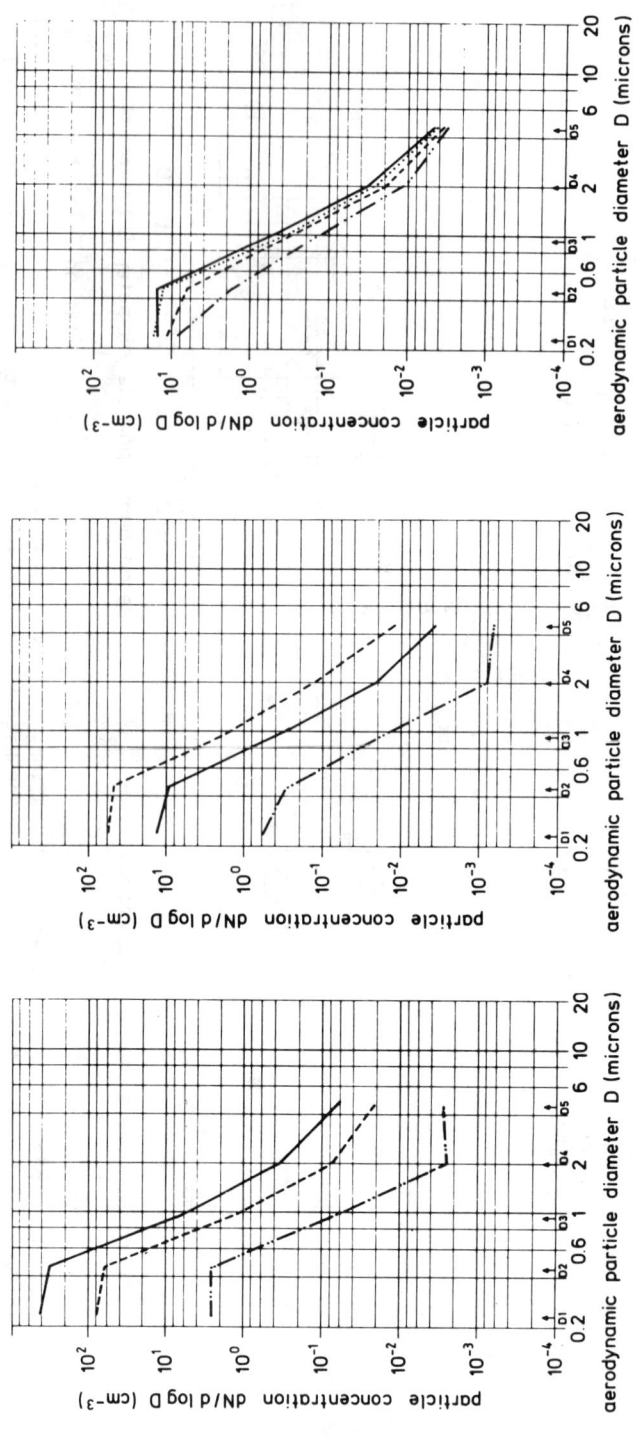

Fig. 7 - The 4 seasons : winter -·-· / spring ···· / summer ——— / autumn ----/ show different spectra on Zugspitze : strong dependence in the range D<1μm, unimportant difference D=1→5μ. valley : practically no dependence

---- mean/ ——— very high/ -·-··- very low particle concentration
Fig. 6 - Station Zugspitze, 2964m The shape of the spectra differs with altitude, concentration drops drastically with altitude

Fig. 5 - Station Garmisch, 740 m, valley

ation step and corresponding computation time is saved if the $N_{i,o}$ are chosen according to the formula:

$$N_{i,o} = \frac{n_i \, D_i^{-3}}{\int f_i \, (D) \, D^{-3} \, dD}$$

This are coordinates of Junge functions $N = dn/d\log D$ $= c_i \cdot D^{-3}$ at the abscissa D_i with constants c_i resulting exactly in the measured number of deposited particles, n_i. For the calculations a program for a Hewlett-Packard 9100 A desk computer with extended memory has been developed. The very complicated integrals $f_i \, (D) D^{-\nu} \, dD$ occurring in the calculations have been approximated as functions of the exponent ν by an interpolation formula. 2 or 3 iteration steps are required on an average, the computation time is in most cases less than on minute.

Since about 5 years the particle deposits on the microscope slides are evaluated by means of an automatic microscope image analyser (Micro-Video-mat, made by Zeiss, W-Germany), consisting of a powerful research microscope with motor-operated scanning stage, a TV camera, a data processing unit, a control panel for manual operation, control electronics for automatic operation of the stage, and a Wang 720 desk computer with typewriter and interface, allowing a completely automatic operation of the whole system.

2. SOME RESULTS, TOTAL MEANS 1970 - 80

All results are given in the form (grafically):
part.conc. $dN/d \log D$ (cm^{-3}) as function of Diameter D (μ), $D_1, D_2 \ldots$ (Fig.2).

The 5 grafs as examples of the in total 51 spectra are speaking for themselves, a detailed description is so far not necessary, some remarks however may be sufficient: *(see legends under figures)*.

References

(1) Jaenicke, R. (1971): Der Doppelstufenimpaktor, eine weitere Anwendung des Impaktorprinzips. Staub-Reinh. Luft 31, 229

(2) Carnuth, W. and Dorn, D. (1977): Ein Doppelstufenimpaktor mit photoelektrischer Meßautomatik. Staub-Reinh. Luft 37, 335

CONCENTRATION CHIMIQUE ELEMENTAIRE DE L'AEROSOL ATMOSPHE-RIQUE PRELEVE SUR LES ALPES FRANCAISES

C. ELICHEGARAY[1], R. VIE LE SAGE[1], M. DARZI[2], P. KUTRAKIS [1]
S. VOLIOTIS[3]

(1) : Laboratoire de Chimie Minérale des Milieux Naturels
              Paris E.R.A. - C.N.R.S. - n° 889 - FRANCE

(2) : Florida State University Dept of Oceanography -
              Tallahassee (U.S.A.)

(3) : Laboratoire de Chimie Instrumentale
              University de PATRAS (GRECE)

Abstract

Atmospheric aerosols have been sampled during two summers
at the "Aiguille du Midi" (3613 m. above sea level).
Time variation concentration and size profile have been
determined for 10 elements.
A relation is established between Enrichment factors and
mean geometrical diameter. Element as S, Br, Zn, Pb, are
very enriched in a fine size fraction at the opposite of
Fe, Ca, Si, Al non enriched and mainly abundant in coarse
size fraction.
Very similar relationships with the Jungfraujoch aerosols
meteorologicals condition and time variation of the concentration suggests also that we are representative of a
West European Background aerosol without significative
local contribution during our sampling.

1. INTRODUCTION

De nombreuses études sont menées actuellement dans le but
d'évaluer l'apport des contributions anthropogéniques au cycle
géochimique des éléments, l'estimation des budgets s'effectu-
ant à partir d'enquêtes d'émissions ou de données caractéri-
sants un "fond atmosphérique" libre de pollution industrielle
(1), (2).
Pour l'Europe du Nord, le massif du Mt Blanc situé dans
les Alpes françaises nous a semblé répondre aux critères de
stations dites "blanches" et notre laboratoire y a effectué
durant deux années consécutives une série de campagnes de
mesure de l'aérosol atmosphérique (juin-septembre 1978 ;
juillet-aout 1979).

2. ECHANTILLONNAGE ET ANALYSE

Les échantillons ont été prélevés sur filtres Nucléopore

(0.4 μm de diamètre de pore) au moyen d'impacteurs en cascade
(type Batelle) et de streakers, ces derniers permettant un
suivi de 24 heures en 24 heures de la variation dans le temps
de la composition chimique de l'aerosol atmosphérique.

Les conditions climatiques rigoureuses ont nécessité
l'installation de l'appareillage de collecte à l'intérieur
d'un abri en duralumin assurant une protection contre la neige
et le vent tout en permettant un échantillonnage représentatif de l'atmosphère ambiante.

La station de prélèvement proprement dite a été installée
près du refuge des "cosmiques" dépendant du laboratoire de
glaciologie de Grenoble, et situé à 3613 m. d'altitude à l'é-
cart de contributions anthropogéniques parasites.

Les analyses ont été effectuées par P.I.X.E.A. (Proton
Induced X Ray Emission Analysis) dans les laboratoires de la
Florida State University (U.S.A.) (3).

3. RESULTATS ET DISCUSSIONS

Les impacteurs type Batelle permettent de séparer l'aérosol atmosphérique en six classes granulométriques.

Ces données nous ont permis de calculer un Diamètre Mé-
dian Pondéré (D.M.P.), qui rend compte du domaine granulométrique dans lequel sont rencontrés préférentiellement les
éléments

$$D.M.P. \ (\mu m) = \frac{\sum_{j} D_j^i \cdot C_j^i}{\sum_{j} C_j^i}$$

D_j^i : Diamètre de coupure de l'étage j

C_j^i : Concentration de l'élément i à l'étage j.

Par ailleurs nous avons calculé les facteurs d'enrichissement moyens des éléments par rapport à la croûte terrestre,
et à l'aluminium comme élément de référence.

Le tableau I fournit les concentrations moyennes, les
facteurs d'enrichissement et les D.M.P. des dix éléments
majoritairement présents dans l'aérosol d'altitude du massif
du Mt Blanc, les autres éléments analysés (P, Ti, V, Cr, Mn,
Ni, Cu, Sr, Bi) n'étant qu'exceptionnellement détecté et à des
teneurs extrèmement faibles. Il permet de distinguer deux
familles d'élements :

Si, K, Ca, Fe, dont les facteurs d'enrichissement sont
proches de l'unité et situés dans le domaine des particules
supérieures au micron.

S, Cl, Zn, Br, Pb très enrichis ou enrichis, majoritairement présents dans le domaines des particules submicroniques.

ELEMENT	X (ng/m^3)	F.E. croûte (Al)	D.M.P. (µm)
Al	165	- ref -	2.4
Si	163	0.3	1.3
S	680	1370	0.4
Cl	45	170	1.2
K	23	0.4	1.3
Ca	28	0.4	3
Fe	23	0.2	2.3
Zn	7	50	0.5
Br	0.4	80	0.25
Pb	1.3	50	0.9

Tableau I : Concentrations moyennes, facteurs d'enrichissement
et D.M.P. des dix éléments majeurs de l'aérosol
alpin.

De nombreuses mesures effectuées au Pôle Sud montrent
une corrélation étroite entre composition chimique des neiges
et glaces, et celle de l'atmosphère environnante. Un phénomène analogue est observé ici, en normant nos résultats à
ceux relatifs à la composition chimique des glaces prélevées
sur le massif du Mt Blanc par Briat, les facteurs d'enrichissement "fictifs" ainsi calculés étant voisins ou inférieurs
à dix. (Tableau II)

X	$\dfrac{(X/Al)_{\text{aérosol}}}{(X/Al)_{\text{glace}}}$
Al	référence
Cl	2.5
K	2.2
Ca	11
Fe	3.7
Zn	0.5
Pb	2.5

Tableau II : Concentrations relatives normées à l'Aluminium
de la composition chimique de l'aérosol alpin
par rapport à la composition chimique des glaces sur le Dôme du Goûter (4210 m.) (4).

Les chiffres du tableau III permettent de comparer ces
données à celles relatives à deux sites particuliers :
- d'une part la Jungfraujoch située dans les Alpes
Suisses et dont les caractéristiques géographiques sont pro-
·ches des notres (5).
- d'autre part l'aérosol Parisien caractéristique d'une
atmosphère urbaine (6).
Les résultats obtenus sur le Jungfraujoch, pour les

éléments dont on dispose de données sont en général du même ordre de grandeur que ceux fournis par le massif du Mt Blanc et attestent d'une relative stabilité de la composition chimique de l'aérosol sur les deux versants des Alpes à quatre années d'intervalles.

Les teneurs en Pb, Br, et Zn et leur facteur d'enrichissement associés sont plus élevés cependant à la Jungfraujoch ou les auteurs signalent des sources de pollution locale.

Malgré tout, tant pour ce qui concerne le massif alpin que Paris, chaque élément demeure dans la même famille granulométrique et la même classe d'enrichissement précédemment définie en dépit de concentrations élémentaires différentes.

Cette similitude sur des sites aussi différenciés rend compte du rôle prépondérant des processus de formation sur la nature des sources émettrices.

On constate également que pour un même élément, les Diamètres médians sont plus faibles dans le massif du Mt Blanc, ceci étant à mettre en relation avec l'altitude du prélèvement.

	PARIS			Mt BLANC			JUNGFRAUJOCH		
	X ng/m^3	F.E	DMP (m)	X ng/m^3	F.E	DMP (m)	X ng/m^3	F.E	DMP (m)
Al	660	ref		165	ref	2.4	51	ref	-
Si	630	1		163	0.3	1.2			
S	1017	480	1.1	680	1370	0.4			
Cl	114	106	1.7	45	170	1.4	7.2	570	-
K	192	1		23	0.4	1.2	20	1.3	-
Ca	350	1.2	5.1	28	0.4	3			
Fe	221	1	3.1	23	0.2	2.3	36	1.1	-
Zn	76	132	0.9	7	50	0.5	10	240	-
Br	37	1840		0.4	80	0.2	1.3	820	-
Pb	218	2040	1.2	1.3	50	0.9	4.4	540	-

Tableau III : Comparaison des concentrations moyennes, facteurs d'enrichissement et taille des particules sur trois sites de prélèvement.

Il s'agit ici d'un déficit relatif en grosses particules due à la sédimentation sèche ou humide.

La figure 1 illustre les variations de concentration de quatre éléments (Al, S, Fe, Pb) caractéristiques des deux familles définies précédemment. Le fait marquant est le parallélisme accentué des profils de variations même pour des éléments dont les facteurs d'enrichissement et la taille traduisent des matériaux d'origine et des processus de formation différents.

D'une façon générale, les variations de concentrations des aérosols dans le temps résultent de la fluctuation propre au flux des diverses contributions, à laquelle il convient d'ajouter celle provenant de la dispersion des polluants au cours du transport et conditionnée notamment par les caractéristiques thermodynamiques de l'atmosphère du moment.

Il parait peu plausible, sur une période de dix neuf jours consécutifs, d'invoquer des sources naturelles et anthropogéniques dont les budgets émissifs et leurs contributions à 3600 m. d'altitude auraient affectés simultanément et dans une proportion analogue le versant français du massif alpin.

Notons à cet égard que l'on peut exclure des contributions locales significatives :

- couverture neigeuse encore importante pour ce qui concerne un apport détritique rapproché,

- notre station était située en permanence, lors des épisodes nuageux, au dessus des couches de nuages à faibles développement vertical recouvrant la vallée de Chamonix, et le plafond d'inversion d'une altitude généralement inférieure à la notre pour les apports anthropogéniques locaux.

L'explication la plus probable parait être un mixage à grande échelle et en amont de notre station des apports de diverses nature et leur homogénéisation à l'intérieur des masses d'air provenant sur les Alpes à cette altitude. Il en résulte un effet "tampon" sur les différents flux émissifs, les éléments étant par la suite affectés d'une manière analogue au regard de leur histoire atmosphérique.

Il demeure bien entendu difficile d'évaluer ici l'importance des contributions anthropogéniques lointaines.

Elles sont présentes, si l'on considère les niveaux d'enrichissement élevés constatés pour Pb, Br, Zn et S, et un rapport Br/Pb égal à 0.31, valeur proche de celle caractéristique des antidétonants additionnés dans les carburants automobiles (Br/Pb = 0.39) (7), mais limitée au vue d'éléments tels que Cr, V, Hg.., rarement décélés.

Le soufre pose un problème particulier, son niveau de concentration étant élevé comparé à celui observé pour les autres éléments sur site urbain et isolé. Identifié qualitativement sous la forme de sulfate particulaire par spectrométrie de photon électrons (E.S.C.A.) et à la différence des autres éléments provenants de la condensation de composés volatils, son abondance s'explique ici en grande partie par l'importance du taux de conversion $SO_2 - SO_4^{--}$ en atmosphère humide et ensoleillée.

4. CONCLUSIONS

L'ensemble de ces résultats montrent que les concentrations de l'aérosol atmosphérique sur le massif du Mt Blanc sont proches de celles observées sur le versant Suisse des Alpes dont les ordres de grandeurs sont 1 à 2 fois inférieurs à celles des zones industrialisées ou polluées. La variation simultanée des concentrations ne s'explique qu'en admettant un brassage à grande échelle des contributions naturelles et anthropogéniques et un apport minimum des sources locales ou situées dans les vallées Alpines.

Ces caractéristiques suggèrent qu'au cours de notre période d'échantillonnage notre site s'est trouvé proche des caractéristiques du bruit de fond d'un aérosol atmosphérique Ouest Européen, peu soumis à des apports industriels.

Support

*Cette étude a ete effectuée avec le support financier du
C.N.R.S. (A.T.P. Nuisances Chimiques n° 3346)*

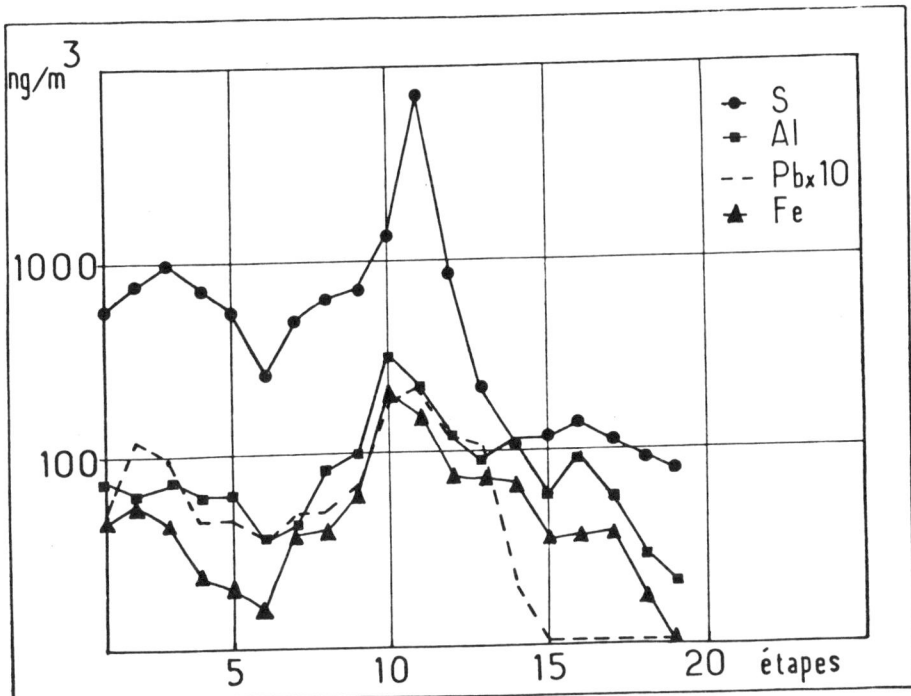

Figure 1 : *Variation des concentrations du S, Al, Pb, Fe,
dans l'aérosol atmosphérique du massif du Mt.
Blanc du 20 juin 1979 au 8 juillet 1979.
(une étape représente 24 heures de prélèvement)*

REFERENCES

(1) ROBINSON E., ROBBINS R.C.
Emissions, concentrations and fate of particulate atmos-
pheric pollutants, Rep. P.R. 6755, Stanford Research
Institute, Suppl. 23-34 (1968)

(2) JOHNSON W.B., WOLF D.E., MANCUSO R.L.
Atm. Env. 12 511-527 (1978)

(3) JOHANSON S.A.E., JOHANSON T.B.
Nuclear Instruments and Methods 137 473-516 (1976)

(4) BRIAT M.
Atm. Pollution, Proceding of the 13th International Col-
loquium, Paris, France
M. Benarie (ed.) Studies in Environmental Sciences, 1
225-228 (1978)

(5) DAMS R., DE JONGE J.
Atm. Env. 10 1079-1084 (1976)

(6) VIE LE SAGE R., ELICHEGARAY C., Ministère de l'Environnement
Symposium sur la recherche en matière de pollution atmosphérique Bordeaux (1978)

(7) BIGGINS P.D.E., HARRISSON R.M.
Env. Sci & Technol, 13 558-564 (1979)

GRANULOMETRIE DE L'AEROSOL MARIN

A. RENOUX[1], G. TYMEN[2], G. MADELAINE[3]

(1) Laboratoire de Physique des Aérosols et de Transfert des Contaminations,
Faculté des Sciences, Université PARIS XII, Av. du Gal De Gaulle,
94000 CRETEIL FRANCE

(2) Laboratoire de Physique des Aérosols et de Radioactivité Atmosphérique,
Faculté des Sciences, U. B. O., Av. Le Gorgeu, 29283 BREST Cédex FRANCE.

(3) Laboratoire de Physique de l'Atmosphère, SPT, DPR, IPSR, CENFAR,
92260 FONTENAY AUX ROSES, FRANCE.

Summary

As ocean covers a large part of the surface of our globe, marine aerosol for its effect on climate or by its reference role in the global
contamination study by anthropogenic sources, is particularly important
and it is fundamental to know its major characteristics, specially its
size distribution that occurs in ocean-atmosphere exchange processes,
in long distance transport phenomena like in cloud-bed formation and
in earth radiative budget.
This paper summarizes results obtained in Brest, in our KERDALAES and
GUISSENY coastal laboratories, and during three oceanic campaigns.
Sampling were carried out by means of experimentally calibrated impactors (CASELLA and ANDERSEN) and Nuclepore filters. For these ones the
collection efficiency has been evaluated by an experimental and theoretical procedure. Samples analysis was made by optical and electron
(transmission and Scanning) microscopy.
Our results have shown that the marine aerosol size distribution is
composed of three log-normal components, corresponding to modes pointed out by WHITBY : Nucleation mode ($D < 10^{-2}$ μm), Accumulation mode
($D = 0.64$ μm) and Coarse Particle mode ($D = 5.4$ μm). These mean values can be modified by anthropogenic influences ; some examples are
mentionned.
In the 0.1 μm < D < 2 μm size range, ammonium sulfate particles, sodium chloride particles, pollutants and mixed particles (sea salt +
pollutants in Brest, NaCl + $(NH_4)_2$ SO_4 in the Atlantic Ocean) correspond to 90 % of the marine aerosol number. Each percentage and mean
particle diameter is indicated. Finally, variations of coefficient B
of the JUNGE' law ($dN/dLogD$ = $C D^{-B}$, $D > 0.2$ μm) according to particle size are studied.

1. INTRODUCTION

L'importance de l'aérosol marin tient d'abord dans le fait qu'il constitue le bruit de fond des particules atmosphériques, et ensuite dans le
rôle qu'il joue dans les processus de transfert de matière océan-atmosphère.
Rappelons également que les particules d'origine marine interviennent dans
le bilan radiatif de la Terre, dans la formation des couches nuageuses,
ainsi que dans les processus de transport de matière à longue distance (1).

On conçoit aisément que l'aérosol marin sert d'élément de référence dans l'étude de la contamination par les sources anthropogènes (2). Il est donc fondamental de connaître ses caractéristiques principales, et sa distribution granulométrique en constitue l'un des éléments essentiels.

Dans cet article nous résumons les principaux résultats obtenus à BREST, dans nos laboratoires annexes de KERDALAES et de GUISSENY, et à la suite de trois campagnes océanographiques.

2. SITES DE MESURE ET APPAREILLAGE

La plupart de nos expériences a été réalisée à Brest (influence maritime et urbaine peu polluée), ainsi que dans deux laboratoires marins annexes situés à KERDALAES et GUISSENY, éloignés de toute pollution. En plus, nous avons participé à trois expéditions océanographiques (MIDLANTE 1974, ROMAN-CAP 1977, THERMOCLINE 1977) localisées sur la fig. 1.

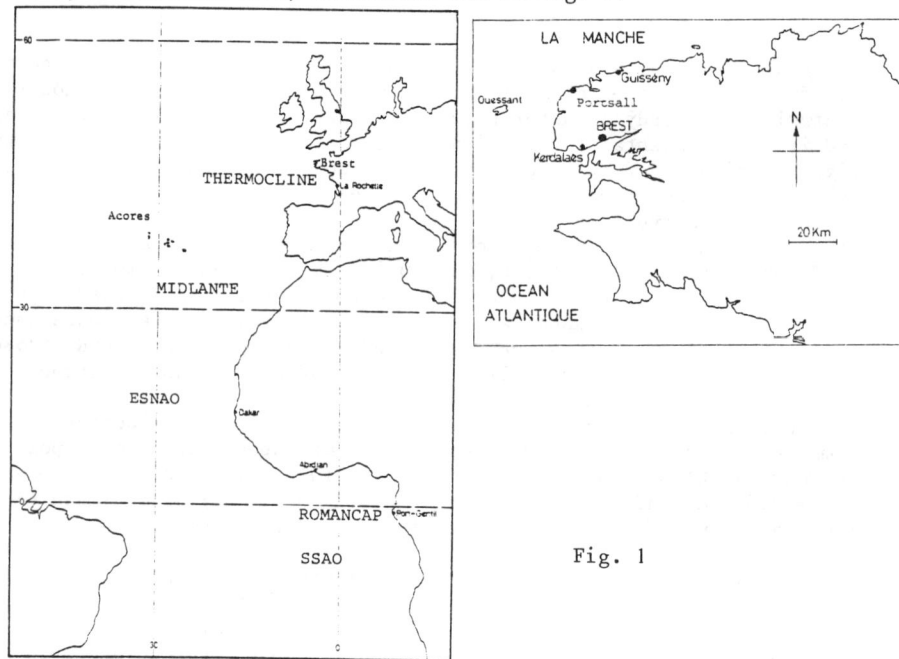

Fig. 1

Notre appareillage a été décrit dans un précédent article (3). Rappelons qu'il comprend des impacteurs en cascade (ANDERSEN et CASELLA) associés à des filtres membranes "NUCLEPORE" (diamètre des pores 0,4 μm). Les particules de dimensions supérieures au micron sont étudiées par microscopie photonique (4), les plus petites par microscopie électronique (transmission et balayage)(5).

3. GRANULOMETRIE DE L'AEROSOL MARIN

3.1. De toutes les études granulométriques que nous avons effectué sur l'aérosol marin, il ressort que ce dernier est formé de trois distributions lognormales correspondant aux modes définis par WHITBY (6) : nucléation, coagulation et grosses particules. Le Tableau I indique la distribution granulométrique moyenne, en nombre, de l'aérosol atmosphérique dans l'Atlantique Nord, proposée par J.F. BUTOR (5) et illustrée par la fig. 2 (7).

TABLEAU I

Mode de Nucléation	Mode de coagulation	Mode des grosses particules
DGN < 10^{-2} µm	DGS = 0,64 µm σ_gS = 1,6	DGV = 5,4 µm σ_gV = 1,7

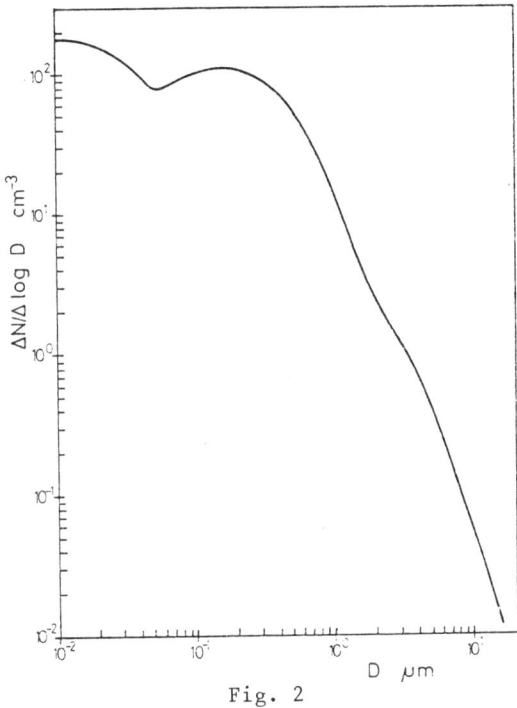

Fig. 2

DGN, DGS, DGV représentent respectivement les diamètres géométriques moyens en nombre, en surface et en volume, les σ_g étant les écarts types correspondants. En effet, le mode de nucléation apparait dans la répartition en nombre $\Delta N/\Delta \log D = f(D)$, le mode de coagulation apparait dans la distribution en surface $\Delta S/\Delta \log D = \phi(D)$ (sa contribution en volume peut aussi être importante, mais surtout dans le cas d'un aérosol urbain), le mode des grosses particules étant lié à la courbe en volume $\Delta V/\Delta \log D = \psi(D)$ (5).

Une comparaison entre les résultats obtenus sur les différents sites de mesure situés sur l'Océan Atlantique montre que les caractéristiques des trois modes varient très peu : l'aérosol marin semble donc beaucoup plus stable que l'aérosol continental.

La Fig. 3 regroupe les résultats de nos trois campagnes océanographiques, ainsi que ceux des campagnes SSAO (8) et ESNAO (9). On notera que les différentes distributions ont la même amplitude, bien qu'elles aient été obtenues par des méthodes différentes. Remarquons également la similitude des courbes MIDLANTE et ESNAO, ainsi que celle des courbes SSAO et ROMANCAP, correspondant à des lieux d'expérimentation relativement voisins.

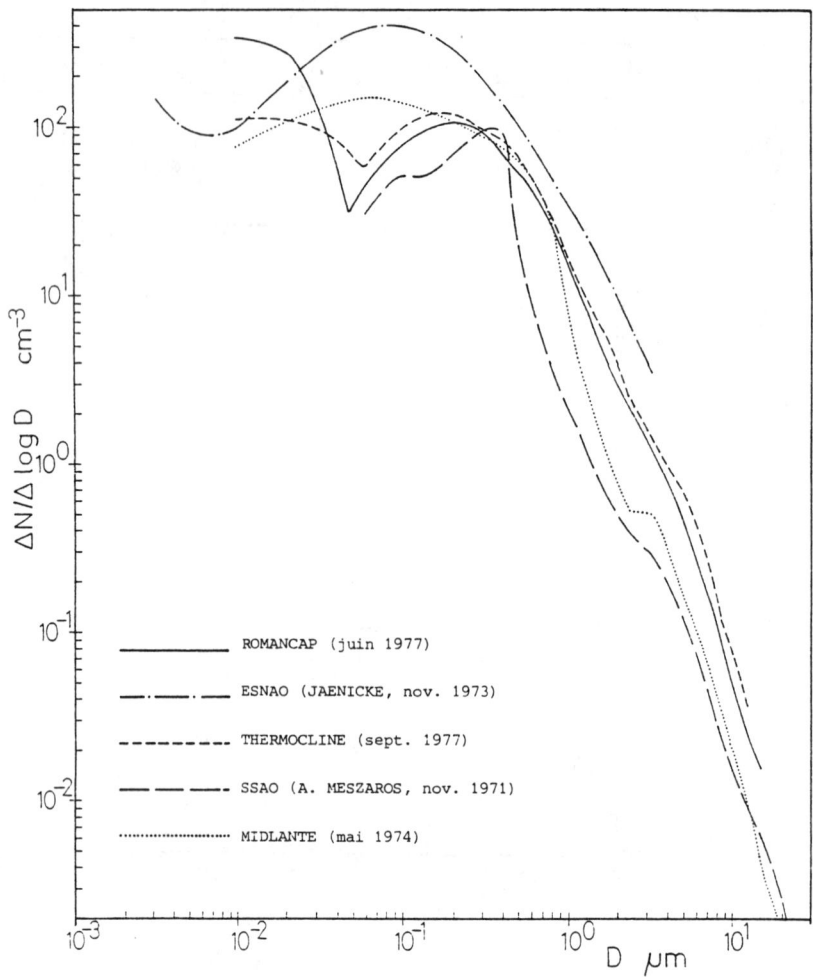

Fig. 3

3.2 Afin de caractériser la distribution granulométrique des aérosols de diamètres supérieurs à 0,2 μm, on utilise souvent la loi de JUNGE (10) :

$$\frac{dN}{d\log D} = CD^{-B} \qquad \text{ou} \qquad \frac{dN}{dD} = CD^{-(B+1)}$$

C et B étant des constantes. On a longtemps pris B = 3. En fait, l'expérience montre que B varie, non seulement en fonction des conditions météorologiques, mais aussi des dimensions des particules (4), ce que montre le tableau II.

TABLEAU II

BREST (influence maritime)	0,6 μm < D < 40 μm B = 2,8	D > 40 μm B = 5,2
KERDALAES (Rade de BREST)	D > 1 μm B = 2,8	
MIDLANTE	D > 2 μm B = 3,1	
ROMAN CAP	0,15 μm < D < 15 μm B = 2,5	
THERMOCLINE	0,15 μm < D < 15 μm B = 2,5	

4. NATURE DE L'AEROSOL MARIN

Le tableau III indique les proportions numériques (%) et le diamètre D des particules identifiées dans le domaine 0,1 μm < D < 2 μm (5). Dans cet intervalle, on constate que les particules de sulfate d'ammonium, de chlorure de sodium, les polluants et les particules mixtes (sel marin + polluants à Brest, NaCl + $(NH_4)_2 SO_4$ en Atlantique et à Brest) forment plus de 90 %, en nombre, de l'aérosol marin, les particules de sulfate d'ammonium étant les plus nombreuses.

TABLEAU III

	$(NH_4)_2 SO_4$	NaCl	Particules mixtes	Polluants
MIDLANTE 1974	86 % D = 0,6 μm	4, %	10 % D = 0,8 μm	0
ROMANCAP 1977	85 % D = 0,6 μm	1,5 %	13,5 % D = 0,8 μm	0
THERMOCLINE 1977	64,5 % D = 0,4 μm	5,5 %	30 % D = 0,8 μm	0
BREST	60-86 % D = 0,6 μm	0-5 %	1-16 % D = 0,8 μm	2-34 % D = 0,5 μm
AÇORES	70 %	20 %	7 %	0
AEROSOL MARIN MOYEN PUR	78,5 % D = 0,53 μm	3,7 %	17,7 % D = 0,8 μm	0

5. ALTERATION DE L'AEROSOL MARIN PAR LA POLLUTION

5.1 Pollution urbaine de faible amplitude.

Comptant plus de 200 000 habitants, assez peu industrialisée, l'agglomération brestoise produit une pollution relativement faible et pourtant, son influence sur l'aérosol marin est certaine, comme on peut le constater sur la fig. 4 tracée en $\Delta V/\Delta \log D$.

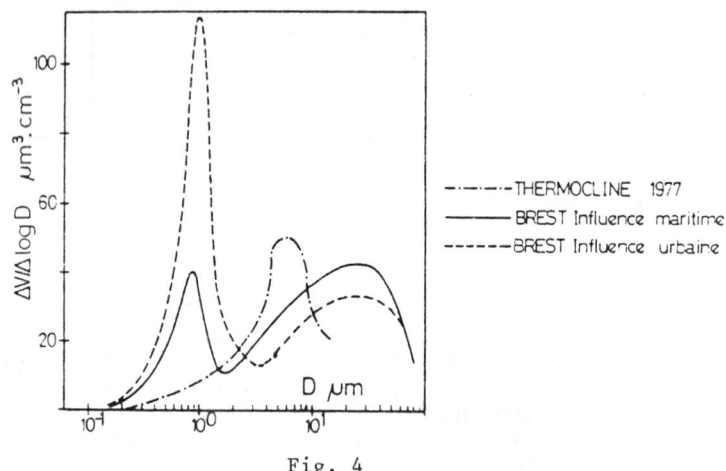

Fig. 4

On remarque que la pollution de la ville se traduit par l'apparition, en plus du mode des grosses particules qui existe seul dans le cas de l'aérosol marin pur (Thermocline, D = 5 µm), du mode de coagulation dont la contribution au volume total de l'aérosol est d'autant plus forte que la pollution est plus marquée. En même temps, le mode des grosses particules diminue d'importance et se déplace vers les grandes dimensions (D = 20 µm).

5.2. Pollution localisée et très importante de la mer.

Le 16 mars 1978, le pétrolier géant "AMOCO CADIZ" s'échouait à la suite d'une violente tempête, sur la côte Nord de la Bretagne, produisant une "marée noire" qui, s'étalant sur plusieurs kilomètres de côtes, représentait plus de 200 000 tonnes de mazout.

Entre mars et juin 1978, nous avons pu, grâce à l'aide financière du Ministère Français de la Santé et de la Famille, effectuer plusieurs études de l'aérosol atmosphérique, à 500 m environ du navire échoué, et sous un vent de mer assez fort (11). La fig. 5 montre, en $\Delta V/\Delta \log D$, les courbes moyennes que nous avons obtenues, la courbe Thermocline servant de référence. La masse totale de l'aérosol est, du fait de cette pollution artificielle de la mer, multipliée par un facteur 20 et, contrairement à ce qu'on estime habituellement (6), le mode de coagulation et celui des grosses particules ne sont pas indépendants. Bien entendu, le spectre dimensionnel de l'aérosol marin est très fortement modifié. Enfin, on constate que les effets de cette pollution sur l'atmosphère disparaissent au bout de trois mois.

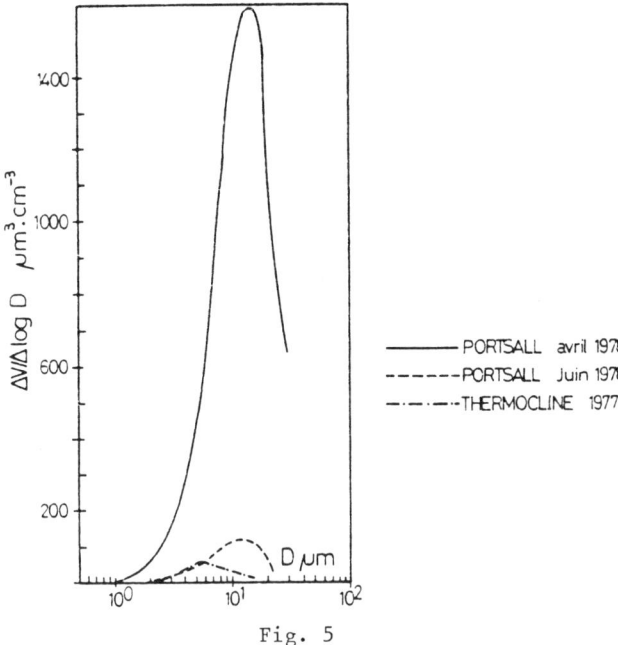

Fig. 5

6. <u>CONCLUSION</u>

L'aérosol marin est donc constitué par trois modes. Les caractéristiques de ces modes sont beaucoup plus stables que celles des modes de l'aérosol continental, mais s'avèrent très sensibles à une pollution, même modérée. A l'avenir, il pourrait être intéressant d'étudier, de façon systématique, l'altération de cet aérosol de fond par différents types de pollution, ainsi que sa composition chimique en fonction des dimensions des particules qui le composent. L'aérosol marin pourrait également servir de support à l'étude du transport particulaire à longue distance dans l'atmosphère.

Pour toutes ces raisons, il nous parait important d'envisager la réalisation d'une station de contrôle de l'aérosol marin, pour laquelle les installations existant déjà à GUISSENY nous paraissent bien adaptées.

<u>BIBLIOGRAPHIE</u>

(1) RENOUX (A.), MADELAINE (G.). L'aérosol marin. Son altération par des influences anthropogènes. Colloque Météorologie et Environnement. Publication MECV, EERM, SMF, Paris, 1980.

(2) TYMEN (G.), BUTOR (J.F.), RENOUX (A.), MADELAINE (G.). Influence of natural or anthropogenic sources on the marine aerosol size spectrum. Congrès du GAF, SCHMALLENBERG (RFA), pp. 8-14, 1980.

(3) RENOUX (A.), TYMEN (G.), BUTOR (J.F.), MADELAINE (G.). Granulometric Spectrum of aerosol particles in slightly polluted seaside. Atm. Environ. 11, pp. 1127-1132, 1977.

(4) TYMEN (G.). Répartition granulométrique de l'aérosol naturel et des particules radioactives issues du radon, en atmosphère maritime et urbaine peu polluée. Thèse BREST, 1978. Rapport CEA R 4965, 1979.

(5) BUTOR (J.F.). Contribution à l'étude de l'aérosol atmosphérique en zones urbaine, maritime et océanique. Thèse BREST, 1980, Rapport CEA R 5080 1981.

(6) WHITBY (K.T.). On the multimodal nature of atmospheric size distributions. VIII International Conference on Nucleation. LENINGRAD, U.R.S.S.

(7) BUTOR (J.F.). Contribution to the study of the atmospheric aerosol in urban, maritime and oceanic areas. IDOJARAS, Vol. 85, n° 3, pp. 117-125, 1981.

(8) MESZAROS (A.), VISSY (K.). Concentration, size distribution and chemical nature of atmospheric aerosol particles in remote oceanic areas. J. Aerosol Sc. 8, pp. 31-38, 1974.

(9) JAENICKE (R.). Aitken particle size distribution in the atlantic north east trade winds. Meteor Forsch. Ergebnisse, 13, pp. 1-9, 1978.

(10) BRICARD (J.). Physique des Aérosols. Rapport CEA R 4832, 1977.

(11) RENOUX (A.), BUTOR (J.F.), JOUAN (M.). Influence, sur l'aérosol marin, d'une pollution accidentelle de la mer. Environ Sc. and Health. à paraître. 1981.

A COMPREHENSIVE PHYSICAL AND CHEMICAL CHARACTERIZATION OF AN URBAN AEROSOL BY WAY OF EXAMPLE BERLIN

G.W. ISRAEL and B. HEITS
Technische Universität Berlin

Summary

Since January 1980 the FG-Luftreinhaltung is investigating comprehensively the physical and chemical nature of the urban aerosol with particular emphasis on high pollution episodes. The aerosol size distribution ($0,01 < d_p < 20$ µm) is measured by an electrical and optical analyzer with a time resolution of about 10 minutes. Simultaneously the total aerosol mass concentration, the light scattering coefficient, the condensation nuclei concentration and the concentration of SO_2, CO, NO, NO_2, O_3 are measured together with pertinent meteorological parameters. For the chemical characterisation the aerosol is collected by Hi-Vols, Low-Vols and impactors. The samples are subsequently analyzed for the metals Pb, Cd, Mn, Ni; the ions NH_4^+, SO_4^{2-}; NO_3^-; and six PAH's: Fl, Py, Per, BaP, BghiP, BkF. These investigations yielded the following first results:

(1) The ratio of SO_2-concentration to the light scattering coefficient is greatly influenced by the source and/or the residence time in the atmosphere for these components.

(2) The mean size of the ambient particulates is smaller in the summer time than in the winter time.

(3) It exists a good linear correlation between the sulfate content of the ambient aerosol and the ambient SO_2- concentration, namely

$$[SO_4^{2-}] = (0,26 \pm 0,03) \cdot SO_2 - (12 \pm 7)$$

(4) The light scattering coefficient, b_{SCAT}, is highly correlated to the ambient sulfate content:

$$b_{SCAT} [10^{-4} \cdot 1/m] = (0,10 \pm 0,01) \cdot SO_4^{2-} [\mu g/m^3] + (1.6 \pm 0.5)$$

(5) The (SO_4^{2-}/NH_4^+) ratio increases with increasing ambient sulfate levels as well as with increasing sulfate proportion of the ambient particulates. The data suggest the presence of H_2SO_4 for sulfate proportions larger than 25 %. The results also strongly indicate that in the presence of H_2SO_4 NO_3^- -ions are lost from the filters. (6) The mean benzo-a-pyrene content of the particulates correlates well with the ambient SO_2 concentration.

1. INTRODUCTION

The continuing concern over the effects of the atmospheric aerosol on man and the environment mandates an understanding of the urban and natural sources of aerosols and the distribution of aerosols within the

local, regional and global atmosphere. In January 1980 we started a still
ongoing project to comprehensively study the physical and chemical nature
of the Berlin aerosol in order to provide a better understanding of the
composition and behavior of the urban aerosol under middle European climatic conditions. Because of the potential health hazards involved particular emphasis is placed on the investigation of high pollution episodes
during the winter months. It is the aim of this paper to briefly describe
our aerosol characterization program and to present first selected results of our studies.

2. AEROSOL CHARACTERIZATION PROGRAM

Table I summarizes the aerosol measurements that have been carried
out through January 1981. The automated system for the in-situ measurement of the particulate size distribution has been described in detail(1).
The analysis for the heavy metal content of the particulates was carried
out by AAS based on the method described by Seifert et al (2). The polycyclic aromatic hydrocarbons (PAH) were analysed by high performance liquid chromatographie (HPLC) according to the analytical procedure described by Fechner (3). Sulfate and nitrite were measured by the photometric methods suggested by Wolfson (4) and Fadrus (5) respectively.
Nitrate and ammonia were analysed with ion-sensitive electrodes (6,7).
The aerosol measurements were complemented by the recording of the gaseous
constituents SO_2 (Monitor Labs 8450), NO, NO_2 (Monitor Labs 8440), O_3
(Monitor Labs 8410), and CO (Ecolyser 2600).
So far the measurements have all been conducted at fixed sites in the
downtown area of Berlin. The main station, FAS, is located at our Institute near the central railway station Zoo. Since May 1981 a fully
equipped mobile van is placed for two weeks at a time at various downtown and suburban locations to study urban gradients as well as the
aerosol influx from the rural invirons. It also measures the pertinent
meteorological data such as wind speed, wind direction, temperature,
global radiation, humidity and pressure.

3. RESULTS

At present we are in the process of analysing the data. Some first
results are presented in the following sections:

3.1 Size distribution of the urban aerosol
Fig. 1 shows a typical number ($\Delta N/\Delta \log D_p$) distribution of the
ambient aerosol on a winter day as measured with the automated in-situ
aerosol measuring system. The shown surface ($\Delta S/\Delta \log D_p$) and volume
distributions ($\Delta V/\Delta \log D_p$) were calculated from the number distribution assuming spherical particles. The surface distribution shows
clearly the three commonly found modes in the particle distribution,
i.e. the nuclei mode ($D_p \leq 0.1 \mu m$), the accumulation mode ($0.1 < D_p < 1 \mu m$)
and the coarse mode ($D_p > 1 \mu m$).
The month of January 1980 was marked by generally high SO_2 and particulate concentrations which resulted in two smog alerts (one of them
on January 17th). The SO_2-concentrations and the light scattering coefficients, b_{SCAT}, for the period January 14 through January 18, are shown
in fig. 2a. Fig. 2b gives the NO and NO_2 concentrations for the same
period. Measurements of the aerosol surface and volume concentrations
which were measured at the times indicated by the arrows are shown in
figures 2c and 2d.

Table I: Berlin aerosol study measurements (July 79 – Jan. 81)

Measured parameter	Sampling method	Collecting material	Time resolution	Sampling frequency	Sampling period	Sampling station
aerosol mass Pb,Cd,As,V,Fe, Fl, Py, Pe, BaP, BghiP, Cor;SO_2	staplex HiVol 83 m³/h	glass fiber filter	24 h	every 6th day	Jul.-Dec.79	5 stations in downtown Berlin(incl. FAS)
aerosol mass, Pb, Cd, Mn, Ni, Fl,Py, Pe, BghiP, BKF, DBacA, DBahA, NO_4^-, NO_3^-, NO_2^-, NH_4^+	staplex HiVol	glass fiber filter Verewa 227/1/60	24 h	daily every 6th day every 10th day every day	1.3.-2.2.80 2.6.-3.31.80 4.1.-12.31.80 1.1.-1.31.81	FAS
aerosol mass $SO_4^=$, NO_3^-, NO_2^-, NH_4^+	KFG-GS 050/3 3 m³/h	Pallflex	24 h	daily daily	9.23-10.18.80 Jan. 81	FAS
aerosol mass distribution Pb,Cd,Mn,Ni, PAH	Hi-Vol Impactor BGI 30 50 m³/h	glass fiber filters	24 h	every 3rd day every 6th day every 10th day	1.3-2.2.80 2.6.-3.31.80 4.10.-7.10.80	FAS
aerosol mass distribution TSP,NO_3^-, NH_4^+	Anderson n.v. impactor, Mark II 1,68 m³/h	Schleicher + Schüll,Nr.10	24 h	every 3rd day daily daily	1.3.-2.2.80 7.10.-8.4.80 9.23-10.18.80	FAS

Also thirty minute averages of the aerosol mass (FH 62I) of the scattering coefficient (Nephelometer MRI 1550 B with heated intake) and of the aerosol size distribution (combination of EAA - TSI 3030; OPC-Roy-co 245, LAS - ASASP 300) were measured during the stated sampling periods at FAS.

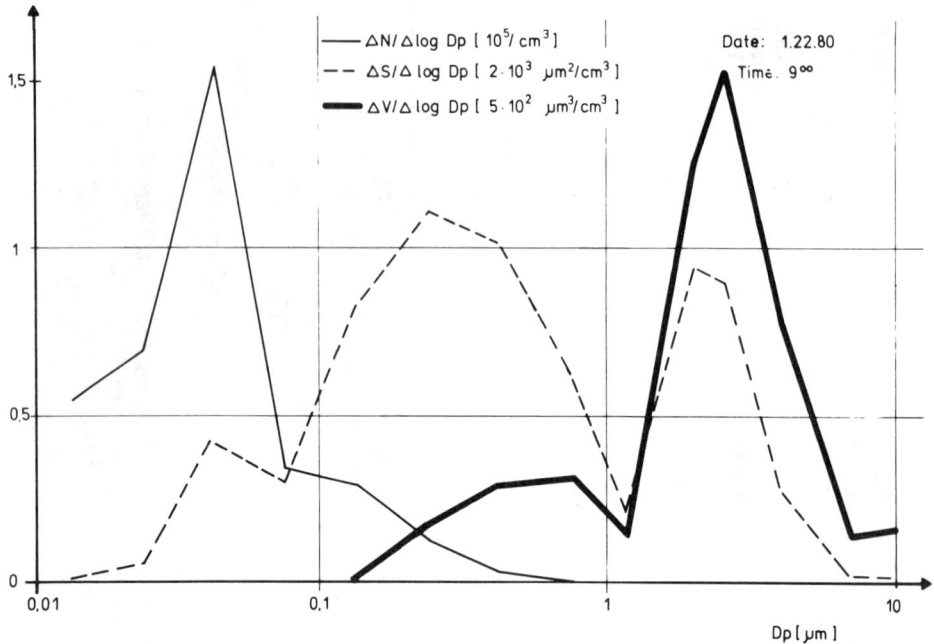

Fig. 1 Typical wintertime size distributions of the aerosol number
($\triangle$ N/$\triangle$ log D_p), surface ($\triangle$ S/$\triangle$ lög D_p) and volume
($\triangle$ V/$\triangle$ log D_p^p)

There are two very different meteorological situations which can
lead to very high SO_2 levels in the city and thus eventually trigger
a smog alert (8). One of these is characterized by anticyclones centered
in the southern DDR, CSSR or southern Poland which result in a rather
strong inversion aloft and southerly winds of 1 - 3 m/s at ground level.
This situation which prevailed on January 14 leads to advection of high
SO_2 concentrations from the industrial centers south of Berlin.
       The other meteorological situation is characterized by very low
wind velocities (0-1m/s) and fog. During such stagnation conditions
the SO_2 concentration is believed to be largely due to the emissions
by the city proper. Such a situation prevailed on January 17.
       The behavior of the scattering coefficient and the NO, which we
believe to be mainly due to vehicular emissions, support the assumption
of two different sources for the SO_2. The SO_2/b_{SCAT} ratio is much larger
on the 14th than during the following days indicating SO_2-sources which
contribute proportionally fewer particulates than do the SO_2-emissions
of the city proper. The general trend of the light scattering coefficient
trails closely the trend of the ambient NO concentrations from one day
to the other, both reaching a maximum on the 17th of January. Thus the
contribution of the city to these pollution levels reached its maximum
during this day.

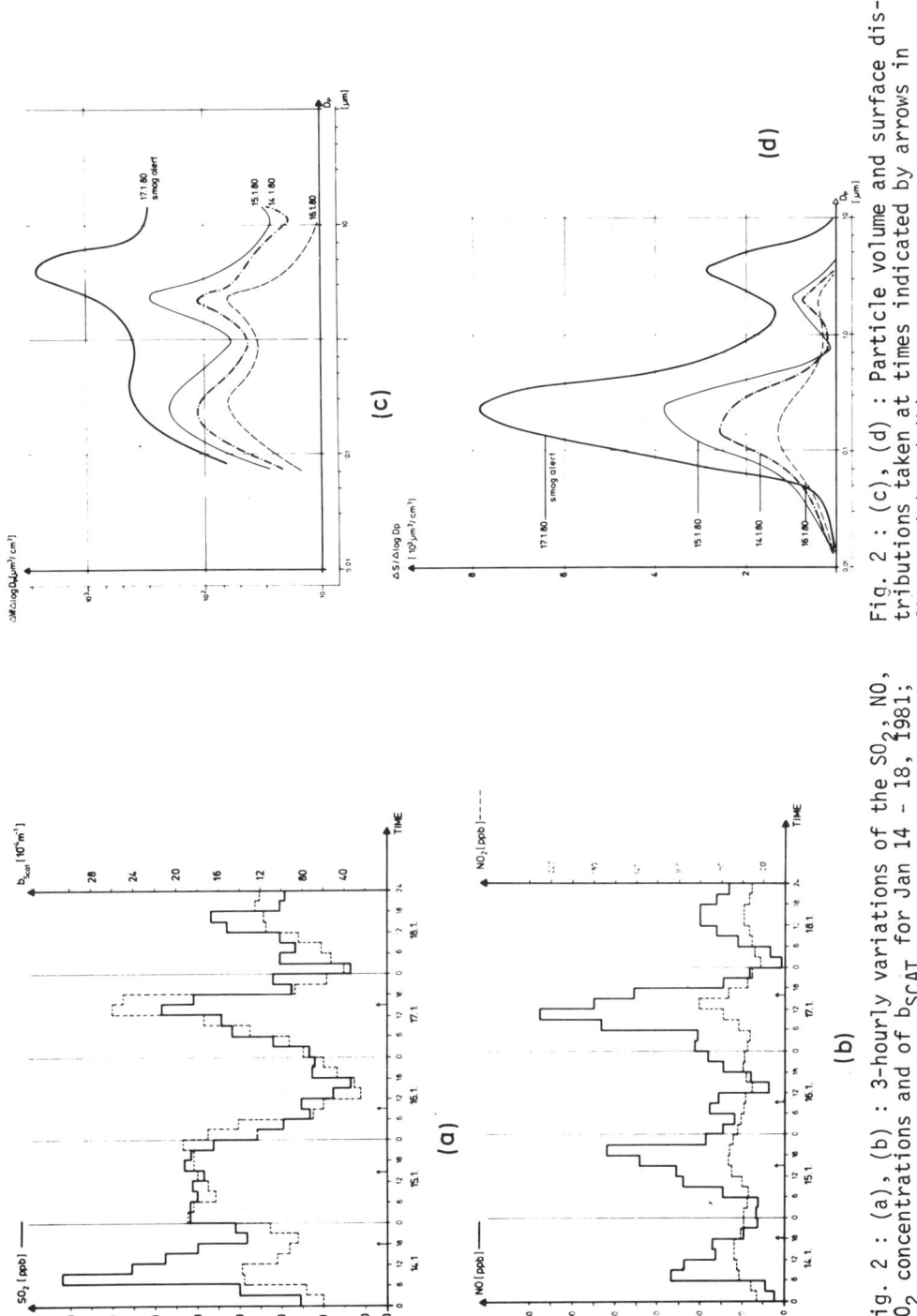

Fig. 2 : (c), (d) : Particle volume and surface distributions taken at times indicated by arrows in figures (a) and (b).

Fig. 2 : (a), (b) : 3-hourly variations of the SO_2, NO, NO_2 concentrations and of b_{SCAT} for Jan 14 - 18, 1981;

- 261 -

The surface and volume distributions of the particulates are very similar for the 14th through 16th, and the total surface and volume is proportional to b_{SCAT}. During the high pollution episode on the 17th, however, a marked shift to larger particles occurs in both distributions, and the usual gap in the volume distribution between the accumulation mode and coarse mode almost disappears.

Fig. 3 shows a comparison between typical aerosol surface distributions for winter and summer. The summer data show a decrease in the absolute values of the particle volume and a marked decrease of the ratio between the total surface in the coarse mode and the surface in the accumulation mode as compared to the winter measurements. This shift of the particle distribution toward smaller particle sizes was also confirmed by aerosol mass distribution measurements with cascade impactors. They revealed that from January through March 44 % and from April through June 62 % of the ambient aerosol mass was due to particulates with d_p < 1.8 μm.

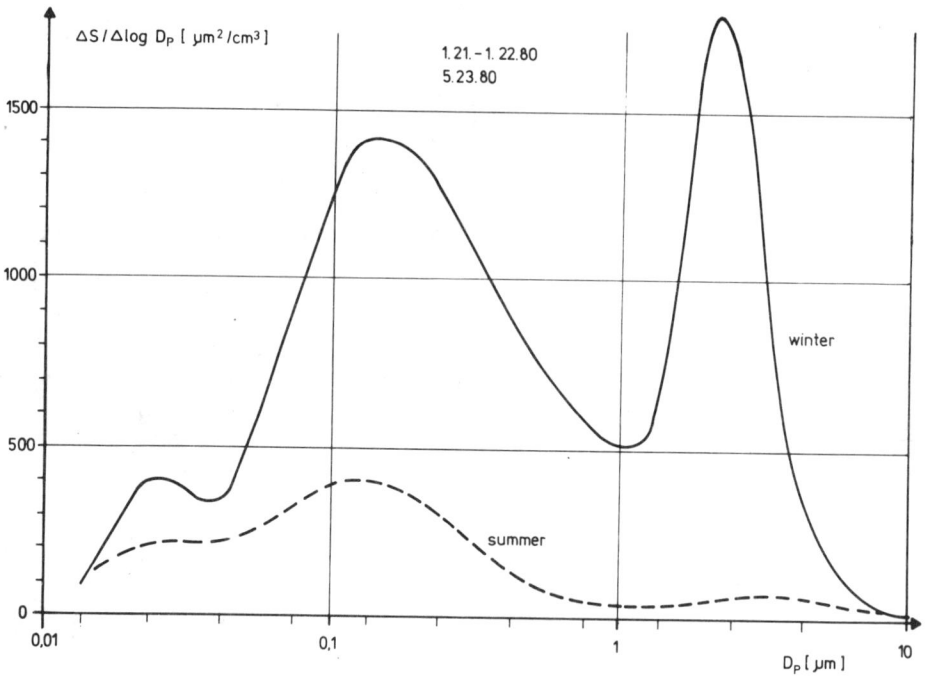

Fig. 3 : Typical aerosol surface distributions for winter and summer time in Berlin

3.2 Sulfur and nitrogen compounds

The following analysis is based on 24 hour measurments during the months of January 1980, October 1980 and January 1981.

The sulfate content of the suspended particulates is clearly related to the ambient SO_2 concentration, as is demonstrated by Fig. 4 and the linear regression analysis of the data yielding:

$$SO_4^= \; [\mu g/m^3] \; = \; (0.26 \overset{+}{_-} 0.03) \; SO_2 \; \mu g/m^3 \; - \; (12 \overset{+}{_-} 7) \tag{1}$$

$$n = 40, \; r = 0.82$$

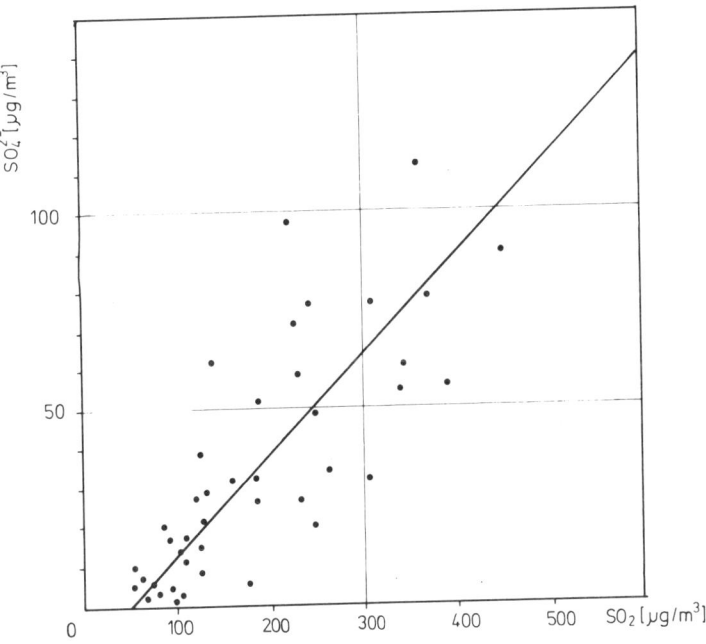

Fig. 4: Relation between particulate sulfate concentration and SO_2 level

The strong correlation between the light scattering coefficient and the sulfate content of the aerosol is evident from Fig. 5. The regression lines are described by

Jan´ 80 : $b_{SCAT} \; [10^{-4}/m] \; = \; (0.10 \overset{+}{_-} 0.01) \; SO_4^{2-} \; [\mu g/m^3] + (0.5 \overset{+}{_-} 1) \tag{2}$

$$n_4 = 12; \; r = 0.92$$

Jan´ 81 : $b_{SCAT} \; [10^{-4}/m] = \; (0.17 \overset{+}{_-} 0.01) \; SO_4^{2-} \; [\mu g/m^3] + (0.6 \overset{+}{_-} 0.4) \tag{3}$

$$n = 25; \; r = 0.95$$

The slope of the regression lines evidently changed from one year to the next. It is possible that this might be due to changes in the chemical composition of the particulates. E.g. the mean ratio of (SO_4^{2-}/NH_4^+) which is indicative of the acidity of the particulates was 2.0 in January 1980 and only 0.51 in January 1981. The slopes of the two regression lines group around a slope of 0.14 which Pierson (9) reported for aerosol measurements in rural Pennsylvania.

Assuming that the NO_3^--ion content of the samples is neutralized by NH_4^+-ions one can estimate the amount of NH_4^+ available to neutralize the SO_4^{2-}-ions by taking the difference between the molar content of NH_4^+ and NO_3^- in the sample, i.e. $(NH_4^+ - NO_3^-)$. Fig. 6 shows that the $SO_4^{2-}/(NH_4^+ - NO_3^-)$ molar ratio increases with the SO_4^{2-} proportion of the

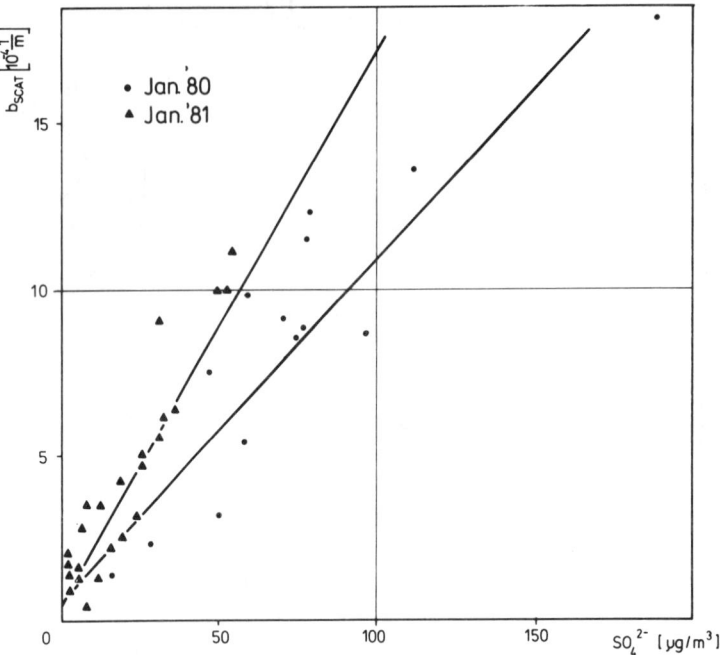

Fig. 5: Relation between light scattering coefficient and sulfate content
at ambient particulates

particulate mass and that there is not sufficient ammonia available to
neutralize the sulfate at high SO_4^{2-} - proportions.

Fig. 6 suggests, under the assumption that the NH_4^+ deficit is compensated for by H^+ ions, that for SO_4^{2-} -contents up to about 15% the sulfate
exists as $(NH_4)_2SO_4$, that for SO_4^{2-} proportions between 15 and 25 % the
sulfate is composed of a mixture of $(NH_4)_2 SO_4 + (NH_4)HSO_4$ and that for
SO_4^{2-} contents larger than 25 % the sulfate consists of a rather acidic
mixture of $(NH_4)H SO_4$ and $H_2 SO_4$. This presence of sulfuric acid is supported by the apparent loss of NO_3^- from the filter samples if the molar
ratio (SO_4^{2-}/NH_4^+) exceeds 1 which is discussed below. During the high
pollution episodes of January 1980 the SO_4^{2-} content of the aerosol mass
reached up to 50 % (190 $\mu g/m^3$) with a sulfuric acid proportion of up to
70 %.

Several investigators have suggested that NO_3^- collected on filters
is metathesized by H_2SO_4 to HNO_3 which evaporates[9]. Evidence of this
reaction and subsequent loss of NO_3^- is seen in fig. 7 which shows a plot
of the (NO_3^-/TSP) ratio vs. the (SO_4^{2-}/NH_4^+) molar ratio. For (SO_4^{2-}/NO_4^+)
molar ratios smaller than 1 there is no correlation between this ratio
and the NO_3^- proportion of the particulate mass as one would expect. Here
the NO_3^- content of the particulate mass ranges from 5 to 15 % with an
average value of 10 %. However, for (SO_4^{2-}/NH_4^+) molar ratios larger than 1,
i.e. with the presumed presence of H_2SO_4 aerosol on the filter, the

(NO_3^-/TSP) ratio decreases rapidly with increasing (SO_4^{2-}/NH_4^+), suggesting a loss of NO_3^- from the filter.

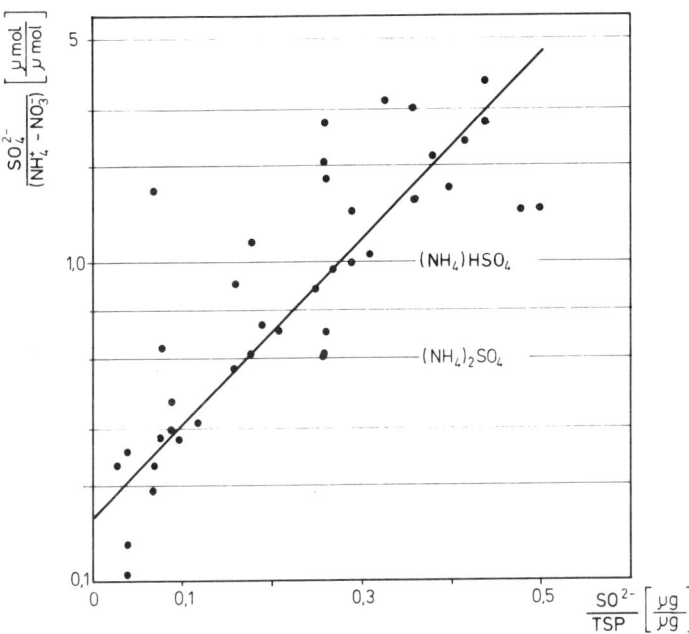

Fig. 6: Comparison of the molar sulfate to ammonia ratio corrected for NO_3^- with the sulfate proportion of the ambient particulates

3.3 Polycyclic aromatic hydrocarbons (PAH)

Ambient particulates were collected on glass fiber filters by high volume samplers every 6th day on 5 stations in downtown Berlin during July through December 1979. Subsequently, they were analysed for the six PAH´s: benzo-a-pyrene, BaP; benzo-ghi-perylene, BghiP; coronene, Cor; fluranthene, Flu; pyrene, Py; and perylene, Per.

The mean concentrations of these compounds for all five stations are summarized in table II.

Table II: Mean PAH concentration of 5 downtown stations in Berlin, July - Dec. 79

ng/m^3	BaP	BghiP	Cor	Flu	Py	Per
	1.3	1.8	1.4	3.2	2.6	0.14

These concentrations and the resulting profiles fall in the range of measurements reported for other urban and industrial centers in Germany and abroad (see e.g. 10).

Impactor measurements revealed that the PAH´s are concentrated in the smaller particle fraction. The particle fraction with D $<$ 2.5 μm contained in the average 65 % of the aerosol mass but 92 % of the PAH mass.

The monthly average BaP concentrations of all stations is plotted

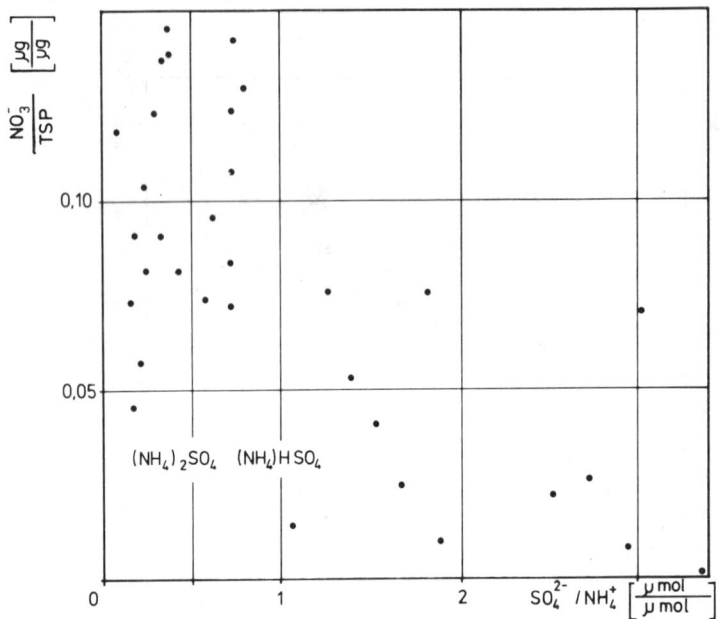

Fig. 7: Comparison of the nitrate proportion of the suspended particulates with the sulfate to ammonia molar ratio of those particulates

against the SO_2 concentration in Fig. 8. the graph shows a very good correlation between both parameters suggesting the same major sources for
SO_2 and BaP, namely fossil fuel combustion. However, wind direction dependence of the BaP/SO_2 ratios by more than a factor of 4 indicate that
the BaP/SO_2 ratio is source and fuel dependent in our area.
      Cor and BghiP showed neither a significant correlation with SO_2 nor
with Pb suggesting that neither fossil fuel combustion nor vehicular
traffic dominated their emissions. But we hope that further analysis of
the data by e.g. multiple linear regression analysis will provide us with
a clue to their possible sources.

Acknowledgement

This research is supported by the Umweltbundesamt der Bundesrepublik
Deutschland.

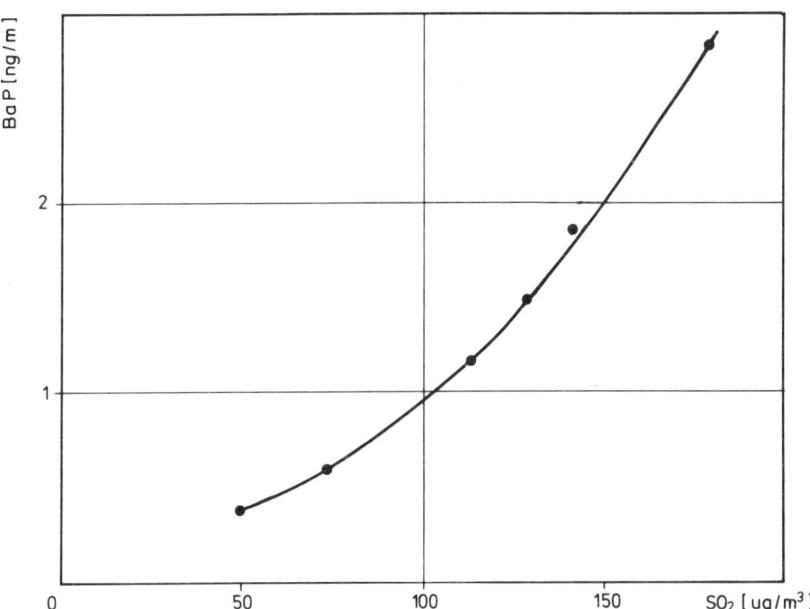

Fig. 8: Comparison between the monthly average BaP concentration of 5 downtown stations with the corresponding mean SO_2 levels. July - Dec. 1979.

REFERENCES

1. B. Heits, G. W. Israel: Ein Schwebstaubmeßsystem zur in-situ Erfassung der Korngrößenverteilung; accepted for publication in "Staub", October 1981.

2. B. Seifert, M. Drews: Atomabsorptionsspektrometrische Bestimmung von Blei, Cadmium, Kupfer, Vanadium, Zink und Arsen in Schwebstaub und Staubniederschlag; Wa-Bo-Lu-Berichte 1/1978 Berlin (1978).

3. D. Fechner: Polyzyklische aromatische Kohlenwasserstoffe in atmosphärischem Staub; Wa-Bo-Lu-Berichte 5/1980 Berlin (1980).

4. J. M. Wolfson: Determination of microgram quantities of inorganic sulfate in atmospheric particulates; J. APCA 30(1980)688.

5. H. Fadrus, J. Maly: Photometrische Bestimmung von Nitrat und Nitrit in Schwefelsäure mit Brucin; Z. Anal. Chemie 246, 230 (1969).

6. Operating manual for the Nitrate Electrode, model 93-07; Orion Research Inc. (1979).

7. Operating manual for the Amonia Electrode, model 95-10; Orion Research Inc. (1978).

8. Monatsbericht Januar 1980 über das Berliner Luftgüte-Meßnetz.
    Herausgeb.: Der Senator für Gesundheit und Umweltschutz, Berlin.

9. W. R. Pierson, W. W. Brachaczek, T. T. Truex, J. W. Butler, T. J.Kor-
    niski: Ambient Sulfate Measurements on Allegheny Mountain and the
    Question of Atmospheric Sulfate in the Northeastern United States.
    In "Aerosols": Anthropogenic and Natural, Sources and Transport -
    T. J. Kneip, P. J. Lioy edit. NY Academy of Sciences 1980.

10. W. Funke, J. König. E. Balfanz, T. Romanowski: The PAH-profiles in
    5 German cities; Atmospheric Environment 15 (887-890) 1981.

THE INTERPRETATION OF PARTICLE SIZE DISTRIBUTIONS OF ORGANIC POLLUTANTS : PHYSICOCHEMICAL AND TOXICOLOGICAL IMPLICATIONS

L. VAN VAECK and K. VAN CAUWENBERGHE
Chemistry Department, University of Antwerp (U.I.A.), B-2610 Wilrijk,
Belgium

Summary

Particle size distribution data for organic pollutants in ambient
aerosols provide a good basis for the study of the fundamental physicochemical processes, which determine the incorporation of these compounds into the particulate phase as well as the effect of aging
during further transport through the atmosphere.

While organic compounds in aerosols of anthropogenic origin are
mostly combustion related, and hence of the condensation type, enrichment of the dispersion mode in organics may often point to natural
contributions. This is illustrated by the particle size distributions of the higher odd-carbon numbered paraffins, e.g. in background
stations with dense vegetation. The seasonal and particle size dependence of the carbon preference index reflects the contribution of
plant waxes.

Aging of the aerosol is most pronounced for remote sites, where
input of fresh condensation aerosol is negligible. For PAH, aging
involves a clear depletion of the submicron fraction and an enrichment of the fraction between 1.5 and 3.0 μm. The phenomenon can also
be described by the increase of the mass median equivalent diameter
within the accumulation mode.

Using the ICRP Task Group on Lung Dynamics model for deposition
of particles in the respiratory tract, total mass concentrations deposited in the pulmonary, tracheobronchial and nasopharyngeal regions
can be calculated. The differences in particle size distribution between remote and urban environments are clearly reflected in the fractions deposited. A rough estimation of the effective intake of a
pollutant (i.e. the amount resorbed by the tissues) is based on the
preponderance of pulmonary resorption (70 %) and will level out almost completely the differences in particle size distributions between sampling stations. Within narrow limits, the resorbed fraction of organic pollutants amounts to about 20 % of the total concentrations, irrespective of the sampling site.

1. INTRODUCTION

The interest in particle size distribution studies of trace pollutants in ambient aerosols is twofold : not only will the obtained data yield
essential information for the assessment of health hazards, related to
the inhalation of toxic aerosol constituents, but also they will provide
a good basis for the investigation of the dynamics of atmospheric aerosols,
such as the physicochemical processes, involved in the incorporation of pollutants into the particulates. In contrast to the inorganic trace elements, of which the particle size distributions are well documented now,

there is a surprising lack of similar data for organic pollutants. Only
a few studies were reported, most of them being limited to only one or a
few polyaromatic hydrocarbons (1-5).

During the five last years we focussed our research efforts on a systematic study of the particle size distributions of about sixty organic
pollutants in ambient aerosols from different areas. For chemical analysis, we used gas chromatography-mass spectrometry (GC-MS). Extensive qualitative studies, carried out before at our laboratory, supported the
choice of the pollutants we quantitated in the size fractionated samples
(6). These include a series of aliphatic hydrocarbons, carboxylic acids,
polycyclic aromatic hydrocarbons (PAH) and their N-analogs. It can be
stated that these compounds characterise adequately the organic solubles,
as far as their volatility and/or polarity does not preclude gas chromatographic separation.

In view of the low concentrations of organic pollutants on the particulates in the atmosphere, a Hi-Vol cascade impactor had to be used and
even then prolonged sampling during one or several weeks was necessary.
In addition to the well known disadvantages of cascade impactors, concerning particle selection (7), we found that for some aerosol constituents
the use of Hi-Vol techniques can induce important sampling artifacts,
physicochemical or chemical in nature. We have treated this subject in
detail elsewhere (8,9).

In previous reports we have already described in detail the sampling
and analytical procedures we elaborated (10,11). Therefore, the purpose
of this paper is to illustrate which methods we used for the analysis of
our particle size distribution data as well as the major conclusions we
could derive from our measurements.

2. RESULTS AND DISCUSSION

2.1. Data analysis

A major challenge in particle size distribution studies remains the
interpretation. The conversion of the results into data, which are directly related to health hazards, is rather straightforward in contrast in the
difficulties, encountered when the interpretation turns to the physicochemical processes. The reason lies not only in the limited resolving power
of Hi-Vol cascade impactors, but also in the fact that particle size distributions of mass concentrations are measured, whereas most physicochemical mechanisms are rather related to number or area concentrations. It is
our experience that a complementary use of different representations of
the particle size distribution data is absolutely necessary to draw correct
and detailed conclusions. In the discussion we will use mainly the cumulative and the Lundgren distributions.

Cumulative distributions are especially useful to relate measurements
from different cascade impactors and to derive relevant data, such as the
MMED (Median mass equivalent diameter) and σ_g (standard deviation). These
two quantities give only meaningful information if the distribution is
unimodal and log-normal in nature. However, most pollutants in ambient
aerosols follow a bimodal distribution and therefore we preferred to calculate partial cumulative distributions within the accumulation range to
derive the corresponding MMED' and $\sigma g'$.

In the *Lundgren distribution*, $\Delta M.(\Delta \log Dp)^{-1}$ is plotted versus
$\log D_p$ (D_p = particle diameter) (12) : the pollutant concentrations in the
particulate fractions are then normalised on the actual size ranges of the
corresponding impactor stages. This yields an histogram approach of the
continuous particle size distribution, as it exists in the atmosphere.

The Lundgren distribution is an excellent tool for the qualitative analysis of particle size distributions.

The *characteristic profiles* are obtained by plotting the basic data, obtained from the chemical analysis, versus the stagenummer. Usually, the discrete points are connected, but the resulting plot has no real physical meaning. However, this approach is commonly used by chemists.

For toxicological assessment ,the pollutant concentrations present in the particles, which are deposited in the respiratory tract, have to be calculated. The model for particle retention, proposed by the ICRP Task Group on Lung Dynamics (13) as a synthesis of literature data, is adequate for this purpose (8) : the deposition probability of inhaled particles of a given size in each of the three major regions of the human respiratory tract (pulmonary, tracheobronchial and nasopharyngeal organs)is directly related to particle size.

However, health hazards will depend primarily on the fraction of the amounts of toxic aerosol constituents, which will be effectively resorbed by the tissues along the respiratory tract. The resorption of a pollutant from a deposited particle is a very complicated process. Until now, only rough approximations can be made for the resorption efficiency : according to Natusch and Wallace (14), we assumed that about 70% of the material deposited in the pulmonary region, will be resorbed, whereas in the nasopharyngeal and tracheobronchial tissues it only amounts to 10 %.

2.2. Incorporation processes

About all organic pollutants under study were found to be almost exclusively (80 % or more) present in the accumulation mode. This reflects the occurrence of the condensation process as the major incorporation mechanism. Only a few exceptions were observed and will be discussed later.

This general conclusion is not unexpected in view of the known anthropogenic sources for organic aerosol constituents : in particular the PAH, but also the aliphatic hydrocarbons and carboxylic acids are combustion related (15-18). The particle size distributions of the PAH in the suburban winter samples are particularly suitable for the illustration of this condensation process as a result of incomplete combustion of fossil fuels in the immediate vicinity (domestic heating).

Table I summarises some data, derived from the cumulative distributions. Submicron particles contain about 60 % of the total non-volatile PAH concentrations in ambient aerosols and the whole accumulation mode accounts for at least 90 %. Consequently, the cumulative distribution for the total particle size range will practically correspond with the one established for the particles below 2.5 µm, and hence, MMED and MMED' values will show good agreement.

According to Natusch and Wallace (14), the divergence between the MMED for the total suspended particulates (TSP) and the one for an individual aerosol constituent reflects the relative importance of condensation versus dispersion mechanisms : indeed, since condensation always involves a preferential enrichmentof the smaller particles, because of their larger specific surface, the MMED of the pollutant will remain at much lower particle size than the MMED for the total aerosol. The data in Table I confirms this idea.

The characteristic profiles for a young condensation aerosol show major contributions on the stages 6 and 5 (< 0.5 µm : 45 %, 0.5 - 1 µm : 30 %); and the profiles sharply decrease for stages collecting larger particles. However, from these plots, one can only conclude that the submicron fractions account for the largest contributions to the total concentrations of organics in the aerosol, but not that these particles have the highest degree of enrichment.

TABLE I : Cumulative distribution data for PAH in suburban winter samples.

Compound	Benzo(b+k)fluoranthenes			Benzo(a+e)pyrenes		
sample no.	1	2	3	1	2	3
Fraction < 2.5 μm (%)	97	90	87	96	90	87
MMED (μm)	0.58	0.74	0.72	0.58	0.58	0.70
MMED' (μm)	0.59	0.64	0.63	0.60	0.60	0.62
MMED (TSP)	1.0	1.1	1.1			

Indeed, the Lundgren distributions, shown in figure 1 indicate that the maximum concentration for particles of a given size is reached between 1 and 2 μm. Using a cascade impactor, providing a supplementary stage to further resolve the submicron particles at a cutoff of 0.5 μm, it is found that the concentration on the particles between 0.5 and 1 μm reaches about the same level as the one between 1 and 1.5 μm. Thus, depending on the cutoffs of the different stages, the Lundgren distributions can look quite different, in spite of the normalisation on the particle size range. Therefore , this representation method, standing alone, does not provide a reliable basis for conclusions.

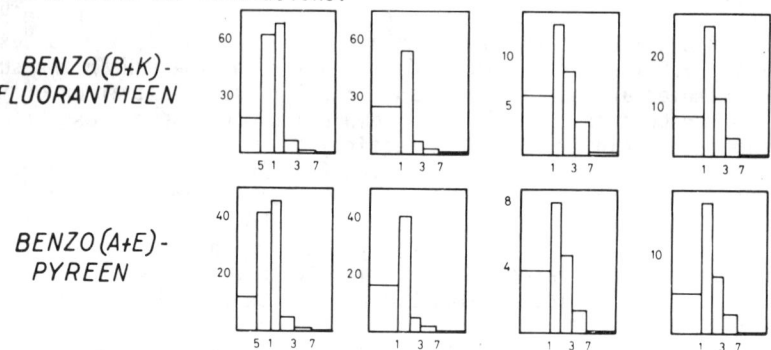

BENZO(B+K)-
FLUORANTHEEN

BENZO(A+E)-
PYREEN

Figure 1 : Typical Lundgren distributions for PAH at the suburb during winter : incorporation by condensation.

Significant enrichment of the particulates by dispersion mechanisms is only observed during summer for the long chain aliphatic hydrocarbons, especially for these with an odd number of carbon atoms (n > 27). However, an important fraction of the total concentration (50-60 %) still remains in the accumulation mode. Some typical Lundgren distributions are shown in figure 2.

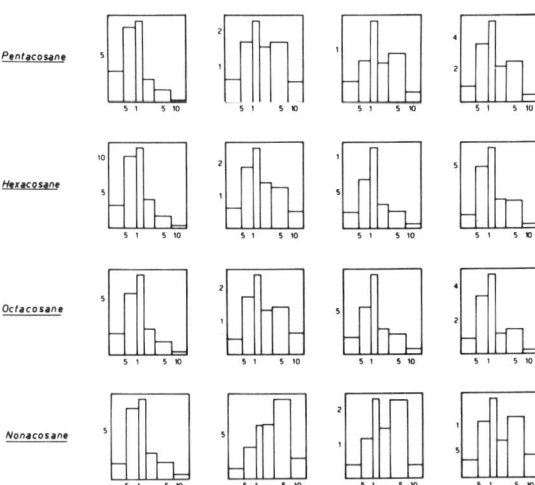

Figure 2 : Particle size distributions of the long chain aliphatic hydrocarbons : incorporation by condensation and dispersion.

The enrichment of the coarse particles with these odd alkanes is
strongly related to the presence of vegetation in the vicinity. It is well
known that waxes at the surface of leaf and plants contain considerable
amounts of these compounds. As a consequence of the biogenesis process,
the odd carbon numbered paraffins reach higher concentrations than the
even ones (19-25). This preference for the odd homologs is very specific
and is used by geochemists to determine the origin of a sample by simple
calculation of a Carbon Preference Index (CPI), i.e. the average of the
concentration ratios of the odd over the even homologs (26). High CPI
values (> 3) point to plant material. Table II summarises the CPI, calculated separately within different particle size ranges. The low values,
observed within the accumulation mode in all samples as well as in the dispersion mode during winter, confirm the anthropogenic origin of the
n-alkanes in these fractions. In contrast, significantly higher values
are reached during summer for the coarse particles and this is most pronounced at the rural and seashore stations.

For toxicological assessment, it is important to know the total deposited and resorbed fractions, as well as the particle size distribution of
the retained material. For all organic pollutants under study, it follows
that :
- the fraction of the total pollutant concentration, which is resorbed,
  remains surprisingly constant : between 18 and 22 %;
- the total deposition is subjected to more important variations : if condensation mechanisms prevail, a fraction of about 55 % is retained, but
  if also dispersion processes become important, significantly higher
  values, up to 75 %, are not unusual.
- the total deposition is mainly determined by the retention in the nasopharyngeal tissues, the pulmonary deposition being nearly constant.

TABLE II : Carbon Preference Index as a function of particle size (μm).

Sample	CPI for particles				
	< 1 μm	1-1.5 μm	1.5-3 μm	3-7 μm	> 7 μm
Winter suburban	1.6	1.6	1.7	2.4	3.8
	1.5	1.5	1.4	1.5	2.0
rural	1.8	1.7	1.6	1.9	2.1
Summer suburban	2.8	2.5	3.9	4.9	2.9
rural	3.5	2.7	3.1	5.0	9.0
sea shore	2.9	2.8	3.7	4.5	8.9

In figure 3 the particle size distribution of the pollutant concentrations in the deposited material for the pulmonary and the nasopharyngeal organs are compared by means of the Lundgren representation method. The distribution of the pollutant concentrations on the particles, deposited in the nasopharyngeal organs is directly related to the occurrence of the dispersion processes. In contrast, these will not affect the histograms for the pulmonary region, which mainly follow the total distribution within the accumulation mode and hence, directly depend on the condensation mechanisms.

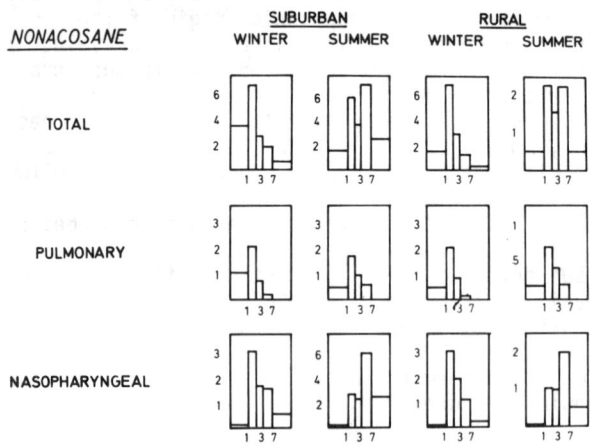

Figure 3 : Particle size distribution of the total and deposited concentrations.

2.3. Aging effects

From the cumulative distribution data, it follows that aging processes do not affect the dominant presence of most organics in the accumulation mode. This is in agreement with the general idea that no mass will
be transferred from the accumulation to the dispersion mode. However,
there are clear indications for a limited shift of the distribution toward larger particle size within the accumulation mode. This is best
illustrated by the MMED' values, in which the contributions of the coarse
particles are filtered out. For the suburban summer samples, the MMED'
are found consistently at somewhat larger particle size than during winter
(cfr. Table III). This is due to the decrease of the combustion related
emissions in the immediate vicinity of the sampling point. The major contributions are now from the industry and road traffic at larger distances.
It is quite surprising that even short term aging can be observed in the
particle size distributions. In the remote areas (30-40 km away from the
major emission sources), the MMED' shift to larger particles is more convincing.

Aging is also reflected in the Lundgren distributions. Figure 4 summarises some examples for the PAH. In contrast to the maximum enrichment
of the particles between 0.5 and 1.5 μm in the freshly emitted condensation aerosol (suburban winter), the PAH become now more sharply distributed around the maximum for the particles between 1 and 1.5 μm, even when
limited aging has occurred. Because of the long term aging at the remote
stations, this trend becomes more clear.

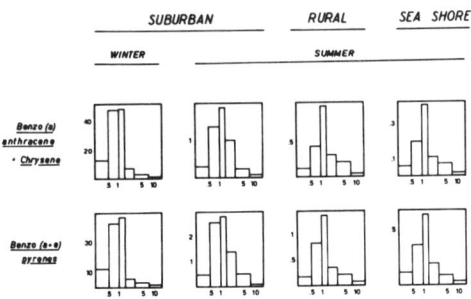

Figure 4 : Survey of Lundgren plots for PAH in suburban and background
stations.

TABLE III : Survey of MMED' values.

	MMED' (μm)			
sample	suburban		rural	
	winter	summer	winter	summer
Compound				
n-heptacosane	0.56	0.63	0.74	0.76
n-octacosane	0.58	0.67	0.69	0.92
n-nonacosane	0.54	0.66	0.74	0.80
benzo(b+k)fluoranthenes	0.59	0.64	0.81	0.82
benzo(a+e)pyrenes	0.60	0.68	0.80	0.80
benzo(ghi)perylene	0.56	0.63	0.85	0.85

    The toxicological implications of the aging processes are related to the limited shifts of the distributions toward the larger particles. In general, as far as no significant dispersion processes occur, the total pulmonary deposition tends to lower somewhat at the background stations, but the increased retention in the nasopharyngeal organs overcompensates for this effect, and hence, causes that the overall deposition also reaches higher values at the background versus the suburban station.

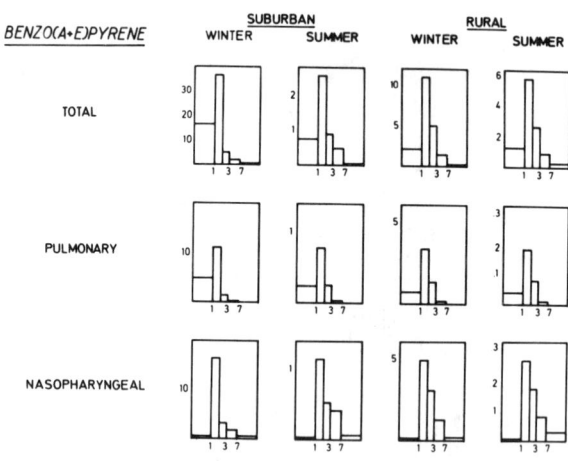

Figure 5 : Particle size distribution of the total and deposited concentrations.

The influence of the aging process on the deposition is clearly illustrated in the Lundgren plots, given in figure 5. The concentrations on
the particles, retained in the pulmonary tissues, again follow the total
distribution within the accumulation mode and reflect the depletion of
the submicron fraction in favour of the particles between 2 and 3 µm at
the rural site. This explains the somewhat lower total pulmonary retention. But the particle size distribution shift also affects the nasopharyngeal deposition : the enrichment of the fraction 2-3 µm is clearly
reflected in the corresponding histograms. Thus, the health hazards in
the remote areas will depend on the final result of two conflicting phenomena : the concentrations are lower, but the overall deposition significantly increases, merely because of the nasopharyngeal retention.

If the effective intake of the pollutants is used as a more realistic
parameter for health hazard assessment, it is found that the resorbed fraction at the background station remains exactly within the same range as for
the suburban site : 20 + 2 %. This is not unexpected in view of the assum-
tions involved : although a dominant role is attributed to the pulmonary
resorption, also the contribution from nasopharyngeally retained material
cannot be neglected. For the pollutants in this study, the nasopharyngeal
resorption compensated for the lower one in the pulmonary tissues.

3. CONCLUSIONS AND RECOMMENDATIONS

Although both analytical and sampling techniques still require further
refinements, in our opinion the study of the particle size distributions
is still justified, in spite of some severe limitations.

An important fraction of the organic aerosol solubles consists of compounds with low volatility and high polarity, which require the use of
LC-techniques. Recent mutagenicity studies attribute an important role
to these polar compounds, some of which are believed to originate, at
least in part, from heterogeneous reactions of the measured compounds, e.g.
PAH, and gaseous trace components in the atmosphere e.g., O_3, NO_2,OH-radi-
cals (27-31).

The occurrence of volatilisation losses for several aerosol constituents upon long term exposure of the already collected particles to the
ambient gas phase, has been shown experimentally, and hence, the particle
size distribution data for volatile organics are not reliable. Indeed,
these volatilisation losses are likely to be related to the actual particle surface, exposed to the air flow, and thus a selective depletion of
the smaller particulates can be expected.

Until now, no precise data are available about the occurrence and the
importance of chemical transformations of aerosol constituents during
sampling. However, from several model studies it follows that the reactivity of PAH can be strongly different. Since we quantified a large number
of compounds for each class, it can be assumed that our measurements covered the whole reactivity range. Nevertheless, by GC we did not detect significant indication for this phenomenon. Further investigations in this
respect should be needed.

Finally, the weakest link in our interpretation concerns the toxicological approach : until now, any method used for relating the particle size
distribution data to health hazard assessment is subjected to major criti-
cisms. Thus, the ICRP model does not take into account the effects of certain obviously important parameters, such as the high relative humidity
in the respiratory tract on the aerodynamic diameter of hygroscopic particles. Furthermore, there is no doubt that the resorption efficiency
of organic pollutants from deposited particles should deserve increased
attention.

REFERENCES

1. M. Kertész-Sáringer, E. Mészáros, R. Várkonyi (1971)
   Atmospheric Environment, 5, 429.
2. R.C. Pierce, M. Katz (1975)
   Environmental Science and Technology, 9, 347.
3. F. De Wiest, H. Della Fiorentina (1977)
   The Science of the Total Environment, 8, 275.
4. M. Katz, R.C. Pierce (1976)
   "Quantitative Distribution of Polynuclear Aromatic Hydrocarbons in
   Relation to Particle Size of Urban Particulates" in "Carcinogenesis",
   Vol. I,"Polynuclear Aromatic Hydrocarbons Chemistry, Metabolism and
   Carcinogenesis", ed. R.I. Freudenthal,P.W. Jones, Raven Press, New
   York, 413.
5. A.H. Miguel, S.K. Friedlander (1978)
   Atmospheric Environment, 12, 2407.
6. W. Cautreels, K. Van Cauwenberghe (1976)
   Atmospheric Environment, 10, 447.
7. V.A. Marple, K. Willeke (1976)
   "Inertial Impactors : Theory, Design and Use" in "Fine Particles", ed.
   BYH Liu, Academic Press, New York, 411.
8. L. Van Vaeck, G. Broddin, K. Van Cauwenberghe (1980)
   Biomedical Mass Spectrometry, 7, 473.
9. K. Van Cauwenberghe, L. Van Vaeck (1981)
   "Toxicological Implications of the Organic Fraction of Aerosols : a
   Chemists View", Mutation Research, submitted.
10. L. Van Vaeck, K. Van Cauwenberghe (1980)
    "Measurement of the particle size distribution and gas phase concentra-
    tion of organic pollutants in ambient air by preprogrammed mass frag-
    mentography" in "Advances in Mass Spectrometry", Vol. VIII, ed.
    A. Quayle, Heyden & Son Ltd., London, 436.
11. L. Van Vaeck, K. Van Cauwenberghe (1977)
    Analytical Letters, 10, 467.
12. D.A. Lundgren, H.J. Paulus (1975)
    Journal of the Air Pollution Control Association, 25, 1227.
13. ICRP - Task Group on Lung Dynamics, II (1966)
    Health Physics, 12, 173.
14. D.F.S. Natusch, J.R. Wallace (1974)
    Science, 186, 695.
15. G.M. Badger, R.W.L. Kimber, J. Novotny (1964)
    Australian Journal of Chemistry, 17, 778.
16. E. Sawicki, J.E. Meeker, M.J. Morgan (1965)
    International Journal of Water and Air Pollution, 9, 291.
17. T.R. Hauser, J.N. Patterson (1972)
    Environmental Science and Technology, 6, 549.
18. D. Hoffmann, E.L. Wynder (1977)
    "Organic Particulate Pollutants : Chemical Analysis and Bioassays for
    Carcinogenicity" in "Air Pollution", ed. A.C. Stern, Academic Press,
    New York, 361.
19. G. Eglinton, A.G. Gonzales, R.J. Hamilton, R.A. Raphael
    Phytochemistry, 1, 89.
20. G. Eglinton, R.J. Hamilton (1963)
    "The Distribution of n-Alkanes" in "Chemical Plant Taxonomy", ed.
    T. Swain, Academic Press, London, 187.

21. G. Eglinton, M. Calvin (1967)
    Scientific American, 216, 82.
22. P.E. Kolattukudy (1968)
    Plant Physiology, 43, 1466.
23. P.E. Kolattukudy, T.J. Walton (1972)
    "The Biochemistry of Plant Cuticular Lipids" in "Progress in Chemistry
    of Fats and Lipids", Vol. 13, ed. K.I. Holmann, Pergamon Press, 121.
24. P.E. Kolattukudy (1975)
    Phytochemistry, 6, 963.
25. A.B. Caldicott, G. Eglinton (1973)
    "Surface Waxes" in "Phytochemistry", Vol. III, ed. L.P. Miller,
    Van Nostrand Reinhold, New York, 1962.
26. E.E. Bray, E.D. Evans (1961)
    Geochimica and Cosmochimica Acta, 22, 2.
27. J.N. Pitts, D. Grosjean, T.M. Mischke, V.F. Simmon, D. Poole (1977)
    Toxicological Letters, 1, 65.
28. K. Teranishi, K. Hamada, H. Watanabe (1978)
    Mutation Research, 56, 276.
29. M. Möller, I. Alfheim (1980)
    Atmospheric Environment, 14, 83.
30. J.N. Pitts, D.M. Lokensgard, P.S. Ripley, K.A. Van Cauwenberghe,
    L. Van Vaeck, L.D. Schaffer, A.J. Thill, W.L. Belser (1980)
    Science, 210, 1347.
31. K. Van Cauwenberghe, L. Van Vaeck, J.N. Pitts, (1980)
    "Physical and Chemical Transformations of Organic Pollutants during
    Aerosol Sampling" in "Proceedings of the First European Symposium on
    the Physicochemical Behaviour of Atmospheric Pollutants", ed. B. Versino
    H. Ott, Commission on the European Communities, Luxembourg, 194.

Acknowledgments

This research was supported by the National Foundation for Scientific
Research of Belgium (NFWO) by means of a grant to L. Van Vaeck as
"research assistant".

THE RELATIVE CONTRIBUTIONS OF SCATTERING AND ABSORPTION OF LIGHT TO VISIBILITY DEGRADATION BY AEROSOLS

A.G. CLARKE and K.J. MORRIS
Department of Fuel and Energy, Leeds University, Leeds LS2 9JT, U.K.

Summary

Simultaneous measurements of the light scattering coefficient b_s and the light absorption coefficient b_a have been made in an urban area. b_s was measured by integrating nephelometer and b_a using the method of Lin, Baker and Charlson. This involves collection of the aerosol on a Nuclepore filter and measuring the reduction in light transmittance using an opal glass diffuser to effectively average out the angular variation of the forward scattered light intensity.

At rooftop level away from the immediate influence of local traffic emissions b_a was found to average 35% of the total extinction coefficient b_e ($= b_a + b_s$). At roadside level where there are significant diesel smoke emissions b_a was found to be up to 10 times higher than the simultaneously measured value on the roof although b_s was increased by less than 50%. Measurements of the mass concentration were made by weighing the filters. The absorption efficiency for freshly emitted particles from traffic b_a/M was estimated to be about 6 m^2 g^{-1} and the corresponding ratio for scattering b_s/M about a factor of 10 lower. The scattering efficiency for the rooftop aerosol was about 2.5 m^2 g^{-1}. The integrating nephelometer thus significantly underestimates the atmospheric extinction coefficient and correspondingly leads to an overestimate of the atmospheric visibility, especially in those situations where significant proportions of carbonaceous aerosols may be anticipated.

1. INTRODUCTION

Atmospheric visibility is inversely related to the atmospheric extinction coefficient which in turn is a summation of the effects of light scattering by gases and particles. Except in very clear atmospheres the contributions from particles are dominant. Interest into the effect of man made pollutants on atmospheric visibility led to the development of instrumentation to measure aerosol optical properties. An instrument to measure the aerosol scattering coefficient had been developed by Beuttell and Brewer (1) and later versions of this integrating nephelometer were constructed by workers in several countries including Charlson and co-workers in the U.S.A. (Ahlquist and Charlson (2)) and Garland and Rae (3) at A.E.R.E. Harwell in the U.K. The U.K. version was specifically developed to study an industrial haze problem at Teesside and has not been widely used elsewhere. The U.S. version has been actively marketed and there are now over 500 instruments in use world wide. Continued interest is likely as air quality standards for visibility have been proposed in the U.S.A. In the absence of any convenient instrument to measure the aerosol absorption effect this has tended to be ignored or assumed to be negligible even though there was early evidence to the contrary (Waldram (4)). Optical absorption measurements over long path lengths in the atmosphere are not very convenient. If the aerosol is collected on a transparent substrate and the reduction in light transmittance measured this does not permit the calculation of the absorption coefficient because no allowance can be made for light scattered by the particles and hence not detected. It was Lin, Baker and Charlson (5) who pointed out that if the particulate layer was placed next to a disk of opal glass the angular distribution of transmitted light plus forward scattered light was made uniform so that measurements over the whole forward hemisphere were unnecessary. Initially the "integrating plate method" was considered to be accurate only to within a factor of two. Subsequently more careful calibration indicated an accuracy of ± 20% (Weiss (6)). Studies in a number of U.S. locations have shown that the aerosol absorption contribution to the extinction is in fact not negligible especially in urban areas.

In order to confirm the U.S. results and to study in particular the effect of traffic emissions on the extinction coefficient we have constructed an absorption meter. It has been used in conjunction with two integrating nephelometers at Leeds University and the preliminary results are reported in this paper.

2.   EXPERIMENTAL DETAILS

Two Meteorology Research Inc. integrating nephelometers were used, one Model 1550 and one Model 1561. These have slightly different optical systems which measure at different peak wavelengths. Model 1550 is a broad band instrument centred round 500 nm. Model 1561 has its peak wavelength at 550 nm which corresponds to the maximum sensitivity of the human eye. In practice after the instruments have been appropriately calibrated with Freon 12 as a standard Rayleigh scatterer the response to atmospheric aerosols was found to be identical. Both instruments were operated with air preheaters to avoid the effect of high humidity on deliquescent particles.

For absorbance measurements particles were collected on 25 mm Nuclepore filters, 0.2 μm pore size. These polycarbonate filters are chosen partly because they are transparent and partly because the collected particles lie on the surface of the filter in contrast to most other types with which particle penetration into the filter occurs. In general, samples of about 0.5 m³ of air were taken over 1 - 2 hours. The collected mass was normally less than 100 μg and was measured using a microbalance when required. Low filter loadings are necessary to avoid multiple scattering effects in the layer of particles. The absorption coefficient is calculated from the reduction in intensity of light transmitted through the filters and the volume of air sampled:-

$$b_a = - (A/V) \ln(I/I_0)$$

where V is the volume of air sampled onto filter area A, 1 is the transmitted light intensity with particles and I_0 that of a clear filter with no particles.

The absorbance measurements were made using a specially constructed instrument following the general principles outlined by Lin, Baker and Charlson (5). The major difference to their design was the use of a large area (100 mm²) photodiode in place of a photomultiplier to detect the light. The lay-out of the instrument is shown in Fig. 1.

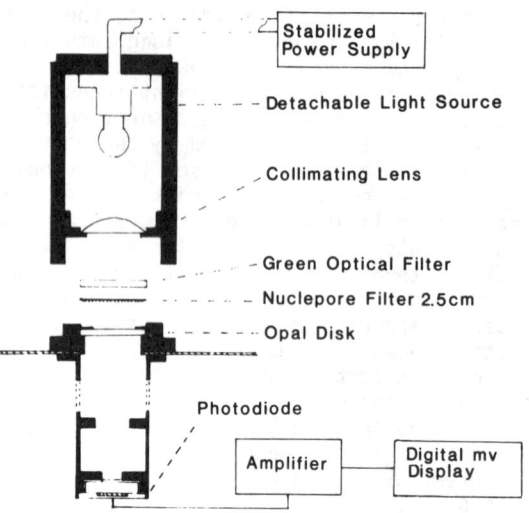

Fig. 1.   The absorbance meter

The light source consisted of a small tungsten filament bulb with a collimating lens.   These were mounted in a turret which could be removed from the rest of the apparatus to allow the introduction of the filters. The bulb was linked to a stabilised power supply unit via a flying lead. The Nuclepore filter was placed, particles downward, on the opal disk and held in position by placing the green glass optical filter on top of it. The photodiode was mounted 85 mm below the opal disk and stray light reflections were reduced with two baffles as shown in the Figure.   The optical filter was chosen to have a peak transmittance at 550 nm with > 5% transmittance in the range 350-620 nm.   The response of the photodiode increased with increasing wavelength in this range so that the combined filter + detector response peaked at 550 nm and had $\geq$ 20% of this peak response over the range 400-600 nm.

The sampling site was at Leeds University ca 1.5 km from the city centre.   The building is immediately adjacent to a busy main road apart from which there is no significant traffic movement or other source of particulate emission within 0.5 km in any direction.   Two sampling points were used, one on the rooftop ca 30 m high and ca 50 m from the road and the other next to the road at a height of 5 m and 3 m from the kerb.

3. RESULTS AND DISCUSSION

Figure 2 shows the results of rooftop measurement of b_s and b_a taken during spring and early summer 1981. b_s is consistently larger than b_a, the slope of the best-fit line corresponding to $b_s = 1.96 \ b_a$ and the correlation coefficient being 0.92. b_a is found to contribute 20 - 50% to the total extinction coefficient b_e (= $b_a + b_s$) the average being 35%. The small contribution of Rayleigh scattering by air (1×10^{-5} m^{-1}) will be neglected as will absorption by gases such as NO .   This finding is in good agreement with the results of Waggoner et al. (7) who found b_a to contribute 35 - 50% of b_e in urban industrial areas of the U.S., 15 - 25% in urban residential areas and 5 - 10% in remote areas.

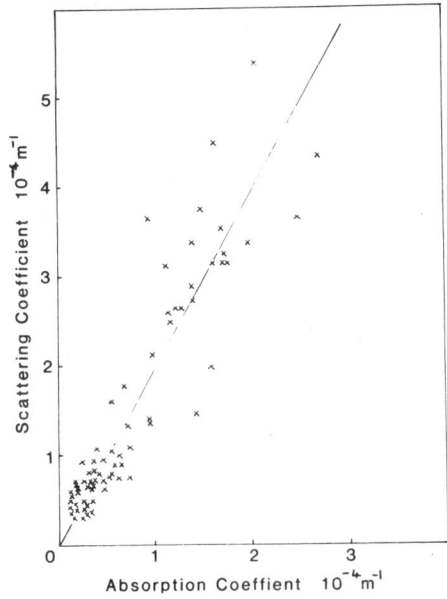

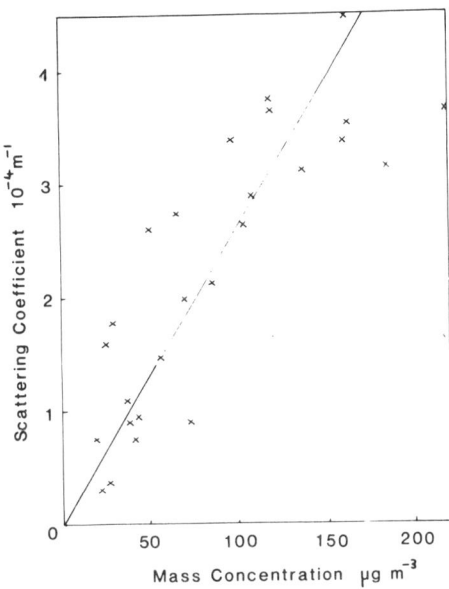

Fig. 2.   Scattering and absorption
          coefficients at rooftop
          level

Fig. 3.   Scattering coefficients
          and mass concentrations
          at rooftop level

Figure 3 shows the relationship of b_s to mass concentration on the
rooftop.   There is significant scatter in the results, the correlation
coefficient being 0.84.   Better correlation might be expected with the
fine fraction of the aerosol using a sampling system such as a dichotomous
sampler (Waggoner and Weiss (8)).   The scattering coefficient to mass
concentration ratio averages 2.5 m² g⁻¹ which is in good agreement with
previous work, e.g. Charlson et al. (9) - 1.7 - 3.6, mean 2.6 m² g⁻¹;
Ettinger and Royer (10) - 3 ± 1 m² g⁻¹; Kretzschmar (11) - 3.4 m² g⁻¹ in
a 2 parameter fit; Clarke et al. (12) - 2.2 m² g⁻¹ also in a 2 parameter
fit.

Figure 4 shows a comparison of the absorption coefficients at rooftop
level and by the roadside.   The correlation is poor (c.c. 0.66) which is
expected since the rooftop sample will rarely be influenced by the traffic
density immediately adjacent to the building.   The roadside absorption
coefficients are from 2 - 10 times higher than at rooftop level for the
daytime samples shown in the figure.   A few additional overnight readings
indicated approximately equal readings for b_a by the roadside and on the
roof in the absence of traffic.   Most of this additional absorption is
thought to arise from carbonaceous diesel engine emissions to which buses
probably contribute the major part.

Figure 5 shows the differences between the roadside and rooftop values
for b_s and b_a.   There are large differences in b_a as previously noted but
very small differences in b_s, i.e. the traffic emissions are strong
absorbers but poor scatters of light.   The data are uncertain but suggest
that the ratio b_a/b_s is of the order of 10.   For carbonaceous particles
with complex refractive indices in the range 1.5 - 0.4 i to 1.5 - 0.7 i
this implies that the average particle size must be about 0.1 μm.   Larger

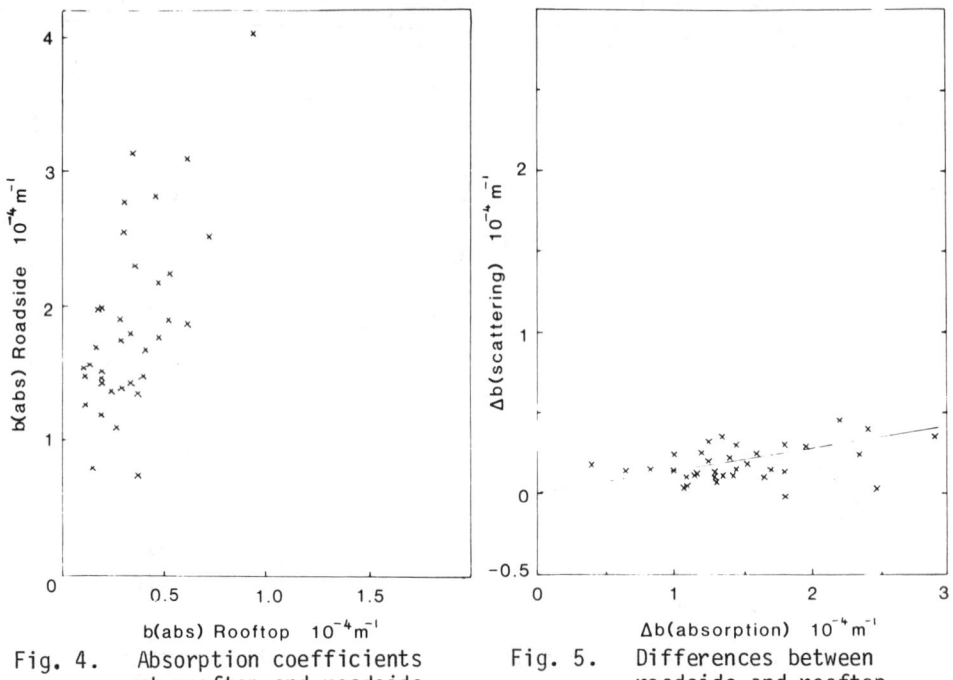

Fig. 4.   Absorption coefficients
           at rooftop and roadside
           levels

Fig. 5.   Differences between
           roadside and rooftop
           coefficients

particles would give a significantly higher proportion of scattering.

We have not yet carried out paired measurements of b_a and mass at both sampling points but a preliminary estimate of the absorbing efficiency (b_a/M) of the traffic emissions can be estimated from the net absorption coefficient (roadside-roof) and the roadside mass reduced by a mass equivalent to the measured b_s on the roof. The approximate value of b_a/M for the traffic emissions is 6 m^2 g^{-1}, with an uncertainty of about a factor of two. This is within the range of 5 to 11 m^2 g^{-1} found by Heisler et al. (13) for inorganic carbon particles in Denver, Colorado and roughly consistent with the results of Sadler et al. (14) in a highway tunnel ($b_a/M = 3$ for M = 30 μgm^{-3} from their regression line).

Pierson and McKee (15) measured b_s in a highway tunnel and compared b_s/M values for traffic emissions with ambient air. They also found particles from traffic to be relatively poor light scatterers, b_s averaging 1 - 2 m g$^-$ compared to their ambient values of > 6 m^2 g^{-1}. (The latter seem higher than found elsewhere). By monitoring the mix of diesel and gasoline powered vehicles they derived b_s for gasoline vehicles to be < 1.4 m^2 g^{-1} and for diesel engines 3.1 m^2 g^{-1}. In the absence of any absorption data they concluded that "light scattering by vehicle aerosol does not substantially degrade atmospheric visibility along the roadway". This conclusion can now be seen to be unwarranted.

The above results all refer to dry particles. As the relative humidity increases water soluble atmospheric particles such as sulphates, nitrates and chlorides absorb progressively more water and the scattering coefficient increases. In contrast the highly absorbing carbonaceous particles may be expected to be unaffected until very nearly 100% R.H. The relative importance of absorption to the overall extinction coefficient

therefore decreases at high humidity. In a sample calculation
Moghadassi (16) took a realistic size distribution of fine particles in
the range 0.05 - 2.5 μm consisting of 40% water soluble salts, 30%
carbonaceous and 30% insoluble, non-absorbing particles with refractive
indices 1.5, 1.55 - 0.66i, and 1.5 respectively. At high humidities the
refractive index and size distribution of the water soluble fraction were
recalculated to allow for increasing water absorption. In the dry state
the carbonaceous particles contributed 57% to b_e and the soluble salts 24%
but by 85% R.H. the relative contributions were almost reversed -
carbonaceous particles 31%, water soluble 58%.

4. CONCLUSIONS

The extinction coefficient of urban air has a highly significant
contribution from absorbing particles. For well mixed air away from very
local sources the absorption contribution can be up to 50% and averaged
35% in our experiments. At high relative humidities the percentage would
be reduced as the total extinction increases but the absorption remains
constant. At sites severely affected by local traffic, especially diesel
vehicles, absorption can dominate the total extinction. The scattering
coefficient which is largely determined by secondary pollutants such as
sulphates may be expected to be less spatially variable within an urban
area than the absorption coefficient.
The integrating nephelometer records only b_s and therefore underestimates b_e. Derived atmospheric visibilities would correspondingly be
overestimated. In remote rural areas the effect is less significant.

ACKNOWLEDGEMENT

The financial support of the Science Research Council in the form of
a studentship for K.J.M. is gratefully acknowledged.

REFERENCES

1. R.G. Bentell and A.W. Brewer.   J. Scientific Inst. 26, 357 (1949).

2. N.C. Ahlquist and R.J. Charlson.   J. Air Poll. Cont. Ass. 17, 457
    (1967).

3. J.A. Garland and J.B. Rae.   J. Physics E: Scientific Inst. 3, 275
    (1970).

4. J.M. Waldram.   Q.J. Met. Soc. 71, 319 (1945).

5. C-I. Lin, M.B. Baker and R.J. Charlson.   App. Opt. 12, 1356 (1973).

6. R.E. Weiss, A.P. Waggoner, D.L. Thorsell, J.S. Hall, L.A. Riley and
    R.J. Charlson.   Proc. Conf. on Carbonaceous Particles in the
    Atmosphere, Paper No. 41, Lawrence Berkeley Lab., California (1978).

7. A.P. Waggoner, R.E. Weiss, N.C. Ahlquist, D.S. Covert, S. Will and
    R.J. Charlson, "Optical Characteristics of Atmospheric Aerosols",
    Atmos. Env. (1981) In press.

8. A.P. Waggoner and R.E. Weiss.   Atmos. Env. 14, 623 (1980).

9. R.J. Charlson, N.C. Ahlquist and H. Horvath.   Atmos. Env. 2, 455
    (1968).

10. H.J. Ettinger and G.W. Royer.   J. Air Poll. Cont. Ass. 22, 108 (1972).

11. J.G. Kretzschmar.   Atmos. Env. 9, 931 (1975).

12. A.G. Clarke, M.A. Moghadassi and A. Williams.   J. Aerosol Sci. 8, 73 (1977).

13. S.L. Heisler, R.C. Henry, J.G. Watson and G.M. Hidy, "The Denver Winter Haze Study, Vol. II.".   Document P-5417-1, ERT, Westlake Village, California (1980).

14. M. Sadler, R.J. Charlson, H. Rosen and T. Novakov.   Atmos. Env. 15, 1265 (1981).

15. W.R. Pierson and D.E. McKee.   J. Air Poll. Cont. Ass. 28, 604 (1978).

16. M.A. Moghadassi.   Ph.D. Thesis, Leeds University (1980).

THE FATE OF LEAD, ZINC AND CADMIUM PARTICLES EMITTED FROM A LEAD SMELTERY STACK

J. HRŠAK and M. FUGAŠ
Institute for Medical Research and Occupational Health, Zagreb
Yugoslavia

Summary

The relationship between lead, zinc and cadmium content of suspended particulate matter (SPM) and depositions was studied in the vicinity of a lead smeltery and related to the relationship of these metals in the ore concentrate, filter and stack dust. In order to get more information about possible changes in particles during their residence in the atmosphere the relative solubility of the three metals in water and in EDTA solution was also investigated.
The results show that a constant enrichment of zinc in particles in relation to lead occurs on the route of emitted dust to the place of deposition. The water soluble fraction of lead and cadmium is smaller in SPM than in depositions. The EDTA soluble portion of zinc increases with the distance of the smeltery stack. Lead solubility in EDTA increases with the decreasing particle size. Further investigations are in progress.

1. INTRODUCTION

The analysis of the relationship between lead, zinc and cadmium in the environment of a lead smeltery for the period 1972-1976 has shown that the Cd/Pb ratio, although varying from day to day, remains in the average the same in both airborne particles and depositions, while the Zn/Pb ratio is higher in depositions indicating possible changes in Zn particles during the transport (1).
In 1978 a new efficient bag filter system was installed in the lead smeltery reducing airborne lead levels, expressed as annual mean, from 20.8 to 1.6 $\mu g/m^3$ and lead in the depositions from 355 to 55 $mg/m^2/month$ (2).
After installation of a new control system the samples of airborne particles and depositions collected continuously during a one-year period were again analysed for lead, zinc and cadmium and the mutual relationships among the concentrations of the three metals were studied. In order to get more information about the possible chemical changes in particles during their residence in the atmosphere the relative solubility of the three metals in H_2O and EDTA solution (against that in HNO_3) was also investigated. The solubility in the EDTA solution was at the same time an indication of their biological availability.

The obtained relationships were related to those in ore
concentrate, filter dust and flue gases.

2. COLLECTION AND ANALYSIS OF SAMPLES

Samples of suspended particulate matter (SPM) were collected at four sites along the smeltery valley: two north and
two southwest from the smeltery. Samples of depositions were
collected at three sites: one close to the smeltery and two
next to the southwest air sampling stations. Dust samples from
flue gases were collected from the duct leading to the stack.
Samples of ore concentrates and bag filter dust were obtained
in the smeltery.
Weekly samples of SPM were collected continuously by
means of low volume sampler on 10 cm Ø membrane filters at a
10 l/min flow rate. A limited number of samples fractionated
by size was collected by a modified Andersen cascade impactor
(3).
Monthly samples of depositions were collected in a
Bergerhof type deposit gauge (4).
The flue gas dust samples were collected on 10 cm Ø membrane filters at a 130 l/min flow rate under isokinetic conditions.
Dust samples and samples collected on filters were divided into two parts, one extracted by 0.5% EDTA solution (pH 8)
and the other by HNO_3. A limited number of samples were divided into three parts and the third part was extracted by redistilled water.
Samples of depositions were filtered and the insoluble
part was extracted with EDTA solution. After the extract had
been decanted the residue was destroyed with HNO_3.

The volumes of EDTA extracts were reduced. The water extracts were evaporated to reduce the volume and concentrated
ammoniacal solution of EDTA was added to give a final solution
with 1% EDTA and pH 8. The HNO_3 extracts were evaporated to
dryness and after the acid vapours had been expelled, they
were redisolved in 1% EDTA solution. Thus all final solutions
contained 1% EDTA at pH 8, which was shown to prevent adsorption of trace metals by walls and to improve atomization in
the presence of phosphates.
The analysis was performed on a UNICAM SP 90 atomic absorption spectrophotometer.

3. RESULTS

The relative proportion (%) of Zn to Pb and Cd to Pb in
all samples as well as the percentage of Pb in the dust (if
such data were available) are shown in Table 1.

The solubility of the three metals in EDTA solution and
redistilled water expressed as percentage of total Pb, Zn
and Cd (HNO_3 soluble) is shown in Table 2.

Table 1 - The relationship Pb/total dust, Zn/Pb and Cd/Pb in the samples

Sample of		N	% Pb in dust	Zn/Pb $\times 10^2$	Cd/Pb $\times 10^2$
Ore con-centrate		2	75	4.74	0.10
Filter bag dust		2	60	8.5	0.80
Stack dust		8	2.4	4.3	0.7
SPM	I	52	7.9	19.0	0.61
	II	50	5.8	18.5	0.60
	III	48	2.9	18.9	0.61
	IV	49	6.3	15.7	0.60
Deposi-tions	V	11		21.4	0.66
	III	12		31.8	0.85
	IV	10		42.3	1.28

Table 2 - Relative solubility of Pb, Zn and Cd in the samples

Sample of		Pb % soluble in EDTA	Pb % soluble in H_2O	Zn % soluble in EDTA	Zn % soluble in H_2O	Cd % soluble in EDTA	Cd % soluble in H_2O
Ore con-centrate	I	11.7	8.0	9.83	7.87	52.9	45.2
	II	28.5	14.5	4.68	2.31	7.1	4.3
Filter bag dust	I	97.8	62.4	90.9	52.1	100.0	79.4
	II	96.2	60.5	91.4	75.3	92.5	75.3
Stack dust		94.5		83.4		83.4	
SPM	I	82.7		71.8		89.0	
	II	84.3		71.8		88.1	
	III	87.2		83.3		89.8	
	IV	84.6		68.0		94.0	
Composite			36.8		61		65.1
Deposi-tions	V	60.3	22.0	61.3	48.4	69.2	50.3
	III	83.3	66.8	71.2	60.8	88.7	80.0
	IV	86.3	66.2	86.5	76.9	90.0	79.0

4. DISCUSSION

The relative proportion of Zn to Pb in the SPM samples in-creased from about 10% before the new filter bag system was in-stalled to 16-19% after installation. This may be explained by a stronger influence of background zinc levels after the concentrations in the flue gases have decreased. The correlation between Pb and Zn in airborne particles, however, has remained high (r > 0.8) suggesting that most particles containing Pb and Zn come from the same source.

The ratio of Zn to Pb (%) in depositions which was before the installation of new filter bag system higher than in the SPM samples (18%) remained high and increased with the distance from the smeltery stack.

The relative proportion of Cd to Pb in the SPM samples is remarkably constant (0.60%) and only a little higher than before introduction of new control measures (0.50%). The correlation between Pb and Cd in the SPM remained very high (> 0.8).

In depositions where the relative proportion of Cd to Pb was practically the same as in SPM (0.52%) a tendency of increase with the distance from the smeltery stack was observed at a rate similar to that of the Zn/Pb ratio.

The change in Zn/Pb and Cd/Pb ratio from ore concentrate to the dust retained by filter bag and emitted through the stack should be attributed to the technological process which is adjusted to produce lead. The results of stack dust measurements are only informative and represent conditions during the period of sampling (30 to 60 min samples within 6 days) at the bottom of the 100 m high stack. Further changes may occur within the stack before the flue gases reach the top. The relative proportion of Cd/Pb in the stack dust is not very different from that in the SPM, but the Zn/Pb ratio is lower. It seems that a constant enrichment of zinc in relation to lead occurs on the route of the emitted dust to the place of deposition.

The relative solubility of Pb, Zn and Cd in EDTA cannot explain the shift in the Zn/Pb ratio. The relatively low extractability of the metals from ore concentrate samples by EDTA and H_2O was expected since they are mostly in the form of sulphides.

The lower solubility of Pb, Zn and Cd in depositions close to the lead smeltery (V) than at the more distant sites (III and IV) is most probably a consequence of low sources emitting ore dust (crushing).

Water extracts of a limited number of composite SPM samples (7 from each site) show a lower solubility of airborne than of deposited lead (37% against 66%) and Cd (65% against 80%) which may indicate that water soluble Pb and Cd are more readily washed out by rain. On the other hand there is practically no difference in the percentage of water soluble fraction of Zn in SPM and depositions at site III, but the EDTA soluble fraction decreases in SPM and increases in depositions from site III to site IV.

There was not much difference in the equivalent aerodinamic mass median diameter (MMD) of particles containing Pb, Zn or Cd (1.4-1.2 µm) but MMDs are generally smaller after the new bag filter system had been introduced. The portion of EDTA chelatable lead tends to increase with the decrease in particle size as shown in Fig. 1.

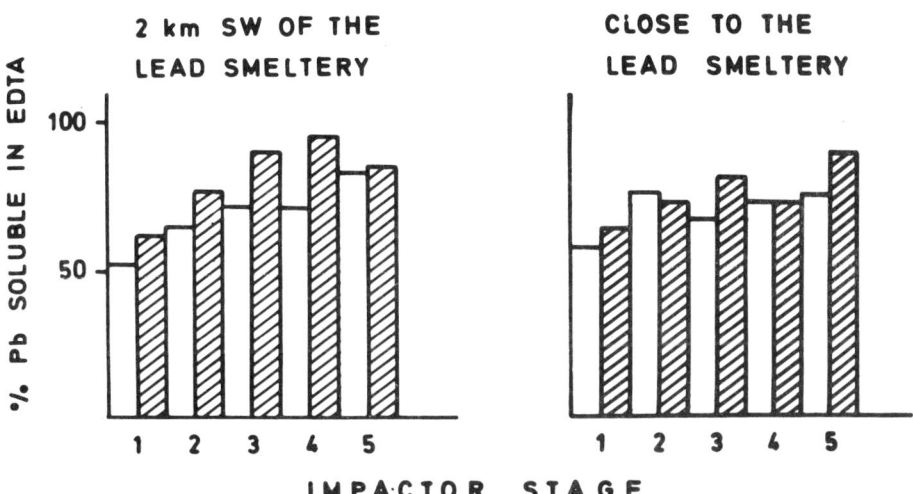

Fig. 1. Percentage of lead chelatable with EDTA by particle
size

Further investigation is in progress with the aim to better define behaviour and fate of Pb, Zn and Cd particles aro-
und the lead smeltery.

References

1. Hršak, J., Fugaš, M.: Distribution of Particulate Lead,
   Zinc and Cadmium Around a Lead Smeltery, Proceedings of the
   International Conference "Management and Control of Heavy
   Metals in the Environment", London 1979, CEP Consultants
   Ltd., Edinburgh 1979, p. 584-587.

2. Fugaš, M., Hršak, J., Souvent, P.: Effect of a New Emission
   Control System on Lead, Zinc and Cadmium Concentrations in
   the Environment of a Lead Smeltery, 5th International Congress for Clean Air, Buenos Aires 1980, Abstracts,
   pp. 221-222.

3. Lee, R.E. Jr. and Flesh, J.P.: A Gravimetric Method for
   Determining the Size Distribution of Particulates Suspended
   in Air Pollution Control Association, New York 1969.

4. VDI - Richtlinie 2119 Bl. 2 (Juni 1972).

A STUDY OF AEROSOLS IN SMOKE

J.A. SCOTT

Physics Department
University College Dublin.

Summary

Experiments have been carried out to determine the physical parameters of aerosols (Condensation Nuclei) in smoke. The measurements were made using the Nolan-Pollak Photoelectric Nucleus Counter together with other ancilliary apparatus which enabled the concentrations, diffusion coefficients and radii of the aerosols to be determined.

Aerosols from two materials - cotton wick and tobacco - were examined. The experiments were carried out in a test chamber which was designed to produce a homogeneous smoke mixture. Using a fan the smoke could be circulated at different flow rates under conditions of laminar flow. The measurements enables studies to be made on the variation of aerosol concentrations and size with various parameters such as circulation air flow rates, time and combustible material.

INTRODUCTION

It has been known for some considerable time that condensation nuclei are produced in combustion processes. However, with the advent of reliable nucleus counters many studies have been carried out on the condensation nuclei produced in combustion. In the Atmospheric Physics Laboratory in University College Dublin, nuclei produced, for example, from bunsen burners and incandescent metals have been used in many studies ranging from the calibration of nucleus counters (1) to the measurement of coagulation coefficients (2).

As a result of renewed interest in atmospheric pollution and indeed in the use of smoke detectors as devices from the early detection of fires (3), work is again being undertaken to study the condensation nuclei produced by combustion. The experiments being carried out in U.C.D. are designed to investigate the properties of these nuclei by measurements made under controlled laboratory conditions.

APPARATUS

(i) Aerosol Measurement: The nuclei were measured using the Nolan-Pollak Photoelectric Nucleus Counter (1). The measurements described in this report were made using the original manual version of the counter rather than the later automatic versions (4) and (5). Because of the operating procedure with the manual counter it is not possible to exceed one measurement every two minutes.

As the initial concentrations of aerosols often exceeded 10^6 nuclei per cm^3 it was found necessary to dilute the samples using pure condensation nucleus free air. By carefully adjusting the volume flows of the pure air and the sample it was found possible to obtain dilutions in the required ratio. In all experiments described here the ratio concentration used was 10:1. The technique was checked at the start of each experimental run and was found to give very good agreement, well within the accuracy of the counter. The experimental arrangement is shown in figure 1.

The diffusion coefficients of the nuclei were measured using the technique of Nolan and Guerrini (6). This makes use of a diffusion box which consists of a number of glass plates formed into a compact pile, each plate being separated from the adjacent plate by thin spacers. The box when completed consists of a number of reactangular channels with the glass plates mounted in the vertical plane so that losses of aerosols under gravity can be assumed to be negligible.

When air containing the nuclei under investigation is passed at a fixed velocity through the diffusion box, and the concentration measured at the entry (Z) and exit (Zv) of the box, it can be shown (7) that the ratio of Zv/Z is given by

$$\frac{Zv}{Z} = 0.9099e^{-x} + 0.053e^{-11.369x}$$

with

$$x = \frac{3.77 \; blc \; D}{aQ}$$

where b, l and 2a are the breadth, length and separation of the plates respectively, c is the number of channels, Q the volume flow per second and D the diffusion coefficient. Having measured Zv/Z, the value of x is obtained from the above equation and hence the value of D can be calculated from the expression:-

$$D = \frac{a}{3.77blc} \; .X.Q.$$

Having determined D, the radius r of the nuclei is obtained from a modified Einstein equation:

$$D = \frac{RT}{N} \; . \; \frac{1 + (\lambda/r) \quad A + Be^{-(cr/\lambda)}}{6\pi \eta r}$$

where R is the gas constant, T the absolute temperature, N is Avogadro's number, λ the mean free path and A, B and C are constants. The actual practical procedure used to determine the diffusion coefficient is to make alternate measurements of the nucleus concentration measured directly (Z) and after passing through the diffusion box (Zv). A measurement of Z or Zv was taken every two minutes, the results were plotted and the corresponding values of Z and Zv were read from the graph. In the experiments a number of diffusion boxes with different constants (k = a/3.77blc) were used with a particular box chosen so as to try to ensure that the ratio of Zv/Z fell within the range 0.7 to 0.3 (7).

(ii) Test Chamber: This consisted of a wooden cabinet 1.67 m long, 0.46 m wide and 0.49 m deep. The interior is divided into two compartments separated by timber 1.06 m long covering the entire width of the chamber (figure 2). A circulating fan is mounted beneath the centre board at one end and an exhaust fan is similarly mounted at the other end. Deflector plates and honeycomb channels are incorporated into the chamber to ensure that the air flow is streamlined and that homogeneous mixing of the smoke takes place. The smoke generator is introduced through a door at one end and sliding door at the other end enables exhaust to be pumped through a metal duct to outside the laboratory.

The circulating fan, which is controlled by a variac enables the air flow in the chanber to be varied. The air flow in the chanber was measured using an "Alnor" hot wire anemometer.

(iii) Smoke Generator: In these experiments the smoke was produced by either a cotton wick or tobacco. The cotton wick 3.2 mm in diameter was lit and the smoldering wick was mounted in a perforated metal cylinder and placed in the chamber by means of the door at the end of the test chamber. The tobacco, in the form of a commercially obtained cigarette, was similarly mounted in the chamber. It should be noted that after each experiment the test chamber was thoroughly ventilated by means of the exhaust before a new experiment was commenced.

EXPERIMENTAL RESULTS

(i) Cotton Wick: The measurements were made in the test chamber using circulating air velocities of 0.16, 0.25 and 0.36 ms^{-1}. In each case the cotton wick was lit and left smoldering in the chamber for three minutes. The nucleus concentration was measured just before the wick was lit, in order to determine the background level, and then at regular intervals until the background concentration was again reached. Checks were made before and after each experiment to ensure that the dilution ratios were constant.

The results of the experiments are shown in figure 3, in which the nucleus concentration is plotted as function of time for the three air velocities. There are a number of points worth noting. In each case the maximum concentration occurs at approximately the same time into the experiment, with the peak concentration increasing with increase in air velocity in the chamber. Another interesting feature is the rate of decrease in concentration with time, with the higher flow rates showing the faster fall off.

(ii) Tobacco: The procedure followed here was exactly the same as that in the case of the cotton, with measurements being made at the same intervals and with the same circulating air velocities in the chambers. The results are shown in figure 4. The features observed with results from nuclei produced from the cotton wick are again evident in this case. One difference is that the peak concentrations at the three air velocities are higher than the corresponding peaks observed with nuclei from the cotton wick.

AEROSOL SIZE

The radii of the nuclei were obtained by measuring the diffusion coefficients. Because of the very rapid changes in concentration in the initial stages of the experiments it was not possible in this present investigation to measure the diffusion coefficients during the first five minutes of each experiment. Thus the first determination of the size is made three minutes after the removal of the smoke source. The results obtained are summarized in the following tables.

(i) Cotton Wick:-

Time after removal of smoke source (min)	Aerosol radius (10^{-8} m) at flow rate (ms^{-1}) of		
	0.16	0.25	0.36
3	3.8	3.8	3.9
5	4.85	4.9	4.8
10	5.1	5.6	7.15
15	5.7	6.7	8.6
20	6.05	7.3	
25	6.25		

(ii) Tobacco:-

Time after removal of smoke source (min)	Aerosol radius (10^{-8} m) at flow rate (ms^{-1}) of		
	0.16	0.25	0.36
3	5.45	5.45	5.45
5	5.6	5.6	5.65
10	6.1	6.15	6.9
15	6.8	7.1	7.5
20	7.5	7.7	
25	7.65		

The results show that for both materials under test the radii of the aerosols measured at the three flow rates are virtually equal for the same material, with the initial size of the aerosols, determined three minutes after removal of the smoke source, being larger from the tobacco. However as can be seen from the tables, the rate of growth in size is bigger for aerosols derived from the cotton wick.

EFFECT OF AIR CIRCULATION ON AEROSOLS

As was stated earlier the test chamber was built to specifications which indicated that the air circulation would remain laminar up to flow rates of 0.76 ms^{-1}, which is twice the max flow rate used in this study. However it was decided to investigate what effect, if any, the air circulation had on both the aerosol concentrations and sizes. The procedure followed was to operate the system exactly as before up to the time of removal of the smoke source. In this set of experiments when the source (i.e. smouldering wick or tobacco) was removed from the test chamber after burning for three minutes, the circulating fan was switched off. Thus from a time of three minutes into the experiments the aerosols were measured as before, but under static (i.e. zero circulation flow) conditions. The results for the cotton wick are shown in figure 5. Comparing these results with those shown in figure 3, it can be seen that the decrease in concentration is much more rapid, in all cases, when the air is circulating. Holub et al (8) when investigating the reduction of airborne radon daughter concentration by plateout on an air mixing fan, found that although there was a variation in condensation nucleus concentrations, they found no correlation of the variation with fan motion.

Experiments were also carried out using tobacco smoke and here again the same trend, as with aerosols from cotton wick was found. The

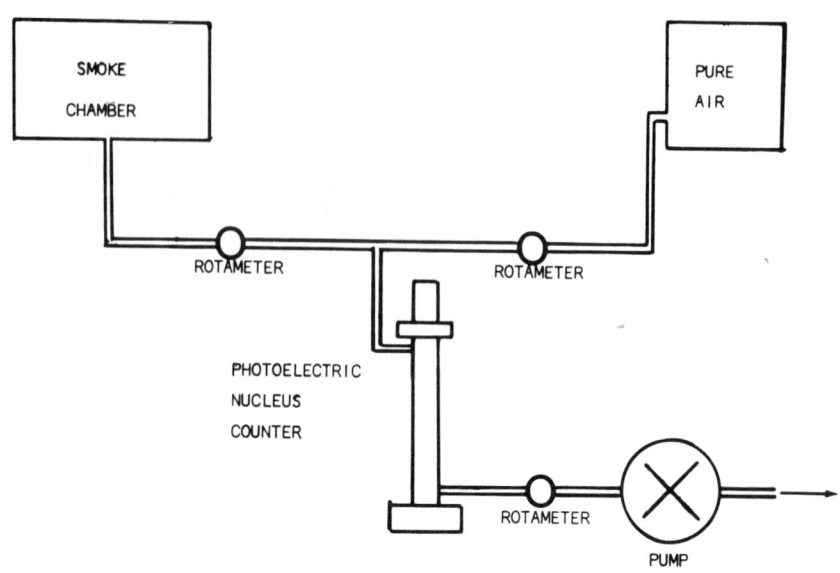

FIGURE 1 - EXPERIMENTAL ARRANGEMENT

(iii) Cotton Wick with static air:-

Time after removal of smoke source (min)	Aerosol radius (10^{-8} m) at flow rate (ms^{-1}) of		
	0.16	0.25	0.36
3	6.05	5.8	5.45
5	6.1	5.9	5.5
10	6.5	6.05	5.9
15	7.2	6.3	6.4
20	7.5	6.6	6.65
25	7.65	6.9	7.3

(iv) Tobacco with static air:-

Time after removal of smoke source (min)	Aerosol radius (10^{-8} m) at flow rate (ms^{-1}) of		
	0.16	0.25	0.36
3	5.7	5.7	5.7
5	5.75	5.85	5.85
10	5.85	6.5	6.15
15	6.45	7.0	6.5
20	7.4	7.5	7.0
25	7.9	7.85	7.55

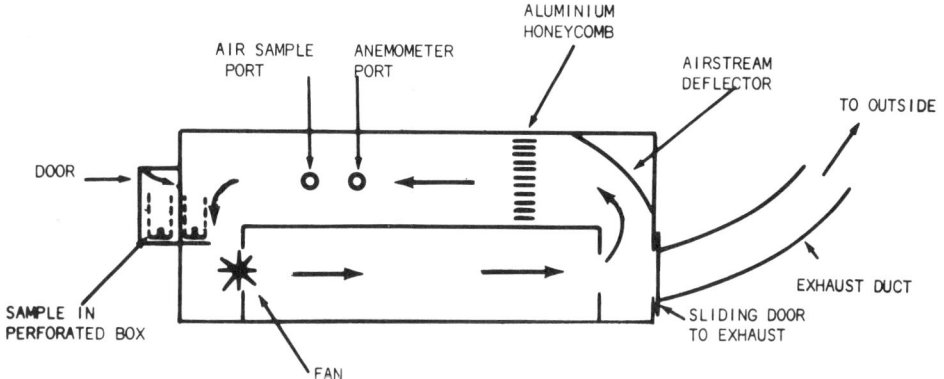

LENGTH = 1.67 m.;   WIDTH = 0.46 m.;   DEPTH = 0.49 m.

FIGURE 2 - TEST CHAMBER

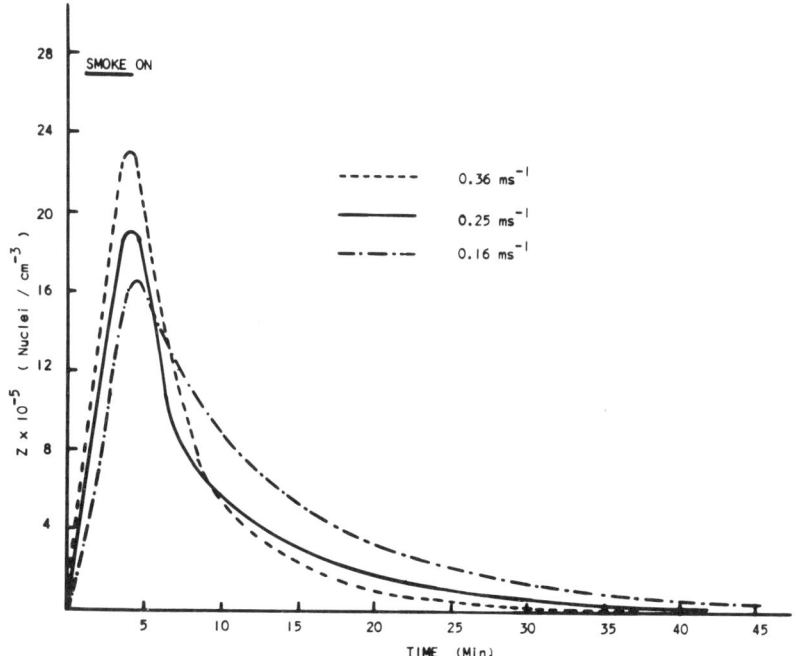

FIGURE 3 - Variation of aerosol concentration with time (COTTON WICK)

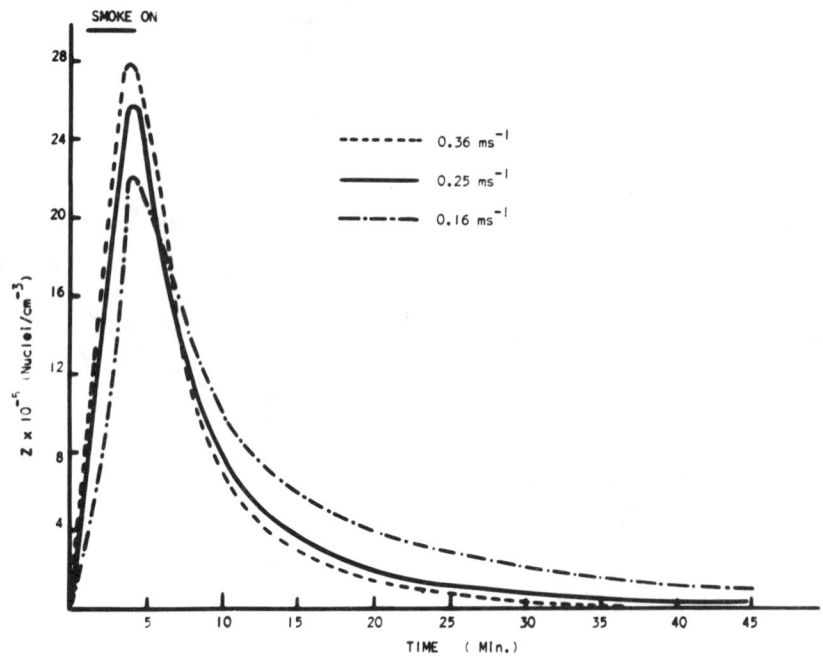

FIGURE 4 – Variation of aerosol concentration with time (TOBACCO)

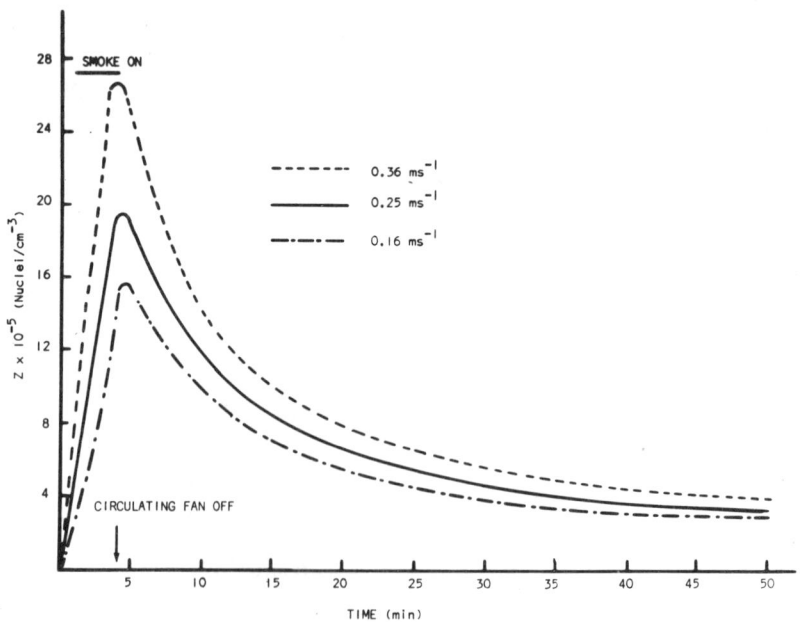

FIGURE 5 – Variation of aerosol concentration with time in static air
(COTTON WICK)

form of the curves exactly correspond to those shown in figure 5.

The radii of the aerosols, from both sources were measured using a method similar to that described earlier. The fan was switched off at the same time as the smoke source was removed from the test chamber, and the initial measurement was made three minutes later. The results of these determinations may be summarized in the tables on page 296.

Comparing these results with those determined with the air circulating in the test chamber (Tables (i) and (ii)) show that the initial size of aerosol measured under still air conditions are in general larger, with the difference in radius, determined at three minutes period, being much greater in the case of aerosols obtained from the cotton wick. The growth in size of the particles with time from both sources is smaller in the static air case.

The results of the experiments discussed now must be considered as an initial investigation. An extended extensive series of experiments is planned to study further the nature and properties of aerosols obtained from combustible materials.

REFERENCES

(1) Nolan, P.J. and Pollak, L.W. 1946. *The calibration of a photoelectric nucleus counter.* Proc. R.Ir. Acad. 51A (2), 9-13.

(2) Nolan, P.J. and Kennan, E.L. 1949. *Condensation nuclei from hot platinum; size, coagulation coefficient and charge distribution.* Proc. R.Ir. Acad. 52A, 171-190.

(3) Helsper, C., Fissan, H.J., Muggli, J. and Scheidweiler, A. 1980. *Particle number distributions of aerosols from test fires.* J. Aerosol Sci. 11, 439-446.

(4) Hayes, E.I. and Scott, J.A., 1969. *An automatic photoelectric condensation nucleus counter.* Proc. R. Ir. Acad. 68A (3), 33-39.

(5) McMahon, G.G. and Scott, J.A. 1979. *An automatic photoelectric condensation nucleus counter with digital recording.* Proc. R.Ir. Acad. 79A (8), 73-80.

(6) Nolan, J.J. and Guerrini, V.H. 1935. *The diffusion coefficients and velocities of fall in air of atmospheric condensation nuclei.* Proc. R.Ir. Acad. 43 (a) 5-24.

(7) Nolan, P.J. and Scott, J.A. 1963. *Observations on the heterogeneity of condensation nuclei.* Proc. R. Ir. Acad. 63(a) 2, 35-47.

(8) Holub, R.F., Droullard, R.F., Wu-Lieh Ho, Hopke, P.K., Parsley, R. and Stukel, J.J. 1979. *The reduction of airborne radon daughter concentration by plateout on an air mixing fan.* Health Phys. 36, 497-504.

THE SIZE AND CHEMICAL COMPOSITION OF SUSPENDED PARTICLES IN THE AMBIENT ATMOSPHERE

P. CLAYTON and S.C. WALLIN
Air Pollution Division, Warren Spring Laboratory,
Stevenage, UK

Summary

The size distributions and chemical compositions of the suspended particles in the ambient atmosphere were determined at five sites in the United Kingdom chosen to represent areas of different industrial, commercial and residential activity. Typical particle size distributions were bimodal with the major modal value in the range 0.6 to 0.8 μm. Average particle concentrations ranged between 40 and 205 μg m^{-3}. Chemically, carbon and sulphur accounted for more than 60 per cent of the total particulate material in many cases.

1. INTRODUCTION

Much information (1,2) has been published describing the physical and chemical characteristics of suspended particulates in various parts of the world. With the exception of surveys of localities with respect to particular industries (3,4,5) and studies of specific pollutants such as sulphate (6) little published work (7) is available regarding the general physical and chemical characteristics of particles in the ambient atmosphere in the United Kingdom.

The investigations described in this paper were therefore intended to provide preliminary information regarding the concentration, size distribution and chemical composition of particles in the ambient atmosphere at a number of sites in the United Kingdom. It was of particular interest to obtain data from areas having different types of industrial, commercial or residential premises and consequently the following locations were chosen;

(a) St Albans - a site representing a residential area.

(b) North Fleet - a site representing a residential area with cement manufacture as the major industrial activity.

(c) London, Vauxhall Bridge Rd - a site representing a commercial area in a large city.

(d) Stoke-on-Trent - a site representing an area of high density housing where solid fuel is used predominantly and where the main industry is pottery production.

(e)　Sheffield　　　　– a site representing the industrial area of a
　　　　　　　　　　　　　large city having steel production as its major
　　　　　　　　　　　　　activity.

2.　EXPERIMENTAL

2.1 Particle Size Distribution and Particle Concentration

　　The particle size distribution was determined using a Washington
University Cascade impactor having nine collection stages with effective
50% cut off diameters (D_{50}) ranging from 0.28 to 21.5 μm.
　　In addition to cascade impactor measurements, the concentrations of
particles were determined in four size ranges (<50 μm, <5 μm, <2 μm and
<0.8 μm) using two basic sampling methods namely:

　　(a)　The United States Environmental Protection Agency high volume
　　　　sampler (8,9).

　　(b)　A low/medium volume sampler consisting basically of a filter
　　　　holder, pump and gasmeter.

The high volume sampler was used to collect particles in the <50 μm size
range whilst variations of the low/medium volume sampler, with suitable
single stage impactors, were used to sample the other size ranges.
　　The sampling periods for the determination of particle concentrations
were nominally of 48 hours duration.
　　A range of filter media were used to facilitate chemical analysis of
the particulate material.

2.2 Chemical Analysis

　　The metals lead, iron, zinc, manganese, copper, cadmium, magnesium and
calcium were determined by atomic absorption spectrophotometry whilst
beryllium, cobalt, chromium, molybdenum, nickel, titanium and vanadium were
determined by atomic emission spectroscopy.
　　A range of techniques, including X-ray fluorescence, colourimetric
analysis and ion specific electrodes, was used to determine other
components of the particulate material such as sulphate, ammonium ion and
chloride ion.　A typical analysis scheme is shown in Fig. I.

3.　RESULTS AND DISCUSSION

3.1 Particle Size Distribution

　　Particle size distributions determined using the cascade impactor were
similar for all sites.　There are several features common to all
distributions, namely:

　　(a)　A peak at approximately 0.6 μm.
　　(b)　A peak or evidence of a peak at 6 μm.
　　(c)　A trough in the region of 15 μm.
　　(d)　In most cases a trough at about 1-2 μm.

　　A typical distribution is shown in Fig. II.　The two peaks correspond
roughly with the trimodal distribution model (10) which has been suggested
for the particulates in the ambient atmosphere and the model also suggests

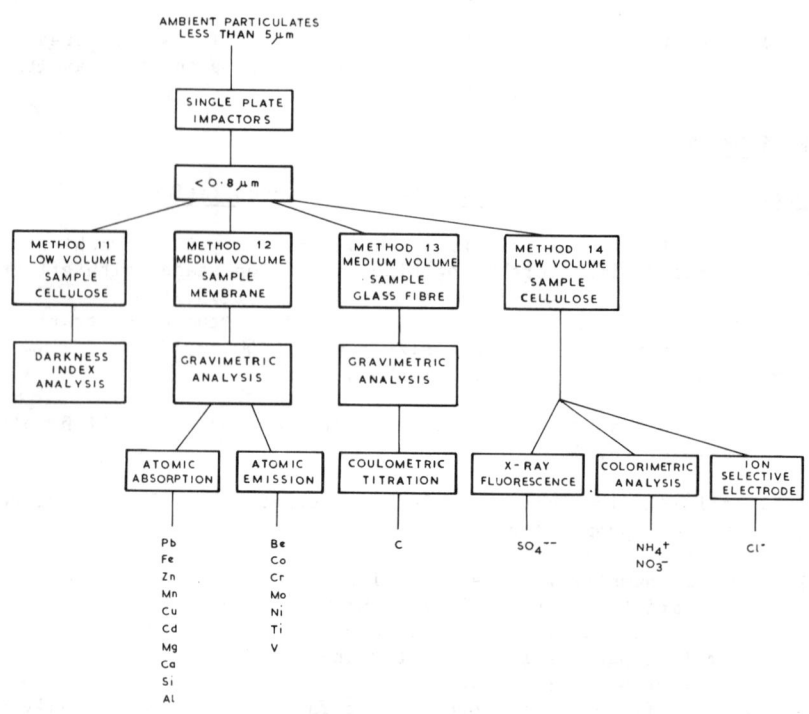

FIG. 1 - SAMPLING AND ANALYSIS PROGRAMME FOR PARTICULATES LESS THAN 0.8 μm

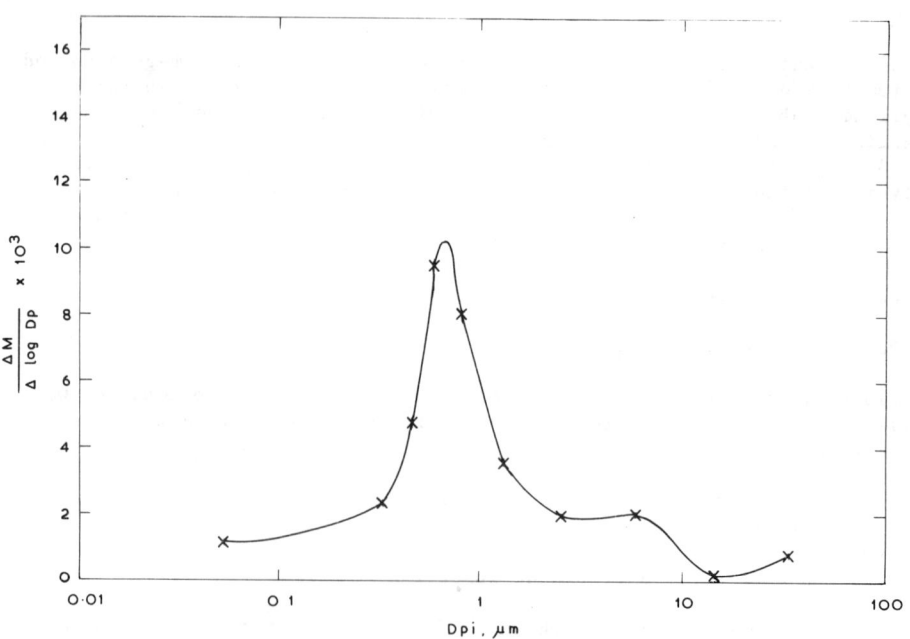

FIG. 2 - MEAN PARTICLE SIZE DISTRIBUTION AT THE ST ALBANS SITE

a minimum in the 1-2 μm region. The minimum at about 15 μm is probably an artefact of the Washington University impactor and the sampling inlet configuration. Other investigators (11) have shown that the first stage of the impactor (i.e. <21.6 μm size fraction) actually collects particles down to about 13.5 μm in size. Thus the second stage (9.6 μm - 21.6 μm) is likely to give low results because particles which should have been collected on this stage are preferentially collected by the preceding stage, giving rise to the minimum at about 15 μm in the distribution.

The nucleation mode of the trimodal distribution model occurs in the <0.1 μm size range and thus the cascade impactor, with a final stage collecting particles down to 0.28 μm in size, is incapable of detecting this mode.

3.2 Particle Concentrations

Forty eight hour average total particulate concentrations measured using the high volume sampler ranged from a minimum 18.1 μg m^{-3} at the St Albans site to a maximum of 512.5 μg m^{-3} at the Sheffield site. Arithmetic mean total particulate (<50 μm) concentrations for a six week period at the St Albans, Northfleet, Vauxhall Bridge Rd, Stoke-on-Trent and Sheffield sites were 40, 79, 114, 148 and 206 μg m^{-3} respectively. The particle concentration results are summarised in Table 1.

TABLE 1. - Summary of Gravimetric Particulate Concentrations at all Sites

Site	Particulate Concentration μg m^{-3}			
	< 50 μm		<5 μm	
	Range	Mean	Range	Mean
St Albans	18.1- 68.6	40.1	10.5- 53.2	28.0
Northfleet	50.7-148.4	78.8	23.8- 63.9	41.4
Vauxhall Br Rd	61.6-286.5	113.9	24.9-180.5	80.3
Stoke	68.7-256.2	148.4	53.0-198.8	130.0
Sheffield	59.2-512.5	205.5	42.1-371.4	131.9

Site	<2 μm		<0.8 μm	
	Range	Mean	Range	Mean
St Albans	9.9- 30.9	18.0	3.9- 31.0	16.0
Northfleet	18.7- 40.8	34.0	6.9- 27.2	14.7
Vauxhall Br Rd	22.6-123.5	71.6	13.7- 29.7	23.2
Stoke	39.9-147.6	95.0	37.6-103.0	69.4
Sheffield	26.2-168.1	64.8	14.1- 83.5	41.1

3.3 The Metals

The contribution of metals to the total particulate material collected in each of the size ranges was small (<10 per cent) for samples from the sites at St Albans, Northfleet, Vauxhall Bridge Rd and Stoke-on-Trent. On the other hand, at the Sheffield site metals accounted for over 20 per cent

of the particulate material in all size ranges. At all the sites the major
metallic component was either iron (Sheffield, Vauxhall Bridge Rd) or
calcium (St Albans, Northfleet and Stoke-on-Trent).

In general, the proportions of metals such as calcium, iron, magnesium
and aluminium in the particulates decreases with decreasing particle size,
indicating that these metals are associated with the coarse mode of the
trimodal distribution model. In contrast, the proportions of lead and zinc
tend to increase with decreasing particle size indicating that they are
associated with the accumulation mode of the trimodal distribution model
(10).

Considering the more toxic, heavy metals, lead was the most abundant
at all sites. The highest concentrations were found at the Sheffield and
Vauxhall Bridge Rd sites, lead being derived primarily from scrap
processing and motor vehicle exhausts respectively. The concentrations of
lead were not high and, with the exception of one site, did not exceed the
US and proposed EC standards of 1.5 μg m^{-3} and 2 μg m^{-3} respectively.
When comparisons are made between these limits and concentrations presented
in this paper it should be borne in mind that the proposed EC standard
refers to a twelve month exposure period and the US standard to a 3
calendar month period.

In Table 2 the metal concentrations obtained during the present survey
have been compared with the concentrations obtained during a Multi-Element
Survey (7) which was carried out as part of an Extension to the UK National
Survey of smoke and sulphur dioxide. Averaging times for the concentrations are similar and therefore the results are directly comparable. The
influence of the steelworks source close to the site at Sheffield is
clearly demonstrated particularly with respect to the maxima of the
concentration ranges. If the results from Sheffield are excluded the
resulting concentrations are similar to those obtained in the Multi-Element
Survey.

TABLE 2. - Comparison of Results with the Multi-Element Survey

Metal	Concentration Range ng m^{-3}		
	Present Survey All Sites	Multi-Element Survey (7)	Present Survey Excluding Sheffield
Lead	250 - 2690	83 - 940	250 - 1590
Iron	70 - 16000	210 - 3300	70 - 3230
Zinc	70 - 7970	41 - 3200	70 - 790
Manganese	2 - 1380	6.8 - 130	2 - 60
Copper	10 - 360	4.9 - 110	10 - 150
Cadmium	<1 - 70	1.0 - 16	<1 - 10
Beryllium	<1	0.06 - 0.44	<1
Cobalt	<1 - 5	0.86 - 4	<1 - 4
Chromium	3 - 340	3.9 - 44	3 - 20
Molybdenum	<1 - 90	0.83 - 7.7	<1 - 10
Nickel	2 - 220	3.3 - 31	2 - 50
Titanium	20 - 230	15 - 86	20 - 230
Vanadium	3 - 100	4.8 - 56	3 - 100

3.4 Carbon and Silicon

In general, silicon analyses show that silicon is associated with the larger particles and accounts for less than 5 per cent of the particles in any size range. Silicon is probably present in the particles as silica or silicates and is likely to be of a terrestrial origin rather than from industrial process or combustion sources.

The percentage of carbon in the particles varies from site to site and from size range to size range. In general the mean percentage of carbon in the particles increases with decreasing particle size. The highest average carbon concentrations were found at the Stoke-on-Trent site, where, in the <2 μm and <0.8 μm size ranges carbon accounted for more than 50 per cent of the particles. The high   proportion of carbon in the particles at Stoke-on-Trent is probably attributable to the emissions from high density housing around the site which relies mainly on coal for heating purposes. The fact that the carbon percentage increases with decreasing particle size is consistent with combustion source emissions.

An investigation into the determination of particulate concentrations using the reflectance method of smoke stain measurement was carried out simultaneously with this investigation (12). The results showed that whilst the smoke shade method is, in general, no longer an accurate measure of total particulate concentrations in gravimetric terms it is an acceptable method for the measurement of carbon concentrations.

Carbon data reported during the present investigations relates to total carbon i.e. the sum of elemental carbon, organic carbon and carbonate carbon. No attempt was made to identify or quantify specific organic compounds such as polynuclear hydrocarbons.

3.5 Sulphate

Most of the analyses indicated sulphate as the second most abundant constituent of the <5 μm particle size range (carbon being the most abundant). In the same size range, 48-hour average sulphate concentrations ranged from 4.6 μg m^{-3} to 33.0 μg m^{-3}. Daily average sulphate concentrations obtained during work carried out as part of the Extensions to the National Survey of Air Pollution (6) ranged from 5 μg m^{-3} to 60 μg m^{-3}.

The proportion of sulphate in the particles has a tendency to increase with decreasing particle size indicating that sulphate is associated with the accumulation mode of the trimodal distribution model. Other workers (10) have also shown that the distribution of sulphate particles measured using impactors approximates to the accumulation mode of the trimodal distribution.

4. CONCLUSIONS

4.1 Taking the five sites as a whole, the particle size distributions, measured using the cascade impactor, exhibit two peaks at about 0.6 μm and 6 μm which approximately correspond respectively to the accumulation and coarse modes of the trimodal distribution model.

4.2 Forty-eight hour average total particulate concentrations (<50 μm) ranged from a minimum of 18.1 μg m^{-3} at the St Albans site to a maximum of 512.5 μg m^{-3} at the Sheffield site. Arithmetic mean total particulate concentrations for the St Albans, Northfleet, Vauxhall Bridge Rd, Stoke-on-Trent and Sheffield sites were 40, 79, 114, 148 and 206 μg m^{-3} respectively.

4.3 With the exception of the Sheffield site, metals account for a small proportion of the total particle concentration, the most abundant being calcium and iron.

Calcium, iron, aluminium and magnesium were associated with larger particle sizes whereas lead and zinc were associated with smaller particles.

At the Sheffield site iron was the most abundant metal and metals as a whole accounted for more than 20 per cent of the total particulate concentration.

4.4 The concentration of lead (6 weeks averages) did not exceed the US standard of 1.5 μg m^{-3} (calendar quarter) nor the proposed EC standard of 2 μg m^{-3} (annual average) at four of the five sites. At the fifth site the average lead concentration was 2.7 μg m^{-3} but this was within the perimeter fence of a steelworks.

4.5 Carbon is a major constituent of the particles accounting for up to 50 per cent of the total concentration mass. In general the mean percentage of carbon in the particles increases with decreasing particle size, indicating that the carbon is associated with the accumulation mode of the trimodal distribution model.

4.6 Sulphate is also a major constituent of the particles and the percentage tends to increase with decreasing particle size, indicating that the sulphate is associated with the accumulation mode of the trimodal distribution model.

5. REFERENCES

1. Lioy, P.J., Kneip, T.J. and Wolff, G.T. Toxic Airborne Elements in the New York Metropolitan Area. J. Air Pollut. Control Assoc., 1978, 28 (5).

2. Kretzschmar, J.G., Delespaul, I., De Ruck, Th. and Verduyn, G. The Belgian Network for the Determination of Heavy Metals. Atmos. Environ., 1977, 11, 263-71.

3. Turner, A.C. The Midlands Metal Survey: Concentrations of 15 Elements in the Air Around Factories in the Birmingham Area and the Black Country - Part I. Stevenage: Warren Spring Laboratory, 1979, Report LR 303 (AP).

4. Turner, A.C. The Midland Metal Survey: Part 2. Stevenage: Warren Spring Laboratory, 1979, Report LR 304 (AP).

5. Clayton, P., Killick, C.M. and Potter, C.J. An Investigation of the Lead Works and their Effect on Ambient Lead Concentrations. Stevenage: Warren Spring Laboratory, 1977, Report LR 269 (AP).

6. McInnes, G. Sulphate in Particulate Survey: Report of the First Years Results. Stevenage: Warren Spring Laboratory, 1978, Report LR 275 (AP).

7. McInnes, G. Multi-Element Survey: Analysis of the First Two Years Results. Stevenage: Warren Spring Laboratory, 1979, Report LR 305 (AP).

8. Federal Register, 1971, 36, 84.

9. Selected Methods of Measuring Air Pollutants. Geneva: World Health
    Organisation, 1976, WHO Offset Publication Number 24.

10. Whitby, K.T. The Physical Characterisation of Sulphur Aerosols.
    Atmos. Environ., 1978, 12 (1-3), 135-59.

11. Proceedings of a Seminar on In-Stack particle Sizing for Control
    Device Evaluation. EPA Report 600/2-77-060, Washington: Office of
    Research and Development, 1977.

12. Bailey, D.L.R. and Clayton, P. The Measurement of Suspended
    Particulate and Carbon Concentrations in the Atmosphere using Standard
    Smoke Shade Methods. Stevenage: Warren Spring Laboratory, 1980,
    Report LR 325 (AP).

INFLUENCE OF PARTICLE PROPERTIES ON HETEROGENEOUS SO$_2$ REACTIONS

R. Dlugi, S. Jordan and E. Lindemann

Kernforschungszentrum Karlsruhe GmbH
Laboratorium für Aerosolphysik und Filtertechnik I
Bundesrepublik Deutschland

Summary

The importance of the transformation of sulfur dioxide in sulfate on catalytic active particles in the atmosphere is still a point of discussion. With respect to heterogeneous reactions on particles the physico-chemical properties of fly ash particles, cement dust and soots were investigated. The chemical and elemental composition of dust samples and of single particles was determined. Physical-chemical properties of particle systems as size distribution, surface area, acidity were measured. The influence of these particle properties on the heterogeneous SO$_2$-oxidation process is discussed.

1. Introduction

According to the present state of knowledge some heavy metal compounds, especially manganes, iron /1,2/ and also soot /3/, have an important catalytic effect on SO$_2$-oxidation. Catalytic SO$_2$-oxidation on the surface of particles produces sulfuric acid or sulfate containing particles. The reaction depends on the concentrations of SO$_2$, H$_2$O, other gases like NH$_3$, NO$_2$, heavy metals, carbon and water soluble compounds on the particle surfaces. Results of former publications /14/ have shown that some particle properties like specific surface area, surface structure and pH on the surfaces exert an important influence on the catalytic SO$_2$-oxidation. Depending on the chemical composition of the particles the reaction is limited by the pH-value of the aerosol. Usually, the reaction stops at pH $\leq$ 2 as shown by Junge and Ryan /5/. Reaction capacities — that is the mass of sulfate or sulfuric acid on the particle per mass of aerosol — of about 0.1 g SO$_2$/g$_{Ae}$ /4/ to 10^{-4} g SO$_2$/g$_{Ae}$ /4, 10, 11,16/were measured for fly ash, soot and various materials with different experimental facilities.

Simulation experiments were carried out to investigate catalytic oxidation of SO$_2$ by airborne natural particles from power plants and cement factories and by soot under conditions which allow to transfer the results to atmosperic conditions /14/. The experiments were carried out in a 4.5 m^3 vacuum tight, temperature and humidity controlled reaction chamber. The chamber had proved its worth in former experiments and has been described in detail by Haury et al. /4/

The chamber was filled with SO_2 polluted, temperature and humidity
controlled air. Experiments were performed with fly ash from coal fired
power plants, sampled from the stack gas behind the electro-filter, from
the electro-filter itself and from cement factories. Different kinds
of artificial soot from Cabot Corp. were also tested. The dust samples
were reheated to the stack temperature, fluidized and dispersed by
a jet of dry air. Before entering the reaction chamber, the aerosol
was stored in a buffer vessel to allow deposition of clustered particles.

Samples of particles were taken after different time intervals to
determine the sulfate and sulfuric acid concentration, respectively,
on particle surfaces. From both types of experiments the reaction
rates and capacities were determined. While total sulfate was determined by isotope dilution analysis /6/, the sulfuric acid was measured
by a gas chromatographic method developed by Penzhorn and Filby /7/.

2. Particle Analysis

The physical properties of the different aerosol systems used in
the experiments can be characterized by the material density, dynamic
shape factor $\varkappa$ and the surface area of the particles. The dynamic
shape factor describes the deviation of the mobility of a non-spherical
particle from the mobility of a sphere with the same volume. Both
paramters and the size distribution of aerosols describe the dynamic
behaviour of particles in the chamber and in the atmosphere. In Tab. 1
measured densities, dynamic shape factors $\varkappa$ for a volume equivalent
radius of about 1 μm and surface areas for different dusts have been
entered.

Table 1:

Material	ρ (g cm^{-3})	$\varkappa$	A(m^2 g^{-1})
Fly Ash	2.5 - 2.8	1 - 1.1	4 - 8
Cement Dust	2.9 - 3.2	1.18 - 1.45	7 - 15
Soot (before reaction)	1.8 - 2	1.5 - 2.9	7 - 220
Soot (after reaction)	1.8 - 2	1.1 - 1.4	7 - 30

The shape of particles their specific surface and, consequently, $\varkappa$
differ substantially, depending on the process of particle formation.
Coal firing in power plants produces temperatures up to 2000 °C. Most
of the elements are gaseous and condense during the cooling process,
depending on their vapour pressure on nuclei. Consequently, the
particles are nearly spherical and have a shell structure (figure 1).
The median particle number diameter of these particles is between
0.5 - 3.0 μm, depending on the power plant operation. A second particle
mode with diameter smaller than about 0.5 μm is observed /15/ but not
studied during these experiments.

The median volume equivalent diamter of particles from cement
factories was between 0.8 and 1.0 μm. The physical structure of these
particles is irregular according to their process of formation in
stone-mills (figure 2). Therefore, their dynamic behaviour ($\varkappa \sim$ 1.3)
differs substantially from that of spheres. The synthetic soot particles
(Vulcan XC-72 R, Elftex 5, Sterling MT) differ in diamter of primary
particles and aggregates, specific surface area and carbon content,

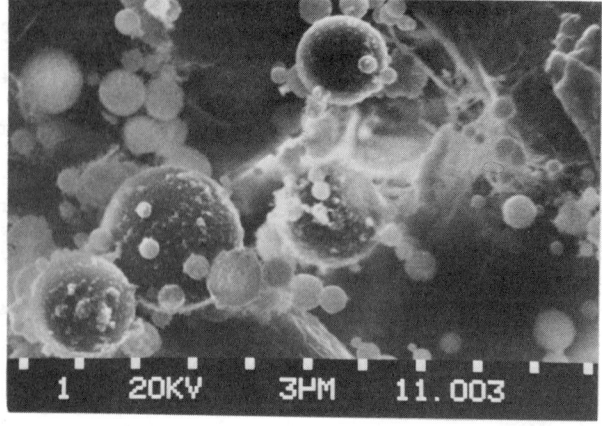

Fig. 1

Fig. 2

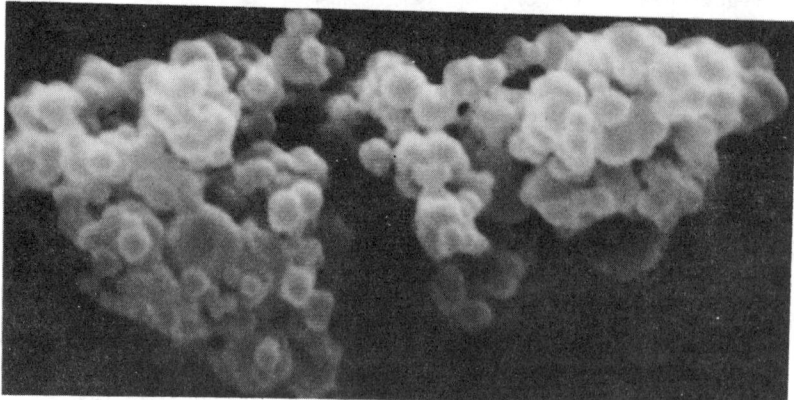

Fig. 3

Fig. 1 - 3: Coal fly ash particles, cement dust and soot

depending on their process of formation. The primary soot particles have a median diamter of about 0.03 - 0.3 µm and agglomerate to clusters of 10 - 10³ particles (Fig. 3). Their carbon content is between 98 and 99.5 %. Figure 4 shows the size distribution of fly ash particles in the reaction chamber compared with a particle distribution measured in the flue gas of a coal fired power plant /8/. The mode of particles behind the electro filter is comparable to the mode in the chamber.

There is only little information on measurements of the size distribution of dust from cement factories and soot particles in the free atmosphere. The particle diameters published in the literature are similar to ours.

The chemical composition of the aerosol was determined by neutron activation analysis. The results of this analysis for particle systems used for simulation experiments are shown in Tab. 2. This analysis was focussed on the detection of catalytic active elements such as Mn, Cr, Ni, V etc. It can be learned from the comparison of coal analysis and fly ash analysis that the catalytic active transition elements are enriched up to a factor of $10^2 - 10^3$ in the dust samples. Particles from coal fired power plants and cement factories contain the elements Na, K, Mn, Fe and Mg at comparable concentrations.

The distribution of elements in each particle, in particular on the surface, is important for SO_2-oxidation. The microstructure of single particles and samples was investigated by microanalysis combined with a scanning electron microprobe and photoelectron spectroscopy (ESCA). For example 350 single dust particles from a power plant were analyzed before and after the reaction with SO_2. It was found that Fe, Ti and Ca are equally distributed over the whole volume of the particles, whereas the elements Mn, V, Cr and also K, Na, Cl and S are enriched on the particle surface. The nucleus of particles consists of aluminium and silicium compounds. A schematic diagram of the element distribution in a fly ash particle is shown in Fig. 5. In 8 % of the particles manganese is detectable. Therefore, it can be concluded that traces of Mn are sufficient for SO_2-catalysis or the much higher Fe-content of particles contributes substantially to sulfate formation.

Before the reaction 30 % of particles contained small amounts of sulfur. Sulfur was found as SO_4^{-2}. On the surface of the particles sulfur compounds such as K_2SO_4, NA_2SO_4, $CaSO_4$, $MnSO_4$ and Fe_2SO_4 were found after the reaction with SO_2.

For particles from cement factories no obvious shell structure was found. $CaCO_3$ and alkali sulfates as well as chlorides were found mainly on particle surfaces. On the surface of soot particles only traces of Al, Ca, Cl and S were detected.

3. SO_2 Transformation on Particle Surfaces

The main results of the simulation experiments in the SO_2-particles-H_2O-air system at temperatures between 2 and 30 °C and relative humidities up to 90 % are: the mass of sulfate on the particles increases with time. It is possible to describe the 'overall reaction', according to a Langmuir-Hinselwood kinetic for a process being

Table 2: Element Analysis of Aerosol Materials and pH-Values

	pH	Na °/oo	K %	Mg %	Ca %	Al %	Ti °/oo	V ppm	Cr ppm	Mn °/oo	Fe %	Ni ppm	Zn °/oo	Cl °/oo
Fly ash														
Sample I	4.7	1.0	3.2	2.9	3.6	13.6	3.1	930	710	0.9	4.9	1600	13.3	1.9
II	5.8	9.0	2.5	1.9	2.5	14.5	1.3	850	660	0.85	4.7	600	5.3	1.4
III	5.1	11.1	3.6	1.1	2.3	13.6	8.8	790	560	0.15	5.7	460	8.1	2.3
IV	5.6	8.0	2.7	1.3	2.5	17.2	1.4	850	270	0.81	2.0	100	4.0	1.1
V	4.2	7.8	3.5	1.4	3.0	17.5	6.7	1063	790	0.14	5.6	500	6.2	0.05
Coal (III)	-	0.4	0.9	0.04	0.01	0.9	0.07	6	9	0.001	2.3	80	0.1	1.4
Cement dust														
PZ 35	10.8	0.87	0.22	1.2	37.3	1.9	1.1	35	44	0.62	1.8	300	2.8	0.01
PZ 45	11.6	0.81	0.99	1.1	42.1	2.3	1.6	41	41	0.65	1.9	500	2.5	0.01
Soot	pH	°/oo	°/oo	°/oo	°/oo	°/oo	ppm	ppm	ppm	ppm	ppm	ppm	ppm	%
Vulcan XC 72	7.5	0.77	0.05	0.3	0.4	0.3	50	2	< 1	0.5	50	—	0.5	2.3
Sterling MT	9.0	0.1	0.05	0.05	0.2	0.03	10	0.1	0.4	1	35	—	0.5	0.035
Elftex	7.0	0.05	0.05	0.1	0.4	-	10	1	0.3	0.5	50	—	0.5	0.28

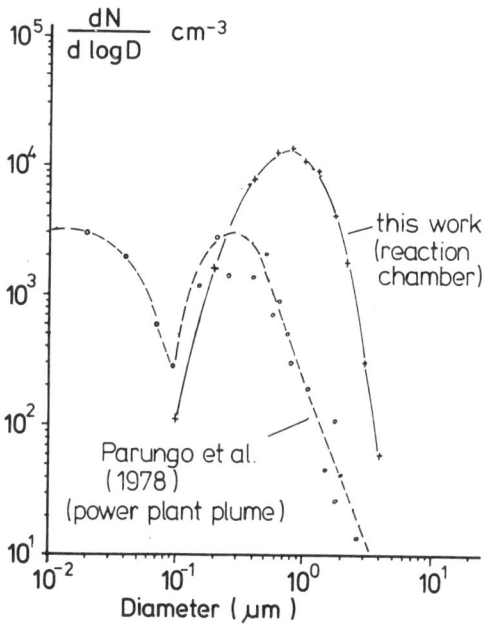

Fig. 4: The size distribution of coalfly ash particles in the
reaction chamber compared with a size distribution from
a plume

Structure of Particles

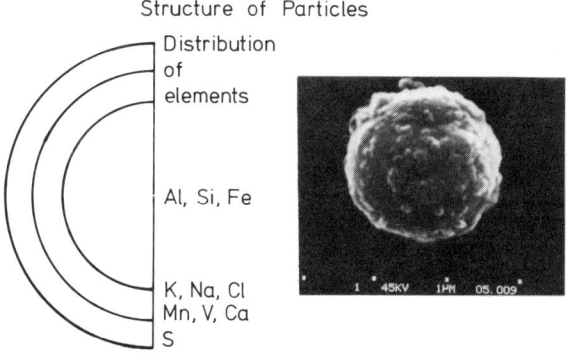

Fig. 5: The schematic structure of coal fly ash particles
after reaction with SO_2

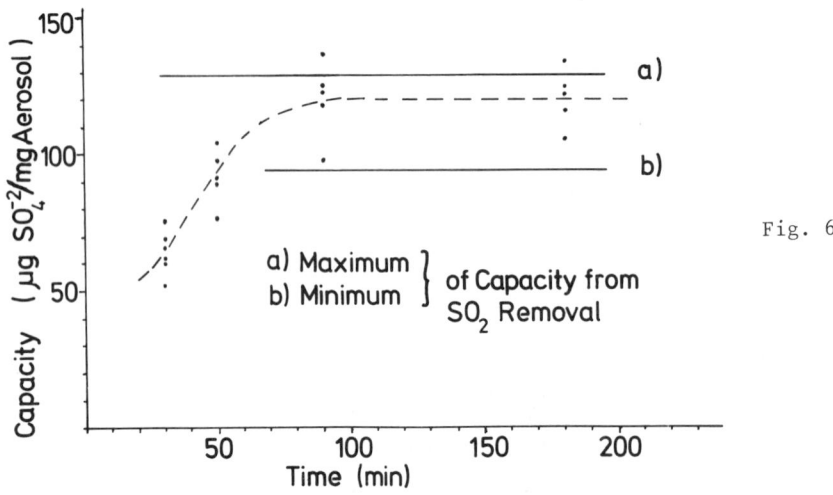

Fig. 6

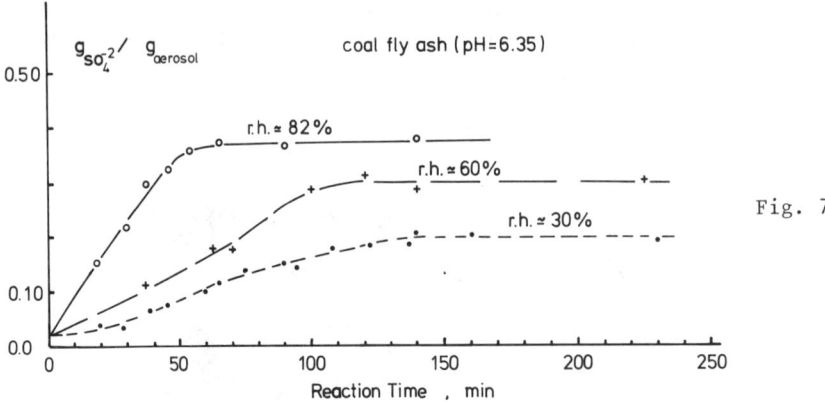

Fig. 7

Fig. 6 - 7: The sulfate formation on coal fly ash particles as a
function of time

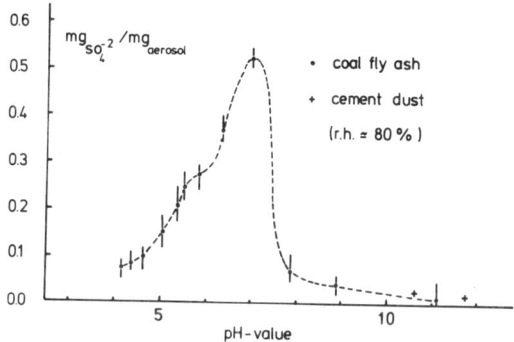

Fig. 8: The capacity as a function of the pH-value

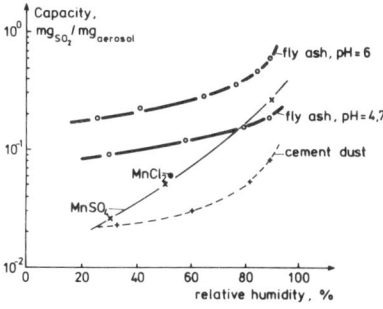

Fig. 9: The capacity for different aerosols as a
function of relative humidity

controlled by surface reactions as done also for example by Britton and Clarke /16/:

$$\frac{d}{dt} \; [\; SO_4^{2-} \;] \; = a_o (1 - \frac{M_{SO_4^{2-}}^t}{M_{SO_4^{2-}}^e})^2$$

were $M_{SO_4}^t{}^{2-}$ is the sulfate formed during the time t and $M_{SO_4}^e{}^{2-}$ is the total mass of sulfate at the end of the reaction. a_o is the rate of sulfate formation which is not a constant but depends e.g. on the relative humidity and the pH-value. The increase with time of total sulfate on particles is shown for fly ash (pH = 4.7) at different relative humidities and 4.5 mg m^{-3} of initial SO$_2$-concentration in Fig. 6 and Fig. 7.

The total sulfate mass on particles reaches a limit after about 100 minutes and a reaction capacity of 0.12 mg$_{SO4}$/m$_{ae}$. Calculations of the reaction capacity from the measurements of SO$_2$ disappearance in the reaction chamber for the same fly ash (see /4/) gives values between the two solid lines a) and b). The transformation rate for this fly ash is $a_o \cong 690$ µg SO$_4 \cdot$ g$^{-1} \cdot$ Min^{-1}. Some experiments give evidence that a_o decreases with decreasing relative humidity (Fig. 7). The reaction capacity depends strongly on the pH-value on the surface of the particles. The reaction capacities measured for different samples of fly ash with different initial pH-values on the particle surfaces are shown in Fig. 8. In all experiments the reaction stopped for pH $\leq$ 2.7.

The reaction capacity increases with the relative humidity. In Fig. 9 the r. h. dependence of capacities for two samples of fly ash with different pH-values are compared with the behaviour of artificial MnSO$_4$ and MnCl-aerosols. MnSO$_4$ - particles have a stronger r. h.-dependence, which is probably due to the higher solubility of the material.

About 7 - 9 % of fly ash is soluble at pH = 2, while about 2 - 6 % are water soluble. A simple correlation between total sulfate and the concentration of a catalyst (Mn, Fe, ...) was not found.

Fig. 9 contains also the capacities measured for cement dust. These particles have an initial pH-value of 11-12 and a pH $\cong$ 8 - 9 at the end of the reaction. This reaction limit (pH $\cong$ 8) might be due to the formation of insoluble sulfates on the particle surfaces.

The surface of soot particles studied was found to be slightly basic. A dependence of the reaction capacity on the relative humidity (0 $\leq$ r.h. $\leq$ 80 %) was not measured. For different soot samples capacities were determined in the order of 10^{-2} mg$_{SO2}$/mg$_{ae}$. The specific surface area of soot particles changes from e. g. 200 m^2/g before the reaction to about 30 m^2 after the reaction.

4. Discussion

Comparison of the results with similar investigations published in the literature shows that the capacities measured here are higher by up to a factor of 10^2 - 10^4. For artificial aerosols of MnCl$_2$

Cheng et al. /9/ measured reaction capacities smaller by a factor of about 10. For fly ash of the same origin and composition as used in this paper, Liberti /10/ found capacities smaller by a factor of 10^3. Judeikis et al. /11/ used fly ash from electro-filters and found also smaller values. All of the authors mentioned have used experimental set-ups which were quite different from the facility described in this paper. Mostly samples were deposited either on filters or on the inside walls of reaction pipes. It might be possible that only the outer layer – that is a fraction of total particles member – of the deposited particles reacted completely with the SO_2.

Therefore, additional first tests were performed with an experimental set-up where particles were deposited on filters. Reaction capacities smaller by a factor of 1,5 – 12 were found for the same thermodynamic conditions as during tests with airborne particles. From these results it seams possible that mass transfer conditions also control the reaction capacities and rates. The analysis of samples from plume measurements under conditions of low r. h. (< 30 %) shows that comparable amounts of sulfates are formed on fly ash particles /12/ as shown in figure 8. For high relative humidities – especially if cooling tower plume or a cloud and stack plume merged – an increase of sulfate content in the particle range D < 0.4 μm is found /13/. This increase is in agreement with our measurements (Fig. 9); but as most sulfate production mechanisms show an increasing rate with increasing relative humidity the contribution of different reaction types to the total SO_2-oxidation is still unknown.

On the basis of the reaction capacites the oxidation of SO_2 on the surface of catalytic active particles in a power plant plume can be estimated, because capacities are not found to be a function of the SO_2-concentration. Assuming the usual source strength of industrial off-gases of about 40-300 mg of aerosols per m³ and 0.5-5 g of SO_2/m^3, a medium total SO_2-removal of 6 % SO_2 can be calculated at high relative humidities (80-92 %) and of 2 % at lower r. h. For polluted air in urban areas up to 15 % of the total SO_2 emitted may be oxidized on the surface of catalytic active particles. For higher r. h. – expecially in clouds or fog the oxidation rate may increase remarkably.

5. References

/1/ Cheng, R. T. et al.: Atmospheric Environment, 5, 987-1008(1971)

/2/ Chun, K.C. and Quon, J.E.: Environ. Sci. Technol., 7, 532-538(1973)

/3/ Novakov, T., Chang, S.G., Harker, A.B.: Science 186, 259 (1974)

/4/ Haury, G., Jordan, S., Hofmann, C.: Atm. Environ. Vol. 12, pp. 281-287 (1978)

/5/ Junge, C.E., Ryan, T.G., (1958): Q.J.R. Met. Soc., 84, 46-55

/6/ Klockow, D., Denzinger, H., Rönicke, G. (1974): Chemie-Ing. Technik 46, 831

/7/ Penzhorn, R.-D., Filby, W.G. (1976): Staub 36, 205-207

/8/ Jockel , W., Abschlußbericht des Forschungsvorhabens "Radio-aktive Emissionen aus konventionellen Kraftwerken", TÜV-Rhein-land e. V.

/9/ Cheng, R.T., Corn, M., Frohlinger, J.O. (1971): Atm. Eviron. 5, 987-1008

/10/ Liberti, A., Brocco, D., Possanzini, M. (1978) Atm. Eviron. 12, 255-261

/11/ Judeikis, H. S., Stewart, T.B., Wren, A.G. (1978): Atm. Environ. 12, 1633-1641

/12/ Mamane, Y., Püschel, R.F. (1979): Geophys. Research Lett. 6, 109-112

/13/ Dittenhoefer, A.C., De Pena, R.G. (1978): Atm. Eviron. 12, 297-306

/14/ Dlugi, R., Jordan S., Lindemann E. (1981) KfK 3187

/15/ Ondov J. M., Biermann, A.H. (1980) Physical and chemical Characterization of Aerosol Emission from Coal-Fired Power Plants; in:Environmental and Climatic Impact of Coal Utilization,Academic Press.

/16/ Britton, L. G. Clarke, A. G. (1979): Atm. Environ. 14, 829-839

SO$_2$ CONVERSION IN A MARINE ATMOSPHERE

H.M. ten Brink, R.K.A.M. Mallant, G.P.A. Kos, J.M. Gouman and
J.F. van de Vate
The Netherlands Energy Research Foundation (ECN), Petten,The Netherlands

Summary

The composition of size-fractionated marine aerosol has been measured
at a coastal site in the Netherlands. The bulk seawater components
dominate in the hypermicron aerosol fraction. Chloride deficit is
observed in almost all samples in the submicron range. The overall value
of the deficit is low however. Excess sulfate up to a concentration of
10 µg.m^{-3} is preferently found in the smallest aerosol sizes. Substitution of chloride is caused by sulfate and probably nitrate which
is also present in polluted marine aerosol. Laboratory measurements
show that the photochemical formation of sulfuric acid from sulfur
dioxide could both explain the presence of sulfate as well as the
substitution in the submicron aerosol. The reaction of hydroxyl-
radicals and sulfur dioxide was found to be the dominant oxidation
path. Nitrate was found to substitute chloride as wel in the simulation experiments. The observed small value of the chloride deficit
in the ambience is due to the fact that condensation of sulfuric acid
occurs on the smallest particles, whereas the mass of the seasalt is
concentrated in the larger size-ranges. Another reason for the low
value of the substitution is the presence of ammonia as a
neutralizing agent.

1. INTRODUCTION

In the aerosol in marine air-parcels often substitution of the
chloride by sulfate is observed [1,2]. The results in our laboratory were
obtained using analysis techniques of a relative nature [1] (electronmicro-
scopy and affiliated methods), however showing the substitution to occur
in individual particles in the submicron size range. The substitution results in the release of the corrosive HCl-gas. To gather absolute values
for this substitution recently quantitative chemical analysis of the aerosol
was used with the additional advantage of the detection of nitrate.

Sulfate formation in marine air is probably caused by the emission of
sulfur-dioxide into marine air parcels and subsequent oxidation of this
component. It was theoretically domonstrated on the last conference [1]
that the condensation of sulfuric acid, formed fotochemically from sulfur
dioxide, onto the marine aerosol could explain both the preferent presence
of sulfate in the submicron aerosol as well as the substitution. Recently
detailed laboratory experiments have been performed to verify this hypothesis and results are reported here.

2. EXPERIMENTAL

Field observations on marine aerosol

In addition to the methods used before [1] ion-chromatography and atomic absorption spectrometry have been applied for the aerosol analysis. Aerosol was sampled on a 8-stage Anderson cascade impactor (aerodynamic range 0.4 - 11 μm). Polyethylene discs were used as the sampling plates because of low contamination properties. Lower detection limits per stage in an average run of 5 hours are 5 ng-m^{-3} for the elements Na, K, Mg and 20 ng-m^{-3} for the anions, Cl$^-$, SO$_4^=$, NO$_3^-$, while initially the detection limit for NO$_3^-$ was higher due to contamination.

Total number concentrations, size distribution, scattering coefficient i.e. the physical properties of the aerosol were determined with the appropriate techniques [1]. Air mass back-trajecttories were provided by the Royal Netherlands Meteorological Institute.

Laboratory experiments

Experiments were carried out in sealed smogchambers of glass (1 m^3), using SO$_2$ (0.1 - 1 ppm) in filtered air and NaCl-aerosol as a well-defined simulation aerosol obtained by nebulizing NaCl-solutions. Atmospheric photochemistry was simulated using HNO$_2$ (10 ppb-100 ppm) as the best-known OH$^{\cdot}$-precursor [4] or a mixture of NO (0.1 ppm) and toluene (0.2 ppm) in air was used [3]. HNO$_2$-gas was generated after the method of Cox [4] mixing NaNO$_2$- and H$_2$SO$_4$-solutions. The reaction was probed by analysis of the aerosol using all the earlier mentioned methods for sampling and analysis; because of lower air volumes detection limits were an order of magnitude higher for the wet-chemical techniques. In addition total filter sampling was performed using teflon membranes. The concentration of the major gas-phase components were followed using the standard techniques [3] with UV d^2-spectrometry for the specific detection of HNO$_2$ [5].

3. RESULTS AND DISCUSSION

As yet the results for nineteen sampling days are available. On these days the total Na-concentration ranged from 0.5 to 20 μg.m^{-3}. In the following the results for two sampling days were disregarded, where the back trajectories showed that the air mass was not of an oceanic origin. For the other days the minimum Na-concentration was 3 μg.m^{-3} while less than 5% of the Na was found in the submicron aerosol. The Na/Cl molar ratio for the total aerosol is 1.07 ± 0.12 against 1.17 in bulk seawater. On the stages 6 and 7 (1.1 and 0.65 μmad) however this ratio is significantly less than the seawater ratio. On one day this value was as low as 0.1. On that particular day the analysis of individual particles, using EDS and Electron Diffraction, showed that particles of about 1 μm consisted of Na$_2$SO$_4$. The absolute value of the Cl$^-$-dificit in our samples is not exceeding 2 μg.m^{-3} and is mostly less than 0.5 μg.m^{-3}.

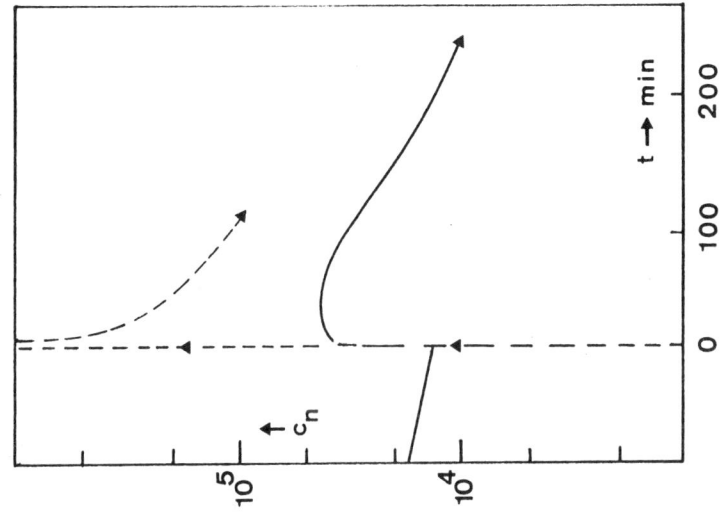

Fig. 2 – Aerosol concentration profiles in an
irradiated mixture of HNO_2 and SO_2 in the presence
and absence of primary NaCl–aerosol.
——— NaCl–aerosol present
----- gas mixture only

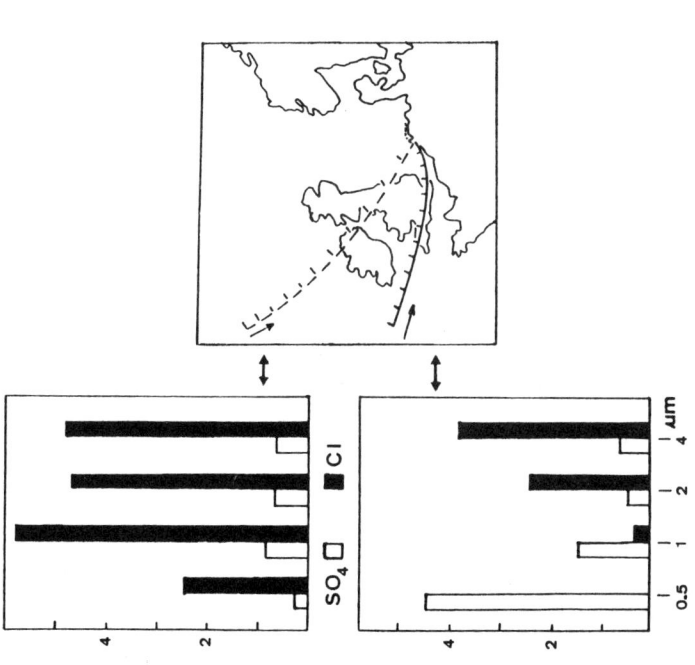

Fig. 1 – Composition of size-fractionated marine
aerosol on dec. 17, 1979 above, and june 19, 1980
below, concentrations in $\mu g.m^{-3}$, together with the
24 hour air mass back-trajectory.

Not only for Cl$^-$ but for all major bulk components the ratio to Na in the larger aerosol fraction (>2 μmad) is that of seawater, e.g. for $SO_4^=$. Excess $SO_4^=$ is observed in all samples in the submicron aerosol from as low as 0.1 $\mu g \cdot m^{-3}$ to 10 $\mu g \cdot m^{-3}$. It was found that in polar air masses behind a cold front the excess $SO_4^=$ concentration was always less than 1 $\mu g \cdot m^{-3}$. The higher excess values, from 4-9 $\mu g \cdot m^{-3}$, occurred in the warm zone accompanying a low with south-westerly air movements. Typical examples of these sampling differences are shown in fig. 1. Significant amounts of (excess) NO_3^- are found both in the submicron and in the larger-sized aerosols depending on the particular day. In the latest five samples NH_4^+ was measured and it was found that NH_4^+ was present in the submicron aerosol proportional to the excess $SO_4^=$-concentration.

In the simulation experiments with SO_2 and NaCl-aerosol, as before [1,3] no significant $SO_4^=$-formation was observed in the dark *. Under irradiated conditions, in the presence of HNO_2, SO_4 was found up to concentrations of 200 $\mu g \cdot m^{-3}$. From the fact that new particle formation is sharply reduced in the presence of NaCl-aerosol (as compared to the experiments without NaCl-aerosol addition; see fig. 2) it was concluded that condensation of the newly formed SO_4 particles was a major process. By size-discrimination of the aerosol it was shown (fig. 3) that $SO_4^=$ is predominantly present in the smallest aerosol fraction in accordance with the expectations [1]. Cl$^-$ appears to be substituted by $SO_4^=$ as well as by NO_3^-, however NO_3^- is found only when the aerosol is deliquiescent (r.h. > 40%).

Even in cases of total excess H_2SO_4 (e.g. fig. 3a) sulfate substitution is restricted to the smallest size-range. However when drawing such an aerosol mixture over a filter complete volatilasation of Cl$^-$ as well as NO_3^- is observed. due to particle-particle interaction on the filter. Using sea-salt aerosol generated from seawater by the bursting-bubble-method [1] identical results were obtained for the substitution.

In a rather better simulation of ambient photochemical conditions a mixture of NO and toluene in air was used instead of HNO_2. On the average 50 $\mu g \cdot m^{-3}$ of SO_4 was formed after 6 hours of irradiation both in the presence and absence of NaCl-aerosol, with analogous results for the substitution.

Addendum

In the experiments with HNO_2 OH$^\cdot$-radicals are formed [4]. The reaction with SO_2 eventually leads to H_2SO_4. Using the HNO_2-method the OH$^\cdot$-concentration can be calculated from the reaction kinetics and the measured photolysis rate constant for HNO_2 in our smogchambers [3]. It appears then that the amount of $SO_4^=$ observed is equal to the calculated concentration using literature values for the rate constants of the respective reactions. In the irradiated mixture of NO and toluene the OH$^\cdot$-concentration is derived from the relative rate of decrease of the toluene concentration, since it is speculated that oxidation of toluene proceeds via OH$^\cdot$-attack only with a known reaction rate constant [6]. Again the $SO_4^=$-production, calculated

---

*
Using a polythelyne reaction vessel which had prior to this been used with high concentrations of NO_2 indeed $SO_4^=$ was found after 8 hours of interaction

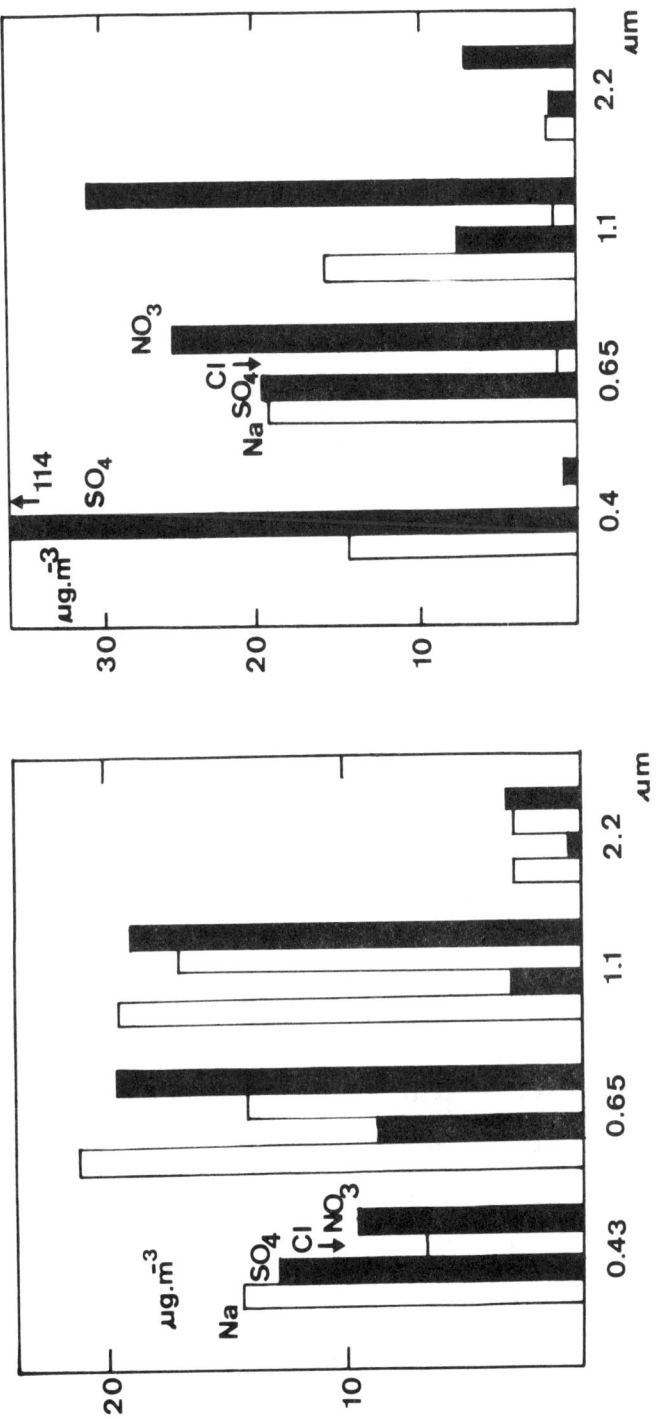

Fig. 3a – Composition of NaCl-aerosol after 2 hours of irradiation in the presence of SO_2 (0.1 ppm) and HNO_2 (50 ppb), on the stages 4 to 8 of an Anderson compactor

Fig. 3b – Idem as in fig. 3a., however increased gas-concentrations : SO_2 (0.95 ppm), HNO_2 (0.8 ppm)

on the basis of the reaction of OH· and SO_2 and the literature value of the rate constant, and the experimental $SO_4^=$-formation are equal within the experimental error. The foregoing shows that the reaction of OH· and SO_2 leads to H_2SO_4 and is indeed the dominant oxidation path under irradiated conditions.

4. GENERAL DISCUSSION

The observed absolute value of the Cl^- deficit ($\lesssim 1$ $\mu g.m^{-3}$) is rather small as compared to the excess $SO_4^=$-concentration. This can be explained as follows. The photochemical oxidation of SO_2 -via the reaction with OH·- in the ambience leads to H_2SO_4 condensing preferently on the smallest aerosol particles. The mass of the seasalt aerosol containing Cl^- however is concentrated in the hypermicron range and it is only in the overlapping range of aerosol sizes, i.e. around 1 μm diameter, that the capacity for the substitution is high. It was further found in the laboratory that substitution of Cl^- by NO_3 can take place in the presence of HNO_3-gas. Also in the ambience NO_3^- is found in the aerosol size-ranges with a Cl^- deficit. This is certainly indicative of an ambient substitution. New measurements are needed to further proof this indication.

The low value of the Cl^--deficit can be further explained by the apparent presence of NH_4^+ in the samples, as the neutralizing agent for H_2SO_4. The presence of NH_4^+ is a proof of the incorporation of continental air into the marine air-parcels since NH_2 is definitely not of a marine origin [7].

ACKNOWLEDGEMENT

The present study was financed by the Netherlands Ministry of Health and Environmental Hygiene. We thank Dr. Slanina and co-workers for performing the wet-chemical analyses.

5. REFERENCES

[1] H.M. ten Brink et al. in "Proceedings of the First European Symposium on Physico-Chemical Behaviour of Atmospheric Pollutants", B. Versino ed; Joint Research Center, Ispra 21020 Italy (1980), 298.
[2] D.R. Hitchcock et al., Atm. Environ. 14(1980), 165 and references therein.
[3] H.M. ten Brink et al., Proceedings of the Fifth International Clean Air Congress, Buenos Aires October 1980, in press; preprints available as ECN-report ECN-80-105.
[4] R.A. Cox, Int. J. Chem. Kinet., Symposium no. 1 (1975), 379 see also ref. [1] pg. 91 c.f..
[5] H.M. ten Brink et al. in "Studies in Environmental Science 1, Atmospheric Pollution 1978", Elsevier Amsterdam (1978), 239.
[6] R. Atkinson et al. Int. J. Chem. Kinet 12(1980), 779.
[7] G. Gravenhorst, Atm. Environ. 12(1978), 707

DISCUSSION ON THE PARTICLE FORMATION PROCESSES IN THE PO VALLEY

M. PAYRISSAT, H. STANGL, G. OTTOBRINI, B. NICOLLIN, B. VERSINO

Joint Research Center Ispra
Commission of the European Communities

Summary

The aerosol size distribution spectrum measured at several sites of the Po Valley has been resolved into its components on the assumption that any single formation process resulted in a log normal distribution. Studying the variation of distribution parameters during summer haze episodes in Ispra, an attempt has been made to identify the main particle formation processes responsible for the build-up of haze. The different particle formation processes, i.e. gas-to particle conversion reactions, mainly through photochemical reactions, coagulations of products and gaseous condensations are discussed, including the problem of possible precursors.

1. INTRODUCTION

Following general measurements made in Ispra and other sites of the Po Valley, experiencing severe haze episodes (1), we have focused our attention on aerosols, trying to explain the presence of particles in this region. Measurements made in winter demonstrated that domestic heating is an important source of particles. The Aitken particles (0,001 - 0,1 μm size), measured in Ispra could be correlated with major local sources, the haze particles with regional and more extended sources since a correlation between total sulphur, scattering coefficient and wind direction showed a south-north transport from industrial areas. On the other hand, sunny and rather windless winter weather shows better visibilities compared to summer.

During summer, where the SO_2 and NO_x levels were 5 - 10 times lower, the Aitken particles reached 60 % of the winter values and the haze particles at least the winter levels without showing on the whole an appreciable difference from place to place. As the relatively low level of summer Aitken nuclei does not correspond with the relatively high level of haze particles, direct transport from industrial areas can be excluded and a formation in rural areas must be considered. The fundamental question is obviously: Are we able to demonstrate the in-situ formation and the origin of summer haze particles?

In the semi-rural area of Ispra, we have then studied the build-up of haze during slightly windy periods of summertime, when the area is only dominated by local valley breezes.

2. EXPERIMENTAL SET-UP AND CALCULATION METHODS

The particles were simultaneously measured with 3 instruments: an integrating nephelometer MRI, a condensation nuclei counter (General Electric C-2), having a sensitivity maximum in the Aitken range (0,001 - 0,1 μm) and an Electrical Aerosol Analyser EAA (TSI Mod 3030).

This latter, used with an auxiliary microprocessor INTEL Mod SBC 660 for data elaboration, was able to size the aerosol concentrations in the range 0,01 - 1 μm every two minutes. The calculation method should determine, for a set of 8 data analysed in this way, the particle diameter present with the highest probability. The growing in size of such a diameter by in situ processes would mean the local build-up of haze.

The variation of the size distribution spectrum was then followed with time. This latter fitted, for the size range studied, on a bimodal log normal model consisting of two distribution modes and was resolved by the Henry straight line method (2) (3) which can be summarized as follows: The cumulative concentration probabilities $\emptyset$ (di) calculated up to each i class of diameter was compared to the integrated values $\emptyset$ (u) versus u of the unit gaussian distribution given by the statistical tables. From each $\emptyset$ (d) or $\emptyset$ (u) the corresponding u values were extracted from the tables and fitted to the regression: $u = a \log d_i + b$.

This latter enabled us to calculate, for each distribution mode, the following log-normal parameters:

- the geometric mean diameter d_g as $d_{50}\%$: $d_g = 10^{-b/a}$
- the geometric standard deviation σ_g as $d_{84}\% / d_{50}\%$: $\sigma_g = 10^{-1/a}$

- the mode m, most probable diameter given by the maximum of each size distribution mode, calculated as:

$$m = e^x \left(\text{with:} \quad x = 2,303 \frac{b}{a} - \ln^2 \sigma_g \right)$$

The calculations were made on a mean distribution, called "resultant distribution" in order to be distinguished from its single components, representative of the particle behaviour during selected time intervals (10 minutes every hour).

The percentage of two minutes single monomodal components (distributions with only one maximum) was simultaneously calculated every hour. This method should follow in a better way the fine structure of the resultant distribution (which could be smoothed for too long time intervals) and was necessary for a better interpretation of the results.

3. RESULTS AND DISCUSSION

The summer haze cycles in the Po Valley are characterized by three successive stages:

- Stage I: one or a few days starting period of excellent weather, often noticed after instable conditions, with local breezes and good visibility.

- Stage II: Consecutive periods of hot and sultry weather often due to continuous days of stagnant air where the build-up of haze and ozone are at a maximum.

- Stage III: the breakdown of the previous situation due to different events, e.g. instable local conditions (thunderstorms), Föhn events or large scale meteorological changes causing rainfalls.

3.1. Aerosol size distributions during the cycle of haze

With respect to the above stages, 3 resultant size distribution types with a more or less stressed bimodality were found (fig. 1). The less pronounced bimodality meant a greater percentage of two minutes MC single Monomodal Component distributions having the maximum (mode m) below the detection limit of the EAA instrument (10^{-2} µm).

Fitting these data, one could determine and characterize the nucleation mode distribution, not properly resolved by EAA measurements when the resultant distribution is bimodal.

The values of the modes m, the percentages of single mono-modal components MC were plotted versus time together with the Aitken nuclei concentrations CN and UV solar energy. At each stage of the haze cycle, 3 typical examples have been taken into consideration:

- At stage I (fig. 2), the concentration of haze particles (greater than 0,1 µm) was very low and the resultant distribution was resolved as a single monomodal component (case fig. 1 A, between 9 a.m. to 16 p.m.)

- At stage II (fig. 3) the spectrum resolved into its two components (case fig. 1 C) showed a continuous shift of the accumulation mode towards the haze particles as seen in the transition between fig. 2 and fig.3, but the significance of "nucleation" mode is discussed (see under 3.2).

- At stage III and during a Föhn event (fig. 4), the resultant distribution showed the decrease of the accumulation mode (due to the progressive dilution and removal of preexisting haze particles) and the beginning of a new cycle.

The distribution parameters, only valid in relative values, are given in the following table.

Distribution parameters	Nucleation mode		Accumulation mode	
	minimum	maximum ⊕	minimum	maximum
Geometric mean diameter d_g (µm)	0,0055	0,022	0,051	0,143
Geometric standard deviation (σ_g)	3,11	1,28	2,12	1,69
Mode m (µm)	0,0015	0,021	0,023	0,12

⊕ To be considered as a coagulation mode (see below)

3.2. Discussion on the related particle formation processes

- Build-up of the nucleation mode distribution

Resolving the monomodal resultant, the formation process was found in a quite low size range. Correlated with UV solar energy, the process can be explained as photochemical gas to particle conversion reactions (gpc) in the homogeneous gas phase.

Similar monomodal configurations explained as in situ formations have been detected by other authors in forest areas during early noon hours (4).

The monomodal size distribution might be compared with the first

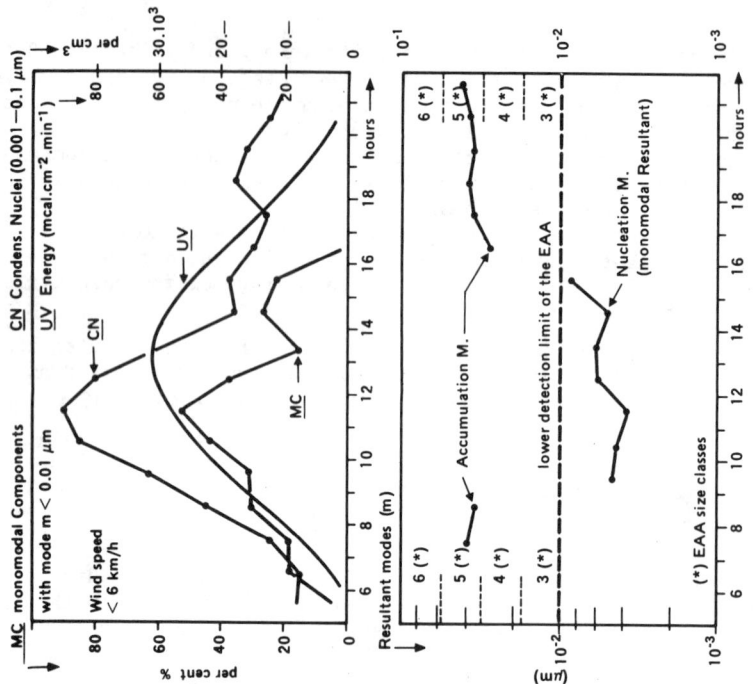

Fig. 2 – Excellent visibility conditions (01.06.80)
Stage I of the haze cycle

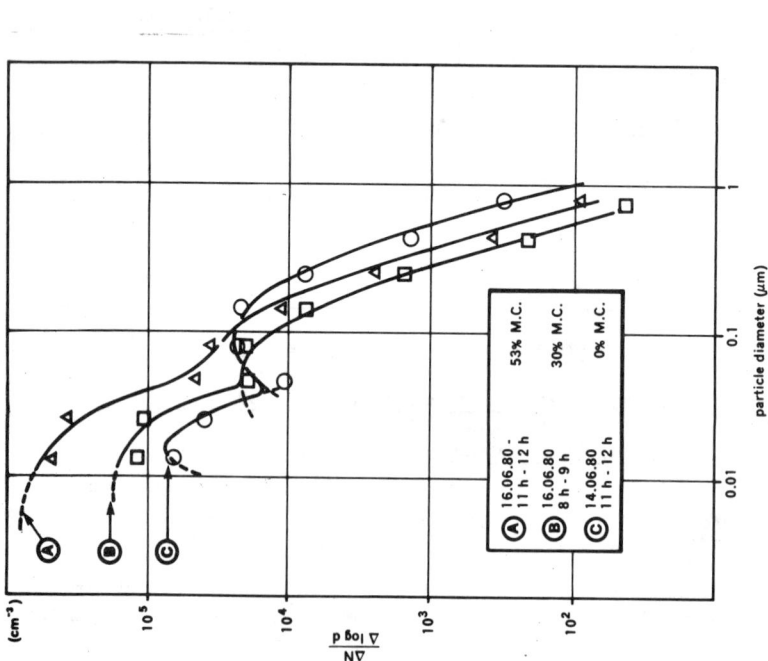

Fig. 1 – Resultant size distributions for different
percentages of monomodal components (M.C.)

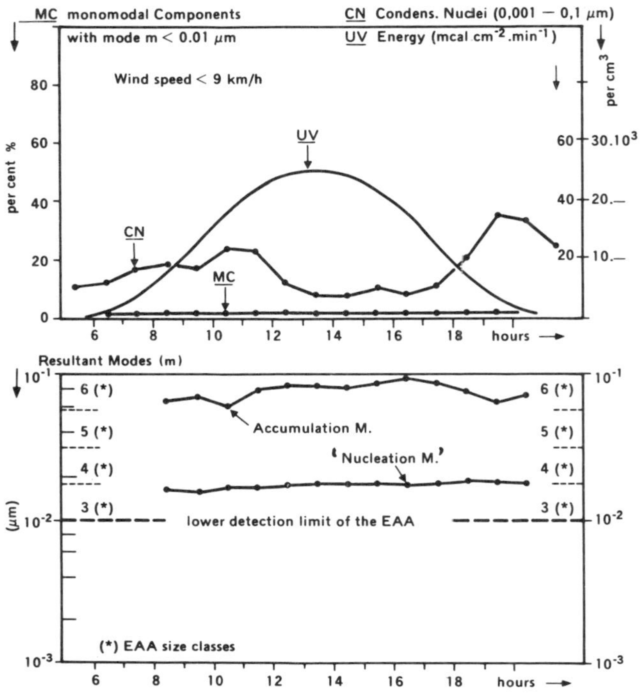

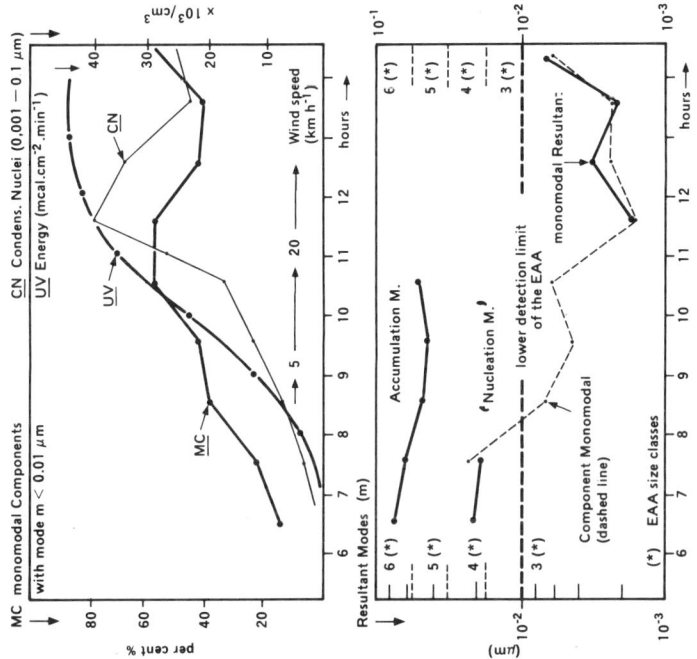

Fig. 3 - Haze conditions (14.06.80)
Stage II of the haze cycle

Fig. 4 - Changing conditions caused by a föehn event
(9.09.77) - Stage III of the haze cycle

component of the well-known trimodal spectrum related in the literature and easily explained: when the concentration of particles greater than 0,1 μm size is very low, the first component mode remains alone.

Resolving the bimodal resultant, measured during night and hazy conditions, the "nucleation mode" was found in a too high size range to be in agreement with a gpc process. Doubtful results could be caused by the lack of data (only 3 for this mode) or by the low efficiency of the EAA instrument, as explained below, since all MC had m higher than 0,01 μm.

The lower Aitken nuclei concentrations during night and in hazy conditions meant lower percentages of MC having a mode m lower than 10^{-2} μm size. The coagulation of generated nuclei during the two minutes response time of the instrument decreased also this percentage and increased, in the same time, the percentage of MC having higher values of m. By averaging too many size distributions during the measuring time, the fine structure of the aerosol could be lost (5) and instead of a nucleation mode, the analyser would measure clustered nuclei and therefore find a coagulation mode. Looking at fig. 4 between 7 - 12 a.m., the decreasing mode of the component monomodal (average distribution found after counting only the single monomodal components) give evidence of successive coagulation mode distributions.

- Build-up of the accumulation mode distribution

The progressive transition from beginning gpc processes, to this mode, correlated with the simultaneous decrease of Aitken nuclei concentrations, explains the in situ build-up of haze by successive coagulations and gaseous condensation processes on preexisting particles.

As soon as the initially generated nuclei have shifted to the coagulation mode, this preexisting aerosol may act now as scavenger for newly formed nuclei (coagulation process) or for gaseous compounds (condensation process), these two processes being more intense at high particle concentrations (6) (7). In this way, the Aitken nuclei concentration decreases as the particle size and the concentration of the preexisting aerosol increases.

Newly formed nuclei could be scavenged onto bigger particles before they grow sufficiently to be seen on the condensation nuclei counter. This type of precursor condensation, in the size range of the nucleation mode, could explain, in stage II and at low relative humidity, the relatively high increase of the mode at low concentrations of Aitken nuclei (8). At relative humidity higher than 80 % encountered mostly during the night, the water vapour condensation is probably the main process of aerosol growth.

3,3. Discussion on gaseous precursor investigations

As the summer concentrations of SO_2 are very low, eventual organic precursors were considered for gpc reactions. Unfortunately, our present GC-MS analysis was not able to detect obvious correlations.

We have then tried to correlate the build-up of haze with the build-up of ozone, indicator for photo-oxidation reactions after having used in bag experiments, some natural and anthrogenic organic precursors.

Unfortunately, the same compound was not always a precursor for ozone but could be simultaneously a sink for O_3 and a source for particles (9).

In fig. 5, where hourly data of visibility and Aitken nuclei concentrations were averaged every 3 hours for 18 selected days of summer 1980, the identified particle formation processes could be generalized. In fact, replacing the most probable sizes by light scattering coefficients (b_s), the build-up of b_s was simultaneously correlated with the decrease of Aitken nuclei concentrations. The build-up of summer haze as in situ processes was then supposed caused by very diffused and unknown precursors and was distinguished from the winter sulphurous smog other source of particle in the region.

4. CONCLUSION

- Aerosol size distribution analysis performed during summer haze episodes in Ispra have shown in the size range lower than 10^{-2} μm two probable homogeneous gas to particle conversion processes through photochemical and non photochemical reactions;

- Consecutive coagulation and gaseous condensations up to 1 μm size cause the build-up of the accumulation mode distribution, the major component of haze (fig. 6);

- The particle growth as given by the increase of m, well correlated with the increase of the light scattering coefficient and with the simultaneous decreases of Aitken nuclei concentrations, evidences the in situ development of the described processes.

- Excluding the local primary sources of particles, not very important in summer, the in situ build-up of haze might be caused by the presence of very diffused unknown precursors in the atmosphere;

- Organic compounds from vegetation likely to be involved in the processes (terpene) have been shown to be of minor importance for direct particle formation. However, the formation of intermediate involatile compounds above forest areas and their removal by preexisting aerosols could be discussed (10).

- The excess concentrations of Aitken nuclei in winter and of haze particles in summer could have the following explanations: The more important primary source from domestic heating in winter and the more important photo-oxidation processes in summer, dependent on UV energy, and temperature.

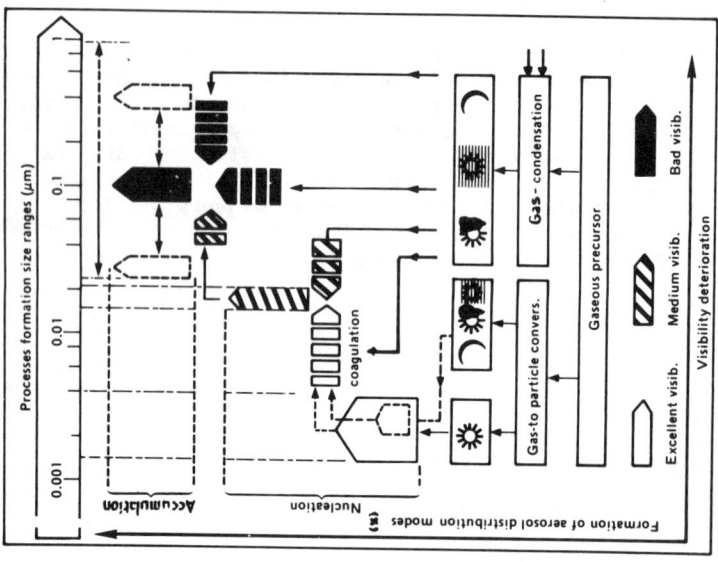

Fig. 6 – Scheme of particle formation processes during the build up of haze (x) EAA Analyser

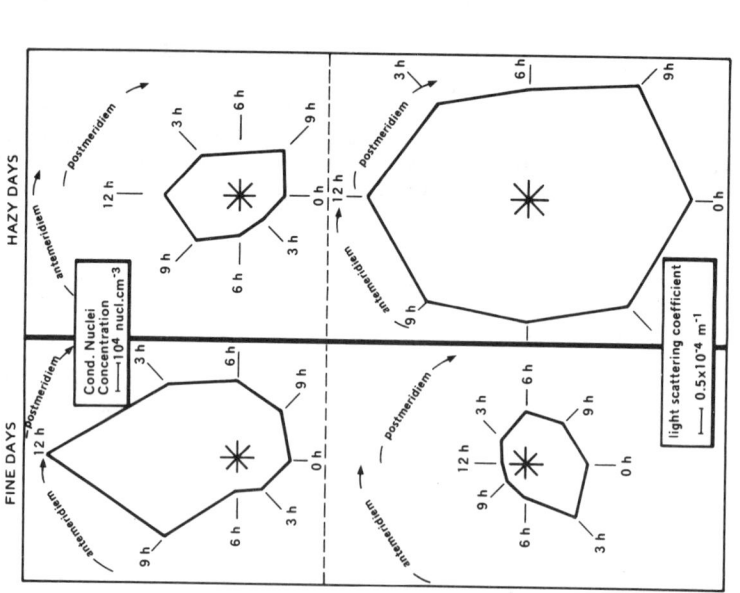

Fig. 5 – Comparisons between concentration of condensation nuclei and light scattering coefficient

REFERENCES

(1) H. STANGL et al.
Discussion of field measurements of pollutants made at Ispra and
other sites of the Po Valley
Proceedings of the 1st European Symposium on "Physico-chemical
behaviour of atmospheric Pollutants", Ispra, 16-18 October 1979

(2) A. RENOUX, J.F. BUTOR, G. MADELAINE
Application des méthodes statistiques à l'étude de la répartition
granulométrique des aérosols atmosphériques
Chemosphère n° 3 pp. 145 - 150 (1975)

(3) M. PAYRISSAT et al.
Measurements and data analysis of atmospheric aerosols
Unpublished work (1980)

(4) W. HAAF and R. JAENICKE
Results of improved size distribution measurements in the
Aitken range of atmospheric aerosols
Journal Aerosol Science, vol. 11, pp. 321 - 330 (1980)

(5) S.H. SUCK, P.B. MIDDLETON and J.R. BROCK
On the multimodality of density functions of pollutant aerosols
Atmosph. Environment, vol. 11, pp. 251 - 255, 1977

(6) P.H. McMURRY and S.K. FRIEDLANDER
New particle formation in the presence of an aerosol
Atmospheric Environment, vol. 13, pp. 1635 - 1651 (1979)

(7) M.L. PERRIN, G. MADELAINE, A. RENOUX
Formation de particules nouvelles en présence de l'aerosol atmosphérique - Proceedings of the 1st European Symposium, Ispra,
16 - 18 Octobre 1979

(8) Programme Progress Report Protection of the Environment.
JRC Ispra Establishment - July - December 1980

(9) C. LOHSE et al.
Photochemical experiments under simulated atmospheric conditions
Proceedings of the 1st European Symposium on "Physico-chemical
behaviour of atmospheric pollutants", Ispra, 16 - 18 October 1979

(10) J. HEICKLEIN
The Removal of atmospheric gases by particulate matter
Atmospheric Environment, vol. 15, n° 5, pp. 781 - 785, 1981.

POLLUTANT CYCLES

Chairman: S. BEILKE

CYCLES OF ORGANIC GASES IN THE ATMOSPHERE

REVIEW

P. BRUCKMANN

Landesanstalt für Immissionsschutz, Essen, W. Germany

Summary

On the basis of recent literature values, it is attempted to describe the elements which are necessary to define the cycles of organic compounds in the atmosphere. Because of their widely differing reactivity, this has to be done for the single compounds. Data on their industrial emissions are available from emission inventories, but data on natural source strengths are still insufficient. Concentrations of many organics have been extensively measured in habitated and remote areas, examples are given. The most important sink processes are attack by reactive intermediates (OH, O_3), whereas dry and wet deposition only play a subordinate role. On the basis of known rate constants, rough lifetimes are estimated. Complete cycles have been described only for long living halocarbons, but the data basis seems to be sufficient now to construct also regional or local budgets for the more reactive ones.

1. INTRODUCTION

The number of existing organic compounds in the atmosphere is legion. A complete description of their cycles seems therefore to be impossible. On the other hand, a mere summary consideration, not taking into account the strongly different reactivities of the single compounds, does not seem to be of great value. We will therefore confine this review to few examples of organic gases with suspected environmental impact. In order to describe the cycles of pollutants, we should know the distribution and effective strengths of their sources and sinks as well as the pathways and products of their removal (1). The knowledge of the rates of the first physical or chemical conversion is not sufficient, because the primary attack on organics is often followed by a cascade of chemical reactions. Even more toxic intermediates than the parent compounds may be formed, until the decay process finally ends up in mineralization or molecules like H_2O or CO_2. Examples are the reaction of dimethylamine and nitrous acid, which gives rise to carcinogenic N-nitrosodimethylamine (2), or the suspected epoxidation of benzo/a_7pyrene by ozone (3). Ultimate aim must therefore be the understanding of the complete cycle from the emission to the ultimate mineralization.

In the following we will try to shed some light on the islands already known in this huge ocean of unknown quantities.

2. EMISSION OF ORGANIC COMPOUNDS

The sources of organics are nearly as numerous as the compounds itself. Information on the man made emission has recently been collected in local or countrywide emission inventories (4) - (7). The survey in table I shows that in highly industrialized regions man made emissions are out-weighing natural emissions of terpenes and isoprene (8). With regard to a whole country, the natural emissions seem to reach the same order of magnitude as the emissions originating from mobile sources. Worldwide, man made emissions are estimated to reach from 5 to 23 % (9), (10), (56) of the total emissions of organic compounds.

The biogenic input of organics is still strongly debated. Estimates of the global input of terpenes and isoprene range from 175×10^6 tons yr^{-1} (11) to $420 - 1330 \times 10^6$ tons CO yr^{-1} (57). Recent estimates of the global natural input necessary to balance the CO cycle give 2100×10^6 tons CO yr^{-1} (56). The actual emission of organic material must be higher, because part of it will not end up as CO, but will pass into the aerosol phase (10) and will be separated from the CO cycle. Besides CH_4, isoprene and terpenes, many other organics are emitted by the biosphere. It has been shown (12) that the ocean acts as a source for light hydrocarbons and methyl chloride (13). Marine aerosols in remote areas contain a variety of higher alkanes and fatty acids with supposed natural origin (14). Vegetation is a source of hundreds of hydrocarbons, organics containing hetero atoms (15) and high molecular waxy components (16). Source strengths for the natural emission of ethene have been given for the U.S.. A portion of 5 % natural emission from the overall ethene emission in the U.S. has been estimated. (17) Quantitative data on most other naturally emitted organics,

Table I.   Estimated emissions (10^3 tons yr^{-1}) in selected areas.
[1]CH_4, C_2H_2 excluded. [2]calculated as CO.

Source	UK,1975(8)	BRD,1975(6)	Rhein-Ruhr area(7)	world
Mobile	660[1]	760	37,8	640.000[2]
Industrial domestic activities	665[1]	1050	301	
natural, terpenes, isoprene	483		20	2.100.000[2](56)
natural, other hydrocarbons	?		?	

however, are lacking, and their source strengths are not known.

Our knowledge of man made emissions is better. Worldwide estimates have been made on the emissions of anthropogenic halocarbons (18). Detailed emission inventories have been established for highly industrialized regions, where the emission is further decomposed into single compounds (7), (4).

As can be seen by a comparison of different local inventories (e.g.(7)), the emitted amounts of single compounds are extremely variable, depending on the type of industry. To give an example, the coal mining and processing industry of Duisburg is emitting 124×10^3 tons methane yr^{-1} and $4,9 \times 10^3$

tons ethene yr^{-1}, respectively. The figures for the petrochemical district around Köln (comparable size and population density) are $4,8 \times 10^3$ tons yr^{-1} and $17,3 \times 10^3$ tons yr^{-1}, respectively. Also the spatial distribution of the source strengths is extremely inhomogeneous [7]. This can be generalized at least for regions with strong industrial point sources (refineries, etc.). For organic compounds with short lifetimes (vide infra), these fine structures have to be taken into account for budget considerations.

A general caveat should be expressed concerning the accuracies of emission data. Error limits are mostly missing, not because of the high precision of the data, but because they are not known. Almost no figure in emission inventories has been measured, but they have been estimated from emission factors and processed substances. Emission factors sometimes change by a factor 2 from one year to another. (example: traffic emissions estimated on the basis of IR are lower by a factor 2 than those based on FID measurements) [7]. At least in the BRD, only factories passing a certain size are obliged to report emission data. Consequently, data from smaller factories are often missing and are not included in the estimates. Emission inventories may therefore be quite uncertain.

3. CONCENTRATIONS OF ORGANIC COMPOUNDS

A lot of work has been published on the concentrations of organics in the atmosphere. Some examples are given in table II. As can be expected from the already discussed inhomogeneous source structure and from the short Lifetimes of many organics in table II (vide infra), the concentrations in polluted areas and rural sites differ up to three orders of magnitude, clearly showing the dominant influence of man made emissions in densely populated areas. This even holds for long living compounds like C_2Cl_4. That the concentrations of terpenes are surprisingly low even in woods [19] has already been mentioned (table II), in contrast to the results from box models based on estimated emissions and known sink processes [20]. The occurance of highly reactive olefines (propene, ethene), together with the relative small latitudinal variation [12] are strong indications for the natural production of light hydrocarbons by the ocean.

While hydrocarbon concentrations in unpolluted air may be sufficiently uniform to define "representative" concentrations for budget considerations, this definition becomes problematic in polluted areas, where concentrations differ by orders of magnitude in time and space. Furthermore, nearly all measurements published so far are ground measurements, which are not representative for the mixing layer. Airborne measurements [21] give strong indications for important fluctuations of the concentration of hydrocarbons, at least in the neighborhood of source areas. Additional data, especially vertical concentration profiles, are needed for budget considerations.

4. SINKS OF ORGANIC COMPOUNDS IN THE ATMOSPHERE

In principle, there are four general sink mechanisms in the atmosphere for organic compounds as well as for other pollutants [1]: dry deposition, rain out and wash out, chemical conversion by homogeneous or heterogeneous reactions, and removal by flux through the boundaries of the considered reservoir (e.g. flux into the stratosphere of long living halocarbons).

4.1 Several authors have determined deposition velocities for organics (table III) and compared them with those for SO_2 under the same experimental conditions. In most cases they have found that deposition velocities of organics were small compared with the uptake of SO_2 by the same soils

Table II. average concentrations (ppb) and main sources of selected organics.

T: traffic, I: industry and heating, N: natural, O: ocean.

compounds of predominantly anthropogenic origin are marked with an asterisk.

a) taken from (5); b) taken from (24); c) taken from (25); d) taken from (13).

compound	city background (22) (23)		traffic (23)	rural (12)	ocean (12)		main sources (15)
ethane	7,5	2,6	4,7	3 -3,5	0,7	-3,3	T, I, N, O
ethene	7,9	3,4	25	0,1-1,5	0,02	-0,5	T, N, I, O
propane	4,0	1,5	2,6	0,8-2	0,1	-2,5	T, N, I, O
propene	1,6	0,9	7,5	0,1-0,5	0,02	-0,2	T, N, I, O
n-butane	4,6	2,2	11,3	0,2	0,03	-1,3	T, N, I, O
n-pentane	1,4	1,7	22,4	0,2	0,01	-1,3	T, N, I, O
benzene*	1,7	2,2	14,7				T, I
toluene*	2,4	2,8	24,9				T, I
m-/p xylenes*	1,3	1,3	16,2				T, I
vinylchloride*	-	0,14	0,14				I
trichlorethene*	0,5[b]	0,6	2,3		0,016[d]		I
tetrachloroethene*	0,8[b]	0,6	1,6		0,04[d]		I
chlorobenzene*	0,2[a]	-	-				I
dichlorobenzene*	0,01[a]	0,5	1,3				I
2-pinene				0,05-1,3[c]			N
limonene		1,2	0,5	0,03-0,27[c]			N
acetaldehyde		4,5	5,1				T, N, I
acetone		8,4	6,0				T, N, I

and materials. These results suggest that dry deposition generally is too slow to compete with chemical reactions (vide infra), with the exception of highly halogenated ones. The situation is different for compounds with low vapor pressures,

Table III.   Deposition velocities v (cm/sec) of selected organics.

compound	material	v(cm/sec)	author
SO_2	grass	0.6 - 0.8	(26)
Lindane, DDT, dieldrin	grass	0.04	(26)
ethene	soil	very small compared	(27)
ethine	soil	with SO_2	(27)
ethene	soil	small compared with SO_2	(17)
Aroclor 1016		0.07	(28)
Chlordane		0.04	(28)
Aroclor 1254		0.11	(28)
Toxaphene		0.24	(28)
dimethyl-sulfide	soil	0.064-0.28 (reversible)	(29)
dimethyl-sulfide	ocean	0.005 (estimated)	(29)

which may be effectively deposited through attachment on particles (e.g. Aroclor 1254 in table III). Otherwise, dry deposition can be neglected unless the chemical conversion times are exceeding one month.

4.2 Almost the same as for dry deposition can be said for rain out and wash out. The uptake of gases into water droplets is governed by Henry's law (26), which describes the partitioning between the two phases. Many organic compounds, especially hydrocarbons and halocarbons, have high vapor pressures and low aqueous solubilities, so that the transition into the water phase becomes negligible. Even the wash out of pesticides is thought to proceed predominantly via their attachment to particles, and not via solution in the water phase (28). This sink mechanism can be quite effective for high boiling PCBs (28). Wet deposition becomes effective, however, for highly polar compounds, especially if the vapor-solution partitioning is strongly affected by chemical reactions in the water phase, as in the case of CH_2O in the presence of HSO_3^- .

4.3 Oxidation reactions with reactive intermediates in the atmosphere, as $OH\cdot$, O_3, $HO_2\cdot$ and O, are by far the most effective sink mechanisms for most organics. Photolysis plays a significant role for some classes of compounds, as lower aldehydes, ketones and nitrites.
    These reactions will purge the atmosphere, if the organic molecules

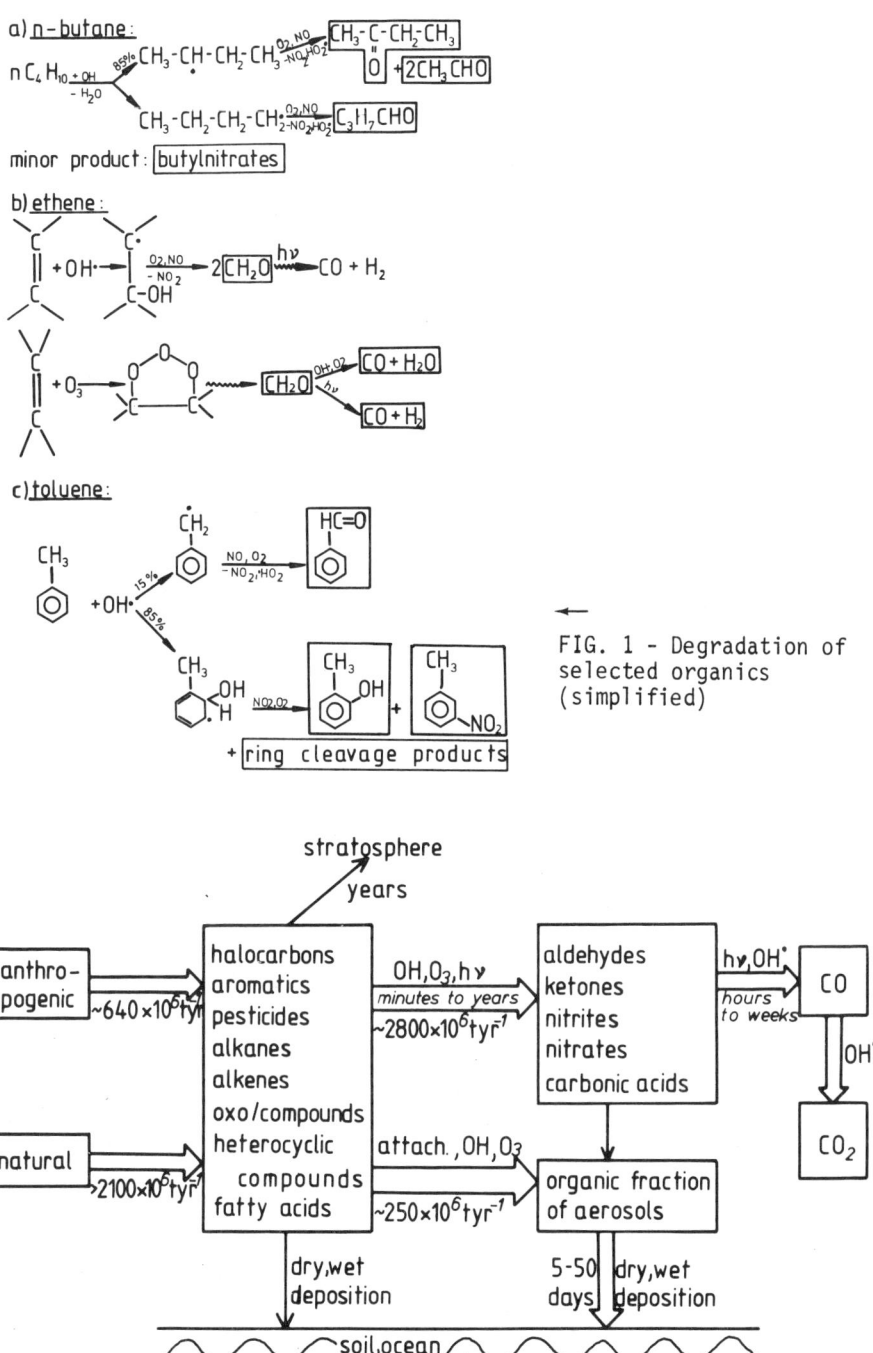

FIG. 1 - Degradation of selected organics (simplified)

Fig. 2 - Fate of organic compounds (simplified)

are either completely oxidized to CO_2 and H_2O, transformed into faster reacting compounds (which in turn will be completely oxidized), or at least
be transformed into more polar compounds or aerosols, so that wet and dry
deposition will become more effective. As excellent reviews dealing with
photooxidation reactions of organics are available (31) - (35), (15), we
will confine ourselves to some basic principles.

The most effective sink process for alkanes is the reaction with the
OH radical. The reactions with other intermediates (O_3, O) are slower by
several orders of magnitude. A brief outline of the reactions are given in
figure I, taking n-butane as example (36), (37). Observed products of reactions performed under simulated tropospheric conditions are aldehydes,
ketones, PAN, PPN, organic nitrates, carbonic acids, hydroxy ketones, alcohols, CO and CO_2. Most of them are quickly broken down further by photolysis (37) and attack by OH radicals. Lower carbonic acids and alcohols, however, are quite stable with lifetimes of one to several weeks (38) and may
survive longer than the parent alkanes.

Olefinic hydrocarbons are generally quickly removed by OH attack and
ozonolysis (figure I), except those substituted by electron drawing substituents (e.g. tetrachloroethene, table IV). For some highly reactive double
bonds as in d-limonene, reaction with ozone becomes more important than attack by OH radicals. At night, only ozonolysis is operative for all olefines. The reactions render mainly aldehydes, ketones, carbonic acids, CO,
alcohols, and finally CO_2. An unsettled question is the amount of organic
aerosols formed in the course of the oxidation reactions of terpenes and
isoprene. It has been reported (39) that these compounds yield principally
aerosol products, whereas recent work (37) has suggested the main products
of OH radical attack to consist of formaldehyde and other gaseous compounds.

Aromatic hydrocarbons substituted by electron donor groups (-OH, $-NH_2$,
$-CH_3$) are quite reactive towards the oH radical. The attack by other intermediates plays only a subordinate role. Products are aromatic aldehydes,
phenols, aromatic nitro compounds, and ring cleavage products (figure I).
As these degradation products generally are less stable than the parent
compounds, rapid further degradation takes place to CO, CO_2, PAN, etc.. An
undissolved question is the amount of products transitioning in the aerosol
phase. As mass balances ovserved in smog chamber studies generally are quite poor (40), considerable particle formation with deposition at the chamber walls may occur.

The chemistry of aldehydes and ketones is of great importance, since
they are intermediates in the degradation of most organics. The attack of
the OH radical is the main reaction path, except for lower aldehydes and
ketones (37), where photolysis is the more efficient sink process. Estimated lifetimes (table IV) are generally shorter than a day. An example of
the reaction sequence is given in figure I.

Degradation products of halogenated alkenes and alkanes can be found
in the literature. (41) - (43).

5. ESTIMATION OF TROPOSPHERIC LIFETIMES

The results presented so far have shown that photochemical reactions
are dominating sink processes for most organics.

If the concentrations of reactive species ($/\overline{O}H/$, $/\overline{O}_3/$, photons) and
their rate constants K_{OH}, K_{O_3}, $\emptyset$ etc. are known, lifetimes τ in the at-

mosphere can be roughly estimated according to the simple relationship

$$\underline{/}1\underline{/} \quad \tau \quad = \quad \frac{1}{K_{OH} \underline{/} OH\_\underline{/} + K_{O_3} \underline{/} O_3\-\underline{/} + \ldots}$$

(τ stands for the time necessary to decrease the concentration to one over e). Rate constants have been determined with an accuracy sufficient for our purpose by several authors and methods.

For a detailed discussion, we refer to the literature (35), (32), (15); values in this work (table IV) are taken from smog chamber studies performed in our laboratory (44) and from the book of Graedel (15), if not stated otherwise.

Problems arise with the definition of timely and spatially averaged concentrations of the reactive species:
- the concentrations of OH etc. depend on the diurnal cycle of the sun.

- in plumes the rate determining step is often not reaction, but transport to the edge of the plume, because reactive intermediates are quenched in the center.

- Altitudinal, seasonal and latitudinal differences in the concentrations as well as temperature dependences of rate constants have to be taken into account for long living compounds (45). Thus, the definition of averaged concentrations heavily depends on the size of the considered reservoir.

Whereas frequency distributions of ozone concentrations have extensively been measured (e.g. (46), (47)), additional data on OH concentrations are still badly needed. OH concentrations of up to 7×10^6 molec. cm^{-3} have been reported in middle europe (48), but it is assumed that such high concentrations describe rather single events than representative concentrations. Average concentrations of $0.5 - 2 \times 10^6$ radicals cm^{-3} have been estimated for a typical summer day (daylight hours only) from model calculations (49), relative disappearance rates in outdoor air (50) and measurements (51). During the night, the concentrations are zero, whereas winter averages (daylight hours) seem to be in the order of 1×10^5 radicals cm^{-3}. Average global OH concentrations of $3 - 5 \times 10^5$ radicals cm^{-3} have been reported from budget considerations ((52), (13)), but it is unclear whether global average values do make sense in the respect to the already discussed variations in time and space.

Nevertheless, neglecting all the factors discussed above, one can obtain at least the order of magnitude of lifetimes of organics, taking the simple expression $\underline{/}1\underline{/}$. This has been done in table IV, tentavely taking concentrations of 1×10^6 molecules cm^{-3} for OH and 50 ppb for ozone, respectively (summer conditions).

We can conclude that the lifetimes of organics cover a wide range from less than one hour (limonene) up to two months (tetrachloroethene). The life times of haloalkanes (not included here) are with several years even considerably longer (45), (13). Organics without strong electron withdrawing substituents, however, have relatively short residence times, especially during summer time. The products of the primary attack have very often even shorter lifetimes, as electron donor groups (OH) or photosensitive groups (aldehydes, keto groups) are introduced. There are exceptions, however, as the formation of quite stable products as trichloroacetylchloride and CCl_4 from the oxidation of tetrachloroethene (41), or the formation of oxirane and methanol from olefines (33). It can be seen from table IV that OH radical attack is not always the dominant sink process. For lower aldehydes and ketones photolysis is more efficient, whereas highly substituted, cyclic

Table IV: Estimated atmospheric lifetimes ($[O_3] = 50$ ppb, $[OH] = 1 \times 10^6$ molec cm^{-3})
and size of reservoir (55) for selected organics.
1) cm^3 molec^{-1} sec^{-1}. 2) for zenith angle of $30°$

compound	$K_{OH}^{1)} \times 10^{12}$	τ_{OH} [h]	τ_{O_3} [h]	$\tau_{hv}^{2)}$ [h]	$\tau_{ges.}$ [h]	reservoir [km]
ethane	0,28 (35)	992			992	world
ethene	8,0	35	122		27	1100
propane	1,9 (30)	146			146	6300
propene	26,5	10	18		6	260
n-butane	2,4	116			116	5000
n-pentane	5,7	49			49	2100
benzene	1,04	267			267	11500
toluene	5,7	49			49	2100
o-xylene	12,5	22			22	950
vinylchloride	6,8	41			41	1800
trichloroethene	2,0	139			139	6000
tetrachloroethene	0,17 (35)	1634			1634	world
chlorobenzene	0,75	370			370	16000
o-dichlorbenzene	0,4	694			694	world
acrylonitrile	2 (38)	139			139	6000
isoprene	74 (30)	4	28		3,5	150
Limonene	142 (35)	2	0,4		0,3	10
formaldehyde	9 (34)	31		5	4,3	180
acetaldehyde	16 (34)	17		7	5,0	200
acetone	0,5 (37)	555		22	21,5	900

Table V. Cycles of halocarbons according to Derwent et al. (45) and Singh et al. (13).

    1) first figure: Northern Hemisphere second: Southern Hemisphere.

    2) S: Stratosphere

compound	τ (yr)	background concentrations (ppt)[1]	source	sinks
$C_2 HCl_3$	0,04	16, < 3	anthr.	OH
$C_2 Cl_4$	0,4	40, 12	anthr.	OH
$CH F Cl_2$	2,9	5, 4	anthr.	OH, 5 % S[2]
$CH Cl_3$	1,0	14,3	anthr.	OH
$CH_3 Cl$	2 – 3	611, 615	ocean, 3×10^6 t yr^{-1}	OH, S
$CH_3 C Cl_3$	5,4 (45) 8 – 10 (13)	113, 77	anthr., $0,7 \times 10^6$ t yr^{-1}	OH, 15 – 20 % S
$CH F_2 Cl$	14			OH, 44 % S
$C Cl_2 F_2$	65 – 70	230, 210	anthr., inventory	S
$C Cl_2 F C Cl F_2$	several	19, 18	anthr.	S, OH
$C Cl F_2 C Cl F_2$	decades	12, 10	anthr.	S, OH
SF_6		0.31, 0.27	anthr.	
$C Cl_4$		122, 119	3×10^6 t yr^{-1}, mainly anthr.	S, ocean

olefines are more readily attacked by ozone.

Concluding, it seems appropriate to disturb the relatively clear picture emerging from table IV. There are some indications that the primary role of the OH radical as "cleanser" is not the whole story. The influence of aerosols, especially in the polluted atmosphere, might have been underestimated. In two recently performed studies (53, 54) of the relative disappearance of hydrocarbons in outdoor air it was found that the alkanes reacted up to three times faster relative to the alkenes, as should have been exspected from their rate constants. One can speculate that there may exist additional sink mechanisms, e.g. the attack by reactive species formed at the surface of aerosols. Further studies are necessary to clarify this point.

6. CYCLES OF ORGANIC COMPOUNDS

In the last row of table IV, Rohde's guideline (55) has been used to estimate the optimum size of the reservoirs necessary for the studies of cycles and budgets. Only for halogenated compounds and ethane, the whole atmosphere or one hemisphere seems to represent an appropriate reservoir. The other cycles have to be studied in a continental (e.g. benzene), regional (e.g. toluene, substituted alkanes) or even local scale (substituted alkenes).

Complete cycles, however, have only been given for long living, anthropogenic halocarbons in a world scale, because of their importance for the stratospheric ozone, and because their emissions are comparably well known. Table V summarizes the main results; for details, the reader is referred to the literature (45, 13).

Surprisingly, no complete cycles of faster reacting organics on a regional or local scale have been published, as far as we know. The material presented so far is on our opinion sufficient to try such a step in the near future, in order to check the single elements - source strengths, concentrations and sinks - for consistency. The results could give new insights in hitherto unresolved questions as the amount of natural emissions of organics besides the terpenes or the role played by heterogeneous reactions and gas to particle conversion.

Concluding, we have tried to give a qualitative picture of the fate of organics in figure II, which may serve as a summary of the results presented so far. Figure II is largely self explanatory and does not need much comment. It should be noted, however, that the figures are taken from different authors (8, 10) and are perhaps not self consistent.

Dispite the remaining "missing links" in the definition of complete cycles it can be stated that great progress has been made in recent years in defining important features - especially the sink processes and the concentration of organics are quite well known.

REFERENCES

(1) S. Beilke and G. Gravenhorst, Proceedings of the first European Symposium of physico-chemical behaviour of atmospheric pollutants, Ispra, 1979, 331 - 353.

(2) D. H. Fine and D. P. Roubehler, Environ. Sci. Technol. 11, 577 - 580 (1977).

(3) J. N. Pitts, Jr., D. M. Lokensgard, P. S. Ripley, K. A. van Cauwenberghe, L. van Vaeck, S. D. Shaffer, A. J. Thill and W. L. Belser, Jr., Science 210, 1347 - 1349(1980).

(4) K. A. Brice and R. G. Derwent, Atmos. Environ. 12, 2045 - 2054(1978).

(5) H. B. Singh, L. J. Salas, A. J. Smith and H. Shigeishi, Atmos. Environ.
15, 601 - 612(1981).
(6) Materialien zum Immissionsschutzbericht 1977 der Bundesregierung an
den Deutschen Bundestag, Hrsg. P. Davids, Erich Schmidt Verlag Berlin,
1977.
(7) Luftreinhalteplan Rheinschiene Süd 1977 - 1981 (Köln), Ed. by Minister
für Arbeit, Gesundheit und Soziales des Landes NW, Kölnische Verlags-
druckerei GmbH, Köln 1977; compare Luftreinhaltepläne Ruhrgebiet West,
Ost, Mitte.
(8) R. G. Derwent and Ö. Hov, Proceedings of the first European Symposium
of physico-chemical behaviour of atmospheric pollutants, Ispra, 1979,
367 - 382.
(9) S. Budiansky, Environ. Sci. Technol. 14, 901 - 903(1980).
(10) G. Ketseridis and R. Jaenicke in: Organische Verunreinigungen in der
Umwelt, K. Aurand, ed., Erich Schmidt Verlag, Berlin 1978, pp. 379-390.
(11) R. A. Rasmussen, J. Air Pollut. Control Assoc. 22, 537 - 543(1972).
(12) J. Rudolph, D. H. Ehhalt and G. Gravenhorst, Proceedings of the first
European Symposium of physico-chemical behaviour of atmospheric pollutants, Ispra, 1979, 41 - 51.
(13) H. B. Singh, L. J. Salas, H. Shigeishi and E. Scribner, Science 203,
899 - 903(1979).
(14) C. L. Weschler, Atmos. Environ. 15, 1365 - 1369(1981).
(15) T. E. Graedel, Chemical Compounds in the Atmosphere. Academic Press,
New York, 1978.
(16) W. A. Hoffman et al., Environ. Sci. Technol. 14, 999(1980).
(17) F. B. Abeles, L. E. Craker, L. E. Forrence and G. R. Leather, Science
173, 914 - 916(1971).
(18) H. B. Singh, L. J. Salas, H. Shigeishi, A. J. Smith and E. Scribner,
Report EPA-600/3-79-107, Project 4487.
(19) Y. Yokouchi, Y. Ambe and K. Fuwa, Chemosphere 10, 209(1981).
(20) E. W. Peterson, Atmos. Environ. 14, 79 - 81(1980).
(21) E. Neuber, H. W. Georgii and J. Müller, Staub-Reinhalt. Luft 41, 91 -
97(1981).
(22) P. Bruckmann and W. Mülder, Schriftenreihe der LIS, in press.
(23) W. Dulson, Schriftenreihe des Vereins für Wasser-, Boden- und Luft-
hygiene 47, Gustav Fischer Verlag, 1978.
(24) U. Bauer, Ruhr Universität Bochum personal communication.
(25) Y. Yokouchi, T. Fujii, Y. Ambe and K. Fuwa, J. Chromatography 209,
293 - 298(1981).
(26) D. E. Glotfelty, J. Air Pollut. Control Assoc. 28, 917 - 921(1978).
(27) K. A. Smith, J. M. Bremner and M. A. Tabatabai, Soil Science 116,
313 - 319(1973).
(28) T. F. Bidleman and E. J. Christensen, J. Geophys. Res. 84, 7857(1979).
(29) H. S. Judeikis and A. G. Wren, Atmos. Environ. 11, 1221 - 1224(1977).
(30) R. A. Cox, Proceedings of the first European Symposium of physico-
chemical behaviour of atmospheric pollutants, Ispra, 1979, 91 - 109.
(31) R. A. Cox and R. G. Derwent, Environ. Sci. Technol. 14, 57 - 61(1980).
(32) B. J. Finlayson and J. N. Pitts, Jr., Science 192, 111 - 119(1976).
(33) J. N. Pitts, Jr. and B. J. Finlayson, Angew. Chem. 87, 18 - 33(1975).
(34) H. G. Wagner and R. Zellner, Angew. Chem. 91, 707 - 718(1979).
(35) R. Atkinson, K. R. Darnall, A. C. Lloyd, A. M. Winer and J. N. Pitts,
Jr., Adv. Photochem. 11, 375(1979).
(36) W. P. L. Carter, A. C. Lloyd, J. L. Sprung, and J. N. Pitts, Jr., Int.
J. Chemical Kinetics, 11, 45 - 101(1979).
(37) R. A. Cox, K. F. Patrick, and S. A. Chant, Environ. Sci. Technol. 15,
587 - 592(1981).

(38) C. Zetsch, Proceedings of the 'Internationale Arbeitstagung über Prüfungsmethoden und Bewertungsverfahren zur Bestimmung des photochemischen Abbauverhaltens von chemischen Stoffen, 2. - 4.12.1980, Berlin',
in press.

(39) B. W. Gay, R. R. Arnts, Report No. EPA-600/3-77-001 b, EPA, Research
Triangle Park, N.C., 1977, p. 745.

(40) R. M. van Aalst, A. C. Besemer and N. Nieboer, Proceedings of the
first European Symposium of physico-chemical behaviour of atmospheric
pollutants, Ispra, 1979, 136 - 149.

(41) B. W. Gay, Jr., P. L. Hanst, J. J. Bufalini, and R. C. Noonan, Environ.
Sci. Technol. 10, 58 - 67(1976).

(42) W. L. Dilling, C. J. Bredeweg, and N. B. Tefertiller, Environ. Sci.
Technol. 10, 351 - 355(1976).

(43) T. Woldbeak and P. Klaboe, Spectrochem. Acta 34 A, 481 - 487(1978).

(44) P. Bruckmann, Proceedings of the 'Internationale Arbeitstagung über
Prüfungsmethoden und Bewertungsverfahren zur Bestimmung des photochemischen Abbauverhaltens von chemischen Stoffen, 2. - 4.12.1980, Berlin',
in press.

(45) R. G. Derwent and A. E. J. Eggleton, Atmos. Environ. 12, 1261 - 1269
(1978).

(46) Photochemical smogformation in the netherlands, Ed. R. Guicherit, TNO
report, 1978.

(47) Photochemische Luftverunreinigungen in der Bundesrepublik Deutschland,
Proceedings, VDI, Düsseldorf, 1980.

(48) D. Perner, D. H. Ehhalt, H. W. Pätz, U. Platt, E. P. Röth and A. Volz,
Geophys. Res. Letters 3, 466 - 468(1976).

(49) R. G. Derwent and Ö. Hov, A. E. R. E. Report R 9434, H. M. Stationery
Office, London, 1979.

(50) R. Guicherit, K. D. van den Hout, C. Huygen, H. van Duuren, F. G.
Römer and J. W. Viljeer, Proceedings of the 11th NATO/CCMS conference,
Amsterdam, 1980.

(51) D. H. Ehhalt, personal communication in the VDI working group 'Luft-
chemie'.

(52) P. Warneck, Planet. Space Sci. 23, 1507 - 1518(1975).

(53) H. B. Singh, J. R. Martinez, D. G. Hendry, R. J. Jaffe, and W. B.
Johnson, Environ. Sci. Technol. 15, 113 - 119(1981).

(54) E. Neuber, personal communication.

(55) H. Rodhe, Atmos. Environ. 12, 671 - 680(1978).

(56) A. Volz, D. H. Ehhalt and R. G. Derwent, J. Geophys. Res. 86, 5163 -
5171(1981).

(57) P. R. Zimmerman, R. B. Chatfield, J. Fishman, P. J. Crutzen and P. L.
Hanst, Geophys. Res. Lett. 5, 679 - 682(1978).

PRODUCTION OF GASEOUS HYDROCARBONS IN SOIL

O. VAN CLEEMPUT, A.S. EL-SEBAAY and L. BAERT
Faculty of Agriculture, University of Ghent,
Coupure 533, 9000 Ghent, Belgium

Summary

The production of methane, ethane, propane and ethylene has been followed in a series (24) of soil samples with different characteristics. Therefore, the soil samples were incubated under closed waterlogged conditions at 25°C and analysed for the hydrocarbons, oxygen and carbon dioxide after 15, 30, 60 and 90 days.
It has been found that in all soils the production of hydrocarbons decreases in the following order : methane $>$ ethylene $>$ ethane $>$ propane. Addition of nitrate to the soil samples retards the hydrocarbon production. After 90 days of incubation about 6 Kg of methane is produced per hectare. Concerning the other hydrocarbons, at maximum concentration during the incubation, a total amount of about 20 g per hectare is produced. The soils with a high organic matter content and a pH value around neutrality or higher have an important oxygen consumption and a pronounced methane and ethylene production. A high or low methane production is not always accompanied by a high or low ethylene production and vice versa. The production of ethane and propane is very low throughout the experiment and not clearly influenced by one or another soil characteristic.

1. INTRODUCTION

The production of gaseous hydrocarbons in soil is important as indication of organic matter transformation under anaerobic conditions. It can cause air pollution and also influence the crop production.
Smith and Russell (11) showed that the production of ethylene is highest in soils with a high content of organic matter. In most cases ethylene is accompanied by a wide range of other hydrocarbons such as methane, ethane, propylene and propane. Concerning the effect of hydrocarbons on plants, Smith and Jackson (9) reported that higher olifins have a smaller effect than ethylene; and saturated hydrocarbons, also formed in soil, are physiologically inactive. It is shown that ethylene is the gaseous hydrocarbon produced by the soil, having the most significant effect on plant growth.
The comparative evolution of ethylene and other hydrocarbons is illustrated by Smith and Restall (10). They found that when soils with different percentages of organic matter (ranging from 1.4 to 38.0 %) were maintained under anaerobic conditions, methane and ethylene were observed in all cases and usually also propane, propylene, n- and iso-butane, and butene-1; no acetylenic hydrocarbons were found.
Besides the influence of organic matter and anaerobic conditions, it is also shown by Smith and Dowdell (8) that in sandy loam soils a clear relationship exists between high moisture content and both the production

of ethylene and the depression of oxygen. There results show, however, a great variability between replicate sampling points.

The factors which have the greatest influence on the production of hydrocarbons e.g. ethylene are considered to be temperature, depression of oxygen, the availability of substrates for microbial activity and the effect of soil moisture content on the air-filled porosity of the soil. It is also indicated that concentrations of oxygen in soil must fall considerably before significant increases in ethylene production occur. Laskowsky and Moraghan (4) showed that the presence of KNO_3 and N_2O reduced the rate of accumulation of methane under anaerobic conditions at 30°C, while Goodlass and Smith (2) demonstrated that no significant relationship could be found between NO_3 concentrations in fresh soils and the quantities of evolved ethylene; the quantity of ethylene, however, was significantly increased with decreasing pH.

It is the aim of this paper to present results on production of the most important gaseous hydrocarbons in a series of belgian soils with different characteristics.

2. MATERIALS AND METHODS

In order to study the production of gaseous hydrocarbons, twenty-four soil samples of different characteristics were incubated at 25°C. The soil samples were collected from the arable layer, air-dried, ground to pass a 2 mm sieve and stored in air-dry conditions. The incubation is carried out in closed flasks of 150 cc, in which 30 g of soil and 60 ml of distilled H_2O is mixed. The incubation flasks were closed by using a rubber septum, allowing also to sample the gas phase by syringe. The soil characteristics are given in table 1. The samples were selected on the basis of their differences in soil texture, carbon content and pH. The gas phase above the suspensions was analysed after 15, 30, 60 and 90 days for oxygen, carbon dioxide and the following hydrocarbons : methane (CH_4), ethane (C_2H_6), propane (C_3H_8) and ethylene (C_2H_4).

Gas chromatographic analysis was carried out on 1 cc gas samples, splitted into one part for the F.I.D. detector (hydrocarbons) and one part for the conductivity detector (permanent gases). The hydrocarbons were separated on 100-120 mesh alumina desactivated with sodium iodide, while O_2 and CO_2 are separated from other permanent gases on a combination of porapak Q and molecular sieve. Concentrations were calculated after comparison of peakhights from the unknown samples with samples of known concentration. The technique has been described by Smith and Dowdell (7). The results are given as percentage of the air space above the soil suspensions.

In order to study the influence of nitrate on production of hydrocarbons, also a series of suspensions was incubated to which 100 ppm NO_3^--N was added.

3. RESULTS AND DISCUSSION

The results of the air composition after 15 and 90 days of incubation are given in table 2 and 3. Out of these tables it can be seen that, irrespective of addition of nitrate, the production of hydrocarbons decreases in the following order : $CH_4 > C_2H_4 > C_2H_6 > C_3H_8$. After 15 days of incubation (table 2) the CH_4 production is about 10-100 times more important than the other hydrocarbons. Addition of 100 ppm NO_3^--N to the suspensions lowers the amount of CH_4 by a factor of about 1.5 to 2; while the amount of the other

Table 1

Characteristics of the different soils

Nr:	Texture (U.S.D.A.)	pH (H$_2$O)	pH (KCl)	%C	%CaCO$_3$	%Clay 0-2μ	%Silt 2-50μ	%Sand 50μ-2mm
1:	Clay Loam	7.76	7.31	2.23	15.8	33.0	29.0	38.0
2:	Silt Loam	6.66	6.16	1.28	0.0	17.6	70.6	11.8
3:	Silt Loam	6.15	5.52	1.15	0.0	13.2	77.0	9.8
4:	Clay Loam	7.74	7.17	2.15	5.6	33.0	30.0	37.0
5:	Sand	4.51	3.70	2.82	0.0	5.8	5.7	88.5
6:	Loamy Sand	4.51	3.77	6.97	0.0	8.7	11.9	79.4
7:	Loamy Sand	4.78	3.60	1.38	0.0	4.1	15.5	80.4
8:	Sand	6.58	5.40	2.11	0.0	2.8	9.3	87.9
9:	Loamy Sand	5.12	3.95	1.29	0.0	3.5	20.0	76.5
10:	Silty Loam	6.82	6.33	0.97	0.0	14.6	74.2	11.2
11:	Sandy Loam	6.00	4.64	0.72	0.0	5.3	38.6	56.1
12:	Clay Loam	7.72	7.35	2.01	8.4	29.0	35.0	36.0
13:	Sandy Loam	5.06	4.25	1.68	0.0	11.3	8.7	80.0
14:	Sandy Loam	6.09	5.62	4.37	0.0	10.1	12.0	77.9
15:	Silty Loam	6.39	5.59	1.17	0.0	10.1	74.3	15.6
16:	Silt	6.95	5.98	1.51	0.0	10.3	81.2	8.5
17:	Sandy Loam	5.11	4.57	9.75	0.0	12.5	14.4	73.1
18:	Clay	7.80	7.35	0.96	5.4	71.5	25.5	3.0
19:	Clay	8.00	7.53	0.90	6.9	55.5	27.1	17.4
20:	Clay	8.10	7.76	0.66	5.9	70.5	24.7	4.8
21:	Loamy Sand	7.41	7.14	2.49	0.0	5.3	8.4	86.3
22:	Sand	4.20	3.35	2.79	0.0	2.0	6.8	91.2
23:	Clay	7.60	6.87	2.10	4.3	41.9	37.5	20.6
24:	Clay Loam	7.92	7.28	0.90	0.0	30.7	37.7	31.6

hydrocarbons remains about the same, with a tendency to be lower. The carbon dioxide content is not very much influenced by addition of nitrate, while the oxygen consumption is somewhat retarded. One reason for the lower production of hydrocarbons after addition of nitrate, can be the fact that the addition of nitrate prevents the redox potential from dropping to low values, retarding the production of methane. It is known that methane is only produced at low redox potentials, while oxygen and nitrate are already reduced at fairly high redox potentials (6). Another reason is that the added nitrate can act as an alternative oxygen substrate for the organisms producing the hydrocarbons (10). It can be seen that in the soils with a high organic matter content, associated with a pH near neutrality or higher, the methane production is important. The results also show that a high or low methane production is not always accompanied by a high or low ethylene production and vice versa. The results after 90 days of incubation (table 3) show the same order of importance for the different hydrocarbons. The methane production is increased by more than ten times, while the amount of the other hydrocarbons is decreased by at least five times. In table 4, the average values of the gaseous products are summarised for the treatments with and without added nitrate and after the four incubation

Table 2 : The percentage of the hydrocarbons and the permanent gasses in different soils (after 15 days)

Number: soil sample	without nitrogen						with 100 ppm NO_3^--N					
	CH_4 %x10^{-2}	C_2H_6 %x10^{-5}	C_3H_8 %x10^{-5}	C_2H_4 %x10^{-5}	O_2 %	CO_2 %	CH_4 %x10^{-2}	C_2H_6 %x10^{-5}	C_3H_8 %x10^{-5}	C_2H_4 %x10^{-5}	O_2 %	CO_2 %
1	0.085	1.91	1.94	31.39	9.31	2.20	0.074	1.53	1.94	26.52	10.54	1.36
2	0.103	3.82	2.59	21.11	12.47	1.73	0.081	3.44	1.94	20.57	14.05	1.73
3	0.050	1.60	1.35	6.71	12.43	2.30	0.040	1.63	1.28	8.81	14.65	2.94
4	1.255	2.28	2.88	21.96	10.46	3.50	2.088	4.29	2.56	21.94	12.88	4.87
5	0.059	1.34	1.29	36.80	13.97	2.89	0.034	0.76	1.29	23.27	13.18	2.36
6	0.069	2.29	1.94	43.30	13.35	2.57	0.042	1.15	1.94	36.26	12.65	3.04
7	0.024	0.76	0.65	10.28	11.24	3.15	0.025	1.15	0.65	14.07	10.01	3.67
8	0.053	1.72	1.94	23.81	6.85	2.89	0.055	1.15	1.94	20.57	9.66	2.62
9	0.046	1.15	1.29	15.15	7.38	2.52	0.049	1.15	0.97	13.53	13.53	2.99
10	0.102	3.05	1.94	10.82	13.00	1.42	0.090	2.29	1.62	9.74	14.05	1.26
11	0.072	0.76	0.97	9.20	13.70	1.63	0.057	0.76	0.97	9.20	14.41	1.94
12	0.080	0.76	1.29	6.49	15.81	1.16	0.040	0.76	0.65	3.25	15.81	0.74
13	0.016	1.53	0.00	6.49	13.00	1.84	0.012	0.76	0.00	2.71	12.30	1.89
14	0.028	0.38	0.00	2.71	14.05	1.73	0.018	0.38	0.00	0.00	13.70	2.20
15	0.001	2.29	1.29	14.61	13.00	1.57	0.001	1.91	0.65	11.91	14.41	1.63
16	0.303	0.76	0.65	14.61	9.84	2.15	0.233	0.76	0.65	12.45	12.30	1.89
17	0.001	1.15	0.97	21.65	11.95	4.99	0.001	1.15	0.00	18.40	13.00	4.93
18	0.014	0.38	0.00	3.25	14.76	0.42	0.012	0.38	0.00	1.62	14.05	0.32
19	0.905	0.00	0.65	0.00	17.22	0.84	0.713	0.00	0.65	0.00	17.92	0.42
20	0.013	0.00	0.00	0.54	19.68	0.03	0.012	0.00	0.00	0.00	20.03	0.03
21	8.841	16.40	6.47	35.70	4.92	1.44	4.962	11.80	4.50	30.31	13.18	0.87
22	0.019	0.76	1.82	25.23	10.89	3.37	0.016	0.76	1.65	33.22	10.54	3.62
23	0.122	1.15	0.00	8.12	16.87	2.36	0.049	0.76	0.00	4.33	17.92	1.89
24	0.061	1.53	0.65	5.95	16.87	1.68	0.045	1.53	0.65	7.58	17.22	1.78

Table 3 : The percentage of the hydrocarbons and the permanent gasses in different soils (after 90 days)

Number: soil sample	without nitrogen						with 100 ppm NO_3^--N					
	CH_4 $\%\times10^{-2}$	C_2H_6 $\%\times10^{-5}$	C_3H_8 $\%\times10^{-5}$	C_2H_4 $\%\times10^{-5}$	O_2 $\%$	CO_2 $\%$	CH_4 $\%\times10^{-2}$	C_2H_6 $\%\times10^{-5}$	C_3H_8 $\%\times10^{-5}$	C_2H_4 $\%\times10^{-5}$	O_2 $\%$	CO_2 $\%$
1	27.034	0.34	0.10	0.59	0.40	4.46	6.477	0.15	0.10	0.49	3.43	4.72
2	7.075	0.34	0.13	0.70	3.16	4.20	2.015	0.17	0.10	0.49	4.22	4.20
3	1.914	2.00	0.33	2.53	3.78	3.73	1.823	1.51	0.85	2.77	4.92	4.57
4	35.369	1.33	0.29	1.08	1.23	4.01	33.930	0.95	0.19	0.50	1.10	4.46
5	6.364	0.11	0.06	1.35	1.85	6.61	0.271	0.08	0.06	1.38	5.71	6.77
6	1.373	0.15	0.13	1.67	3.43	5.56	0.176	0.08	0.06	1.46	5.45	5.67
7	3.041	0.10	0.13	1.32	1.49	6.72	1.071	0.10	0.13	1.46	1.85	7.09
8	8.941	0.12	0.16	1.51	1.45	4.15	36.326	0.72	0.19	0.97	1.10	4.72
9	3.745	0.08	0.13	1.30	1.32	6.30	5.153	0.08	0.13	1.23	1.32	6.56
10	4.647	0.23	0.16	0.73	2.90	3.25	4.506	0.21	0.13	0.68	3.34	3.78
11	6.336	0.10	0.20	0.86	3.16	3.99	7.885	0.05	0.06	0.85	4.74	4.30
12	0.524	0.00	0.06	0.00	6.32	4.20	0.062	0.00	0.06	0.00	5.18	4.41
13	0.101	0.80	0.08	1.43	8.61	5.88	0.204	0.40	0.11	0.85	10.89	5.25
14	0.031	0.07	0.16	0.00	6.32	4.62	0.008	0.05	0.16	0.00	6.85	4.93
15	15.206	2.08	0.13	0.11	1.32	3.88	11.968	2.08	0.13	0.11	1.32	3.88
16	61.952	1.43	0.13	0.41	1.23	3.31	50.688	1.27	0.19	0.22	1.23	3.52
17	0.046	0.21	0.35	3.05	5.27	7.56	0.021	0.11	0.05	2.49	2.11	4.36
18	0.019	0.04	0.06	0.00	6.15	3.73	0.007	0.04	0.00	0.00	8.43	2.73
19	0.003	0.00	0.00	0.00	3.87	2.68	0.002	0.00	0.00	0.00	4.57	2.41
20	0.001	0.00	0.00	0.00	12.47	0.23	0.001	0.00	0.00	0.00	12.74	0.10
21	103.066	1.62	0.70	2.80	1.21	1.65	62.515	1.60	0.90	2.59	1.21	1.68
22	4.576	1.50	3.20	3.24	1.25	9.18	1.442	1.15	0.90	2.59	1.05	8.40
23	15.206	0.08	0.13	0.00	1.30	5.14	5.745	0.06	0.13	0.00	1.76	5.41
24	18.811	0.09	0.27	0.11	1.98	4.04	3.576	0.06	0.10	0.00	1.63	4.72

Table 4

Average amounts of gaseous products after different periods of

incubation and in the presence (+) and absence (-) of nitrate

Gaseous product	Nitrate addition	Days of incubation			
		15	30	60	90
CH_4 $(\% \times 10^2)$	-	0.513	5.112	9.457	13.558
	+	0.364	2.882	7.901	9.828
C_2H_6 $(\% \times 10^5)$	-	1.990	3.364	0.907	0.534
	+	1.677	2.725	0.695	0.455
C_3H_8 $(\% \times 10^5)$	-	1.357	1.986	0.591	0.291
	+	1.104	1.749	0.410	0.201
C_2H_4 $(\% \times 10^5)$	-	15.661	19.433	4.913	1.033
	+	13.760	18.125	4.553	0.880
O_2 $(\%)$	-	12.625	9.960	5.963	3.394
	+	13.833	12.074	8.254	4.006
CO_2 $(\%)$	-	2.099	3.590	4.299	4.544
	+	2.125	3.547	4.256	4.526

periods. Out of this table it can be seen that except for the methane production, the concentration of all hydrocarbons is highest after 30 days of
incubation, whereafter it importantly decreases. It is clear that all hydrocarbons are produced while the oxygen concentration is still fairly
high. It should, however, considered that the gas phase composition is
measured in the space above the suspension. This means that the suspension
can be depleted of oxygen while still an important amount of oxygen remains
in the gas phase. It is known that the diffusion of oxygen in water is
10,000 times slower than in air (3, 5). The important methane production
also indicates that the soil suspension is under strict anaerobic conditions (1). Taking into account the space volume (70 ml) above the suspension, it can be calculated that, in the absence of nitrate, after 90 days
of incubation about 2 mg methane per Kg of soil is produced or about
6 Kg/Ha; concerning the other hydrocarbons, at maximum concentration, a
total amount of about 7 µg/Kg of soil or 21 g/Ha is produced. In the presence of nitrate, the methane production is decreased by more than 25 % and
the sum of the other hydrocarbons by about 10 %.

A correlation study between the different hydrocarbons after 15 days
of incubation showed that methane, ethane and propane are positively correlated to one another at 0.01 level of significance. At this level of significance, ethylene is positively correlated only with propane; it is also
positively correlated with ethane but at 0.05 level of significance.
Ethane, propane and ethylene are negatively correlated (0.01 level) with

the oxygen content in the gas phase. A correlation study between the hydrocarbons and the different soil characteristics showed that the ethylene production is positively correlated with the organic matter content at 0.05 level. In the used soils, the organic matter content is also positively correlated with the sand content (0.05 level), making that the ethylene production is positively correlated with the sand fraction of the soil and consequently negatively correlated with the clay content. The other hydrocarbons are not clearly correlated with the studied soil characteristics.

Out of this study it can be concluded that under waterlogged conditions, the following gaseous hydrocarbons are produced in decreasing order : methane, ethylene, ethane and propane. The presence of nitrate in the soil suspension retards the hydrocarbon production. Soils with a high organic matter content and a pH value around neutrality or higher have an important oxygen consumption and a pronounced methane and ethylene production.

4. REFERENCES

1. Bell, R.G.. 1969. Studies on the decomposition of organic matter in flooded soil. Soil Biol. Biochem. 1:105-116.
2. Goodlass, G., and K.A. Smith. 1977. Effect of pH,organic matter content and nitrate on the evolution of ethylene from soils. Soil Biol. Biochem. 10:193-199.
3. Greenwood, D.J.. 1961. The effect of oxygen concentration on the decomposition of organic materials in soils. Plant Soil, 14:360-376.
4. Laskowsky, D., and J.T. Moraghan. 1967. The effect of nitrate and nitrous oxide on hydrogen and methane accumulation in anaerobically incubated soils. Plant Soil 27:357-368.
5. Lemon, E., and J. Kristensen. 1960. An edaphic expression of soil structure. Trans. 7th Intern. Congr. Soil Sci. 1:232-240.
6. Patrick, W.H. Jr., and D.S. Mikkelsen. 1971. Plant nutrient behavior in flooded soil. In "Fertilizer Technology and Use", 2nd edition, Soil Sci. Soc. Amer., Madison, Wisc. :187-215.
7. Smith, K.A., and R.J. Dowdell. 1973. Gas chromatographic analysis of the soil atmosphere : automatic analysis of gas samples for O_2, N_2, Ar, CO_2, N_2O, and C_1-C_4 hydrocarbons. J. Chromatog. Sci. 11:655-658.
8. Smith, K.A., and R.J. Dowdell. 1974. Field studies of the soil atmosphere. 1. Relationships between ethylene, oxygen, soil moisture content, and temperature. J. Soil Sci. 25:217-230.
9. Smith, K.A., and M.B. Jackson. 1973. Ethylene, waterlogging and plant growth. Ann. Rep. Agric. Res. Council, Letcombe Lab., Wantage, England.
10. Smith, K.A., and S.W.F. Restall. 1971. The occurrence of ethylene in anaerobic soil. J. Soil Sci. 22:430-443.
11. Smith, K.A., and R.S. Russell. 1969. The occurrence of ethylene and its significance in anaerobic soil. Nature, Lond. 222:769-771.

DISTRIBUTION D'ALCANES ET DE COMPOSES AROMATIQUES C_6 ET C_7 DANS L'ATMOSPHERE DE REGIONS SEMI-RURALES BELGES *

M.TERMONIA
Institut de Recherches Chimiques
Ministry of Agriculture, Museumlaan, 5
1980 Tervuren, Belgium

Abstract. (occurence of alkylbenzenes and alkanes immissions in Belgium)

Quantitative analyses of benzene, benzene derivatives and alkanes in urban and rural air were performed by high resolution gas chromatography (HRGC) and HRGC-mass spectrometry-data system.
The data generally fit the distribution of these compounds determined in exhaust gases from gasoline motors. These results parallel those obtained in other countries. From the immission data, it can be deduced that a C7/C6 ratio higher than unity corresponds to emissions from automotive traffic, while a C_7/C_6 ratio lower than unity could be related to a direct source of benzene evaporation.
Deviations were observed when the relative humidity was high, probably due to the difference of solubility of benzene and toluene in water. The results concerning alkane immissions indicate reduced concentration levels in comparison with the aromatic fraction.

Key words

Alkylbenzenes and alkanes in air - Distribution and origin of C_6 and C_7 aromatics in air. HRGC/MS/DS.

Résumé

L'analyse quantitative des aromatiques C_6 et C_7 et d'alkanes C_7-C_{20} en atmosphère urbaine et rurale a été effectuée par chromatographie à haute résolution (HRGC) et par HRGC-spectrométrie de masse-système informatique. Les concentrations mesurées coïncident généralement avec la distribution de ces composés dans les gaz d'échappement des moteurs à essence. Ces résultats sont à comparer avec ceux obtenus dans d'autres pays. Des données à l'immission, il peut être déduit qu'un rapport C_7/C_6 plus petit que l'unité peut être révélateur d'une source directe d'évaporation de benzène. Des écarts sont observés quand l'humidité relative est élevée, probablement dus à la différence de solubilité du benzène et du toluène dans l'eau. Les résultats concernant les alcanes font état de niveaux de concentration plus faibles que ceux de la fraction aromatique.

Mots clés

Alkylbenzènes et alcanes dans l'air - Distribution et origine des composés aromatiques C_6 et C_7 - HRGC/MS/DS.

----

* : une partie de ce travail a été réalisée dans le cadre du programme national R-D Environnement-Air (Services de la politique scientifique).

INTRODUCTION

La participation, maintenant établie (1, 2), d'hydrocarbures volatils à la formation d'oxydantsphytotoxiques tels que l'ozone ou le nitrate de peroxyacétyl a fait croître l'intérêt de ceux qui se préoccupent de la protection de la végétation, notamment pour l'étude de l'origine et des voies de disparition de ces hydrocarbures.
Les réactions de ces composés avec les radicaux hydroxyles peuvent, suivant la valeur de la constante de vitesse d'ablation de l'atome d'hydrogène (3), conduire plus ou moins efficacement à la formation de radicaux hydrocarbonés qui, réagissant à leur tour avec l'oxygène de l'air, peuvent produire des radicaux peroxydiques :

$$RH + OH \xrightarrow{\quad(1)\quad} R^{\cdot} + H_2O$$
$$R^{\cdot} + O_2 \xrightarrow{\quad(2)\quad} RO_2^{\cdot}$$

Ces derniers peuvent alors entrer en compétition avec l'ozone naturellement présent dans l'air selon les étapes suivantes :

$$RO_2^{\cdot} + NO \xrightarrow{\quad(3)\quad} RO^{\cdot} + NO_2 \qquad k_3$$
$$O_3 + NO \xrightarrow{\quad(4)\quad} O_2 + NO_2 \qquad k_4$$

Généralement, l'efficacité de la réaction (3) est beaucoup plus grande que celle de (4) (k_3 k_4) et on peut donc assister à une accumulation d'ozone. La nature de l'hydrocarbure (RH) (étape 1) qui subit l'ablation d'hydrogène règle, entre autres facteurs, le taux de cette accumulation.
A cet égard, il apparaît donc nécessaire de réunir des données au sujet de la composition de la fraction hydrocarbonée de l'air, surtout en milieu peu pollué, c'est à dire où les seules réactions de l'ozone à considérer sont 3 et 4.
Des études en chambre de simulation ayant montré que les aromatiques légers (benzène, toluène et autres alkylbenzènes) possèdent une réactivité photochimique non négligeable, nous présenterons ici les résultats de mesures à l'immission de ces composés en même temps que ceux des composés aliphatiques les plus fréquemment rencontrés.

PARTIE EXPERIMENTALE

Les techniques d'échantillonnage et d'analyse ont été décrites précédemment (4). Les échantillons sont enrichis en pompant l'air à analyser au travers d'adsorbants de charbon actif selon une méthode décrite par Grob (5). Les éluats (CS_2) sont analysés par chromatographie gazeuse (Carlo Erba) sur colonnes capillaires couplée à un spectromètre de masse (MAT 311A). Les spectres de masses sont acquis et dépouillés par ordinateur (INCOS 2000).
Les chromatogrammes obtenus sont semblables à celui repris dans la figure 1.

RESULTATS ET DISCUSSION

Une série de substances est omniprésente dans les analyses de composés organiques de l'air, indépendamment de la nature du site étudié. Il s'agit du benzène, toluène, de l'éthylbenzène, des xylènes, des éthyl-toluènes, du cumène, des mésitylènes et dans une moindre mesure des alcanes volatils C_7 à C_{12}.
Ces substances sont typiques de l'analyse de l'essence et du mazout utilisés comme combustible des moteurs équipant les véhicules automobiles et

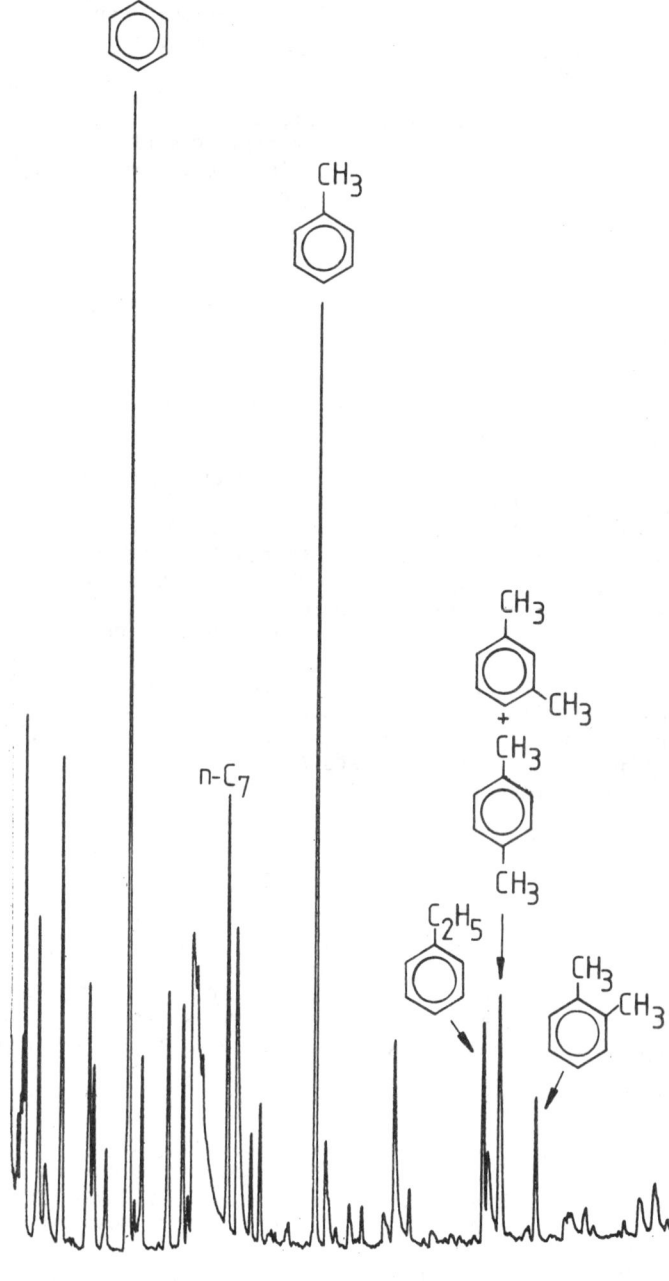

figure 1 : Chromatogramme de l'analyse d'air à
à Tervuren. Volume d'air échantillon-
né : 100 1. Conditions analytiques :
voir ref. 4

et comme combustible des brûleurs des chaudières à mazout des installations de chauffage.

Le tableau I reprend les données du présent travail et celles de la littérature qui sont directement comparables. La composition des gaz d'échappement est donnée au bas de ce tableau. On peut constater que la distribution des aromatiques de l'air coïncide avec celle des gaz d'échappement. Ce résultat est à mettre en parallèle avec ceux obtenus par d'autres laboratoires (5-14), confirmant dans certains cas l'origine de ces substances dans l'air.

La distribution des aromatiques et l'ordre dans lequel leurs concentrations peuvent se classer sont considérés comme indicateurs de la nature de l'origine de l'émission. Un rapport toluène/benzène (C_7 / C_6) plus grand que l'unité et de l'ordre de 1,5 à 4,5 est représentatif d'une pollution par les gaz d'échappement de véhicules automobiles (7).

Au vu des mesures faites à Vilvoorde, Feluy et Diksmuide, on peut donc conclure que ces sites sont effectivement pollués pour les gaz d'échappement. D'après Pilar et al. (8) des rapports $(C_7)/(C_6)$ plus élevés indiqueraient qu'une évaporation directe d'essence serait en cause.

Par contre, Burghardt et al. (7) émettent des doutes au sujet de cette hypothèse en faisant remarquer que la tension de vapeur du benzène étant plus élevée que celle du toluène, si l'évaporation directe a lieu, il faut au contraire s'attendre à observer un rapport $(C_7)/(C_6)$ plus faible.

Nous pensons en effet qu'il faut tenir compte de la tension de vapeur saturante de chacun des constituants des gaz d'échappement, de l'essence, mais aussi de facteurs extérieurs tels que la température et l'humidité de l'air. Ce dernier paramètre en particulier pourrait avoir une grande influence sur le rapport $(C_7)/(C_6)$, la solubilité du benzène dans l'eau étant beaucoup plus élevée que celle du toluène.

Ce raisonnement nous invite à croire qu'un rapport $(C_7)/(C_6)$ faible peut indiquer soit la présence d'une source d'évaporation directe d'essence par temps sec, soit simplement la pollution par les gaz d'échappement par temps très humide.

Analysés à la lumière de ces considérations, les résultats obtenus à Baudour, Tervuren et dans une moindre mesure à Diksmuide peuvent être expliqués de la manière suivante :

Dans le cas de Baudour $(C_7)/(C_6)$ = 0,40), et de Tervuren $(C_7)/(C_6)$ = 0,65) les échantillonnages ont été effectués par temps sec, il faut donc admettre la présence de source(s) directe(s) d'évaporation d'essence ou de sources indépendant de benzène.

A Tervuren, aux échantillons prélevés en mai 1979 par temps pluvieux, correspondent un rapport $(C_7)/(C_6)$ = 0,4 ce qui confirme l'hypothèse d'une source d'évaporation directe d'essence ou de benzène.

D'une manière générale, si l'on désire se servir de valeurs de rapport $(C_7)/(C_6)$ pour estimer la nature de la source de pollution, nous pensons qu'il convient d'être prudent, et de considérer le problème en termes de compétition entre la tension de vapeur du benzène et du toluène et leur solubilité dans l'eau.

Les alcanes sont aussi partiellement témoins d'émissions dues à la combustion de combustibles fossiles (6). Ils apparaissent effectivement en plus grandes concentrations l'hiver, tout comme les aromatiques, mais les plus volatils d'entre eux (C_6-C_{12}) en plus faible quantité que les alkylbenzènes. Cette constatation a également été rapportée par Grob (5); il suppose que les alcanes de masse molaire plus élevée proviennent uniquement des moteurs. Diesel et des brûleurs de chauffage à mazout, sans pouvoir opérer de distinctions analytiques entre ces 2 sources.

Tableau I : Alkylbenzènes dans l'air (concentrations moyennes en $\mu g/m^3$)

Site	Benzène	Toluène	m+p-xylène	o-xylène	éthyl-benzène	1,2,3-triméthyl-benzène	1,3,5-triméthyl-benzène	Nombre de mesures	Ref.
Delft	2 (1,7)	12 (10)	3 (2,5)	0,4 (0,33)	0,9 (0,75)	-	-	90	7
Den Haag	42 (4,9)	86 (10)	85 (9,9)	21 (2,4)	22 (2,6)	9,3 (1,1)	9,3 (1,1)	24	7
Rotterdam	80 (5,6)	144 (10)	117 (8,1)	30 (2,1)	26 (1,8)	71 (4,9)	55 (3,8)	12	7
Amsterdam	19 (3,9)	49 (10)	30 (6,1)	13 (2,7)	13 (2,7)	-	-	3	9
Los Angeles	208 (3,4)	610 (10)	416 (6,8)	35 (0,6)	112 (1,8)	-	14 (0,2)	126	10
Zürich	173 (11,8)	147 (10)	130 (8,8)	40 (2,7)	38 (2,6)	-	10 (0,6)	1	5
Zürich	112 (5,9)	189 (10)	-	-	-	-	-	-	11
Etterbeek	4 (5,4)	7,4 (10)	3,5 (4,7)	1,1 (1,5)	1,7 (2,3)	-	0,6 (0,8)	1	*
Vilvoorde	5,3 (3,7)	14,4 (10)	10,7 (7,4)	1,9 (1,3)	3,4 (2,4)	1,1 (0,8)	0,9 (0,6)	2	*
Feluy	1,8 (6,9)	2,6 (10)	0,92 (3,5)	0,06 (0,2)	0,08 (0,3)	-	-	2	*
Baudour	2,3 (25,2)	0,91(10)	0,72 (7,9)	0,31 (3,4)	0,35 (3,9)	-	0,09 (1,0)	5	*
Tervuren	3,2 (16)	2,0 (10)	0,6 (3,0)	0,35 (1,8)	0,41 (2,1)	-	0,06 (0,3)	8	*
Diksmuide	0,16 (9,4)	0,17(10)	0,08 (4,7)	0,04 (2,4)	0,05 (2,9)	-	0,01 (0,6)	4	*
Floride	-	23,0 (10)	17,0 (7,4)	12,1 (5,3)	6,5 (2,8)	-	1,0 (0,4)	4	12
Floride	-	24,5 (10)	11,7 (4,8)	12,3 (5,0)	5,0 (2,0)	-	2,0 (0,8)	3	12
Grosser Feldberg Taunus	1,2 (3,6)	3,32(10)	1,69 (5,1)	0,43 (1,3)	0,74 (2,2)	0,15 (0,5)	0,15 (0,5)	1	13
Gaz échappem. (g %)	2,4 (4,4)	5,5 (10)	2,2 (4)	-	-	-	-		6 et *
Essence (g %)	2,2 (3,7)	5,9 (10)	5,3 (9,0)	2,2 (3,7)	1,6 (2,7)	-	1,0 (1,7)		7
Production de solvants industriels (10^8 kg)	33,5 (13,6)	24,6(10)	19,1 (7,8)	-	20,1 (8,2)	-	-		14

* : ce travail

REFERENCES

1. B.DIMITRIADES - Effects of hydrocarbon and nitrogen on photochemical smog formation. Environ. Sci. Technol. 6, 253 (1972).

2. J.N.PITTS Jr., B.J.FINAYLSON - Mechanisms of photochemical air pollution. Angew. Chem. Internat. Edit. 14, 1 (1975).

3. R.A.COX, R.G.DERWENT, M.R.WILLIAMS - Atmospheric photooxidation reactions. Rates, reactivity and mechanism for reaction of organic compounds with hydrowyl radicals. Environ. Sci. Technol., 14, 1 (1980).

4. M.TERMONIA, X.MONSEUR, G.ALAERTS, A.DEMEYER, P.DOURTE and J.WALRAVENS - Proc. First Europ. Sympos. Physico-chemical Behaviour of Atm. Pollutants ISPRA, 1979, p. 52; B.VERSINO, H.OTT, eds, CEC, Brussels, 1980.

5. K.GROB, G.GROB - Gas liquid chromatographic mass spectrometric investigation of C_6-C_{20} organic compounds in an urban atmosphere. An application of Ultra-Trace Analysis on Capillary Column. J.Chromatogr., 62, 1 (1971).

6. L.S.TUESDAY, cité dans "Atmospheric Chemistry", J.Heicklen, Acad. Press Inc., 1976, p. 169.

7. E.BURGHARDT, R.JELTES - Gas chromatographic Determination of Aromatic Hydrocarbons in Air Using a Semi-Automatic Preconcentration Method. Atm. Env. 9, 935 (1975).

8. S.PILAR, W.F.GRAYDON - Benzene and Toluene Distribution in Toronto Atmosphere. Environ. Sci. Technol. 7, 628 (1973).

9. Ozon und Begleitsubstanzen im Photochemischen Smog. VDI Berichte 270, VDI Verlag GmGH, Dusseldorg, 1976, p. 75.

10.W.A.LONNEMAN, T.A.BELLAR, A.P.ALTSCHULLER - Aromatic Hydrocarbons in the atmosphere of the Los Angeles Basin. Environ. Sci. Technol., 2, 1017 (1968).

11.K.GROB, G.GROB - Neue Züriches Zeitung aug. 7. Cité dans : E.MERIAN, Some environmental programs in Zwitzerland, Chimica, 27, 521 (1972).

12.W.A.LONNEMAN, R.L.SEILA, J.J.BUFALINI - Ambiant Air Hydrocarbons concentrations in Florida. Environ. Sci. Technol., 12, 459 (1978).

13.K.H.BERGERT, V.BETZ - Erfahrungen beider quantitativen Analyse von flüchtigen organischen Mikroveruntreinigungen in Luft. Chromatographia, 7, 681 (1974).

14.J.GARNER, cité dans "Atmospheric Chemistry", J.Heicklen, Acad. Press Inc. 1976, p. 172.

ETUDE DE LA SOURCE D'OZONE DANS LA COUCHE LIMITE PLANETAIRE

A. LOPEZ, S. PRIEUR, J. FONTAN, P.S.KIM
Laboratoire de Physique des Aérosols et Echanges Atmosphériques
ERA n°378, C.P.A., U.P.S. 118, route de Narbonne, Toulouse
France

Summary

The study of the ozone source in the planetary boundary layer presented here has two aims.

Experimentaly, in urban and forest areas, to establish a budget of the ozone in the planetary boundary layer under an anticyclonic period with a low air exchange between the P.B.L. and the free atmosphere.

The production or destruction rate is obtained by the measurements of ozone concentration near the ground and by the variations of vertical exchange measured with a natural radioactive tracer : the radon.

The first results have show that for a urban area, a good negative correlation between ozone and pollution levels is noted, for a forest are we noted a photochemical production of ozone, naking good loss due to chemical reactions and ground deposition.

Numerical simulation, to consider the O_3 level variations with a box model in the P.B.L., the variations of the mixed layer -deduced from the variations of the radon concentrations- and the chemicals reactions are studied. The photochemical reactions include the NO_x, CH_4, CO, isoprene . The O_3 destruction at the ground is also considered.

The first results have shown that the daily variations of O_3 in the air arise from that of the intensity of the vertical exchanges in the P.B.L.. The photochemical production remains low, less than 10 per cent. The introduction of the isoprene oxidation chain in the conditions considered (low polluted zone) give a chemical destruction of O_3 over a period of two days.

1. INTRODUCTION

On a attribué pendant longtemps la présence d'ozone dans les basses couches de l'atmosphère à des phénomènes de transports verticaux avec les hautes couches (1, 2, 3, 4, 5).

Beaucoup plus récemment des mesures expérimentales ont mis en évidence au niveau du sol, dans des zones très étendues, de fortes teneurs en ozone. Ceci, principalement par période anticyclonique, c'est-à-dire lorsque les échanges entre la C.L.P. et l'atmosphère libre sont les plus faibles (6,7).

L'hypothèse d'une source d'ozone photochimique, mettant en jeu les oxydes d'azote et les hydrocarbures a été émise pour interpréter ces épisodes (8, 9, 10).

Le problème que nous avons entrepris d'étudier est de préciser l'influence de ces différentes composantes.

Cette étude est abordée de deux façons :

- expérimentalement :

Par la mesure du bilan d'ozone dans la CLP, en période anticyclonique lorsque les échanges entre CLP et atmosphère libre sont faibles ;

- par simulation :

Avec un modèle à boite de CLP. Les variations de la hauteur de mélange sont mesurées à l'aide d'un sodar et d'un traceur radioactif naturel, le radon. Les réactions chimiques et photochimiques avec les oxydes d'azote, le méthane, le monoxyde de carbone, l'isoprène et les terpènes sont considérées ainsi que la destruction de l'ozone par déposition en surface.

I. MESURES EXPERIMENTALES

Les teneurs en ozone, mesurées au niveau du sol dans la CLP en différentes zones:rurales, forestières, urbaines, mettent en évidence une variation annuelle ainsi qu'une variation journalière caractéristique (11).

Les mesures sont faites à l'aide de deux analyseurs, l'un à absorption UV (Environnment S.A., type 100 3AH), l'autre à chimieluminescence avec l'éthylène (MELOY).

En ce qui concerne la variation annuelle (figure 1) on note des valeurs maximums durant les mois d'été. Cette forme de variation a été interprétée par certains auteurs comme résultant de l'intensité des échanges entre troposphère et stratosphère.

En ce qui concerne la variation journalière (figure II), on note une valeur minimum la nuit, maximum le jour. L'interprétation de ces courbes est délicate car les teneurs sont soumises à la variation des échanges verticaux à l'intérieur de la couche limite planétaire.

Ces échanges verticaux, tout au moins par période de beau temps peuvent être schématisées à partir d'une structure stratifiée (figure III).

On définira une couche de hauteur Ho, appelée couche limite planétaire qui se maintient en général par période anticyclonique le jour et la nuit. Cette couche est délimitée par une inversion de température à son sommet. Elle se caractérise durant la journée par un mélange important par turbulence d'origine thermique conduisant à des profils verticaux de répartition homogène pour des éléments dont le temps de vie est supérieur à quelques heures. Sa hauteur de l'ordre de 1000 m dans nos régions varie en fonction de la saison, elle est plus faible l'hiver que l'été.

La nuit, par suite du refroidissement radiatif du sol, on assiste au développement d'une zone stable limitée à quelques centaines de mètres qui freine les échanges avec les couches supérieures. Cette couche stable évolue le matin avec le réchauffement du sol. La hauteur et la fréquence d'apparition de la couche nocturne d'inversion Hi varie en fonction de la saison.

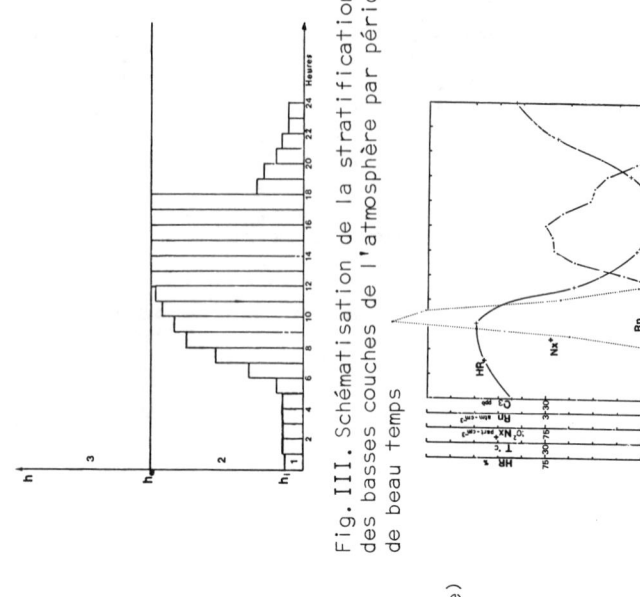

Fig. I. Variation annuelle des concentrations en ozone mesurées au
niveau du sol à Colomiers (zone semi-rurale du Sud-Ouest de la France)

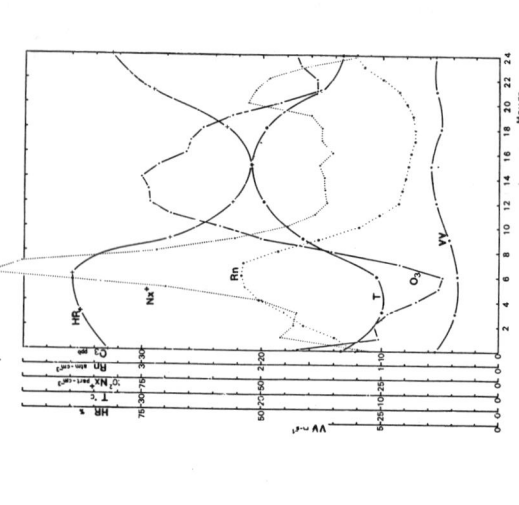

Fig. III. Schématisation de la stratification
des basses couches de l'atmosphère par péric
de beau temps

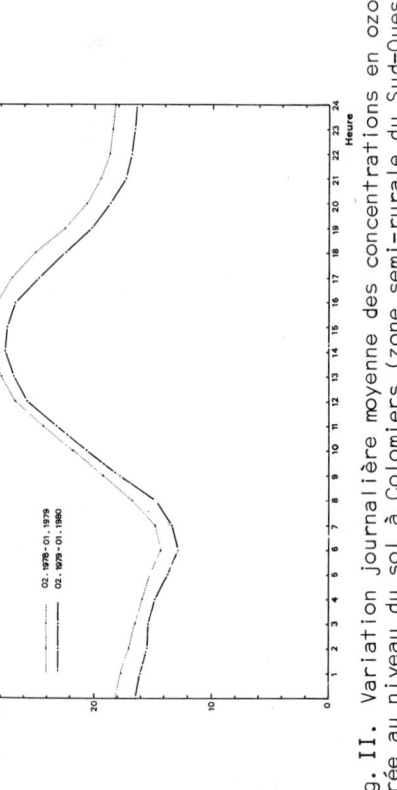

Fig. IV. Variation journalière moyenne des
teneurs en ozone mesurées sur l'agglomération toulousaine

Fig. II. Variation journalière moyenne des concentrations en ozone mesurée au niveau du sol à Colomiers (zone semi-rurale du Sud-Ouest de
la France)

C'est ainsi que l'on peut définir trois zones :
- zone 1 :
Adjacente au sol, elle correspond la nuit à la couche nocturne d'inversion stable, le jour à la couche limite planétaire instable.
- zone 2 :
Apparait la nuit entre le sommet de l'inversion nocturne de température et le sommet de la couche limite planétaire. Par suite des deux inversions de température délimitant cette zone, elle peut être considérée localement comme non influencée ni par le sol, ni par les couches supérieures.
- zone 3 :
Correspond à l'atmosphère libre. Par période anticyclonique, les échanges entre l'atmosphère libre et la zone 1 le jour et la zone 2 la nuit sont faibles. En fonction des périodes de la journée les mécanismes d'évolution de l'ozone dans les différentes couches varient :
- Période nocturne :
Dans la zone 1, nous avons une disparition de l'ozone par dépôt au sol et réactions chimiques qui doit conduire à une diminution des concentrations d'autant plus grande que l'épaisseur de la couche stable près du sol sera faible. C'est ce que l'on note par période de forte stabilité.
Dans la zone 2, les seuls processus de disparition de l'ozone correspondent aux réactions chimiques. Leur importance dépend des teneurs en monoxyde d'azote que l'on avait durant la journée précédente dans la couche limite planétaire. En dehors des zones de pollution, les teneurs en NO sont considérées comme faibles.
On peut penser, bien que nous n'ayons que peu d'expériences pour le prouver que les teneurs en ozone évoluent peu au cours de la nuit dans la zone 2.
- Période diurne :
Le matin avec le réchauffement du sol, la hauteur de la zone 1 augmente. Les teneurs en ozone sont influencées par l'apport de la zone 2 correspondant à la variation de la hauteur qui va se traduire par une augmentation des concentrations. En outre, les déperditions par dépôt au sol et réactions chimiques subsistent. Une production photochimique pourrait intervenir si les concentrations en polluants (oxydes d'azote + hydrocarbures) sont en quantités suffisantes.
La dynamique journalière de la hauteur de ces différentes zones peut être obtenue expérimentalement à partir de la mesure d'un traceur radioactif : le radon, comme l'ont déjà montré des études effectuées par notre laboratoire (12).
L'examen des premiers résultats obtenus au niveau de différents sites amène certaines remarques.
a. Zone urbaine (figure IV) :
On retrouve le cycle caractéristique décrit précédemment. L'augmentation, le matin des teneurs en ozone intervient avec la diminution du radon, c'est-à-dire avec l'augmentation de la hauteur de mélange équivalente due au réchauffement du sol.
On note une anti-corrélation très marquée entre ozone et noyaux d'Aitken. La source de pollution particulaire dans la zone considérée (agglomération toulousaine) est essentiellement due au trafic automobile (13). Il est vraisemblable que l'augmentation des teneurs en N.A., le matin s'accompagne d'une

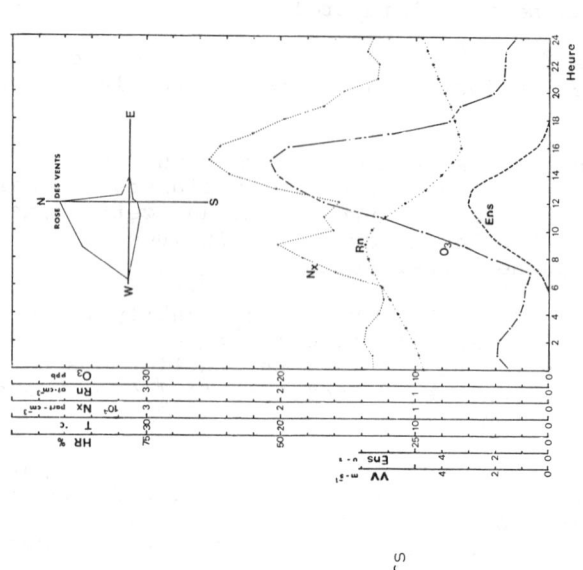

Fig. VI. Variation journalière moyenne de l'ozone, du radon, des noyaux d'Aitken et de l'ensoleillement mesurée dans la zone forestière des Landes (octobre 1978)

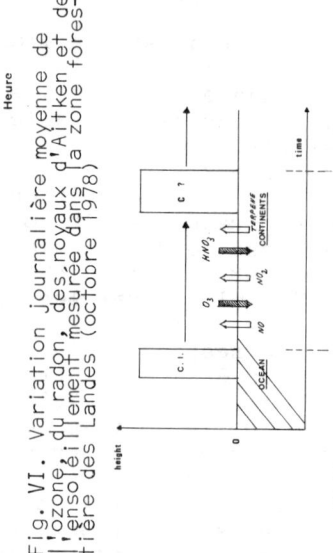

Fig. VII. Modèle d'évolution des teneurs en ozone dans une masse d'air d'origine océanique se déplaçant sur le continent.

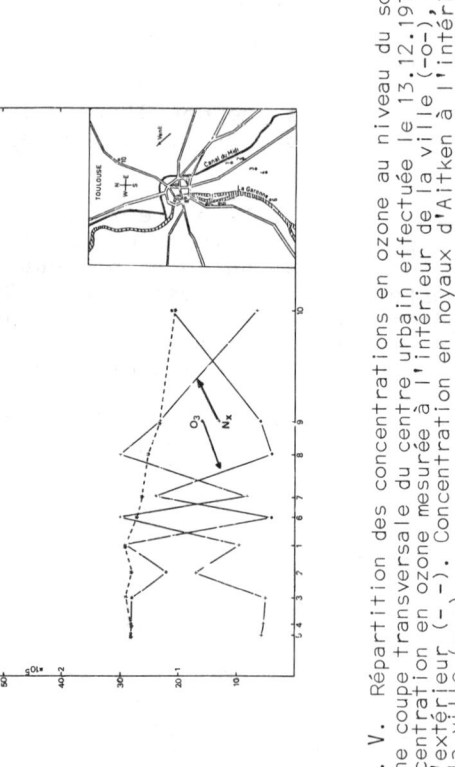

Fig. V. Répartition des concentrations en ozone au niveau du sol lors d'une coupe transversale du centre urbain effectuée le 13.12.1979. Concentration en ozone mesurée à l'intérieur de la ville (-o-), à l'extérieur (- -). Concentration en noyaux d'Aitken à l'intérieur de la ville (——).

Fig. VII. Variation journalière de la source d'ozone dans la zone forestière des Landes (période octobre 1978).

augmentation du NO. La destruction de l'ozone résulterait alors d'une réaction chimique entre ces deux éléments. Cette destruction de l'ozone en zone urbaine est mise en évidence sur la figure(V)qui montre l'évolution des teneurs lors d'une coupe transversale de la zone.

b.Zone forestière (figure VI).

On retrouve ici une variation caractéristique. On note cependant que l'augmentation des concentrations le matin intervient deux heures avant que ne décroisse le radon indicateur des échanges verticaux. Ceci semble montrer l'existence d'une source d'ozone dans les basses couches de l'atmosphère.

Nous avons déterminé, moyennant certaines hypothèses à partir des variations journalières d'ozone et du radon, la source ou le puits d'ozone dans la couche limite (figure VII). Elle est déterminée pour une colonne d'air de 1 cm^2 de section droite.

Les hypothèses de calcul sont les suivantes :
- Les teneurs volumiques d'ozone dans la couche 2 ne varient pas pendant la période nocturne.
- Le flux au sol de radon est constant en fonction du temps. Une mesure quantitative de la production d'ozone demande une mesure quantitative du flux de radon qui peut être obtenue par une étude comparative entre sondeur acoustique et radon (14).

On trouvera le détail du calcul dans la thèse de P.S.KIM (11). On note une production qui augmente le matin à partir de 7 h, qui devient maximum vers 12-13h, lorsque l'ensoleillement est maximum, qui décroit à partir de 13 h pour atteindre une valeur négative vers 17 h. La nuit la valeur reste légèrement négative résultant principalement de la destruction de l'ozone par déposition de surface.

2. SIMULATION

Il s'agit d'un modèle de CLP a structure stratifiée selon le schéma que nous avons déjà présenté. On étudie dans ce modèle l'évolution des teneurs en ozone dans une masse d'air d'origine océanique se déplaçant sur le continent (figure VIII).

Les différents processus intervenant sur l'évolution des teneurs en ozone sont pris en compte :
- transports verticaux
- réactions chimiques et photochimiques
- déposition en surface.

Différentes conditions initiales sont envisagées.
. Ce modèle repose sur différentes hypothèses :
- relative aux transports convectifs : stratification de la couche limite planétaire comme nous l'avons explicité précédemment.
- relative aux réactions chimiques et photochimiques.

Nous avons considéré la photochimie d'une atmosphère non polluée correspondant au passage d'une masse d'air sur une zone rurale.

Nous avons introduit les chaines de réaction indiquées dans le tableau 1. Les mécanismes réactionnels à l'intérieur de ces chaines sont encore très incertains principalement pour les terpènes.

O-H : Chimie inorganique

$O_3 + h\gamma \rightarrow O_2 + O\ (1D)$ $J = 2,6.10^{-5}$ (15)
$O(1D) + M \rightarrow O + M$ $K = 2,9.10^{-11}$ (16)
$O(1D) + H_2O \rightarrow 2\ \dot{O}H$ $K = 2,1\ 10^{-10}$ (15)
$O + O_2 + M \rightarrow O_3 + M$ $K = 1,6\ 10^{-34}$ (15)
$\dot{O}H + O_3 \rightarrow H\dot{O}_2 + O$ $K = 5,15\ 10^{-14}$ (15)
$H\dot{O}_2 + O_3 \rightarrow \dot{O}H + 2O_2$ $K = 1,4\ 10^{-15}$ (15)
$H\dot{O}_2 + \dot{O}H \rightarrow H_2O + O_2$ $K = 3,0\ 10^{-11}$ (17)

O-H-N : Chimie inorganique

$NO + O_3 \rightarrow NO_2 + O_2$ $K = 1,5.10^{-14}$ (15)
$NO + \dot{O}H \rightarrow HNO_2$ $K = 2,0\ 10^{-12}$ (15)
$NO + H\dot{O}_2 \rightarrow \dot{O}H + NO_2$ $\begin{cases} K = 3,5\ 10^{-13} & (17) \\ K = 8,1\ 10^{-12} & (15) \end{cases}$

$NO_2 + h\gamma \rightarrow NO + O$ $K = 3,8^{-3}$ (15)
$NO_2 + \dot{O}H \rightarrow HNO_3$ $K = 1,0\ 10^{-11}$ (15)
$NO_2 + O_3 \rightarrow NO_3 + O_2$ $K = 2,4^{-17}$ (15)
$NO_2 + HO_2 \rightarrow HNO_2 + O_2$ $K = 3,0\ 10^{-14}$ (15)
$NO_3 + NO \rightarrow 2NO_2$ $K = 8,7\ 10^{-12}$ (15)
$NO_3 + h\gamma \rightarrow NO_2 + O$ $J = 2,3\ 10^{-3}$ (15)
$NO_3 + h\gamma \rightarrow NO + O$ $J = 2,3\ 10^{-3}$ (15)
$HNO_2 + h\gamma \rightarrow \dot{O}H + NO$ $J = 1,5.10^{-4}$ (15)
$HNO_2 + \dot{O}H \rightarrow NO_2 + H_2O$ $K = 2,1.10^{-12}$ (16)
$HNO_2 \rightarrow$ Rain out $K = 1,0\ 10^{-6}$ (15)
$HNO_3 + h\gamma \rightarrow \dot{O}H + NO_2$ $J = 1,1\ 10^{-7}$ (15)
$HNO_3 + \dot{O}H \rightarrow NO_3 + H_2O$ $K = 1,5.10^{-13}$ (15)
$HNO_3 \rightarrow$ Rain out $K = 1,0\ 10^{-6}$ (15)

Méthane : chimie organique

$CH_4 + \dot{O}H \rightarrow \dot{C}H_3 + H_2O$ $K = 7,0\ 10^{-15}$ (15)
$\dot{C}H_3 + O_2 + M \rightarrow CH_3\dot{O}_2 + M$ $K = 1,7.10^{-32}$ (15)
$O(1D) + CH_4 \rightarrow \dot{C}H_3 + \dot{O}H$ $K = 3,6.10^{-10}$ (15)
$CH_3\dot{O}_2 + NO \rightarrow CH_3\dot{O} + NO$ $K = 5,9\ 10^{-13}$ (15)
$CH_3\dot{O} + O_2 \rightarrow H_2CO + H\dot{O}_2$ $K = 1,3\ 10^{-17}$ (15)
$H_2CO + h\gamma \rightarrow H + HCO$ $J = 7,1\ 10^{-6}$ (15)

$H_2CO + h\nu \rightarrow H_2 + CO$ $J = 2,2 \ 10^{-5}$ (15)

$H_2CO + OH \rightarrow \overset{\bullet}{H}CO + H_2O$ $K = 1,4 \ 10^{-11}$ (15)

$H\overset{\bullet}{C}O + O_2 \rightarrow CO + H\overset{\bullet}{O}_2$ $K = 1,10 \ 10^{-13}$ (15)

$H_2CO \rightarrow$ Rain out $K = 1,1 \ 10^{-6}$ (16)

$\overset{\bullet}{H} + O_2 + M \rightarrow H\overset{\bullet}{O}_2 + M$ $K = 1,4 \ 10^{-32}$ (15)

$CO + \overset{\bullet}{O}H \rightarrow CO_2 + \overset{\bullet}{H}$ $K = 3,1 \ 10^{-13}$ (15)

Isoprène : chimie organique

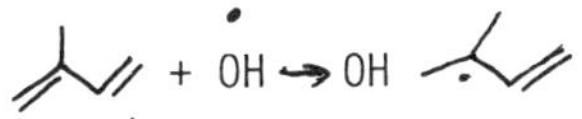

     $K = 7,8 \ 10^{-11}$ (18)

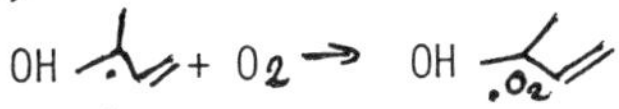

     $K = 6,2 \ 10^{-12}$ (19)

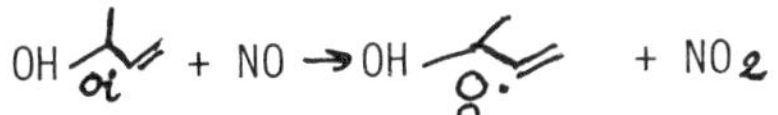

     $K = 3,0 \ 10^{-13}$ (19)

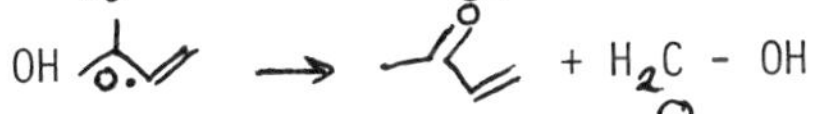

     $K = 2,5 \ 10^{3}$ (19)

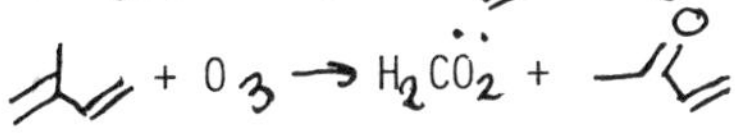

     $K = 1,18 \ 10^{-17}$ (20)

     $K = 1,4 \ 10^{-11}$ (21)

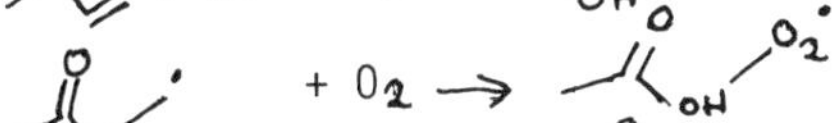

     $K = 6,2.10^{-12}$ (19)

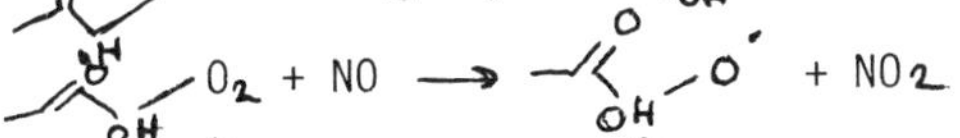

     $K = 3,0 \ 10^{-13}$ (19)

     $K = 2,5.10^{3}$ (19)

     $K = 6,2 \ 10^{-12}$ (19)

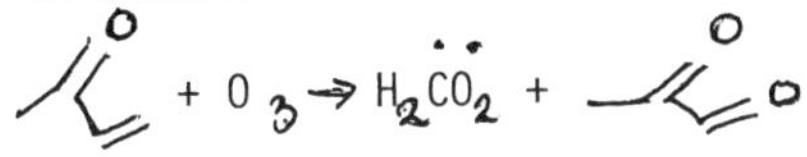

     $K = 1 \ 10^{-17}$ estimé

     $J = 8,3 \ 10^{-4}$ (18)

$H_2\overset{\bullet}{C}OH + O_2 \rightarrow CH_2(\overset{\bullet}{O}_2)OH$ $K = 6,5 \ 10^{-12}$ (19)

$CH_2(O_2)OH + NO \rightarrow CH_2(\overset{\bullet}{O})OH + NO_2$ $K = 5,5.10^{-12}$ (19)

$CH_2(\overset{\bullet}{O})OH + O_2 \rightarrow HCOOH + H\overset{\bullet}{O}_2$ $K = 2,4 \ 10^{-15}$ (19)

$CH_2(O_2)OH + CH_3\overset{\bullet}{O}_2 \rightarrow CH_3\overset{\bullet}{O} + O_2 + CH_2(\overset{\bullet}{O})OH$ $K = 2,6.10^{-14}$ (19)

$H_2\overset{\bullet}{C}O_2^{\bullet} + O_2 \longrightarrow CH_2(O_2^{\bullet})(\overset{\bullet}{O_2})$ $\qquad$ K = 6,5 10^{-12} $\qquad$ (19)

$CH_2(\overset{\bullet}{O_2})(\overset{\bullet}{O_2}) \longrightarrow O_2 + CH_2(\overset{\bullet}{O})(\overset{\bullet}{O})$ $\qquad$ K = 4,7 10^{3} $\qquad$ (19)

$CH_2(\overset{\bullet}{O})(\overset{\bullet}{O}) + O_2 \longrightarrow HO_2 + HCO_2$ $\qquad$ K = 4,0 10^{-16} $\qquad$ (19)

$HCO_2 \longrightarrow H + CO_2$ $\qquad$ K = 2,5 10^{8} $\qquad$ (19)

$CH_2(O)(\overset{\bullet}{O}) + NO \longrightarrow NO_2 + H_2CO$ $\qquad$ K = 3,15 10^{-12} $\qquad$ (19)

$\overset{O}{\underset{\bullet}{\diagup\!\diagup}} + O_2 \longrightarrow CH_3 - \overset{\overset{O}{\diagup\!\diagup}}{C} - O\text{-}\overset{\bullet}{O}$ $\qquad$ K = 6,7 10^{-12} $\qquad$ (17)

$CH_3 - \underset{\underset{O}{\diagdown\!\diagdown}}{C} - O\text{-}\overset{\bullet}{O} + NO_2 \longrightarrow CH_3 - \underset{\underset{O}{\diagdown\!\diagdown}}{C}\text{-}O\text{-}O\text{-}NO_2$ $\qquad$ K = 1,4 10^{-12} $\qquad$ (19)

$CH_3 - \underset{\underset{O}{\diagdown\!\diagdown}}{C}\text{-}O\text{-}\overset{\bullet}{O} + NO \longrightarrow NO_2 + CH_3 - \underset{\overset{\bullet}{O}}{C} = O$ $\qquad$ K = 3,3 10^{-12} $\qquad$ (17)

$CH_3 - \underset{\underset{O}{\diagdown\!\diagdown}}{C} - \overset{\bullet}{O} \longrightarrow \overset{\bullet}{CH_3} + CO_2$ $\qquad$ K = 2,4 10^{10} $\qquad$ (17)

$CH_3 - \underset{\underset{O}{\diagdown\!\diagdown}}{C} - O\text{-}O\text{-}NO_2 \longrightarrow CH_3 - \underset{\underset{O}{\diagdown\!\diagdown}}{C}\text{-}O\text{-}\overset{\bullet}{O} + NO_2$ $\qquad$ K = 1,6 10^{-4} $\qquad$ (19)

- Dépôt au sol : Nous avons choisi une vitesse de déposition constante d'une valeur moyenne de 0,40 cm/s (21)
. Equation régissant le modèle :
- Couche 2 :

$$\frac{dC_{i,2}}{dt} + C_{i,2} \cdot \frac{1}{H_i - H_o} \frac{dH_i}{dt} = P_r - P_e \qquad (Ci,2)$$

- Couche 1 :

$$\frac{dC_{i,1}}{dt} + C_{i,1} \cdot \frac{1}{H_i} \cdot \frac{dH_i}{dt} = Pr - Pe \ (Ci,1) + Prs - Pes \ (Ci,1)$$

H_o : hauteur de la couche d'inversion de température caractérisant la couche limite planétaire
H_i : hauteur de mélange équivalente
Ci,j : concentration du composent i dans la boite j
Pr : terme de production volumique d'origine photochimique
$Pe(Ci,j)$: terme de destruction volumique d'origine photochimique.
Prs : taux d'émission ramené à l'unité de volume dans la couche de mélange pour le composé considéré.
$Pes(Ci,j)$: terme de destruction hétérogène ramené à l'unité de volume du constituant i (déposition au sol, lessivage)
$1/H_i \ dH/dt$; $1/H_i - H_o \cdot dH_i/dt$: termes décrivant le mélange vertical due à une variation de la hauteur de mélange équivalante.
. Résultats
Les résultats obtenus par le modèle pour chacun des cas étudiés sont présentés sous forme graphique.
a. Tracé donnant l'évolution des concentrations pendant 2 cycles journaliers
b. Histrogramme donnant la proportion des différents processus intervenant dans l'évolution des teneurs en ozone (dépôt au sol, photochimie, transport vertical) dans la couche de mélange adjacente au sol
c. Estimation des quantités d'ozone formées ou détruites par photochimie et chimie dans la couche de mélange pour une colonne de section unité.
La figure IX a, correspond a un cas sans émanation gazeuse d'origine végétale.
La variation journalière est caractérisée par une décroissance la nuit, la valeur minimale est obtenue à 6 h au moment du lever de soleil, une augmentation durant la matinée qui conduit à un maximum de concentration vers 14h.
Le NO_2 s'accumule durant la nuit par suite de l'absence de photodissociation. Le NO augmente juste après le lever du soleil correspondant à la photolyse du NO_2 qui s'est accumulé la nuit dans une couche de hauteur faible.
Au niveau des deux cycles journaliers, on note une diminution de l'ozone qui se traduit globalement comme un appauvrissement de la masse d'air au fur et à mesure qu'elle pénètre sur le continent. La figure IX(b) montre que 80% des pertes sont dues au dépôt au sol. La photochimie n'intervient que pour environ 5%. On peut noter que l'augmentation matina-

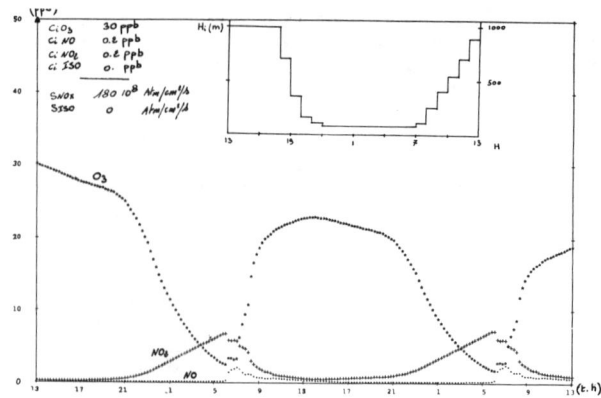

Fig. IXa. Evolution des teneurs en ozone et en oxydes d'azote dans une masse d'air océanique se déplaçant sur le continent en l'absence d'émanations gazeuses d'origine végétale.

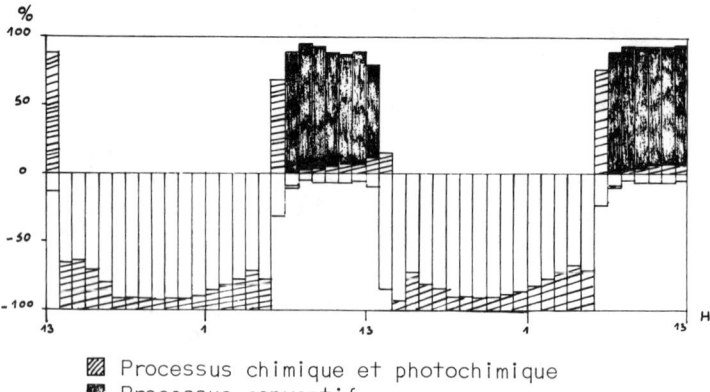

    ▨ Processus chimique et photochimique
    ▤ Processus convectif
    ☐ Processus par déposition sec et humide

Fig IXb. Histogramme décrivant la part des différents processus dans la variation de la concentration d'ozone au niveau du sol pour la couche 1.

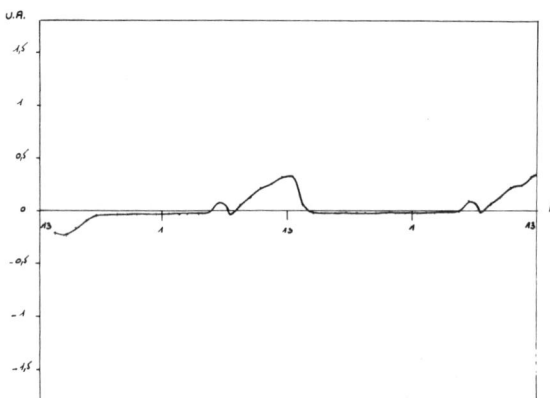

Fig. IXc. Variation de la production et de la perte d'ozone d'origine chimique et photochimique dans la couche de mélange près du sol (couche 1) pour une surface unité.

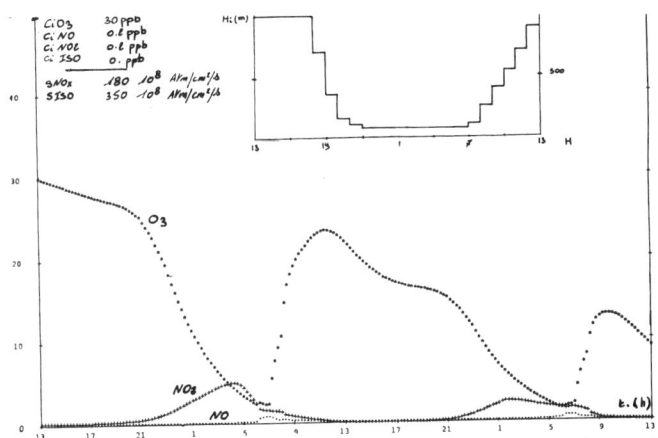

Fig. Xa. Evolution des teneurs, près du sol, en ozone et oxydes d'azote dans une masse d'air se déplaçant sur le continent en présence d'une source continue d'isoprène

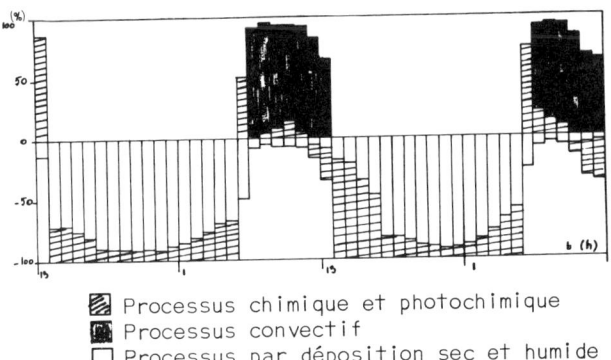

▨ Processus chimique et photochimique
■ Processus convectif
☐ Processus par déposition sec et humide

Fig. Xb. Histogramme décrivant la part des différents processus dans la variation de la concentration d'ozone au niveau du sol pour la couche 1 en présence d'une source continue d'isoprène

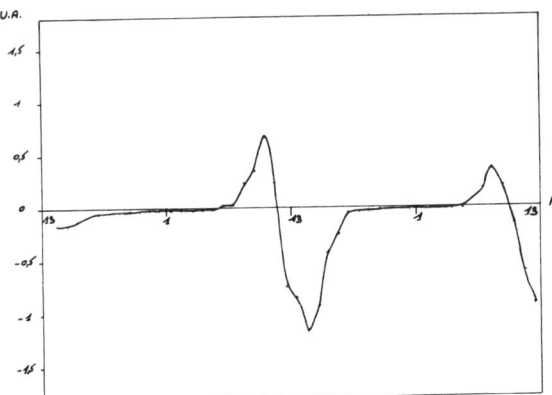

Fig. Xc. Variation de la production et de la perte d'ozone d'origine chimique et photochimique dans la couche de mélange près du sol pour une colonne de section unité en présence d'une source continue d'isoprène

le est due essentiellement au transport vertical. La figure
IX c) représente le terme source ou puits d'origine chimique et
photochimique. La valeur positive entre 8h et 16h correspond à
une légère formation photochimique. Elle est négative pendant
tout le reste du temps.

La figure (X a) correspond à un cas par lequel la chaine
d'oxydation de l'isoprène est prise en considération avec une
émanation continue au cours de la journée. Au niveau des deux
cycles journaliers on note une décroissance des teneurs en
ozone mettant là aussi en évidence globalement un appauvrissement de la masse d'air. Il est plus important que dans le
cas précédent par suite d'une destruction d'origine chimique
plus grande.

Au niveau du cycle journalier, on observe une nette amplitude de variations. Durant la nuit, l'accumulation de NO_2
est moindre due à la formation de PAN.

La figure (X b) montre que le pourcentage de formation photochimique reste faible, par contre la destruction chimique
de l'ozone devient prépondérente l'après-midi. Durant la nuit
la destruction résulte essentiellement du dépôt au sol.

La production d'origine chimique et photochimique est positive le matin entre 7h et 12 h (figure X c), négative à partir de 13 h. On note que le maximum de formation photochimique
intervient à 11h résultant de l'oxydation de l'isoprène. La
disparition d'origine chimique est maximum vers 15h-16h. Elle
est due à l'accroissement des radicaux (principalement HO_2)
provenant de l'oxydation de l'isoprène et à une destruction
du NO_2 par le PAN résultat final de la chaine d'oxydation.

Il est intéressant de noter que cette forme de variation
correspond à celle que nous avons mesurée dans la zone forestière des Landes bien que les réactions à prendre en compte ne
soient pas forcément les mêmes. En effet les émanations gazeuses d'une forêt de résineux sont principalement des terpènes
avec une source dont l'intensité doit vraisemblablement varier
avec le cycle journalier.

3. CONCLUSION

Les premiers résultats expérimentaux obtenus ont montré
une disparition d'ozone au niveau d'une zone urbaine. Ce puits
s'accompagne d'une anticorrélation très nette entre les teneurs
en ozone et noyaux d'Aitken utilisés dans notre étude comme
indicateur de pollution. La réaction chimique primaire de
l'ozone sur le NO, spécifique d'une pollution automobile, caractéristique du site étudié est vraisemblablement responsable
de cette perte. Une formation mettant en jeu les oxydes d'azote et les hydrocarbures pourrait intervenir dans le panache
hors de la zone urbaine.

Au niveau d'une zone forestière de résineux, les premiers
résultats obtenus montrent l'existence d'une source d'ozone
d'origine photochimique dans la couche de mélange dont la contribution à la couche limite planétaire reste faible. Les
mesures effectuées se traduisent par un bilan à l'échelle de
la journée nul, c'est-à-dire une compensation des pertes par
la source.

Ces mesures expérimentales effectuées dans des conditions
météorologiques pas toujours très favorables ne nous permet-

tent pas de tirer une conclusion définitive. Il serait en particulier intéressant d'obtenir des résultats dans des périodes de l'année ou l'émanation d'essences végétales est plus importante et de vérifier par ailleurs dans cette zone l'influence d'oxydes d'azote d'origine naturelle ou anthropogénique. ·

L'étude sur modèle, montre un bilan négatif à l'échelle de la journée. Cette étude doit être continuée et complétée au niveau des mécanismes réactionels, d'une part, au niveau des conditions aux limites fixant l'intensité des sources des composés pouvant intervenir d'autre part. Au niveau des mécanismes réactionnels, il serait intéressant d'y introduire les réactions sur les terpènes. Par ailleurs, la connaissance précise de certaines vitesses de réaction en particulier NO + HO$_2$ → OH + NO$_2$ s'avère nécessaire.

BIBLIOGRAPHIE

(1) REGENER "Neue messungen der vertikalen verteilung des ozons in der atmosphäre" Z. Phys., 1938, 109
(2) JUNGE "Global ozone budget and exchange between stratosphere and troposphere" Tellus, 1962, 14
(3) FABIAN "A theoretical investigation of tropospheric ozone and stratospheric - tropospheric exchange processes
(4) DANIELSEN et al. "Observed distribution or radioactivity, ozone and potential vorticity associated with tropopause golding" J. Geophys. Res., 1970, 75
(5) DANIELSEN, MOHNEN "Project dustorm report : ozone transport in situ measurements, meteorological analyses of tropopause golding" J. Geophys. Res., 1977, 82
(6) GUICHERIT, VANDOP "Photochemical production of ozone in western Europe (1971-1975) and its relation to meteorology" Atm. Env., 1977, 11
(7) COX et al. "Long-range transport of photochemical ozone in north-western Europe" Nature 1975, 255
(8) HAAGEN-SMIT "Chemistry and physiology of Los Angeles smog" J. Ind. Eng. Chem., 1952, 44
(9) RIPPERTON "Ozone and ozone precursors in the atmosphere of chapel Hill, North Carolina" J. Geophys. Res., 1965, 70
(10) CHAMEIDES, WALKER "A photochemical theory of tropospheric ozone" J. Geophys. Res., 1973, 78
(11) KIM "Contribution à l'étude des sources et des puits d'ozone dans la couche limite planétaire" Thèse 3e cycle UPS, 1980 Toulouse
(12) FONTAN et al. "Une méthode de mesure de la stabilité verticale de l'atmosphère près du sol" Boundary Layer Meteo. 1979, 17
(13) LOPEZ A., FONTAN J., BOULARD P. "Mesure de l'intensité de la source de noyaux d'Aitken en site urbain. Influence des différentes composantes" Atmos. Environ., 1981 , sous presses
(14) GUEDALIA D., NTSILA A., DRUILHET A., FONTAN J. "Monitoring of the atmospheric stability above an urban and suburban site using sodar and radon mesurements" J. Appl. Meteorol., 1980, t.19, n°7
(15) CHAMEIDES, STEDMAN " Tropospheric ozone , coupling transport and photochemistry "J. Geophys. Res., 1977, 82, n° 12
(16) HAMEED, PINTO, STEWART "Sensitivity of the predicted (OH-CH$_4$ perturbation" J. Geophys.Res., 1979, 84, n°2

(17) GRAEDEL, SCHIATONE, "2D studies of the kinetic photochemis-
     try of the urban troposphere" Atm. Env., 1981, vol.15
(18) WINIR et al., "Relative rate constants for the reaction of
     the hydroxyl radical with selected ketones, chloroethenes,
     and monoterpenes" J. Phys. Chemistry, 1976, vol.80, n°14
(19) HENVEDT et al. "Quasy-steady state approximations in air
     pollution modeling" Int. J. of Chemical Kinetics, 1978,
     vol.X
(20) E.P.A. ARNTS et al. "Photochemistry of some naturally emi-
     tted hydrocarbons" EPA 1979, 600-3-79-081
(21) GARLAND, DERWENT "Destruction at the ground and the diurnal
     cycle of concentration of ozone and others gazes" Quart.
     J.R. Met. Soc., 1979, 105.

VERTICAL PROFILES OF FORMALDEHYDE IN THE TROPOSPHERE

U. SCHMIDT and D.C. LOWE
Kernforschungsanlage Jülich GmbH
Institut für Chemie 3: Atmosphärische Chemie
Postfach 1913, D-5170 Jülich, F.R.G.

Summary

Tropospheric formaldehyde, HCHO, has been determined for several sets of air samples collected in continental air over the Eifel region of West-Germany (51°N; 6°E) during aircraft flights up to altitudes of about 7 km. The observed mixing ratios show a rather high variability that strongly depends on the meteorological conditions. The vertical profiles obtained during periods with winds from the west to north-west show values generally decreasing from 0.3 ppbv above the boundary layer to less than 0.1 ppbv at 5 to 7 km altitudes. Within the boundary layer and during periods with easterly winds, high values of 4 to 5 ppbv have been observed.
Comparison with a 1-D model of tropospheric HCHO shows that these values are more than an order of magnitude higher than expected from the photochemical oxidation of ambient levels of methane. The additional sources of HCHO are probably due to anthropogenic HCHO transported from the surface during periods of intense vertical mixing in the lower troposphere or the direct photochemical production of HCHO from NMHC's. However, in view of the results of a few parallel profiles of C_2 to C_5 alkenes and alkanes the latter possibility does not seem likely.

1. INTRODUCTION

The distribution of formaldehyde, HCHO, in both clean air as well as in polluted air is of current interest in air chemistry for several reasons. HCHO is formed as an intermediate compound during the photochemical oxidation of methane by the following basic series of reactions (1)

$$CH_4 + OH \quad \rightarrow \quad CH_3 + H_2O \qquad (R\ 1)$$

$$CH_3 + O_2 + M \quad \rightarrow \quad CH_3O_2 + M \qquad (R\ 2)$$

$$CH_3O_2 + NO \quad \rightarrow \quad CH_3O + NO_2 \qquad (R\ 3)$$

$$CH_3O + O_2 \quad \rightarrow \quad HCHO + HO_2 \qquad (R\ 4)$$

These reactions form the first past of a sequence of reactions in which the carbon in methane, CH_4, is oxidized in the atmosphere and eventually is transformed into CO since HCHO is photochemically destroyed through,

$$HCHO + h\nu \quad \rightarrow \quad HCO + H \qquad \qquad (R\ 5)$$

$$+ h\nu \quad \rightarrow \quad H_2 + CO \qquad \qquad (R\ 6)$$

$$HCHO + OH \quad \rightarrow \quad HCO + H_2O \qquad \qquad (R\ 7)$$

and $\quad HCO + O_2 \quad \rightarrow \quad HO_2 + CO \qquad \qquad (R\ 8)$

CO is finally oxidized to CO_2 by

$$CO + OH \quad \rightarrow \quad CO_2 + H \qquad \qquad (R\ 9)$$

There is a possibility that this sequence of reactions might be interrupted with a subsequent reduction in the yield of HCHO and CO. For example, due to low concentrations of NO, CH_3O_2 might react with other species. It is beyond the scope of this paper to discuss the oxidation of methane in full detail. However, we would like to refer to the papers by Calvert (2), Logan et al. (3), in which - among many others - discussion is concentrated on these processes. However, even now it is evident that representative measurements of HCHO will be invaluable in deciding on the overall efficiency of the oxidation of carbon from methane to carbon monoxide a major part of the most important hydrocarbon cycles in the atmosphere.

Adding the reaction

$$H + O_2 + M \quad \rightarrow \quad HO_2 + M \qquad \qquad (R\ 10)$$

to the sequence of reactions listed above, shows that the methane oxidation chain reactions result in a net production of radicals belonging to the HO_x family because the HCHO destruction series (R 5 - R 7) lead to a HO_2-production that is larger than the OH-consumption. The concentration of HCHO in atmospheric air therefore is an indicator for the reactivity of the air mass.

It has been recognized for a long time, that HCHO (and other higher aldehydes) play such a role in the smog-chemistry of sunlit polluted air. In this case the HO_x enhancement is even larger because OH attack of higher hydrocarbons generally results in a net HO_x production during formation of the respective higher aldehydes, the carbon of which will subsequently be present in at least one HCHO molecule formed as an intermediate (4 - 6). The results of recent measurements have shown that non-methane-hydrocarbons (NMHC) are also present in the clean atmosphere (7 - 8) and that their mixing ratios are large enough to enable them to significantly contribute to the photochemical production of carbon monoxide. Again measurements of the mixing ratio of HCHO are of current interest to decide on the relative importance of methane and NMHC compounds. Only a few measurements of HCHO in clean surface air have been reported (9 - 13). In this paper we will present the results of HCHO analyses that have been performed on air samples collected during a series of aircraft flights up to 7 km altitude over the Eifel region in West-Germany.

2. EXPERIMENTAL

During the last two years we have developed a new HCHO sampling and analysis technique (13) which in principle consists of two steps.

a) Stripping of HCHO from about 1 m^3 of air into aqueous 2.4-dinitrophenylhydrazine-solutions, where the reaction product HCHO-2.4-dinitrophenylhydrazone is formed as a derivative. The solution is contained in an all glass sampling tube filled with about 1.5 l of raschig rings. This sampler, when continuously rotated, has a sampling efficiency of 98 % for a minimum load of about 50 ml of blank solution. Figure 1 shows a schematic view of the sampling train.

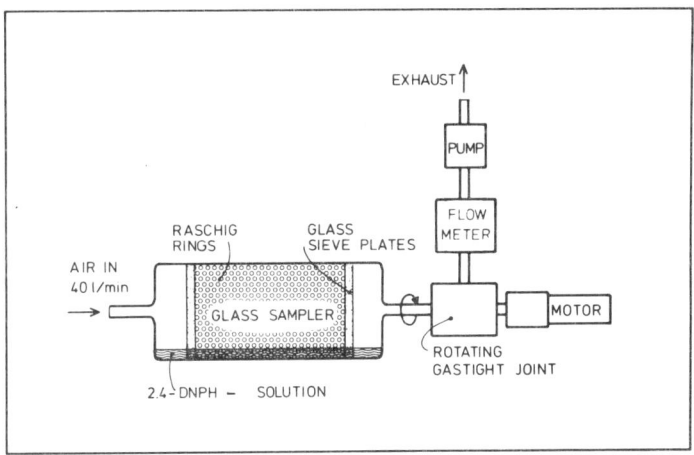

Figure 1: HCHO sampling system (schematic).
The sampling chamber is mounted in a frame with the free end supported by a ball-bearing race and rotated by a motor at about 30 rpm. The motor coupling is combined with a rotating gas tight joint.

b) After sampling the HCHO-2.4-dinitrophenylhydrazone content of the solution is determined by means of HPLC. Calibration is done by comparing the sample with a standard prepared by reacting 2.4-dinitrophenylhydrazine with diluted HCHO solutions derived from a stock solution. A typical chromatogramm is shown in Figure 2.

For the purpose of the aircraft sampling program the sampling system had to be modified. Since the samplers must be rotated during sampling we could not use a closed system. Ambient air was transferred to the inlet part of the sampler through a glass tube the inner diameter of which was large enough to allow the air to be flushed through the tube at very high flow rates and driven by the ram pressure of the aircraft. Only part of this flow was pumped through the sampler and the excess draft allowed to go to waste.

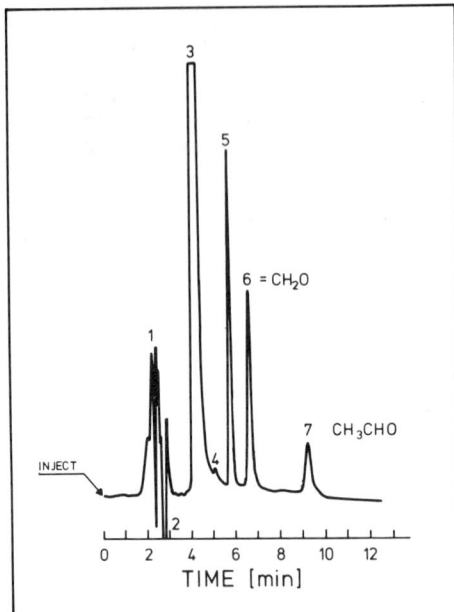

Figure 2: Typical liquid chromatogram of sampling solution used to collect an air sample at the KFA Jülich. Conditions: mobile phase: 65 % MeOH, 35 % H_2O; pressure: 72 bars; flowrate: 1 ml/min, 1. ion peaks; 2. solvent peak; 3. main 2.4-DNPH peak; 4. 5. secondary 2.4-DNPH peaks; 6. formaldehyde-2.4-DNPH; 7. acetaldehyde-2.4-DNPH. This sample had a CH_2O-concentration of 42 ng/ml as CH_2O-2.4.-dinitrophenylhydrazone.

The analysis technique is now developed to the point, where an absolute concentration of 0.1 ng HCHO per ml of solution can be detected with the HCHO in the form of the derivative. Generally the samplers are loaded with about 100 ml of solution and 1 m^3 (STP) of ambient air is sampled. Therefore the lower limit of detection for this technique is 10 ng/m^3, which corresponds to a mixing ratio of 7 pptv or -practically- of about 0.01 ppbv.

However, this limit cannot be reached for all samples obtained under field conditions, because problems such as sample storage and transfer are encountered. Furthermore the actual mixing ratio is derived from the difference of the HCHO contents of the sample and blank solution. For the measurements reported here the lower limit of detection is estimated to amount to 0.03 ppbv and the total error is about 10 % for mixing ratios of about 0.3 ppbv.

3. MEASUREMENTS AND RESULTS

We have recently reported first measurements of the HCHO mixing ratio in the free troposphere over Continental Europe (10). After the first flights in 1979 the sampling system has been considerably improved and another five flights have been made. They were planned on days when the weather conditions

were such that the winds were blowing from either westerly or
easterly directions at all altitudes. The heavy industrialized
area "Rhein-Ruhr valley" is located to the east of the sampling
region, and therefore profiles observed during easterly winds
should reflect the impact of anthropogenic sources. To demonstrate these differences the respective measurements made
during both weather conditions are plotted separately in Figure 3 and Figure 4.

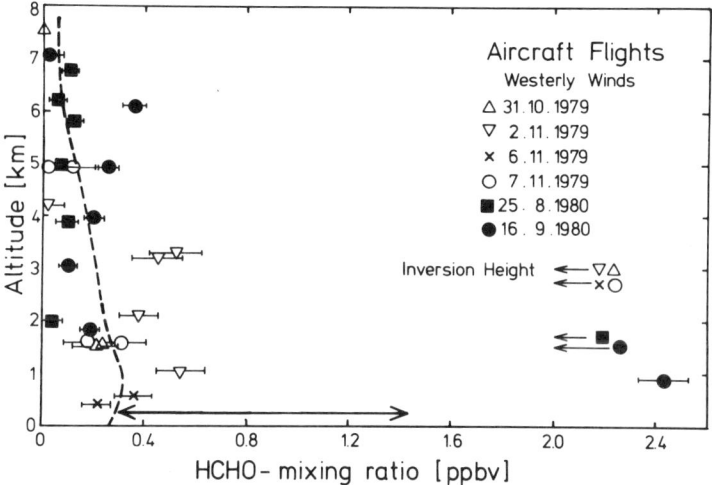

Figure 3: HCHO measurements in the troposphere above the Eifel region, FRG,
during westerly winds. The top of the boundary layer is shown on each day
by the horizontal arrows. The large arrow at the bottom represents 74 measurements made at the surface at Jülich. The dashed curve is a theoretical
HCHO profile computed by the KFA 1-D-model (14).

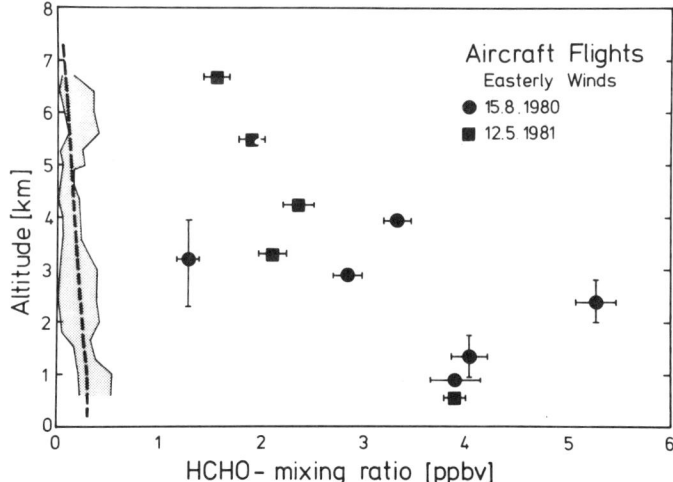

Figure 4: HCHO measurements in the troposphere above the Eifel region, FRG,
during easterly winds. For comparison the shaded area shows the range of all
samples collected during westerly winds. The dashed curve is a theoretical
HCHO profile computed by the KFA 1-D-model (14).

During most of these flights suplementary whole air grab
samples were collected in parallel and analyzed by J. Rudolph
and coworkers at our institute for CO, CH_4, CO_2, and NMHC (C_2-
C_5). Some of these data will be reported along with the discussion of the HCHO results in the next section.

4. DISCUSSION

The data plotted in Figure 4 show that on days when the
air masses had passed over heavy industrialized areas the
HCHO mixing ratios were much higher (by a factor of ~ 10) than
observed when maritime air masses reached the sampling region.
However, even these data (see Figure 3) show considerable
natural fluctuation. This reflects the impact of additional
strong HCHO sources, though their effect appears to be reduced due to dilution by mixing processes. Methane photochemistry can only account for HCHO mixing ratios of about
0.3 ppbv at 1 km altitude decreasing to values around 0.05 ppbv
at 7 km altitude as predicted by a 1-D diurnal averaged model
(14). Further evidence for such sources is due to the frequent observation of much larger HCHO mixing ratios within
the boundary layer. A large series of measurements made at
Jülich gave values of 0.86 ± 0.56 ppbv. These and many measurements reported by various authors (9 - 12) give ample evidence for the identification of the sources as anthropogenic
production processes.
However, the limited profile measurements available do
not allow investigation or identification of specific HCHO
sources. Particularly they do not allow validation of photochemical processes, that might produce HCHO in the free atmosphere.
Simultaneous measurements of abundant hydrocarbons are
one means of obtaining further information on whether the
HCHO has been directly transported from sources on the ground
or results from in situ photochemical production from hydrocarbons. Some of the light hydrocarbons may act as a indirect
source for tropospheric HCHO. The dominant removal of alkanes
and alkenes involves reaction with OH radical and these reactions occur much faster than the analogous reaction with
methane (R 1). If the abundance of these hydrocarbons especially the highly reactive ones as C_2H_4, C_3H_4 reaches the
ppbv-level they may significantly contribute[4] (> 10 %) to the
total photochemical HCHO production (2).
After development of a suitable sampling and detection
technique Rudolph et al. (15) have started an intensive program to investigate the role of NMHC in atmospheric chemistry.
Whenever possible analyses of NMHC (C_2 - C_5) were made on
samples collected either simultaneously or during supplementary flights during the aircraft HCHO sampling program.
In their paper Rudolph et al. report the results of three
profile measurements made at the end of October and in early
November 1979 over the same region (the respective HCHO data
are plotted in Figure 3). During that period winds from north-
west to west predominated and strong inversion layers were
located between 2 and 2.5 km altitude.

Their data are plotted in Figure 5 for comparison with another set of profile data obtained during a sampling flight on May 12, 1981, a day when easterly winds prevailed at all altitudes and no pronounced inversion layer was observed.

The mixing ratios of some NMHC compounds considered were substantially higher on this day, and therefore show the same behaviour as the HCHO mixing ratio. Highest deviations are observed for acetylene, C_2H_2, which has both similar sources and a similar chemical lifetime as CO, and even for CH_4, elevated mixing ratios have been observed. This indicates that both local and distant sources have released the different NMHC, as also suggested for HCHO. Furthermore emission from sources located close to the sampling area must undergo rapid vertical mixing. It appears that the data do not indicate a defined boundary layer.

Although rather high mixing ratios of some NMHC's were observed on the May 12, 1981 their possible contribution cannot account for the large HCHO mixing ratios measured during the same flight. Qualitative information on the effect of photochemical processes may be derived from the fact that one sample collected at 5.5 km altitude showed a significant content of acetaldehyde (CH_3CHO), which is an intermediate product of photochemical oxidation of higher alkanes but has anthropogenic sources, too. We believe that most of the HCHO observed during this flight must be due to direct emissions from anthropogenic surface sources.

5. CONCLUSIONS

The measurements reported here provide only rough information on the vertical distribution of HCHO in continental air. We plan to continue the sampling program for both HCHO and NMHC. The profile flights made to date have been more exploratory in nature. Future measurements will focus on those meteorological conditions that allow the investigation on the relative contribution of photooxidation of hydrocarbons as a source for atmospheric HCHO.

The measurements performed so far nevertheless revealed two basic findings:
a) The impact of surface emissions from anthropogenic sources is not restricted to the boundary layer. Highly elevated mixing ratios of CO, CH_4, HCHO, and NMHC observed during certain weather conditions, demonstrate the effectiveness of turbulent mixing processes up to altitudes as high as several kilometers above the boundary layer.
b) The abundances of light hydrocarbons are large enough that HCHO production through NMHC photooxidation should be considered as an additional HCHO source. However, direct experimental evidence from in situ measurements is not yet available.

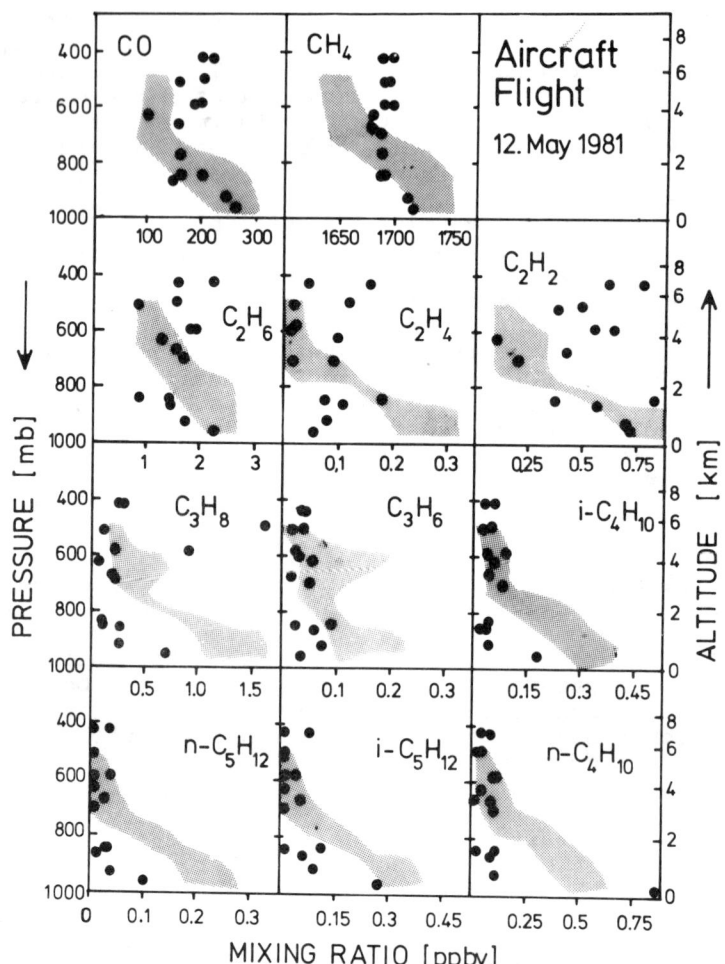

Figure 5: Measurements of NMHC, CO, and CH₄ above the Eifel region, FRG, during easterly winds (Rudolph, personal communication). The shaded envelopes show the distribution of data collected during westerly winds above the same region (15).

ACKNOWLEDGEMENTS

Some of the aircraft flights have been sponsored by the Federal Ministry for Research and Technology (grant FKW O6). We are grateful to J. Rudolph for making unpublished data available.

REFERENCES

(1) Levy, H. II. - Normal atmosphere: large radical and formaldehyde concentrations predicted. Science, 173, 141 (1971)

(2) Calvert, J.G. - The homogeneous chemistry of formaldehyde generation and destruction within the atmosphere. Proceedings of the NATO Advanced Study Institute on Atmospheric Ozone, Aldeias das Acoteias, Portugal, October 1-13, 1979, pp. 153-190, FAA Report No. EE-80-20, US-DOT, Washington, D.C., USA

(3) Logan, J., M.J. Prather, S.C. Wofsy, M.B. McElroy - Tropospheric chemistry: a global perspective. J. Geophys. Res., 86, in press (1981)

(4) Graedel, T.E., L.A. Farrow, T.A. Weber - Kinetic studies of the photochemistry of the urban atmosphere. Atmospheric Environment, 10, 1095-1116 (1976)

(5) Isaksen, S.A., O. Hov, E. Hesstvedt - Ozone generation over rural areas. Environ. Sci. Technol., 12, 1279-1284 (1978)

(6) Cox, R.A. - Rates, reactivity and mechanisms of homogeneous atmospheric oxidation reactions. Proceedings of the "First European Symposium on Physico-Chemical Behaviour of Atmospheric Pollutants", Ispra, Italy, October 16-18, 1979 (B. Versino and H. Ott, Eds.) Commission of the European Communities, Brussels, Belgium

(7) Rudolph, J., D.H. Ehhalt, G. Gravenhorst - Recent measurements of light hydrocarbons in remote areas. Proceedings of the "First European Symposium on Physico-Chemical Behaviour of Atmospheric Pollutants", Ispra, Italy, October 16-18, 1979 (B. Versino and H. Ott, Eds.) Commission of the European Communities, Brussels, Belgium

(8) Zimmermann, P.R., R.B. Chatfield, J. Fishman, P.J. Crutzen, P.L. Hanst - Estimates on the production of CO and H_2 from the oxidation of hydrocarbon emissions from vegetation. Geophys. Res. Lett., 5, 679-682 (1978)

(9) Platt, U., D. Perner - Direct measurements of atmospheric CH_2O, HNO_2, O_3, NO_2, and SO_2 by differential optical absorption in the near U.V. J. Geophys. Res., 85, 7453-7458 (1980)

(10) Lowe, D.C., U. Schmidt, D.H. Ehhalt - A new technique for measuring tropospheric formaldehyde (CH$_2$O). Geophys. Res. Lett., 7, 825-828 (1980)

(11) Fushimi, K., Y. Miyake - Contents of formaldehyde in the air above the surface of the ocean. J. Geophys. Res., 85, 7533-7536 (1980)

(12) Neizert, V., W. Seiler - Measurement of formaldehyde in clean air. Geophys. Res. Lett., 8, 79-82 (1981)

(13) Lowe, D.C., U. Schmidt, D.H. Ehhalt, C.G.B. Frischkorn, H.W. Nürnberg - Determination of formaldehyde in clean air. Environ. Sci. Technol., 15, in press (1981)

(14) Röth, E.P. - Modellrechnungen zur atmosphärischen Reaktionskinetik zur Untersuchung des Einflusses von Spurenstoffen auf die Troposphäre und Stratosphäre mit Analyse der Schwankungsbreiten. Final Report Project BMFT FKW 21. Federal Ministry for Research and Technology, Bonn-Bad Godesberg, FRG, 1981

(15) Rudolph, J. D.H. Ehhalt, A. Khedim, C. Jebsen - Determination of C$_2$-C$_5$ hydrocarbons in the atmosphere in the lower ppb and upper ppt level. J. Chromatography, in press (1981)

AIRBORNE MERCURY AND ITS ORIGIN

Cyrill BROSSET
Swedish Water and Air Pollution Research Institute,
P.O. Box 5207, S-402 24 Gothenburg Sweden

Summary

Total airborne mercury has been monitored at 14 places
in Sweden. The result shows very small variation of its
concentration with geographical site. On the contrary,
there is a clear seasonable variation giving maxima in
February-March and September-October and a minimum in
June-July. The total airborne mercury in Sweden consists
of up to 80% of a background part, the rest being of
anthropogenical origin. This rest is highly dependent on
wind direction and correlated to black particle concentration. The importance of identification of specific
mercury compounds, especially those watersoluble
in ambient air and in emissions has been stressed.

1. INTRODUCTION

During the last years many lakes in Sweden have been investigated with respect to the content of mercury in fish po-
pulations (1 , 2 , 3).
Statistical evaluation of the results shows primarily a correlation between high mercury content in pike muscle and low
pH. This relation is, however, different in different regions
(4)   indicating a rather complex behavior.
Furthermore, it has been shown that the mercury concentration level in fish is higher in a region (0 - 80 km) around
a chlor-alkali plant than within a reference area without any
local mercury source.
Most important is, finally, the fact that the relevant
high mercury concentrations are often observed in lakes quite
isolated from all kinds of industry. Such lakes may only be
polluted by wet and dry deposition of airborne mercury. However, connection between airborne mercury and its concentration in lakes and soil has until now not been established as

with the exception of concentrations in biota and sediment, comprehensive studies of mercury in the environment have earlier not been undertaken in Sweden.

Consequently, it was necessary to study this problem and an investigation sponsored by Project Coal - Health - Environment, The National Environment Protection Board, The Central Operating Management, and The Power Supply Board, on the origin and transport of different airborne mercury compounds was started in 1980. Part of the result hitherto obtained is presented below.

2. THE SCOPE OF THE TOTAL INVESTIGATION AND THE STATE OF ART TODAY

The whole study comprises the following main parts:

1. Monitoring of total mercury in different parts of Sweden

2. Monitoring at a few places of two groups of airborne mercury compounds; such which are easily soluble in water and such which are almost insoluble.

3. Monitoring of mercury concentration in precipitation.

4. Measurement in different types of lakes of mercury concentrations in the water phase.

Short after the start of this study it was realized that the analytical technique until now available first had to be considerably improved.

Furthermore, some important physical constants for relevant mercury compounds had to be determined. All this was a difficult task and some problems here are not yet solved.

However, a number of routine methods of sampling and analyses have now been developed and also applied in field measurements.

This has been the case for mainly item 1 and partly item 2 above.

3. TOTAL AIRBORNE MERCURY

Total airborne mercury was measured at 14 different places in Sweden as dayly mean concentration. At 7 stations the duration of the measurement was one year. At the other stations shorter.

The method of sampling and analyses adopted was similar to the one proposed by Braman et al.(5,6). At sampling air was sucked with a flow of 1 l/min through a tube packed with glass beads covered with a layer of gold. The gold sorbent was activated and blanked in a stream of N_2 at $\sim 400°$ C.

Braman has indicated that all relevant mercury compounds, i.e. metallic mercury vapor (Hg^O), mercury chloride ($HgCl_2$), methyl mercury chloride (CH_3HgCl) and dimethyl mercury ($(CH_3)_2Hg$), are sorbed and retained by activated gold. However, it turned out that desorption of the polar mercury compounds takes place when large volumes humid air are sampled.

This difficulty was overcomed by sampling for the whole sampling period of 24 h only during 15 minutes an hour.

The analysis were performed according to Braman (l.c.) using He-plasma for excitation of the mercury line (253.65 nm) and measuring its intensity. The detection limit observed was 0.002 ng and 0.01 ng could be measured with an uncertainty of ± 20%.

The results obtained were evaluated in two steps.

In the first one monthly averages were calculated for the different sampling places and compared with each other.

Part of the result is given in Table I and Fig. I.

Table I

Monthly mean values of total airborne mercury (ng/m^3) in three places in Sweden.

Year	1979			1980								
Month	10	11	12	1	2	3	4	5	6	7	8	9
West coast	3.4	3.0	3.0	3.7	3.8	4.1	3.1	2.3	2.3	2.6	3.0	3.6
Baltic Sea	3.5	3.4	2.7	5.3	5.3	5.7	3.7	3.3	2.5	2.8	3.5	5.2
Skåne (south Sweden)				4.7	4.5	3.9	3.0	2.7	2.4	2.1	3.7	4.1

Figure I

Monthly mean values of total airborne mercury.
October 1979 - September 1980.
Station Rörvik represents the Swedish west coast, Hoburg represents the Baltic Sea and Ekeröd represents the southern province of Sweden (Skåne).

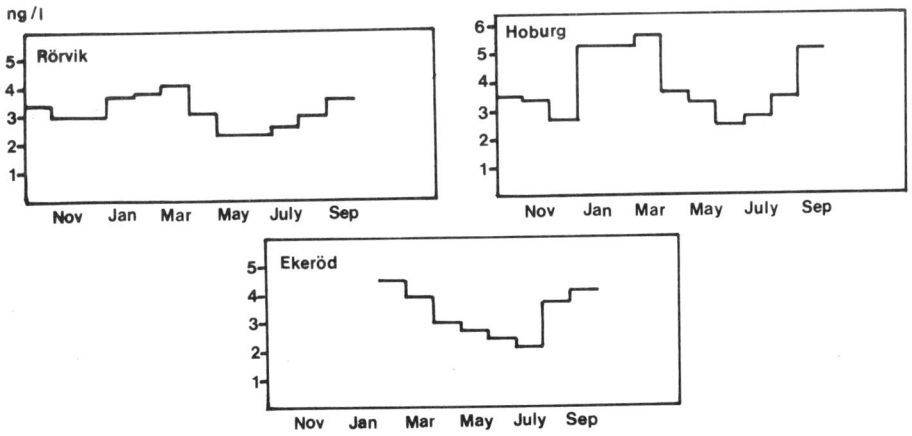

It could be stated that the concentration of total mercury was very evenly distributed over large areas. The seasonal trend observed at the three stations in Table I and Fig. I was also the same at all the other sampling places as well.

It was seen (Fig. 1) that the concentration of total mercury has a maximum in February-March (4-6 ng/m^3) and another one in September-October (4-5 ng/m^3). A minimum is observed in June-July (2-3 ng/m^3). The reason for this trend will be discussed in a later paper.

The second part of the evaluation was aimed at establishing, if possible, the sources of the total airborne mercury. As usual in such cases observed concentration and high level wind directions were combined in windrose diagrams.

Such a one for station Hoburg is shown in Fig.II.

Figure II

Windrose diagram for Hoburg, October 1979-September 1980. Total mercury in ng/m^3.

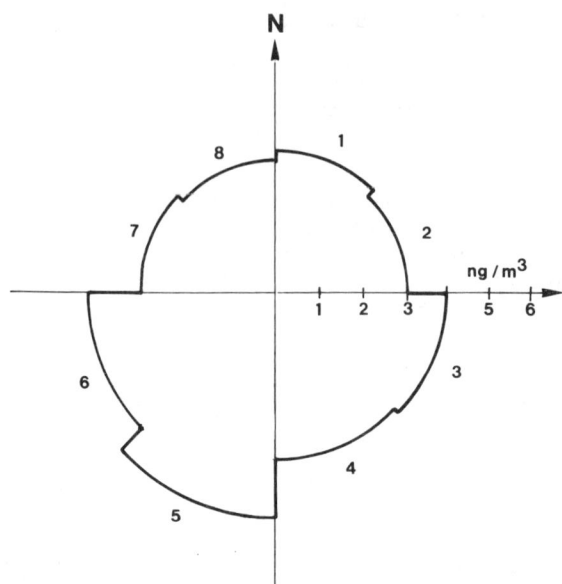

It is seen that this diagram is rather circular in shape meaning that total mercury is mainly independent of wind direction.

However, the slight maxima observed when wind is blowing from south to south west and south west to west is interesting as they are typical for the behavior of black particles.

Hence there was a possibility that part of the total mercury could be transported by the same air masses as soot particles.

To clarify this point regression lines were calculated for dayly averages of total mercury concentration as a function of black particle (soot) concentration.

The result is seen in Table II.

Table II

Correlation between total mercury (Hg_T) and soot for different high level wind directions.

Station	Wind-direction	n	r	m ng Hg_T/ /µg soot	b ng/m^3 Hg_T
Hoburg	N-E	23	0.48	0.15	2.80
Oct-79-	E-S	20	0.67	0.18	2.84
Sept -80	S-W	20	0.65	0.13	3.20
	W-N	26	0.39	0.26	2.67
Ekeröd	N-E	12	0.47	0.063	2.59
Jan -	E-S	17	0.74	0.091	1.98
Sept -80	S-W	17	0.69	0.099	2.68
	W-N	23	0.52	0.15	2.68
Rörvik	N-E	17	0.46	0.10	2.08
Oct -79-	E-S	20	0.77	0.13	1.97
Sept -80	S-W	23	0.72	0.072	2.80
	W-N	29	0.35	0.073	2.53

n = number of data pairs   b = intercept
r = coefficient of correlation
m = slope

This table indicates that of the total airborne mercury (here 2.5 - 5 ng/m^3) on the average 2.6 ng/m^3 is present in the air when soot is absent.

This average seems to constitute the background concentration which is rather independent of place and wind direction. The rest (the difference between the total mercury and the background) is highly correlated to wind direction and to black particle concentration, hense it may be of anthropogenic origin. To be more precise, this part of total airborne mercury seems to have been emitted at the same rate as soot in the respective air masses.

These calculations show that about 80% of the total mercury in Sweden constitute the background. The rest seems to correspond to emissions from distant and local sources, some of which could be identified.

4. INVESTIGATIONS IN PROGRESS

Having got some overview of the behavior of total mercury in air, it was important to elucidate the chemical composition of the background and of the anthropogenic mercury and to establish the origin of the first one.

Ongoing work has shown that a part of the total mercury in air is present as watersoluble compounds, probably $HgCl_2$, and CH_3HgCl.

The seasonal variation of the total watersoluble mercury is now being measured as well as mercury concentration in precipitation.

It has been found that the concentrations in the water and the gas phase measured so far are in agreement with the Henry's law constants that recently have been determined for CH_3HgCl and $HgCl_2$ by Lindqvist et al. (7).

It should finally be pointed out that the chemical composition of airborne mercury and emitted mercury may be imperative for the impact of the emissions on environment.

As metallic mercury seems to be predominant in the air emission of it usually will increase the correspondent level only scarcely. On the other hand, emissions of watersoluble mercury may radically change the relevant concentrations in air with high impact on the environment as result.

Therefore, it seems to be very important not only to measure released quantities of total mercury but also to identify the mercury compounds emitted.

References

(1) Grahn, O., Hultberg, H., and Jernelöv, A.: 1975
     The National Environment Protection Board, Sweden.
     IVL publ. B-292 (In Swedish only)

(2) The National Environment Protection Board, Sweden, 1976
     PM 722.

(3) The National Environment Protection Board, Sweden, 1979
     PM 109.

(4) Hultberg, H., and Hasselrot, B.: Jan. 1981
     Paper pres. at IVA-KHM Symposium on Mercury, Stockholm.
     (In Swedish only)

(5) Braman, R.S., and Johnson, D.L.: 1974
     Selective absorption tubes and emission technique for
     determination of ambient forms of mercury in air.
     Environ. Sci. & Technol. 8, No 12, p 996.

(6) Braman, R.S.: 1971
     Membrane Probe-Spectral Emission Type Detection
     System for Mercury in Water.
     Anal. Chem., Vol 43, No 11.

(7) Lindqvist, O., and Iverfeldt, A.
     (To be published)

SPECTROSCOPIC MEASUREMENTS OF NITROUS ACID AND
FORMALDEHYDE - IMPLICATIONS FOR URBAN PHOTOCHEMISTRY

C. KESSLER, D. PERNER and U. PLATT
Kernforschungsanlage Jülich GmbH
Institut für Chemie 3: Atmosphärische Chemie
Postfach 1913, D-5170 Jülich, FRG

1. INTRODUCTION

In the polluted troposphere night-time concentrations of nitrous acid of several ppb have been observed. The significance of nitrous acid for photochemistry is based upon its rapid photolytic destruction after sunrise, which yields hydroxyl radical production rates of up to 2×10^7 molec cm^{-3}sec^{-1}. Thus the photolysis of nitrous acid gives rise to strong photochemical activity in the morning.

2. EXPERIMENTAL

Nitrous acid, formaldehyde, nitrogen dioxide, sulfur dioxide and ozone are measured by differential absorption spectroscopy in the ultraviolet region. The technique used has been described previously (1). The measurements reported here were carried out with light paths of one to 3.5 km. The species mentioned are identified by their characteristic absorption spectra between 300 and 370 nm. The concentration can be calculated from the strength of the absorption bands according to Beer's law.

In the spectral range from 340 nm to 370 nm nitrogen dioxide exhibits the major absorption structure. In order to detect other absorption bands for example of nitrous acid, formaldehyde and ozone which absorb also in this spectral region, the ambient air spectrum has to be divided by a weighted NO$_2$-reference spectrum (fig. 1, 2).

The weighting factor can be determined from nonoverlapping NO$_2$ bands, thus the NO$_2$-structure can be eliminated completely. In this way around 360 nm nitrogen dioxide and nitrous acid and around 340 nm nitrogen dioxide, nitrous acid, formaldehyde and ozone can be measured simultaneously.

With a time resolution of 20 minutes, a typical level of background noise corresponds to an absorption of 10^{-4} at a 3.5 km light path. Under these conditions, a detection limit of 50 ppt HONO, 300 ppt CH$_2$O, 300 ppt NO$_2$, 40 ppt SO$_2$ and 5 ppb O$_3$ is obtained.

location		path length	HONO night time maximum	NO$_2$
JÜLICH , FRG	FEB·28·79 - SEP·26·80	3.5 km	.05 - 2.2	5. - 45.
RIVERSIDE , CA , USA	AUG·2·79 – SEP·2·79	1 km	1.1 - 4.1	40. - 80.
COLOGNE , FRG	MAR·25·80 - APR·8·80	1.9 km	.4 - 2.	12. - 60.
LOS ANGELES, CA , USA	JUL·5·80 – AUG·8·80	1 km	.4 - 8.	20. - 105.

concentrations given in ppbv

Table 1: HONO-measurements by uv-spectroscopy, concentrations given as ppb

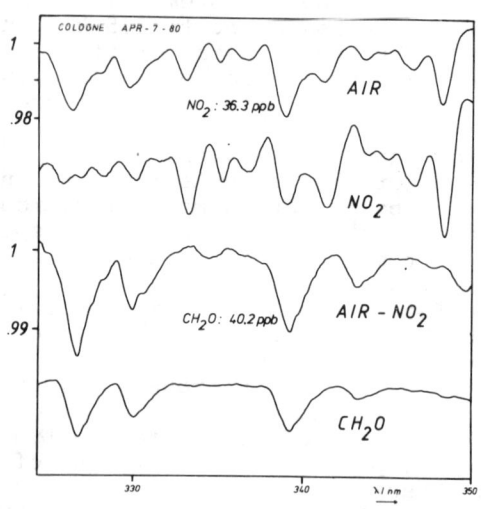

Figure 1: Absorption spectra of NO2 and CH2O

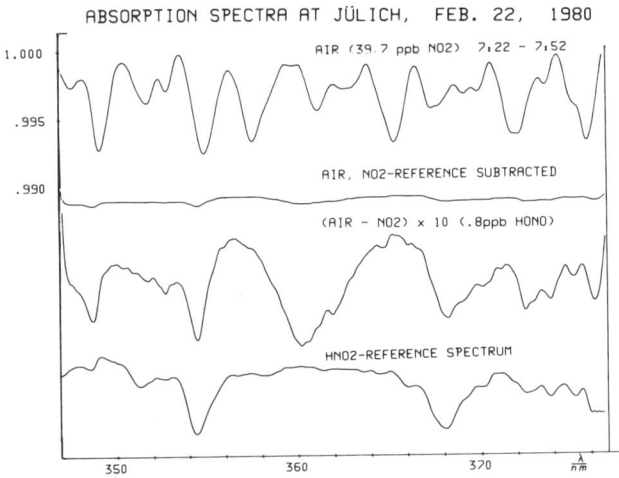

Figure 2: Absorption spectra of NO2 and HONO

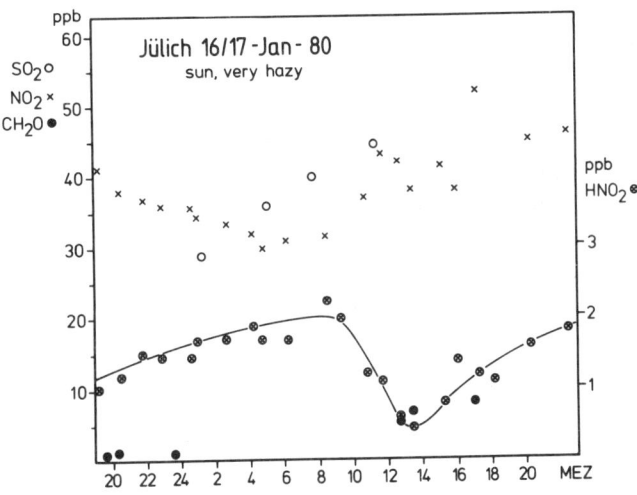

Figure 3: Diurnal variation of HONO:
high equilibrium concentrations during short winter days

3. RESULTS AND DISCUSSION

Using this technique, nitrous acid was detected at four different sites (table 1), which vary from moderately polluted to strongly polluted (2-6). Given is the range of the night-time maximum concentration of nitrous acid along with the range of the nitrogen dioxide concentration measured at the same time.

At two further sites, at the west coast of Ireland and in the Hunsrück, Germany, which both are believed to be rather unpolluted measurements failed, i.e. the concentration of nitrous acid was below the detection limit of 50 ppt.

The temporal behaviour of nitrous acid is composed of a steady increase during the night and a rapid decay after sunrise. In moderately polluted areas, night-time production rates up to 0.25 ppb per hour have been observed (fig. 3, fig. 4).

Homogeneous reactions to form HNO_2 at night are

$$NO + HONO_2 \rightarrow NO_2 + HONO, \quad k_a = 5 \times 10^{-19} \; cm^3 sec^{-1} \quad (7) \qquad (a)$$

and

$$NO + NO_2 + H_2O \rightarrow 2 \; HONO, \quad k_b \quad 4.4 \times 10^{-40} \; cm^3 sec^{-1} \quad (8) \qquad (b)$$

During the day, there may be additional reactions involving photolytic cycles.

Taking the homogeneous rate constants and the NO_x concentrations in question, however, reactions (a) and (b) are too slow by a factor of more than 1000. In terms of the latter reaction a rate constant of about 10^{-35} $cm^3 sec^{-1}$ is required to give the HONO concentrations observed. This urges that nitrous acid is formed heterogeneously, probably at wet walls or wet aerosol particles (fig. 5)

At any rate, it may be expected that the HONO concentration is determined by the concentration of nitrogen oxides available. The data obtained at Jülich and at Cologne are plotted as the ratio of nitrous acid to nitrogen dioxide. The data are slightly correlated with the relative humidity. The correlation with nitrogen dioxide is similar. However, correlation with the concentration of sulfur dioxide is good and the correlation with the product of sulfur dioxide and relative humidity is even better (fig. 6). This suggests, that under conditions in question (i.e. night-time relative humidity above 75 %) the concentration of sulfur dioxide and relative humidity, probably among other parameters as aerosol composition determine the formation rate of nitrous acid.

Several reactions have been suggested to form nitrous acid heterogeneously, partly involving sulfur dioxide or sulfur acids as a catalyst:

$$NO + NO_2 + H_2O \xrightarrow{aerosol} 2 \; HONO \qquad (c)$$

$$2 \; NO_2 + H_2O \xrightarrow{aerosol} HONO + HONO_2 \qquad (d)$$

$$NO + NO_2 + 2 \; H_2SO_4 \rightarrow 2 \; HSO_4NO + H_2O \qquad (e)$$

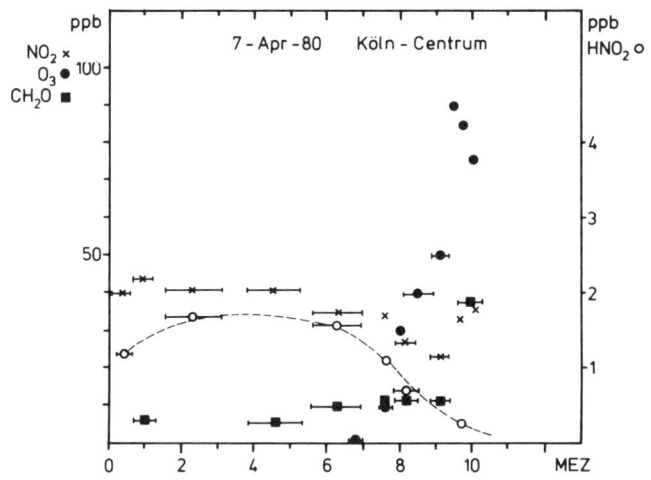

Figure 4: Diurnal variation of HONO:
strong photochemical activity following the photolysis of nitrous acid

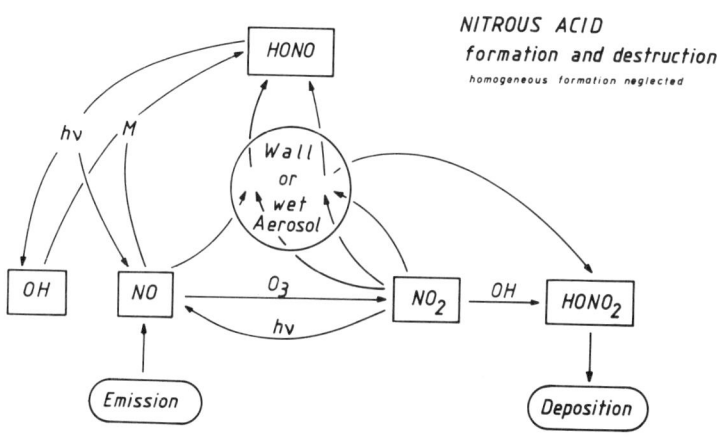

Figure 5: Simplified HONO-NOx-system

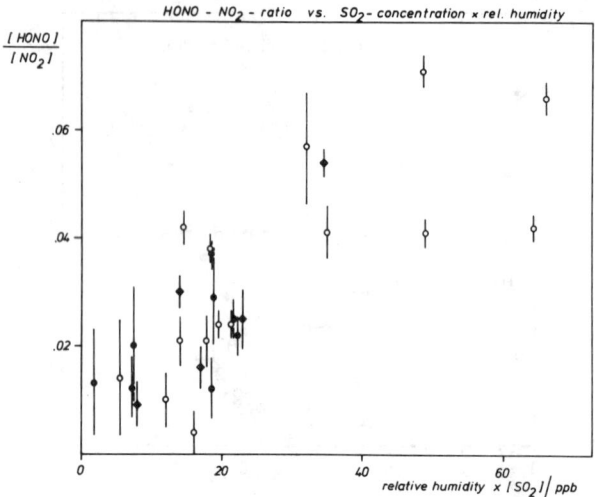

Figure 6: Correlation of nitrous acid night time maxima with
SO2 and relative humidity

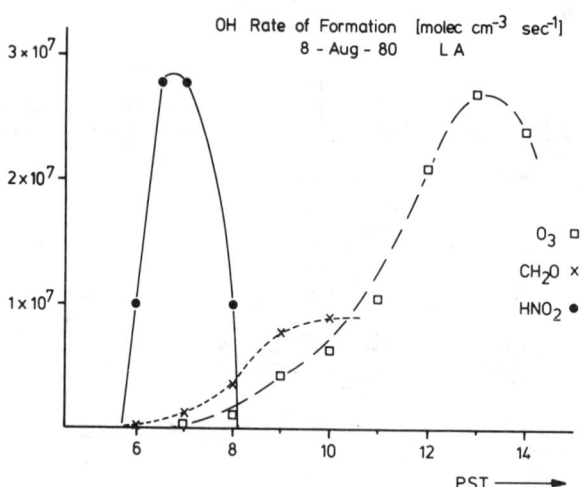

Figure 7: OH formation from photolysis of nitrous acid, formaldehyde and ozone

$$2\ NO_2 + H_2SO_4 \rightarrow HSO_4NO + HONO_2 \qquad (f)$$

$$2\ NO_2 + H_2SO_3 \rightarrow HSO_4NO + HONO. \qquad (g)$$

Nitrosylsulfuric acid reacts rapidly with water to give nitrous acid

$$HSO_4NO + H_2O \rightarrow H_2SO_4 + HONO. \qquad (h)$$

The concentration of nitrous acid reached may be limited by the reaction

$$2\ HONO \qquad \rightarrow NO + NO_2 + H_2O, \qquad (i)$$

which is probably heterogeneous as well. In this way the apparent equilibrium observed before sunrise in most of the nights can be explained.

The measurements carried out at Jülich and at Cologne show average night-time maxima of nitrous acid of about 1 ppb. In few cases, concentrations as high as 2 ppb have been observed.

After sunrise, within about two hours nitrous acid is photolyzed by solar radiation to give nitric oxide and hydroxyl radicals. This photolysis of nitrous acid produces hydroxyl radicals with rates up to 2×10^{-7} cm^{-3}sec^{-1} at a time when radical production from formaldehyde and ozone photolysis is low.

Once there are hydroxyl radicals available in the early morning, a chain photo oxidation is started to give ozone, aldehydes, PAN, and other secondary pollutants. As can be derived from figure (7), the radical production from ozone and aldehydes is low at the time when the photolysis of nitrous acid takes place. This leads to the conclusion, that the photolysis of nitrous acid possibly accelerates the build up of ozone and aldehydes. It may be estimated that the additional hydroxyl source increases the total photochemical turnover during the day.

REFERENCES

(1) Platt, U., D. Perner, and W. Pätz, Simultaneous measurements of atmospheric CH$_2$O, O$_3$, and NO$_2$ by differential optical absorption, J. Geophys. Res., 84, 6329-6335, 1979.

(2) Perner, D., and U. Platt, Detection of nitrous acid in the atmosphere by differential optical absorption, J. Geophys. Res. Lett., 6, 917-920, 1979.

(3) Platt, U., D. Perner, Direct measurements of atmospheric CH$_2$O, HNO$_2$, O$_3$, NO$_2$, and SO$_2$ by differential optical absorption in the near uv, J. Geophys. Res., 85, 7453-7458, 1980.

(4) Platt, U., D. Perner, G.W. Harris, A.M. Winer, and J.N. Pitts, Observation of nitrous acid in an urban atmosphere by differential optical absorption, Nature, 285, 312-314, 1980.

(5) Platt, U., D. Perner, and C. Kessler, HNO$_2$ formation in industrial regions, to be published 1981.

(6) Carter, W., G.W. Harris, A.M. Winer, J.N. Pitts, W. Long, U. Platt, and D. Perner, Observations of nitrous acid in the Los Angeles atmosphere and their implications for prediction of ozone formation and ozone-precursor relationships, Environm. Sci. Techn., to be published 1981.

(7) McKinnon, I.R., J. G. Mathieson, and I.R. Wilson, Gas phase reaction of nitric oxide with nitric acid, J. Phys. Chem., 83, 779-780, 1979.

(8) Kaiser, E.W., and C.H. Wu, A kinetic study of the gas phase formation and decomposition reactions of nitrous acid, J. Phys. Chem., 81, 1701-1706, 1977.

RATES OF SULPHUR DIOXIDE SINK PROCESSES

S. VANDENDRIESSCHE, C. VINCKIER and S. DE JAECERE

Katholieke Univ. Leuven, Lab. anal. anorg. scheikunde
Celestijnenlaan 200F, 3030 Heverlee, Belgium

Summary

Based on literature data, a study was made of processes removing
sulphur dioxide from the lower troposphere. This study aims at
supplying input data to air quality models. Some conceptual aspects
of both empirical rate parameter determination and modelling of sink
processes are dealt with. Typical values of pseudo-first-order rate
constants for wet deposition and atmospheric oxidation of SO_2 are
discussed. Since the rate of pollutant absorption at the surface is
often determined - or at least influenced - by atmospheric transport,
a more sophisticated approach involving 'dry deposition velocities'
is preferred here. Reasonably accurate prediction of these dry deposition velocities is feasible for most combinations of surface and
meteorological situation, and has been set out in full length in a
report (2).

1. INTRODUCTION

Due to rather intensive investigation during the last decade of the
fate of pollutants, and especially SO_2, one might believe SO_2 sink processes to be well known. This belief is disappointed, however, when one is to
advance numerical values of sink rate parameters for use in modelling. It
may be worthwhile here to consider what possible gaps between empirical
and modelling efforts would be due to. While doing so quite in general, we
may at the same time have a look at the state of the art with respect to
SO_2.

The perfect dispersion and transport model should calculate the
concentration distribution $\chi(x,y,z,t)$ by working out the continuity equa-
tion $\partial\chi/\partial t + \partial/\partial x(\chi \cdot V_x) + \partial/\partial y(\chi \cdot V_y) + \partial/\partial z(\chi \cdot V_z) = q(x,y,z,t) -$
$k_{tot}(x,y,z,t) \cdot \chi$ for every volume element $dx \cdot dy \cdot dz$ of the atmosphere in-
fluenced by emissions. V_x, V_y en V_z are the net components of the transport vector from the volume element; q is the source term, and $k_{tot} \cdot \chi$ is
the sink term. Application of this principle would require infinite input
data and infinite calculation effort; therefore, real models have to sim-
plify the problem drastically. If we let x, y and z be, respectively, the
wind direction and the horizontal and vertical cross wind directions, V_x,
V_y and V_z can be considered, in a first approximation, the wind speed and
the lateral and vertical diffusion velocity.

This paper does not review models, it only deals with the sink term.
In models, sink processes will be easily tractable if they are first order
with respect to the pollutant concentration, thus if the parameter k_{tot}
can be considered constant. This is by no means obvious, however, and we
will consider k_{tot} a function of x, y, z and t.

2. GROUPS OF SINK PROCESSES

The total removal rate is usually split up into contributions from three groups of sink processes, dry deposition, wet deposition and conversion, so that $k_{tot} = k_d + k_w + k_r$.
'Dry deposition' means pollutant absorption or retention at the earth's surface (irrespective of whether this is dry or wet), and pollutant transfer from the atmosphere to the surface by vertical turbulence. In most instances, this process is the major one. It is a sink ($k_d > 0$) only at the surface; in the atmosphere above the surface, it causes transport (a downward contribution to V_z) only. For the larger aerosol particles (say, $\geqslant 10$ μm), gravitational settling is to be added to this group.
'Wet deposition' is uptake of pollutant by precipitation; it is by nature discontinuous in time, k_w being 0 in the absence of precipitation. The uptake may occur before as well as after coagulation of cloud droplets to falling raindrops ('rainout' or 'in cloud scavenging' versus 'washout' or 'below cloud scavenging').
Conversion in the atmosphere can include reactions in the gas phase, in suspended droplets and on particles; for sulphur dioxide, it implies the formation of sulphate.
One should not split up k_{tot} before having checked up if no sink process is included twice and if none is overlooked. Both dangers are real. So, k_r is to account for any reaction of SO_2 in cloud droplets; no additional removal is caused by merely the coagulation of part of these dropplets. Therefore, rainout as a sink mechanism doesn't make much sense, and k_w should allow for washout only. On the other hand, the very fast $\frac{k_w}{k_{tot}} = 6 \cdot 10^{-4}$ s^{-1} as found from one of the Danish COST-flights was most probably due to absorption of SO_2 by sea spray. This phenomenon can occur only at rather high wind speed and over water, but may be very fast then, and will often not be kept in mind when estimating the rates of dry deposition and of conversion in the atmosphere.

3. DRY DEPOSITION

The uptake flux at the surface is called $-F$ (the sign allows for definition of a positive diffusion coefficient K_x). This flux is assumed equal to the vertical turbulent flux $F = \overline{w'\chi'}$ through the lowest tens of metres of the atmosphere (the covariance of vertical wind speed component with concentration). It causes a vertical concentration gradient : $-F = K_x \cdot \partial\chi/\partial z$.
The dry deposition velocity from a height z, $v_g(z) = -F/\chi(z)$, usually has the order of magnitude 10^{-2} m s^{-1}; hence units of cm and s are common in this context.
The actual removal of SO_2 occurs in the thin layer of air close to the surface. In this layer, the concentration is called $\chi(o)$; the upper boundary of the layer is called z_x. Enough definitions have been given here to present the resistances formalism describing the dry deposition velocity :

$$\frac{1}{v_g(z)} = \frac{\chi(z) - \chi(o)}{-F} + \frac{\chi(o)}{-F} = r_g(z) + r_c,$$

respectively. $r_g(z)$ is the aerodynamic resistance to pollutant transport from the height z to the surface : $r_g(z) = \int_{z_x}^{z} dz/K_x(z)$; r_c is the kinetic resistance to uptake at the surface. This formalism is the simplest, and most often the best, model to describe the deposition process. For a smooth surface, r_c in s cm^{-1} is equal to $1.3 \cdot 10^{-4}$ times the number of SO_2-to-surface collisions needed for 1 molecule of SO_2 to be absorbed (1). For complex surfaces, r_c is to be found empirically. From surface roughness and meteorological data, $r_g(z)$ can be calculated; it decreases with increasing wind speed (inversely linearly to a first approximation) and with increasing surface roughness. The order of magnitude of r_g and r_c often is the same. In dry deposition to wet and/or alkaline surfaces, r_c becomes negligible compared to r_g, and diffusion from the atmosphere to the surface is rate determining; r_c becomes high and rate limiting, however, in dry deposition to dry, inert surfaces.

An ideal dispersion and transport model would in principle need only r_c as an input parameter to account correctly for dry deposition. Calculation of the dry deposition velocity from a height z (say 1 to 10 m) above the surface has two advantages :
- Many models do not calculate the correct dependence of K_x on z; in the lowest air layer, however, K_x is easily calculated from a surface roughness parameter and meteorological information (2).
- Most surfaces are complex (e.g., vegetated) : pollutant uptake does not occur at one particular height. The parameters in the resistivity formalism can be found for complex surfaces as well, though they may lack explicit physical meaning.
In a model dry deposition can be described as if it occurred in a layer with thickness dz at height z : $k_d(z) = v_g(z)/dz$; at any other height $k_d = 0$. No error is introduced as long as the assumption of equal uptake flux and vertical transport flux remains correct : depletion of the air layer below z should not contribute significantly to the depletion of the polluted air mass.

Accounting for dry deposition by calculating the dry deposition flux to the surface, as was outlined here ('surface depletion' modelling), is quite demanding. A much simpler and often used approach ('source depletion' modelling) is, in fact, conceptually unfounded : k_d is used as independent of height, as if dry deposition would occur uniformly throughout the atmosphere. Not seldom, however, the eventual result will be very much the same. Where r_c is large enough to be rate limiting, v_g becomes $\simeq 1/r_c$, independent of height. Indeed, dry deposition ceases causing a concentration gradient in the lower air layer, since vertical mixing becomes fast compared to absorption at the surface. Higher in the atmosphere, however, this fast mixing dies out, and with a temperature inversion layer present above the polluted air layer, this can happen quite abruptly. In such a case, the height Z where the inversion begins (order of magnitude 1 km) may be considered as a lid upon the lower troposphere. Dry deposition is then apparently correctly described using a k_d independent of height and approximately equal to $(Z \cdot r_c)^{-1}$ or v_g/Z (order of magnitude $\leqslant 10^{-5}$ s^{-1}).

Where r_c is not rate limiting, an apparent dry deposition velocity $v_d = Z \cdot \bar{k}_d$ can be calculated in much the same way by following the concentration decay as a function of travel time (see 6.4). To find the dry deposition velocity $v_g(z)$ empirically, the deposition flux F and the concentration $\chi(z)$ are to be measured. The apparent deposition velocity v_d found using the 'mass balance method' is smaller than $v_g(z)$ if v_g is smaller than $1/r_c$, buth a height Z with distinct transition from fast to slow vertical mixing is present; it has little physical meaning if there is no such height.

The same procedure used the other way round may be found in source depletion modelling. The parameter v_d to be used there has not only the disadvantage of being unpredictable : this drawback could be got round by collecting empirical or data-fitted values. The more serious objection is, that the parameter v_d in source depletion models cannot be tuned to calculate both the local removal (deposition rate) and the fraction of pollutant remaining airborne optimally. The discrepancy can be as large as an order of magnitude (3).

The dry deposition velocity $v_g(z)$ can be calculated for a known terrain and for a known meteorological situation (2). Specification of k_d as a function of x and y requires mapping input parameters such as r_c and the surface roughness. Specification of k_d as a function of time means essentially allowing for r_c to be influenced by weather and season conditions (2, and references therein).

4. CHEMICAL CONVERSION

Since chemical (or, for aerosols, physical) conversion processes will be different for each pollutant , only SO_2 is dealt with here. Any conversion leads eventually to sulphate formation : water soluble sulphate accounts well for all sulphur in aerosols (4).

Gas phase oxidation of SO_2 occurs mainly through reaction with OH and peroxy radicals :

$$SO_2 + OH \overset{(M)}{\rightarrow} HSO_3; HSO_3 \ldots \rightarrow \ldots S(VI)$$

$$SO_2 + RO_2 \rightarrow SO_3 + RO; SO_3 + H_2O \rightarrow H_2SO_4 \text{ or}$$

$$\overset{(M)}{\rightarrow} RO_2SO_2 ; RO_2SO_2 \ldots \rightarrow \ldots S(VI)$$

These reactions contribute $\Sigma k(A+SO_2) \cdot [A]$ to k_r; in the northern hemisphere, the time averaged value is $\leqslant 10^{-6} \text{ s}^{-1}$. Specification in space is hardly worthwhile; in time it is quite simple, however, since the co-reactants are formed photochemically. At night, or with overcast sky, the contribution to k_r is negligible ($< 10^{-7} \text{ s}^{-1}$). In sunny weather, it will be $\simeq 3 \cdot 10^{-6} \text{ s}^{-1}$ during winter and $\simeq 10^{-5} \text{ s}^{-1}$ during summer (references in 2).

The rate of aerosol-catalysed reactions will be pseudo-first order with respect to SO_2 only for background aerosol, the extent of reaction being insignificant (5). The oxidation rate caused by fresh aerosols, released together with the SO_2, will tend to be rather pseudo-second order; it will decrease with increasing dispersion. The extent of reaction will not exceed a few %, and is much easier to account for assuming some primary sulphate emission. Catalytic oxidation of SO_2 on particles is important in producing a very acid and hygroscopic aerosol, but not as a sink for sulphur dioxide.

In spite of the low liquid water content of clouds and fog (order of magnitude tenths of a ml m^{-3}), oxidation in water droplets is the major form of chemical conversion in the atmosphere. It contributes $k_{ox} \cdot f_w \cdot C_{S(IV)}/\chi$ to k_r; f_w is the volume fraction of liquid water in the atmosphere, $C_{S(IV)} = [SO_2 \cdot aq] + [HSO_3^-] + [SO_3^{2-}]$ as equivalent SO_2 with the same units as χ, and k_{ox} is the pseudo-first order oxidation rate constant for the S(IV) pool, $-d \, ^{ox}C_{S(IV)}/C_{S(IV)} \cdot dt$ if no further SO_2 is absorbed:

$$k_{ox} \cdot ([SO_2 \cdot aq] + [HSO_3^-] + [SO_3^{2-}]) = [SO_2 \cdot aq].\Sigma k(A+SO_2 \cdot aq).[A] +$$

$$[HSO_3^-].\Sigma k(A+HSO_3^-).[A] + [SO_3^{2-}].\Sigma k(A+SO_3^{2-}).[A]$$

In clouds and fogs, k_{ox} is rate determining to the contribution to k_r. Part of the oxidation occurs by dissolved O_2. This reaction is extremely sensitive to catalysis and inhibition. Literature values for k_{ox} vary over many orders of magnitude (10^{-6} s^{-1} to $\gg 10^{-2}$ s^{-1}). Both reaction mechanism and kinetics are very complex. A simulation in the laboratory showed oxidation by O_2 to be much slower than oxidation by O_3 in solution (6).

The reactions of ozone with bisulphite and especially sulphite are fast (references in 2): $k(O_3+HSO_3^-) \simeq 1\cdot10^5$ to $4\cdot10^5$ l mol^{-1} s^{-1} and $k(O_3+SO_3^{2-}) \simeq 3\cdot10^8$ to $2\cdot10^9$ l mol^{-1} s^{-1}. With 10^{-9} mol l^{-1} O_3 in solution (0.05 ppm in the air) and at pH $\simeq 5$, the contribution of the ozone reactions to k_{ox} takes the order of magnitude 10^{-2}s^{-1}. This will contribute some 10^{-5} s^{-1} to k_r if f_w is a few 10^{-7}. The reaction rate increases rapidly with increasing droplet pH.

In contrast with this, the contribution to k_{ox} caused by dissolved hydrogen peroxide is essentially independent of pH. The kinetic expressions given in a very recent paper (7) are very simple compared to those in previous studies; the HSO_3^- ion would be oxidized to a HSO_4^- isomer, the reaction being reversible; at pH $\gg 1$, the reaction with H^+ to rearrange the isomer would be rate limiting. At 25° C, and in the wide pH range where HSO_3^- is the dominant S(IV) species, $k_{ox} \simeq 5\cdot10^7[H^+][H_2O_2]$ l^2 mol^{-2} s^{-1}. A constant sulphate production rate results at constant SO_2 concentration. With $f_w \simeq 3\cdot10^{-7}$ and 10^{-3} ppm H_2O_2 in the gas phase, a contribution $\geqslant 10^{-4}$ s^{-1} to k_r is found at 10^5 C. This sort of calculation can be very misleading, however. Mostly, polluted air contains more SO_2 than H_2O_2. The net reaction $H_2O_2 + SO_2 \rightarrow H_2SO_4$ cannot remove more $SO2$ than $H2O2$ is available. Since the reaction is fast, it may be safer to assume that a cloud system will oxidize as much SO_2 as H_2O_2 enters it. This will in turn be cumbersome in modelling; therefore, the use of empirical k_r values seems justified.

Specification of the contribution of reaction in droplets to k_r in space and time would not only require a measure for f_w in models, but also for the presence of oxydants and of atmospheric constituents influencing pH. Normally, insufficient input information will be available to make this possible to any accuracy. Hence, one will confine oneself to discriminating the presence of clouds or fog from their absence.

5. WET DEPOSITION

Diffusion of SO_2 from the gas phase to raindrops sets an upper limit $k_w \leqslant 10^{-4}$ ·(J in mm/hour)$^{0.5}$ s^{-1} to the rate of washout (8). This limit will only be attained when the rain intensity J is high, or when the raindrops have a high pH or contain much H_2O_2. By far most of that potential k_w is due to the smaller raindrops, and exactly these will usually reach saturation to S(IV) early in their fall. The rate at which they will further absorb SO_2 is then limited by the oxidation rate of S(IV). Some ten 'washout models' have been reported in literature (references in 2) to account for this effect in a conceptually correct way. The input data needed to calculate k_{ox} and pH as functions of both fall time and diameter of raindrops is not normally available, however. Therefore, washout models tend to have little use; one will prefer the use of empirical or semi-empirical k_w values, lower than the diffusion rate limited ones (order of magnitude 10^{-5} s^{-1} or so).

6. CONCEPTS OF EMPIRICAL RATE PARAMETER DETERMINATION

6.1 TOTAL REMOVAL RATE

In principle, $k_{tot}(x,y,z,t)$ is to be found empirically through

$$k_{tot} = - \frac{1}{\chi} \cdot [\frac{\partial \chi}{\partial t} + \frac{\partial}{\partial x}(\chi \cdot V_x) + \frac{\partial}{\partial y}(\chi \cdot V_y) + \frac{\partial}{\partial z}(\chi \cdot V_x)].$$

In practice, the problem will be reduced to a one-dimensional steady-state one : the decay of SO_2 will be followed as a function of the travel time t' in the atmosphere. Thus, the effluents from a well defined source will be followed in a Lagrangian way.

Studying k_{tot} already implies that no information on sink mechanisms will be obtained. The one-dimensional approach has additional drawbacks. To correlate a travel distance x to a travel time t' = x/$\bar{u}$, the dependence on height of both wind speed and wind direction is to be ignored. Also, the k_{tot} found is allowed to be dependent on time and travel distance only : any dependence on y and z is ignored; especially this last implication is annoying when dealing with a sink process tied to a particular height.

The most correct method (the 'mass balance method') makes use of the pollutant flow through a plume cross section : Q(x) = $\int\int u.\chi.dy.dz$. Such measurements are demanding : they require horizontal and vertical scanning through plume cross sections. The decay rate constant (averaged over y and z) is then k_{tot} = -dQ/Q·dt'; if decay is first order with respect to SO_2 (if k_{tot} is independent of x), k_{tot} is found as the slope of a plot of - ln Q(t') versus t', since k_{tot} = [ln Q(o) - ln Q(t')]/t'.

In order to avoid these elaborate measurements, the SO_2 decay is sometimes compared to the concentration evolution of an inert tracer gas (χ_i) subject to the same dispersion. In this 'tracer method', one puts k_{tot} = dχ_i/χ_i·dt' - dχ/χ·dt', so that k_{tot} is the slope of a plot of $ln [\chi_i(t')/\chi(t')]$ versus t'. The assumption of identical dispersion is confusing, however. The tracer method actually assumes V_y and V_z for pollutant and tracer gas to be identical at any point of space. This in turn would require all sink processes to occur uniformly throughout the atmosphere, so that relative concentration gradients, and hence diffusion velocities, remain uninfluenced anywhere. This assumption is patently incorrect whereever depletion by dry deposition, for instance, is significant. Use of the tracer method with ground level concentration data (e.g., 9) may cause tremendous overestimation of k_{tot}.

In a third method ('quasi-Lagrangian cell'), a vast air mass is followed over a long distance, so that dispersion can be neglected (or corrected for by estimation). Then the decay rate constant may be found simply from the decay of average concentration : $k_{tot} \simeq -d\bar{\chi}/\bar{\chi}·dt'$, as the slope of a plot of -ln $\bar{\chi}$ versus t'. Here, however, care should be taken in choosing the 'average' concentration : one should not expect the concentration at one height close to the surface to be representative for the whole air mass.

6.2 OXIDATION RATE

Empirical information on k_r requires measurement of both SO_2 and sulphate. Secondary pollutant will be indicated by the index s (χ_s, Q_s, both given as equivalent SO_2). Here again, measurement of pollutant flows is the most correct method. Assuming no rain, Q and Q_s vary with travel time as dQ/dt' = -(k_d+k_r).Q and dQ_s/dt' = k_r.Q-k_{ds}.Q_s, where k_{ds} is the equivalent of k_d for sulphate. It can be found empirically by following another aerosol constituent (for which index a is used), subject to the same deposition processes as sulphate, but having zero production rate : k_{ds} = - dQ_a/Q_a·dt'.

The most correct formula for empirical determination of k_r reads :

$$k_r = \frac{Q_s}{s}\left(\frac{dQ_s}{Q_s \cdot dt'} - \frac{dQ_a}{Q_a \cdot dt'}\right)$$

As long as Q_s is much smaller than Q, however, deposition processes needn't be considered. Calling $R = Q/(Q + Q_s)$ and assuming k_r independent of t', we find from working out $\int_o^t dR/R \cdot dt'$ that

$$\ln R(t') = \ln R(o) - k_r \cdot t' - (k_d - k_{ds}) \cdot \int_o^{t'} (1-R) \cdot dt'.$$

For $R \simeq 1$, this equation simplifies to $k_r \simeq \ln[R(o)/R(t')]/t'$, so k_r is the slope of a plot of $-\ln R(t')$ or $-R(t')$ versus t'.

The use of concentrations instead of flows causes, in principle, the same error as the tracer method does. χ_s/χ becomes equal to Q_s/Q as soon as $k_d = k_{ds}$, however. This condition will seldom be met, thus an error will usually be made, but in most cases it will be smaller than the one that would be caused by the tracer method in a similar situation.

6.3 WET DEPOSITION RATE

The S(IV) and part of the S(VI) in rain water originate from both rainout and washout. The fraction of the S(VI) due to wet deposition of unoxidized SO_2 cannot be found from wet deposition experiments. Therefore, study of wet removal of SO_2 by analysis of collected rain water is useful only around an isolated source, so that wet deposition in the area under the plume can be clearly distinguished from the background. In such an experiment, about the sum of oxidation and wet deposition of SO_2 will be measured, since washout of sulphate is rather fast. The wet deposition observed will increase rapidly with distance from the source because of both horizontal and vertical dispersion, causing contact of the plume with an increasing number of raindrops for an increasing time. When dispersion approaches completeness, the wet deposition due to the source will normally become indistinguishable from the background. The procedure of Högström (10) to get round this problem does not calculate k_w nor $k_r + k_w$, but (2) sets an upper limit to k_{tot}.

Rain water analysis has also the inherent drawback that no rain collector can avoid dry deposition. Using well designed collectors, set out only for the duration of the experiment, still an error of a factor of 2 can be made due to absorption of SO_2 and sulphate directly from the air (11). Using rain samples from routine networks, the situation may be much worse.

One can also study the decay of SO_2 in the atmosphere rather than the presence of sulphur in rain. Concentration measurements in air are more reliable, of course, but one will only be able to evaluate the difference in k_{tot} between dry and rainy weather. This difference will be due to each of the three groups of sink processes; also, the problem of the dependence of concentration on height will tend to be severe here, since the wet surface is a very good sink for SO_2.

6.4 DRY DEPOSITION RATE

From measurement of Q, Q_s and Q_a at different travel times ('mass balance method'), a plume-averaged k_d can be found as $\bar{k}_d = k_{tot} - k_r$ (assuming dry weather). From the foregoing discussion, one will have understood that such a first order decay parameter has little physical meaning. However, in some Lagrangian models, where dry deposition is described as a first order decay process, the best choice will evidently be the use of an empirical value measured under similar circumstances.

In most instances, measurement of the dry deposition flux will be more useful. This can occasionally be done by determining the quantity of SO_2 absorbed per unit of surface area and of time. There is no severe analytical problem in the case of fresh snow, for instance. For some other surfaces, the use of $^{35}SO_2$ is feasible. It is more common, however, to measure the vertical turbulent flux. The eddy correlation method is the most direct one : F is calculated as the covariance of w with χ. Both are to be measured at the same place, and with the same response time not exceeding a few tenths of a second. This method has been used already (12). Flame photometry, measuring total sulphur rather than SO_2, was used for analysis. Nevertheless, the accuracy attained ($\pm$ 50 %) was quite good. The gradient method is the least demanding one in its experimental set-up, and is the most frequently used one. The vertical flux is calculated as $-F = k^2 \cdot z^2 \cdot (\phi_x \cdot \phi_M)^{-1} \cdot (\partial u/\partial z) \cdot (\partial \chi/\partial z)$ with k = 0.41; ϕ_x and ϕ_M are stability functions ($= 1$ in neutral, <1 in unstable and >1 in stable atmosphere), and can be calculated from $\partial u/\partial z$ and $\partial T/\partial z$ using empirical flux-gradient relationships (references in 2). In diabatic atmosphere, measurement at several heights and determination of gradients is recommended. In neutral atmosphere, measurement at 2 heights, z_1 and z_2, is sufficient : $-F = k^2 \cdot [\chi(z_2) - \chi(z_1)] \cdot [u(z_2) - u(z_1)]/[\ln(z_2/z_1)]^2$; here the accuracy is limited by the measurements. In diabatic atmosphere, the limited knowledge of ϕ_x and ϕ_M adds some uncertainty.

7. EMPIRICAL AND SEMI-EMPIRICAL RATE PARAMETERS

Conceptually valid measurements of k_{tot} are few in number; they have not contributed sufficient knowledge of the SO_2 removal rate under different circumstances to represent a significant source of information relevant to SO_2 removal modelling. More measurements of the rate of conversion to sulphate are available; their major feature is scatter, showing the fitfulness of the ambient atmosphere. Together with laboratory results and literature on radical concentrations, however, they yield a satisfactory knowledge of SO_2 conversion in dry weather conditions; typical values have been given already. Experimental access to wet atmospheric systems is troublesome; data-fitting is the main source of information on removal rate parameters. First order rate parameters of a few 10^{-5} s^{-1} are often found for both conversion in clouds and washout. Little chemical input information is available for predicting in cloud conversion rates; in spite of the many wet deposition experiments, the same holds for washout. One study of the difference $(k_{tot})_{wet} - (k_{tot})_{dry}$ due to rain (13) is worth mentioning here, though actually a statistical evaluation of Eulerian measurements at ground level was made; at rain rates $\lesssim$ 2 mm/hour, some $4 \cdot$ to $5 \cdot 10^{-5}$ s^{-1} was found, increasing to some 10^{-4} s^{-1} at higher rain rates. The first value should represent the increase of k_d and k_r in rainy weather; with J in mm/hour, $k_w \simeq 2.5 \cdot 10^{-5}$ (J-2) s^{-1} is found for $2 < J < 5$.

Dry deposition has intensively been reported on; this mechanism is well documented now for many types of uniform terrain (references in 2). Since terrain parameters, such as r_c and surface roughness, cannot be entered in models with infinite resolution, effective parameters are needed to describe surface grids with non-uniform coverage. Here, much is left to do. Both surface roughness and r_c have been mapped for on extensive area in North-America (14), allowing for variation of these parameters with weather and season conditions. Surface roughness and average dry deposition velocities have recently (15) been mapped for the Benelux area; we intend to do the same for r_c, so that it will be possible in models to calculate the actual dry deposition velocity under the prevailing weather conditions.

8. REFERENCES

(1) H.S. Judeikis and T.B. Stewart, 1976. Atmos. Environ. 10, 769-776.
(2) S. Vandendriessche et al., 1981. Final report to DPWB (see acknowledgment); in Dutch. To be available from DPWB.
(3) R. Berkowicz and L.P. Prahm, 1978. Atmos. Environ. 12, 379-387.
(4) D.A. Hegg and P.V. Hobbs, 1980. Atmos. Environ. 14, 99-116.
(5) A. Liberti et al., 1978. Atmos. Environ. 12, 255-261.
(6) S.A. Penkett and J.A. Garland, 1974. Tellus 26, 284-289.
(7) L.R. Martin and D.E. Damschen, 1981. Atmos. Environ. 15, 1615-1621.
(8) A.C. Chamberlain, 1953. Report AERE HP/R 1261, Atomic Energy Research Establishment, Harwell, U.K.
(9) E. Weber, 1970. J. Geophys. Res. 75, 2909-2914.
(10) U. Högström, 1974. Atmos. Environ. 8, 1291-1303.
(11) T.D. Davies, 1974. WMO Report No. 368, 567-578.
(12) I.E. Galbally et al., 1979. Nature 280, 49-50.
(13) P.R. Maul, 1978. Atmos. Environ. 12, 2515-2517.
(14) C.M. Sheih et al., 1979. Report ANL/RER-79-2, Argonne National Laboratory, U.S.A.
(15) H. van Dop, (1981). To be published in Atmospheric Environment.

ACKNOWLEDGMENT

This research was performed as a part of the 'National R-D Programme on Environment -Air', Services for Science Policy Programming, Belgium (Dienst Programmatie van het Wetenschapsbeleid, Wetenschapsstraat 8, B-1040 Brussel).

INVESTIGATION OF THE REGIONAL DISTRIBUTION OF WET DEPOSITION OF POLLUTANTS

C. PERSEKE, H.-W. GEORGII, E. ROHBOCK
Department of Meteorology and Geophysics
University of Frankfurt/M (FRG)

ABSTRACT

Deposition processes are main sinks for atmospheric pollutants. In order to assess wet and dry deposition a precipitation chemistry network has been established in the
Federal Republic of Germany. For sampling and separating
wet and dry deposition a wet/dry collector has been designed. The following components are analyzed: pH, $SO_4^=$,
NO_3^-, trace metals. In this paper first results on wet deposition of pH, sulfate and nitrate are presented for the
period Aug. 79 to Aug. 80. The concentration pattern shows
differences by a factor of about two between polluted areas
(Ruhr area) and unpolluted areas. Wet deposition is estimated and discussed in relation to meteorological conditions.

To determine the relative importance of sulfate and
nitrate for the acidity of rain the ratio $SO_4^=/NO_3^-$ is discussed.

1. INTRODUCTION

During the last years atmospheric removal processes have
gained growing attention because of some serious effects of
acid rain associated with the sulfur and nitrogen emissions
(1,2). Besides dry deposition, wet deposition is a dominant
sink for atmospheric trace gases (e.g. SO_2, NO_x) and aerosols.
Thus, large amounts of pollutants are removed from the atmosphere and deposited on the ground. A first estimation of
the dry and wet deposition of sulfur in the Federal Republic
of Germany has shown that dry SO_2 deposition is the main sink
in heavy polluted areas. Whereas the relative importance of
wet deposition increases in less polluted regions and regions
with high rainfall amounts (3).

In order to assess wet and dry deposition on a regional
scale, a deposition network has been established in the Federal
Republic of Germany since August 1979 with a measuring period
of two years, only. The network consists of 12 stations, representing polluted areas (e.g. Ruhr area) and less polluted
areas.

2. MEASUREMENTS

For sampling and separating wet and dry deposition a wet/

dry collector has been designed. The collector consists of
two separate sampling units for rain and for dry fall-out.
Rain is collected in a polyethylene funnel and dry deposition
in a high-walled glass-container according to the standard
method by BERGERHOFF (4). By means of a rain sensor the rain-
and dustgauge is covered alternatively. A detailed description of the collector is given elsewhere (5).

Rain is sampled on a daily basis, dry deposition during
a period of 14 days.

The sampling efficiency of the rain gauge was determined
by comparison with the precipitation rate measured by the German Weather Service. It amounts to 84%-94% and depends on the
rain intensity.

The chemical analyses include sulfate, nitrate and trace
metals.

3. RESULTS AND DISCUSSION

Based on first results the temporal variation of the sulfate and nitrate concentration in rain is discussed for the
period August 1979 to September 1980. Furthermore, the regional concentration and wet deposition pattern of sulfate and
nitrate is presented for the winterseason 1979/80 and the
summerseason 1980.

4. TEMPORAL VARIATION

The temporal variation of the sulfate concentration (precipitation weighted monthly means) in rain seems to be similar at all stations during the period August 1979 - September
1980 (Fig. 1). During autumn (1979) a decrease of the sulfate
concentration is observed. Lowest concentration values between 1.4 mg S/l in polluted and 0.4 mg S/l in unpolluted
areas are found in November and December. During spring 1980
an increase of the sulfate concentration to maximum values of
2.6 - 6.6 mg S/l is observed in May. The increase is most
pronounced in polluted regions. In order to explain these
high values the general weather situation in May was taken
into account.

The weather situation in May 80 was predominated by low
gradient with anticyclonic flow-pattern, resulting in a dry
period of about 10 days. Precipitation with increased concentrations of pollutants occurred before and after the dry
period. The precipitation events (showers) were associated
with a postfrontal advection of cold airmasses: both continental airmasses of polar origin from East Europe and aged
maritim polar airmasses, reaching Germany from southwest.

Due to convection with vertical mixing in the cold air
the measured air concentration at ground level is low. Polluted airmasses (from local sources) are mixed into upper
layers and are included into precipitation formation processes.

In showers compared to warmfront rain elevated concentrations are found (6).

Low concentration values in June and July are the result
of a west-weather situation with frontal precipitation systems
reaching Germany in short intervals.

For the nitrate concentration a similar temporal variation with highest values during May is found. On the molar
basis the nitrate concentration is somewhat lower than the
sulfate concentration during most time of the year. Only
during spring the nitrates are more increased than the sulfates.

5. REGIONAL DISTRIBUTION

      The regional distribution of the sulfate concentrations
makes obvious that higher sulfate concentrations in rain are
observed at all stations during summer and not during the
heating period. We have to keep in mind that the SO_2-emis-
sions are higher in winter compared to summer, but that the
homogeneous oxidation-processes during summer lead to a more
rapid formation of sulfate and may be the main factor determining the sulfate concentration in rain. In USA increased
sulfate concentrations are observed during summer, too (7).
During the winterseason the sulfate concentration pattern
shows a maximum with 1.9 mg S/l in the Ruhr area. The concentration decreases to values of 0.7 mg S/l in unpolluted areas.
      During summer the maximum has shifted to the area south-
west of the Ruhr. The gradient from polluted to less polluted
areas is not as pronounced as in winter. In contrast to the
concentration pattern of the precursors of acid substances,
which is strongly determined by the emission pattern, the
regional distribution of acid substances in rain is more uniform. Thus, indicating the integrating effect of frontal precipitation fields. It is interesting to compare the sulfate
concentration in rain in Germany with values measured in the
Netherlands (8). Due to coastal influences higher sulfate concentrations (1.8 - 2.3 mg S/l) especially in winter are found
at Dutch stations.
      The spatial distribution of the nitrate concentration
shows differences with regard to the location of the maximum
area. During winter highest nitrate concentrations of 0.7 -
0.8 mg N/l are observed further northeastwards in the region
of Hamburg and Braunschweig decreasing to values of 0.2 mg
N/l in Deuselbach. During summer the pattern changed. Maximum values of 0.85 mg N/l occur southwest of the Ruhr area.
At the northern stations lower nitrate concentrations are
measured in summer compared to the winterseason. Whereas at
the southern stations, especially at Deuselbach higher concentrations are observed in summer. It is assumed that the increased nitrate content in rain at the rural site in Deuselbach is due to natural nitrogen emissions from agricultural
activities. The nitrate concentrations measured in the Netherlands with values between 0.5 and 0.8 mg N/l agree quite well
with our values (8).

6. WET DEPOSITION

      In Fig. 2 the wet sulfur deposition pattern is presented
for the winterseason 1979/80 and the summerseason 1980. During
winter sulfur depositions up to 160 mg S/m^2 month occur in the
highly industrialized Ruhr area and in the region around

Hamburg. Outside the maximum area the deposition decreases
to values less than 60 mg S/m^2 month. A comparison with the
distribution of total precipitation rate during that period
shows that the deposition is higher in areas with high precipitation amounts.

During summer the sulfur deposition is increased by a
factor of two as a result of a long lasting precipitation
period. The general weather situation during June, July was
characterized by a sequence of frontal systems from the atlantic crossing Germany with their precipitation fields.

In winter the regional distribution of the wet nitrate
deposition (Fig.3) shows maximum depositions of 60-70 mg
N/m^2 month in an area extending from Hamburg over the Ruhr
area to the Rhein Main area. In the summer season the nitrate
deposition is increased, too. Maximum values of 100-110 mg
N/m^2 month are found. A decrease by a factor of two to less
polluted areas and areas with less precipitation is to be seen.
On a yearly basis the nitrate deposition amounts to 700-800 mg
N/m^2 year in the Ruhr area and at mountain stations with high
precipitation rates. These values are in accordance with an
evaluation by BONIS and MESZAROS (9) of the wet nitrate deposition in Europe.

An evaluation of the wet deposition as a function of the
winddirection at the surface layer shows that highest deposition is found with advection from southwest for the ground
stations and from northwest for the mountain station Kleiner
Feldberg. This reflects the predominance of westerly winds
during frontal precipitation events.

7. ACIDITY

The temporal variation of the monthly average pH values
and of the monthly precipitation rates is plotted in Fig. 4
for the period October 1979 - September 1980. At all stations
a winter peak of pH-values, indicating low acidity is observed.
At the time of the winter peak high precipitation rates are
measured. Low pH values are found in October, from January
to March and in June. While there seems to be an influence
of rainfall rate on pH in winter, the picture in spring and
summer is not so clear.

The acidity in rainwater is a result of the overall
charge balance of cations and anions in the solution. KLOCKOW
et al. (10) found a good correlation between the H^+ concentration versus the excess sulfate concentration ($SO_4^= - Ca^{++}$),thus
indicating that the acidity is mainly determined by sulfates.
An evaluation of the pH-data in the OECD project "long range
transport of air pollutants" makes obvious that, in addition
to the sulfates, the nitrates play an important role for the
acidity in rain. In order to determine the relative importance of sulfate and nitrate for the acidity of rain (before
neutralizing components are acting) the ratio $2 SO_4^=/(2 SO_4^= + NO_3^-)$ was calculated for yearly depositions, assuming for each
mole sulfate an equivalent of 2 moles H^+ and for each mole
nitrate an equivalent of 1 mole H^+. The contribution of the
sulfates sums up to 62 - 71% depending on the particular sampling site. The sulfate contribution is highest in the in-

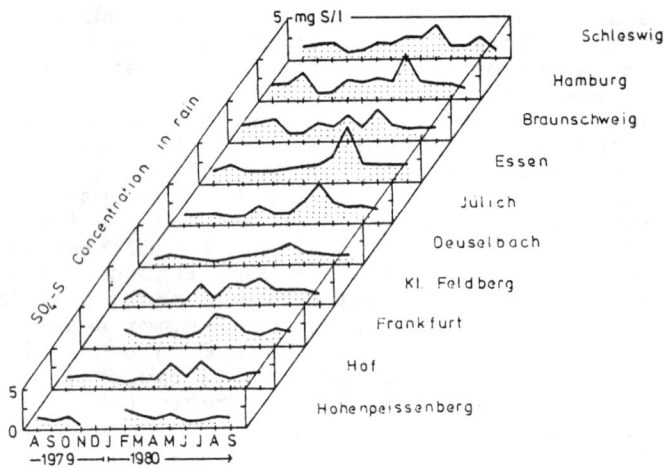

Fig. 1: Temporal variation of the sulfate concentration in rain at ten German stations during the period August 1979 - September 1980

dustrial areas. These values agree with investigations in the United States, where 60-70% of the acidity in rain can be attributed to sulfur compounds, and the remaining 30% to nitrates and chlorides (12).

8. CONCLUSIONS

The evaluation of the first year measuring period in the German deposition network has shown that the wet deposition of acid pollutants is highest during the summer season due to high precipitation rates. The temporal variation of the sulfate and nitrate concentrations in rain indicates that highest concentration values occur during May, at the beginning of the vegetation period, together with elevated dry deposition rates. These increased deposition rates are of particular interest because acid pollutants may result in damages to the vegetation. Further measurements are necessary to find out, whether high wet depositions are typical for German summers. Due to the large variation of the precipitation-rates a measuring period of two years only is too short to obtain representative values.

It should be mentioned that the concentration pattern shows regional differences of a factor 2 only, due to the integrating effect of precipitation. Thus, less polluted areas can receive enhanced wet depositions.

ACKNOWLEDGEMENT

Financial support by the Umweltbundesamt of the FRG is gratefully acknowledged.

Wet SO$_4$-S Deposition (mg S/m^2/month)

WINTER 79/80   SUMMER 80

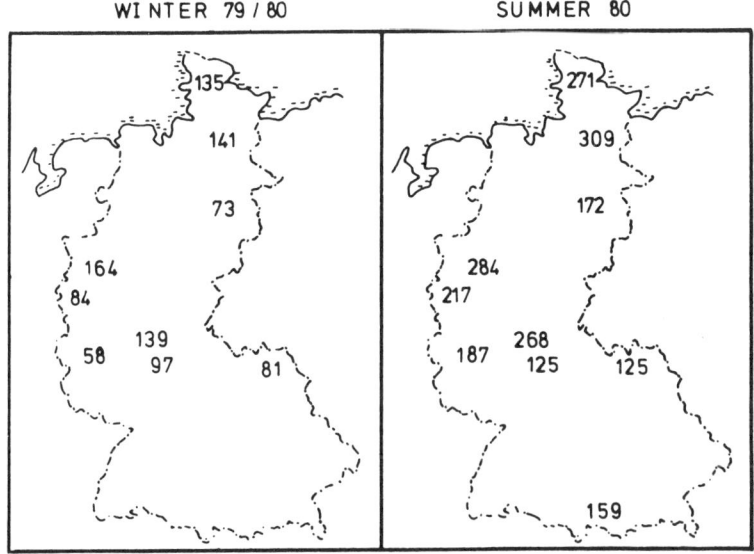

Fig. 2

Wet NO$_3$-N Deposition (mg N/m^2/month)

WINTER 79/80   SUMMER 80

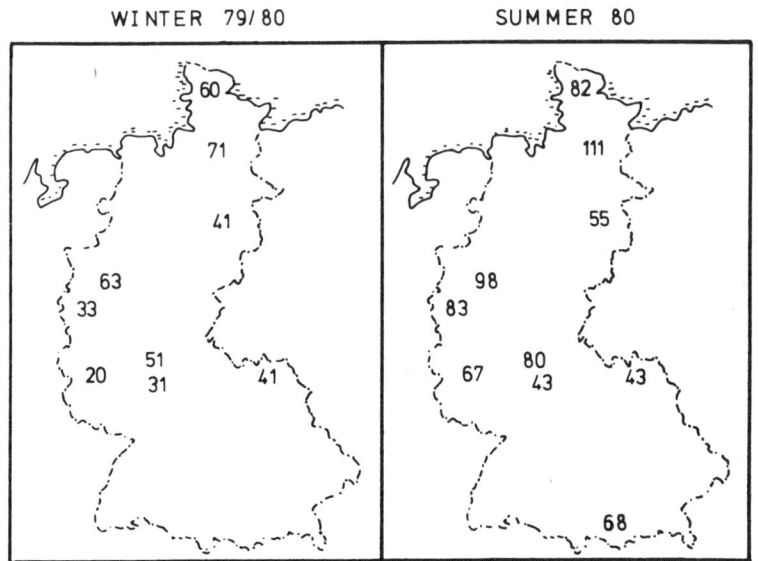

Fig. 3

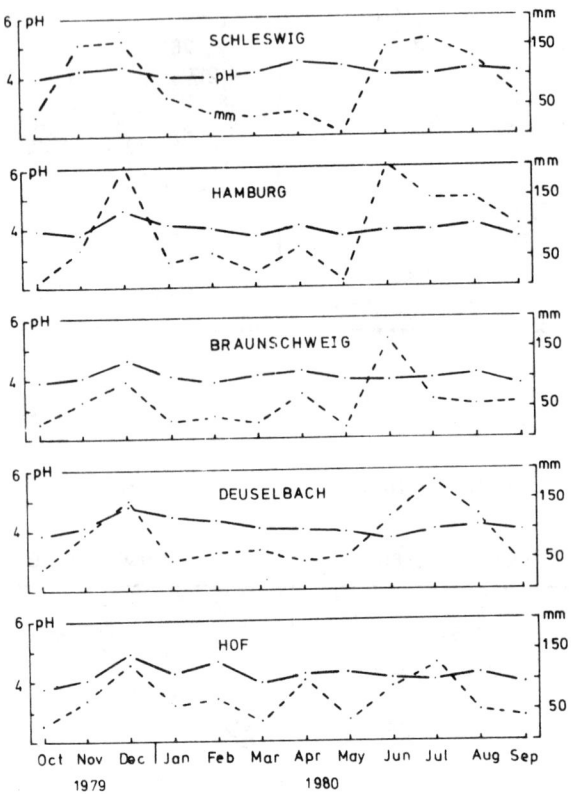

Fig. 4: Temporal variation of monthly averaged pH-values and precipitation rates

9. LITERATURE

1) AMBIO (1976)   Proc. Int. Conf. on the Effects of Acid
Precipitation
Vols 5-6, Telemark, Norway, 14-19 June

2) ODEN, S. (1976)   The acidity problem - an outline of concepts, Proc. first Int. Symp. on Acid Precipitation and the Forest Eco-System

3) GEORGII, H.-W., PERSEKE, C. (1979)   Some results on wet
and dry deposition of sulfur compounds,
Proc. First Europ. Symp.-Physico-Chemical
Behaviour of Atmospheric Pollutants, Ispra,
16-18 Oct. 79, pp 410-418

4) VDI-Richtlinie 2119, Blatt 2 (1971)

5) GEORGII, H.-W., GRAVENHORST, G., PERSEKE, C., ROHBOCK,
E., (1980)   Untersuchung über die trockene und feuchte
Deposition von Luftverunreinigungen in der
Bundesrepublik Deutschland
Bericht im Auftrag des Umweltbundesamtes

6) KINS, L. (1981)   Untersuchung der Spurenstoffkonzentra-
tion im Verlauf von Einzelniederschlägen, Diplomarbeit, Manuskript, Universitätsinstitut für
Meteorologie und Geophysik, Universität
Frankfurt

7) PACK, D.H. (1979)   Acid Precipitation - The Physical
System, Proc. Advisory Workshop on the Formation
of Acid Precipitation
EPRI Rep. No. EA 1074, 3-137

8) Meetnet voor Bepaling van de Chemische Samenstelling van
de Neerslag in Nederland. Jahresbericht 1980,
Royal Netherland Met Institut (KNMI). De Bilt,
Inst. f. Public Health (RIV) Bilthoven

9) BONIS, K., MESZAROS, E., PUTSAY, M. (1980)   On the at-
mospheric budget of nitrogen compounds over
Europe
Idöjaras, Vol. 84, No. 2, 57

10) KLOCKOW, D., DENZINGER, H. RÖNICKE, G. (1978)   Zum Zusammenhang zwischen pH-Wert und Elektrolytzu-
sammensetzung von Niederschlägen
VDI Berichte Nr. 314, pp 21-26

11) OTTAR, B. (1978)   An assessment of the of the OECD-Study
of long range transport of air pollutants.
Atm. Env. 12, 445-454

RESULTATS PRELIMINAIRES D'UNE ETUDE DE

LA COMPOSITION CHIMIQUE DE L'AEROSOL COTIER

par J. MORELLI[1], T. MARCHAL[1], L. GIRARD-REYDET[1], B. REMY[1],

A. DUTOT[1], P. PERROS[2] et P. CARLIER[3]

1. Laboratoire de Chimie Minérale des Milieux Naturels (Equipe de
   Recherche associée au CNRS n° 889), Université de Paris VII,
   2 Place Jussieu, 75 251 Paris Cédex 05, France

2. Laboratoire de Physico-Chimie Instrumentale, Université de
   Paris VII, 2 Place Jussieu, 75 251 Paris Cédex 05, France

3. Laboratoire de Chimie Organique Physique (laboratoire associé au
   CNRS), Université de Paris VII, 1 rue Guy de la Brosse,
   75 005 Paris, France

ABSTRACT

        A study on the chemical composition of atmospheric aerosol star-
ted recently on the coast of Brittany. Presently, two sampling campaings
have been carried out, in strong oceanic winds situations (December 1980)
and in calm weather conditions (June 1981). The samples have been
subjected to multielements analysis by X ray fluorescence spectrometry.
We are presenting herin some prelimary results related to two of their
main components, chlorine and sulfur. They show for these elements clear
variations of atmospheric concentrations corresponding to various classes
of particles size, with a decrease between December 1980 and June 1981
for coarse particles. An opposite effect is observed for submicronic aerosol, only in the case of sulfur, small particles being always very poor
in chlorine. Such results are discussed taking into account formation
processes of the collected particles : bubbles bursting at sea surface,
production of sulfur containing aerosol from gazeous components of both
natural and man-made origins. We eventually underline the interest of
the selected sampling zone, which has yet few industries and forms an
extreme point of the European continent in Atlantic, as a reference for
future comparisons with similar studies performed in other areas.

1. INTRODUCTION

On sait qu'il existe des échanges de matière continuels entre l'océan et les continents, via l'atmosphère (1, 2). Une fraction non négligeable des constituants à l'état gazeux et particulaire produits par la source marine est susceptible d'être entraînée plus ou moins loin à l'intérieur des terres. Inversement, des gaz et des particules émis par des sources continentales naturelles et anthropogéniques sont disséminés au-dessus de l'océan, certains polluants pouvant ainsi contaminer le milieu marin.

De tels échanges de matière impliquent des transits au-dessus de zones littorales. A l'échelle européenne, la façade maritime de notre continent s'étire sur près de 38.000 km. Elle accueille une partie notable de la population et de l'activité économique. Il apparaît donc nécessaire de développer des observations systématiques sur la qualité de l'air en milieu côtier en vue d'une meilleure connaissance des caractéristiques de l'environnement atmosphérique, de son rôle dans les grands cycles biogéochimiques et de ses éventuelles perturbations sous l'effet des activités humaines.

Dans cette perspective, notre groupe a commencé récemment une étude régionale de la composition chimique de l'aérosol atmosphérique sur le littoral du Finistère, en Bretagne. Cette région, encore peu industrialisée, forme une pointe de l'Europe Occidentale s'avançant dans l'Atlantique (figure 1). En raison du régime des vents dominants (circulation d'Ouest), elle est fréquemment soumise à l'influence de masses d'air océaniques peu polluées.

A l'heure actuelle, deux campagnes de prélèvement par impaction ont été effectuées, l'une en Décembre 1980, l'autre en Juin 1981. Nous présenterons ici de premiers résultats relatifs à deux principaux constituants de nos échantillons : le chlore et le soufre. Les fluctuations des concentrations atmosphériques et de la distribution granulométrique de ces éléments seront ensuite discutées en les reliant à différents types de processus de formation de l'aérosol collecté.

2. CONDITIONS D'ECHANTILLONNAGE ET TECHNIQUES MISES EN OEUVRE

Nos échantillons ont été prélevés sur 3 sites dont l'emplacement est indiqué sur la figure 1 (Pointe Saint Mathieu, Pointe du Raz et Penmarc'h), dans des phares ou sémaphores. Les paramètres météorologiques y sont relevés régulièrement. La côte est abrupte dans les deux premiers sites, contrairement à Penmarc'h, où des champs d'algues se découvrent à une marée basse.

La campagne de Décembre 1980 a été marquée par des conditions de forts vents vents océaniques de secteur Ouest, avec des averses fréquentes. En Juin 1981, la situation a été très différente : temps calme ensoleillé, absence de pluie, légère brise venant dans la majorité des cas du secteur Nord, avec surtout sur le site de Penmarc'h des masses d'air ayant souvent transité au-dessus des terres.

Au total, on a effectué une quinzaine de prélèvements couvrant en général des périodes de 2 à 4 jours, à des hauteurs au-dessus de la mer de 22 à 85 m. On a utilisé à cet effet des impacteurs en cascade à 6 étages avec flux dérivé au 6ème étage conçus au Laboratoire de Chimie Minérale des Milieux Naturels de l'Université de Paris 7. Les caractéristiques en ont été décrites lors du premier symposium européen sur le comportement physico-chimique des polluants de l'atmosphère (3).

FIGURE 1

Au niveau de chacun des étages des impacteurs, les particules ont été collectées sur des filtres Nuclépore. Ceux-ci ont été soumis à une analyse multiélémentaire par spectrométrie de fluorescence X. Un des avantages de cette technique tient au fait qu'elle est non destructive, permettant ainsi d'effectuer ultérieurement des analyses complémentaires. On se limitera dans ce qui suit à l'examen de résultats de mesures en chlore et soufre.

3. RESULTATS

A. Résultats relatifs à l'ensemble de la matière particulaire collectée

Le tableau 1 donne les concentrations atmosphériques totales en chlore et soufre à l'état particulaire, obtenues en ajoutant pour chacun de ces deux éléments les contributions des diverses classes de taille, ainsi que le rapport pondéral Cl/S correspondant.

Site	Date de prélèvement	Hauteur de prélèvement (m)	Cl$_{-3}$ ng.m	S$_{-3}$ ng.m	Cl/S
POINTE SAINT MATHIEU	13-17 Déc. 80	22	2 800	330	8,5
	14-18 Juin 81		*2240*	*630*	*3,5*
	18-22 Juin 81	42	*610*	*920*	*0,7*
	22-26 Juin 81		*960*	*1360*	*0,7*
POINTE DU RAZ	13-17 Déc. 80	85	4 250	700	6,1
	17-29 Déc. 80		1 330	300	4,4
PENMARC'H	13-15 Déc. 80	5 70	8 140 5 930	870 550	9,3 10,8
	15-18 Déc. 80	5 70	5 770 4 350	510 380	11,7 11,3
	13-17 Juin 81	*34*	*1570*	*670*	*2,3*
	21-25 Juin 81	*34*	*560*	*890*	*0,6*
		70	*730*	*1000*	*0,7*

TABLEAU I

B. DISTRIBUTION GRANULOMETRIQUE DU CHLORE ET DU SOUFRE

Il apparaît que les fractions déposées sur les étages 1 et 2, 3 et 4, 5 et 6 de nos impacteurs ont des caractéristiques voisines. L'interprétation se trouve simplifiée en considérant les masses d'élément collectées par des étages successifs regroupés deux à deux. On est alors amené à prendre en compte 3 domaines de tailles de particules A, B, C avec des diamètres de coupure à 4 μm et 1 μm. Ces coupures semblent judicieuses pour différencier des grandes familles de particules qui n'ont pas la même aptitude à participer à des transports atmosphériques, les plus gros aérosols (classe A, d> 4 μm) ayant une probabilité élevée d'être éliminés rapidement.

Le tableau 2 donne les contributions de chacune des classes A, B, C, exprimées en % des masses totales de chlore et soufre particulaires collectées.

Site	Date de prélèvement	Hauteur de prélèvement (m)	CHLORE			SOUFRE		
			A	B	C	A	B	C
POINTE SAINT MATHIEU	13-17 Déc. 80	22	72	27	1	52	34	13
	14-18 Juin 81		59	40	1	19	41	40
	18-22 Juin 81	42	68	27	5	7	39	54
	22-26 Juin 81		63	35	2	15	32	53
POINTE DU RAZ	13-17 Déc. 80	85	68	27	5	50	32	18
	17-29 Déc. 80		43	55	2	38	47	15
PENMARC'H	13-15 Déc. 80	5	55	42	3	40	56	14
		70	52	44	3	30	51	19
	15-18 Déc. 80	5	65	26	9	43	42	15
		17	59	39	1	35	45	20
	13-17 Juin 81	34	37	60	3	11	44	45
	21-25 Juin 81	34	40	53	7	8	41	51
		70	50	48	2	7	41	52

TABLEAU 2

4. DISCUSSION

On notera au préalable que des mesures en aluminium, élément généralement pris comme indicateur d'origine crustale, ont permis de vérifier que la contribution terrigène au chlore et soufre de nos échantillons a toujours été négligeable (en moyenne, environ 1°/oo)[*].
Des analyses complémentaires en sodium par spectrophotométrie d'absorption atomique sont par ailleurs en cours en vue d'une évaluation précise de la contribution de la source océanique[**]. D'ores et déjà, nous pouvons cependant dégager un certain nombre de faits majeurs de nos résultats.

[*]*la contribution crustale à la concentration atmosphérique mesurée d'un élément X s'obtient par la relation :* $X_{crustal} = Al_{collecté} \times (X/Al)_{croûte\ terrestre}$
On suppose que tout l'aluminium collecté est d'origine crustale.

[**]*Par analogie, on peut calculer pour un élément X une contribution de la source marine au moyen de la relation :* $X_{marin} = Na_{collecté} \times (X/Na)_{eau\ de\ mer}$
On admet que le sodium collecté est d'origine marine.

Dans les conditions d'échantillonnage des deux premières campagnes sur le littoral breton, la valeur des concentrations en chlore et
soufre particulaires n'a pas été très profondément affectée par la hauteur
de prélèvement au-dessus de la mer. Il est donc probable que l'influence
des sources locales n'a pas été très marquée par rapport à celle de sources plus lointaines, y compris à Penmarc'h. Les conditions de vent défaro-
rables qui ont prévalu sur ce site en Juin 1981 (brise de terre), n'ont
pas permis de mettre en évidence l'impact de l'activité biologique des proches
champs d'algues côtiers. Les concentrations atmosphériques en soufre n'y
ont pas été plus fortes que celles mesurées simultanément à la Pointe
Saint-Mathieu, alors que les champs d'algues découverts à marée basse sont
connus pour être des sources très actives de gaz soufrés et de noyaux de
condensation (4, 5).

En ce qui concerne le chlore particulaire, on a observé des concentrations totales en moyenne nettement plus élevées dans les conditions
de fort vent du large de Décembre 1980 que par temps calme en Juin 1981
-4700 ng.m^{-3} au lieu de 1100 ng.m^{-3}- (tableau 1). Le chlore étant censé
être essentiellement marin, ces variations peuvent être reliées à la production de particules de sels par éclatement de bulles à la surface de
l'océan. On sait en effet que ce phénomène est d'autant plus actif que la
vitesse du vent est plus forte (6, 7, 8). Il apparaît par ailleurs que le
chlore collecté était pour une très large part porté par d'assez grosses
particules dans la gamme de taille des "jet drops" (9), la contribution en
masse de l'aérosol submicronique (classe C) étant toujours faible et inférieure à 10% (tableau 2). On rejoint donc ainsi les conclusions d'autres
observations du même type dans l'atmosphère marine (10, 11). On peut voir
sur la figure 2 des exemples de variations saisonnières des concentrations
en chlore associées aux 3 classes de taille considérées.

Dans le cas du soufre, on note une tendance inverse à celle men-
tionnée pour le chlore, sa concentration totale sous forme particulaire
ayant été dans l'ensemble plus élevée en Juin 1981 qu'en Décembre 1980
-en moyenne 900 ng.m^{-3} au lieu de 500 ng.m^{-3}- (tableau 1). Ces valeurs
recouvrent celles rencontrées par d'autres auteurs très loin des côtes
en diverses régions océaniques (10, 11, 12). On est d'autre part frappé
par l'ampleur du remaniement de la distribution granulométrique du soufre
d'une campagne d'échantillonnage à l'autre (tableau 2). Alors que la contribution en masse de l'aérosol submicronique représentait à peine 15 à
20 % du soufre particulaire total en Décembre 1980, elle atteint 40 à 54 %
en Juin 1981, les concentrations correspondantes passant de 50-120 mg.m^{-3}
à 250-730 ng.m^{-3}. On assiste parallèlement à une diminution de la contribution des plus grosses particules (classe A) -7 à 19 % au lieu de 30 à 50 %
-et des concentrations en soufre qui leur sont imputables -en moyenne 100
ng.m^{-3} au lieu de 220 ng.m^{-3}-. Ces profonds changements sont illustrés par
la figure 2. Ils traduisent un ralentissement de la production d'assez
grosses particules de sulfate marin par effet de pétillement couplé à une
intensification de la formation d'un aérosol fin par conversion gaz/parti-
cules. On sait en effet que des gaz soufrés de l'atmosphère donnent lieu
à de tels processus (15, 16, 17, 18).

Ces composés gazeux ayant de multiples sources naturelles et anthropogéniques, la question de l'origine du soufre incorporé aux particules submicroniques est complexe. En décembre 1980, par vent d'ouest, sa
concentration était proche du "bruit de fond" mesuré au-dessus de zones
océaniques non polluées de l'hémisphère Sud (50 à 100 ng.m^{-3}) (19). Elle
peut donc être attribuée pour une large part à la conversion du SO_2 issu de
précurseurs gazeux biogéniques soufrés tels que le DMS, qui sont

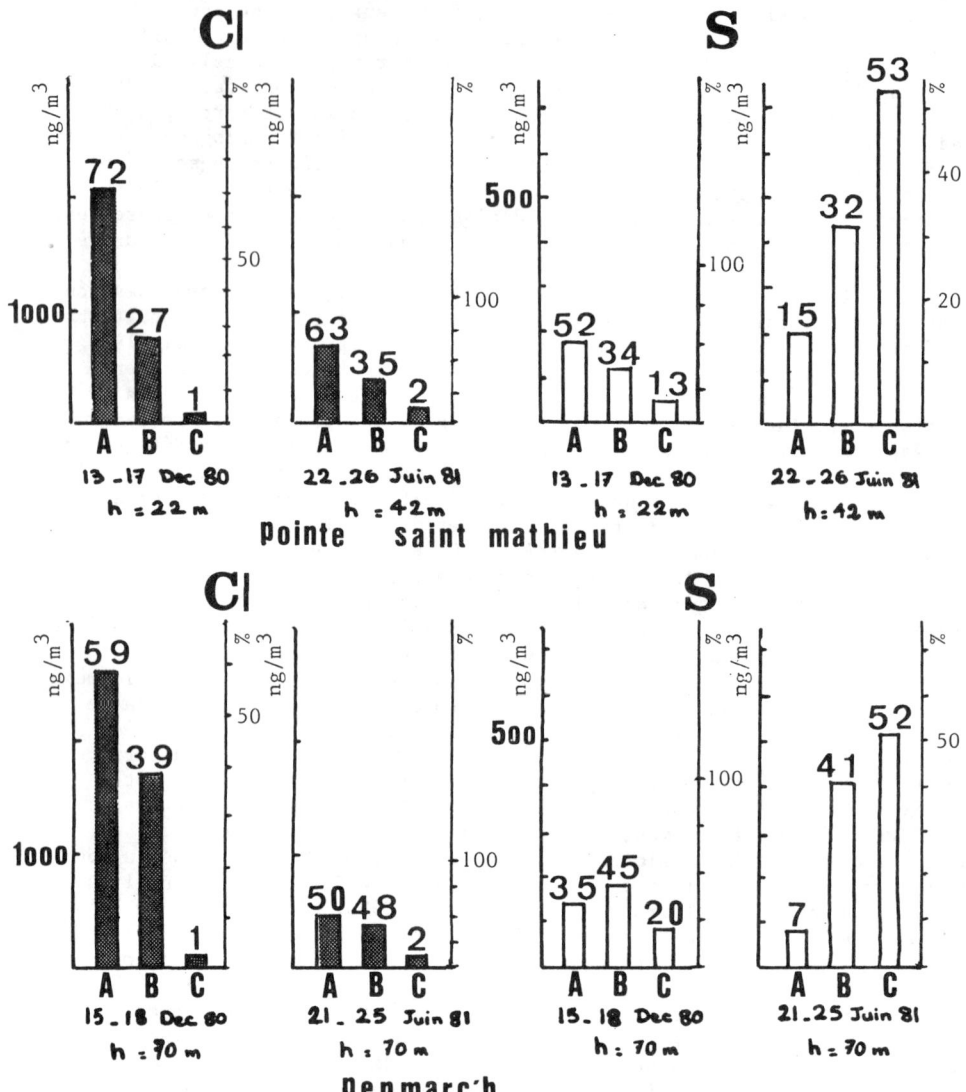

FIGURE 2

émis à la surface de l'océan (20) [*]. L'accélération de ces phénomènes à la fin du printemps, période où l'activité biologique du milieu marin est stimulée et où la longueur du jour favorise l'activité photochimique de l'atmosphère, a pu contribuer aux fortes concentrations du soufre particulaire submicronique observées en juin 1981. Toutefois, étant donné qu'à cette époque les masses d'air avaient souvent transité au-dessus des terres, une contribution prépondérante d'une pollution soufrée reste possible, que celle-ci provienne de sources proches (agglomération de Brest) ou même plus lointaines. On ne peut enfin écarter complètement un effet résultant d'éventuelles émissions végétales de composés gazeux soufrés.

Les variations constatées dans la distribution granulométrique du chlore et du soufre ont eu bien entendu des répercussions au niveau du rapport pondéral Cl/S. Il s'avère que celui-ci a toujours été inférieur au rapport marin de référence -Cl/S = 21,4-, tant pour l'ensemble de la matière particulaire collectée (tableau I) que pour chacune des classes de tailles considérées. D'une manière générale, le rapport Cl/S décroit lorsque la taille des particules diminue. Il présente un minimum très marqué pour l'aérosol submicronique, ce qui recoupe les résultats de travaux antérieurs (11), avec des valeurs plus faibles en Juin 81 -0,03 à 0,2- qu'en Décembre 1980 -0,6 à 6,8-. Les écarts au rapport Cl/S de l'eau de mer doivent être reliés à la fois à une perte en chlore particulaire due à sa transformation sous forme gazeuse (7, 11) et à un gain de soufre de diverses origines résultant des processus de conversion gaz/particules évoquées précédemment. On comprend dès lors que dans la composante fine de l'aérosol marin, du sulfate de sodium remplace le chlorure de sodium (21).

5. CONCLUSIONS ET PERSPECTIVES

Bien que les résultats de notre travail aient encore un caractère préliminaire, on peut en dégager les faits suivants :
- Les mesures sur la qualité de l'air que nous développons sur la côte de Bretagne, région encore très peu industrialisée qui constitue une pointe avancée du continent européen fortement soumise à l'influence de masses d'air océanique, devraient pouvoir servir de référence pour de futures comparaisons avec des études similaires faites en d'autres régions.
- Nous avons observé dans des situations météorologiques bien distinctes de nettes variations des concentrations atmosphériques du chlore et du soufre incorporés aux diverses classes de tailles de particules collectées.
- Pour le chlore comme pour le soufre, les concentrations associées aux plus grosses particules ont diminué de Décembre 1980 (conditions de fort vent océanique) à Juin 1981 (temps calme).
- La contribution en masse des particules submicroniques, qui sont toujours très pauvres en chlore, a été beaucoup plus élevée en Juin 1981 (40 à 54 %, soit 250 à 730 ng.m^{-3}), qu'en Décembre 1980 (15 à 20 %, soit 50 à 120 ng. m^{-3}). Ces dernières valeurs sont proches du "bruit de fond" mesuré au-dessus de zones océaniques très éloignées des sources de pollution.
- Nos résultats peuvent expliquer par une conjonction de divers processus de formation de l'aérosol collecté : phénomène de pétillement à la surface de l'océan, production d'un aérosol fin à partir de gaz soufrés d'origines naturelles et anthropogénique.

---

*On ne doit cependant pas totalement exclure une contribution de fines particules de type "film drop" qui sont produites lors de l'éclatement de bulles à l'interface air/mer (6, 9).

Nous nous proposons à l'avenir d'approfondir l'étude des contributions respectives des diverses sources de l'aérosol côtier grâce à un suivi de la trajectographie des masses d'air à grande et moyenne échelle. Des mesures complémentaires en sodium devraient d'autre part permettre de préciser la part de la composante océanique. Enfin, la compréhension des variations relatives des concentrations en chlore et soufre de l'aérosol côtier nécessitera une étude des échanges de ces éléments entre phases gazeuse et particulaire.

Remerciements

Nous remerçions M. G. TYMEN, qui a facilité sur le plan local l'organisation de notre première campagne de prélèvements atmosphériques ainsi que les personnels des sémaphores de la Marine Nationale et du Service des Phares et Balises pour leur bienveillant accueil. Le travail entrepris s'insère dans la première phase d'un contrat de recherche accordé par la Direction des Etudes et Recherches d'Electricité de France" (Convention AEE 287).

BIBLIOGRAPHIE

(1) Symposium International sur les échanges océan/atmosphère de matière à l'état particulaire, Nice, 4-11 Octobre 1973. Articles dans J. Rech. Atmosph., 8 (3-4), 1974.

(2) Symposium on the role of the oceans in atmospheric chemistry, IAMAP 3rd Scientific Assembly, Hamburg, 24-27 August 1981, Abstracts edited by National Center for Atmospheric Research, Boulder, Colorado.

(3) VIE LE SAGE R., GRUBIS B., BERGAMETTI G., ELICHEGARAY C. et et MALINGRE G., 1980 "Mise au point d'un impacteur en cascade à flux dérivé, Proceedings of the First European Symposium on Physico-Chemical Behaviour of Atmospheric Pollutants (Ispra, Italy, 16-18 October 1979), 322-327.

(4) BONSANG B., NGUYEN B.C. et PAUGAM J.Y., 1976. Sur la formation de gaz sulfurés et de noyaux d'Aitken dans l'atmosphère le long du littoral. C.R. Acad. Sci. Paris, Série D, 283 : 1285-1289.

(5) PAUGAM J.Y., NGUYEN B.C., BONSANG B. et FONGANG S., 1977. Production de noyaux d'Aitken et de composés soufrés dans l'air au-dessus d'une zone littorale. Chemosphere, 6 : 333-339.

(6) BLANCHARD D.C., 1963. The electrification of the atmosphere by particles from bubbles in the sea, in Progress in Oceanography, 1, 71-202, Mary Sears Ed., Pergamon Press, Oxford.

(7) CHESSELET R., MORELLI J. et BUAT-MENARD P., 1972. Some aspects of the geochemistry of marine aerosols. In : "The changing chemistry of the oceans". Proc. 20th Nobel Symp., Aspenäsgarden, 1971, Sweeden, Ed. by B. DYRSSEN et D. JAGNER, Stockholm, 93-114.

(8) MORELLI J., 1977, "Contribution à l'étude du cycle atmosphérique du potassium marin", Thèse d'Etat, Univ. Paris VI.

(9) MACINTYRE F., 1974, Chemical fractionation and sea-surface microlayer

processes, <u>In</u> : The Sea, Vol. 5, Chapter 8, Ed. E.D. GOLDBERG, John
Wiley & Sons, New York, 245-299.

(10) TYMEN G., BUTOR J.F., RENOUX A. et MADELAINE G., 1975, "Quelques
caractéristiques de l'aérosol situé au-dessus de l'Atlantique
Chemosphere, 4, 357-360.

(11) BERG W.W., Jr., 1976, Chlorine Chemistry in the marine atmosphere,
PhD Thesis, Department of Oceanography, Florida State University.

(12) NGUYEN BA CUONG, BONSANG B., PASQUIER J.L. et LAMBERT G., 1974.
Composantes marine et africaine des aérosols de sulfates dans l'Hémisphère Sud. <u>J. Rech. Atmosph.</u>, <u>8</u> (3-4) : 831-844.

(13) GRAVENHORST G., 1978, Maritime sulfate over the North Atlantic.
<u>Atmosph. Envir.</u>, <u>12</u> : 707-713.

(14) MESZAROS E., 1978, Concentration of sulfur compounds in remote continental and oceanic areas. <u>Atmosph. Envir.</u>, <u>12</u> : 699-705.

(15) BRICARD J., BOULAUD D., MADELAINE G. et VIGLA D., 1975, SO_2 transformation in controled atmosphere leading to production of aerosol
particles", Water, Air and Soil Pol., 4, 435-445.

(16) Sulfur in the atmosphere, Proceedings of the International Symposium
of Dubrownik, Yougoslavia, 7-14 September 1977, Pergamon Press, 1978.

(17) MADELAINE G., 1980, "Physique et dynamique de l'aérosol atmosphérique", Proceedings of the First European Symposium on Physico-Chemical Behaviour of Atmospheric Pollutants (Ispra, Italy, 16-18 October
1979), 261-267.

(18) BEILKE S. and GRAVENHORST G., 1980, "Cycle of pollutants in the
troposphere", Proceedings of the First European Symposium on Physico-Chemical Behaviour of Atmospheric Pollutants (Ispra, Italy,
16-18 October 1979), 331-353.

(19) DARZI M. and WINCHESTER J.W., 1981, Marine aerosol Composition in
the Indian Ocean, Symposium on the role of the oceans in atmospheric
chemistry, IAMAP 3rd Scientific Assembly, Hamburg, 24-27 August 1981,
Abstracts edited by National Center for Atmospheric Research,
Boulder, Colorado.

(20) NGUYEN B.C., BONSANG B. and LAMBERT G., "Oxidation processes of sulfur components in the marine atmosphere", Paper presented at the
IAGA/IAMAP Joint Assembly, Seattle, (1977).

(21) TEN BRINK H.M., MALLANT R.K.A.M., GOUMAN J.M., KOS G.A. and
VAN DE VATE J.F., 1980, "SO_2 conversion in a marine atmosphere",
Proceedings of the First European Symposium on Physico-Chemical
Behaviour of Atmospheric Pollutants (Ispra, Italy, 16-18 October
1979), 298-306.

THE INCREASE OF CARBON DIOXIDE AT RURAL SITES OF GERMANY

W. GROSCH , W. FLECK
Umweltbundesamt, Pilotstation Frankfurt
und
D. Jost
Umweltbundesamt, Berlin

Summary

Due to growing concern about the CO_2-problem worldwide
monitoring of atmospheric CO_2 has been increased since
the middle of the fifties.
   The investigation of long-term CO_2-trends provides
an useful element for studying the atmospheric CO_2-cycle.
Up to now most trends are reported for baseline-stations
such as Mauna Loa, South Pole etc. For polluted areas like
Central Europe evaluations of such trends are more
difficult due to strong biogenic and antropogeneous interferences.
   In this paper an attempt is made to elaborate a long
term trend for Germany. Local interferences are reduced by
combining the data of three sites at several hundreds
kilometers distance. Periodic time dependent variations
are suppressed by the use of 24-month running mean values.
First results are presented covering the period 1972 -1980
for five rural stations in the Federal Republic of Germany.
   CO_2 increase was found when eliminating a distortion
of about 2 years period and 3 ppm amplitude from the data.
Seasonal variation was suppressed by the same procedure.
The mean annual increase rate is calculated to be 0.97ppm/y
for the period studied. A similar analysis made with the
data of another rural site showed quite a different result.
At that place within the mixing layer a period of stagnation during the early 70th was followed by an accelerating
increase of CO_2-content in the second part of that decade.

1. EINLEITUNG

   Kohlendioxid ist ein wesentlicher Bestandteil der Erdatmosphäre und eine der Grundlagen für die Photooxidation im
Biotop. Veränderungen des Kohlendioxidgehaltes der Luft werden
heute vorrangig auf den im Zuge der Industrialisierung gestei-
gerten Energiebedarf und die intensivierte Nutzung der Land-
flächen zurückgeführt. Obwohl die Atmosphäre, verglichen mit
anderen biogeochemischen Reservoiren im Kohlenstoffkreislauf,
quantitativ eine eher untergeordnete Rolle spielt, ist es
gerade diese Qualität eines begrenzten Reservoires, die ihre
Schlüsselstellung im Kohlenstoffkreislauf ausmacht.
   Die wissenschaftlichen Aktivitäten in den letzten Jahr-
zehnten zeigen folgerichtig die Tendenz, mit Hilfe spezieller
Meßstrategien zu einem verfeinerten Verständnis des Kohlen-

stoffspeichers Atmosphäre, seines inneren Aufbaus und seiner Wechselwirkung mit biotischen und abiotischen Kohlenstoff-quellen und -senken zu gelangen. Auch die Arbeiten über den Gehalt der Luft mit Kohlendioxid zeigen diese Tendenz.

So war man in den fünfziger Jahren zunächst bestrebt, möglichst präzise Verfahren für die Messung des CO_2 zu entwickeln, mit denen der Gehalt der Luft an Kohlendioxid an wenigen Stellen fernab von großen Quellen und Senken bestimmt werden konnte. Für diese Stationen lassen sich relativ leicht Kriterien fest-legen, die eine Datenauswahl erlauben, aus der konsistente Resultate ableitbar sind. Solche Selektionsverfahren sind in größerer Zahl bekannt (vgl. KEELING (1960),PALES u. KEELING (1965), BROWN u. KEELING (1965)).

Mitte der sechziger Jahre begann man in zunehmendem Maße mit der Einrichtung von Stationen, an denen stärker der regio-nale Kohlendioxidpegel untersucht werden sollte. SAWYER (1971) wies darauf hin, daß zur Abschätzung künftiger globaler Trends mehr Kenntnisse über CO_2-Verteilung in den verschiedenen Reservoiren der Luft, der Vegetation erforderlich sind und darüber, wie die Gleichgewichte zwischen diesen Reservoiren erreicht und gesteuert werden. Über die Auswertung solcher Meß-reihen von regionaler Bedeutung ist bisher wenig bekannt. Immerhin haben WOODWELL et..al. (1973) sowie VERMA und ROSENBERG (1976) interessante Auswertemöglichkeiten aufgezeigt.

Einer der Gründe für diese Ausgangslage dürfte in den biogenen und anthropogenen Interferenzen zu sehen sein, die der-artige Datenkollektive prägen. Die Regionalstationen liegen in einem Quellenfeld, dessen Aktivität weitgehend unbekannt ist, und werden von ebenfalls unbekannten Senkenfeldern beeinflusst. Beide Einflußfaktoren spiegeln sich in einer erheblichen räum-lichen und zeitlichen Variabilität des Meßsignals wider. Mehr als bei den Baseline-Stationen hängen Resultate von Regional-analysen von der Untersuchungsmethodik ab.

In dieser Arbeit werden CO_2-Meßergebnisse von regional-typischen Meßstationen des Umweltbundesamtes aus den Jahren 1972 bis 1980 darauf untersucht, ob sich mit Hilfe einfacher Untersuchungsmethoden Trendaussagen ableiten lassen, die ein-gehende Datenanalysen sinnvoll erscheinen lassen. Das Daten-kollektiv wurde ausgewählt, da an den z.T. bereits in den sech-ziger Jahren von der Deutschen Forschungsgemeinschaft errichte-ten Stationen zunächst methodische Arbeiten durchgeführt wurden, die 1972 in die Festlegung eines einheitlichen Meßverfahrens mündeten, das dann bis heute unverändert geblieben ist.

2. MESSVERFAHREN

Das NDIR-Meßverfahren, das für die Messungen eingesetzt wurde, ist im Detail von der WMO (1978) beschrieben und berück-sichtigt die Erfahrungen, die an den Baseline Stationen gewonnen wurden. Mit Rücksicht auf einen rationellen Meßaufwand sind jedoch eine Reihe von Vereinfachungen eingeführt worden, was zur Folge hat, daß die Qualität der Daten nicht das an Baseline-Stationen angestrebte Maß erreicht. Für eine Interpretation der Daten sind folgende Fakten von Bedeutung:

(1) Wegen der Überlappung der Infrarotspektren von CO_2 und H_2O werden Kalibriergase und Meßluft vor der Meßstrecke in ihrer

Feuchte stabilisiert.

(2) Regelmäßige Kalibrierungen innerhalb des Meßnetzes
(Intercomparison Program) und die Resultate der in ein bis zwei
Wochen Abstand durchgeführten Sekundärkalibrierungen lassen
erwarten, daß innerhalb des Netzes eine Genauigkeit von etwa
1ppm erreicht wird.

(3) Die Kalibrierstandards und Vergleichsgase sind
Mischungen von CO_2 in N_2 und an die Gase des SCRIPPS Institutes
für Ozeanographie angeschlossen (1959 Adjusted Index Scale).
Die in dieser Arbeit angegebenen Meßwerte des Netzes sind daher
in diesem Maßstab angegeben. Hinsichtlich des Vergleichs mit
anderen Meßstationen sind die Hinweise von BISCHOF (1977) und
PEARMAN (1977) zu beachten.

3. LAGE DER MESSSTATIONEN

Die Besonderheiten der Lage sind für die Auswertung von
Meßreihen zu berücksichtigen, bis auf Koordinaten und Höhe jedoch kaum in einfacher Weise darstellbar. Die im folgenden auf-
geführte quantitative Einteilung des Umfeldes (km-scale) stellt
daher auch nur einen Versuch in dieser Richtung dar, der eine
grobe Einschätzung möglicher biogener Einflußfaktoren ermög-
lichen soll.

Schauinsland (S), 47°55'N, 7°54'E ist im Südwesten der
Bundesrepublik Deutschland am Rande des Rheintales in 1205m NN
gelegen. Bedingt durch die Höhe, liegt die Station relativ
häufig über der Mischungsschicht. Das Umfeld besteht zum
größten Teil aus Wald (50%), Weideland (10%) und Wiesen (40%).

Brotjacklriegel (B),48°49'N, 13°13'E, ist ebenfalls eine
Bergstation, die im Südosten des Bundesgebietes liegt. Inmitten
des Bayrischen Waldes gelegen, reicht ihre Höhe von 1016m NN
ebenfalls aus, um Messungen über der Mischungsschicht durchzuführen. Das Umland besteht zu 84,5% aus Wald, 15% Wiesen und
0.5% bebauten Flächen.

Deuselbach (D), 49°46'N, 7°3'E, liegt inmitten einer
Hügellandschaft im Westen der Bundesrepublik in 480m NN. Die
Umgebung ist weitgehend von Landwirtschaft geprägt (42% Äcker,
45% Weideflächen). Hinzu kommen 10% Wald, 3% Wiesen und be-
bautes Land.

Langenbrügge (L), 52°48'N, 10°45'E, liegt im Osten der
Lüneburger Heide in 74m NN. Die Station wird deutlich durch
Ferntransporte von Luftverunreinigungen beeinflußt. Da sich die
Umgebungsbedingungen dieser Station änderten, werden die Daten
dieser Station in einer speziellen Arbeit untersucht werden
und sind daher hier ausgeklammert worden. Gegenwärtig sind
25% des Umlandes landwirtschaftlich, 66% forstwirtschaftlich
genutzt. 6% sind Heide und Gras.

Westerland (W), 54°55'N, 8°18'E, auf der Insel Sylt unweit
der Grenze zu Dänemark, ist die nördlichst gelegene Station des
Netzes. Die Lage der Station in den Dünen der Ostküste der
Nordsee ist für die Beobachtung von See kommender Luftmassen
günstig, da Beeinträchtigungen durch lokale Quellen bei geeigneten Transportbedingungen nicht auftreten.

4. DATENBASIS

In den Tabellen 1 bis 5 sind die arithmetischen Mittelwerte für die einzelnen Monate des Beobachtungszeitraums angegeben, die aus den kontinuierlichen Registrierungen der o.a.
Meßstationen errechnet wurden. In der letzten, mit YR markier-
ten Zeile ist der jeweilige Jahresmittelwert angegeben. Die
Werte dienten als Grundlage für die folgenden Analysen. Wie die
Arbeiten von VERMA und ROSENBERG (1976) und die von WOODWELL
et al. (1973) erwarten ließen, ist die saisonale Variation
außerordentlich hoch, wenn man sie mit den Amplituden der
Jahresgänge von Background-Stationen vergleicht. Eine ent-
sprechende Gegenüberstellung enthält die Übersicht von
MANNING (1981). Die Amplituden der Variation sind für die hier
analysierten Daten ähnlich den Werten, die WOODWELL et al.(1973)
in den industrialisierten Gebieten Nordamerikas feststellte, wo
ähnliche Bedingungen vorlagen. Sie schlossen aus ihren Messungen, daß es möglich sei, Trends im atmosphärischen CO_2-Gehalt
auch dieser Gebiete zu prüfen, wenn die Messungen ausreichend
intensiv sind. Für die in den Tabellen 1 bis 5 genannten Werte,
die auf mehr als 45 000 Meßstunden zurückgehen, wurde unterstellt,
daß sie diese Bedingung erfüllen. Wie man den in diesen
Tabellen angegebenen Jahresmittelwerten entnehmen kann, sind
außer den jahreszeitlichen Schwankungen noch andere Fluktuationen vorhanden, die einen möglichen säkularen Trend überdecken
und ein ungünstiges Signal-Rausch-Verhältnis verursachen.

5. RAUSCHUNTERDRÜCKUNG

Für die Unterdrückung des zeitlichen Rauschens wird daher
die Anwendung eines geeigneten Filterverfahrens notwendig sein.
Daneben gibt es Unterschiede in den zeitgleich festgestellten
Werten verschiedener Stationen, die sich nicht mit den bereits
erwähnten Meßunsicherheiten erklären lassen und diese bei
weitem übersteigen. Um derartige Einflüsse in den Analysen möglichst gering zu halten, bietet es sich an, mehrere Meßreihen
zusamenzufassen, um so eine Kompensation der zufälligen Einflüsse herbeizuführen.
Da eine großräumige Veränderung des CO_2-Gehaltes beschrieben werden soll, müßten die Stationen, die diese Veränderung
aufzeigen können, ähnliche saisonale Schwankungen aufweisen und
untereinander einen horizontalen Abstand haben, der gegenüber
der Ausdehnung der Störeinflüsse groß ist. Da die letzte Bedingung selbst bei einem Abstand der Stationen von mehreren hun-
dert Kilometern nur unvollkommen erfüllt ist, wurde ein orien-
tierender Vergleich von saisonalen Variationen durchgeführt.
Grundlage bildeten die Daten von Baseline Stationen, die von
NOAA/HANSON(1977) angegeben sind.
In Abbildung 1 sind den Meßergebnissen von Key Biscayne
und Barrow die Ergebnisse der deutschen Stationen Schauinsland,
Brotjacklriegel und Westerland gegenübergestellt. Bei den Daten
aus Barrow liegt das sommerliche Minimum im August wie bei den
drei deutschen Stationen, deren Jahresverlauf - abgesehen von
dem bei Barrow fehlenden Frühwintermaximum - sehr ähnlich den
Werten vom nordamerikanischen Kontinent ist. Sehr unterschied-
lich ist hingegen der Verlauf des CO_2-Gehaltes in Key Biscayne,

wo das Sommermaximum wie bei anderen Stationen in südlicheren
Breiten erst im September auftritt und die Amplitude der Variation viel geringer ist. Die Werte der Stationen Deuselbach und
Langenbrügge sind der Übersicht halber in Abbildung 1 nicht
aufgeführt. Man kann jedoch leicht verifizieren, daß diese Werte
erheblich außerhalb der Bandbreite liegen, die für die drei
oben genannten Stationen B,S und W gegeben ist.

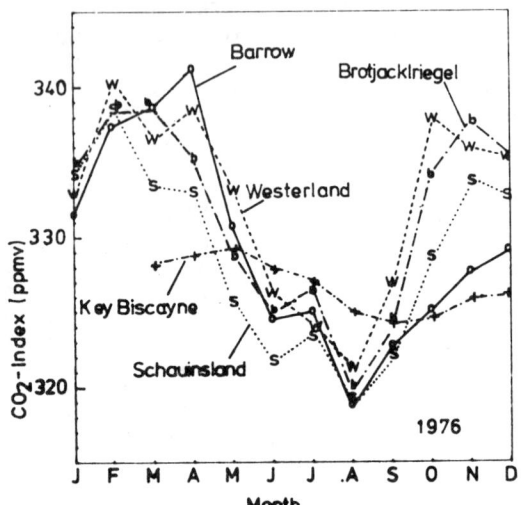

Figure 1: Monthly mean values (1976) for different
sites in the northern hemisphere.

Eine monatsweise Zusammenfassung der Daten dieser drei
Stationen durch die Bildung arithmetischer Mittelwerte wurde
daher als Ausgangsbasis für die weitere Analyse herangezogen.
Diese Daten werden im folgenden als BSW-Index bezeichnet.
Es ist offenkundig, daß unter den Bedingungen, die Abbildung 1 widerspiegelt, nicht zu erwarten ist, daß die saisonale
Variation durch die räumliche Mittelung beeinflußt wird. Wie
Abbildung 2 zeigt, ist die Zeitreihe des BSW-Index von erheblichen saisonalen Einflüssen gekennzeichnet, die durch eine
geeignete zeitliche Filterung unterdrückt werden müssen. Im
Prinzip kann dazu jedes Tiefpaßfilter geeigneter Eckfrequenz
und Charakteristik Verwendung finden. Für eine Übersichtsanalyse, wie sie hier vorgenommen werden soll, bietet sich eine
sehr einfache Methode an, die auch bei den ersten Analysen der
Mauna Loa - Daten bereits angewandt wurde. Wie PALES und KEELING
(1965) damals zeigten, ließen sich saisonale Variationen durch
die Bildung übergreifender Mittel - seinerzeit wurden jeweils
12 Monatmittelwerte zusammengefaßt - sehr gut eliminieren. Die
Methode hat zudem den Vorteil, daß sie nicht voraussetzt, daß
die saisonale Variation einen sinusförmigen Verlauf hat. Selbst
ausgefallenste Variationsmuster werden unterdrückt, vorausge-
setzt, sie wiederholen sich in dem Zeitrhythmus, der durch die
Mittelbildung vorgegeben wird.

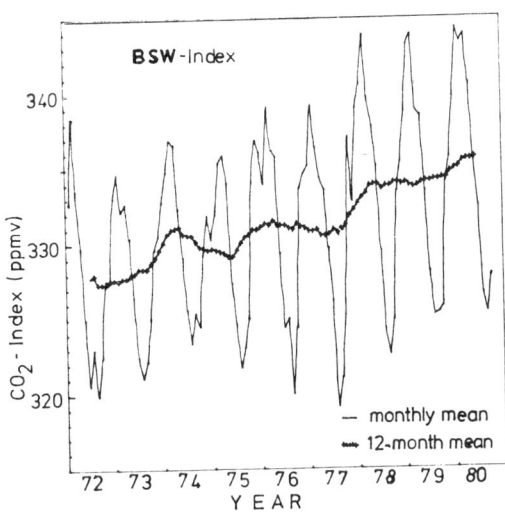

Figure 2: Monthly means and 12-month running means
of BSW-Index. Running means are plotted
versus the sixth month of the appropriate
12-month interval.

Eine Anwendung dieser Verfahrensweise führte zu den in
Abbildung 2 als markierte Kurve dargestellten Werten. Der
Kurvenverlauf ist der besseren Übersicht wegen so dargestellt,
daß Änderungen im Verlauf der gefilterten Kurve synchron zu
denen der Ausgangsfunktion verlaufen. Es ist auffallend, daß
auch die solchermaßen geglättete Kurve noch einen periodischen
Term enthält. Mit einer Amplitude von etwa 3ppm und einer
Wellenlänge von ca. 2 Jahren überdecken diese Fluktuationen
einen langzeitigen Anstieg der CO_2-Konzentration.

Bei dieser Erscheinung muß es sich nicht um echte periodische Variationen handeln, da der analysierte Zeitraum sehr
kurz im Vergleich zur Wellenlänge ist. ANGELL(1977) hat für
die außertropischen nördlichen Breiten Temperaturanomalien
beschrieben, die in ihrer Phase synchron zu den beobachteten
CO_2-Fluktuationen sind. Es dürfte sich daher um einen ähnlichen
Effekt handeln, wie ihn NEWELL und WEARE (1977) für den Pazifik
beschrieben haben. Auf eine nähere Untersuchung wurde hier
verzichtet, da dies eine sehr umfangreiche Analyse erfordert
hätte. Da eine zweijährige Periodizität für die Troposphäre
sehr außergewöhnlich wäre, bedarf dieser Effekt jedoch noch
eingehender Untersuchung.

Für die vorliegende Analyse wurde diese Fluktuation als
eine periodische Störung betrachtet, die mit einer einfachen
Änderung des Mittelungsverfahrens eliminiert werden kann.
Durch Verlängerung des Mittelungszeitraums auf 24 Monate ließen
sich die entsprechenden Wellen unterdrücken und aus der Trendlinie entfernen.

6. TRENDLINIEN

Die auf diese Weise erzeugte Trendlinie der gleitenden
24-Monatsmittelwerte ist in Abbildung 3 dargestellt und in

ähnlicher Weise synchronisiert worden, wie oben bereits erwähnt. Wie man der Abbildung entnehmen kann, verläuft die BSW-Indexlinie über den gesamten Zeitraum nahezu gleichmäßig und wurde daher durch eine lineare Funktion approximiert. Nach dieser in Abbildung 3 gestrichelt eingezeichneten Geraden, die nach der Methode der kleinsten quadrate ermittelt wurde, wurden in dem 1975 abgeschlossenen Zweijahreszeitraum im Mittel 330 ppm erreicht. Die Anstiegsrate liegt bei 0.97 ppm/yr.Dieser Wert für die Anstiegsrate weicht kaum von den Werten ab, die PETERSON et al.(1977) für die Änderungen des CO_2-Gehaltes auf Mauna Loa angegeben haben. Die Konzentrationsangaben dieser Autoren sind um etwa 3 ppm geringer als die hier angegebenen. Es wäre verfrüht, diesen Unterschied in der Höhe der angegebenen Werte zu interpretieren, da die Stationen nicht mit CO_2-in-Luft Kalibriergasen interkalibriert werden.

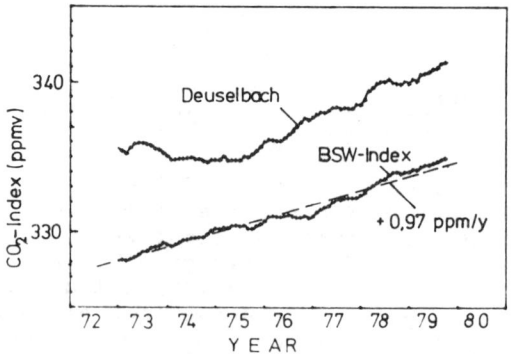

Figure 3: 24-month running mean of the atmospheric CO_2-concentration for Deuselbach and BSW-Index. Means are plotted versus the 12th month of the appropriate interval. The straight line (linear least square fitted) indicates a rate of increase of o.97 ppm/yr for the period studied.

Für ein besseres Verständnis der bei Regionalstationen auftretenden Besonderheiten ist in Abbildung 3 eine Trendlinie für die Station Deuselbach angegeben, die ebenfalls aus der Bildung von gleitenden 24-Monatsmittelwerten resultiert. Der zeitliche Verlauf dieser Trendlinie unterscheidet sich -abgesehen von dem deutlichen Konzentrationsunterschied- auffallend von dem des BSW- Index. Auf eine stagnierende bis rückläufige Entwicklung der CO_2- Konzentration in den frühen 70er Jahren folgte etwa ab 1976 ein Anstieg,der steiler ausfiel als der des BSW- Index im gleichen Zeitabschnitt. Hierin drückt sich aus, daß es lokal durchaus Abweichungen von der großräumigen Entwicklung gibt, die durch den Einfluß von Quellen und Senken geprägt werden und über mehrere Jahre andauern können. Erst in einem längeren Zeitscale setzt sich dann diese extrem niederfrequente Fluktuation nicht mehr gegen den säkularen Trend durch. Da gerade diese Abweichungen der lokalen und regionalen Entwicklung von der großräumig beobachteten neue Informationen über den Einfluß von Quellen und Senken erwarten lassen, ist hier ein weiterer Ansatzpunkt für eingehendere Untersuchungen gegeben.

7. SCHLUSSFOLGERUNGEN

Eine mit einfachen Verfahren durchgeführte Übersichtsanalyse der an den Stationen des Umweltbundesamtes von 1972 bis 1980 ermittelten Meßwerte über den atmosphärischen CO_2-Gehalt hat gezeigt, daß es

(1) möglich ist, an Hand dieser Daten einen langzeitigen Anstieg der CO_2- Konzentration in Westeuropa aufzuzeigen, dessen Anstiegsrate im Untersuchungszeitraum bei 0.97 ppm/yr liegt,

(2) zweckmäßig ist, für diese Belange eine Zusammenfassung verschiedener Meßreihen, z.B. zu einer Indexreihe, durchzuführen,

(3) notwendig ist, die Daten eingehend auf eine nicht saisonale Fluktuation zu untersuchen, die im Meßzeitraum einen Zweijahresrhythmus zeigte und im Zusammenhang mit Temperaturanomalien in den nördlichen gemäßigten Breiten gesehen werden muß,

(4) notwendig ist, die Abweichungen des zeitlichen Verlaufs der CO_2-Konzentration näher zu untersuchen, die die zeitliche Entwicklung im globalen und regionalen Scale unterscheiden,

(5) sinnvoll ist, wenn internationale Organisationen ein Intercomparison Programm mit CO_2-in-Luft Kalibriergasen durchführen würden, damit auch die Unterschiede in den Konzentrationsangaben verschiedener Autoren interpretierbar werden.

8. SCHLUSSBEMERKUNG

Die Autoren danken an dieser Stelle all den Mitarbeitern, die an der technischen Durchführung dieser Langzeitmessungen seit vielen Jahren mitarbeiten und ohne deren Fleiß und Sorgfalt diese Auswertung nicht möglich gewesen wäre.

Table 1 : Monthly mean CO_2-concentration at Schauinsland, F.R.G. (1959 adjusted index scale -300 ppm)

	1972	1973	1974	1975	1976	1977	1978	1979	1980
JAN	30,3	29,7	29,0	28,3	34,4	35,5	38,4	39,9	40,5
FEB	34,0	30,4	35,6	30,3	38,7	32,9	40,1	45,5	39,4
MAR	32,1	30,6	34,9	34,9	33,6	31,3	36,5	36,0	42,7
APR	30,1	30,4	31,0	33,8	33,2	33,8	39,0	36,1	39,5
MAY	24,8	24,1	26,4	28,1	25,9	28,9	34,9	30,0	34,2
JUN	21,0	20,4	22,7	24,0	21,8	25,6	26,5	26,4	30,0
JUL	23,4	18,6	20,3	16,9	23,7	18,6	21,5	20,5	26,7
AUG	17,7	19,9	23,3	21,4	19,0	21,1	20,5	22,4	25,4
SEP	17,4	21,5	22,4	23,6	22,2	23,0	22,6	24,8	26,8
OCT	24,5	26,3	27,9	28,7	28,9	29,3	31,1	32,9	-
NOV	24,1	28,3	28,9	35,5	33,9	32,6	33,3	35,2	-
DEC	26,6	30,4	27,8	32,0	32,7	34,3	36,9	36,9	-
YR	25,5	25,8	27,4	28,1	29,0	28,9	31,7	32,1	33,9[*]

Table 2 : Monthly mean CO_2-concentration at Brotjacklriegel, F.R.G. (1959 adjusted index scale -300 ppm)

	1972	1973	1974	1975	1976	1977	1978	1979	1980
JAN	35,1	34,8	35,1	30,8	34,6	38,6	40,2	42,6	42,2
FEB	34,8	35,8	37,8	32,7	38,4	35,7	43,6	41,0	42,4
MAR	31,7	35,7	35,7	35,0	38,6	35,9	41,2	38,5	43,8
APR	29,5	31,6	30,3	32,1	35,5	34,4	39,2	38,3	41,3
MAY	25,2	27,7	28,3	24,5	29,1	29,0	35,0	32,5	35,5
JUN	20,2	22,5	25,4	22,7	25,1	26,8	31,1	28,5	31,5
JUL	21,4	22,1	23,5	21,9	26,8	19,6	20,8	25,9	26,5
AUG	18,0	20,0	26,0	24,1	19,9	21,0	21,9	24,2	25,3
SEP	19,9	24,7	23,3	27,0	24,1	27,4	25,5	25,2	27,5
OCT	27,5	29,5	30,4	36,3	34,2	34,4	33,2	32,2	-
NOV	30,6	32,0	33,0	45,9	37,8	36,3	34,2	39,3	-
DEC	29,7	34,0	32,3	40,8	35,6	37,8	36,6	37,9	-
YR	26,9	29,0	30,0	31,1	31,6	31,4	33,5	33,8	34,6*

* mean of given values

Table 3 : Monthly mean CO_2-concentration at Deuselbach, F.R.G.
(1959 Adjusted index scale -300 ppm)

	1972	1973	1974	1975	1976	1977	1978	1979	1980
JAN	42,4	43,8	38,3	37,5	37,0	41,2	42,1	49,4	48,7
FEB	44,9	38,8	40,6	39,3	46,7	38,3	50,4	52,2	47,1
MAR	37,3	39,8	47,2	40,6	40,6	39,3	39,6	44,5	50,2
APR	32,0	36,7	38,2	37,0	38,6	37,5	42,2	44,3	43,7
MAY	31,4	32,2	32,6	33,3	32,6	37,5	37,1	40,5	39,8
JUN	30,7	28,6	30,3	30,0	30,0	35,6	31,3	35,3	36,2
JUL	30,3	28,7	28,4	35,3	34,4	30,5	31,2	34,8	32,6
AUG	29,6	28,4	27,6	29,1	26,2	36,4	28,2	30,7	35,6
SEP	30,8	30,3	26,0	26,9	31,3	33,5	30,3	32,3	32,7
OCT	36,4	38,0	34,5	35,9	39,9	42,7	40,0	41,6	-
NOV	42,5	37,7	37,5	42,0	43,5	39,0	49,4	43,4	-
DEC	41,1	41,7	36,6	41,7	45,0	46,4	44,5	43,1	-
YR	35,8	35,3	34,8	34,9	37,2	38,2	38,7	40,9	40,7*

Table 4 : Monthly mean CO_2-concentration at Langenbrügge,
F.R.G. (1959 Adjusted index scale -300 ppm)

	1972	1973	1974	1975	1976	1977	1978	1979	1980
JAN	44,9	49,1	43,3	34,9	36,7	47,9	48,5	53,2	65,3
FEB	46,8	38,5	41,7	40,9	51,5	45,5	52,2	52,0	53,7
MAR	40,7	41,6	40,7	36,9	43,0	40,8	41,3	44,7	54,2
APR	33,6	37,9	35,1	39,1	37,9	36,2	43,2	44,7	47,0
MAY	35,7	39,2	36,4	35,7	37,1	37,6	44,2	46,1	45,9
JUN	39,0	37,6	37,4	32,3	39,0	38,6	45,1	41,0	52,2
JUL	42,3	38,1	35,3	35,2	34,9	37,1	37,8	38,5	42,9
AUG	35,2	39,2	n.m.	38,8	33,8	41,7	40,4	42,8	43,7
SEP	44,7	39,0	36,0	34,7	39,6	40,0	33,5	44,3	51,2
OCT	45,5	44,3	35,6	42,1	49,3	55,3	48,5	45,5	-
NOV	42,9	37,0	42,9	47,2	43,3	38,1	51,6	46,6	-
DEC	47,8	43,3	34,1	41,3	44,2	47,5	48,3	44,9	-
YR	41,6	40,4	30,0*	38,2	40,8	42,2	44,5	45,3	50,7

* mean of given values

Table 5 : Monthly mean CO_2-concentration at Westerland, F.R.G.
(1959 Adjusted index scale -300 ppm)

	1972	1973	1974	1975	1976	1977	1978	1979	1980
JAN	n.m.	39,7	40,3	36,5	32,9	43,4	43,2	47,0	50,1
FEB	46,4	30,3	37,2	43,2	40,3	40,5	48,0	45,2	48,5
MAR	37,0	32,8	39,1	37,7	36,5	36,1	41,4	42,3	45,3
APR	n.m.	29,2	35,8	36,3	38,6	32,5	35,3	41,0	40,9
MAY	n.m.	27,7	32,6	29,6	33,2	31,1	35,8	38,1	37,8
JUN	n.m.	24,5	28,7	27,5	26,2	n.m.	31,0	29,3	36,1
JUL	24,3	22,5	26,4	26,5	24,4	n.m.	30,6	29,1	n.m.
AUG	22,2	21,9	26,8	23,9	21,2	n.m.	25,4	29,6	n.m.
SEP	n.m.	24,0	27,8	24,5	27,0	n.m.	26,5	27,4	n.m.
OCT	n.m.	30,5	29,4	37,4	38,0	47,5	36,2	37,2	–
NOV	28,1	31,2	36,3	38,8	35,9	31,4	39,8	39,6	–
DEC	43,0	34,0	30,0	30,5	35,2	46,5	45,2	43,0	–
YR	33,7*	28,9	32,5	32,7	32,3	38,1*	36,5	37,4	43,4*

* mean of given values

9. LITERATUR

C.D. KEELING (1960): The concentration and isotopic abundances
of carbon dioxide in the atmosphere, Tellus, 12, 200-203
J.C.PALES u. C.D.KEELING (1965): The concentration of atmos-
pheric carbon dioxide in Hawaii, J.Geophys. Res., 70,6053-6076
C.W.BROWN u. C.D.KEELING (1965): The concentration of atmos-
pheric carbon dioxide in Antarctica,J.Geophys.Res.,70,6077-6085
J.S.SAWYER (1971): Possible effects of human activities on the
world climate, Weather, 26, 251 -262
G.M. WOODWELL et al. (1973); Atmospheric CO_2 at Brookhaven,
Long Island, New York;Patterns of variation up to 125 m.
J. Geophys. Res., 78, 932 - 940
S.B.VERMA u. N.J.ROSENBERG (1976): Carbon dioxide concentration
and flux in a large agricultural Region of the Great Plains
of Noth America, J. Geophys. Res.,81,(C), 399 - 405
WMO (1978): International operations handbook for measurements
of background atmospheric pollution, WMO Publ. No. 491, Genf
W. BISCHOF (1977): Comparability of CO_2-measurements,
Tellus, 29, 435 - 444
G.I. PEARMAN (1977): Further studies of the comparability of
baseline atmospheric carban dioxide measurement, Tellus,
29, 171 - 181
M.R.MANNING (1981): An assessment of BAPMon Data currently
available on the concentration of CO_2 in the atmosphere (unpubl)
NOAA/HANSON (1977): Geophysical monitoring for climatic
change, No.5, Summary report 1976, Boulder, Colorado
J.K. ANGELL (1977): Recent climatic trends;in : Proceedings of
the meeting on education and training in meteorological aspects
of atmospheric pollution and related environmental problems,
WMO Publ. No. 493, Genf , 3 - 20

R.E. NEVELL u. B.C. WEARE (1977): A relationship between
atmospheric carbon dioxide and Pacific sea surface temperature
Geophys. Res. Letters, 4, 1-2
J.T.PETERSON et al.(1977): NOAA carbon dioxide measurements
at Mauna Loa Observatory, 1974 - 1976, Geophys. Res. Letters,
4, 354 - 356

MEASUREMENT OF F^-, Cl^-, NO_3^- AND $SO_4^=$-IONS IN RAINWATER AND

PARTICULATE MATTER BY AID OF IONIC-CHROMATOGRAPHY

MÜLLER, J., H. REUVER
Umweltbundesamt, Pilotstation - Frankfurt
D. JOST
Umweltbundesamt, Berlin

Summary

By aid of ionic chromatography fluoride, chloride, nitrate and sulfate were measured in particulate matter and rainwater of ambient air. In particulate matter, the urban samples compared to rural ones had about a factor 2-4 higher concentrations. In rainwater the concentrations varied only up to 20%. Furtheron, particulate matter was sampled with an 8-stage Andersen-impactor and each stage analyzed for the four ions and the pH-value. It was found, that the pH varies in the order of 1-2 scales (pH 4-5,5) between the fine and coarse particle fractions. The smaller the particles the more acid they are. Under consideration of the normalities, between the pH and the mass fractions of $SO_4^=$, NO_3^- and Cl^- a good correlation was found. These ions constitute more than 90% of the pH-value in particulate matter. $SO_4^=$ which is mostly bound in fine particles contributes about 60% and NO_3^- about 30% to the pH-value.
The measured data bring new light into the atmospheric cycles of sulfur and nitrogen. The mass size distributions can be used for determination of the residence times of the particle-bound ions. Under Middle European conditions, a mean residence of 2,2 days for SO_2-sulfur and 1,6 days for NO_x-nitrogen was estimated.

1. INTRODUCTION

The ions F^-, Cl^-, NO_3^- and $SO_4^=$ play an important role in air chemistry as they constitute about one quart of the mass of particulate matter in ambient air. The emitted gases NO_x and SO_2 are to a certain quantity transformed into NO_3^- and $SO_4^=$-ions which can be detected in rainwater and particulate matter. In order to get a profound knowledge about the atmospheric cycles of these substances a reliable measuring method is desirable. Formerly, these ions were measured by different not very efficient methods. Now, ionic chromatography represents a powerful tool for detecting these ions.

2. EXPERIMENTAL

By ionic chromatography the ions can directly be analyzed in aqueous solutions with high precision and sensivity (1). The ions are simultaneously measured and registered by the chromatograph. Rainwater can be analyzed without pre-preparation of the samples. Even in samples collected in clean air the ions can be detected with sufficient high sensitivity. The detection limits are in the order of 0,01 μg/ml for F^-, 0,1 for Cl^-, 0,1 for NO_3^- and 0,1 for $SO_4^=$.

Particulate matter sampled on glassfiber filters is pretreated with an ultrasonic source in order to desintegrate the filters and to suspend the particles in aqueous solution. The samples are centrifugated and the clear solutions can directly be analyzed.

The mass size distributions of particulate matter were measured by use of an 8-stage Andersen-impactor. The particle fractions were captured on glass discs and rinsed off with destilled water. The solutions then were analyzed in the ionic-chromatograph. The pH-values were measured with an electrode. In addition, the water-soluble portions of particulate matter on each stage were gravimetrically determined.

The samples of rainwater and particulate matter were collected in urban and rural air.

3. RESULTS AND DISCUSSION

In Tab. 1 the mean values of samples taken in the course of one year are represented. The concentrations of $SO_4^=$ and F^- in particulate matter of urban air are twice as high as in rural air. In the case of NO_3^- the difference is about 2,5 and for Cl^- even a factor 5. In the case of chloride, therefore, in addition to its air chemical behaviour large urban sources can be assumed.

The differences in rainwater concentrations between urban and rural air are only up to 20%. This can be explained by the fact that rain-out is the dominating process for the accumulation of substances in rain-water (2). Wash-out processes play only an additional, less important role.

In Fig. 1 the measured mass size distributions of Cl^-, NO_3^-, F^- and $SO_4^=$ in particulate matter of urban air are drawn. $SO_4^=$ is almost entirely bound in the fine particle mode whereas Cl^- has also a certain amount of mass in the coarse particle mode. The sum of the four ions represents about one quart of the mass of total particulate matter.

In Fig. 2 the sum of the four ions under consideration of the normalities is compared to the measured pH-value in each size fraction. A parallel behaviour can be noticed. With increasing mass of the ions the pH goes down. The smaller the airborne particles the more acid they become.

The fine particles (diameter $<$ 2µm) are removed from the atmosphere almost entirely by wet removal processes whereas coarse particles ($<$2µm) are mainly taken off by sedimentation. The acid character of the fine particles is transformed into the rainwater. The low pH of the particles induces a good solubility. Measurements showed that about 3/4 of the fine particle mass is water-soluble, whereas in the case of coarse particles only half of the mass is soluble. Substances which are mainly incorporated in fine particles, therefore, have large soluble portions.

The mass size distribution of a substance in particulate matter can be taken as an information to determine its atmospheric residence time. The residence time of fine particles is mainly coupled to the atmospheric water cycle, and the sedimentation of coarse particles can be calculated by Stokes' law. These two processes added determine the particle removal in the range of 0,1 to 100 µm diameter (3).

To each particle size fraction a mean residence time τ_i can be attached. By summation over the mass fractions X_i on each stage of the impactor (Fig. 1), the residence time of the considered substance can be calculated:

$$\tau = \sum_i x_i \tau_i, \qquad \sum x_i = 1$$

By use of the Middle European wet residence time of about 3,5 days we get for the particle-bound ions the following residence times:

	Cl^-	NO_3^-	F^-	$SO_4^=$	
τ:	2,1	2,4	2,6	2,9	days

$SO_4^=$ has the relatively longest residence time because its mass portion in the fine particle mode is the highest. The respective acid H_2SO_4 has the lowest vapour pressure compared to HNO_3, HF and HCl. Molecular H_2SO_4 therefore, has only a very short life time and condensates quicker into the particle phase (gas to particle conversion) than the other acids. On account of this fact, $SO_4^=$ is enriched in the size range of airborne particles which have the longest residence time. Sulfur, to a certain extent, pushes the more volatile acids away from small particles.

By aid of the measured mean gas phase concentrations of SO_2 and NO_2 (Tab. 1) the residence times of SO_2-sulfur and NO_x-nitrogen can be estimated. In the case of sulfur, the emitted SO_2 in the atmosphere builds up a SO_2-reservoir and a particle-bound $SO_4^=$-reservoir. An amount C of the emitted quantity Q is converted into $SO_4^=$, coupled by the factor α : $C = \alpha \cdot Q$. In stationary equilibrium between the emission and the two reservoirs can be written:

$$(1) \quad \tau_{SO_2-S} = \frac{m_{SO_2-S}}{Q} \, , \quad \tau_{SO_4-S} = \frac{m_{SO_4^=-S}}{\alpha \cdot Q}$$

$m_{SO_2-S}, m_{SO_4^=-S}$: mass of the sulfur reservoirs

$\tau_{SO_2-S}, \tau_{SO_4^=-S}$: residence times of SO_2 and $SO_4^=$

By use of 1/4 for the conversion factor (α), $\tau_{SO_4^=}$ determined by the measured mass size distributions and the rural values for the masses of SO_2 and $SO_4^=$ (Tab. 1), the residence time of SO_2 is calculated as 1,5 days. The residence time τ_S of total atmospheric sulfur calculated by the equation

$$(2) \quad \tau_S = \tau_{SO_2-S} + \alpha \, \tau_{SO_4^=-S}$$

becomes 2,2 days.
In the case of NO_x-nitrogen the residence time τ_{NO_x-N} can be determined under equivalent considerations.
By looking on Tab. 1 we notice that sulfur compared to nitrogen is enriched in its particulate concentration. Relatively more SO_2 than NO_2 is converted into the particle-phase whereas in rainwater equivalent amounts can be detected. This means that relatively more nitrogen than sulfur is already removed from the atmosphere in the gas-phase. The high vapour pressure of HNO_3 ($\approx$ 60 Torr) induces that only a fraction is converted into particles. Measurements showed that about 1/3 of atmospheric nitrate remains in the gas-phase. A certain amount is deposited as HNO_3-gas. The higher gas phase portions cause a relatively quicker removal of nitrogen. Nitrogen is by a factor of 1,75 less enriched in particles than sulfur. In the case of nitrogen, therefore, for the conversion factor α the value 1/7 is used. The equations (1) and (2) applied furnish for the residence times of NO_2 1,3 days and for total NO_x-nitrogen 1,6 days.
The shorter residence time of NO_x-nitrogen compared to sulfur limits its importance as a long-range problem.
The cycles of chlorine and fluorine can not yet be sufficiently handled because only sporadic values of the gas-phase concentrations are available.

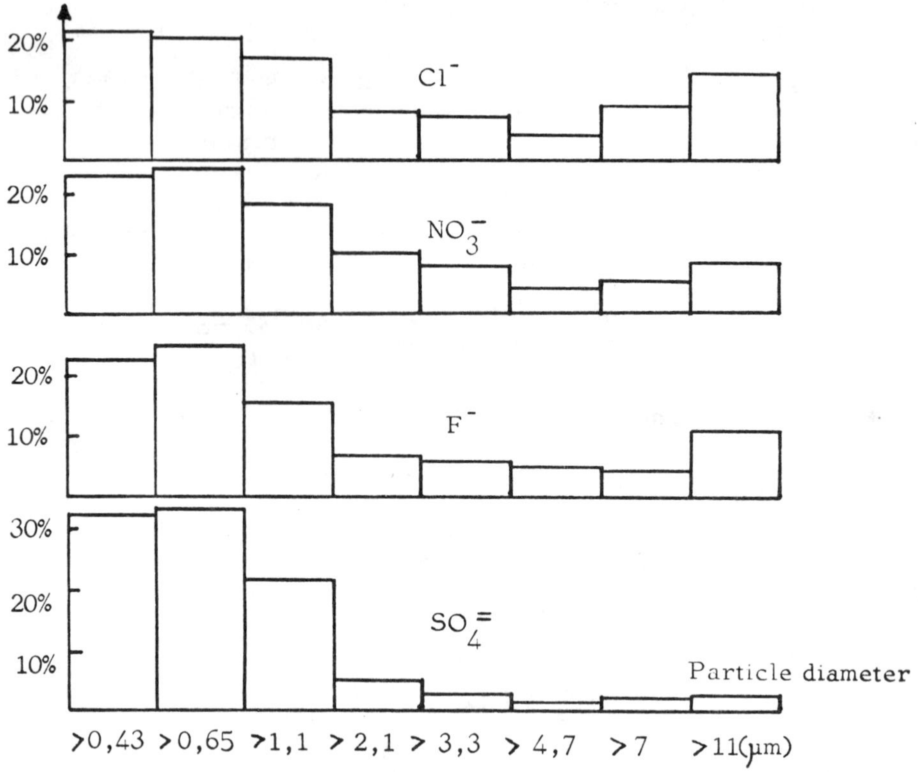

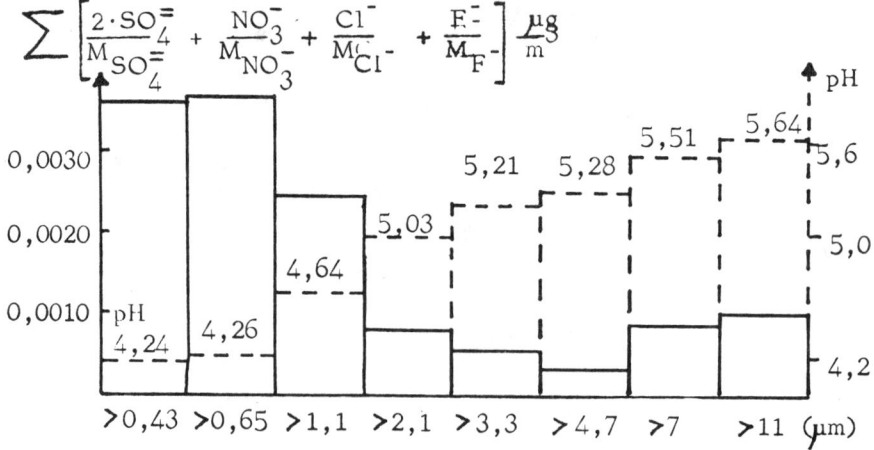

Fig. 1 - Mass size distributions of Cl^-, F^-, NO_3^- and $SO_4^=$ in particulate matter versus particle diameter

$$\sum \left[\frac{2 \cdot SO_4^=}{M_{SO_4^=}} + \frac{NO_3^-}{M_{NO_3^-}} + \frac{Cl^-}{M_{Cl^-}} + \frac{F^-}{M_{F^-}} \right] \frac{\mu g}{m}$$

Fig. 2 - Concentrations of $SO_4^=$, NO_3^-, Cl^- and F^- devided by the molecular weight and pH in particulate matter versus particle diameter

Tab. 1: Concentrations of F^-, Cl^-, NO_3^- and $SO_4^=$ in urban and rural air

Gas concentrations ($\mu g/mg^3$):

	NO_2	NO_2-N	SO_2	SO_2-S
urban	50	15,3	80	40
rural	11,8	3,6	14	7

Concentrations in particulate matter ($\mu g/m^3$):

	F^-	Cl^-	NO_3^-	NO_3^--N	$SO_4^=$	$SO_4^=$-S
urban	0,057	1,28	8,18	2,5	18,07	6,02
rural	0,030	0,27	3,9	0,95	10,4	3,5

Concentrations in wet precipitation ($\mu g/ml$):

	F^-	Cl^-	NO_3^-	NO_3^--N	$SO_4^=$	$SO_4^=$-S
urban	0,15	1,46	4,80	1,08	6,90	2,30
rural	0,13	1,27	4,00	0,90	5,72	1,90

Ref.:
1. Jansen, K. H.: "Chromatographie auch in der anorganischen Analytik durch Ionenchromatographie", GIT-Fachzeitschrift für das Laboratorium, Heft 12, S. 1062-1071 (1979), c/o GIT-Verlag E. Giebler, Darmstadt

2. Müller, J.: "Wet removal of heavy metals from the atmosphere", c/o Proc. Int. Conf. on Heavy Metals, Toronto/Can. pp 987-999 (1975)

3. Jaenicke, R.: "Über die Dynamik atmosphärischer Aitkenteilchen", Ber. Bunsenges. Phys. Chem. 82, 1198 - 1202 (1978)

4. Husar, Rudolf, report: "Atmospheric sulfur and nitrogen budget" c/o Max Planck Inst. für Chemie, Mainz (Mai 1981)

TRANSPORT AND MODELLING   –   FIELD EXPERIMENTS

Chairman: A. J. ELSHOUT

MEASUREMENTS OF NO-OXIDATION IN POWER PLANT PLUMES BY CORRELATION SPECTROSCOPY

S.BEILKE and H.MARKUSCH

Umweltbundesamt-Pilotstation Frankfurt

and

D.JOST

Umweltbundesamt Berlin

Abstract

The Barringer correlation spectrometer(COSPEC) is a remote sensing spectrometer which can measure overhead burdens of SO_2 and NO_2 gas.The use of COSPEC in a moving vehicle permits overhead measurements along the traversing path of the moving laboratory.This remote sensing technique is widely used to measure atmospheric plume dispersion and emission strengths of area,line- and point sources.

Apparently unknown is another application of the COSPEC system:it can be used to determine chemical conversion rates.In this paper a method is introduced which permits a quatitative determination of the oxidation of NO to NO_2 in plumes of point or line sources such as a coal-fired power plant using SO_2 as a quasi-inert tracer. Our first measurements were carried out under different atmospheric stability conditions and have widely confirmed the large variation of NO-oxidation rates reported in the literature.

On the basis of an extrapolation of our results the conclusion can be drawn that for a slightly unstable, neutral and stable atmosphere,50% NO is oxidized to NO_2 at distances greater than 10 km from the power plant investigated.

1. INTRODUCTION

One problem of current interest in the Federal Republic of Germany is the oxidation of NO to NO_2 in plumes of point or line sources such as a coal-fired power plant.

This oxidation is a complicated process depending on a series of chemical and physical parameters.Nitrogen oxide (NO) is oxidized by three reactions:

$$2\ NO\ +\ O_2\ \longrightarrow\ 2\ NO_2 \qquad (1)$$

$$NO\ +\ O_3\ \longrightarrow\ NO_2\ +\ O_2 \qquad (2)$$

$$NO + RO_2 \longrightarrow NO_2 + RO \qquad (3)$$

Reaction(1) is only of importance if the NO-concentrations are sufficiently high.Such conditions can exist within the stacks(a small oxidation is possible due to 3-6% residual oxygen)and in the plume close to the stacks as shown by model calculations of Elshout and Steenkist(1974),Varey et al.(1979) and by Cocks and Fletcher(1979).
According to the model calculations of Elshout and Steenkist (1974) and of Varey,Sutton and Marsh(1979),oxidation of NO with oxygen is the dominant mechanism for plume travel times up to ca. 5 minutes for most situations occuring in the atmosphere.This mechanism contributes up to ca. 10% to total NO-oxidation covering the range of possible NO-concentrations and most conditions for plume dispersion.

For many situations occuring in the lower atmosphere,oxidation of NO by O_3(reaction 2) is by far the most effective oxidation mechanism(Hegg et al.,1976;Ogren et al.,1976;Davis et al.,1974;White,1977;Elshout et al.,1980).
Almost all measurements in plumes have shown that ozone concentrations within the plumes were depressed below their levels in the background air at least in the near-field of the plant suggesting that the NO-conversion is controlled by the rate at which a plume mixes with the ambient air rather than by chemical reactions within the plume.As will be shown in chapter 3,this result is supported by our own measurements.

The oxidation of NO by peroxy radicals(reaction 3) is of minor importance.It only plays a role in a polluted atmosphere containing high concentrations of reactive hydrocarbons in the presence of high UV-radiation.

Measurements in different power plant plumes have shown a wide range of NO-oxidation rates.
A large variation of NO-oxidation rates was also predicted by the model of Varey et al.(1979) taking into account different meteorological conditions and NO-emission levels.

In table 1 an attempt has been made to integrate some of the experimental investigations reported in the literature into a unified picture.In this table integrated 50% NO-oxidation times and/or distances(integrated over the plume cross section) are given along with atmospheric stability data and ozone background concentrations as far as it was possible.The integrated 50% values were either given by the authors cited in column 1 or were extrapolated by the authors of this paper. As can be seen in this table,50% NO-oxidation was found to fall between 5-15 minutes to 2-4 hours plume travel time corresponding to a few kilometers up to ca.100 km.
As measured by Elshout(1980) under unstable atmospheric conditions(Pasquill stability classes B - C,rapid mixing), NO-oxidation increases with increasing ozon concentrations in the ambient air.On the other hand,under more stable conditions(Pasquill stability classes D and E),NO-oxidation is not very dependent on ozone background concentrations which is in qualitative agreement with the model predictions of Varey et al.(1979).

Authors	Background Ozone	Atmospheric Stability	50% NO-oxidation integrated over plume cross sect.	R e m a r k s
Davis,D.D. Smith,G. Klauber,G. (1974)	50-80 ppb	Widely ranging mixing conditions	12 - 60 min	Morgantown plant(1000 MW, 75% gas-fired,25%coal-fired) Washington,DC,USA 11 aircraft flights between Oct.1973-Aug.1974 1.6 - 40 km from the plant
			12 min	1 night time flight
Hegg,D.A. Hobbs,P.V. Radke,L.F. (1976)		Widely ranging mixing conditions	ca. 5 min- 2 hrs	Centralia plant(1200 MW, coal-fired,Washingt.,USA) Four Corners plant(3000 MW, coal-fired,NM,USA) Hobbs plant(300 MW,gas,NM) Longview plant(900MW,gas,Tex 30 aircraft flights 0.8 - 70 km from the plant
Ogren,J.A. Blumenthal Vanderpol (1977)	50-70 ppb	Widely ranging mix.conditions	13 min-3 hrs[1]	Four Corners plant(3000 MW, coal-fired,NM,USA) San Juan plant(coal-fired, New Mexico,USA) Aircraft measurements during July 1976, 0 - 85km (1): Values are determined from Fig.4-7 in Ogren et.al.
Melo,O.T. Lusis,M.A. Stevens,R. (1978)	62-114 ppb	Pasquill class[1] between D and E (very stable; plume trapped in a stable layer)	30min-4 hrs[2] corresponding to 20 km - 100 km	Nanticoke plant(Lake Erie, Canada) 8 helicopter flights Nov.75 NO-oxidation does not depend on O3-background concentr. (1)Pasquill classes determ. by Elshout(1980) (2)50%-NO oxidation extrapolated from table 1 in Melo et al.(1978)
Elshout,A. (1980)	20 ppb	Pasquill class B and C (unstable,very rapid mixing)	ca. 2 hrs[1]	Massvlakte plant(two 540 MW units,gas-fired,Netherlands) 7 aircraft measurements between 0.2-9 km NO-oxidation increases with increasing O3-concentration
	40 ppb	i d e m	ca. 30 min[1]	
	60 ppb	i d e m	ca. 8 min[1]	
	25 - 50 ppb	Pasquill class D (neutral,mixing not very effect.)	ca. 25 min[1]	i d e m 6 flights NO-oxidation does not depend on O3 background concentr. (1) 50% NO-oxidation extrapolated from fig.2 in publ.

TAB. 1 : NO-oxidation in power plant plumes. 50% NO-oxidation as a function of plume travel time or distance from the plant integrated over the plume cross section.The values are compiled on the basis of literature release data.

2. MEASUREMENTS

During the period August 1980 to August 1981 first remote
sensing measurements were conducted at the Staudinger power
plant which is located along the Main river about 20 km east
of the city of Frankfurt/Main.It is a coal-and oil fired
plant with four units feeding into four stacks of different
hights between 195 m(units 1-3) and 250 m(unit 4).The power
plant has an electrical output of 1500 MW at full load but
was operating at less than 900 MW during our measurements
(only coal-fired units 1 to 3 were in operation).

First measurements were carried out during August 1980
using a Barringer remote sensing spectrometer(COSPEC II)
which can measure overhead burdens of SO_2 and NO_2-gas simul-
taneously.Since April 1981 measurements have been performed
using two sophisticated COSPEC-instruments:a COSPEC V for
SO_2 and a COSPEC IV b for NO_2.The use of COSPEC-instruments
in a moving vehicle permits overhead measurements along the
traversing path.Readers who are interested in a detailed
description of the COSPEC-system are ref ered to publications
of Millán and Hoff(1977) and Millán et al.(1976).

The determination of NO-oxidation rates is based on a combination of a tracer technique with mass balance considerations.
Measurements for studying integrated NO-oxidation were performed in the near-field of the plant along traverses at
different distances from the stacks between ca. 30 meters
and 15 km.The duration of each traverse was between 20 seconds
and ca. 3 minutes depending on plume width and traffic conditions.
Sulfur dioxide is used as a quasi-inert tracer which is justified for most situations since SO_2-removal due to oxidation
and dry deposition proceeds much slower than NO-oxidation at
least in the near field of the plant(Husar et al.,1978;
Wilson,1978;Gillani et al.,1981).
As an independent measure to validate the above assumption,
measurements of total SO_2 by COSPEC can be used.For the
measurements carried out up to now,the above assumption was
found to be reasonable.
Another assumption was that the total amount of NO_x(NO + NO_2)
remains constant in the near-field investigated.This assumption seems also to be realistic in the light of experimental
results of different groups(Ogren et al.,1976;Hegg et al.,
1976;Flyger et al.,1977;Davis,1977;Forrest et al.,1979;
Guicherit et al.,1980).
Our measurements of the mass ratio SO_2/NO_2 integrated over
the plume cross section as a function of distance from the
plant are used to determine NO-oxidation rates.The determination of integrated NO-oxidation rates is based on the ratio
of emission factors SO_2/NO_x which depends on both operating
conditions of the plant and sulfur content of the coal(both
are delivered by the plant operating personal).This ratio
had to be determined for each measurement separately.

3. RESULTS AND DISCUSSION

Effluent gases from power plants burning coal contain high
concentrations of sulfur dioxide(SO_2) and NO_x(NO + NO_2)
relative to the ambient atmosphere.Nitrogen oxides(NO_x)
emitted from power plants consist mostly of NO.

According to Elshout(1980),for power plants in the Netherlands,the NO portion is at least 95%,in many cases higher
than 98%.
Our measurements by correlation spectroscopy adjacent to the
stacks of the Staudinger power plant have widely confirmed
the above figures.For example,the ratio SO_2/NO_2 was measured
and the degree of NO-oxidation determined on August 28,1980
at distances between 25-50 meters from the stacks of units
1 and 3.The ratios NO/NO_x were found to fall between 95 % and
98.5 %(10 measurements).The ratio at stack exit should be
slightly higher than the above values taking into account an
NO-oxidation between stack exit and place of measurement.

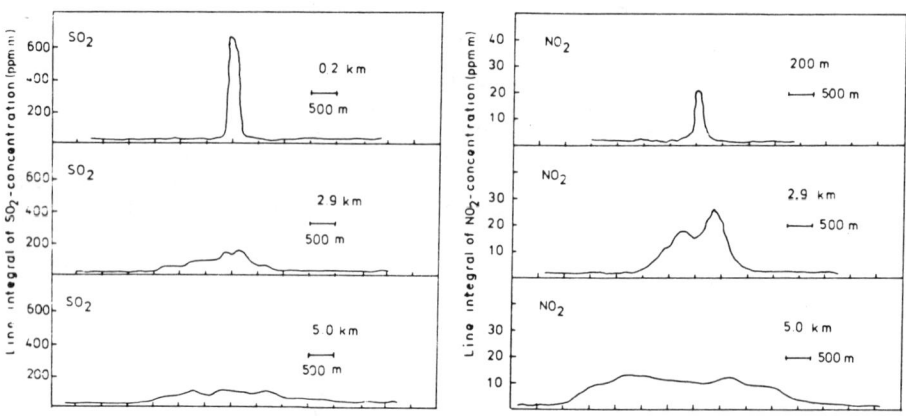

FIGURE 1 : Line integrals of concentrations for SO_2(1a) and
NO_2(1b) at three different traverses downwind:
0.2 km; 2.9 km ; 5 km.(29 Aug.1980,unstable,Pasquill
class B-C).

Our measurements for studying NO-oxidation were carried out
along traverses at different distances downwind of the stacks
between ca. 30 meters and 15 km depending mainly on road-
and traffic conditions,atmospheric stability and wind direction.

One example is given in figures 1a and 1b showing line
integrals of SO_2-and NO_2-concentrations(gas burdens) at three
traverses downwind measured on 29 August 1980.Contrary to
SO_2,the integrated amount of NO_2 increases with increasing
distance from the stacks for most cases investigated up to

now.
Figure 2 shows the decrease of the SO_2/NO_2-mass ratio with
increasing distance from the plant for 4 cases.

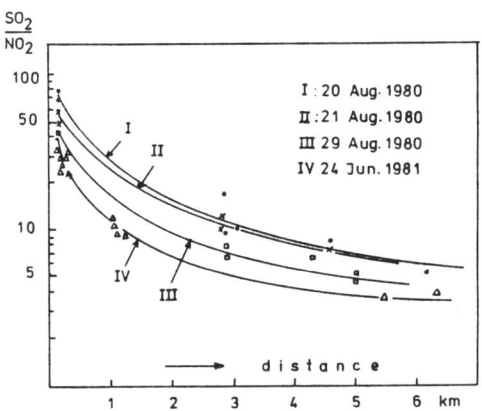

FIGURE 2 : Decrease of the SO_2/NO_2-mass ratio integrated over
plume cross section in the plume of the Staudinger
plant as a function of distance.

The percentage oxidation ratio NO_2/NO_x(as NO_2) integrated
over the plume cross section as a function of distance between plant and plume centerline at different distances downwind is shown in figures 3 to 5.

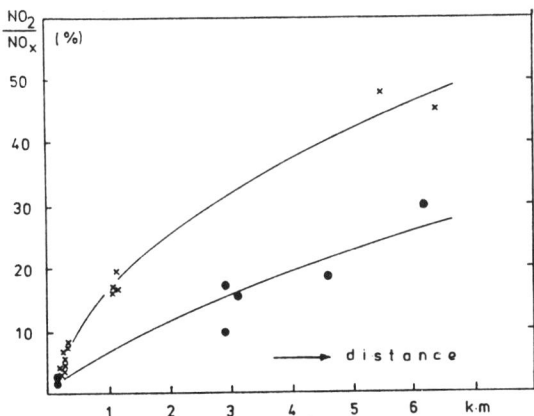

FIGURE 3 : Oxidation of NO in power plant plumes. Percentage
oxidation ratio NO_2/NO_x(as NO_2) integrated over the
plume cross section as a function of distance from
the plant.(××× 24 June 1981:unstable,Pasquill class
B-C, O_3 at Frankfurt: 55 ppb ; ●●● 20 Aug.1980:neutral,
Pasquill class D, O_3 in Frankfurt: 35 ppb)

Figure 3 shows two curves along with the points which are
based on our remote sensing measurements of the SO_2/NO_2-ratio.
The scatter of these points is most likely caused by varying
conditions under which NO is oxidized to NO_2(ozone background
concentrations,mixing of the plume with ambient air) during
the ca. 2 hours lasting measurements rather than by varying
emissions or uncertainties of our remote sensing measurements
or data processing.

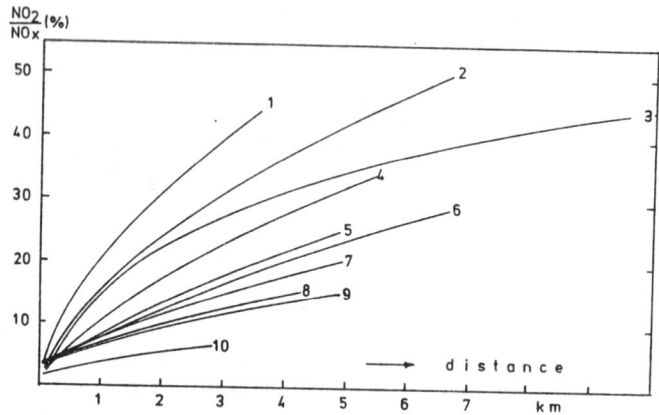

FIGURE 4 : Oxidation of NO in power plant plumes.Percentage
oxidation ratio NO_2/NO_x(as NO_2) integrated over the
plume cross section as a function of distance.

```
1: 30 July 1981 (unstable,Pasquill class B-C,35 ppb O₃)
2: 24 June 1981 (unstable,Pasquill class B-C,55 ppb O₃)
3: 28 July 1981 (unstable,Pasquill class  C ,25 ppb O₃)
4: 29 Aug. 1980 (unstable,Pasquill class B-C,55 ppb O₃)
5: 21 Aug. 1980 ( neutral,Pasquill class C-D, -    )
6: 20 Aug. 1980 ( neutral,Pasquill class  D ,35 ppb O₃)
7: 21 July 1981 ( neutral,Pasquill class  D , 5 ppb O₃)
8: 29 May  1981 ( neutral,Pasquill class  D ,55 ppb O₃)
9: 22 July 1981 ( neutral,Pasquill class  D ,10 ppb O₃)
10: 30 Apr. 1981 ( neutral,Pasquill class  D , 5 ppb O₃)
```

Ozone was measured at 2 sites in Frankfurt ca. 15 m above
ground.These O_3-values are only a rough criterion for the
range of O_3-concentrations near the plume.

In figure 4 the results of 10 measurements are compared
covering a wide range of conditions for NO-oxidation.The
scatter of the points measured is not included to allow a
clearer presentation of the results.As no direct data of ozone
concentrations near the plume were available,ozone concentrations measured at two sites in Frankfurt(ca.15 m above ground)
were given for each curve.These values can only be a rough
criterion for the range of ozone concentrations near the
plume.As seen in this figure,the variation of percentage
oxidation is large.For example,at a distance of 3 km downwind
of the plant between ca. 8 and 40% of NO is oxidized.The
higher oxidation rates are connected with unstable conditions
(Pasquill stability classes B-C,C) in which case mixing of

the plume with ambient air is more effective than for many
situations under Pasquill class D(neutral).
The smallest percentage oxidation values were observed under
overcast conditions connected with slight rainfall(curves
7-10).The main reason is most likely a reduced photochemical
production of ozone.

One example for extremely small NO-oxidation is given in
figure 5 showing our results obtained on 4 August 1981.On this
day the plume was trapped in a stable layer aloft.Plume rise
and dispersion and NO-oxidation occurred within the first
kilometers downwind of the stacks.Estimates of plume dilution
rates can be based upon measured plume widths.Between ca. 1
km and 13 km no further plume dilution occurred as indicated
by the constant plume width(lower figure).As a result,a
further NO-oxidation could not be observed suggesting that
NO-oxidation is diffusion controlled at least in this case.

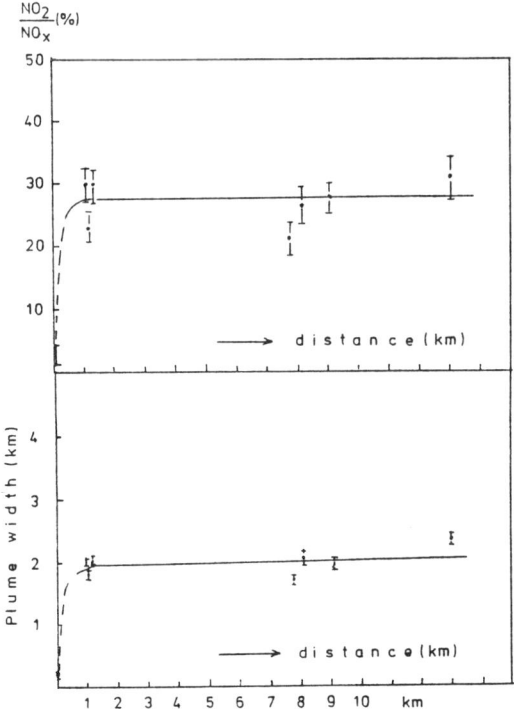

FIGURE 5 : NO-oxidation in power plant plumes.Percentage oxidation
ratio NO_2/NO_x(as NO_2) integrated over the plume cross
section as a function of distance from the plant.
(4 Aug.1981:the plume was trapped in a stable layer aloft,
60-80 ppb O_3).
Ozone was measured at 2 sites in Frankfurt ca. 15 m above
ground.The ozone values are only a rough criterion for the
range of O_3-concentrations near the plume.

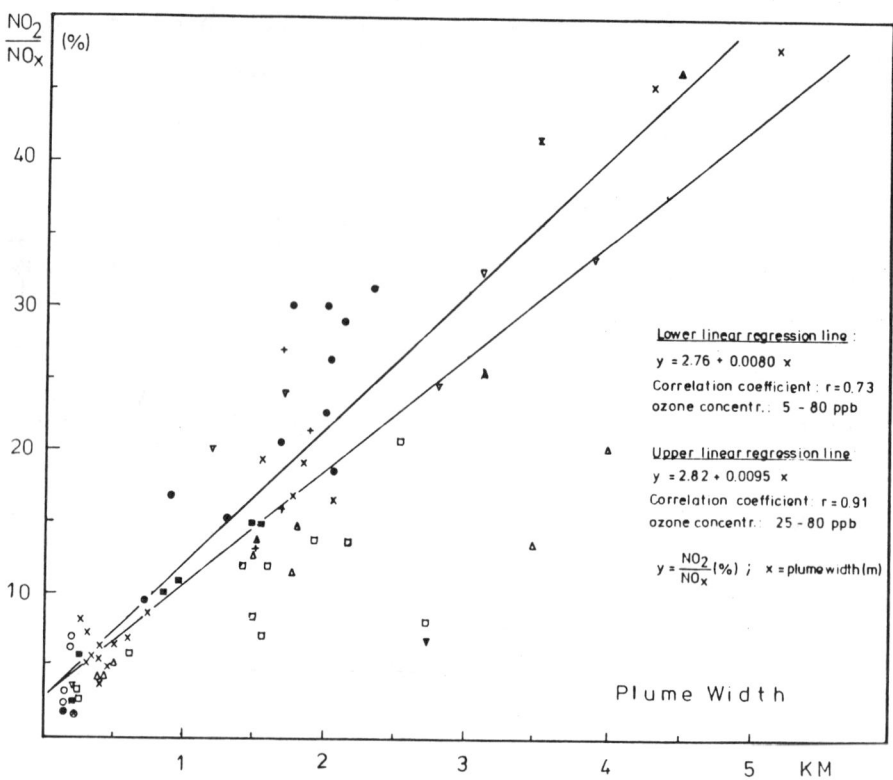

FIGURE 6 : NO-oxidation in power plant plumes.Percentage oxidation
ratio NO_2/NO_x (as NO_2) integrated over the plume cross
section as a function of plume width.

⊠	30 July 1981	(unstable,Pasquill class B-C,35 ppb O_3)
×	24 June 1981	(unstable, PC B-C , 55 ppb O_3)
▲	28 July 1981	(unstable, PC C ,25 ppb O_3)
▽	29 Aug. 1980	(unstable, PC B-C , 55 ppb O_3)
+	21 Aug. 1980	(neutral , PC C-D , -)
⊛	20 Aug. 1980	(neutral , PC D ,35 ppb O_3)
△	21 July 1981	(neutral, PC D , 5 ppb O_3)
■	29 May 1981	(neutral, PC D ,55 ppb O_3)
☐	22 July 1981	(neutral, PC D ,10 ppb O_3)
▼	30 Apr. 1981	(neutral, PC D , 5 ppb O_3)
○	12 June 1981	(neutral, PC D , -)
●	4 Aug. 1981	(stable , PC D-E ,60-80 ppb O_3)

Ozone was measured at 2 sites in Frankfurt ca. 15 m
above ground.The ozone values are only a rough criterion
for the range of O_3-concentrations near the plume.

In figure 6 the percentage oxidation NO_2/NO_x is given as a
function of plume width for 12 measurements regardless of
background ozone concentrations.As can be seen in this figure,
there is a tendency(correlation coefficient:0.73,lower regression line) of percentage oxidation to increase with increasing plume width(plume dilution) suggesting that the NO-
oxidation rate is generally controlled by the rate at which

the plume mixes with the ambient air rather than by the relevant chemical reaction of NO with ozone within the plume.
The correlation coefficient increases to r = 0.91(upper regression line) if the points for extremely low ozone concentrations are omitted(21 July 1981: 5 ppb O_3 ;22 July 1981:
10 ppb O_3; 30 April 1981: 5 ppb O_3).

Another result is that Pasquill classes are only a rough
criterion for the rate at which a plume mixes with the ambient
air especially for Pasquill class D(neutral conditions).

On the basis of an extrapolation of our results it seems to
be that for a slightly unstable,neutral and stable atmosphere,
50 % NO is oxidized to NO_2 at distances greater than 10 km
from the plant.This result is at least in a qualitative agreement with measurements of Elshout et al.(1980) who concluded
that for ozone concentrations smaller than 50 ppb in a stable,
neutral and slightly unstable atmosphere,50 % NO is oxidized
beyond a distance of 10 km from the stacks of two power plants
in the Netherlands.

Although our percentage oxidation values are in reasonable
agreement with values reported by Elshout et al.(1980),more
measurements are necessary especially during winter to provide
a more general picture of the NO-oxidation in power plant
plumes.

4. ACKNOWLEDGEMENTS

We thank the plant operators of the Staudinger power plant
(Preussen Elektra) for supporting this project.
Special thanks go to E.Simon and G.Stahl for providing plant
operating parameters and for their valuable suggestions and
useful discussions.

5. REFERENCES

Cocks,A.T. and Fletcher,J.S.(1979)
    A model of the gas-phase chemical reactions of power
    station plume constituents.Laboratory report No.RD/L/R 1999
    (Job.No.VF 418).Report of CERL(Central Electricity Research
    Laboratories,Leatherhead,Surrey,England)

Davis,D.D.;Smith,G. and Klauber,G.(1974)
    Trace gas analysis of power plant plumes via aircraft
    measurement:O_3,NO_x,and SO_2 chemistry.Science,Vol.186,
    pp.733-736.

Davis,D.D.(1977)
    OH radical measurements:impact on power plant plume chemi-
    stry.Report EPRI EA-465,project 676,Final report,Dec.1977.
    Prepared for Electric Power Research Institute,Palo Alto,
    CA,USA.Project manager: C.Hakkarinen.

Elshout,A.J. and Steenkist,R.(1974)
    A numerical approach to the effect of the oxidation of NO
    on the dispersion of these pollutants from a point source.

Technical note.Paper presented at Ispra(Italy),October
1974.Project COST 61a.

Elshout,A.J.(1980)
Dispersion and transport-modelling and field experiments.
Proceedings of the First European Symposium on Physico-
Chemical Behaviour of Atmospheric Pollutants.Ispra
16-18 October 1979(Project COST 61a bis),pp.510-515.

Elshout,A.J.;Van Duuren,H.;Römer,F.G. and Viljeer,J.W.(1980)
Messungen aus Flugzeugen:Verbreiterung und Umwandlung
primärer luftfremder Komponenten in Rauchfahnen.
Proceedings of the First European Symposium on Physico-
Chemical Behaviour of Atmospheric Pollutants.Ispra
16-18 October 1979(Project COST 61a bis),pp.451-457.

Flyger,H.;Lewin,E.;Lund Thomsen,E.;Fenger,J;Lyck,E. and
Gryning,S.E.(1977)
Physical and chemical processes of sulfur dioxide in the
plume from an oil-fired power station.Riso Report No.
328,March 1977.Available from:Riso Library,Research
Establ.Riso,DK-4000 Roskilde,Denmark.

Forrest,J.;Garter,R. and Newman,L.(1979)
Formation of sulfate,ammonium and nitrate in oil-fired
power plant plumes.Atm.Env. 13,p.1287.

Gillani,N.V.;Kohli,S. and Wilson,W.E.(1981)
Gas-to-particle conversion of sulfur in power plant
plumes:I.Parametrization of the conversion rate for dry,
moderately polluted ambient conditions.
Paper presented at the Symposium on Plumes and Visibility,
Grand Canyon,USA,10-14 Nov.1980.Accepted for publication
in Atm.Env.,Vol.15,1981.

Guicherit,G.;van den Hout,K.D.;Huygen,C.;van Duuren,H.;
Römer,F.G. and Viljeer,J.W.(1981)
Conversion rate of nitrogen oxides in a polluted atmos-
phere.Paper presented during Technical Meeting on Air
Pollution Modelling and its Application, 11 th NATO-
CCMS,Amsterdam,24-27 Nov.1980.To appear in the proceed.
of this meeting.

Hegg,A.D.;Hobbs,P.V. and Radke,L.F.(1976)
Reactions of nitrogen oxides,ozone and sulfur in power
plant plumes.Report EPRI EA-270(Research project 572-3),
Final Report,Sept.1976.Prepared for Electric Power
Research Institute,Palo Alto,CA,USA.Project manager:
C. Hakkarinen.

Husar,R.B.;Patterson,D.E.;Husar,J.D.;Gillani,N.V. and
Wilson,W.E.(1978)
Sulfur budget of a power plant plume.Atm.Env. 12,
pp. 549-568.

Melo,D.T.;Lusis,M.A. and Stevens,R.D.S.(1978)
Mathematical modelling of dispersion and chemical reactions in a plume-oxidation of NO to NO_2 in the plume of
a power plant.Atm.Env.12,pp.1231-1234.

Millan,M.M. and Hoff,R.M.(1977)
    Dispersive correlation spectroscopy:a study of mask
    optimization proceedures.Appl.Optics 16,pp.1609-1618.

Millan,M.M.;Gallant,A.J. and Turner,H.E.(1976)
    The application of correlation spectroscopy to the study
    of dispersion from tall stacks.Atm.Env.10,pp.499-511.

Ogren,J.A.;Blumenthal,D.L. and White,W.H.(1976)
    Determination of the feasibility of ozone formation in
    power plant plumes.Report EPRI EA-307(Research project
    572-1,2).Vol.II:Final Report,Nov.1976.
    Prepared for Electric Power Research Institute,Palo Alto,
    CA,USA.Project manager:C.Hakkarinen.

Ogren,J.A.;Blumenthal,D.L. and Vanderpol,A.H.(1977)
    Oxidant measurements in Western power plant plumes.
    Final Report:EPRI EA-421(Research project 861-1).
    Vol.1:Technical analysis(July 1977).Prepared for Elec-
    tric Power Research Institute,Palo Alto,CA,USA.
    Project manager:C.Hakkarinen.

Pasquill,F.(1974)
    Atmospheric diffusion. 2nd edition Halsted Press: a
    division of John Wiley and Sons, 429 pages.

Varey,R.H.;Sutton,S. and Marsh,A.R.W.(1979)
    The oxidation of nitric oxide in power station plumes.
    A numerical model.Laboratory note no.RD/L/N 184/79
    (Job.No.VC 401).Report of CERL(Central Electricity
    Research Laboratory,Leatherhead,Surrey,England)

White,W.H.(1977)
    NO_x-O_3 photochemistry in power plant plumes:comparision
    of theory with observations.Env.Sci.Techn.,Vol.11,
    No.10,Oct.1977,pp.995-1000.

Wilson,W.E.(1978)
    Sulfates in the atmosphere:a progress report on project
    MISTT.Atm.Env. 12,1-3,pp.537-548.

MEASUREMENTS BY AEROPLANE OF THE DISTRIBUTION
OF OZONE AND PRIMARY AIR POLLUTANTS

H. van Duuren and F.G. Römer
N.V. KEMA, Environmental Research Department, Arnhem, the Netherlands

H.S.M.A. Diederen, R. Guicherit and K.D. van den Hout
TNO Research Institute for Environmental Hygiene, Delft, the Netherlands

Summary

For several years N.V. KEMA and TNO-IMG have carried out air pollution measurements by aeroplane. The objectives of the project were:
- to study the vertical and horizontal distribution of pollutants (mainly ozone) under different meteorological conditions, including photochemical pollution episodes;
- to study the behaviour of pollutants downwind of industrial complexes;
- to study plume behaviour and (photo)chemical reactions in plumes.
Results of the study are presented and implications for modelling purposes and abatement strategies are discussed.

1. INTRODUCTION

Since 1972, TNO-IMG and N.V. KEMA have carried out measurements by aeroplane of air pollutants in the Netherlands. The aim of these measurements was to study: -photochemical air pollution, particularly the horizontal and vertical distribution of ozone;
-plume behaviour of power plants and the oxidation of nitric oxide;
-the influence of power plants on the emission level of the urban industrial Rotterdam area.
In a co-operative study of TNO and KEMA, some flights were made to gain a better insight into the contribution of industrial and urban areas to the concentration levels of ozone.

This paper deals mainly with results of recent flights which took place under conditions favourable for photochemistry, though typical results of former experiences are used to illustrate the train of thoughts. In this manner descriptions are given of:
- the vertical distribution of ozone in the lower troposphere under various atmospheric conditions to examine in what measure ozone is of natural or antropogenic origin;
- measurements of ozone levels near different sources (urban, industrial areas) to study the contribution from these areas and the contribution from distant upwind sources.

Finally, some remarks are made regarding the horizontal distribution of ozone over larger distances in the Netherlands.

2. EQUIPMENT

The equipment used is surveyed in Table I. In the first period, 1972-1975, only ozone was measured. This was done with an electrochemical cell (Van Dop et al. 1979). The instrument was mounted in an aeroplane, Cessna, of the National Police Force. Since 1975, a specially adapted aircraft, Piper Navajo Chieftain, has been used. This plane is equiped with measuring instruments owned by KEMA. A description of the aircraft and the instruments is given elsewhere (Viljeer and Van Duuren, 1977; Römer et al., 1979; Elshout et al., 1979).

Besides the pollutants ozone, oxides of nitrogen and sulphur dioxide, height, air velocity and temperature are measured.

The output signals of the instruments are recorded using a datalogger (Monitor Labs 9400) and a cartridge recorder (Perex 6300).

Table I.

Period	Component	Instrument	Principle
1972-1975	O_3	TNO-IMG	electrochemistry
1978-1981	O_3	Bendix 8002	chemiluminescence
	SO_2	Teco 43	fluorescence
	NO, NO_2, NO_x	Teco 14 D	chemiluminescence
	air velocity	King Radio Corporation	static pressure
	height	KDC 3800	absolute pressure
	temperature	Rosemount 102 BE+amplifier(KEMA)	resistance

3. THE VERTICAL DISTRIBUTION OF OZONE

During the measurements the conditions were favourable for photochemistry: the weather was characterized by little cloudiness and temperatures during daytime of more than 20 °C and the air masses had been transported over some hundreds of kilometres over polluted areas. For these circumstances, a distinct daily cycle of the mixing layer height exists, which varies from some hundreds of metres at night to about 2 km in the daytime. In figure 1, the results of the flights are shown schematically by the profiles A through F.

PROFILE A is encountered at night or in the early morning in background areas or on days with low quantities of photochemical air pollutants. The build-up of a stable layer by night-time irradiation near the surface prevents vertical exchange. The result is a rather uniform vertical concentration profile (natural ozone) above this layer and a reduction by deposition ("physical quenching") within this layer.

PROFILE B can be met after sunrise when vertical exchange starts. The same profile is found at night in an unstable or neutral atmosphere. Transition from profile B to A takes place when the air is transported over polluted areas with a high density of rather low sources and ozone reacts with NO and/or reactive hydrocarbons ("chemical quenching").

PROFILES C AND D. Ozone profile C is encountered in polluted areas before mixing is complete and also because of horizontal inhomogeneity during the measurements. Profile C is transformed in the D type after mixing is complete. As yet, it is not clear to what measure the measured profile variations in the mixing layer result from horizontal variations encountered during the spiral shaped flight tracks.

PROFILE E. During photochemical episodes at night and in the early morning a stable layer is formed in which ozone is physically and chemically quenched. Between the subsidence inversion and the radiation inversion an aged smog layer with high concentrations of ozone can be maintained.

PROFILE F. This profile can be found sometimes during photochemical episodes after sunrise when photochemical processes and mixing in the radiation inversion layer are starting and before mixing of the aged smog layer and the lower inversion layer is complete. Under these conditions the measured concentration levels of ozone in the near ground layer are sometimes higher than in the aged smog layer. Later on, when mixing is complete, profile F changes into profile D.

4. CONCENTRATION FIELDS AROUND SOURCE AREAS

Two flights were specially directed to the study of the horizontal profiles. To this end, the measurements were performed at one particular height halfway the mixing layer, with a flight trajectory covering an area with dimensions of 100 to 200 km. Elevated ozone levels were found to extend over large areas, with variations on a smaller spatial scale.

In seven flights, the relation between the spatial variations of the ozone levels and the emissions of several source areas was studied in more detail. The source areas studied were: the Rijnmond district (the industrialized region around Rotterdam), the Ruhr area in Germany, and Eindhoven, a medium-sized town (200 000 inhabitants) in a semi-rural environment.

The flights were planned in the afternoon of days when a rather homogeneous and stationary windfield, accompanied by elevated ozone levels, was expected. Each flight was divided into a number of straight, horizontal flight tracks of about 50 km according to the following strategy.

Upwind of the source area, in cross wind direction, various tracks were flown at different heights between the ground and the top of the mixing layer; occasionally a track was flown above the mixing layer. In this way an approximately vertical cross section of the mixing layer, so also of the background concentration field, could be constructed. Next, a similar, detailed cross section was made directly downwind of the source area, from which an estimate of the emission of pollutants and the quenching of ozone by NO could be made. Further downwind, other cross sections were made in order to study the time evolution of the polluted air. The distances between the cross sections were chosen according to the average wind speed, in such a way that the various cross sections each time were taken in approximately the same air mass.

Results of three flights are illustrated in the figures 2-4 and are discussed in some detail. The ozone patterns shown are derived from tracks at one particular, intermediate height, which can be considered representative for the bulk of the mixing layer. The isopleths were determined by interpolation between the profiles measured.

Figure 2 shows the ozone levels found around the Rijnmond district in the afternoon of August 30, 1979. The wind was southeast with a speed, averaged over the mixing layer, of 8 m/s and a mixing height of 1000 m. In the lee of the industrialized harbour situated west of Rotterdam, the quenching of ozone, particularly by NO emission of an isolated source - in this case a large power plant - is clearly visible. Because of the limited mixing of the air above sea the effect of this plume remains visible over

a large distance. Ozone concentrations at the southwest border are clearly
increased. Analyses of the air trajectories show that this air mass had
been transported over the industrialized area of Antwerp. The influence of
the Rijnmond district is bounded more distinctly at the northeast side.
This air is transported alongside the Rijnmond district; the concentrations
hardly change during the measurement period. In the air transported across
the Rijnmond district the ozone levels increase at a rate of 15 ppb/hr.

Figure 3 gives more or less similar results from another flight around
Rotterdam on September 16, 1980. Here the wind is south, with an average
speed of 5 m/s and a mixing height of 1000 m. At the west side, relatively
unpolluted air is present, while air which had passed over Antwerp about
five hours before, now moves across the east part of the Rijnmond district.
The SO_2 measurements show an addition by the Rijnmond sources of about 12
tons/hour, which is comparable with the load of about 9 tons/hour calculated from the plume shaped SO_2 profile in the air from Antwerp. In the narrow east band, the ozone levels increase significantly, partly due to the
emissions of Antwerp alone, but perhaps also due to the interaction of
emissions of Antwerp and Rotterdam. In the air transported over the largest
Rijnmond sources, quenching is seen directly downwind, followed by a slow
increase of the ozone concentrations.

Finally, figure 4 shows the ozone pattern observed around Eindhoven on
September 2, 1980. The wind direction was southsoutheast, the wind speed 5
m/s and the mixing height 1200 m. Apart from a very slight quenching directly downwind of the city, the emissions do not affect the ozone levels significantly. The most conspicuous feature in the figure is the strong quenching
of ozone in a plume east of Eindhoven. This plume arises from two sources
about 50 km upwind and contains NO_x and SO_2, but probably little hydrocarbons. Apparently as a result of this, the contribution to the ozone levels
remains negative, even after a travel time of about five hours.

The general picture arising from these flights is, that the pollution
levels encountered downwind of source areas in the Netherlands, are strongly influenced by the pollution emitted by sources further upwind. Plumes of
increased ozone which can be associated with precursor source areas, are
found at distances of several tens of kilometres. Often, in situations with
complicated wind fields, these plumes cannot be traced back to the sources.
Much closer to the sources, but sometimes at large distances as well, quenching is the predominant effect on ozone. Such air with diminished ozone
concentrations has experienced less dispersion and can usually be traced
back to the source without much difficulty. The travel time after which
the contribution to the ozone levels changes from negative to positive varies considerably, probably from one or two hours to many hours, depending
on the pollutant mixture.

5. CONCLUSIONS

Ozone concentrations in ambient air are often lower than in the bulk
of the mixing layer and are therefore not representative for the concentrations met in the first few kilometres of the atmosphere. Because the concentrations of natural ozone (above the subsidence inversion layer) are much
lower than in the mixing layer it must be concluded that the higher values
for ozone are due to photochemical processes of precursors.

During episodes of photochemical ozone formation, ozone levels are increased over large areas. On a smaller spatial scale important variations

altitude (km.)　　　　　　　altitude (km.)

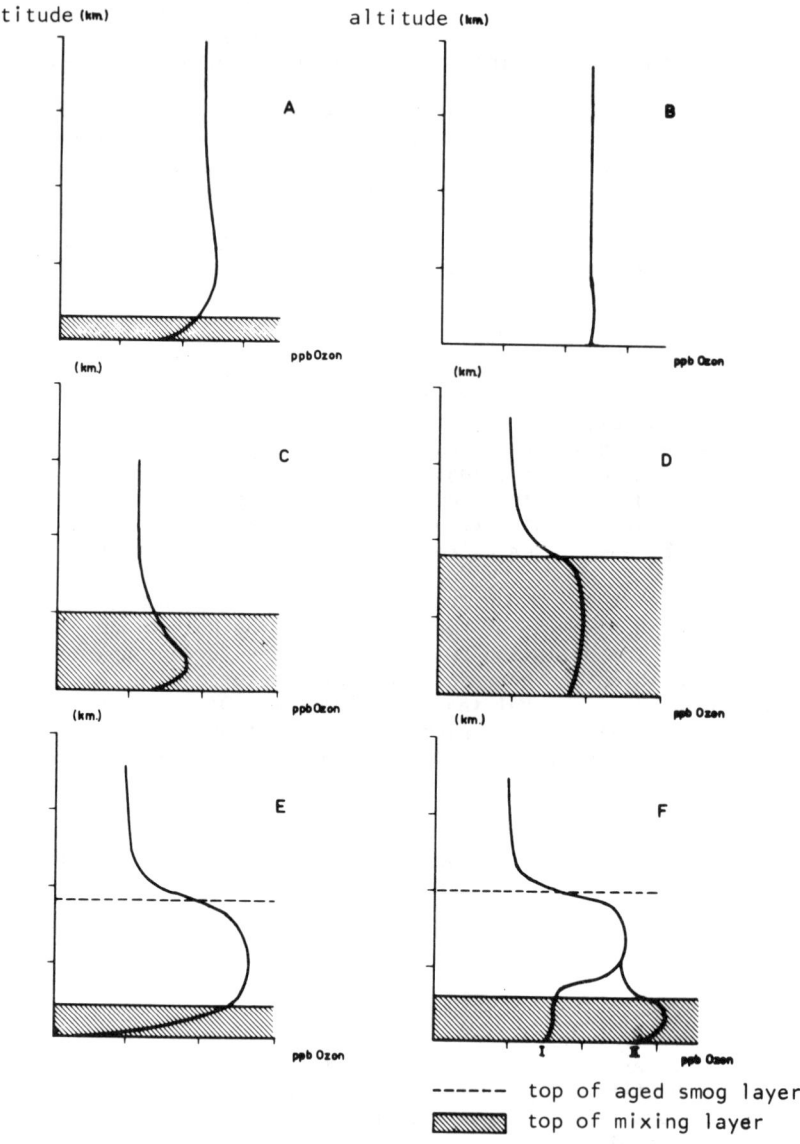

A

B

(km.)　　　　　　　ppb Ozon

(km.)　　　　　　　ppb Ozon

C

D

(km.)　　　　　　　ppb Ozon

(km.)　　　　　　　ppb Ozon

E

F

I　　　II

ppb Ozon

ppb Ozon

------ top of aged smog layer

▧ top of mixing layer

Fig. 1. Vertical profiles of ozone.

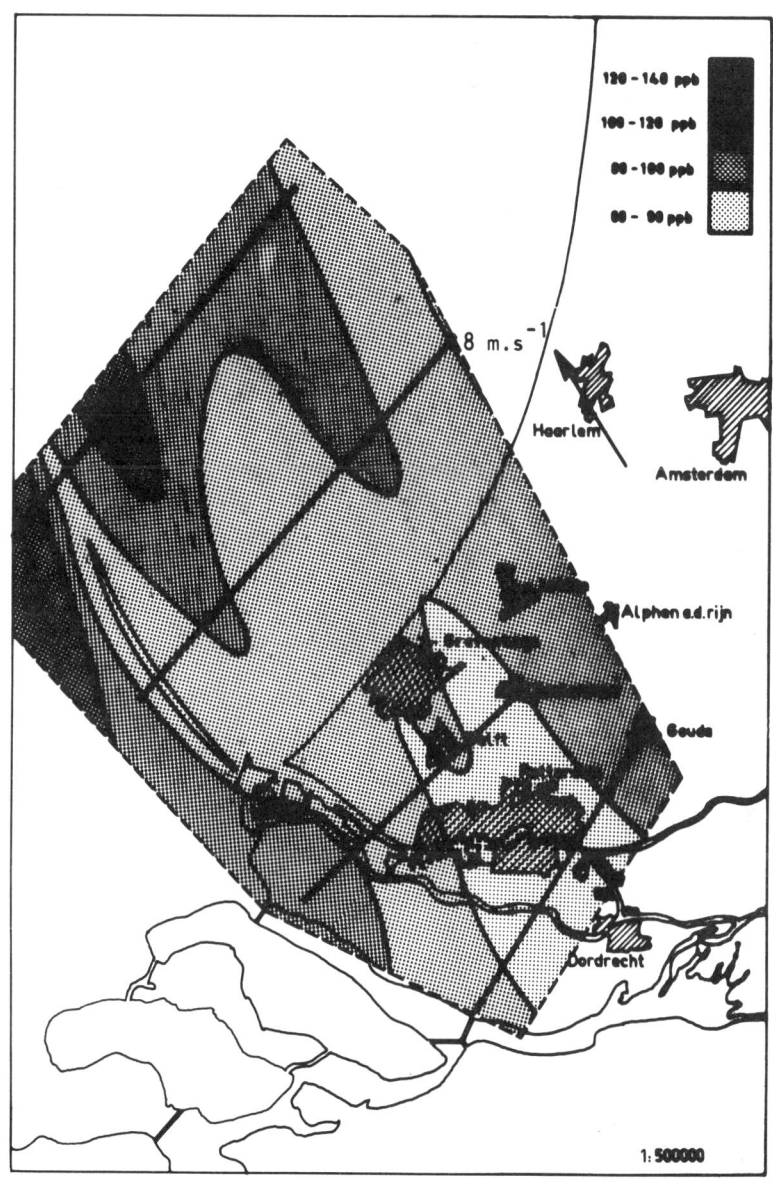

Fig. 2. Development of the ozone concentration on a height of
650 m, as measured on August 30, 1980 around the Rijnmond district.

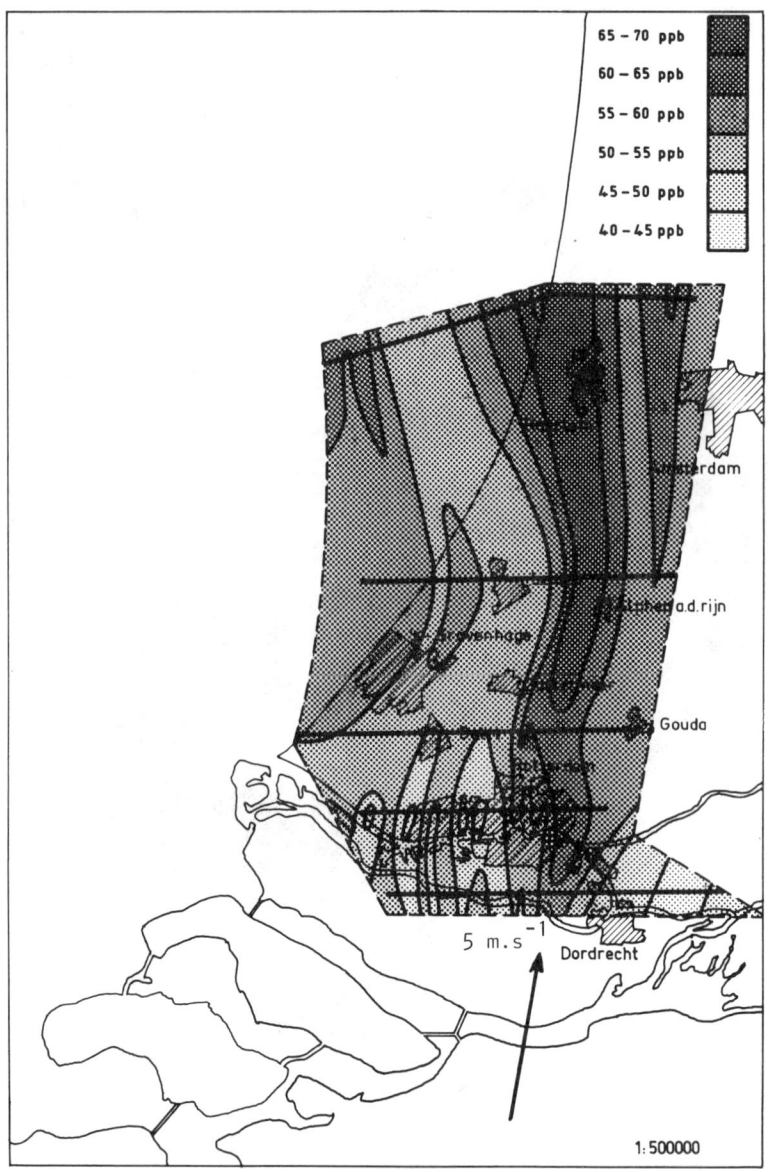

Legend:
- 65 – 70 ppb
- 60 – 65 ppb
- 55 – 60 ppb
- 50 – 55 ppb
- 45 – 50 ppb
- 40 – 45 ppb

Amsterdam

Alphen a.d.rijn

s-gravenhage

Gouda

5 m.s^{-1}

Dordrecht

1:500000

Fig. 3. Development of the ozone concentration on a height of 400 m, as measured on September 16, 1980 around the Rijnmond district.

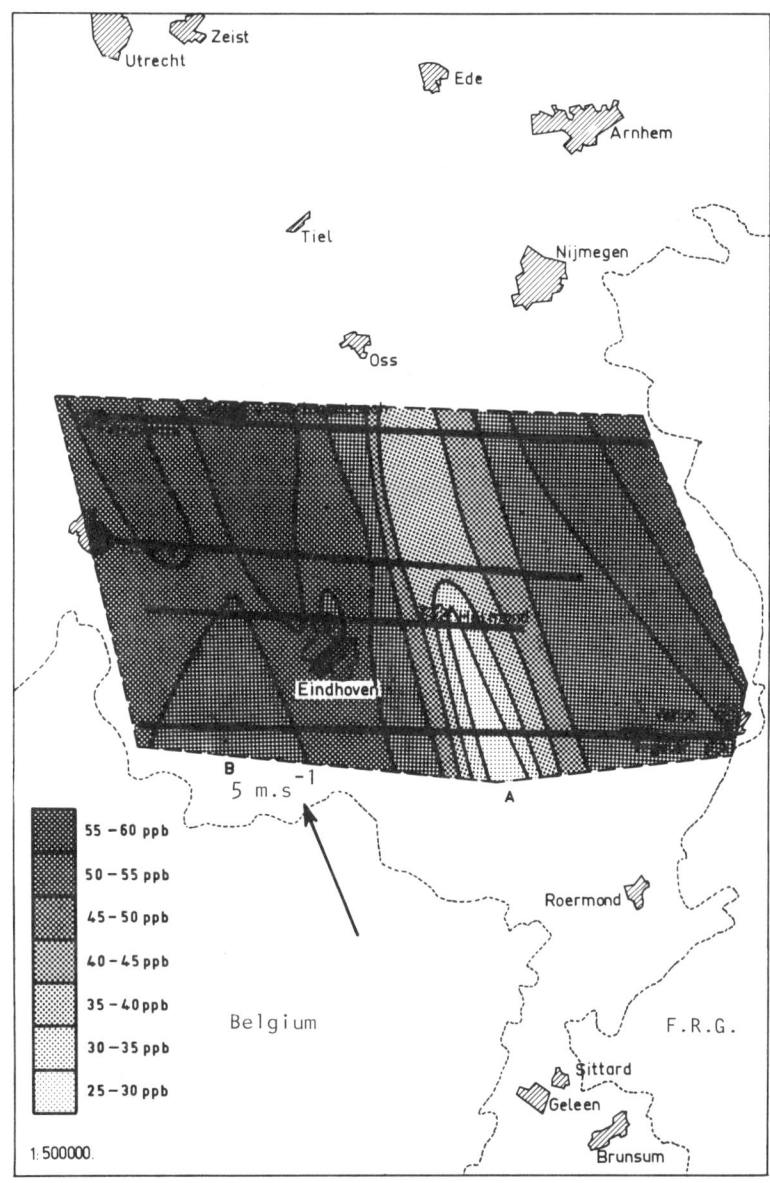

Fig. 4. Development of the ozone concentration on a height of
550 m, as measured on September 2, 1980 around the
town of Eindhoven.

of the ozone levels occur. Close to sources, quenching by NO gives rise to plumes of diminished ozone concentrations; further away the contributions usually become positive. The travel time at which the contribution changes sign varies considerably. The net effect of sources on the ozone levels depends on the mixture of pollutants emitted and the composition of the air in which the plume is dispersed.

Localized actions to diminish precursor emissions do not effectively reduce the ozone levels in the surroundings, and, because of diminished quenching by NO, may even lead to increased ozone concentrations close to the source. Regulatory actions aimed at the reduction of ozone levels should therefore be undertaken on a regional or even larger scale.

6. FURTHER STUDIES

For a better interpretation and understanding of the ozone concentration levels measured under photochemical conditions more comprehensive measurements around industrial areas will be carried out.

In a related program, the composition of the aged smog transported above the nocturnal ground inversion layer during the night and in the early morning will be studied. This aged smog layer containing pollution from foreign sources is known to have a predominant effect during photochemical episodes.

7. REFERENCES

H. van Dop, R. Guicherit and R.W. Lanting (1979): Some measurements of the vertical distribution of ozone in the atmospheric boundary layer. Atmospheric Environment 11, p. 65-71.

J.W. Viljeer and H. van Duuren (1977): The measurement by aeroplane of atmospheric pollution, in particular in smoke plumes (in Dutch). Elektrotechniek 55, p. 540-547.

F.G. Römer, H. van Duuren, A.J. Elshout, J.W. Viljeer (1979): Messungen aus Flugzeugen: Verbreitung und Umwandlung primärer luftfremder Komponenten in Rauchfahnen. VGB Konferenz Kraftwerk und Umwelt 1979, p. 134-140 (Essen, BRD).

A.J. Elshout, H. van Duuren, F.G. Römer, J.W. Viljeer (1979): Messungen aus Flugzeugen: Verbreitung und Umwandlung primärer luftfremder Komponenten in Rauchfahnen. Proceedings First European Symposium: Physico-chemical behaviour of atmospheric pollutants, p. 451-457, Ispra 16-18 October 1979.

AIRCRAFT MEASUREMENTS OF OXIDANTS AND PRECURSORS IN THE

PLUMES OF HEAVY-INDUSTRIALIZED AREAS

NEUBER, E., H.-W. GEORGII
Institute of Meteorology and Geophysics,
University Frankfurt/M

MÜLLER, J.
Umweltbundesamt
Pilotstation Frankfurt/M

Summary

Aircraft measurements of C_2-C_8-hydrocarbons (HC), nitrogen-oxides, ozone and meteorological parameters were carried out above and in the environment of an industrialized area during anticyclonic weather conditions. The behaviour of precursors in the oxidant formation processes as obtained from smog chamber experiments is confirmed in the atmosphere by these field measurements. The ratio of high reactive to low reactive HC's decreases during the transport from urban to rural areas. In contrast, the NO_x/NO ratio increases and the ozone concentrations are enhanced significantly in the downwind areas. In recent measurements ozone concentrations, exceeding 200 ppb, were observed. Estimates of the mean daytime OH-radical-concentrations under "photosmog conditions" in the atmosphere are made by use of the observed olefine removal rates ($[OH]$ = $2 - 18x10^6$ molecules/cm^3).
In the lower 3.5 km of a polluted atmosphere, the actinic solar flux, which is related to the NO_2-photolysis-rate, was attenuated up to 50% at various zenith angels.

I. INTRODUCTION

Photochemical air pollution results from reactions between nitrogen oxides, organic compounds and oxygen in the atmosphere which are initiated by solar radiation. The effects of the compounds formed by those processes (O_3, PAN and others) comprises potential damage on human health, vegetation and materials (1). Analytical data represent the occurrence of photochemical smog episodes in the Rhein-Ruhr area (2). The extent and the abundance of these episodes are closely related to the variability of the synoptic weather patterns (3).
Several attempts were made to develop abatement strategies for photochemical air pollution by use of smog chamber and modelling data (4, 5). Because these data are valid only to a limited extend, aircraft measurements were carried out to

provide input data for a photochemical air quality simulation
model, developed for the Rhein-Ruhr area.

Furthermore the simultaneous measurements of the atmospheric HC-, NO_x-, NO- and O_3-concentrations as well as various
meteorological parameters should give informations about the
abundance and the behaviour of precursor-compounds and oxidants
in a polluted atmosphere.

II. EXPERIMENTAL

The nitrogen oxides and the C_2-C_8-hydrocarbons were
measured as precursor compounds.

The HC were sampled by an enrichment procedure and analysed by a gas-chromatograph which allows the detection of
19 HC-species (6). A sample period of 3 to 5 minutes is required to accumulate a relevant sample in cooled adsorption
tubes ($-78^{\circ}C$). This time corresponds to a distance of 9 to
15 km at usual flight velocities. The NO_x-concentrations were
continously measured during the airborne operations by a
Monitor Labs NO_x-Analyser (ML 8440).

Ozone was taken as an oxidant-tracer and was measured by
a Bendix Ozone Analyzer (8002).

The incoming and outgoing actinic fluxes were measured by
an UV-radiometer (UV 103, MACAM). The detector consists of a
horizontal plate sensor and two filters equipped with cosine-
correction-plates.

The actinic fluxes on a horizontal surface can be measured
in the wavelength range between 330-400 nm and 290-330 nm,
using the different filters. The pressure-temperature dependence of the equipment was tested in the laboratory.

The aircraft data acquisition-system consists of two
4-channel strip chart recorders which allow to record the
trace-gas concentrations, the relative humidity, temperature
and pressure as well as the actinic fluxes.

Horizontal and vertical flight patterns were designed to
determine the spatial and temporal distribution of the measured
parameters. The measurements took place upwind and up to 150 km
downwind of the Rhein-Ruhr area. In order to estimate the time
change of precursor and oxidant concentrations, the measurements
were carried out from 6.00 - 9.00 a.m. above the source areas
and from 2.00 - 6.00 p.m. in the downwind sites. In particular
the flights were arranged in such a way, that the measurements
were carried out in the same air parcel throughout the day. By
knowledge of the wind distribution this can be verified approximately. In addition to the horizontal flight-patterns
vertical profiles and cross sections were carried out to a
height of 3.5 km in order to measure the natural ozone level
above the mixing layer.

III. RESULTS AND DISCUSSION

The aircraft measurements were carried out above the
Rhein-Ruhr area during anticyclonic weather conditions (Fig.1).
In the morning the typical vertical ozone and nitrogen oxides

Table 1: HC-removal rate derived from the temporal change
          of HC-acetylene ratios

		K_{rHC} $[h^{-1}]$	r^2 x)
olefines	(C_2-C_8)	- 0.26	0.50
C_2H_4		- 0.23	0.39
C_3H_6		- 0.41	0.62
alkanes	(C_2-C_8)	- 0.19	0.44
$i-C_4H_{10}$		- 0.11	0.19
$n-C_4H_{10}$		- 0.20	0.40
$i-C_5H_{12}$		- 0.16	0.27
$n-C_5H_{12}$		- 0.19	0.48

x) coefficient of determination, calculated for the pseudo-
first order decay (n=11)

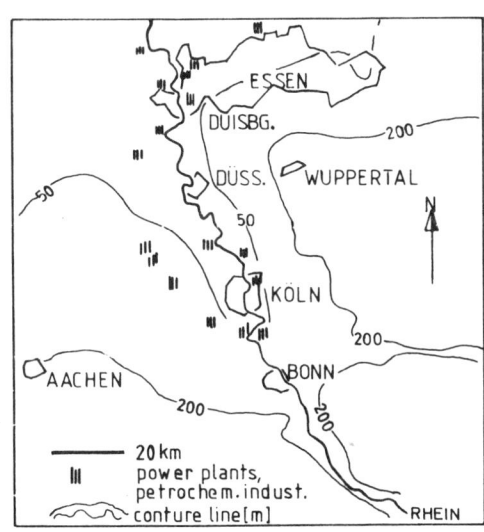

Figure 1 - Map of the Rhein-Ruhr area

distributions show significant gradients in opposite directions up to a height of about 500 m above the ground level (Fig. 2). The corresponding temperature-profiles are characterized by well established inversions in the lowest air layers (Fig. 2). These inversions suppress the vertical mass-exchange in the atmosphere, depending on the inversion strength. This leads to a separation of the ground based air from the upper layers. Therefore, an enrichment of pollutants, emitted at ground level, can take place. The accumulation of anthropogenic gases in the lowest air layers act in addition to the earth's surface as an ozone sink, which explains the observed decay of the ozone concentration in the lowest 5 ͺ٥ m of the atmosphere above the source area.

In the afternoon the ground inversions are dissipated whereas a subsidence inversion in heights from 1500-2000 m is observed. The vertical ozone profile is almost homogeneous in contrast to the morning situation (Fig. 2). No significant variations of the ozone-concentration occured in the middle-tropospheric niveau above 2 km height, in contrast to the increased ozone content in the mixing layer below the subsidence inversion. These profiles demonstrate the influence of atmospheric stability, mixing and dilution processes associated with ozone sink- and generating mechanisms on the vertical ozone distribution in the mixing layer during photosmog episodes. The actinic fluxes in the wavelength-range from 330-400 nm, which are related to the NO_2-photolysis-rate, were attenuated up to 50% in the lowest 3.5 km of a polluted atmosphere.

In Fig. 3 a typical horizontal distribution of the HC-concentrations measured in the downwind site of the Ruhr area at 300 m above ground level is represented. The concentrations of the HC-compounds were summarized as HC-fractions. Strong gradients in the HC-concentration-field and different compositions in the HC-spectra were observed. The horizontal distribution of the nitrogen oxides and ozone concentrations measured simultaneously show strong opposite gradients above the source area (Fig. 4). The ozone maxima occured in a distance of 80 km from the center of the Ruhr-area. In Fig. 5 the vertical cross sections of the ozone concentrations and the HC- and nitrogen oxides-concentrations above the flight tracks from E to F (see Fig. 3) in 80 km distance from the center of the Ruhr area are represented. The ozone maximum of 140 ppb coincided with high HC- and low NO_x-concentrations at a height of 1000 m above ground level. Particularly, the ratio of the very reactive propene to the relatively inert acetylene (C_3H_6/C_2H_2) decreases with increasing height. The observed propene-acetylene ratio is by a factor of 10 to 50 lower in comparison to the measured ratios above the Ruhr area. The less reactive HC-compounds benzene and acetylene become more dominant in the HC-spectrum above the mixing layer.

The C_2-C_8-HC are removed from the atmosphere under "photosmog conditions" by gas-phase photo-oxidation, mainly affected by OH-radicals (8). Furthermore, the olefine-ozone reactions are the most dominant HC-removal processes in "aged" photosmog (9). The involvement of the HC in the oxidant formation pro-

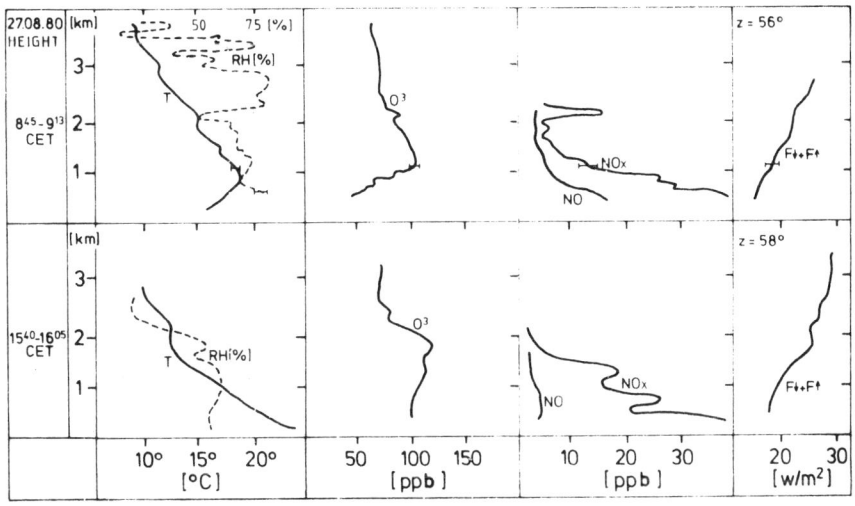

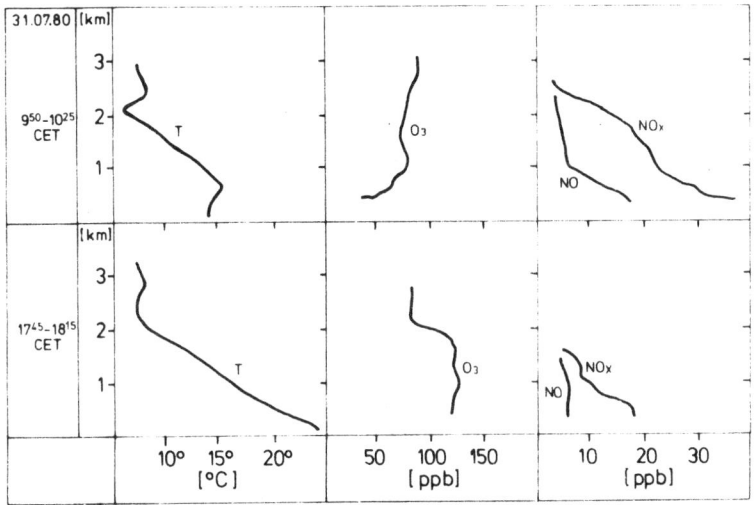

Fig. 2 - Typical vertical profiles of ozone- and nitrogen oxides-concentrations and meteorological parameters at various daytimes during photosmog episodes in the Rhein-Ruhr area. F↑ + F↓ = sum of the incoming and outgoing actinic flux on a horizontal plate (330 - 400 nm)).

cesses cannot be directly derived from their measured spatial and temporal distribution in the atmosphere, because physical and chemical sink mechanisms are superimposed by mixing and dilution processes. Nevertheless, the influence of the HC-removal by chemical reactions can be approximately separated from mixing and dilution processes in the atmosphere by use of the temporal change in the ratios between very reactive and less reactive HC-compounds. Acetylene was used as an inert compound in similar studies because its reactivity referring to OH-radicals and ozone is by factor of 100 to 1000 lower as compared to the olefines (10, 11). The HC-acetylene ratios were measured in the morning above the urban areas and in the afternoon at several distances above the downwind sites. By balloon ascents and radio sonde wind data, the travelling times of the air parcels, in which the HC-sampling took place, were calculated. Under the assumptions, that the measurements were carried out in the same air parcels throughout the day and that no mixing-in of "fresh emissions" or "aged" air essentially affects the HC/acetylene ratio in the air parcels during their transport, HC-removal rates can be estimated using the temporal change of the observed HC-acetylene ratios (10). In Fig. 6 the alkane- and olefine/acetylene ratios are plotted versus the travelling time. The olefine/acetylene-especially the propene/acetylene ratios decrease significantly with the travelling time (Fig. 7). Under the assumptions, mentioned above, the observed decay can be due to the HC-removal by gas-phase photooxidation. Smog chamber experiments and atmospheric studies have shown that the HC-oxidation products are essential participants in the reactions, generating peroxyradicals, which play an important role in the NO to NO_2 conversion (12, 13). In Fig. 8 the NO_x/NO ratios, which were observed during the HC-sampling periods are plotted versus the travelling time of the air parcels. The ratios increase with increasing travelling times throughout the day indicating the occurrence of NO-oxidation processes. The ozone concentration also increases significantly (Fig. 9).

The temporal change in the HC-acetylene ratios can be described by a pseudo-first order decay (8):

$$\frac{[HC]}{[C_2H_2]}_t = \frac{[HC]}{[C_2H_2]}_o \times e^{(-K_{rHC} t)}$$

Under the above mentioned assumptions, K_{rHC} represents the HC-removal rate. In table 1, K_{rHC}- values of the HC-fractions and some single compounds are summarized.

By use of the following equations and the observed HC-removal, the mean daytime OH-radical concentration can be estimated (11):

$$\frac{d\,[olefine]}{dt} = -(\,K_{ol}\,[\overline{OH}]\,[olef] + K_{o2}\,[O_3]\,[olef])$$

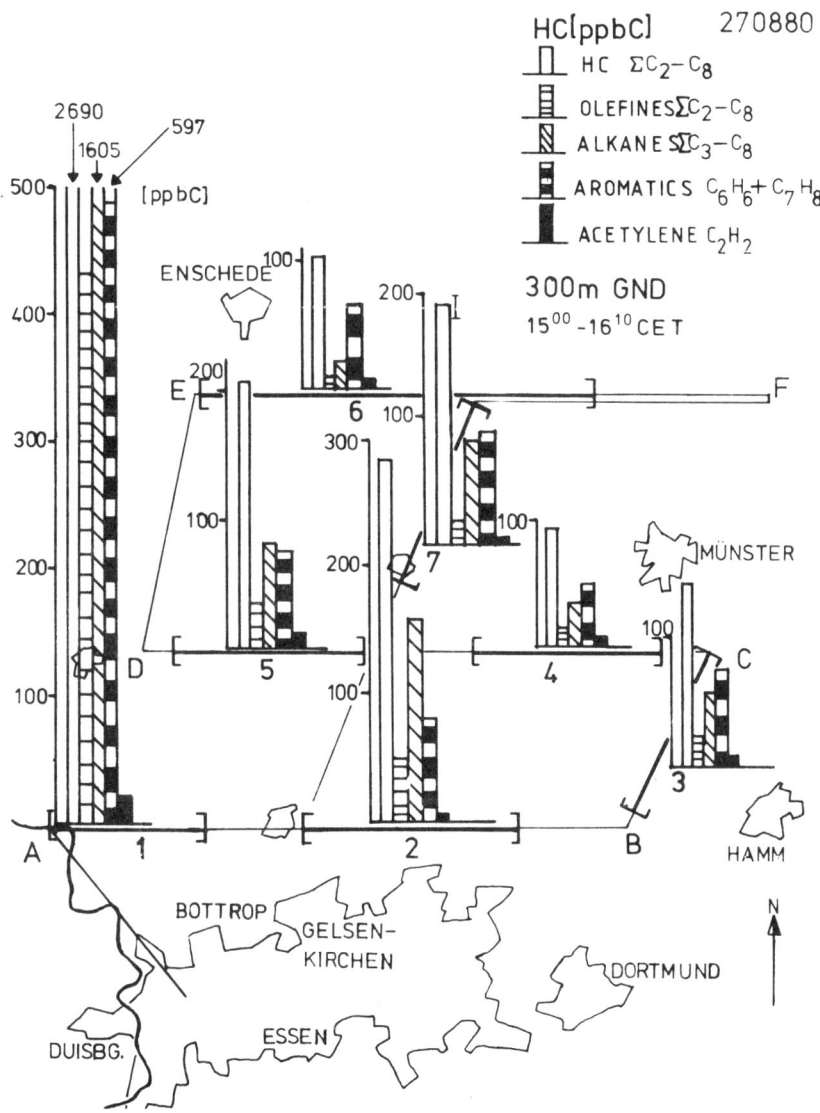

Fig. 3 - Flight pattern and typical horizontal distribution
of the HC-concentration in the downwind site of the Ruhr area
(300 m above ground level; southerly wind directions,
[——] HC-sampling positions)

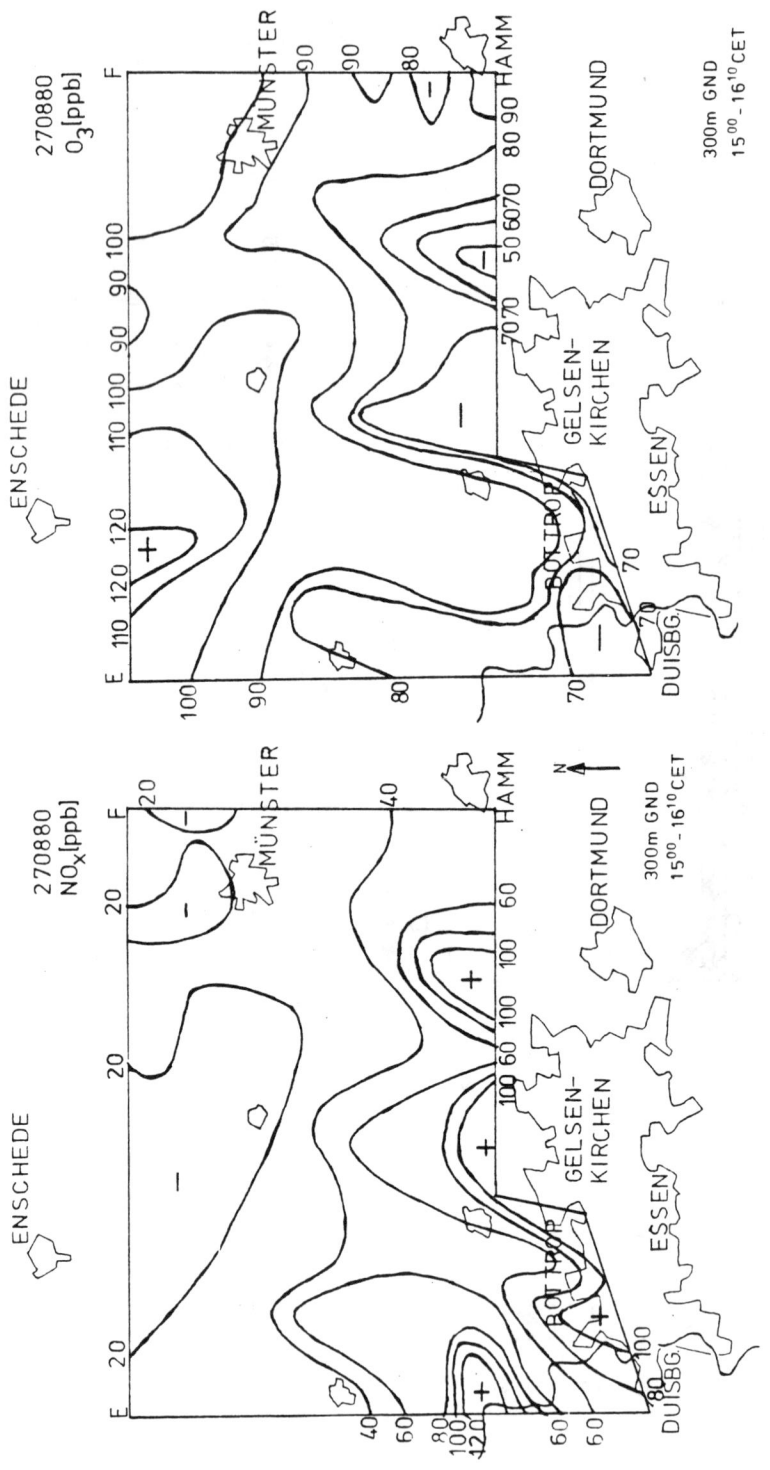

Fig. 4 – Typical horizontal distribution of the ozone and nitrogen oxide-concentrations in the downwind site of the Ruhr area (300 m above ground level, southerly wind directions).

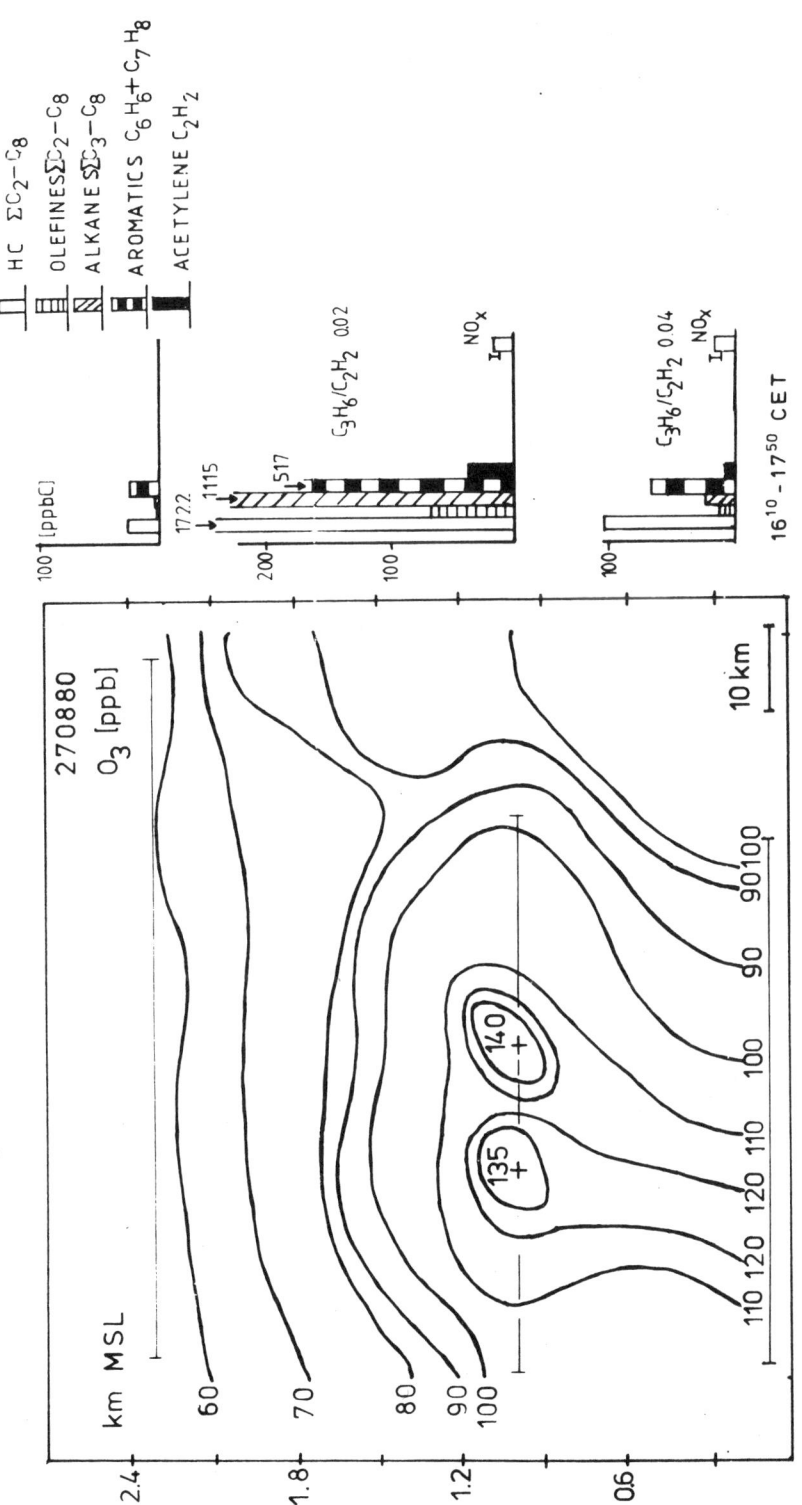

Fig. 5 – Vertical cross section of the ozone-, HC- and nitrogen oxide-concentration above the flight track E →F in 80 km distance from the Ruhr area ([——] HC-sampling positions)

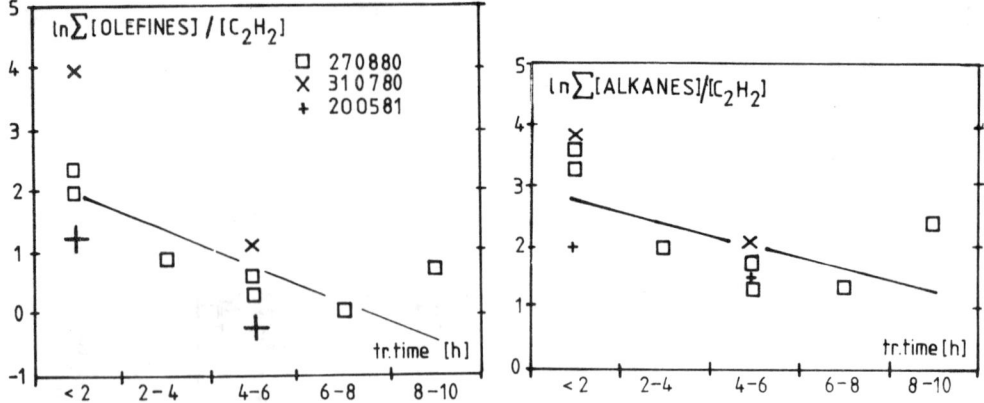

Fig. 6 - Plot of the HC-fractions to acetylene ratio versus the travelling time (tr.time) ppb/ppb

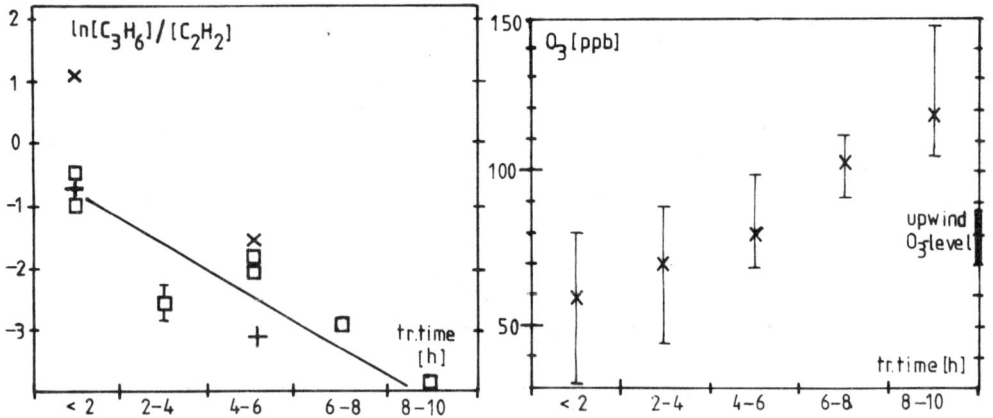

Fig. 7 - Plot of the propene/ acetylene ratio versus the travel time

Fig. 9 - Plot of the ozone concentration versus the travel time measured during the HC-sampling periods

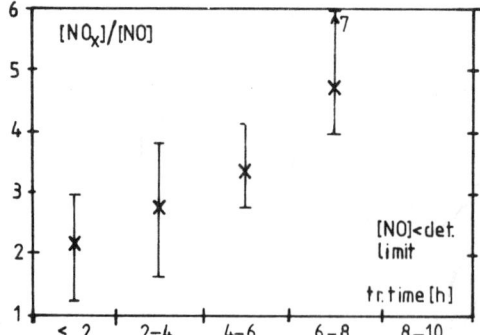

Fig. 8 - Plot of the $[NO_x]/[NO]$ ratios versus the travel time measured during the HC-campling periods

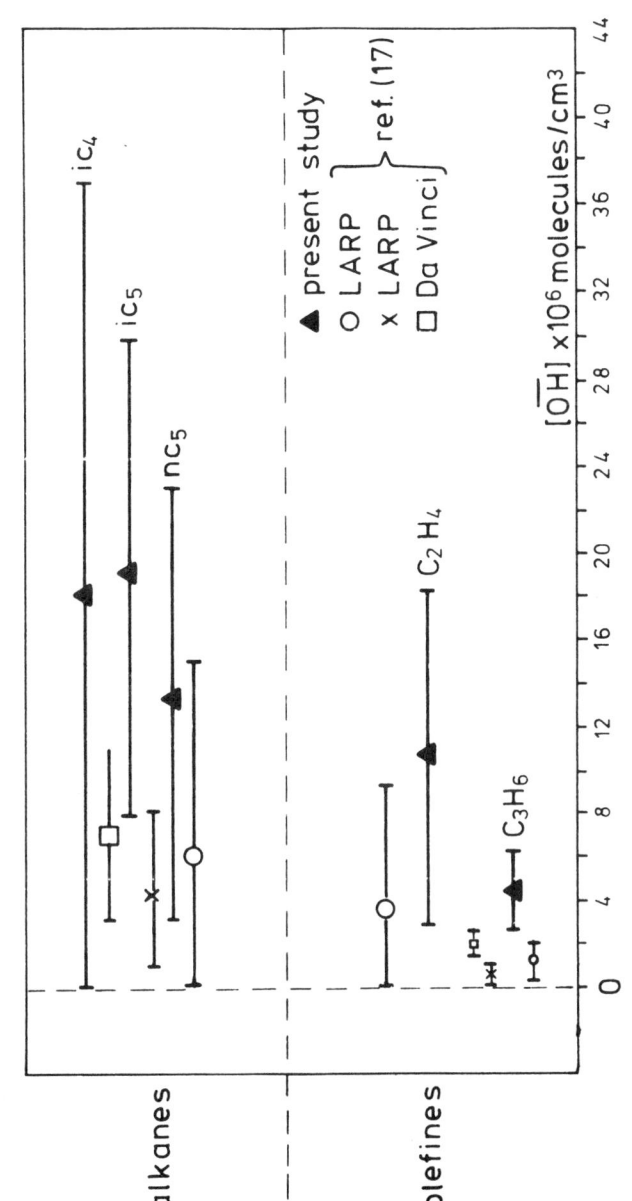

Fig. 10 – Summary of the mean daytime UH-radical concentration calculated by the HC-removal in the mixing layer

$$\frac{d[\text{alkane}]}{dt} = - K_p \ [\overline{OH}] \ [\text{alkane}]$$

(K_p, K_{o1}, K_{o2}: reaction rate constants at 20°C, 1 atm; ref. 14,15,16)

In Fig. 10 the calculated mean daytime OH-radical concentrations under "photosmog conditions" are represented in addition to the results of other studies in which the HC-acetylene ratios were also used for estimating the atmospheric OH-radical concentrations (17). The OH-radical concentrations calculated from the olefine removal vary between 2 - 18×10^6 molecules/cm^3. In particular, Fig. 9 shows that OH estimates are strongly dependent on the hydrocarbon used to obtain them. The estimated OH levels and uncertainties associated with them increase with decreasing reactivity of the HC.

IV. CONCLUSIONS

The aircraft measurements demonstrate that the behaviour of precursor-compounds in oxidant-formation postulated by smog chamber and modelling data, can be observed in the polluted atmosphere under "photosmog" conditions.

ACKNOWLEDGEMENTS

These investigations were sponsored by the Umweltbundesamt of the FRG.

LITERATURE

(1) TNO: Photochemical Smog-Formation in the Netherlands, Editor: R. Guicherit, 1978

(2) FRICKE, W.: Die Bildung und Verteilung von anthropogenem Ozon in der unteren Troposphäre, Berichte des Institutes für Meteorologie und Geophysik der Universität Frankfurt/M, Nr. 44, 1980

(3) MUSCHALIK, B.: Der Einfluß meteorologischer Parameter auf die bodennahe Oxidatienverteilung, UBA-Forschungsbericht 79-10402502/03/04, 1979

(4) DERWENT, G., O. HOV: Computer Modeling Studies of the Impact of Vehicle Exhaust Emission Controls on Photochemical Air Pollution Formation in the United Kingdom, Environmental Science & Technology, 14, 1360-1366, 1980

(5) GUICHERIT, R., P.J. BLOKZIJC, E. PLASS: Some Notes on the Abatement of Photochemical Ozone Production in: Photochemical Smogformation in the Netherlands TNO, 1978

(6) NEUBER, E., H.-W. GEORGII, J. MÜLLER: Verteilung leichter Kohlenwasserstoffe an Orten unterschiedlicher Luftqualität, Staub-Reinhalt. der Luft, 3, 91-97, 1981

(7)   PETERSON, J.T., K.L. DEMERJIAN, K.L. SCHERE: Active
      Solar Flux and Photolytic Rate in the Troposphere, Int.
      Conference on Photochemical Oxidants Pollution and its
      Control, Raleigh, 1976, Proceedings, 763-774, 1977

(8)   BRUCKMANN, P.:   Reaktionen ausgewählter organischer Ver-
      bindungen anthropogener Schadstoffe, Ergebnisse der VDI
      Arbeitsgruppe Luftchemie, Düsseldorf, 1980 (Manuskript)

(9)   COX, R.A.:   Rates, Reactivity and Mechanisms for Homo-
      geneous Atmospheric Oxidation of Atmospheric Pollutants,
      Proceeding, Ispra 1979

(10) CALVERT, J: Hydrocarbon Involvement in Photochem.-Smog
      Formation in Los-Angeles-Atmosphere, Env. Science and
      Technology, 10, No. 3, 256-262, 1976

(11) GUICHERIT, R., K.D. VAN DEN HOUT, C. HYGEN, H. VAN DUUREN,
      F. G. RÖMER, J.W. VILJEER: Conversion Rate of Nitrogen
      Oxides in a Polluted Atmosphere, Paper presented at 11th
      NATO-CCMS International Technical Meeting on Air Pollu-
      tion Modelling and its Applications, Amsterdam, 24-27
      November 1980, appeared in the Proceedings

(12) KERR, J.A. et al.:   The Mechanism of Photochemical Smog
      Formation, in: Adv. in Environm. Science and Technology,
      (Hrsg. J. Pitts u. R. Metcalf), 1-262, John Wiley&Sons,
      N.Y. 1974

(13) SCHURATH, U.:   Physikalisch-chemische Grundlagen der
      Photosmog-Bildung unter Berücksichtigung von Smogkammer-
      Messungen in: UBA-Forschungsbericht Nr. 79-10402502/03/04
      1979

(14) HUIE, R.E., J.T. HERRON: Temperature Dependance of the
      Rate Constants for Reactions of Ozone with some Olefines
      Int. chem. Kinetic. Sympos. I, 165-181, 1975

(15) ATKINS, R., J.N. PITTS: Rate Constants for the Reaction
      of OH-Radicals with Propylene and the Butenes over the
      Temperature range 297-425 K. J. Chem. Phys. 63, 3591-3595,
      1975

(16) DARNALL, K. R., R. ATKINS, J. N. PITTS: Rate Constants
      for the Reaction of the OH-Radical with Selected Alkanes
      at 300 K, J. Phys. Chem. 82, 1581-1584, 1978

(17) SINGH, H.B., R. MARTINEZ, D.G. HENDRY, R.J. JAFFE,
      B. JOHNSON: Assessment of the Oxidant-Forming Potential
      of the Light Saturated Hydrocarbons in the Atmosphere,
      Env. Science and Techn., Vol 15, 1, 113-119, 1981

STUDIES OF THE FATE OF ATMOSPHERIC EMISSIONS IN POWER PLANT PLUMES OVER THE NORTH SEA

A.S. KALLEND, A.R.W. MARSH, G.M. GLOVER, A.H. WEBB, D.J. MOORE
P.A. CLARK, B.E.A. FISHER, D.J.A. DEAR, P. LIGHTMAN, C.K. LAIRD

CENTRAL ELECTRICITY RESEARCH LABORATORIES

Summary

An intial review is presented of a programme of aircraft flights to follow the evolution of power plant plumes over distances of several hundred kilometres. Plumes were successfully tracked in real time using SF_6 tracer for distances of up to 700 km and SO_2 oxidation rates determined. The results of measurements of pH and detailed chemical analysis of cloud water both in clean and polluted air are also reported.

1. Introduction

The transport of pollutants from large point sources over distances of the order of 1000 km has been linked with ecological effects observed in parts of the world normally considered remote from industry and its effects. In particular, the long range transport of sulphur dioxide and, to a lesser degree, of nitrogen oxides is suspected of being directly linked with the acidification of surface water through acid precipitation (e.g. Oden, 1968, Likens and Bormann, 1974). Particular areas of concern are Southern Scandinavia, the N. Eastern United States and Southern Ontario in Canada.

A complex chain of processes links the production of an emission with any subsequent environmental effect. Emissions may be modified by physical dilution, deposition, homogeneous gas phase reactions, heterogeneous reactions on particles or reaction in water droplets and the rates and extent of these processes may vary widely, influenced by such factors as wind speed, solar intensity and trace chemical constituents in the ambient air. In such circumstances it is unrealistic to expect, a priori, a simple association of cause and effect.

Major areas of uncertainty exist in this sequence of processes. In particular most field studies of plume transport have been confined to a maximum distance of ∿ 100 km from source so that there is a need for case studies over longer distances to establish a better data base for modelling long range transport. For the same reason few measurements of oxidation rates of plume constituents exist over long distances under various atmospheric conditions and, although reactions in cloud are thought to be a major route for the oxidation of SO_2 this prediction is based solely on the results of laboratory measurements (Junge and Ryan, 1958, Urone and Schroeder, 1969, Erickson et al, 1977, Penkett et al, 1979) and there have been relatively few measurements even of cloud acidity, no comprehensive chemical analysis reported except from mountain-top sites (Falconer and Kadlecek, 1980) and no measurements at all in the neighbourhood of clearly defineable plumes.

To address some of these uncertainties Central Electricity Research Laboratories, in 1979, initiated a programme of aircraft flights in

collaboration with the British Meteorological Office and with support from
E.P.R.I. to investigate the interactions of power plant plumes with the
atmosphere during transport over distances of several hundred kilometres.
The first stage of this programme has now been completed and we present
here a brief initial review of some of the early results.

2.   Experimental

Details of the experimental equipment carried on board the Meteorological Office Lockheed C130 Hercules and the Cranfield Insitute of
Technology Jetstream aircraft will be described elsewhere. Briefly
each aircraft is equipped with flame photometric SO_2 analysers and
chemiluminescent instruments to measure ozone, nitric oxide and total
nitrogen compounds, all this equipment being modified for aircraft use.
The C130 aircraft carries equipment for particle sizing and droplet
imaging and both aircraft have facilities to take filter samples. These
are analysed for aerosol material by a combination of ion chromatography
and spectrographic techniques. Absorbent filters have also been used to
measure sub ppb background levels of SO_2 (Johnson and Atkins, 1975). The
analytical sensitivity is sufficient to permit the use of 10-15 min
sampling periods in the plume, thus enabling a degree of structure to be
determined. In addition to the chemical instrumentation, the C130
aircraft is also equipped to determine a wide range of meteorological
parameters.
Each aircraft also carries a cloud water collector, based on centri-
fugal separation using a static vane within a 50 mm diameter pipe through
which air flows by ram effect. Wind tunnel tests have shown that >90%
of cloud droplets above 7 μm diameter were collected and the aircraft
experiments showed that cloud could satisfactorily be sampled provided
the liquid water content exceeded $\sim$ 0.3 g/m^3. After separation part
of the water flows to a glass electrode which gives a continuous pH record
and part to a fraction collector for later analysis.
An essential part of the programme has been the successive monitoring
of the same parcel of air at different times and an important aid in
achieving this has been the use of a tracer in the power station plume.
SF_6 was injected continuously, usually at 50 kg/hour and the plume was
also time-marked by injecting 25 kg of perfluoro-methyl-cyclohexane during
a half hour period, repeated at about six hourly intervals. The tracers
are detected in the aircraft continuously by an instrument developed in
these laboratories (Blackburn and Dear, 1981) and the continuous measurements are supported by laboratory chromatographic analysis of bag samples
taken on the aircraft.

3.   Aircraft Flight Programme

Flights were made over and around the U.K. mostly in conditions of
the prevailing south-westerly winds. The objective was to establish
complete histories of air masses in different meteorological conditions so
measurements were made on background air, upwind of the emission sources,
downwind in the near-field of the power station, and the air mass approach-
ed the far side of the North Sea. To achieve continuity of measurement
within the timescale established by the wind-speed, co-ordinated flights
by the two aircraft were required. An idealized flight pattern is shown
in Fig. 1 with the Jetstream making the background flight off Cornwall
and the C130 operating over the North Sea. The location and timing of
individual flights naturally varied with the meteorological conditions

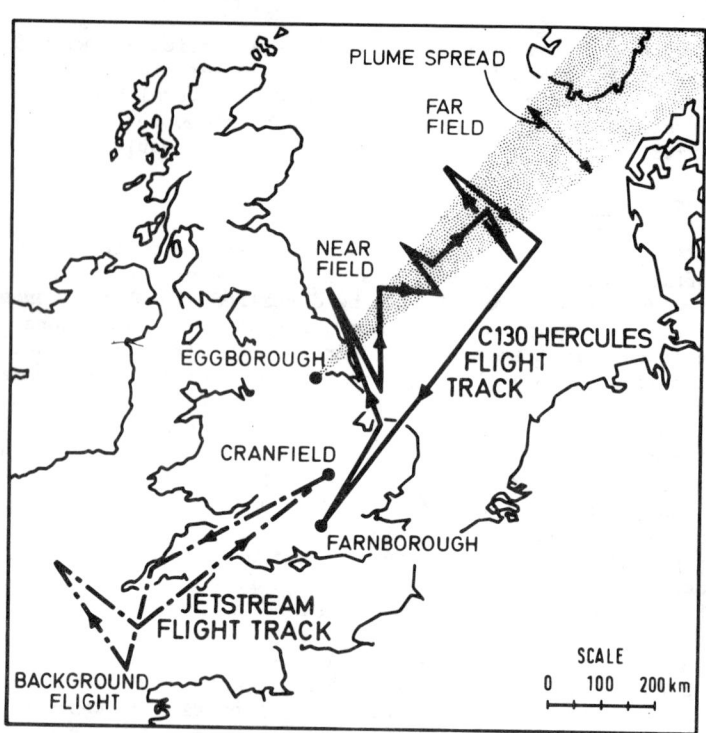

FIG.1 IDEALIZED FLIGHT PLAN

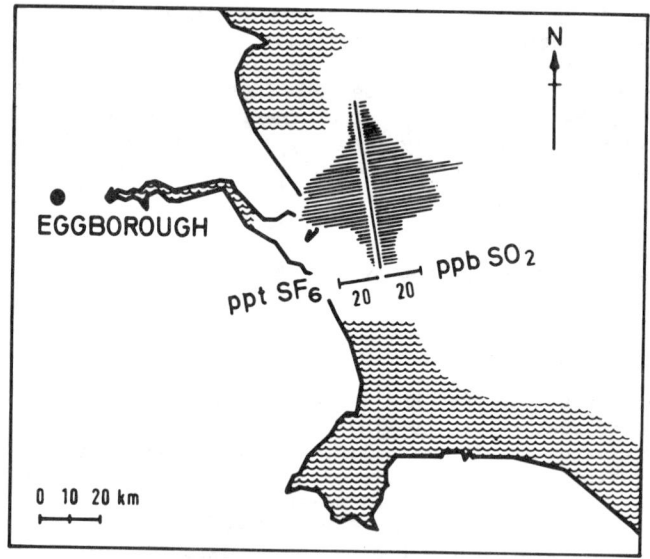

FIG. 2 SO₂ & TRACER MEASUREMENTS IN THE NEAR FIELD DISPLAYED ALONG THE AIRCRAFT TRACK

and, depending mainly on the windspeed, the experimental programmes extended over a period between one and three days. To date the C130 has carried out eleven flights and the Jetstream twenty two, some of the latter being shorter flights for development or other special purposes.

4. Results and Discussion

A complete review of all the results is not possible at this stage. We, therefore, highlight some of the more interesting results, with the reservation that the conclusions should not be generalized without due care.

4.1 Gas and Filter Measurements

Figs 2 and 3 show typical measurements made in the medium-near field about 100 km from Eggborough Power Station on 11/12/81, a clear sunny day in mid-winter. Fig. 2 gives a topographical representation of the SO_2 and SF_6 measurements and Fig. 3 is a corresponding chronological plot of ozone and total oxides of nitrogen obtained during repeated traverses of the plume. The influence of the power station plume is very apparent with SO_2, SF_6 and NO_x rising to maxima, while the O_3 level fell.

As can be seen from Fig. 2, the ratio of SO_2 to SF_6 in the plume at this point is about x1000, very similar to their ratio in the original emission. The implication is that the observed plume is predominately that of Eggborough P.S. with little contribution from other sources or loss of SO_2 by dry deposition and chemical conversion. The reasoning is shown in a different way in Fig. 4 which displays "excess" SO_2,

$$\text{Excess } SO_2 = \text{Measured } SO_2 - k \times \text{Measured } SF_6$$

where k is the ratio os SO_2 to SF_6 in the emission. If the plume simply dilutes physically, by mixture with clean background air, then excess SO_2 will remain at zero. As can be seen from Fig. 4, this is a fair approximation, for the instantaneous concentrations.

An alternative way of considering the tracer data is to calculate total fluxes from a cross-plume integration of the measurements, together with the measured wind speed (about 15 m/s and the mixing depth (about 750m). The results from different runs range from 180 to 270 kg/hour with an average of 226 kg/hour and this, evidently, is much larger than the actual injection rate of 50 kg/hour. The most likely reason is that, the plume had not expanded to fill the whole mixing depth and was still concentrated mainly within a layer about 200m deep. Evidence for this view is also given in Fig. 5 which shows height profiles of SO_2, NO_x and O_3 at a location just south of the main plume. The limit of the mixing layer at 750m is clear, but subsidary layering at about 250m and 600m seems to be present. Since our measurements were made at 300m, it seems probable that the power station plume was restricted mainly to the lower layer, and separated from the plumes of neighbouring major sources.

A further example of plume behaviour under different meteorological conditions is given by the results of the flight on 29/1/81 (Fig 6). The Eggborough plume was identified very clearly by the SF_6 measurements close to the Danish coast, over 600 km from the power station. In the same location, SO_2 concentrations similar to those in Fig. 2 (at 100 km from the station) were observed. These levels both of SF_6 and SO_2 were high and appear to have arisen from a combination of an accelerating air flow with subsidence.

The NO_x and ozone measurements in Fig. 3 for 11/12/80 indicate the low level of photochemical activity which would be expected at this time

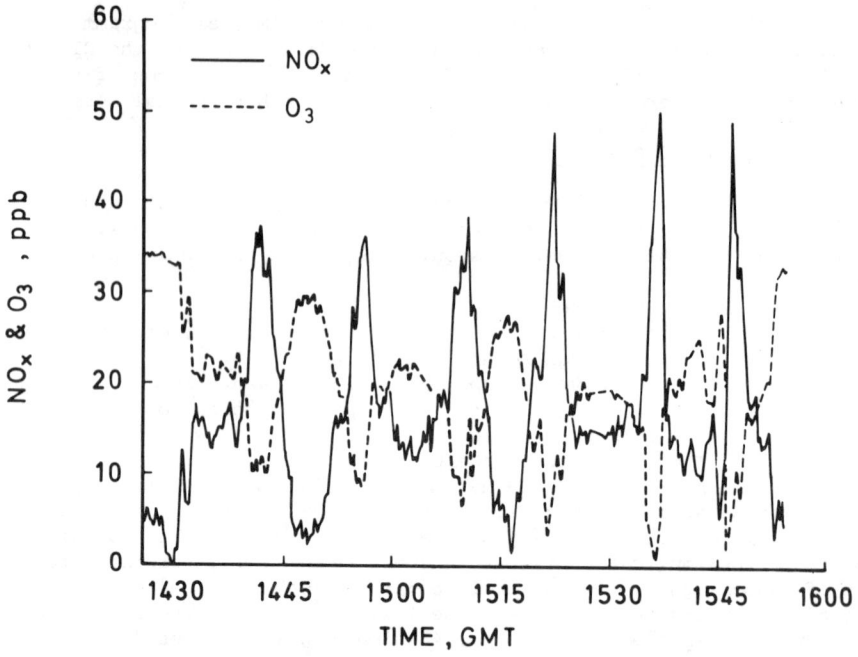

FIG. 3   OZONE & OXIDES OF NITROGEN :-
REPEATED PLUME TRAVERSES

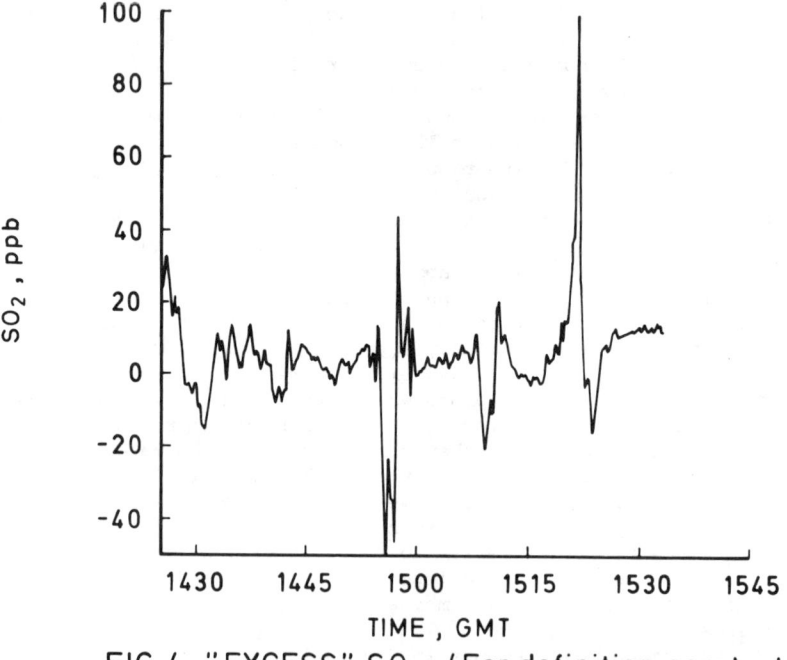

FIG. 4   "EXCESS" SO$_2$ (For definition see text.)

of year in the U.K. At this stage, 1½-2 hours after emission, the NO
had mostly oxidized to NO_2 and the peak levels of NO seen were only about
10 ppb. The sum of $NO_2 + O_3$ remained fairly constant at 35-40 ppb during
the period shown on Fig. 3, as would be expected if NO is being oxidised
predominantly by ozone. The quantity (NO) (O_3/NO_2) which is related to
the balance between the reaction of NO and O_3 and the reverse photolytic
decomposition of NO_2 had an average value of 2.5 ppb. This indicates the
low isolation and is quite consistent with a vlue of 2.6 ppb calculated
from the rate constants and an average solar zenith angle of $80^\circ C$.

Several preliminary estimates of the rate of oxidation of SO_2 have
been made from the measurements of sulphate aerosol in the plume (less
that which came from the background air). For a clear sunny day in
mid-summer (18/6/80) measurements made at around mid-day just over 100 km
downwind of the source, corresponding to a plume travel time of about 5½
hrs, gave an average oxidation rate of $\sim$ 1% hr evaluated from the
application of a single layer diffusion model. This is quite consistent
with predictions from reactive plume models. A similar rate is obtained
for the mid-winter flight (11/12/80) using the aerosol data in Table 1
although a substantially lower rate is predicted for mid-winter by
chemical kinetic models. In contrast with these observations, the results
of 3/3/81 indicated a much higher SO_2 removal rate. On this occasion, with
wind from the NE the Jetstream located the SF_6 tracer in the north-west
Midlands, about 150km from Eggborough. The associated SO_2 concentration was
near the limit of detection ($\sim$ 2ppb) indicating an SO_2 removal rate
of over 30% per hour. Although precipitation had occured, the reasons
for this high overall rate are not yet clear, but an indication may be
given by the aerosol analysis (Table 1, last line). This shows not
merely high sulphate, but also high ammonium leading to an alkaline ion
balance. It seems possible that the high SO_2 removal rate may be
associated with high ammonia levels in the background air mass.

The general pattern of the measurements thus shows great variation.
However, some systematic features can be seen, for example, in the
aerosol analyses in Table 1. The background aerosol sample taken on
11/12/80 is typical of 'clean' oceanic air. Apart from sodium chloride
from sea salt, the only significant constituent is a relatively low level
of sulphate. The overall ion balance appears slightly acid, but as all
the constituents are in such low concentration this may be an artifact of
the analytical uncertainties. The background sample from 22/7/80 is
different in containing much higher concentrations of sulphate and ammonium. Although the local wind was south westerly, a back trajectory indicated that the air mass was of continental origin and the sulphate is
clearly man-made. The aerosol is still fairly neutral, because of the
ammonium level. The background sample from 2/3/81 has intermediate levels
of sulphate and ammonium and also has the near-neutral ion balance which
seems to be typical of background aerosols which have been airborne for
some days.

The plume aerosols for 11/12/80 and 23/7/80 differ from their corresponding background samples (11/12/80 and 22/12/80) mainly in containing
higher sulphate levels and being correspondingly more acidic. Clearly,
the major process to have occurred is the oxidation of SO_2 to sulphuric
acid aerosol. It is notable that, despite the presence of significant
concentrations of oxides of nitrogen in the plume, there was little nitrate
in the aerosol - probably reflecting the volatility of nitric acid and its
derivatives. The plume aerosol from 3/3/81, as discussed above, was
different again in that it had both higher sulphate and ammonium levels
resulting in a slightly alkaline overall ion balance.

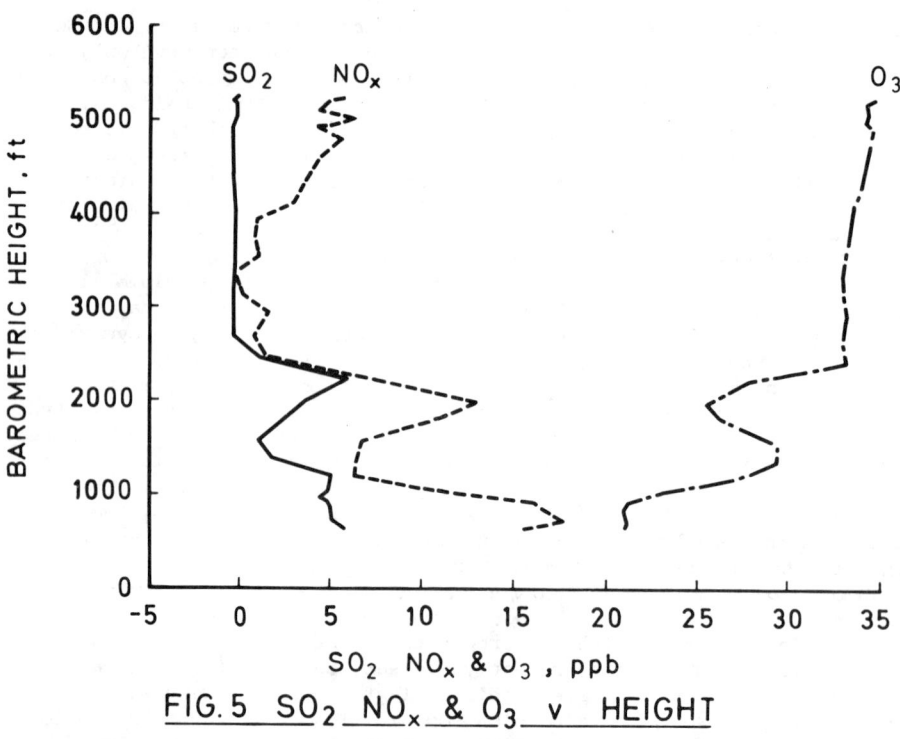

FIG.5 SO₂ NOₓ & O₃ v HEIGHT

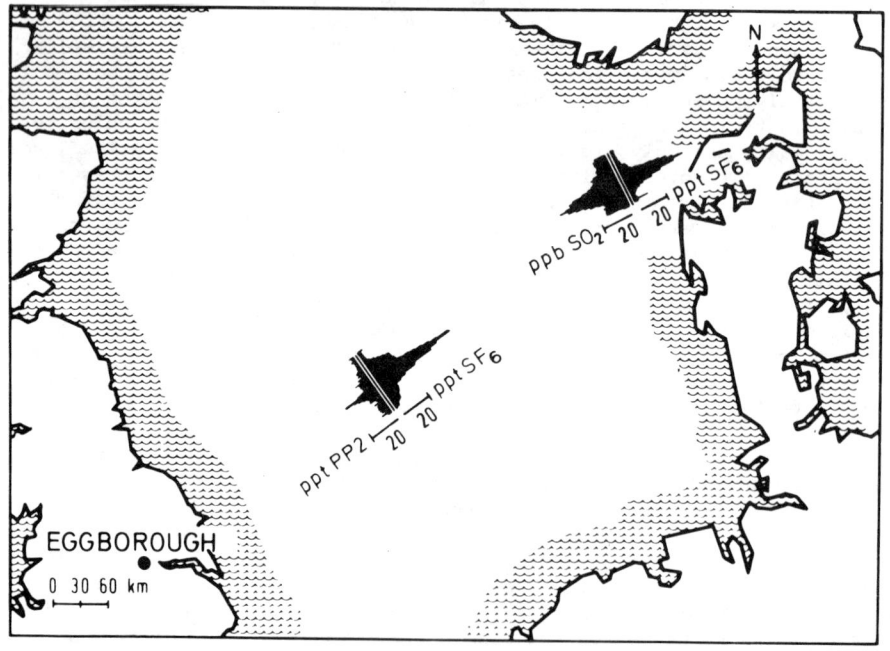

FIG.6 SO₂ & TRACER MEASUREMENTS IN FAR FIELD

Table 1. Aerosol Loadings ($n.eq/m^3$) in Background Air & Plumes

	Cl^-	NO_3^-	$SO_4^=$	Na^+	K^+	Ca^{++}	Mg^{++}	NH_4^+	H^+ Calc
Background 11/12/80 Lancashire Coast	10	ND	9	7	2	ND	ND	ND	11
Background 22/7/80 Start Point towards Cherbourg	7	7	160	9	5	-	-	153	7
Background 2/3/81 Lincolnshire Coast	ND	12	63	12	2	15	3	66	-5
Plume Measurement 11/12/80 Off Humber	19	ND	50	24	ND	7	ND	16	23
Plume Measurement 23/7/80 Yorkshire Coast	11	31	348	24	8	14	-	183	161
Plume Measurement 3/3/81 West Midlands	ND	3	351	13	5	19	4	331	-17

Table 2 Composition of Cloud and Precipitation

Micro eq/litre	$SO_4^=$	NO_3^-	Cl^-	Na^+	K^+	Ca^{++}	Mg^{++}	NH_4^+	pH
Cloud, mean upwind U.K.	261	73	1154	905	112	53	237	126	4.3
Cloud, mean downwind U.K.	558	240	857		576	27	163	169	3.5
Precipitation mean OECD Sites N1-N10	60	27	112	100	-	12	22	30	4.35

4.2 Cloud Water Chemistry

Conditions were suitable for collecting significant quantities of cloud water on eight flights and over one hundred individual samples have been analysed. Although the majority were from non-precipitating layer cloud it is useful as a first step to compare the measured concentrations with typical observations of precipitation composition. Such a comparison is presented in Table 2 from which it is seen that most solute ions are more concentrated in cloud water than in precipitation by about an order of magnitude, although the ratios are similar. Falconer and Kadlecek (1980) report similar high values (40-1206 μ equ l^{-1} SO_4) at the relatively remote site at Whiteface Mountain and, more recently, Hegg and Hobbs (1981) observed sulphate levels up to 400 μ equ l^{-1} over Los Angeles and Western Washington. Observations in precipitating cloud give lower values, much more comparable to precipitation itself.

The difference between the cloud water upwind and downwind of the U.K. lies primarily in the higher levels of SO_4^-, NO_3^- and NH_4^+ observed in the downwind samples. Sulphate and nitrate clearly derive from man's activities and the ammonium level is probably mechanistically related to these. Sodium and chloride are likely to derive mostly from sea salt, although the downwind cloudwater also contains excess chloride, probably from coal-burning emissions. The land origin of calcium is clearly reflected.

The cloudwater results also present features which are less easy to quantify. For example, comparing cloudwater with the aerosol analyses in Table 1 it can be seen that the cloudwater contains much higher levels of sodium chloride and nitrate, relative to sulphate, than are found in aerosols. This probably reflects the dominance of giant sea salt particles in cloud nucleation processes, but the effect cannot yet be modelled quantitatively.

This brief review is necessarily preliminary in nature. Each single flight of the C130 aircraft recorded over one million data points. Further analysis especially in the area of coordinating chemical and meteorological data, will continue to give valuable results.

References

A.J. Blackburn and D.J.A. Dear, 1981, A Continuous Monitor for Sulphur hexafluouride and perfluouromethyl cyclohexane in air, (to be published).

D.A. Johnson and D.H.F. Atkins 1975 An Airborne system for the sampling and analysis of sulphur dioxide and atmospheric aerosols. Atmos. Environ. $\underline{9}$, 825-9

Oden, S., 1968, Swedish Nat. Sci. Res. Council Ecology Committee Bul. 1, p. 68

Likens, G.E. and Bormann F.H., 1974, Science, $\underline{184}$, 1176

Junge, C.E. and Ryan, T.G., 1958, Q.J. Roy. Met. Soc. $\underline{84}$, 46.

Urone, P. and Schroeder, W.H., 1969, Envir. Sci. Tech. $\underline{3}$, 436.

Erickson, R.E., Leland, M.Y., Clark, R.L. and McEwen, D., 1977 Atmos. Env., $\underline{11}$, 813

Falconer, D. and Kadlecek, J.A., 1980, A.S.R.C.

Publication No. 748, A.S.R.C., Albany, New York

Penkett, S.A., Jones, B.M.R., Brice, K.A., and Eggleton, A.E.J., 1979
Atmos. Env. $\underline{13}$, 123

Hegg, D.A. and Hobbs, P.V., 1981, Atmos. Env., $\underline{15}$, 1957,

Acknowledgement

The authors acknowledge the helpful participation of many of their
colleagues at CERL and, in particular, that of Mr D. Hattan and
Dr J. Crabtree of the Meteorological Office.

This paper is published by permission of the Central Electricity
Generating Board.

PHOTOCHEMICAL OZONE TRANSPORT IN AN INDUSTRIAL COASTAL AREA

G. GIOVANELLI (INST. FISBAT-CNR, BOLOGNA, ITALY),
F. FORTEZZA, L. MINGUZZI, V. STROCCHI and W. VANDINI
(Public Health and Prevention Lab., RAVENNA, ITALY)

Abstract

During summer months the entire coast near Ravenna experiences
photochemical smog typical of coastal industrial area. Local air
circulation can bring about transport and accumulation of precursors
of ozone over the sea during the morning hours while the land breeze is
blowing; during the afternoon hours, when the sea breeze sets in, one
finds ozone levels over the land which remain high until late evening.
On days with W or NW circulation, one finds a regular diurnal variation
of ozone well correlated with solar radiation. With winds from E-SE,
the variations of ozone are irregular and can mantain high levels (30-
60 ppb) even during the nightime hours. Horizontal transport apperars
to be the most realistic explanation for these nocturnal ozone levels.
Here, we report the examination of data from a network of gas pollution
analysers, one of which was placed on an offshore drilling rig at 20
Km from the coast.

1. INTRODUCTION

The lower tropospheric ozone concentration is controlled by different
mechanisms of production and destruction. The intrusions of stratospheric
ozone connected with events of cyclogenesis and tropopause folding seem to
be the main source of global tropospheric ozone. On the other hand, if we
take into account the remarkable concentration levels and diurnal variations
in ozone content on a local scale, there seems to be a dependence on the
nitrogen oxide and hydrocarbon reactions.

Recently several authors [1,2,3 and 4] poinded out that the transport
phenomena of ozone even with trajectories of hundreds of kilometers.

In the industrial area of Ravenna (see Fig.1) are found petrochemical
facilities (ANIC and SAROM), three thermal power plants and a number of
middle-size and small factories. In september and october 1978 we carried
out the first measurements of ozone . The concentration levels (up to 90
ppb / hourly average) showed that the entire coastal area was affecded by
photochemical reactions [5],(Fig.2).

This paper reports the results of some measurement campaigns intended
to show clearly the ozone production and destruction phenomena, as well as
its possible recirculation mechanisms under breeze conditions.

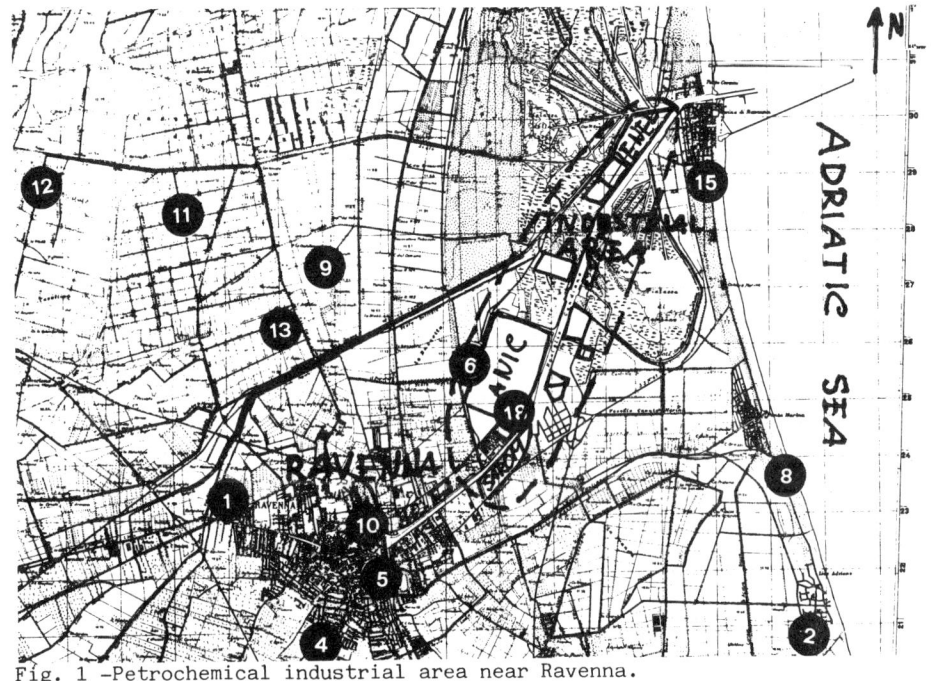

Fig. 1 –Petrochemical industrial area near Ravenna.

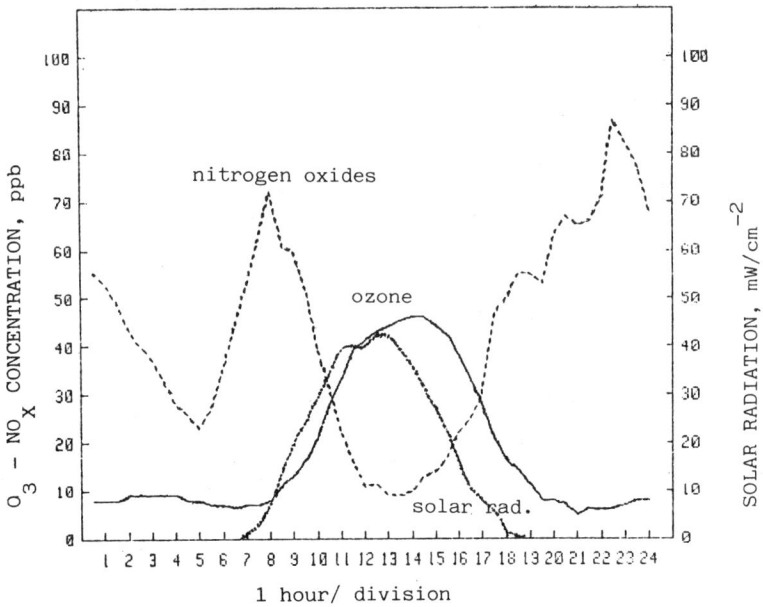

Fig. 2 –Monthly average (October 1978) diurnal variation of O_3, NO_x and
solar radiation referred to the AMGA station (point 6, see Fig.1).

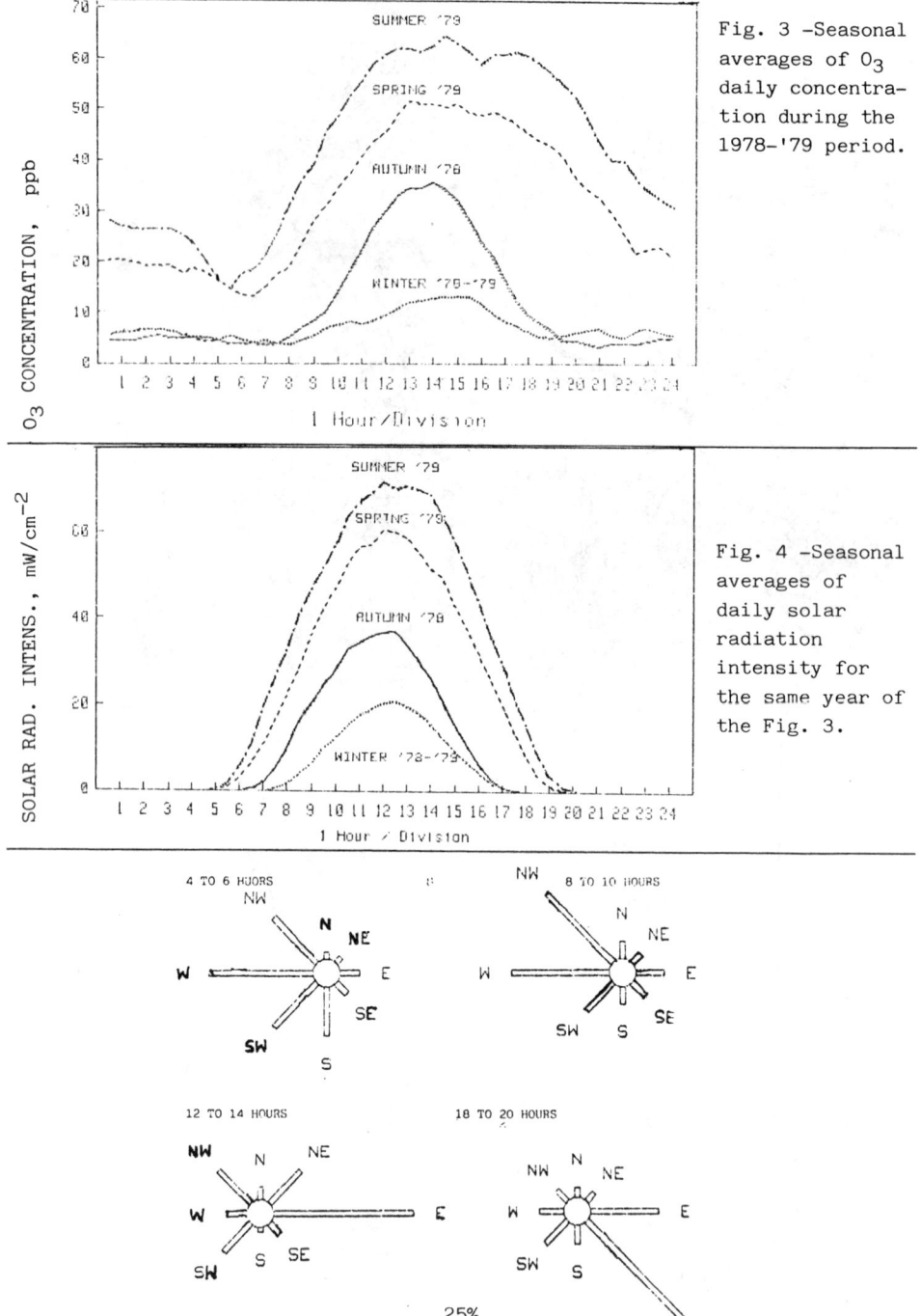

Fig. 3 —Seasonal averages of O_3 daily concentration during the 1978-'79 period.

Fig. 4 —Seasonal averages of daily solar radiation intensity for the same year of the Fig. 3.

Fig. 5 - Roses of the wind frequencies (July 1978).

2. MEASUREMENT SITES AND DATA ANALYSIS

The seasonal averages of ozone daily concentration are reported in Fig. 3. The corresponding values of the total solar radiation intensity are shown in Fig. 4. It can be seen that the curves of ozone diurnal variation in spring and summer (Fig. 3) show a maximum at noon hours and a prolonged high level into the late afternoon.

Taking into account that the spring-summer period is generally characterized by breeze circulation, the diurnal variation of ozone concentration has been calculated according to the four main wind directions (W, NW, E, SE, see also the Fig. 5). Dividing the data into the two classes W-NW (land breeze) and E-SE (sea breeze) gives two very different curves of the ozone diurnal variation |6|, (see Fig. 6, 7).

In the case of W-NW winds the ozone concentration is well correlated with the solar radiation intensity and the nitrogen oxide concentration, but a second peak in the ozone curve deos not appear in the late afternoon and the nocturnal levels of ozone are low.

In the case of E-SE vinds the correlation becomes worse, the second peak in the ozone curve is very marked and the nocturnal levels are high.

High nocturnal ozone levels were also measured by Steinberger and Ganor |7| in Tel-Aviv area. The authors suggest three possible mechanisms justifying the phenomenon: i) horizontal transport, ii) stratospheric descent, iii) ozone trapped under an inversion layer. In the case of the coastal area near Ravenna only the i) may justify the high ozone levels in the late afternoon and in the night.

But the horizontal transport hypothesis involves the formation of an ozone-rich air-mass over the sea which is suggested as being trasported by the sea breeze to the land during the afternoon. The process can last during the night too if the wind direction remains E-SE.

To verify the ozone accumulation and recirculation hypothesis, three measuremant points have been chosen for ozone and nitrogen oxides along a line, that was also the prevaling direction of the land or sea breeze during the summer months (E-SE, W-NW line). The first station (AMGA) is included in the gas pollution monitor network of the Public Health and Prevention Laboratory of Ravenna and is near the industrial area (about 6 Km from the coast, see Fig. 8). The second station (COOP 3) is on the coast in a camping area near Punta Marina, while the third (PCB) is on an offshore drilling platform at 22 km from Punta Marina.

The typical diurnal variations of ozone concentrations for the three monitor points are shown in Fig. 9. These are got plotting the half-hour averages of the ozone data for the period between the 18th and 31st of July 1980. From the analysis of Fig. 9, one can draw the following evaluations:

i) during the day the air over the sea in front of the industrial area is rich with photochemical ozone. In fact the ozone values measured on the offshore drilling platform (PCB station) are higher on the average than the values measured in the other two land stations.

ii) Also during the night the ozone levels on the sea station remain rather

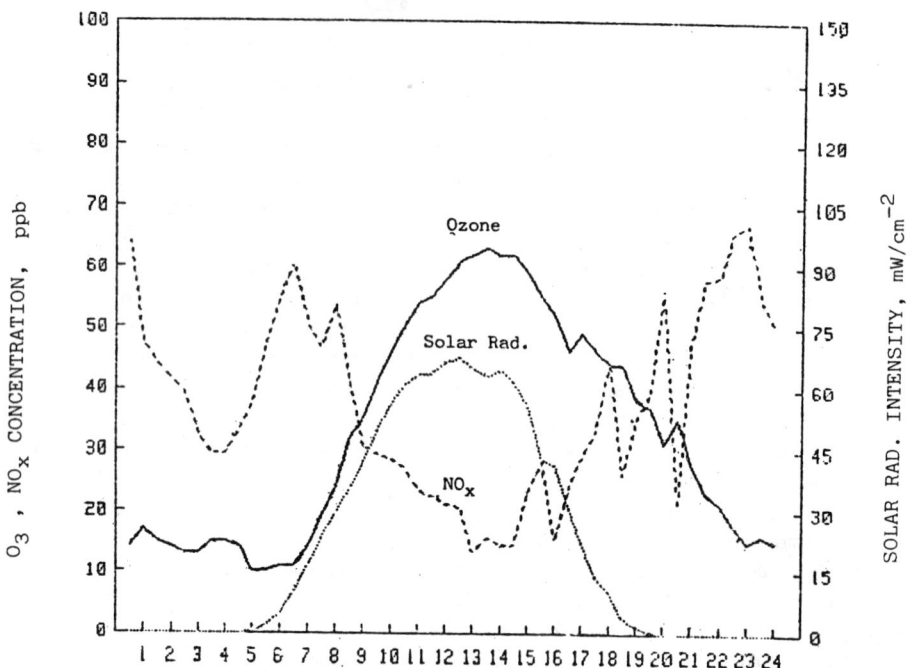

Fig.6 - Diurnal variation of O_3, NO_x and Solar Rad. with wind from W+NW.

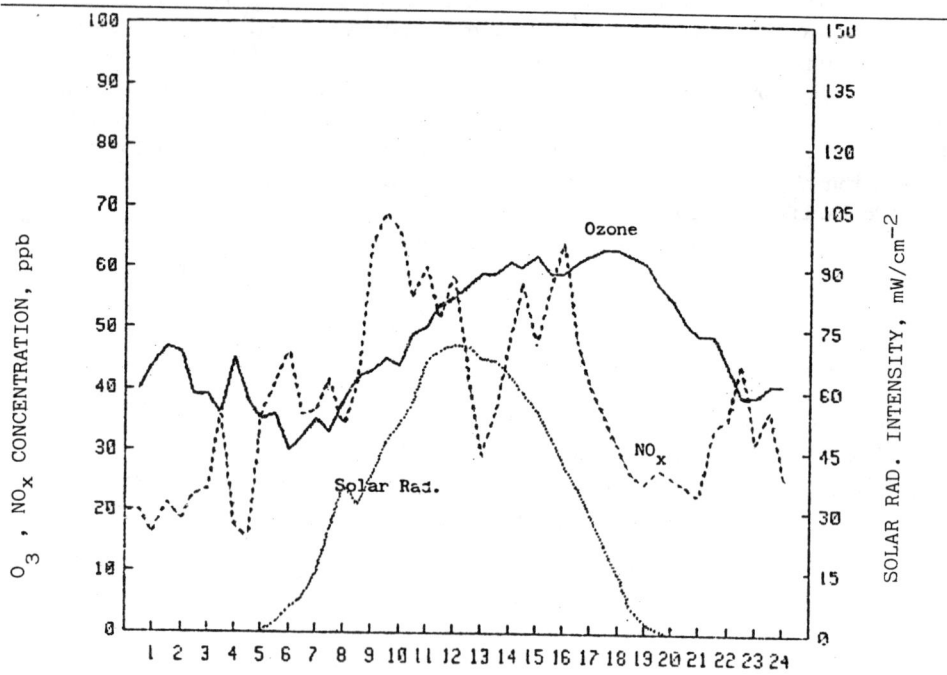

Fig.7 - Diurnal variation of O_3, NO_x and Solar Rad. with wind from E+SE.

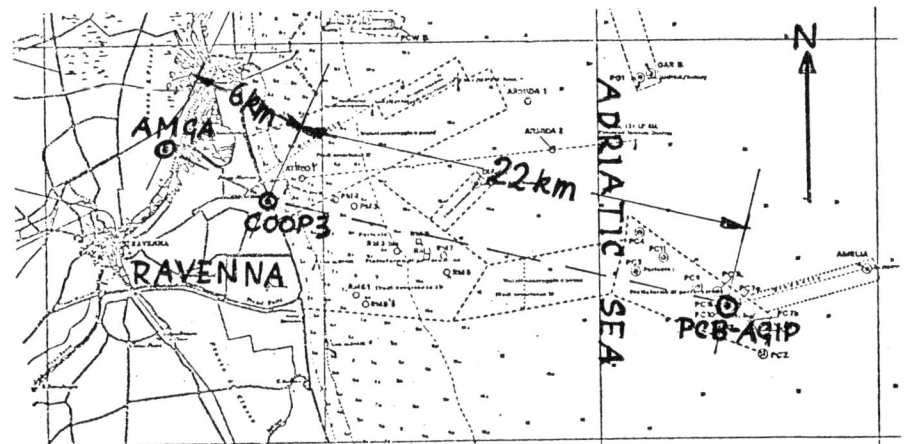

Fig. 8 – Mapping of the three measurement points along the prevaling
direction of the land and sea breeze.

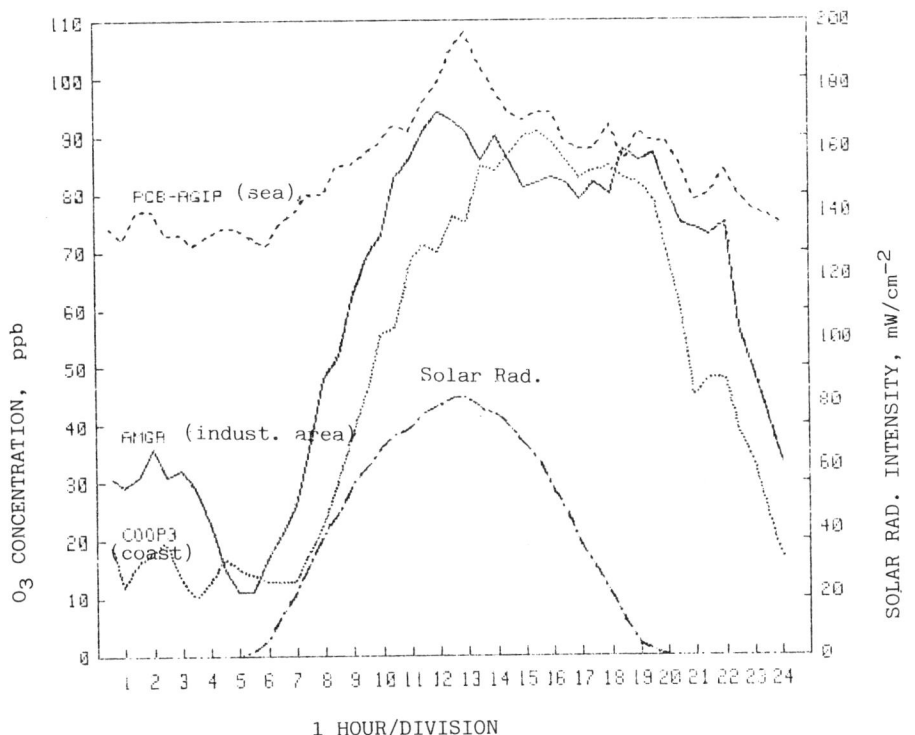

1 HOUR/DIVISION

Fig. 9 – Typical diurnal variations of O_3 and Solar Rad. for the three
monitoring points shown in Fig.8 (data of the period between
18th and 31st of July 1980).

high (70-80 ppb) and are always higher than the corresponding values
measured on the land. These nocturnal ozone levels on the sea are con-
siderably above the average background values (20-40 ppb), |8|. This
may be caused by a combination of weak vertical mixing due to the high
atmospheric stability on the sea at night and slower velocities in the
destruction processes (physical and chemical quenching) of O_3 on the sea.
iii) The three O_3 curves in the growing part (in the morning) follow well
the solar radiation curve. In fact the sea area in front of the industrial coast is affected by nitrogen oxides under the land breeze. Thus
the ozone due to the solar radiation simultaneously forms on the entire
sea area. On the contrary during the afternoon the ozone levels in the
three stations are affected by transport phenomena due to the sea breeze.
The descending parts of the concentration curves show different temporal
slopes. In fact the time lag (at equal concentration) between two curves
is comparable to the time interval that an air mass takes to run the
distance between the two stations.

Figs. 10, 11, 12 show clearly what has been said above about the correlation between the ozone concentration values to those of the total solar
radiation intensity.[*] We found the best correlation line between O_3 concentration and solar radiation in the morning (increasing radiation and land
breeze): at the AMGA station with 0.5 hour delay, at the COOP3 station 1.5
hour, and PCB-AGIP 1 hour. In the afternoon (decreasing radiation and sea
breeze) the best correlation is: at PCB-AGIP -0.5 hour delay (in this case
the ozone reduction precedes that of solar radiation), at COOP3 4.5 hours,
and at AMGA 5.5 hours. The difference between the latter two delays (1
hour) is the time necessary for an air mass to run the distance between the
COOP3 and AMGA stations. The negative delay found at the PCB-AGIP station
in the afternoon suggests that the maximum fall-out ozone area is farther
from the PCB-AGIP platform.

(*) The ozone and solar radiation data in these correlations have been
divided into two classes: morning and afternoon.

3. CHEMICAL AND PHYSICAL DESTRUCTION PROCESSES OF OZONE AT GROUND LEVEL

The data presented in the above paragraph show that, in a coastal area,
the photochemically produced ozone is subject to phenomena of local circulation. It is then interesting to see what part the processes of destruction
of the ozone play along the sea trajectories.

On the areas near to the ground, the destruction processes of ozone
are principally: i) physical quenching at contact with the surface,
and ii) chemical quenching in the lowest layers of the troposphere.
i) Physical quenching: the dimensions of these processes can be estimated
taking a column of air of unitary width, length L and height H, that
contains a concentration of O_3 of uniform density C. If q is the reaction constant for the process of destruction, the concentration of O_3
in the air column is:

$$C = C_o \exp^{(-q/H \cdot t)}$$

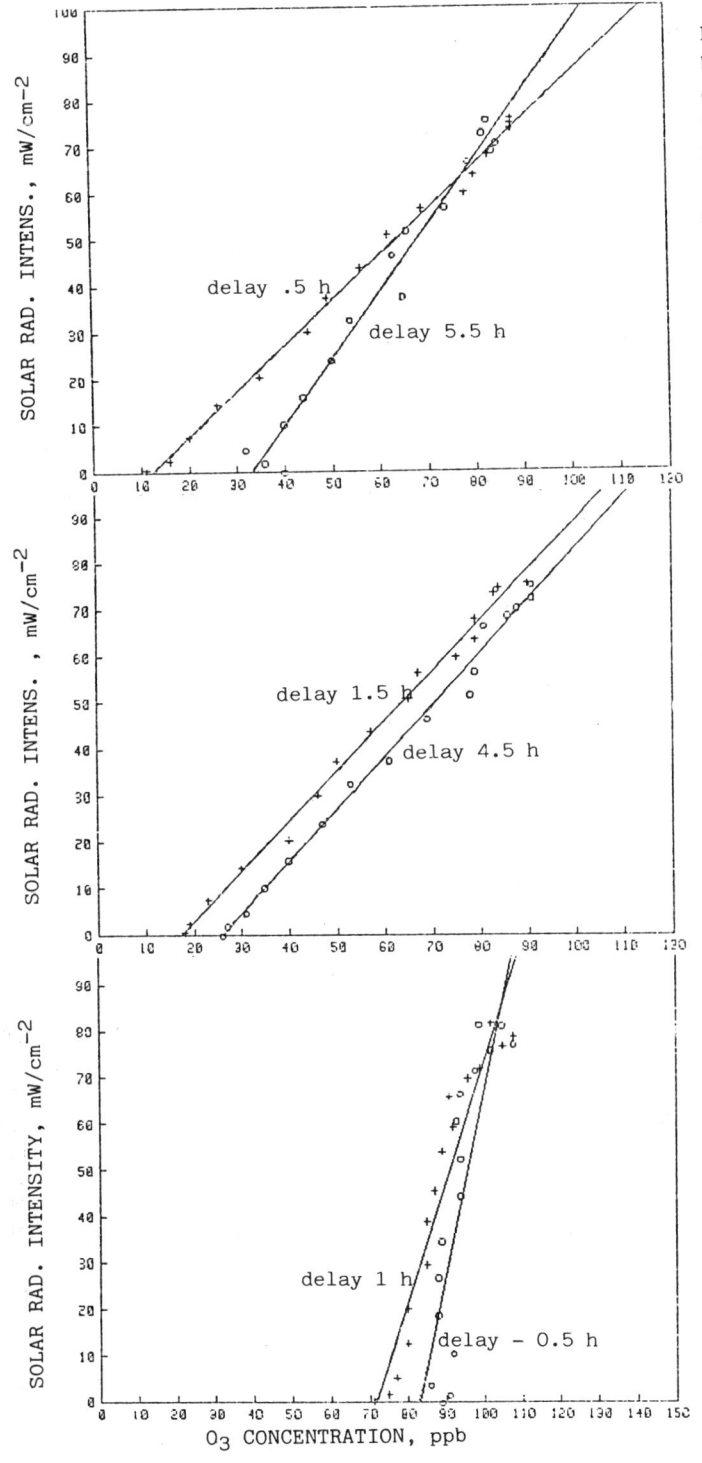

Fig.10 -Correlation
between the Ozone
of AMGA monitor and
the Solar Radiat.:
1) in the morning
   (+) line;
2) in the afternoon
   (o) line.

Fig;11 -Correlation
between O_3 of
COOP3 monitor and
Solar Rad. :
(+) morning line;
(o) afternoon line.

Fig.12 -Correlation
between O_3 of
PCB-AGIP monitor
and Solar Rad. :
(+) morning line;
(o) afternoon line.

where C_0 is the initial concentration of O_3 and H/q is the "lifetime" of the ozone, |9|.

For our case one can take H as 200 m (for example with an inversion layer at this level). Below we give H/q for different values of q for surfaces of earth, sea and snow |10|.

	q (cm/sec)	H/q (hours)
earth	0.6	9
sea	0.04	140
snow	0.02	280

Over the sea ozone has a lifetime 15 times longer than on dry land surfaces.

ii) Chemical quenching: in particular, in polluted areas, different chemical reactions have an important role in ozone destruction. The reactions with NO and olefinic hydrocarbons are the most important:

$$O_3 + NO \longrightarrow O_2 + NO_2 \qquad k = 23 \ ppm^{-1} min^{-1}$$

$$O_3 + HC \longrightarrow products \qquad k = 5 \cdot 10^{-2} \ ppm^{-1} min^{-1} \ (for \ olefins)$$

These two processes are, however, important only at high industrial and/ or vehicular concentrations. Over the sea for lack of NO and hydrocarbons, these processes can be neglected (when the wind is blowing from the sea).

Ozone can then be involved in recirculation processes from the sea with breezes and may also be transported to large distances without considerable destructions.

4. CONCLUSIONS

The Po Valley can be considered as a large natural reactor for chemical transformations in the atmosphere. It is, in fact, closed on three sides by mountains, while the fourth side is coast with reducest dimension breeze cells. In the summer, in anticyclonic conditions, the weak local circulations, the temperature inversions and the high solar radiation can create periods of high photochemical smog. Moreover, in the coastal petroleum centres, as at Ravenna, these phenomena can be prolonged in time because of the recirculation in the breeze cells.

Over the sea ozone and its precursors can find particular conditions, such as a greater atmospheric stability, limiter vertical exchanges, and lower destruction processes for which they can be transported for tens of kilometers without undergoing a particular dilution or transformation.

5. ACKNOWLEDGEMENTS

This work was suppored by a collabotation between the CNR and the Regional Government of Emilia-Romagna. The author thank the ENI for the instrunent assistence in the PCB offshore drilling platform.

6. REFERENCES

(1) Cox R.A., Eggleton A.E.J., Derwent R.G., Lovelock J.E. and Pack D.H.
    (1975) Long-range transport of photochemical ozone in the north-
    western Europe : Nature, 255, 118-121.

(2) Lyons W.A. and Cole H.S. (1976) Photochemical oxidant transport: meso-
    scale lake breeze and synoptic scale aspect: J. Appl. Met., 15,733 .

(3) Cleveland W.S. and Kleiner B. (1975) Transport of photochemical air
    pollution from Camden-Philadelphia urban complex: Env. Sci.Thecnol.,
    9, 869-871.

(4) Karl T.R. (1978) Ozone transport in St. Louis area: Atmos. Envir., 12,
    1421-1431.

(5) Fortezza F., Minguzzi L., Stocchi V. and Vandini W, (1980) Ossidi di
    azoto, idrocarburi ed ozono nell'area industriale ravennate: Acqua &
    Aria, 8, 1023-1026.

(6) Vandini W.,Minguzzi L.,Fortezza L. and Strocchi V. (1980) Assessment
    of nitrogen oxides,hydrocarbons and ozone in the industrial area of
    Ravenna through one year of measurements: Acts of V International
    Clean Air Congress: IUAPPA, October 20-26, Buenos Aires.

(7) Steinberger E.H. and Ganor E. (1980) High ozone concentrations at night
    in Jerusalem and Tel-Aviv, 1975 : Atmos. Envir., 14, 221-225.

(8) NAPCA (1970) Air quality criteria for photochemical oxidants: Report
    no. AP 63.

(9) Guicherit R. and Van Dop H. (1977) Photochemical production of ozone
    in western Europe (1971-1975) and its ralation to meteorology: Atmos.
    Envir., 11, 145-155.

(10) Aldaz L. (1969) Flux measurements of atmospheric ozone over land and
    water: J. Geophys. Research, 74, 6943-6946.

(11) Carassiti V., Chiorboli C., Bignozzi C.A., Maldotti A. and Minghuzzi L.
    (1978) Fotoinquinamento in aree industriali chimiche: effetti di mi-
    scele inquinanti: Report of Photochemical Institute, CNR, Universaty
    of Ferrara (Italy).

HERKUNFTSANALYSE DES ATMOSPHÄRISCHEN AEROSOLS IN WIEN

H. MALISSA, H. PUXBAUM UND B. WOPENKA

Institut für Analytische Chemie
Technische Universität Wien, Österreich

(Source analysis of the vienna atmospheric aerosol)

Summary

In this paper a first report about the work on source
analysis of the Vienna atmospheric aerosol is presented.
A set of 37 aerosol samples was collected with 5-stage
cascade impactors. The mass-size distribution of the aerosol was evaluated gravimetrically, covering the size range
0.1 - 25 µm AD. The elements Ca, Mg, Al, Fe, Mn, Cr, Cu,
Pb, Zn, V and Sr were analyzed by ICP-AES. Using factor
analysis 5 factors representing 93.5 % of the variance
of the data set were extracted. Three factors with simple
loading structures were associated with the source categories traffic, oil-combustion and mineralic abrasion
products. Two factors with complex factor loadings could
not be attributed to individual sources.

Zusammenfassung

Die gegenständliche Arbeit bildet einen Teilbericht über
die geplante Herkunftsanalyse des Wiener Aerosols. Sie
umfaßt die Probenahme von 37 Aerosolproben im Sommer 1981
mit 5-stufigen Kaskadenimpaktoren, die gravimetrische
Analyse der Massen - Korngrößenverteilung des Aerosols
im Korngrößenbereich 0,1 - 25 µm AD, die Analyse der
Elemente Ca, Mg, Al, Fe, Mn, Cr, Cu, Pb, Zn, V und Sr
durch ICP-AES und die Durchführung einer Faktorenanalyse
mit dem Element - Datensatz. Es wurden 5 Faktoren
extrahiert, mit welchen 93,5 % der Varianz des Datensatzes erklärt werden konnten. Drei Faktoren mit einfacher Ladungsstruktur wurden mit den Quellengruppen
Verkehr, Ölverbrennung und mineralischer Abrieb in
Verbindung gebracht. Zwei Faktoren wiesen eine komplexere Ladungsstruktur auf und konnten nicht direkt Einzel-
quellen zugeordnet werden.

1. EINLEITUNG

    Bei Untersuchungen der Massen - Korngrößenverteilung von
Aerosolproben aus dem Wiener Stadtgebiet wurden stets hohe
Anteile an feinen (< 2 µm AD) als auch an groben (> 2 µm AD)
Teilchen festgestellt (1). Um eine wirksame Reduktion der
Aerosolbelastung zu erzielen, wird Information über die
Beiträge der wichtigsten aerosolverursachenden Quellen benö-

tigt. Dies führte zum gegenständlichen Projekt der Herkunftsanalyse des atmosphärischen Aerosols in Wien.

Die Herkunftsanalyse soll an 2 umfangreichen Datensätzen
auf folgende Weise durchgeführt werden: Aerosolproben werden
in Wien in der Nähe des Stadtzentrums mit Impaktoren gesammelt
- 37 Tagesproben im Sommer 1980, 20 Halbtagesproben im Winter
1980/81. In vier Korngrößenfraktionen im Größenbereich
0,1 - 25 µm AD (Sommer 1980), in 5 Fraktionen im Bereich
0,04 - 25 µm AD (Winter 1980/81) werden folgende Komponenten
analysiert: C, SO_4, NO_3, Br, Cl, F, Ca, Mg, Al, Fe, Mn, Pb,
Zn, Cu, Cr, V und Sr. Zur Herkunftsanalyse wird zunächst durch
Faktorenanalyse die Zahl und Art der Haupteinflußfaktoren
ermittelt und Verursacherquellen bzw. -gruppen zugeordnet.
Durch ein Regressionsmodell soll hierauf der Beitrag bestimm-
ter Aerosolquellen zur Zusammensetzung des atmosphärischen
Aerosols festgestellt werden.

Über die faktorenanalytische Herkunftsanalyse von Aerosolen sind bereits Übersichtsartikel erschienen (2-5), sodaß
eine weitere Diskussion der literaturbekannten Arbeiten nicht
erforderlich erscheint. Im folgenden sollen jedoch kurz die
Besonderheiten des geplanten Projekts hervorgehoben werden:
      - Zur Analyse gelangt ein umfangreicher Datensatz
      von 57 Proben.
      - Während bisher maximal 2 Größenfraktionen untersucht
      wurden, gelangen in der gegenständlichen Arbeit 4 bzw. 5
      Fraktionen zur Analyse.
      - Die chemische Analyse umfaßt Haupt- und Nebenkomponenten
      sowie die wichtigsten Markerelemente.
Im folgenden wird über den 1. Teil der Arbeit, umfassend die
Probenahme und ICP-AES Analyse der Proben vom Juni - Aug. 1980
und eine Faktorenanalyse, berichtet.

2. PROBENAHME UND ANALYSE

Die Probenahme wurde in Wien an der Meßstelle Getreidemarkt in 12 m Höhe mittels 5-stufiger Kaskadenimpaktoren,
Typ "TU 80", im Zeitraum Juni 1980 - August 1980 vorgenommen.
Die Probenahmeintervalle betrugen in der Regel 24 h, die
Sammelrate lag bei 80 l/min. Die Abscheidung des Aerosols
erfolgte auf Aluminiumfolien in vier Korngrößenfraktionen,
logarithmisch geteilt im Größenbereich 0,1 - 25 µm AD. Die
eingesetzten Impaktoren wurden von A.BERNER konstruiert und
beschrieben (6).

Zur Bestimmung der Massen-Korngrößenverteilung wurden die
Proben ausgewogen. Für die Elementanalyse wurde ein Aliquot
der Probe in Aufschlußgefäße (Teflon) übertragen, mittels
HF - HNO_3 bei 190°C aufgeschlossen, mit H_3BO_3 komplexiert und
hierauf durch ICP-AES (Induktiv gekoppelte Plasmaatomemissions-
spektroskopie) einer simultanen Multielementanalyse unterzogen.
Die Analyse erfolgte auf einem 3 kW - Argon/Stickstoff - ICP -
Gerät, bestehend aus einem leistungsstabilisierten LINN -
Generator, Frequenz 27,12 MHz, einem GREENFIELD - Plasmabren-
ner, MEINHARD - Zerstäuber und einem 1 m Vakuumspektrometer
(BAIRD Spectrovac 1000) mit simultaner Erfassung von 24 Linien.
Details der Analysenmethode wurden bereits beschrieben (7).

AEROSOL WIEN SOMMER '80

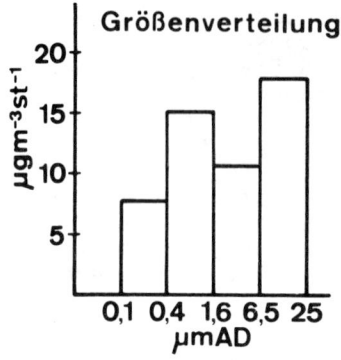

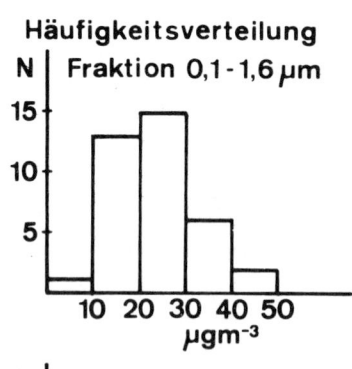

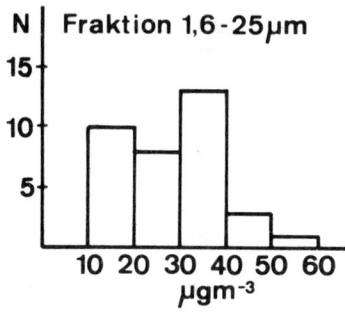

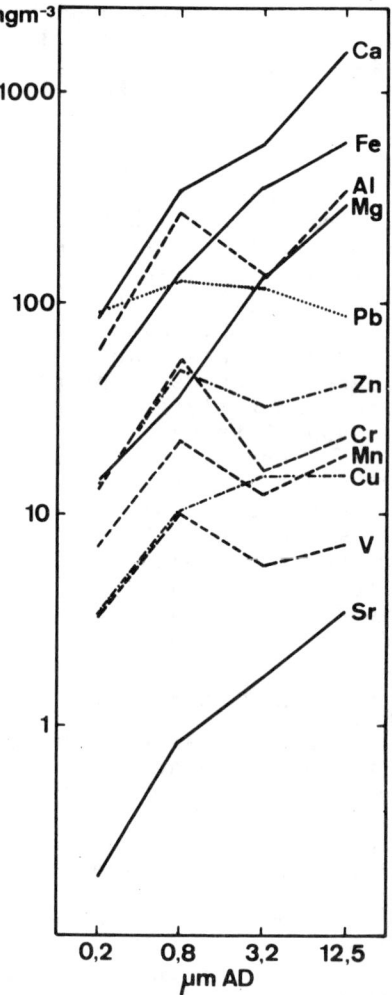

Abb. 1 - Massengrössenverteilung der Aerosolkonzentration Messtelle Wien, Getreidemarkt - Probenserie Juni-Aug. 1980, Mittelwerte, n=37. Häufigkeitsverteilung der Fein- und Grobteilchenkonzentration. st .. Impaktorstufe

Abb. 2 - Massengrössenverteilung der Konzentration der Elemente, Messstelle Wien, Getreidemarkt - Probenserie Juni-August 1980, Mittelwerte, n=37

Tabelle 1: Korrelationsmatrix (n = 18)

	M-F	Al-F	Ca-F	Cr-F	Cu-F	Fe-F	Mg-F	Mn-F	Pb-F	Sr-F	V-F	Zn-F	M-C	Al-C	Ca-C	Cr-C	Cu-C	Fe-C	Mg-C	Mn-C	Pb-C	Sr-C	V-C	Zn-C
M-F	1.00																							
Al-F	-.24	1.00																						
Ca-F	-.39	.51	1.00																					
Cr-F	.16	.52	.04	1.00																				
Cu-F	-.10	.78	.06	.58	1.00																			
Fe-F	.32	.51	-.18	.84	.76	1.00																		
Mg-F	.27	.69	-.02	.86	.75	.93	1.00																	
Mn-F	-.40	.20	.86	.10	-.03	-.16	-.13	1.00																
Pb-F	.33	-.12	-.51	.75	.24	.67	.51	-.23	1.00															
Sr-F	.29	.72	.39	.49	.42	.57	.70	.06	-.02	1.00														
V-F	-.24	.55	.77	.52	.23	.22	.40	.77	-.01	-.40	1.00													
Zn-F	.19	-.41	-.31	.14	.06	.04	-.15	.06	.50	-.60	-.14	1.00												
M-C	.48	.02	-.61	.49	.31	.59	.63	-.56	.60	.07	-.04	.17	1.00											
Al-C	-.13	-.09	-.35	.11	.31	.13	.14	-.18	.27	-.48	-.03	.49	.58	1.00										
Ca-C	.03	-.03	-.50	.24	.39	.31	.32	-.35	.39	-.36	-.07	.44	.77	.95	1.00									
Cr-C	.13	.51	-.06	.53	.72	.67	.76	-.06	.27	.27	.39	.06	.72	.61	.73	1.00								
Cu-C	-.39	-.08	-.74	.53	.43	.76	.66	-.70	.73	.18	-.31	.07	.80	.26	.49	.48	1.00							
Fe-C	.60	.10	-.59	.45	.29	.70	.70	-.65	.53	.40	-.17	-.18	.85	.15	.40	.55	.89	1.00						
Mg-C	.64	.15	-.40	.48	.16	.64	.64	-.50	.47	.54	-.04	-.27	.63	-.26	.03	.32	.75	.88	1.00					
Mn-C	.74	-.08	-.56	.12	.04	.44	.43	-.65	.24	.34	-.28	-.24	.70	-.03	.22	.38	.68	.90	.86	1.00				
Pb-C	.49	.11	-.56	.61	.27	.76	.69	-.55	.71	.42	-.13	-.13	.69	-.08	.18	.33	.91	.62	.90	.92	1.00			
Sr-C	.55	-.06	-.45	.27	-.01	.41	.51	-.53	.33	.32	-.06	-.28	.81	.24	.40	.47	.62	.87	.90	.92	.74	1.00		
V-C	.77	.12	-.49	.41	.38	.64	.66	-.54	.44	.34	-.10	.12	.86	.32	.53	.68	.71	.86	.71	.83	.70	.67	1.00	
Zn-C	.40	.20	-.51	.45	.38	.57	.63	-.55	.45	.15	.01	.03	.90	.36	.62	.71	.77	.80	.73	.86	.73	.72	.65	1.00

Tabelle 2: Rotierte Faktorenmatrix, Ladungen > 0.6 sind
unterstrichen, Fine: Fraktion 0,1 - 1,6 µm AD,
Coarse: Fraktion 1,6 - 25 µm AD.

Variable	Faktor 1	Faktor 2	Faktor 3	Faktor 4	Faktor 5
Mass-Fine	.81	-.20	-.12	.00	.24
Al-Fine	-.07	.89	-.03	.26	-.30
Ca-Fine	-.41	.19	-.27	.79	-.30
Cr-Fine	.27	.71	.03	.25	.56
Cu-Fine	-.05	.89	.32	-.04	.05
Fe-Fine	.44	.78	.07	-.02	.38
Mg-Fine	.49	.84	.12	.13	.16
Mn-Fine	-.48	.02	-.14	.83	.09
Pb-Fine	.33	.26	.12	-.17	.86
Sr-Fine	.40	.65	-.47	.29	-.23
V-Fine	-.05	.38	.03	.88	.02
Zn-Fine	-.20	-.27	.44	-.01	.75
Mass-Coarse	.73	.22	.56	-.16	.22
Al-Coarse	-.03	.02	.97	-.08	.13
Ca-Coarse	.20	.13	.94	-.17	.16
Cr-Coarse	.39	.57	.67	.21	-.03
Cu-Coarse	.59	.44	.18	-.55	.34
Fe-Coarse	.87	.34	.12	-.31	.08
Mg-Coarse	.86	.32	-.25	-.17	.13
Mn-Coarse	.94	.05	.01	-.27	-.11
Pb-Coarse	.72	.43	-.15	-.35	.33
Sr-Coarse	.89	.06	.21	-.08	-.09
V-Coarse	.84	.24	.34	-.09	.11
Zn-Coarse	.69	.34	.43	-.19	:08

Tabelle 3: Vergleich Faktoren - Quellenzuordnung
(v% ... relativer Anteil an der Gesamtvarianz)

Faktor	HOPKE et al. (9)	v%	Quellenzuordnung LEWIS et al. (8)	v%	Diese Arbeit	v%
1	Erdkruste und Kohleverbrennung	53	Bodenabrieb, Straßenstaub und Kohleverbrennung	35	Straßenstaub, unbekannte Quellen	46
2	Seesalz	11	Unbestimmte anthropogene Quellen	24	Unbestimmte anthropogene Quellen	20
3	Heizöl	11	Konversion	13	Abrieb	14
4	Verkehr	9	Verkehr	11	Heizöl	7
5	Unbestimmt	9			Verkehr	6
6	Müllverbrennung	6				

3. ERGEBNISSE

Durch Auswaage der Aerosolabscheidungen auf den Aluminiumfolien erhält man die fraktionierte Massen-Größenverteilung des Aerosols. Bei linear-logarithmischer Darstellung wird im untersuchten Korngrößenbereich von 0,1 - 25 µm AD eine bimodale Verteilung ersichtlich. Abb. 1 zeigt die Verteilung als Mittelwert über alle Ereignisse. Die Abbildung enthält weiters die Häufgkeitsverteilung der Massenkonzentration in zwei Korngrößenbereichen (Feinteilchen < 1,6 µm AD und Grobteilchen > 1,6 µm AD).

Die mittlere Verteilung der untersuchten Elemente in doppelt logarithmischer Darstellung zeigt Abb. 2. Nach dem Verlauf der Verteilungsfunktion lassen sich zumindest drei "Verteilungstypen" unterscheiden: 1) Elemente, welche in der groben Korngrößenfraktion angereichert sind (6,5 - 25 µm AD) wie Ca, Mg, Sr, Fe. 2) Elemente mit bimodaler Verteilung, wie Al, Zn, Cr, Mn, V. 3) Elemente mit relativ gleichförmigen Konzentrationsverlauf über den gesamten Korngrößenbereich, Pb und Cu.

Die Faktorenanalyse wurde auf der Großrechenanlage der TU Wien (Cyber 74) mit dem Programm "Factor" des Statistikprogrammsystems SPSS 7 durchgeführt. Durch Zusammenziehen der Massen- und Elementdaten der beiden Fein- und Grobteilchenfraktionen wurden 24 Variable gebildet. Diese wurden einer Hauptkomponentenanalyse mit anschließender Varimax - Rotation unterzogen. Den Ausgangspunkt der Faktorenanalyse bildet die Matrix der linearen Korrelationskoeffizienten, welche in Tab. 1 für 18 Fälle mit kompletten Datensätzen angegeben ist. Durch die Faktorenanalyse erfolgt eine Reduktion der Korrelationsmatrix auf eine Anzahl von Faktoren, welche den Hauptanteil der Varianz des Datensatzes verursachen.

Tab. 2 zeigt das Ergebnis der Faktorenanalyse. Die resultierenden Faktoren werden aufgrund der Faktorenladungen bestimmten Quellengruppen zugeordnet. Die Faktoren 3 - 5 weisen relativ einfache Ladungsstrukturen auf und sind den Quellengruppen "mineralischer Abrieb" (hohe Ladung der Elemente Ca und Al in der Grobfraktion), "Ölverbrennung" (hohe Ladung für V in der Feinfraktion) und "Verkehr" (hohe Ladung für Pb in der Feinfraktion) zuzuordnen. Die Faktoren 1 und 2 weisen eine komplexere Ladungsstruktur auf, sodaß eine eindeutige Zuordnung derzeit nicht möglich ist. Der Faktor 1 zeigt Ähnlichkeit mit dem 1. Faktor in der Analyse des Aerosols in Charleston (8). Dieser Faktor ist in beiden Fällen mit den Variablen Masse-Feinteilchen, Masse-Grobteilchen, Fe, Sr und Pb als Grobteilchen hoch geladen. LEWIS und MACIAS (8) ordnen diesem Faktor Bodenabrieb, Wiederaufwirbelung von Straßenstaub und Staub aus Kohleverbrennung zu. Die genannten Autoren fanden ebenfalls keine Erklärung für die Ladung der Masse-Fein in diesem Faktor. Faktor 2 zeigt ebenfalls Ähnlichkeit mit dem Faktor 2 der genannten Autoren, welche diesem Faktor eine "komplexe Kombination anthropogener Einflußfaktoren" zuordnen.

4. DISKUSSION

Die in dieser Arbeit extrahierten Faktoren und deren Quellenzuordnung sind in Tab. 3 zwei bereits publizierten Ergebnissen gegenübergestellt. Obwohl aus gänzlich unterschiedlichen Datensätzen abgeleitet, lassen sich Ähnlichkeiten bei 4 Faktoren ableiten. Eine Trennung der komplexeren anthropogenen und natürlichen Einflüsse (Faktor 1 bei allen 3 Arbeiten, Faktor 5 bei (9) und die Faktoren 2 bei (8) und dieser Arbeit) ist derzeit noch nicht möglich. Faktoren mit einfacher Ladungsstruktur und hohen Ladungen für Tracerkomponenten (z.B. Pb, V, SO_4) konnten in den genannten Arbeiten spezifischen Quellengruppen zugeordnet werden (Tab. 3). Wie bereits HOPKE et al. (9) feststellten, ist die Einbeziehung von Tracerelementen und von Elementen, welche die Hauptmasse des Aerosols bilden, für die Faktorenanalyse vorteilhaft.

Daher wird die weitere Arbeit die Einbeziehung von Komponenten umfassen, welche die Hauptmasse des feinteiligen Aerosols bilden (C, SO_4), als auch solcher, die Tracerfunktion ausüben (SO_4, Br, Cl).

Danksagung

Die Arbeit wird durch das Bundesministerium für Gesundheit und Umweltschutz, Projekt III-430.007, unterstützt.
Die Autoren danken Herrn Prof. Dr. Laqua für die Ermöglichung, Herrn Dr. Broekaert für die Mithilfe bei den ICP - Messungen am Institut für Spektrochemie und angewandte Spektroskopie, Dortmund.

LITERATUR

1) H. Puxbaum, Fresenius Z.Anal.Chem. 298,110 (1979)

2) G.E. Gordon, Envir.Sci.Technol. 14,792 (1980)

3) J.A. Cooper, J.G. Watson jr., J.Air.Pollut.Control Ass. 30, 1116 (1980)

4) B. Wopenka, Mathematische Methoden zur Erfassung atmosphärischer Aerosolquellen, in H. Malissa, H. Puxbaum (Hsg.) Analytische Chemie und Luftschadstoffe, Inst. Anal. Chem. TU Wien 1981

5) B. Wopenka, H. Puxbaum, Use of receptor models for the evaluation of aerosol sources, 14th European Symp. on Computerized Control and Operation of Chemical Plants, IIASA, Laxenburg, Sept.8-11, 1981

6) A. Berner,Chem.-Ing.-Techn. 50, 399 (1978)

7) J.A.C. Broekaert, B. Wopenka, H. Puxbaum, Application of ICP-AES to the analysis of aerosol samples collected by cascade impactors (in Vorbereitung)

8) C.W. Lewis, E.S. Macias, Atmos. Environ. 14, 185 (1980)

9) P.K. Hopke, E.S. Gladney, G.E. Gordon, W.H. Zoller, A.G. Jones, Atmos. Environ. 10, 1015 (1976)

SIZE-SEGREGATED MEASUREMENTS OF PARTICULATE ELEMENTAL
CARBON BY OPTICAL METHODS AT URBAN AND REMOTE SITES

J. HEINTZENBERG

Department of Meteorology, University of Stockholm
Arrhenius Laboratory, S-106 91 Stockholm, Sweden

Summary

Size-segregated aerosol samples were taken during 2 winter pollution
periods and in clean summer air at different remote locations in
the European Arctic >74°N. At 2 urban sites in Hamburg and Stockholm
reference samples were collected. By means of a newly developed
integrating sphere photometer these filter samples have been analysed
for aerosol light absorption coefficients and particulate elemental
carbon (PEC). The relatively high PEC concentrations in winter confirm other findings about the Arctic winter atmosphere having an
aged continental aerosol burden. In summer very small light absorption coefficients of $4.5 \cdot 10^{-8}$ m^{-1} were measured, similar to upper
tropospheric background values. For the climatically important months
of March through May the key optical aerosol properties extinction
coefficient, single scattering albedo and absorption to backscatter
ratio were determined. The arctic haze aerosol is found to contribute
to atmospheric heating, even in the summer. A first PEC size distribution was determined in clean polar summer air. The results show
systematic variations in PEC size distribution from urban to remote
locations and seasonal variations in the sink region which may be
exploited to quantify aerosol removal process in long distance
transport studies.

1. INTRODUCTION

Both urban and non-urban aerosols exhibit a non-selective absorption
throughout the solar spectrum (1). In many urban aerosols and at one
remote site this absorbing aerosol component has been identified as
particulate elemental carbon (PEC) or soot (2). Soot therefore causes one
of the major climatic effects of the aerosol, namely the transfer of part
of the incoming solar energy into sensible heating of the atmosphere. Excluding bush and forest fires with natural causes, PEC is a typical
anthropogenic aerosol component produced in carbonaceous fuel combustion
processes.

Besides its effect on the energy balance of the atmosphere, its properties as a tracer for long distance transport of anthropogenic air
pollution provides an interesting aspect of the problem of soot in the
atmosphere. PEC is emitted in particulate form and its chemical stability
excludes chemical transformations during its lifetime in the atmosphere.
Primary soot particles have sizes generally below .05 μm radius. Through
rapid coagulation in the vicinity of the sources and slower cloud condensation and coagulation processes these primary particles will spread
throughout the aerosol size distribution. It is likely that they will
accumulate, as do many other materials, in the size range between 0.05

and 0.5 µm radius, where atmospheric residence times have their maximum (3). Therefore, the quantitative exploitation of PEC data in long distance transport studies requires knowledge about the size distribution of soot and its change with distance from the source area.

An assessment of the role of soot as an atmospheric tracer also requires knowledge about the chemical composition of the particulate matter associated with it. The ratio of PEC to other chemical species in the aerosol with known anthropogenic sources should yield information about the relative contribution of different elimination processes along the pathway of air pollution. It may also help to identify certain source areas.

Results of chemical analysis of aerosol samples from the European Arctic suggest that this region during the winter season is exposed to anthropogenic air pollution from Eurasian sources (4). Hence the arctic sink region appears to be a good test case for using soot as an aerosol tracer in atmospheric long distance transport studies. In this study light absorption and PEC results from size-segregated aerosol samples are presented. Aerosol samples taken at different arctic locations during winter and summer season have been analysed with a newly developed soot photometer. Through complementary gravimetric and chemical analysis, mass fractions and relations to other important chemical species in the arctic aerosol have been determined. Together with corresponding results from urban sampling locations in central and northern Europe, these measurements provide a first data base for incorporating size segregated soot data into long distance transport studies for the investigation of pollution pathways from Europe to the Arctic.

2. EXPERIMENTAL PROCEDURES

The PEC sampling was done with single stage high volume impactors with cut-off radii between 0.1 and 0.5 µm radius (5) plus a total filter sampler. Particles not removed by the impactor are collected on 47 mm diameter Delbag Microsorban-98 filters. For the optical analysis of the aerosol samples a double beam, single detector integrating sphere photometer has been developed (6).

The PEC results for atmospheric samples are reported as absorption-equivalent amounts of M71 soot according to a mass calibration of the photometer with a well characterized channel type carbon black (6) (Monarch 71, Cabot Corp.). This interpretation relies on the assumption that under most circumstances PEC is the single aerosol component dominating light absorption.

From a large number of tests with M71 hydrosols a 10% accuracy in the mass determination with the soot photometer was found. However, for atmospheric aerosol samples this value has to be increased if the size and composition of soot containing particles is not well known. The size segregated soot sampling of the present study reduces this uncertainty. Thus an accuracy on the order of 50% for the PEC values in atmospheric samples seems to be a reasonable figure.

As complementary gravimetric analysis the total suspended aerosol mass (TSP) was determined from separate filter samples. Chemical analysis of the major element sulphur and iron which may cause interfering light absorption as hematite were determined through particle induced X-ray emission at the institute of nuclear physics in Lund, Sweden.

Table I. Physical and chemical properties of arctic aerosol samples; grand average winter and summer results compared to results from urban samples.

Location	CNC cm^{-3}	σ_{SP} (0.55μm) m^{-1}	σ_{SP} (0.55μm) m^{-1}	TSP	C_{total}	$C_{<0.1\mu m}$ $ng\ m^{-3}$	S	Fe
Hamburg, May 1981	1)	1)	$1.6 \cdot 10^{-5}$	$43 \cdot 10^3$	$1.5 \cdot 10^3$	630	1)	1)
Stockholm, Feb. 1981	$1.2 \cdot 10^4$	$4.7 \cdot 10^{-5}$	$4.9 \cdot 10^{-6}$	$19 \cdot 10^3$	550	230	1500	25
Arctic winter grand average	380	$1.5 \cdot 10^{-5}$	$8.2 \cdot 10^{-7}$	4700	70	33	900	18
Arctic summer grand average	100	$1.7 \cdot 10^{-6}$	$4.4 \cdot 10^{-8}$	1200	3	0.8	90	1.1

1) data not taken

3. ARCTIC FIELD SAMPLING CAMPAIGNS

Samples were taken in the arctic region in a series of two late winter experiments and one summer experiment during the Swedish arctic expedition Ymer-80. Details about instrumentation and measuring sites can be found in (6) and (11).

To avoid local contamination all sampling was done with brushless motors and with filters on the pump exhausts while the samplers were controlled by high sensitivity condensation nuclei counters (CNC). No aerosol samples were taken at CNC-values significantly raised (~2x) over the background level as experienced at the particulate site during the sampling period.

4. RESULTS AND DISCUSSION

Grand average results for winter and summer experiments are compared to PEC results from urban locations in Table I.

The winter data provide the possibility of a direct comparison with other arctic observations near Barrow, Alaska. Both the light scattering coefficients and the concentrations of the major ion SO_4^{2-} are in close agreement with corresponding Alaskan results. However, light absorption coefficients σ_{AP} are about 35% lower than Alaskan results of Rosen et. al. (2) for the same period.

The winter PEC and light absorption data add more support to Rahn's hypothesis of the arctic region being generally polluted by anthropogenic aerosols during the winter season. Despite the three to five thousand kilometers travel distance from the Eurasian source areas, light scattering and absorption coefficients are consistently close to continental background values (7) and are only 3-6 times lower than the urban results from Stockholm. Besides these elevated aerosol burdens in winter – quite unexpected for a pure sink region as the Arctic – direct injections of highly polluted European air into the Spitsbergen region were observed, producing events lasting from hours to a few days. One winter event (NYÅ-81, sample 7) is shown in Fig. 1. Both aerosol light scattering and absorption properties of that particular air mass were similar to urban values found in Scandinavia. Particulate sulfur was even higher than in the Stockholm samples which can be explained by most of the SO_2 present in the source region being converted to sulfate by the time the air reaches Spitsbergen.

The mass median radius for PEC in arctic haze was found to be around 0.15 μm which is close to the value of 0.14 μm found by Countess et. al. (8), for urban aerosols in Denver. The substantial amounts of PEC found in both winters in the Aitken size range indicate a significant fraction of combustion aerosols in arctic haze. This is one more piece of evidence for the anthropogenic pollution character of the arctic winter aerosol.

While these winter results mainly confirm previous findings about arctic winter aerosol burdens being close to continental background values, the summer experiment on the icebreaker Ymer represents the first systematic study of PEC in very clean tropospheric background air. According to Rahn's coupling - decoupling hypothesis (9) the Arctic is separated from the midlatitude source regions by the polar front in summer. Indeed, the summer results are typified by extremely low levels of the physical and chemical aerosol characteristics analysed in our samples.

The general level of the light scattering coefficient σ_{SP} again was in close agreement with summer results from Barrow, Alaska (10). The

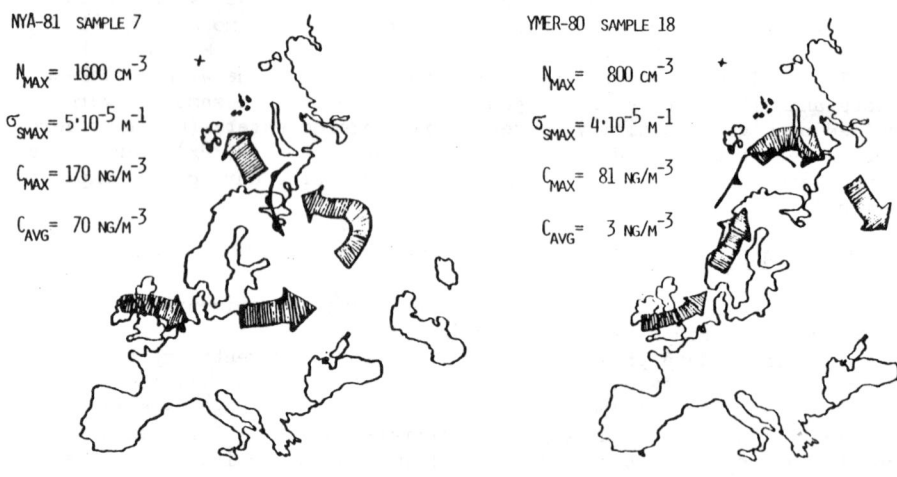

NYA-81 SAMPLE 7

N_{MAX} = 1600 cm^{-3}

σ_{SMAX} = 5·10^{-5} m^{-1}

C_{MAX} = 170 ng/m^{-3}

C_{AVG} = 70 ng/m^{-3}

YMER-80 SAMPLE 18

N_{MAX} = 800 cm^{-3}

σ_{SMAX} = 4·10^{-5} m^{-1}

C_{MAX} = 81 ng/m^{-3}

C_{AVG} = 3 ng/m^{-3}

Fig. 1 – Examples of winter (NYÅ-81, sample 7) and summer (Ymer-80, sample 18, 74°N, 25°E) PEC injections into the European Arctic. The flow patterns shown are typical for 1-2 days before and during the sampling periods.

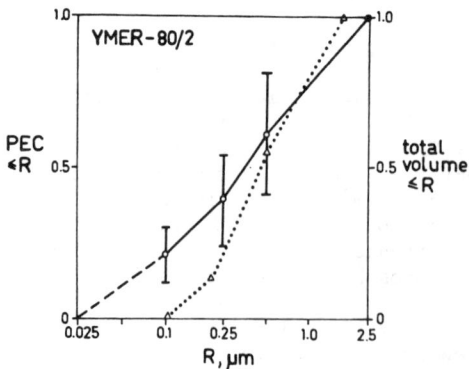

Fig. 2 – Ymer-80, leg 2 grand average cummulative PEC size distribution (O) and grand average volume distribution (Δ) from impactor measurements (13). The error bars on the PEC distribution refer to ± 1 standard deviation. The total aerosol volume was 0.6 μm^3 m^{-3} and the total PEC 3 ng m^{-3}.

grand average σ_{AP} of $4.4 \cdot 10^{-8}$ m^{-1} is almost identical with the only upper tropospheric background value measured on Mauna Loa, Hawaii (7).

In winter 2% of the fine particle mass or about 1.5% of TSP was found to be elemental carbon. The aerosol in the clean polar summer air on the other hand has a ten times smaller mass fraction which is occupied by strongly light absorbing material. Thus the climatic implications of soot in the Arctic, which need to be considered mainly during the polar day, are less drastic than the late winter data would indicate.

Additional perspective on the climatic importance of soot erosols in the Arctic is provided by evaluating the aerosol single scattering albedo ω_o and the absorption to backscattering ratio. The median ω_o is 0.95 in winter and 0.97 in summer in the Arctic, and is comparable to the value reported by Weiss et. al. (7) for Manua Loa, this value is significantly greater than the values given therein for urban and rural continental aerosols. In spite of the relatively high values for ω_o the net effect of the arctic aerosol is to heat the atmosphere, as can be determined by an evaluation of the absorption/backscatter ratio. The average ratio σ_{BSP}/σ_{SP} in first experiment was 0.16 (11). When combined with the single scattering albedo of 0.95 an average value of 0.33 is obtained for the absorption to backscatter ratio. Based on the approach of Mitchell (12), a ratio above 0.05 would result in atmospheric heating above a snow covered surface, even in the summer.

The generally clean air conditions prevailing during the arctic summer months do not exclude the possibilities for short-term injections of polluted air from mid-latitudes, such as the event in early August 1980. As shown in Fig. 1, the air mass reaching the Barents Sea had passed over the British Isles and Western Scandinavia. As a consequence, light scattering and absorption coefficients went up by a factor of 20-30, approaching the winter levels found on Spitsbergen. However, the pathway of the polluted air mass did not allow for a deep intrusion which might have raised the general aerosol burden of that sector of the Arctic as evidenced by the return of the aerosol levels to the pre-event values.

Combining the data from the 3 impactors and the total PEC filter samples a grand average cummulative PEC mass size distribution was constructed. The resulting PEC distribution is compared in Fig. 2 to the grand average volume distribution determined by Jaenicke et. al. (13) during the Ymer-80 expedition. The lower and upper size limits are set somewhat arbitrarily. The most important feature is the median radius being about 0.5 µm for both PEC and the total volume distribution. Comaped to the arctic haze in winter this means a shift of both PEC and total particle volume towards larger sizes.

In the source regions PEC which is predominantly emitted as Aitken particles will contribute strongly to the aerosol volume found below 0.1 µm. This is reflected in both urban experiments where roughly 50% of the PEC was found in the Aitken size range. On the other hand, an aerosol reaching the arctic sink region after several thousand kilometers transport will exhibit an Aitken size range which is strongly depleted by coagulation, condensation and precipitation processes. With particle concentrations decreasing during the transport, coagulation inside the Aitken range becomes less effective. At the low remaining concentration levels wet removal through cloud condensation and precipitation seems to be the dominating elimination process. Our summer CNC-observations strongly support this hypothesis: During snow-showers and even when cruising through extensive fog banks the total particle count went from the general level of about 100 cm^{-3} down to or not more than 1-5 cm^{-3}. According to the annual variation of precipitation in the Arctic this net removal is expected to be least effective in late winter and most effective in summer. This may explain the changes of the ratio $C_{<0.1\mu m}$ to C_{total} from 0.3 for the winter 1981 experiment to the value 0.2 for the summer data.

The relatively high PEC concentrations in winter confirm other findings about the Arctic winter atmosphere having an aged continental aerosol burden. The climatically important combination of sunlight and elevated PEC concentrations occurs during the months of March through May. For this period the optical properties of aerosols i.e. the extinction coefficient, the single scattering albedo, and the absorption to backscatter ratio, were determined for the arctic haze. A heating of the arctic atmosphere is infered from these optical aerosol properties.

There are systematic variations in the PEC size distribution from urban to remote locations and seasonal variations in the sink region. As PEC is a chemically inert primary component of anthropogenic aerosol, it may be possible the use of PEC size distribution data to quantify the effects of coagulation and were removal process in long distance transport studies such as the investigation of the origin of arctic haze.

REFERENCES

1. Fischer K. (1980) Contrib. Atm. Phys. 43, 244.
2. Rosen H., Novakov T., and Bodhaine B.A. (1981) Atm. Env. 15, 1371.
3. Jaenicke R. (1978) Berichte Bunsenges. Phys. Chem. 82, 1197.
4. Heintzenberg J., Hansson H.Ch., Lannefors H. (1981) Tellus 33, 162.
5. Winkler P. (1975) Geophys. Res. Lett. 2, 45.
6. Heintzenberg J. (1981) Paper submitted to Atm. Env.
7. Weiss R.H., Waggoner A.P., Charlson R.J., Thorsell D.L., Hall J.S., and Riley L.A. (1979) Proc. Conf. on Carbonaceous Particles in the Atmosphere, 257.
8. Countess R.J., Cadle S.H., Groblicki P.J., and Wolff G.T. (1981) J. APCA 31, 247.
9. Rahn K.A., Borys R.D., and Shaw G.E. (1977) Proc. 9th International Conference on Atmospheric Aerosols, Condensation and Ice Nuclei. Galway, Ireland, 21.
10. Bodhaine B.A., Harris J.M., and Herbert G.A. (1980) Atm. Env. 15, 1375.
11. Heintzenberg J. (1980) Tellus 32, 251.
12. Mitchell J.M. (1971) The effect of Atmospheric Particles on Radiation and Temperature. In Man's Impact on Climate. Edited by W.H. Matthews, W.W. Kellog, and G.D. Robinson. MIT Press, Cambridge, Mass.
13. Jaenicke R., and Schütz L. (1981) Paper presented at the European Geophysical Society Meeting. Uppsala, Aug. 24-29, 1981.

ON THE USE OF SF_6-TRACER RELEASES FOR THE DETERMINATION OF FUGITIVE EMISSIONS

B. VANDERBORGHT, J. KRETZSCHMAR and T. RYMEN
Nuclear Energy Research Center, B-2400 Mol, Belgium
F. CANDREVA, R. DAMS
Institute for Nuclear Sciences, B-9000 Gent, Belgium

Summary

The principle and applicability of tracer releases for the accurate measurement of fugitive emissions is discussed and illustrated with some practical examples. At the emission source SF_6-tracer gas is released at a well-known constant rate. With proper adaption of the tracer emission device to the fugitive emission source thorough mixing and homogeneous dispersion of tracer and pollutant can be obtained. In this way the tracer to pollutant concentration ratio is the same in the dispersed plume as in the emission and the pollutant emission rate can be calculated by measuring the tracer to pollutant ratio at any place in the plume.

1. INTRODUCTION

Although dispersion calculation for emissions from high stacks predict a maximum ambient air concentration at a certain distance from the source, the highest concentrations are very often measured very close to the factory. Control of point-source emissions has in certain problem areas not produced the anticipated improvement in ambient air quality. The reason appears to be fugitive emissions at low- or ground-level altitude. Fugitive emissions seem to be especially important for aerosol emissions from metallurgical plants. For the complete understanding of air pollution situations - through emission-inventories, mathematical modelling etc. - it is necessary to be able to quantify the fugitive emissions.

Since the emission area is usually not well defined, and since volume flow rates are usually unknown, the normal emission measurement techniques cannot be applied. The measuring methods actually described in the literature can be classified into "quasistack" sampling or roof monitoring in which the emission area is by some means physically restricted or into "reversed modelling" where emission is calculated from immission measurements and mathematical dispersion models.(1)

In this paper the principle and possibilities of a tracer technique for the quantification of fugitive emissions from a metallurgical plant is described.

A common feature of all mentioned procedures is the discontinuous character preventing routine continuous emission measurements.

2. PRINCIPLE

The principle of the procedure is as follows :
A tracer component is discharged at a constant, well-known rate at the emission site of the pollutant. When the tracer and the pollutant are well mixed in the emission plume and when they are dispersed in the same way,

then the concentration ratio of both components will remain constant all over the plume and equal to the emission rate ratio :

or $\dfrac{\chi_p}{\chi_t} = \dfrac{Q_p}{Q_t}$ (1)

where χ = concentration $\left[M.L^{-3}\right]$
Q = emission rate $\left[M.T^{-1}\right]$
p = pollutant
t = tracer

Knowing the tracer emission rate and measuring the concentration of the tracer and the pollutant in the plume give the possibility to calculate the unknown pollutant emission rate Q_p. It is obvious that the quality of the results will depend on the validity of equation (1) for which a good homogenization and equal dispersion of tracer and pollutant is necessary. The characteristics of the tracer release (place, time, flow-rate, temperature, shape of nozzle) and the choice of the measuring points for the simultaneous determination of pollutant and tracer concentrations must be optimized as a function of this homogeneity.

When the fugitive emissions out of a factory workplace are to be measured, different sampling strategies are possible. Suppose a building in which two furnaces are the sources of uncontrolled emissions of metal fumes (figure 1). Aerosol is evacuated from the inside atmosphere of the workplace by means of natural draft through rooftop ventilation openings and by means of a ventilator with the hood a few meters above one furnace and the exhaust opening in the wall or the roof of the building. These emissions are easily entrained in the wake on the lee of the building and the source cannot be considered as a point source.

For the determination of these fugitive emissions — by the method of tracer release and measuring tracer and pollutant concentration — three different sampling locations are possible.

1) Tracer and pollutant are sampled outside at distances from a few meters up to a few hundreds meters downwind the building. With this configuration the total emission of the building is measured.
2) Sampling units are placed along the ventilation opening in the rooftop.
3) Isokinetic sampling is performed in the exhaust tube of the ventilator.

In the two latter cases homogenization of tracer and pollutant becomes very critical. By proper adjustment of the tracer release and sampling places the emissions of the different sources are discerned and the emissions through the rooftop and through the ventilator can be calculated separately. In this paper the properties of the three methods will be compared.

3. EXPERIMENTAL

In an antimony (Sb) metallurgical plant the technique has been used to quantify the fugitive emissions emanating from a 18m high 30mx20m long workplace in which a convertor and a refinery furnace were installed (figure 1). One of the experiments, in this workplace, will be described here as an example. Uncontrolled Sb dust emissions originated mainly from the filling and the slag removal from the convertor as well as from the same actions on and the emptying of the refinery furnace. Minor leak emissions on filter installations did also occur. Sb aerosol had a mean aerodynamic diameter of 1 to 2 μm.

Sulfurhexafluoride (SF_6) gas was used as a tracer. SF_6 is an inert, non-toxic gas, stable up to about 500°C, routinely detectable up to 50 $ng.m^{-3}$ and normal background concentrations are below the detection limit. SF_6 was discharged at the convertor (main emission place) at a rate of about 27 $g.min^{-1}$.

SF_6 and Sb were simultaneously sampled at three places in the roof of the workplace (W_1 to W_3 in figure 1), at one place (W_0) in the exhaust tube of the hood above the convertor and 8 places outside at distances between 15 and 180m from the hall (R_1 to R_8 in figure 2).
Up to this distance sedimentation and deposition of Sb aerosol is neglectible, and there is no basic difference between the dispersion of Sb-aerosol and SF_6 gas.
The sampling places outside were chosen in such a way that interferences from other parts of the factory were minimal. Four sampling periods of each 30 minutes were performed. During sampling the local wind direction at the 30m-level was between 200° en 205° with a windspeed between 7 and 8,5 $m.s^{-1}$ and neutral atmospheric stability. For the sampling points in the roof (W_1 to W_3) the aerosol was sampled on the same filter for the 4 periods as the access to these places was difficult.

Aerosol samples were taken using LIB-type low volume samplers with Whatman 41 filters. Sb was determined by neutron activation and X-ray fluorescence. SF_6 samples were collected by filling plastic bags and were gas chromatographically analysed. For further details on the sampling and analysis procedures see reference (2).

4. Results and discussion

The results of the emission calculation are summarised in table I.

TABLE I : Sb emission from a workplace as determined by the concentration
ratio Sb/SF_6 in : ambient air downwind of the building ($R_1 \rightarrow R_7$)
ventilator exhaust tubo (W_0)
rooftop of the building ($W_1 - W_3$)

sampling period	Sb emission in kg/h based on sampling points			Activities of	
	$R_1 \rightarrow R_7$	r.s.	W_0	Convertor	Refinery furnace
1	1.6 ± 0.4	23 %	2.5	8' filling	17' slag
2	1.1 ± 0.4	33 %	1.3	6' filling 4' slag	short emissions
3	2.6 ± 0.5	19 %	2.7	12' filling	3' slag
4	1.0 ± 0.2	22 %	0.7	9' filling	
Average	1.6	24 %	1.8		

r.s. relative standard deviation in procent on 7 measurements

- 519 -

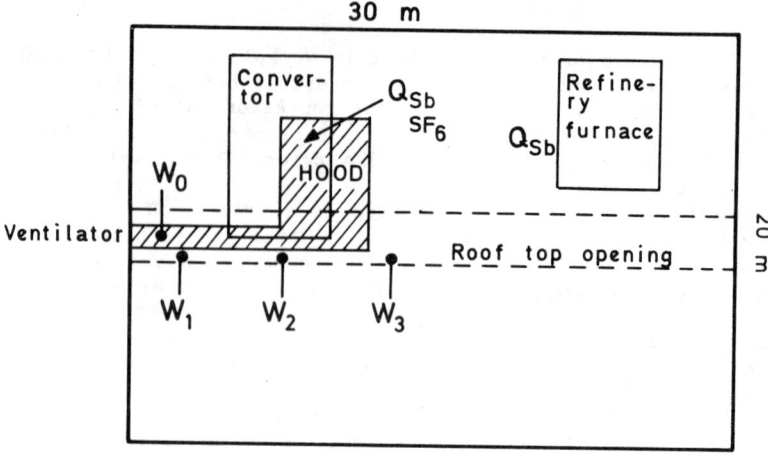

Sb$_2$O$_3$ aerosol   Ø 1-2 µm

Fig. 1 - Layout of emission and sampling points in the workplace

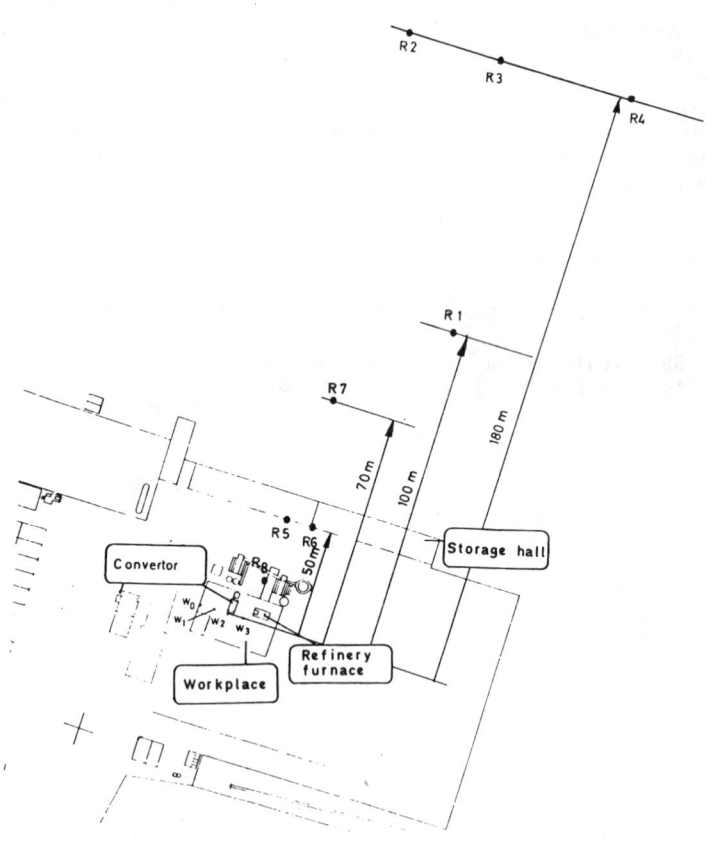

Figure 2 - Layout of emission and sampling points

sampling place	Sb emission in kg/h averaged over 4 periods
W_0	1.8
W_1	1.4
W_2	1.4
W_3	0.9
average	1.4 ± 0.37

The relative standard deviation on the emission determination for each sampling period through the 7 sampling points ($R_1 \rightarrow R_7$) outside the building and on a distance larger than 30 m ranges between 19 and 33 % with an average of 24 %, and no sampling place gave a systematic deviation from the average. In sampling point R_8, 15 m downwind of the roofcenter, on the other hand the Sb/SF_6 ratio was systematically higher than in the other points. At this short distance, the dispersed plumes from the two Sb sources (convertor and furnace) were not yet well homogenized, resulting in a higher Sb/SF_6 ratio in the plume of the refinery furnace than in the plume of the convertor, where the SF_6 emission was performed. At larger distance the plumes of the two installations seem to be well mixed since the Sb/SF_6 remains from place to place constant between acceptable limits. The considerable aerodynamic turbulence in the wake at the leeside of the building, is responsible for this good result.

The calculated Sb emission for the consecutive sampling periods is compatible with the emission causing activities on the plant apparatus, except that the emission during period 4 is lower than expected in comparison with the three former periods.

The sampling points W_1 to W_3 and the hood of the ventilator, and consequently also point W_0, are placed above the convertor. Visual observation of the emission plumes of the convertor and of the refinery furnace indicates that emission results obtained by these points can in a first approximation be equalized to the emission from the convertor and that the contribution of the refinery furnace to the measurements in those points is neglectible.

With the measurements inside the workplace – respectively in the rooftop (W_1 to W_3) and in the exhaust tube (W_0) – two different methods to calculate the emissions can be used. With the first method the global pollutant emission is obtained, with the second it is possible to determine what fraction goes through the hood, respectively the roof.
a. By using the total SF_6 emission in equation (1) and the SF_6 and Sb concentration respectively in the points W_0, W_1, W_2 and W_3, the total Sb-emission from the convertor is obtained. Averaged over the four sampling periods, the emission results obtained by the points W_0 to W_3 are respectively 1,8 – 1,4 – 1,4 and 0,9 kg Sb/h (table I) and averaged over the four points : 1,38 ± 0,37 kg/h or 1,38 kg/h ± 27 %. If SF_6 and Sb were well homogenized in the plume above the convertor the four results should be the same.
Considering the relatively long sampling time (4x30'), the 27 % relative standard deviation on the emission calculation indicates that the homogenization of Sb and SF_6 right above the convertor is not completely attained.

b. The measurement of the flow rate in the exhaust tube enables to calculate the mass flow of SF_6 and Sb through the hood and the ventilator
(Q^H).

Subtraction from the total SF_6 emission $(Q^T_{SF_6})$ gives the SF_6 emission
through the roof $(Q^R_{SF_6})$ by draft ventilation. Substitution of this
value in equation (1) gives the Sb emission through the roof (Q^R_{Sb})
instead of the total Sb emission, this of course with the condition of
sufficient homogenization of tracer and pollutant. The results of this
exercise are summarized in table II.

TABLE II : Sb and SF_6 emission through the ventilator and the roof of the
workplace

sampling period	emission in kg/h			sampling place	emission in kg/h
	$Q^H_{SF_6}$ (1)	$Q^R_{SF_6}$ (2)	Q^H_{Sb} (1)		Q^R_{Sb} (3)
1	0.24	1.48	0.34	W_1	1.08
2	0.39	1.18	0.32	W_2	1.07
3	0.41	1.20	0.68	W_3	0.70
4	0.39	1.23	0.17		
average	0.36	1.27	0.38	average	0.95

(1) through the volume flow rate measurment in the exhaust tube
(2) $Q^R_{SF_6} = Q^T_{SF_6} - Q^H_{SF_6}$
(3) $Q^R_{Sb} = Q^R_{SF_6} \cdot (\frac{X_{Sb}}{X_{SF_6}})_R$

The sum of the Sb-emission through the roof and through the ventilator
should be the same as the total Sb emission from the convertor determined
by the previous procedure (sub a). Averaged over the four sampling
periods this gives :

$$Q^T_{Sb} = Q^R_{Sb} + Q^H_{Sb} = 0.95 + 0.38 = 1.3 \text{ kg Sb/h}$$

which is only a few procents less than the 1.4 kg/Sb found by the other
procedure (table I).
It can easely been shown mathematically that both results should be
identical with complete mixing of tracer and pollutant, i.e. when

$$(\frac{X_{Sb}}{X_{SF_6}})_R \equiv (\frac{X_{Sb}}{X_{SF_6}})_H$$

The total Sb emission averaged over the four sampling periods, is
slightly higher when determined through places outside the building (Q_{Sb}
= 1.6 kg/h) than through the sampling points in the roof and the ventilator (Q_{Sb} = 1.4 kg/h). This was to be expected since the places
outside are influenced by the emissions of the complete building while
the points inside were focussed on the emissions of the convertor only.
Moreover measurements outside can be positively interfered by the background caused by other parts of the factory. Measurements of the Sb concentration upwind the building showed that this was a minor problem.

5. Conclusions and epiloge

Fugitive emissions can be quantified by the method of tracer release and measuring the tracer to pollutant ratio in the dispersed plume. The critical factor of the method is the homogenization of the tracer with the pollutant. This can be optimized by proper adjustment of the tracer emission device, the sampling time and the selection of the sampling places.

With concentration measurements a few tens or hundreds meter away from the source, in the example of the paper this is outside the building, the tracer to pollutant concentration ratio remains constant from place to place within acceptable limits. The total pollutant emission from the building can be determined.

With concentration measurements close to the source, for example in the rooftop and the ventilation of a workplace, sufficient homogeneity is more difficult to obtain. Nevertheless can the emission be determined with reasonable precision, even in presence of another source close to the main source. The efficiency of an exhaust hood can be evaluated.

Within the sampling time, the emission of the pollutant can fluctuate. If during this time the tracer emission and the dispersion conditions remain constant, the time averaged pollutant emission can nevertheless be determined.

The procedure has been used to determine emission factors for fugitive emissions. These data have subsequently been used for the mathematical modelling of the Sb concentration in the environment around the plant (2,3,4). The procedure has also been used to determine the fugitive emissions of a metal refinery installation before and after construction of a system for reduction of the fugitive emission. With this information the reduction of the pollution level in the environment, due to this investment, has been estimated using a bi-gaussian disperion model.

This research was carried out in the "National R & D Programma, Leefmilieu-Lucht" of the DPWB (Ministery of Science Policy).

References

(1) "Wanted : fugitive emissions"
     S. Budiansky. Environmental Science and Technology V14, N8 (1980) 904-905.
(2) "Luchtverontreiniging door een metallurgisch bedrijf". Eindrapport Nationaal R-D Programma Leefmilieu-Lucht
     Diensten voor de Programmatie van het Wetenschapsbeleid, Brussel, 1981.
(3) "Influence of the meteorological input data on the comparison between calculated and measured aerosol ground level concentrations and depositions"
     I. Mertens, J. Kretzschmar and B. Vanderborght
     Proceedings of the "12th ITM on air pollution modelling and its application"
     NATO/CCMS Palo Alto August 1981
(4) "Depositie rondom een non ferro bedrijf"
     B. Vanderborght, I. Mertens, J. Kretzschmar
     Extern N4 (1981) in press

UTILISATION D'UNE EQUATION POUR L'ENERGIE CINETIQUE

TURBULENTE DANS LES MODELES NUMERIQUES DE MESO-ECHELLE

APPLICATIONS ET COMMENTAIRES

C. BLONDIN et G. THERRY
Direction de la Météorologie
Etablissement d'Etudes et de Recherches Météorologiques
73-77 rue de Sèvres - 92100 BOULOGNE BILLANCOURT
France

Summary

An equation for the spatio-temporal evolution of the turbulent
kinetic energy is added to the classical equations set of a numerical mesoscale model to improve the description of the turbulent
processes in the atmospheric planetary boundary layer. The relia -
bility of the mathematical and numerical framework is assessed by
comparison with high order closure model results. A two-dimensional
numerical simulation of a land-sea breeze cycle is achieved using
two different methods for the exchange coefficients computation :
analytically and based on energy budget. The discrepancies between
the two runs are analysed according to the thermodynamical properties of the boundary layer and interpreted through trajectories
pattern studies. Further developments with a view to pollution
diffusion studies are then presented and discussed.

1. INTRODUCTION

     Les effets orographiques ou thermiques influencent de façon déter-
minante la structure à méso-échelle de la couche limite planétaire dans
laquelle s'effectuent généralement le transport et la diffusion des
nuages de polluants issus de sources industrielles ou domestiques. L'accès
à cette structure est impossible au moyen de simples renseignements
météorologiques de routine. L'apport de la modélisation numérique ou
physique des écoulements atmosphériques semble donc irremplaçable.
     On trouvera dans BLONDIN (1979) la description d'un modèle de simu-
lation numérique des écoulements atmosphériques à méso-échelle. Ce modèle
a été développé afin de rendre compte, outre des influences orographiques,
des contraintes qu'imposent les flux énergétiques de chaleur au sol sur
la structure moyenne et turbulente de la couche limite planétaire.
Nota : Dans la suite, toutes les notations non expliquées sont classiques.

2. L'ENERGIE CINETIQUE TURBULENTE

     2.1. La turbulence dans les modèles numériques de méso-échelle

     Le passage de l'écriture des équations de l'hydrothermodynamique

d'un fluide atmosphérique à l'échelle du laboratoire à celle de ces
mêmes équations à méso-échelle s'effectue par application d'un opérateur
de moyenne approprié (vérifiant les axiomes de Reynolds et pour lequel on
applique l'hypothèse ergodique). Les processus atmosphériques d'échelle
inférieure à la dimension spatiale sur laquelle on effectue la moyenne
sont décrits en terme de perturbation α' par rapport à la grandeur
moyenne $\bar{\alpha}$ correspondante et leur contribution apparaît au travers des
équations par la divergence de leur flux $\vec{\nabla}.\overline{\alpha'\vec{v}'}$.

Dans la mesure où on ne veut pas calculer explicitement l'évolution
de ces flux, qui fait appel à la connaissance de corrélations d'ordre
supérieur (problème de la fermeture des équations de la turbulence), on
doit paramétrer ces termes, c'est à dire les estimer en fonction des
caractéristiques moyennes de l'écoulement.

Dans le cadre de ce formalisme, la proposition la plus simple et la
plus communément utilisée consiste à relier les flux turbulents au gradient de la quantité moyenne par :

$$\overline{\alpha'u'_i} = -K^\alpha_i \frac{\partial\bar\alpha}{\partial x_i}$$

K^α_i s'appelle le coefficient d'échange et la spécification de son
mode de calcul permet de fermer les équations.

L'idée fondamentale dans cette démarche consiste à concentrer l'information sur les propriétés turbulentes de l'atmosphère dans ce seul
coefficient d'échange ; en respectant le critère suivant : moins l'atmosphère est stable, plus le mélange turbulent est important et donc plus
le coefficient d'échange doit être grand.

Pour arriver à cette fin, plusieurs degrés de complexité et plusieurs
démarches existent. La plus simple consiste à définir un critère de
stabilité pour toute la couche limite planétaire que l'on classera par
exemple en stable, neutre et instable et à donner à K une valeur constante
pour chacune de ces classes, typiquement 1, 10 et 100 m^2/s. Il est clair
que si quelques applications pratiques, en particulier en matière de
pollution (ex : études à caractère statistique) s'avèrent possibles dans
un contexte aussi simple, il devient hors de question d'essayer de simuler
avec tant soit peu de réalisme l'évolution diurne de la couche limite
planétaire et d'espérer traiter conjointement de manière satisfaisante
le problème de la dispersion des polluants dans les basses couches de
l'atmosphère.

A un stade plus élevé, on cherchera bien évidemment à calculer K
au moyen de formules analytiques. On peut classer ces formules en deux
catégories :

Dans la première, on applique des formules différentes suivant la
stabilité globale de la couche limite et pour chaque classe de stabilité,
généralement au plus trois : stable, neutre, instable, on obtient directement K par K(ζ). Un exemple type se trouve dans O'BRIEN (1970)
utilisé par BLONDIN (1979).

Dans la deuxième, on cherche à relier K à l'état de turbulence locale
de l'atmosphère. On aura alors par exemple K = K (Ri) où Ri est un nombre
de Richardson local

$$Ri = \frac{g}{\theta_o}\left(\frac{\partial\theta}{\partial z}\right)/\left(\frac{\partial|\vec{V}|}{\partial z}\right)^2$$

BLONDIN (1980) emploie une telle formulation tirée de J.F. LOUIS (1979).

Ces deux approches sont complémentaires, en ce sens que si la
turbulence, au niveau de couches stables dans l'atmosphère, a un caractère local, il existe au contraire une organisation structurée dans
une couche limite convective dont il est important de rendre compte
dans une simulation numérique. Nombre de modélisateurs adopte d'ailleurs

cette vision des choses et calcule leur coefficient d'échange suivant l'une ou l'autre des formulations en fonction de la stabilité globale de la couche limite, ou du signe du flux cinématique de chaleur sensible Q_o.

On conçoit qu'une harmonisation de ces deux approches doit pouvoir se réaliser si l'on devient capable de produire une information sur l'ensemble de la structure turbulente de la couche limite planétaire. Nous allons montrer que la gestion d'une équation de l'évolution de l'énergie cinétique turbulente permet d'atteindre cet objectif.

2.2. Equation de l'énergie cinétique turbulente

L'énergie cinétique turbulente, $\bar{e}$, est la quantité :

$$\bar{e} = \frac{1}{2}\left(\overline{u'^2} + \overline{v'^2} + \overline{w'^2}\right)$$

On peut formaliser son équation d'évolution par

$$\frac{\partial \bar{e}}{\partial t} = A + D_E + P_M + P_T + D_P$$

où A =   advection   = $-\vec{V}\cdot\vec{\nabla}\bar{e}$

D_F =   diffusion   = $\vec{\nabla}\cdot\left(K_e\cdot\vec{\nabla}\bar{e}\right)$

P_M =   production mécanique   = $K\left[\left(\frac{\partial \bar{u}}{\partial \eta}\right)^2 + \left(\frac{\partial \bar{v}}{\partial \eta}\right)^2\right]$

P_T =   production thermique   = $-K\alpha_T\beta\left(\frac{\partial \bar{\theta}}{\partial \eta} - \gamma\right)$

D_P =   dissipation   = $-C_E\dfrac{\bar{e}^{3/2}}{L_E}$

L_E est une longueur, caractéristique des tourbillons les plus énergétiques, γ, un pseudo-gradient de température, β, le coefficient de flottabilité g/θ_o et α_T une constante. On remarquera que seul le cisaillement vertical du vent horizontal a été retenu dans le terme de production mécanique.

Deux difficultés apparaissent pour gérer cette équation : la définition d'une condition initiale sur le champ $\bar{e}$ et la formulation d'une condition limite au sol satisfaisante. Dans un premier temps, une solution simple a été adoptée à savoir : à l'instant initial, $\bar{e}$ est supposé avoir partout la même valeur, non nulle mais très faible, et on applique la condition $\partial\bar{e}/\partial\eta = 0$ à la limite inférieure de l'atmosphère. Ces deux points méritent d'être revus, en cherchant un état initial solution d'une équation stationnaire simplifiée pour $\bar{e}$ par exemple et en reliant la valeur de $\bar{e}$ au sol, c'est à dire en fait à z_o, hauteur de rugosité, aux vitesses caractéristiques u_* et $w_* = (\beta h Q_o)^{1/3}$ où h est la hauteur de la couche limite.

Suivant l'idée originale de KOLMOGOROV (1942), on postule que les coefficients d'échange appliqués dans les équations des grandeurs moyennes ($\bar{u}$, $\bar{v}$, $\bar{\theta}$, $\bar{q}$, $\bar{e}$) au niveau des termes de diffusion verticales, et qui sont censés être théoriquement reliés à l'état turbulent de l'atmosphère, s'expriment en fonction de $\bar{e}$ par :

$$K = C_K L_K \bar{e}^{1/2}$$

Les flux verticaux sont alors calculés suivant les formules suivantes :

$$\overline{u'w'} = -K\frac{\partial\bar{u}}{\partial\eta} \quad ; \quad \overline{v'w'} = -K\frac{\partial\bar{v}}{\partial\eta} \quad ; \quad \overline{\theta'w'} = -K^\theta\left(\frac{\partial\bar{\theta}}{\partial\eta} - \gamma\right);$$

$$\overline{e'w'} = -K_e\frac{\partial\bar{e}}{\partial\eta}$$

avec

$$K = C_K L_K \bar{e}^{1/2} \quad ; \quad K^\theta = \alpha_T K \quad ; \quad K_e = C_e K$$

2.3. Longueurs caractéristiques et pseudo gradient

Dans sa formulation initiale, KOLMOGOROV faisait l'hypothèse que les coefficients d'échange K n'étaient fonction que d'une seule vitesse, $\bar{e}^{-1/2}$, et d'une seule longueur, $L = L_K = L_E$.

On a peu de moyens expérimentaux pour accéder à une estimation de L_K, en particulier dans les couches convectives car les gradients verticaux des grandeurs moyennes sont très faibles. Par contre, il est plus aisé d'évaluer L_E.

Les simulations numériques effectuées en utilisant une seule longueur de mélange calée sur les mesures de la longueur dissipative ne reproduisent pas correctement le développement de la couche limite convective, et c'est en vue de l'étude de ce genre de situations que certains aménagements ont été entrepris. Ce n'est pas pour autant que le problème des couches stables puisse être considéré comme résolu, bien au contraire, mais, devant le manque de connaissances théoriques et expérimentales de ce cas, on se satisfait facilement d'un formalisme qui aboutit à une valeur de coefficient d'échange faible dans de telles conditions.

La difficulté de simuler l'évolution de la couche limite convective avec une seule vitesse caractéristique $\bar{e}^{-1/2}$ et une seule longueur L_E provient fondamentalement du fait que dans ce cas les variances de vitesses horizontales et verticales n'ont ni la même amplitude ni surtout les mêmes processus de production. On serait donc tenté de gérer séparément $\overline{w'^2}$ et $(\overline{u'^2} + \overline{v'^2})$. Pour ne pas entrer dans un degré de complexité supplémentaire, on peut justifier (BLONDIN et THERRY, 1981) de résoudre ce problème en différentiant les longueurs L_K et L_E, de telle sorte qu'en régime convectif, on ait :

$$L_K / L_E \propto \left(\overline{w'^2} / \bar{e} \right)$$

L'introduction d'un terme γ pseudo gradient de température, dans le calcul du flux de chaleur sensible et de la production thermique d'énergie cinétique se réfère à l'aspect suivant de la couche convective : il est admis qu'une couche convective n'est pas complètement bien mélangée et neutre mais présente une stratification légèrement instable dans sa moitié inférieure, légèrement stable dans la moitié supérieure. Dans une partie de celle-ci, le flux de chaleur sensible est encore ascendant, et donc impossible à simuler avec la formulation classique du flux de chaleur sensible :

$$\overline{w'\theta'} = - K^\theta \frac{\partial \bar{\theta}}{\partial y}$$

Afin d'obtenir des résultats plus réalistes, un artifice commode consiste donc à écrire (DEARDORFF, 1972) :

$$\overline{w'\theta'} = - K^\theta \left(\frac{\partial \bar{\theta}}{\partial y} - \gamma \right)$$

avec
$$\begin{cases} \gamma = 5 \, \dfrac{Q_o}{w_* h} & \text{si} \quad Q_o > 0 \\ \gamma = 0 & \quad Q_o \leqslant 0 \end{cases}$$

2.4. Test de validation

Le réalisme de ce formalisme a été validé en comparant les résultats d'une simulation de la situation du jour 33 de la campagne de Wangara (CLARKE et all, 1971) réalisée sur la version uni-dimensionnelle de notre modèle avec la même simulation présenté par ANDRE et all (1978) qui utilise un modèle de turbulence d'ordre élevé.

Outre le comportement tout à fait similaire de la température moyenne, on a pu vérifier la concordance des profils succesifs dans le temps du flux $\overline{\omega'\theta'}$ et des différents termes "puits" et "source" dans l'équation adoptée pour $\overline{e}$. Une idée de la valeur de cette simulation est donnée par la figure 1, où sont reportés les profils verticaux de $\overline{e}$ dans le modèle ANDRE (à gauche) et BLONDIN-THERRY (à droite) à 12, 14 et 16 heures TU.

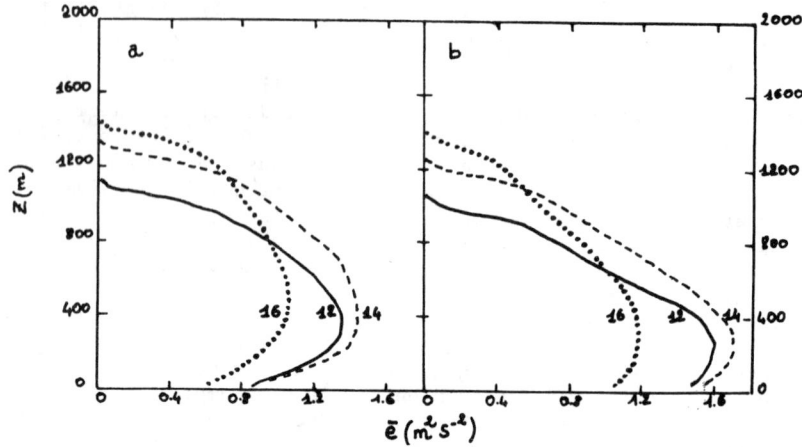

Figure 1 : Profils verticaux d'énergie cinétique turbulente
à 12, 14, 16 h TU le jour 33 de Wangara dans ANDRE et all
(a) et BLONDIN et THERRY (b)

3. SIMULATION BI-DIMENSIONNELLE D'UN CYCLE DE BRISE - ETUDE DE TRAJECTOIRES

Pour illustrer et vérifier les corrections apportées par le calcul des coefficients d'échange et des flux (en particulier $\overline{\omega'\theta'}$) suivant les formules précédentes, nous avons réalisé la simulation d'un cycle de brise terre-mer purement thermique à l'aide de la version bi-dimensionnelle de notre modèle, une première fois avec le formalisme de J.F. LOUIS (référencé ci-dessous avec le modèle K), une seconde fois avec le formalisme décrit au paragraphe 2 (référencée ci-dessous comme expérience avec le modèle $\overline{e}$). Naturellement, les contraintes extérieures, à savoir paramètres astronomiques du soleil, paramètres physiques relatifs à la terre et à la mer, sont les mêmes dans les deux simulations. Par contre, l'évaluation des flux radiatifs, dans l'atmosphère et au sol est issu d'un calcul, réalisé au moyen d'un modèle relativement sophistiqué (GELEYN et all 1979).

Les différences observées sur les champs thermodynamiques calculés peuvent être analysées et interprétées de la façon suivante :
La brise de mer s'établit un peu plus tôt, mais surtout avec une plus grande intensité dans le modèle K que dans le modèle $\overline{e}$. Cette tendance se renverse en milieu d'après-midi et on constate l'extinction de la brise de mer deux heures plus tôt dans le modèle K que dans le modèle $\overline{e}$. Durant cette dernière période, l'extension verticale du

front de brise est plus importante que dans le modèle $\overline{e}$. La nuit, une brise de terre, faible, s'installe dans les deux modèles, mais l'écoulement devient plus rapidement faiblement turbulent dans le modèle $\overline{e}$.

Ces grands traits répondent très bien aux améliorations théoriques apportées dans le modèle $\overline{e}$ et sont à relier à la différence de comportement dans la propagation verticale du flux de chaleur sensible entre les deux modèles, dans la mesure où les flux de surface présentent quant à eux, une évolution extrêmement similaire. Dans le modèle K , le flux de chaleur se propage dans une couche instable au départ moins épaisse que celle dans laquelle se manifeste cette propagation dans le modèle $\overline{e}$ par suite de l'influence du pseudo-gradient γ . La température des premières couches augmente donc plus vite dans le modèle K , créant un gradient horizontal de pression au niveau de la côte plus fort. Au fur et à mesure que la convection se développe, et que toute la couche limite tend à être bien mélangée, la divergence du flux de chaleur sensible dans le modèle K diminue par suite de la très faible valeur des gradients. Par contre l'air au dessus de la terre se réchauffe plus fortement dans le modèle $\overline{e}$ et sur une plus grande épaisseur. Quand le flux Q_0 devient négatif, l'instabilité dans les basses couches du modèle $\overline{e}$ est encore suffisante pour entretenir une convection organisée pendant plus d'une heure. Par contre, dès que les termes sources dans l'équation de $\overline{e}$ deviennent négligeables, l'énergie tombe presque immédiatement à une valeur très petite, entrainant le calcul de coefficients K plus faibles que ceux obtenus par la formule analytique, à conditions égales de champs moyens.

Afin de rendre plus parlantes ces différences et pour montrer l'importance qu'elles revêtent pour certaines applications pratiques, telles les études de pollution sur sites côtiers, nous avons développé dans notre modèle un algorithme permettant de suivre la trajectoire de particules. Le pointage de la position de chaque particule se fait tous les pas de temps du modèle dynamique (ici 30 secondes).

La figure 2 montre l'ensemble des trajectoires (modèle K en haut, modèle $\overline{e}$ en bas) de particules émises toutes les demi-heures entre 07h31 (une heure après le lever du soleil, instant initial de la simulation) et 17h31 (la fin de la simulation correspond à 5h31 le lendemain), et ce à 200 m d'altitude à la verticale de la côte.

Les numéros apparaissant sur ces figures correspondent à l'ordre chronologique d'émission. L'intervalle séparant deux croix sur une même trajectoire correspond à un transport d'une heure.

On notera la meilleure cohérence temporelle dans la simulation du modèle $\overline{e}$, due à une bonne organisation de la circulation dans la couche limite, caractéristiques importantes des situations convectives comme rappelé en 2.1. D'autre part, si on assimile l'enveloppe de ces trajectoires à une zone sensible à la pollution, on constate l'importante différence d'extension verticale observée à l'intérieur des terres par suite du maintien d'une activité convective en fin d'après midi (trajectoires 14 et 15).

4. APPLICATIONS AUX ETUDES DE DIFFUSION DE POLLUANTS

L'énergie cinétique turbulente constitue un des paramètres de base pour décrire l'état turbulent de l'atmosphère. Il est donc légitime de vouloir en tirer parti pour mettre sur pied une méthode de calcul visant à simuler la diffusion des polluants.

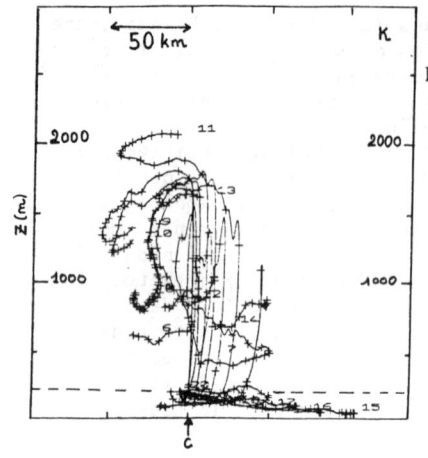

Modèle K

Figure 2 : Ensemble des trajec-

toires des particules émises
toutes les demi-heures à 200 m
(ligne pointillée) à la verticale
de la côte (point C)

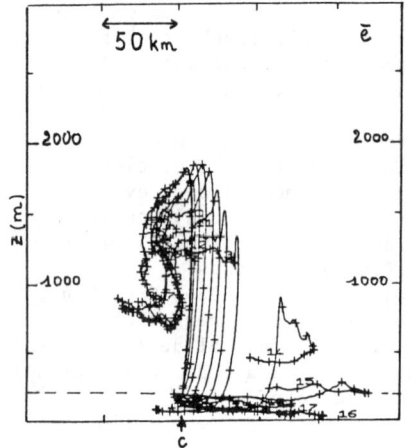

Modèle $\bar{e}$

Une approche lagrangienne semble bien appropriée puisqu'on peut espérer au travers de $\bar{e}$ remonter aux variances séparées $\overline{w'^2}$ et $\left(\overline{u'^2} + \overline{v'^2} \right)$, quoiqu'un important travail de recherche fondamentale et d'exploitation de données expérimentales soit à envisager pour ce faire.

On peut cependant déjà mettre en place un algorithme de calcul adéquat. Pour étudier la dispersion d'un nuage de polluants, on simulera la dispersion d'un flux de particules solides émises de la même source, en ayant tenu compte d'une éventuelle sur-hauteur.

La position de chaque particule sera calculée par

$$\overline{X}^{m+1} = \overline{X}^m + \left(\overline{u}^m + u'^m \right) \Delta t$$

$\overline{X}^{m+1}$: position de la particule à l'instant $(n+1) \Delta t$
$\overline{X}^m$: position de la particule à l'instant $n \Delta t$

$\overline{u}^m$: vitesse moyenne à l'instant $n\Delta t$ au point $\overline{X}^m$

u'^m : perturbation de vitesse à l'instant $n\Delta t$

$\overline{u}^m$ est calculé par interpolation linéaire entre les valeurs des vitesses aux points de grille les plus voisins de $\overline{X}^m$ du modèle dynamique.

u'^m est généré par un processus aléatoire dont l'ensemble des réalisations répond à une distribution centrée de Gauss de variance $\overline{u'^2}(\overline{X}^m)$ estimée à partir de $\overline{e}(\overline{X}^m)$

La méthode exige évidemment la gestion d'un grand nombre de particules. Pour vérifier le bon comportement numérique de cette méthode, une première expérience à deux dimensions a consisté à émettre des particules dans un champ de vent constant ($\overline{u}(\overline{z}) = u_o$; $\overline{w}(\overline{z}) = 0$) en spécifiant les variances $\overline{u'^2} = \overline{w'^2} = 4\,m^2/s^2$

On a pu ainsi vérifier que, dans ce cas, les écarts types de positions vérifiaient au premier ordre les relations théoriques :

$$\sigma_x^2 \propto \overline{u'^2}\,t \quad ; \quad \sigma_z^2 \propto \overline{w'^2}\,t$$

Pour illustrer maintenant de façon plus parlante les possibilités de cette méthode, nous avons simulé le transport et la dispersion d'un nuage de particules dans un champ de vent présentant un cisaillement vertical : $u(\overline{z}) = u_*/\chi \, Log(\overline{z}/\overline{z}_o)$ et un champ de température

présentant une zone de forte stabilité surmontant une zone de forte instabilité, la source se trouvant aux 2/3 supérieurs de celle-ci. On différencie les variances suivant la valeur du Richardson local.

La figure 3 montre l'allure du champ de concentration calculée sur une grille régulière ($\Delta x = 100\ m$, $\Delta z = 10\ m$) à l'instant $t = 300\ s$. (Le module du vent au niveau de la source vaut 6.8 m/s).

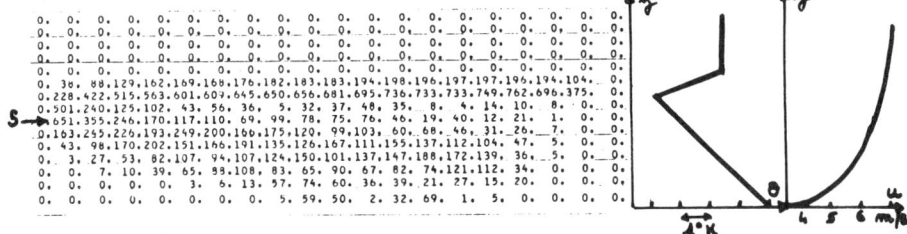

Figure 3 : Concentration (unité arbitraire) La source est au point S
A droite, profils verticaux de température potentielle et module du vent

On constate que le panache se divise en deux lobes, un dans la couche stable, l'autre dans la couche instable. La ligne inférieure donne une idée des concentrations au sol, à mettre en relation avec l'intensité de la source.

5. CONCLUSION

On a su prendre en compte dans un modèle de simulation numérique des écoulements atmosphériques à méso-échelle l'information apportée par la gestion d'une équation d'évolution de l'énergie cinétique turbulente. On s'est ainsi doté d'un outil permettant de reproduire l'influence des effets orographiques et thermiques sur l'écoulement synoptique. On a démontré la faisabilité et l'intérêt d'études de trajectoires dans des champs à forte variabilité spatio-temporelle et tenté de mettre en place un algorithme permettant de simuler la diffusion des panaches de polluants.

Une confrontation avec des données réelles s'impose. L'expérience de la CEE à Fos sur Mer pourrait en être l'occasion.

BIBLIOGRAPHIE

BLONDIN C. (1979) : Simulation numérique de l'atmosphère à méso-échelle. Proceedings du ler Symposium sur le comportement physico-chimique des polluants atmosphériques. Ispra (Italie) Octobre 1979, p 479-491.

BLONDIN C. (1980) : Projet CYBELE. Exploitation de la campagne "Vallée du Rhône 77". Note technique de l'EERM N° 65 (nouvelle série).

BLONDIN C. et G. THERRY (1981) : Analysis of particles trajectories during a land-sea breeze cycle using two-dimensional numerical meso-scale models. Proceeding du 12ème I.T.M. "On air pollution modelling and its applications". PaloAlto (USA). Août 1981.

CLARKE R.H. et all (1971) : The Wangara Experiment. Boundary layer data. Paper N° 19, Division Meteorological Physiko, CSIRO, Australia

DEARDORFF J.W. (1972) : Theorical expression for the counter gradient vertical heat flux. J. Geophys. Res. Vol 77. p 5900-5904.

GELEYN J.F. et A. HOLLINGWORTH (1979) : An economical analytical method for the computation of the interaction between scattering and line absorption of radiation. Contribution to atmospheric Physics Vol 52 N° 1.

KOLMOGOROV A.N. (1942) : Equations of turbulent motion of an incompressible turbulent fluid. IZV, S ci . Phys. Vol VI N° 1-2, p 56.

LOUIS J.F. (1979) : A parametric model of vertical eddy fluxes in the atmosphere. Boundary layer Meteorology N° 17, p 187-202.

O'BRIEN J.J. (1970) : A note on the vertical structure of the eddy exchange coefficient in the planetary boundary layer. J. Atm. Sci. Vol 27 p 1213-1215.

TAYLOR G.I. (1921) : Diffusion by continuous movements. Proceedings London Math. Society, Vol 20, p 196-211.

INTERCOMPARISON OF LIDAR AND SPECTROSCOPIC DATA FOR THE STUDY OF POWER PLANT EMISSIONS

P. Camagni[+], E. De Blust[+], M. De Groot[*], C. Koechler[+], R. Michelon[+], A. Pedrini[+], S. Sandroni[*]

Chemistry[(*)] and Electronics[(+)] Divisions
J.R.C.- Ispra (Varese) Italy

Summary This paper describes a general method for the combination of Lidar and Cospec techniques, with their respective abilities to detect particulate matter or specific molecular components. It is shown that by simultaneous monitoring of a plume and proper exploitation of the absorption and scattering data, one can afford a consistent intercomparison of particulate behaviour vs. gaseous behaviour in terms of integral burden distributions. Application of these concepts is demonstrated, with reference to some in-field experiments. While the interpretation of the phenomena can only be qualitative at the moment, the information obtained shows already a very interesting potential for mass balance and mass flow evaluations.

Introduction

Knowledge of the pathways of combustion products in the lower atmosphere is of great concern for the planning, regulation and control of strong localized sources, such as power plants. From this arises a continuous demand for methods, apt to improve the study of pollutant emission, transport and fallout. In this connection, important aspects of any monitoring technique are: 1) detection specificity and sensitivity for one or more pollutant substances; 2) ability to give a synoptic view, at least on a local scale; 3) complementarity with other techniques. Point sampling, employing chemical sensor networks, is often unequalled as far as concerns the first point. However, the other benefits are better afforded by remote sensing techniques. Among these the applications of light scattering (by means of LIDAR) and molecular absorption (by means of COSPEC) are particularly developed (1,2). The two techniques, as is known, are specifically sensitive to the particulate and to the gaseous components of emissions. So far, work has been centred on the development of their respective capabilities; much remains to be done in order to exploit them in joint operation, so as to achieve fast, simultaneous monitoring of multicomponent pollu-

tant structures. The relative behaviour of particle vs. gaseous
emissions from power plants would be a major field for such applications; it could give in particular a great advance in the
knowledge of physicochemical processes due to the complex interactions of gases and aerosols, thus helping also a substantial refinement of dispersion models.

A special effort to define problems and related methodologies in this field has been developed by the J.R.C. groups during recent monitoring campaigns (3). The present paper describes a typical exercise concerning the comparison of particle
vs. gaseous behaviour in a stack plume and related determinations of pollutant mass-flow and mass-balance.

Methodological survey

The principles of Lidar and Cospec observations are schematically compared in Fig. 1. Here we sketch a situation where
a simplified plume is analyzed by repetitive scanning along the
same transversal section with the two instruments. Complete
analysis of the plume will obviously imply the extension of similar concepts to a number of other sections, at different
heights and distances from the stack. Within the given section,
we fix our attention on a particular direction of observation
(say, trace $\underline{c}$) characterized by common values of elevation ϑ and
azimuth φ for LIDAR and COSPEC.

The Lidar response is schematized by a profile of backscattered light intensity vs. range (continuous line) which is
spatially resolved along the trace of observation. From the distribution $I \equiv I(R)$ one can work out two integrals of special
operative interest. They are:

(1)
$$\vec{P_{\underline{c}}} = \int_{Rmin}^{Rmax} \vec{R} \cdot I(\vec{R}) dR \bigg/ \int_{Rmin}^{Rmax} I(\vec{R}) dR \qquad \vec{R} \equiv (R, \vartheta, \varphi)$$

giving the centre of mass of scattering matter along trace $\underline{c}$,
and

(2)
$$M_{\underline{c}} = \iint_{\Delta\alpha}^{Rmax}_{Rmin} R I(\vec{R}) dR d\alpha \qquad d\alpha \equiv d\vartheta \text{ or } d\varphi$$

which represents on a relative scale the scattering mass-densi-
ty, integrated over the element of section area $RdRd\alpha$ on which
it insists, $\Delta\alpha$ being the angular step between successive traces;
in this way $M_{\underline{c}}$ is a direct (relative) measure of the integrated
scattering burden along the trace, weighted by the areal frac-

tion of the total section that pertains to the given trace, in the discrete succession of equally spaced traces a, b, c etc. The sum $\sum_j M_j$ will represent the total section burden as far as scattering matter (i.e. particulate) is concerned. A corresponding quantity for absorbing matter (SO_2 or NO_2) may be obtained from the Cospec data along the same trace. In fact the Cospec signal is a direct measure of the integral absorption burden A_c, in terms of concentration x distance. This is schematized by the shaded rectangle of Fig. 1.

What lacks to the Cospec data is the range-resolved distribution, hence the centre of mass of the gaseous species. This information can be borrowed from the Lidar data, assuming that the centre of mass of particles coincides with that of the gaseous pollutants, over any given trace.

If $\vec{P}_a$, $\vec{P}_b$, ...$\vec{P}_k$ etc. is the locus of baricentral points for the various traces belonging to a given section, we may now treat this as an abscissa (conveniently projected onto coordinate planes) and plot with respect to it the corresponding trace burdens of scattering matter and absorbing matter, M_k and A_k. We obtain in this way representations such as those of Fig. 2, showing the distribution of burden across the section as a function of a transversal parameter (for instance, the azimuthal angle φ). The areas under these curves are a measure of total burdens $\sum_k M_k$ and $\sum_k A_k$ of the concerned species; they represent, on a relative scale, the respective masses of pollutants over the entire section. In addition, taking a set of arbitrary levels B_1, B_2, ... B_j on the ordinate, one can identify on the abscissa sets of points corresponding to these levels, thus spanning the entire section into subsections of increasing burden intensity.

As an outcome of this analysis, we finally obtain two basic groups of data for each of the concerned species:
a) total particulate burden and total gas burden, to be correlated with the centre of mass of each section;
b) spatial coordinates of equal (relative) burden levels along the baricentral traverse.

From these data an overall picture of plume evolution can be finally established, in terms of geometrical distribution as well as of total mass balance. In order to achieve that, we need at this point a coherent spatial scheme for the description of a series of sections. A convenient picture will be obtained by converting the data for the actual sections into consistent data for proper cross-sections, defined around the same centre of mass. Fig. 3 illustrates the procedure, with reference to an actual experiment. The distribution of a number of sections,taken above the stacks of the Turbigo power plant, is shown here in two coordinate planes N-E (ground) and H-E (vertical). The common location of LIDAR and COSPEC in the ground plane was out

of scale to the North. In each plot the line connecting the centres of mass of successive Lidar sections allows us to define
an average plume trajectory and consequently the corresponding
cross-sections at any point (see dotted projections). After evaluation of the respective direction cosines, any segment on the
traverse of the experimental section can be projected into a
corresponding segment in the cross-sectional plane. Likewise
the measured total burden can be normalized to cross-sectional
area. By this procedure we are able finally to produce ensemble
plots, where the respective evolution of burden profiles and
total burden is schematically represented (Figs. 4 and 5).

Discussion of a typical experiment

The above described methodology was applied to investigate
the plume behaviour over the ENEL power plant of Turbigo (Italy)
during recent campaigns. Figs. 4 and 5 summarize the results of
an exercise performed on October 6,1980, on the following chimneys of the eastern group: L/1 (95 m, 250 MW) ; L/2 (150 m, 330
MW) ; L/3 (150 m, 540 MW) . These are aligned in approximately
a NW direction. The two monitoring units were placed close to
each other about 1300 m to the North of the stacks. The LIDAR,
with its 2 J Q-switched Ruby laser and associated collector
optics, as well as the correlation spectrometer (COSPEC III,
Barringer Res.) have been described elsewhere (3,4). Both units
are capable of programmable motions in azimuthal and vertical
directions and are assisted by data acquisition systems.

Soundings of the atmospheric parameters by pilot balloons
indicated a mean wind direction to NE, with appreciable direc-
tional shear, during the entire exercise, i.e. from 15.30 to
17.00 LMT; however, the average wind speed, with respect to
ground, was found to fall from $\sim 1,7ms^{-1}$ to $< 0,7ms^{-1}$ (ultrasonic anemometer data) after 16.00 LMT. For this reason, the data
have been analyzed as two separate experiments, referring respectively to the first and to the second part of the above quoted period. Fig. 3 depicts the arrangement used in the second
part, when vertical sectioning at increasing distance from the
stacks was performed. Fig. 4 summarizes the results for this
case: here the abscissa L is the length at ground along the
plume axis, i.e. the unfolded succession of segments connecting
the centres of mass of Fig. 3b, while the ordinates are the
heights of these centres. The rectangular areas, with different
shades as shown for particles and SO_2, are oriented as the appropriate projections of local cross-sections at any point.
Their linear dimensions express (in ordinate units) the lateral
spread, i.e. the width corresponding to trace burdens greater
than or equal to a given reference level (see definitions of
Fig. 2); this level is fixed for convenience at 10 ppmxmetre
for SO_2 and at 10xM_0 for particles (M_0 being the background

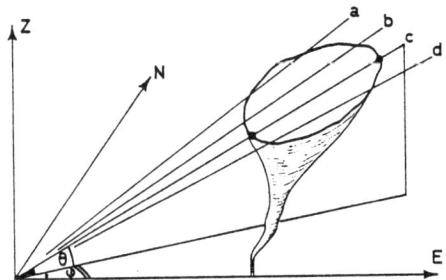

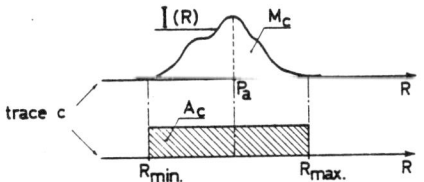

Fig. 1 - Schematics of Lidar and Cospec sectioning across a plume.

atmospheric burden measured by LIDAR, in arbitrary units). The thickness of the rectangular areas is normalized to represent total burden, expressed in terms of the above units multiplied by length; the same quantity is also quoted in the numerical insert.

The main aspects of our observations may be summarized as follows: 1) satisfactory overlap of spatial SO_2 and particle distribution in any given section, in spite of sizeable variations of lateral width and local plume trajectory between successive sections; 2) good conservation of total SO_2 burden and particle burden along the plume, as shown by the insert. Such conservation is all the more remarkable if one considers the vertical wind components which were acting during the measuring period (as revealed by ultrasonic anemometer) and which are visualized in the vertical shears of the trajectory.

An earlier view of the same plume, taken by a series of azimuthal and vertical sections, is illustrated in Fig. 5 with similar criteria. The same observations of points 1) and 2) above are seen to hold; in particular the individual conservation of SO_2 and particle burdens is well demonstrated. Combining the evidence of Figs. 4 and 5 the good consistency of mass-balance determinations for the two species is confirmed, at least for distances up to 1 Km. and for time intervals characterized by a reasonable uniformity of meteorological parameters. This gives further support to previous determinations (3).

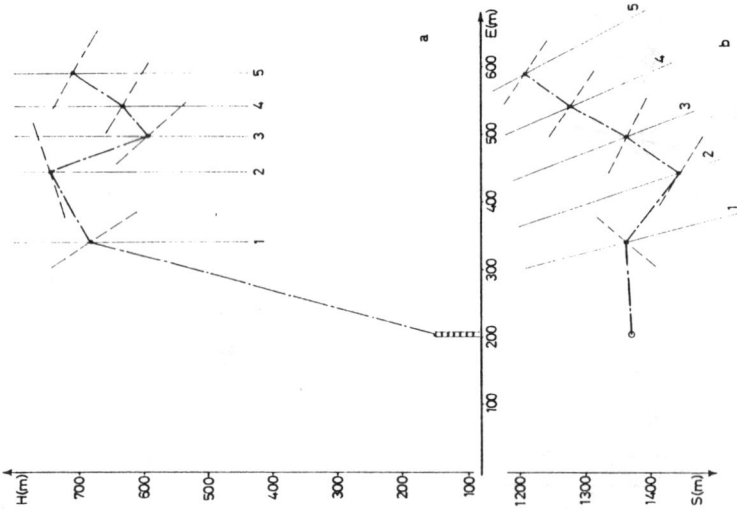

Fig. 3 - Geometrical characterization of an ensemble of vertical sections with respect to coordinate planes N-E (ground) and H-E (vertical).

—·—·— Average plume trajectory
——————— Trace of experimental sections
— — — — Trace of cross-sectional planes

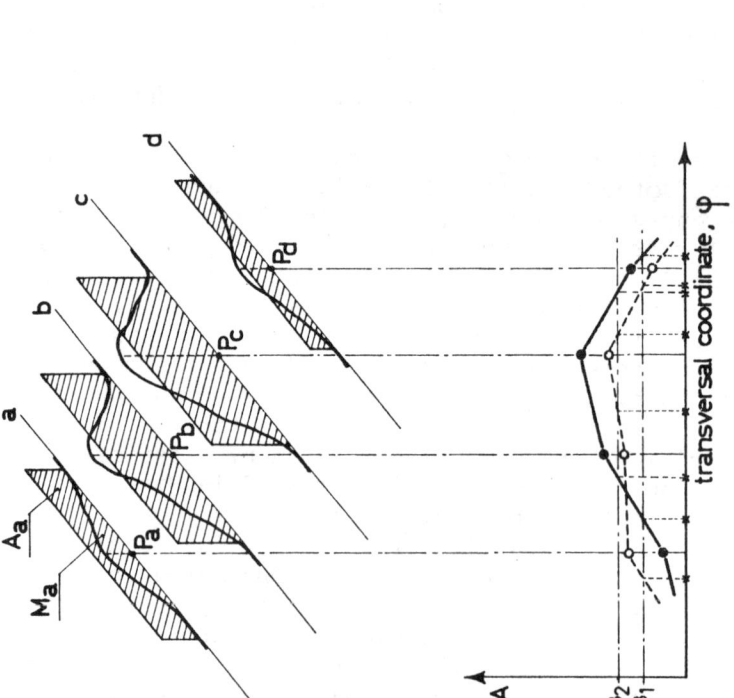

Fig. 2 - Construction of burden distribution for gas and particles, along the baricentral traverse of a given section.

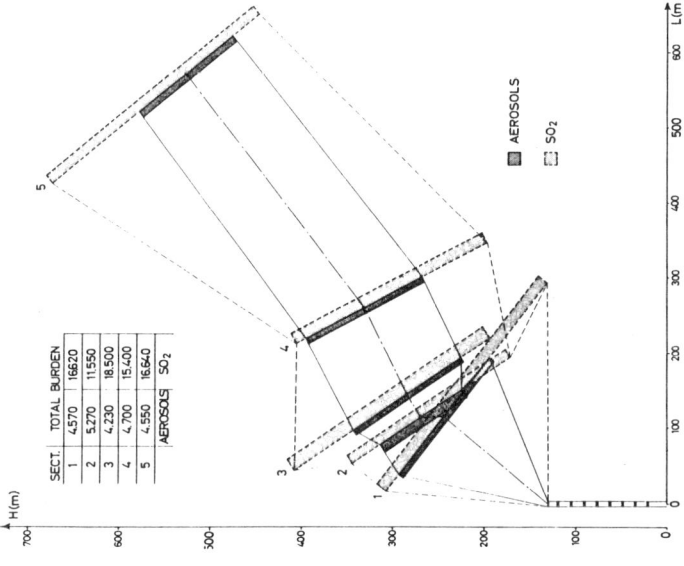

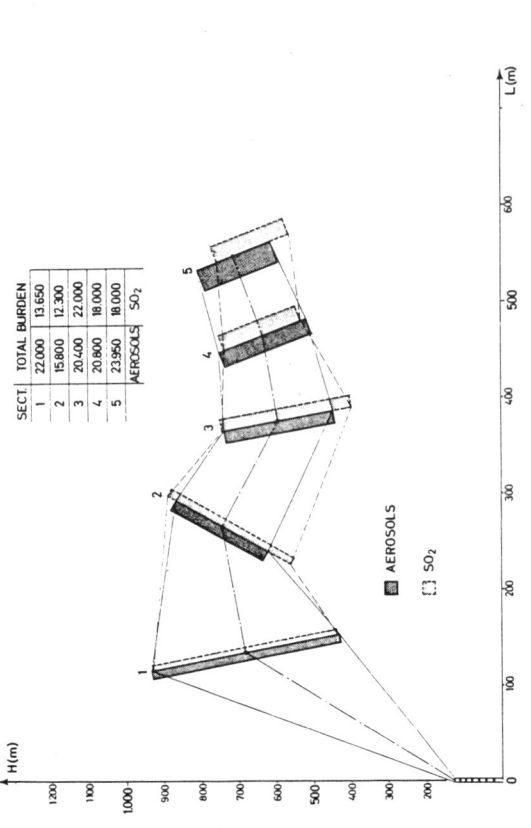

Fig. 4 - Evolution of SO₂ and particle burdens along the plume, according to the experimental arrangement illustrated in Fig. 3. The monitored chimney group is graphically represented by the equivalent chimney.

Time of measurements : 16.10 to 17.00 LMT, 6/10/1980.
Average wind speed : 0,7 ms⁻¹ hor.; 0,8 ms⁻¹ vert.
For graphical symbols and numerical inserts, see text.

Fig. 5 - Same as Fig. 4, illustrating plume evolution at an earlier time (15.30 to 16.00 LMT). Average wind speed : 1,7 ms⁻¹ hor.; 1,6 ms⁻¹ vert.

A remarkable difference between the situations of Figs. 4 and 5 is represented by the fact that absolute figures of particle burden are not conserved from one experiment to the other, while gas burdens approximately are. The reasons for such a variation might by complex; one certainly is the significant difference of the average horizontal wind speed, which as noticed before dropped from 1,7 to < 0,7ms^{-1} on passing from the experiments of Fig. 5 to those of Fig. 4. In terms of total flux conservation, this fact accounts already for an increase of more than 2 in absolute burden, i.e. more than half the total variation observed. To complete the explanation, however, it is necessary to include the possibility of physico-chemical processes capable to enhance the mass and/or the size of "visible" aerosols, when their residence time is increased (as in the case of Fig. 4). Hints of such phenomena were already invoked to explain earlier experiments on a vertical plume, in the presence of a stagnant atmosphere. Anyhow , one cannot exclude at this time that a substantial difference in the diffusion of gas and aerosols may play a role in the observed discontinuity of mass-balance.

It will be interesting to extend the present experiments to the study of plumes from coal-fired power plants, where a higher particle-to-gas ratio holds and the absence of water vapour simplifies the physico-chemical conditions.

Acknowledgements

The authors are grateful to the Direction of the ENEL power plant for technical support during the exercise, and to Mr. H. Hasenjäger of J.R.C. - Ispra for his valid support with meteorological data.

References

1) Collis, R.T.H., Russell, P.B.: Lidar Measurement of Particles and Molecules by Elastic Back-scattering and Differential Absorption in: Laser Monitoring of the Atmosphere (D.E. Hinkley, Editor) p. 71, Springer-Verlag, 1976.

2) Hamilton P.M., R.H. Varey, and M.M. Millan: Atm. Environm. 12, 127 (1978).

3) Camagni P., E. De Blust, C. Koechler, R. Michelon, A. Pedrini, M. De Groot and S. Sandroni - Nuovo Cimento C (in press).

4) Preliminary Report (September 1979) and Final Report (in press) of the IV C.E.C. Campaign on Remote Sensing of Atmospheric Pollution, Turbigo, 1979; C.E.C. and Enel Editors.

THE REAL MEANING OF DRY DEPOSITION VELOCITIES FOR DISPERSION CALCULATIONS

I. MERTENS, J.G. KRETZSCHMAR and B. VANDERBORGHT

Nuclear Energy Research Centre, B-2400 Mol, Belgium

SUMMARY

Different schemes for the calculation of dry deposition of gases and aerosols have been reviewed and the possibilities for their incorporation into a bi-Gaussian dispersion model have been analized. At least for short range calculations the source depletion approach seems to remain an acceptable approximation. The implications for practical impact assessment studies have been evaluated by means of a practical case involving fugitive emissions and chimney emissions of a specific metal. As field measurements of concentration- and deposition-levels were available over a period of more than a year the model results were verifiable. The sensitivity of the model calculations, as well as the deviations between measured and calculated values, are analized as a function of the available (measured or theoretically derived) set of values for the deposition velocity.

1. INTRODUCTION

Within the framework of the Belgian National Research and Development Program Environment-Air of the Ministry of Science Policy a detailed investigation of the environmental behaviour of Sb-particulates emitted by a nonferrous metal industry situated in an open and flat region has been carried out over a period of approximately two years (1979-1980). The main purpose of this testcase was the development and verification of appropriate methodologies to deal with the different aspects of the impact of nonferrous metal industries upon the environment. The specific choice of Sb as testcase is due to the fact that the involved factory is the only Sb-emitter in the region, so that problems with background-levels were avoided, that the factory was willing to cooperate and that no direct health hazards were existing.

The purpose of this paper is to present some of the results of the modelling undertaken within the context of the testcase. The required information on the emissions (three chimneys, many fugitive emissions), the local atmospheric conditions and the ground-level concentrations and depositions have been obtained by different teams of the Gent University (R.U.G.), the University of Antwerpen (U.I.A.), the Nuclear Energy Research Centre (S.C.K./C.E.N.) and the National Data Bank (I.H.E.). The lay-out of the monitoring network around the Sb-plant is given in Fig. 1. In each monitoring site 21 h-average ground-level concentrations and monthly deposition values were obtained. Meteorological data are measured half-hourly averages while the emission data are best estimates based on selective measurements, tracer releases and a systematic inventory and registration of the activities in the plant.

The model calculations have been restricted to a period of 14 consecutive months during which the required input information was as complete as possible.

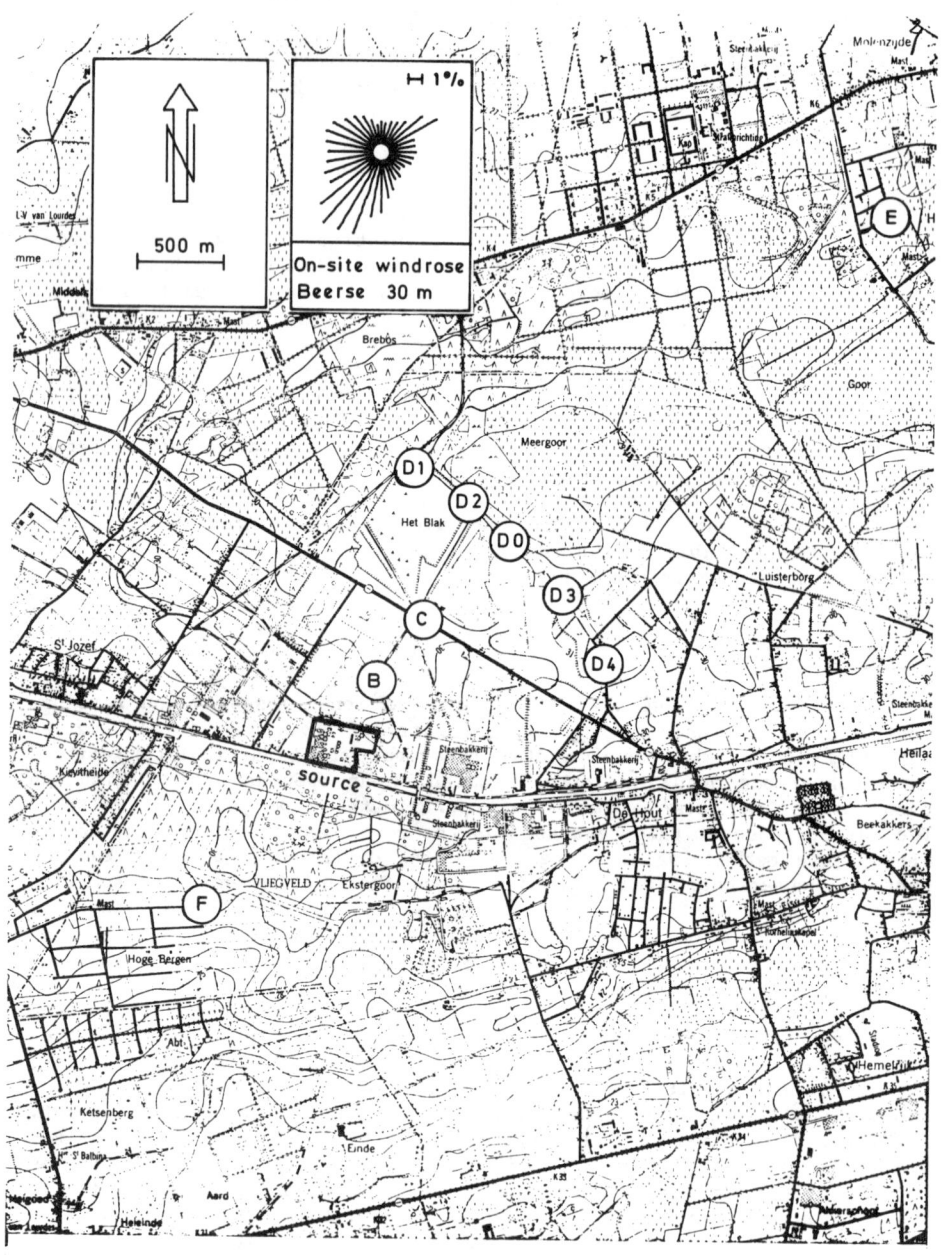

Figure 1 : Lay-out of the Sb-monitoring network (concentration and deposition).

2. DESCRIPTION OF THE MODEL

All calculations have been carried out by means of the Immission Frequency Distribution Model (IFDM), a bi-Gaussian dispersion code developed at the S.C.K./C.E.N. (1,2) Mol, Belgium. The most important features of the model version actually used for this study are as follows :
- half-hourly averages as input data for the emissions and the meteorological parameters;
- turbulence typing scheme and corresponding dispersion parameters based on the ratio of the potential temperature gradient and the square of the average wind speed measured at 69 m above ground-level as described by Bultynck and Malet (3);
- daily- and monthly averages obtained by taking the arithmetic average over consecutive half-hourly values;
- plume rise formula of Stümke (4) for the chimney releases;
- all fugitive emissions released at approximately 15 m above ground-level (no plume rise);
- transport speed in the Gaussian formula is the average wind speed at the effective release height obtained by means of the power law wind speed profile (m = 0,53 for very stable to m = 0,10 for very unstable conditions);
- all sources taken into account as point sources without any initial dispersion due to building effects;
- deposition calculated by means of source depletion;
- no sedimentation was used as granulometric investigations indicated that almost all particulates were smaller than 1 μm;
- mixing height unlimited.

3. INFLUENCE OF THE DRY DEPOSITION VELOCITY ON THE CALCULATED CONCENTRATIONS

As mentioned before wet and dry deposition were calculated by means of the source depletion approach. For wet deposition a constant washout-rate of 2.10^{-4} s^{-1} was used while for the dry deposition different numerical values for the dry deposition velocity v_d were tried out. In a recent review paper by Sehmel (5) values of 0,06 cm/s and 0,4 cm/s are given for the dry deposition velocity of Sb-particulates while $v_d = 1$ cm/s is a quite frequently used value for particulates in general. With each of the three v_d-values concentrations and depositions were calculated in each of monitoring sites A to F where simultaneously measured values were available (LIB high volume samplers with NAA for the concentrations, NILU deposit gauges with AAS for the depositons).

Since averaging time for measuring ground-level concentrations was much less than for measuring depositions, the sensitivity of the model calculations will be tested for the former.

In Fig. 2 cumulative frequency distributions of measured daily Sb-concentrations are compared with the corresponding cumulative frequency distributions of the calculated values obtained respectively with $v_d = 0$, 0.06, 0.4 and 1 cm/s in three points at different distances from the factory (Fig. 1).

In site B, close to the source measured values are smaller than calculated values and vice versa in site E at greater distance. The calculated concentrations in B are for 99 % due to the fugitive emissions. These

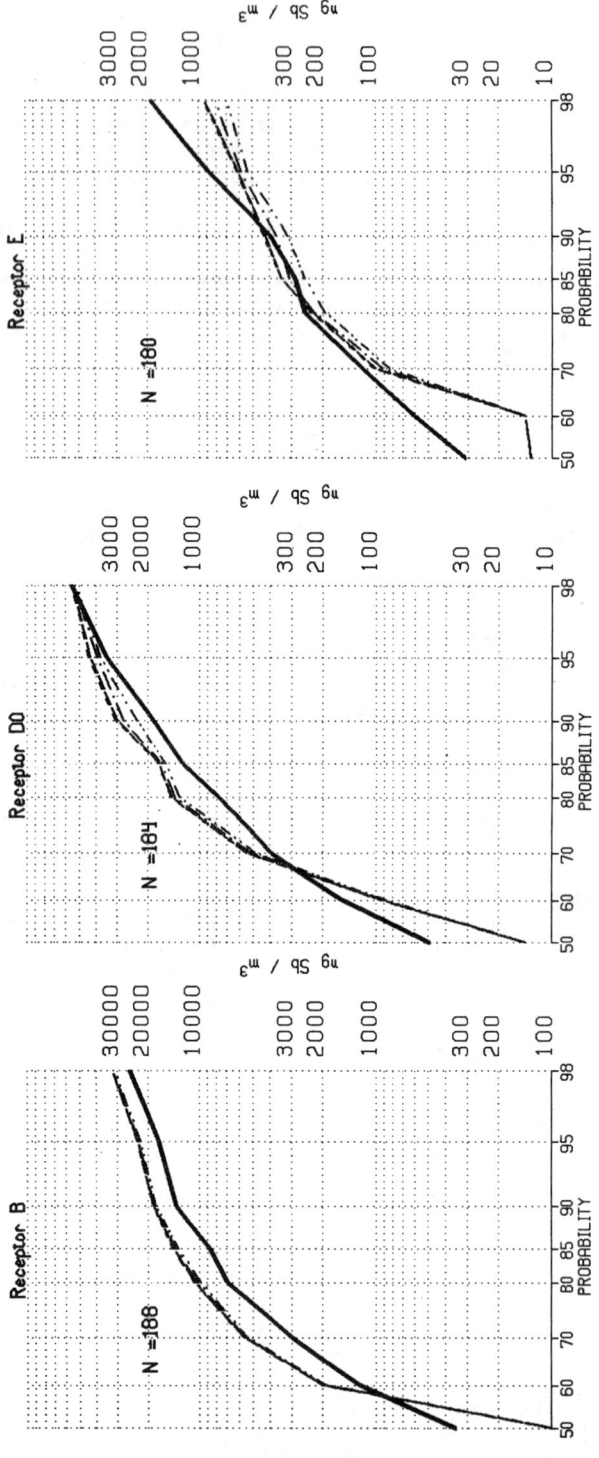

Figure 2 - Cumulative frequency distributions of Sb-levels in Beerse - Period : June 1979 - July 1980

measured values

calculated values; no depletion
calculated values; v_d = 0.06 cm/s
calculated values; v_d = 0.4 cm/s
calculated values; v_d = 1 cm/s

almost ground-level sources are in reality initially dispersed due to building effects while this has not been taken into account in the model calculations. The underestimation at greater distance could be due to an underestimation of the impact of the chimney emissions or to resuspension phenomena which were not included in the model.

The influence of the choice of v_d increases with distance from the source (B at 385 m, Do at 1445 m, E at 4120 m). The average daily Sb-concentration decreases with 4 % in B, 12 % in Do and 19 % in E when taking account v_d = 1 cm/s instead of no depletion. The value of v_d = 1 cm/s gives the best agreement between calculated and measured values in all points except for the most distant point E where the model already underestimates the concentrations for v_d = 0.

4. INFLUENCE OF THE CHOICE OF THE DISPERSION PARAMETERS

The sensitivity of the bi-Gaussian model for the dispersion parameters $\sigma_y(x)$ and $\sigma_z(x)$ has been discussed by many different investigators. The purpose of the actual exercise is not to repeat well-known facts but to illustrate the previous findings since the correction factor for dry deposition, using a source depletion model, is also influenced by the dispersion parameters.

In order to do so the measured half-hourly average temperature gradient along the meteorological tower in Mol was used to determine the corresponding Pasquill stability class as described in the U.S. Nuclear Regulatory Guide (6). Note that the same meteorological measurements were used in the previous calculations based on the turbulence typing scheme of the S.C.K./C.E.N. The correspondence between the frequency of occurrence of the stability categories in respectively Pasquill's and S.C.K./C.E.N.'s system is illustrated in Table 1.

TABLE 1 : Relative Frequency of Occurrence of Pasquill's and S.C.K./-C.E.N.'s Stability Classes during the Test Period (14 m)

	E1	E2	E3	E4	E5	E6	E7	Σ
A	0	0	0	1,6	0,3	0	0	1,9
B	0	0	0,1	1,4	0,6	0	0	2,1
C	0	0	0,5	3,2	1,1	0	0	4,8
D	1,0	13,4	16,0	6,0	0,8	0	0	37,2
E	10,3	25,8	0,5	0	0	0	0,3	36,9
F	10,6	0,4	0	0	0	0	0,4	11,4
G	3,5	0	0	0	0	0	0	3,5
Σ	25,4	39,6	17,1	12,2	2,8	0	0,7	97,8

The calculated concentrations, obtained by using this data set and the corresponding dispersion parameters as given by Gifford (7), are compared with the measured ones as illustrated in Fig. 3. The influence of v_d on the calculated concentrations increases with increasing distance but not in the same way as for the S.C.K./C.E.N. stability classification scheme. The average daily Sb-concentrations decrease with 1 % in B, 12 % in Do and 29 % in E when switching from no deposition to dry de-

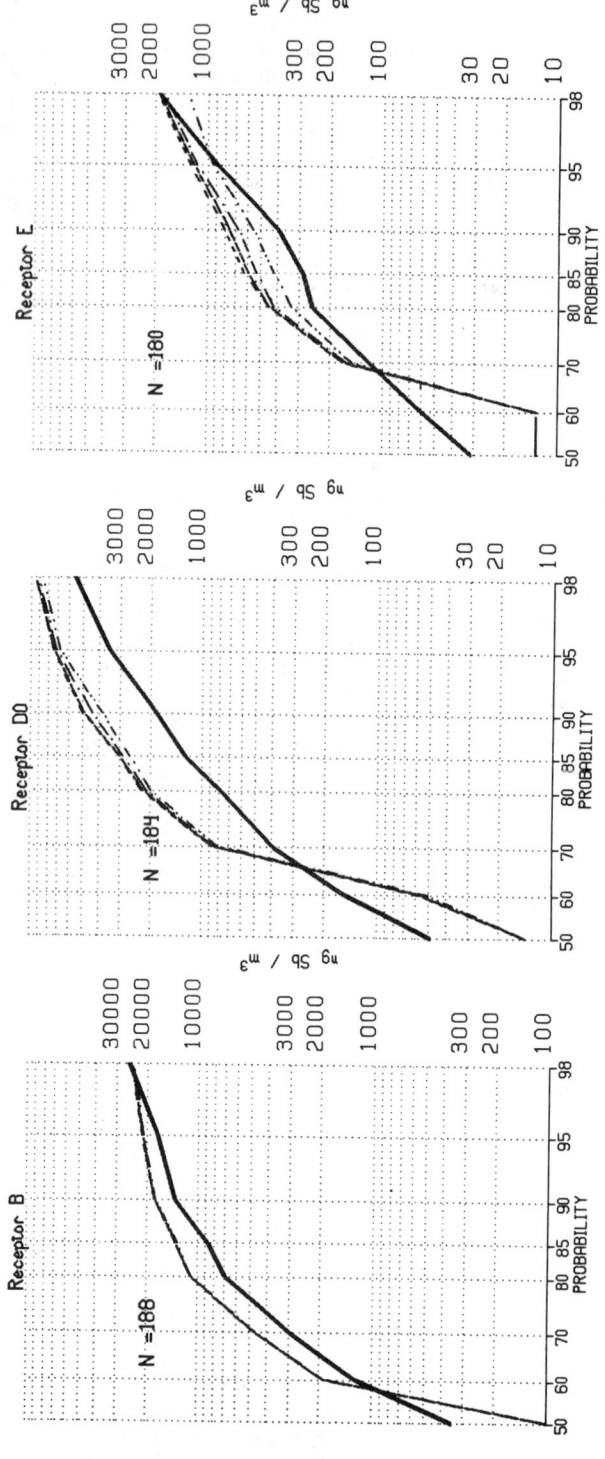

Figure 3 - Cumulative frequency distributions of Sb-levels in Beerse - Period : June 1979 - July 1980
measured values
calculated values; no depletion (Pasquill scheme)
calculated values; v_d = 0.06 cm/s (Pasquill scheme)
calculated values; v_d = 0.4 cm/s (Pasquill scheme)
calculated values; v_d = 1 cm/s (Pasquill scheme)

position with v_d = 1 cm/s. The value of v_d = 1 cm/s gives again the best agreement between calculations and measurements.
Comparing the Pasquill results for v_d = 1 cm/s to the corresponding ones calculated in the S.C.K./C.E.N. scheme and to the measured ones shows (Fig. 4) no great difference close to the source, while farther away (Do) the results based on the Pasquill input show larger deviations from the measured values.
At the most distant point E there isn't much difference except for the higher percentiles where Pasquill gives better agreement. These phenomena are due to the fact that the Pasquill parameters for the dominant stable and neutral conditions give narrower plumes than the S.C.K./C.E.N. dispersion parameters.

5. INFLUENCE OF THE DRY DEPOSITION VELOCITY ON THE CALCULATED DEPOSI-TIONS

Using the same data sets as before monthly depositions (dry plus wet) were calculated in the same receptors. Time series of measured and calculated monthly values are given in Fig. 5 for the S.C.K./C.E.N. diffusion typing scheme and in Fig. 6 for the Pasquill one.
The influence of v_d will naturally be more pronounced on the deposition calculations than on the concentration calculations, even in B. On the other hand the influence is of the same order in the three receptors at different distances : the linear relation between deposition and v_d dominates the distance dependent relation between concentration and v_d.
The patterns calculated for v_d = 1 and 0,4 cm/s are completely the same and the ratio of 2,5 between the two v_d's can be found in the calculated depositions. This doesn't hold for v_d = 0,06 cm/s. This means that for v_d > 0,06 cm/s the dry deposition is responsable for the major part of the calculated total deposition.
Only for very small v_d influence of the washout-rate can be seen. From Fig. 5 and 6 it is also clear that v_d = 1 cm/s gives the best agreement between measured and calculated depositions as well for the S.C.K/C.E.N. scheme as for the Pasquill scheme as this was the case for the calculated concentrations. The value v_d = 1 cm/s is also supported by some experimental findings as the ratio between the measured average dustfall and the measured average concentrations varies between a low 0,7 ± 0,3 for sites C and D2, and a high 2 ± 1,1 for D3 (Fig. 1).
Although some quite acceptable results were obtained here with a dry deposition velocity of 1 cm/s this doesn't proof at all that this is the magic number. A detailed analysis of all the available data from the different experiments carried out during the Sb-testcase showed that major uncertainties still exist as to the real physical meaning of dust-fall measurements in general and the accuracy of actual dustfall measurements under field conditions.
Somewhat related problems are the physics and the mathematical treatment and quantification of resuspended particulate matter and its influence upon the measured and calculated concentration and deposition values.

6. CONCLUSIONS

By means of the available data of a detailed study over fourteen months of the Sb-levels around a nonferrous metal industry it was pos-sible to demonstrate :

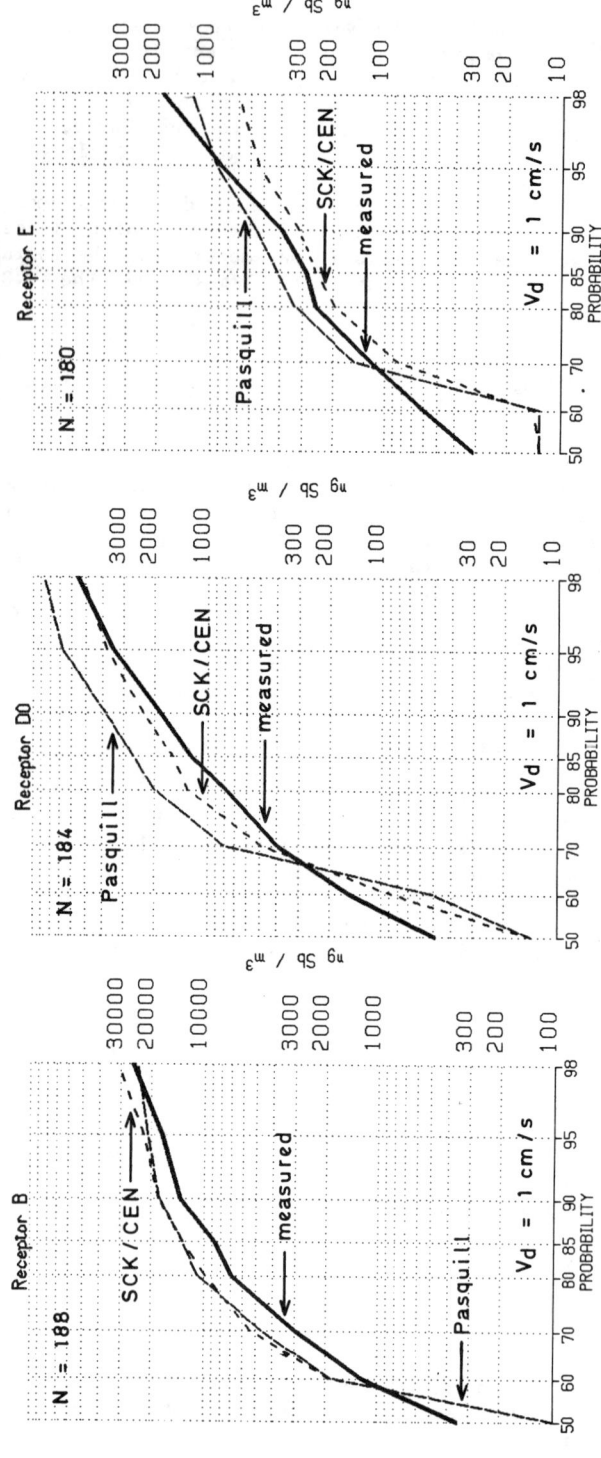

Figure 4 - Comparison between cumulative frequency distributions of measured Sb-levels and calculated ones by means of respectivelw Pasquill's and SCK/CEN's stability classification scheme.

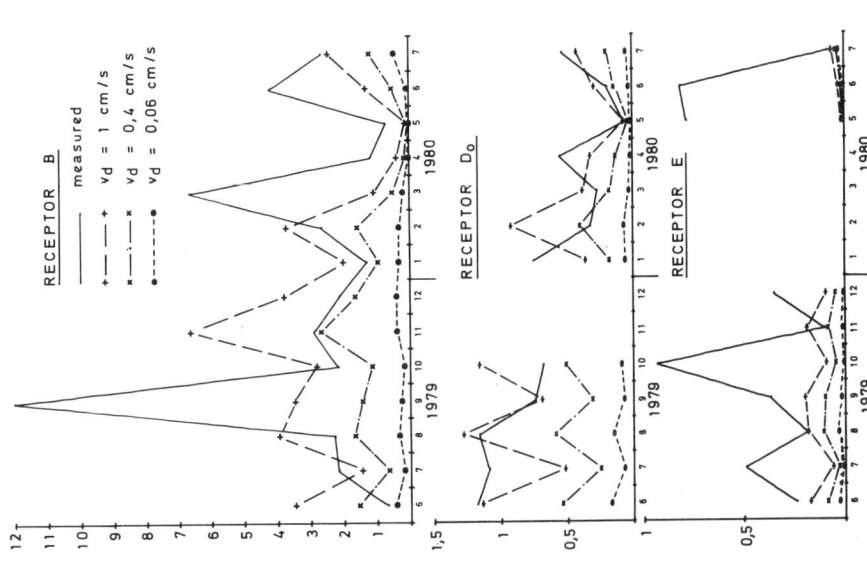

Fig. 5 - Comparison between measured Sb-depositions and calculated ones by means of the SCK/CEN's stability classification scheme.

Fig. 6 - Comparison between measured Sb-depositions and calculated ones by means of the Pasquill's stability classification scheme.

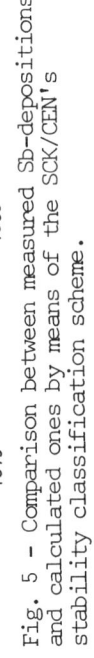

- that acceptable agreement between calculated and measured concentration and deposition values in monitoring sites at different distances from the source can be achieved by means of a simple bi-Gaussian dispersion model using the source depletion approach;
- that major uncertainties still exist as to the physics and the real meaning of the measurements as well as the calculation of deposition and resuspension.

7. ACKNOWLEDGEMENT

Within the framework of the Sb-testcase of the National R & D Programme Environment-Air, under the direction of C. De Wispelaere (DPWB), it has been possible to collect the sufficient amount of reliable data for the proper application of the IFDM dispersion and deposition simulation model.

As modellists we greatly appreciate the efforts of the teams of R. Dams (RUG), F. Adams (UIA), A. Cottenie (RUG) and M. Legrand (IHE) as well as the support of our scientific, technical and administrative colleagues at the Nuclear Energy Research Centre in Mol.

REFERENCES

1. J.G. Kretzschmar, G. Cosemans, G. De Baere, I. Mertens and J. Vandervee, Some practical examples of the impact of individual sources upon the cumulative frequency distributions of the daily SO_2-concentrations in an urban and industrial area, in : "Proceedings of the Eight International Technical Meeting on Air Pollution Modeling and its Application", Nato CCMS (1977).

2. J.G. Kretzschmar, G. De Baere and J. Vandervee, The Immission Frequency Distribution Model of the S.C.K./C.E.N., Mol, in : "Modeling, Identification and control in Environmental Systems", Vansteenkiste, ed., North-Holland Publ. Co. (1978).

3. H. Bultynck and L.M. Malet, Evaluation of the atmospheric dilution factors for effluents diffused from an elevated continuous point source, Tellus, 24 : 455 (1972).

4. H. Stümke, Vorschlag einer empirischen Formel Für die Schornstein-überhöhung, Staub, 23 : 549 (1963).

5. G.A. Sehmel, Particle and gas dry deposition : a review, Atm. Env., 14 : 983 (1980).

6. Anon., Regulatory Guide 1.23. Onsite meteorological programs, U.S. Nuclear Regulatory Commission (1972).

7. F.A. Gifford, Turbulent diffusion-typing schemes : a review, Nucl. Saf., 17 : 68 (1976).

REASONS FOR SEASONAL AND DAILY VARIATIONS OF CO_2

AND O_3 AT 0.7, 1.8, AND 3.0 KM ALTITUDE RECORDED

SINCE 1977

R. REITER (Director) and
H.-J. KANTER
Fraunhofer-Institut für Atmosphärische Umweltforschung
Garmisch-Partenkirchen, FRG

Summary

Since 1977, recordings are taken of the O_3 and more recently also of the CO_2 at the 3 neighoring mountain stations Garmisch (0.7 km), Wank peak (1.8 km), and Zugspitze peak (3.0 km altitude). By means of the Zugspitze cable car telemetry system hourly profiles of the O_3 are measured on many days. The seasonal and daily variations are para- meterized and largely clarified through meteorological and ecological data. Aside from vertical transport processes, photochemical processes in the boundary layer play an important role for the O_3 - even under pure air conditions - while the CO_2 level is dominantly determined by the activity of different kinds of the biomass as a function of weather and climate where a considerable influence of the altitude level on density and kind of vegetation is observed. At 3 km altitude, strato- spheric intrusions prevail as far as O_3 is concerned, the CO_2 is al- ready there coupled to the global level only.

1. OBJECTIVES

The purpose of this research is the study of time variations of CO_2 and O_3 and their reasons in the lower troposphere through simultaneous re- cordings at different levels (e.g. by mountain stations and cable car tele- metry) to elucidate the dependence of the vertical profile of the concen- tration of both gases on the diurnal and annual variation, on meteorologi- cal parameters, on the vertical exchange in both directions; it includes further derivation of anthropogenic sources and photochemical processes (O_3). Through parameterization of CO_2 and O_3 data obtained by meteorological influence factors, the above objective shall be achieved step by step. This paper can only give a few rare examples of results. Nevertheless they show that the proposed schedule has fully been met and that the facilities and methods are appropriate to our purposes.

2. MATERIALS AND METHODS

2.1. Ozone

The 3 recording stations, see Fig. 1 with small horizontal distances in the valley -No.1- (740 m), on the Wank -No.2- (1780 m) and on the Zug- spitze -No.3- (2964 m) have been equipped with chemoluminescence devices and automatic calibration systems so that the quality of recordings meets the present highest demands. Further, an ozone radiosonde (ECC) was electro- nically adapted and calibrated in our lab such that it delivers from summer

'80 along with our other facilities aboard the Zugspitze cable car currently profiles of ozone from 1.0 to 2.96 km altitude. In all, about1000 individual profiles of ozone have been taken by this manner. The total of meteorological and many further parameters are recorded at the stations.

2.2. Carbondioxide

At the stations mentioned under 2.1., the CO_2 has been continuously
recorded. Innovations during 1980: Installation of a fully-automatic recording station closely below the Zugspitze peak; this work proved very time-
consuming so that the station could be put into operation only from the end
of fall 1980. All 3 stations are equipped with automatic calibration systems
based on the Keeling CO_2 scale.

2.3. Evaluations

All evaluations are made by computer. Parameterization by means of the
simultaneously known meteorological conditions is in progress.

3. RESULTS

In this short paper only some few results can be discussed. Essentially
more information is available which will be combined with data currently
obtained during 1981.

3.1. Ozone

Fig. 2 shows that in compiling data from 1977 - 80, incl., the earlier
published (1) extreme daily variation is found at the valley station where
the O_3-conc. exceeds in the afternoon the values from Wank and Zugspitze
during intense insolation (rel. sunshine duration SD >80%). This holds
equally for spring and fall. If we have only diffuse light (sunshine
duration <1 h), we note in Fig. 2 (right half) that a daily variation does
still exist at the valley station but the maximum values do no longer exceed those at the mountain stations. Hence this phenomenon is statistically
established and the question arises as to the thickness of this layer with
extremely high ozone concentration. The question is answered by using the
Zugspitze cable car telemetry for the derivation of vertical ozone profiles.
Fig. 3 shows on a typical single day in summer the evaluation of 17
O_3-profiles. Line A includes the successively detected vertical O_3-profiles.
We observe the nocturnal deficit in the boundary layer which extends to an
altitude of about 2 km ASL and is filled-up in the course of the day. In
the afternoon we find even in the lowest layer an O_3-conc. higher than that
at Zugspitze peak which remained constant throughout the day. The simultaneously measured temperature gradients (line B in Fig. 3) reveal from morning till noon a temperature inversion at about 2.2 km alt. which dissolves
through vertical exchange in the afternoon. At the end we have an adiabati-
cally mixed layer. Yet, no appreciable O_3-transport from the valley to the
Zugspitze is observed. The inversion in line B is also apparent till early
afternoon from the kink in the O_3-profile (line A). The change of the O_3-
profile suggests photochemical O_3 production in the lower layer between
ground and inversion but by no means an O_3-transport from higher levels to
the lower troposphere. Line C gives in addition the vertical profiles of
the positive electrical conductivity showing a light kink at 2.2 km alt.
with higher values above. This, too, indicates that no transport from Zugspitze level down to the valley took place. Line D in Fig. 3 contains the
O_3 daily variations, calculated from the vertical O_3-profile, at the differ-

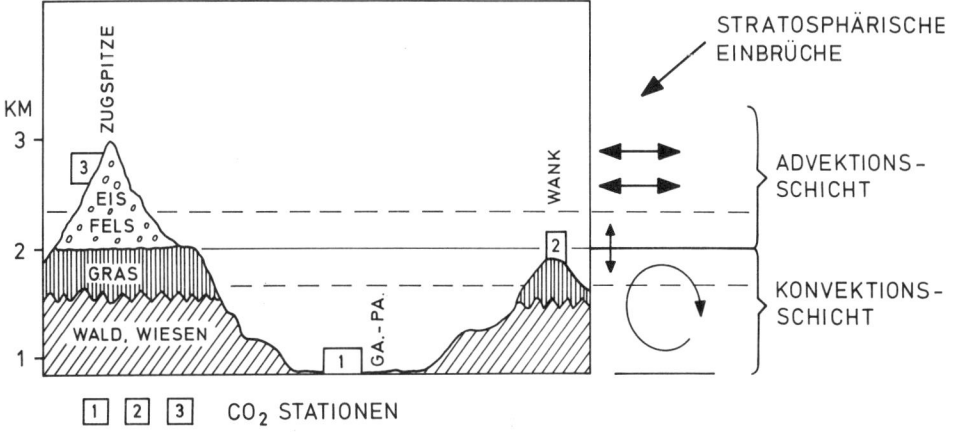

Fig. 1

The CO_2 and O_3 recording net

1: Valley, 740 m a.s.l. (Garmisch) - mainly meadows and forests

2: Wank Peak, 1780 m a.s.l., above timber line, meadows

3: Zugspitze Peak, 2964 m a.s.l. - in rocks, partly covered
with ice and snow

ent altitudes (km) from 100 to 100 m. We clearly note how the O_3-conc. increased currently during the day in a layer thickness to about 2.2 km where the increase dropped however slowly with height. From 2.3 km alt. and above the O_3-conc. remained constant. It can also be seen that in late afternoon in the valley (1 km) higher concentrations have been reached than at the Zugspitze (3 km). Hence, this case points definitely to an afternoon photochemical production of ozone in the lower troposphere.

Fig. 4 shows another type. It is distinguished by the fact that a daily variation of O_3 exists in the near-ground air layer but remained small (see lines A + D), nevertheless this increase prevailed up to 2.2 km. In a mean-level layer (Fig. 4, line D) the O_3-conc. remained constant up to 2.7 km altitude. Above, we find however a steep increase in the O_3-conc. which leads to essentially higher concentrations as have been measured in the valley (contrast to Fig. 3!). This behavior of the O_3-profile - clearly evident from Fig. 4, line A + D - suggests that we have in the present case a subsidence of stratospheric air into the troposphere. This is confirmed by the high $Be7$ concentration and the temperature profiles in line B. From 14.00, it is obvious that adiabatic warming takes place. If we view in line C the positive electrical conductivity, we recognize by the steep increase of conductivity at higher altitude from 14.00 that extremely pure stratospheric air subsided. But this air could neither through subsidence nor through mixing processes penetrate to the valley floor. From this we may conclude that stratospheric O_3 injections are less important to the O_3 behavior at lower altitudes, and that 2 independent O_3 layers exist, one above, one inside the boundary layer (comparing Figs. 3 and 4).

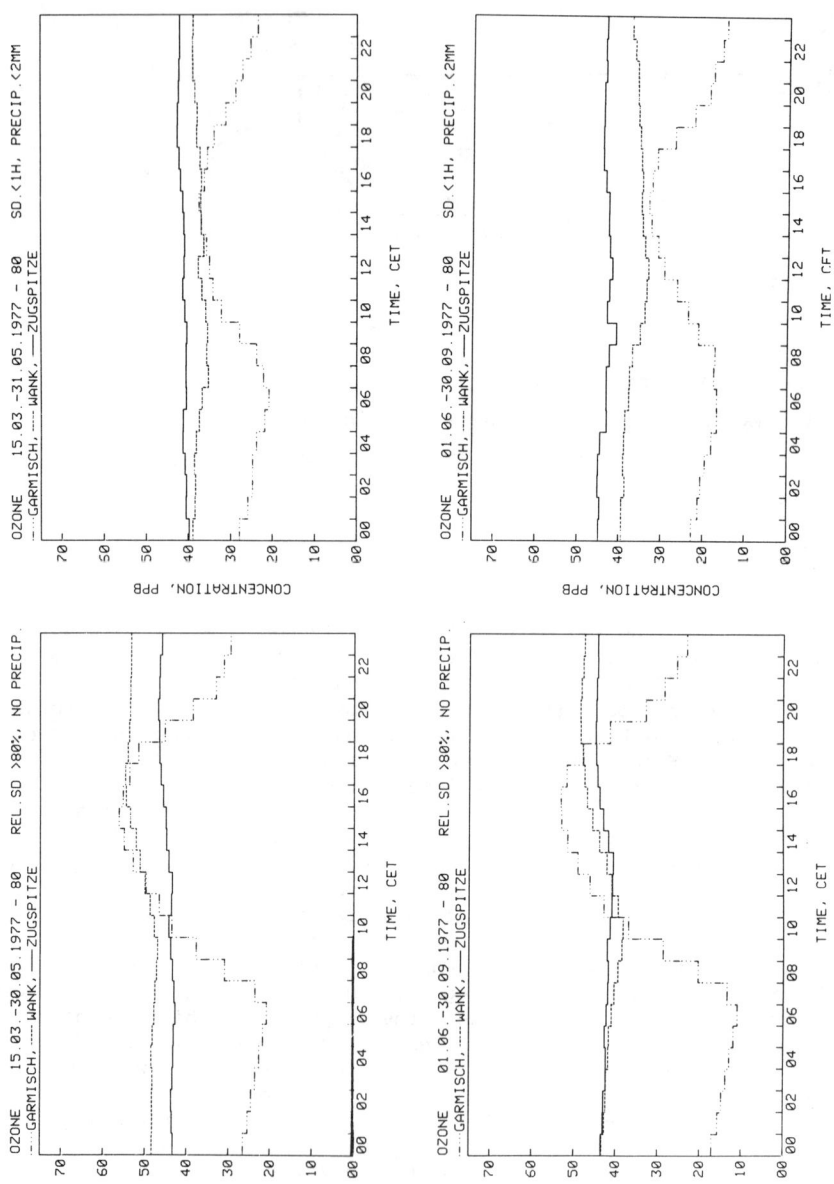

Fig. 2 - Daily variations in the valley (Garmisch) depending on sunshine duration (SD)
On the mountain stations Wank and Zugspitze only weak variations

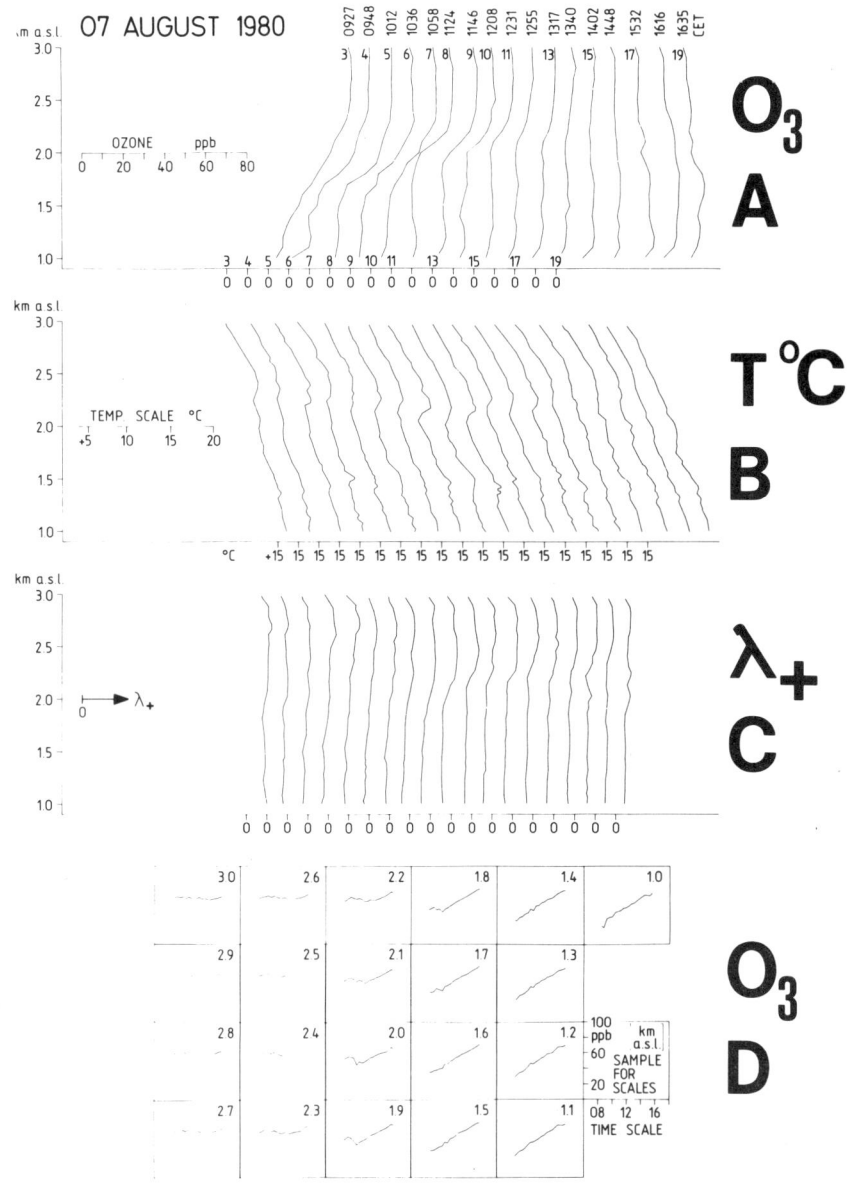

Fig. 3 : Soundings by using cable car telemetry system

7 Aug.   A: Ozone profile / B: Temperature profile /
1980   C: Profile of electric air conductivity

           D: O_3 variations with time between 1.1 and 3.0 km a.s.l.

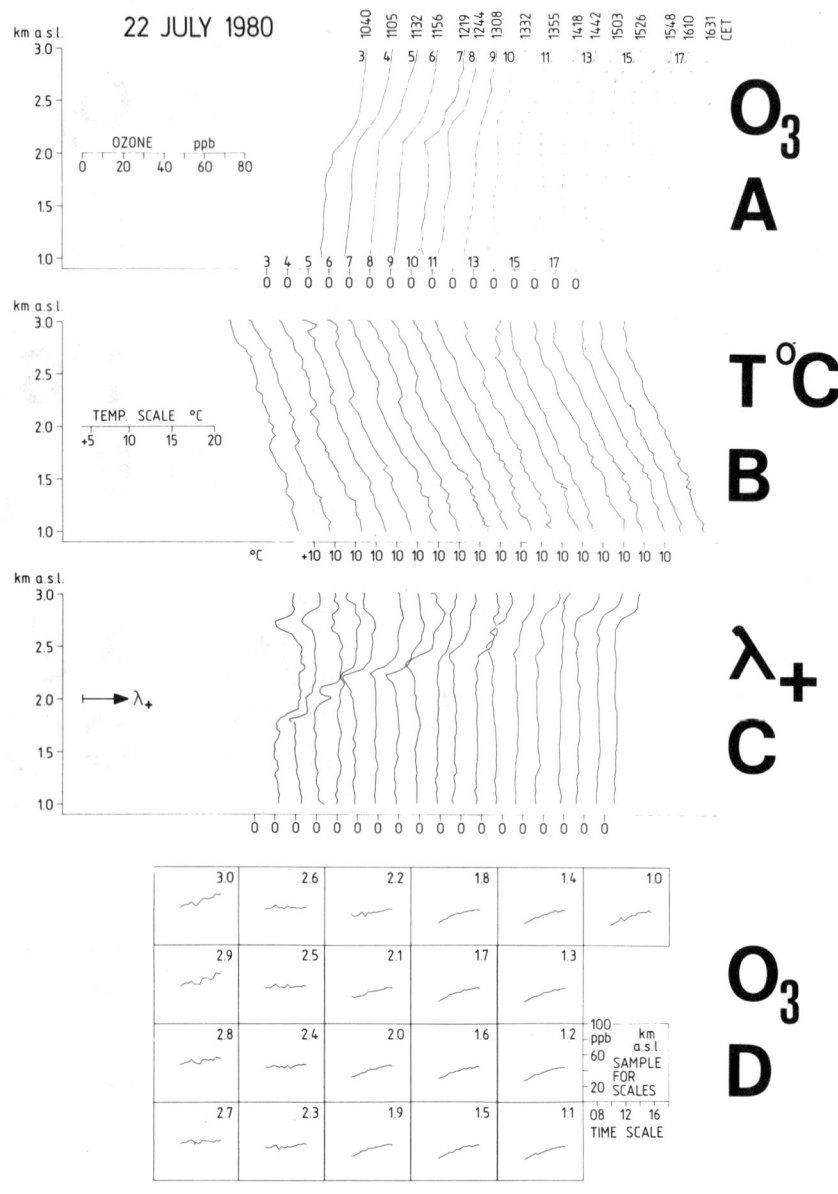

Fig. 4 : Same as Fig.3, however on 22 July 1980 with a strong intrusion of O₃ from the stratosphere down to 2 km a.s.l.

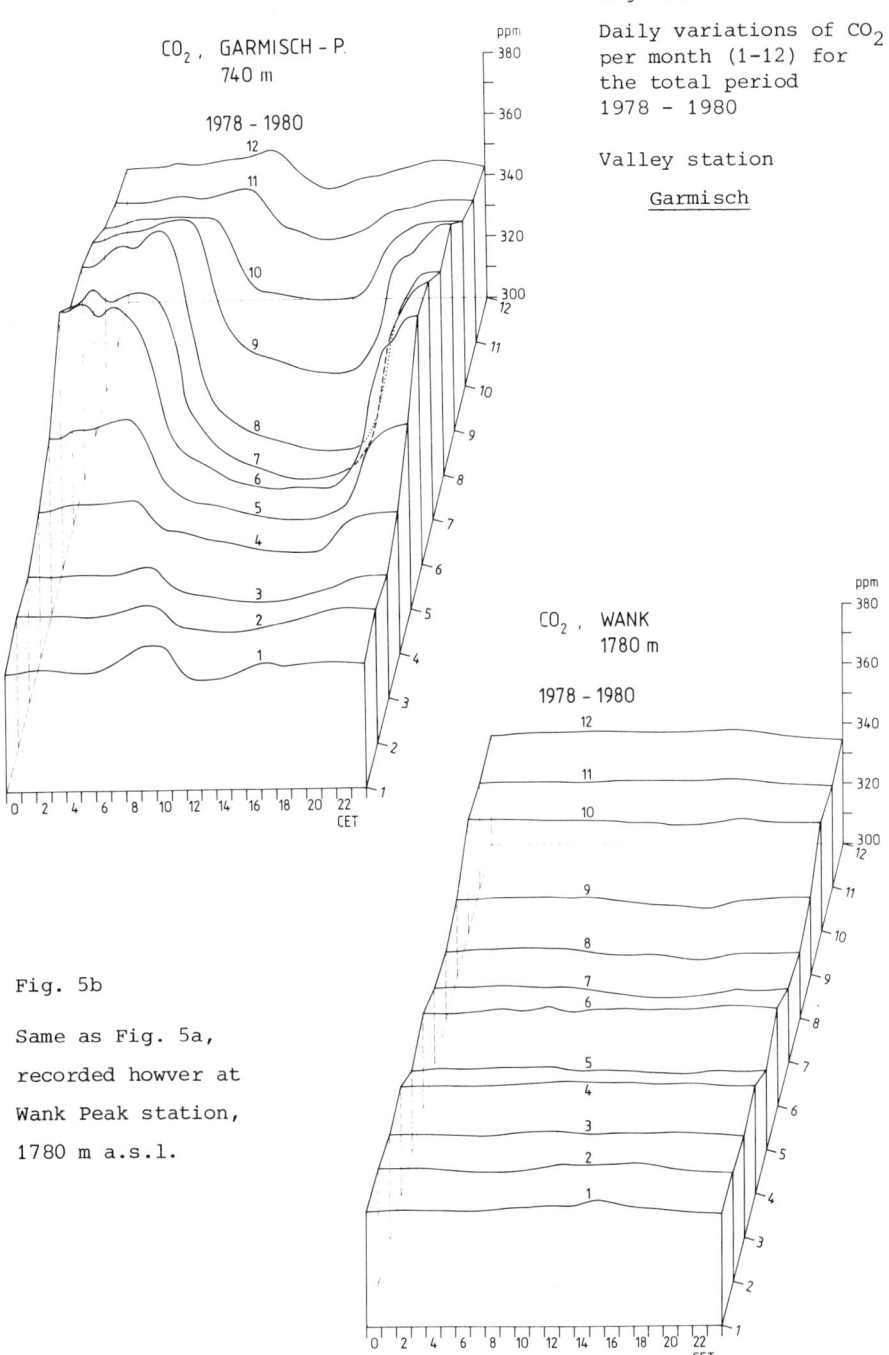

CO₂ , GARMISCH - P.
740 m

1978 - 1980

Fig. 5a

Daily variations of CO_2
per month (1-12) for
the total period
1978 - 1980

Valley station

Garmisch

CO₂ , WANK
1780 m

1978 - 1980

Fig. 5b

Same as Fig. 5a,
recorded howver at
Wank Peak station,
1780 m a.s.l.

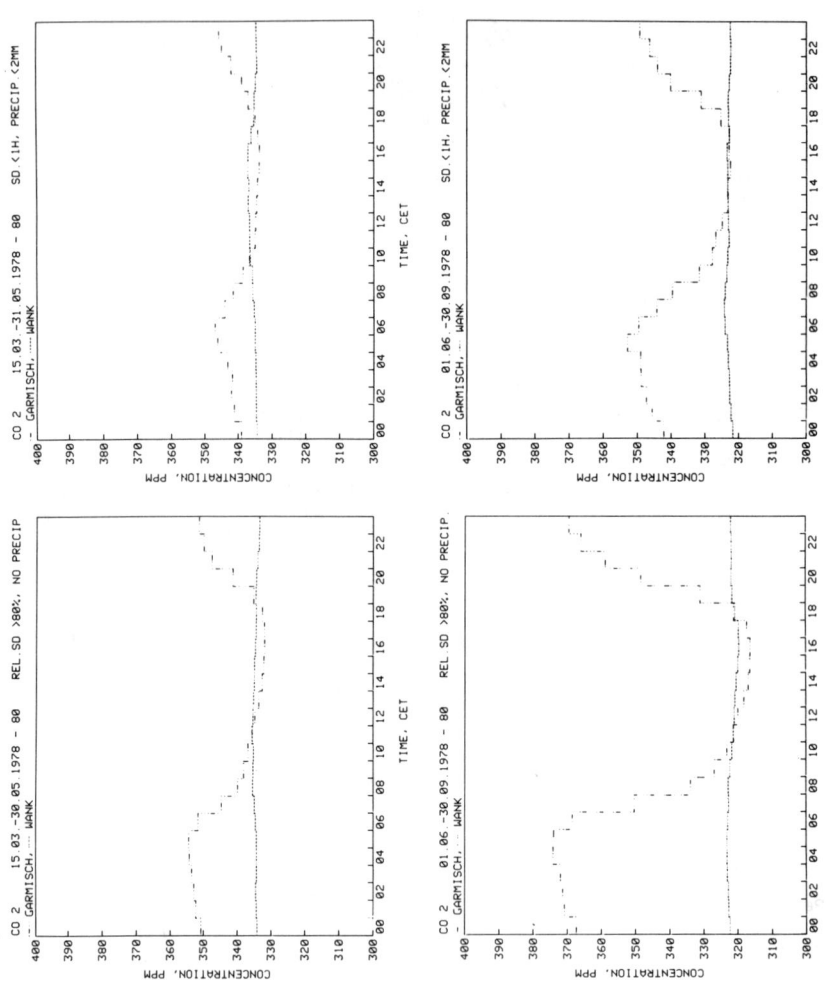

Fig. 6 – Daily variations of CO_2 in the valley (Garmisch) depending on sunshine duration and season. Mean values 1978) 1980. Practically no variations at Wank Peak station.

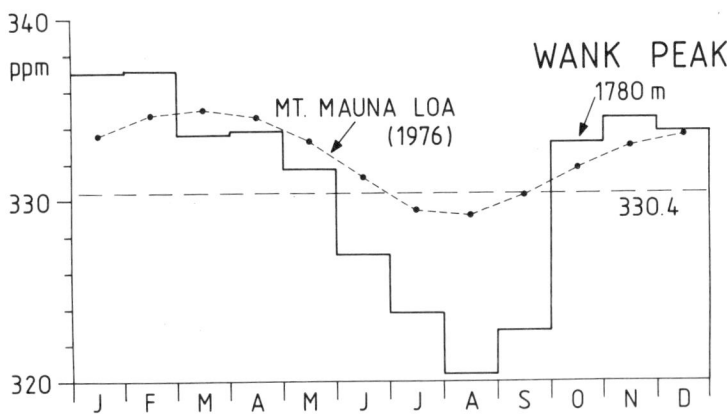

ANNUAL VARIATION CO$_2$
1978 - 1980

Fig. 7 : Annual variation of the CO$_2$ concentration at Wank Peak
at 1780 m a.s.l. Mean values 1978 - 1980.

Good agreement with the phase of the CO$_2$ recordings
on Mt. Mauna Loa, Hawaii

3.2. Carbondioxide

The results of CO$_2$ recordings, see also (2), at 740 and 1780 m are
summarized in Fig. 5. In graphs a - b we clearly observe the season-depen-
dent daily variation in the valley (5a) which is governed by the following
components:
During exposure to light and higher temperatures net-photosynthesis prevails
through CO$_2$ decomposition and in warm seasons at night, CO$_2$ production dominates by respiration of plants and decay processes in humus. The small maxima in the winter months 0800-1100 in the valley are - identified by the behavior of SO$_2$ gas - definitely of anthropogenic origin while the influence
of the biomass plays practically no longer a role in these seasons. It
clearly follows from the graph in Fig. 6 that from spring to fall the intensity of insolation (sunshine duration >80% or <1 h) has indeed a decisive
influence on the net photosynthesis by day in the valley, covered with vegetation. Fig. 6b shows that the variations from day to night at an altitude
of 1 km above the valley floor (1780 m a.s.l.) and above the timber line are
very low at all seasons (slightly indicated in the summer months). However,
there is a well pronounced variation from month to month (see Fig. 7) which
is well in phase with the variation at Mt. Mauna Loa, Hawaii.
That means from an altitude, higher than 1 km above the ground local
influences on CO2 conc. are becoming less important and global influences
begin to prevail.

4. CONCLUSIONS

Thus, it can be concluded that the installed facilities are suitable in every respect for studying the research subject posed. At the same time it is however beyond doubt that essentially more data must be gathered in order to arrive through an in-depth parameterization at a final solution.

References

(1) R. Reiter and H.-J. Kanter: Daily and Annual Variation of Tropospheric Ozone Under Pure Air Conditions at 740, 1780, and 2964 m ASL and its Possible Causes.
     Quadrennial Ozone Symposium of the International Ozone Commission, Boulder, Colorado, 4-9 August 1980

(2) R. Reiter and H.-J. Kanter: First Results of Simultaneous Recordings of the CO_2-Concentration From a Valley Station and a Neighboring Mountain Station at an Altitudinal Difference of About 1 km.
     Arch. Met. Geophys. Biokl., Ser. B., 28, 1-13 (1980)

ANALYSIS OF A PHOTOCHEMICAL SMOG EPISODE AND PREPARATION OF THE
METEOROLOGICAL INPUT DATA FOR A THREE DIMENSIONAL AIR QUALITY
DISPERSION MODEL

B. SCHERER and R. STERN
Institut für Geophysikalische Wissenschaften der Freien Universität
Berlin, 1000 Berlin 33, Thielallee 50, FRG.

ABSTRACT

A photochemical smogepisode during June 1976 in the Cologne/Bonn area
is analysed using standard meteorological observations and air quality
measurements. Preparation of the meteorological input data for the
three dimensional photochemical dispersion model involves the utilisation of two mesoscale models. Computed meteorological parameters,based
on different grid sizes, are presented and compared with measurements.

1. INTRODUCTION

     Air quality monitoring in the Rhine-Ruhr area has shown that oxidant
levels can be substantial during specific weather conditions (1). Thus, in
order to improve air quality during such photochemical smog episodes, emission control strategies must be employed. One of the greatest potential for
assessing the effectivness of specific oxidant control strategies involves
the use of a deterministic photochemical dispersion model. In order to
apply such a model to the Rhine-Ruhr area, a photochemical smog episode is
analysed with regard to the following points:
- meteorological conditions
- local and regional aspects of transport and dispersion processes
- distribution of precursor and oxidant levels.
The result of this analysis serves to determine data requirements for the
SAI airshed model (2).
     One of the main features of the modeling area is the complicated wind-
field induced by the complexity of the terrain. Since the density of wind-
measurement stations inside the modeling area is too low to create gridded
wind data as needed by the airshed model, numerical simulation of the
spatial and temporal variing windfields is carried out using two different
meteorological mesoscale models: REWISIM (3), a one-layer model for the
simulation of near ground windfields and the three dimensional UVMM-model
of MAHRER and PIELKE (4). The two models are used with different gridsizes
in order to get some information about subgrid-effects which may be important for the assessment of the model performance by comparing the measured
and calculated concentration levels.

2. DESCRIPTION OF THE AREA

     The Cologne basin is a highly industrialized region in the western
part of the FRG (Fig.1). The emissions in that area are predominantly from

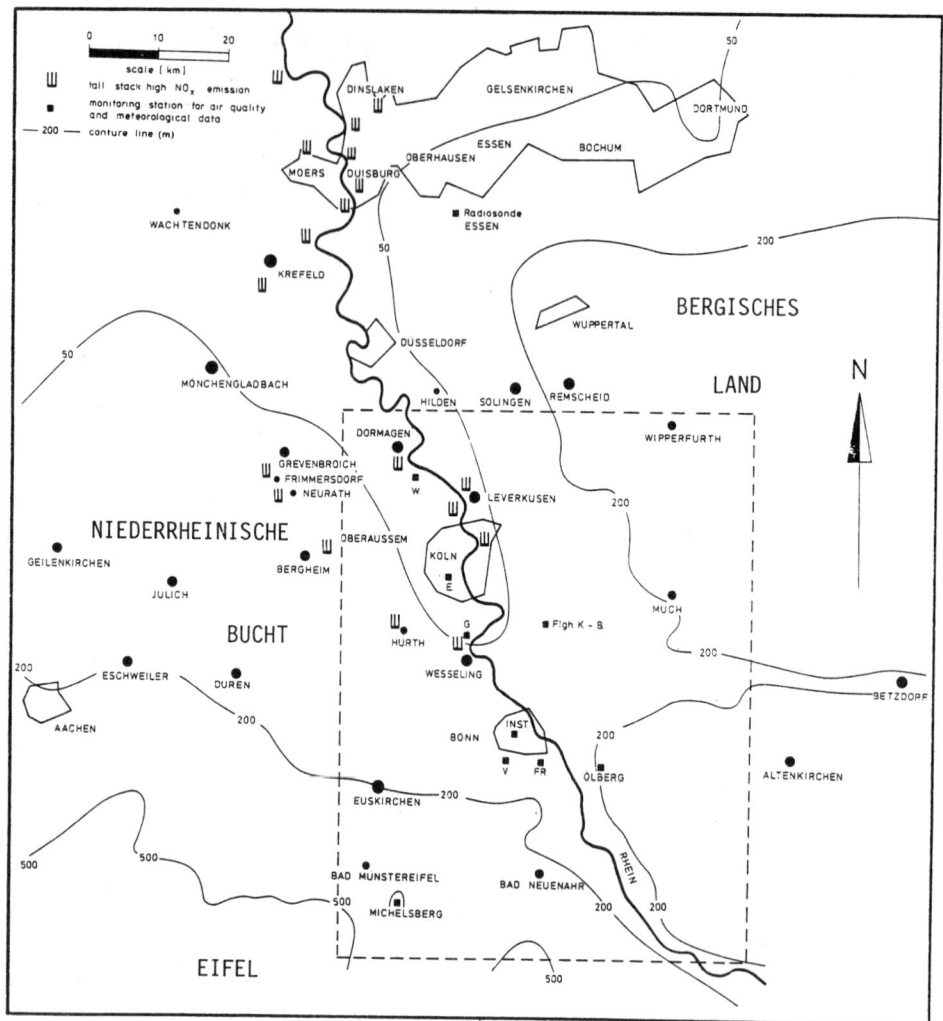

Fig.1 : The Rhine-Ruhr area. Symbols are described in the text

chemical and petrochemical industries situated primarily north and south of Cologne (5). The next adjoining large urban-industrial area is Düsseldorf situated about 40 km to the north, and which itself lies on the southern edge of the Ruhr area, the largest industrial area in Europe.

At several stations within the region shown in Fig.1 half-hourly average air quality data have been monitored. Two rural stations, Michelsberg and Ölberg, are situated in the south of Bonn on tops of hills 580 m and 450 m high, respectively. Two urban monitoring stations, one located inside Bonn (Inst), and the other inside Cologne, Eifelwall (E). Two additional stations are situated in rather industrialized areas to the north and south of Cologne. Godorf (G), to the south, is located near an oil refinery and a petrochemical plant. The northern station, Worringen (W), is located in the neighborhood of a chemical plant. All stations provide ozone data and most of them NO$_x$ data as well. Extensive flight measurements upwind and

HEIGHT : 900 m

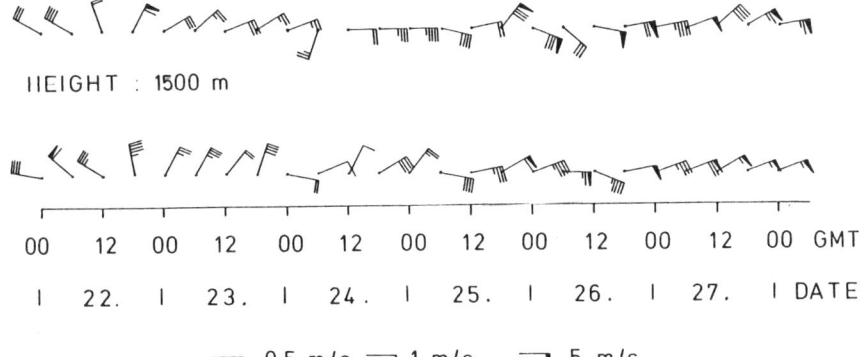

IIEIGHT : 1500 m

```
00   12   00   12   00   12   00   12   00   12   00   12   00 GMT
  I  22.  I  23.  I  24.  I  25.  I  26.  I  27.  I DATE
```

⌐ 0.5 m/s ⌐ 1 m/s   ⌐ 5 m/s

Fig.2 : Windmeasurements, Radiosonde Essen, June 22-27, 1976

downwind of the urban-industrialized areas provide additional data of ozone and NO_x during this episode (1).

3. METEOROLOGICAL SITUATION

During the episode a high pressure system is situated over northern Europe. Its center moves rapidly eastward and a high pressure ridge develops which extends from the Azores over England and middle Europe to Russia. The location of the Cologne/Bonn area with respect to the pressure system changes from a front to a backside position on the southern edge of the ridge. This leads to a change of the flow over the area from west to northeast on June 22, followed by very weak winds on June 23 and 24, and a steady easterly flow with moderate winds from June 25 on as illustrated in Fig.2 for the levels 900 and 1500 m.

Insolation of more than 15 hours results in a strong heating of the lower atmosphere during the episode. Rising dayly maximum near-ground temperatures at the stations Cologne, Düsseldorf and Essen as shown in Fig.3 demonstrate this heating. Fig.3 also illustrates the inhomogeneous horizontal temperature distribution in the modeling region with the most extrem daily variation at Cologne airport. A comparison of Cologne airport and Cologne city air temperature measurements (not shown) also reveals large differences between urban and rural areas (6-8°C) during the night, indicating heat island effects.

The variation of the winddirection near ground at Worringen and in 700 mb over Essen is shown in Fig.4. While the ground level wind shows a daily pattern, the upper flow turns uniformly from west to east. This nearly total disconnection of near ground and upper level winds indicates a strong local windsystem.

SODAR measurements in Cologne (6) and temperature soundings at Essen show that the top of the night time inversion appears at about 400 m height throughout the episode. During the day the maximum height of the mixing layer rises from 1500 m at the beginning of the episode up to 2500 m at June 25.

4. OBSERVATIONAL RESULTS AND ANALYSIS

The observed variations of ozone concentrations at five stations in

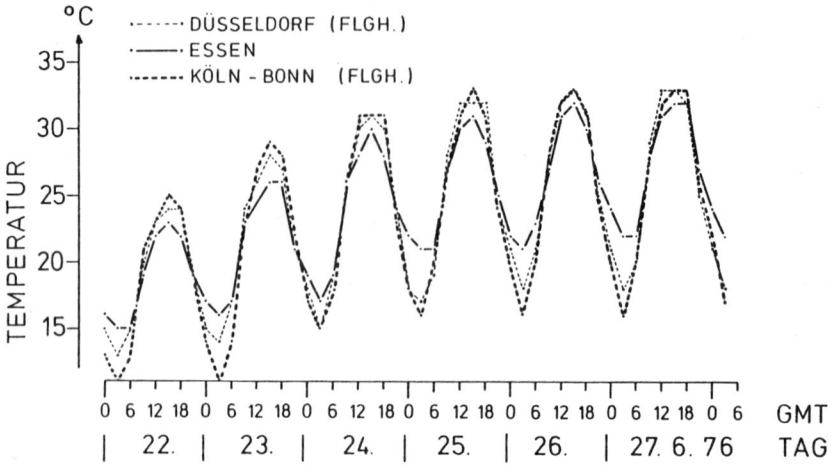

Fig.3 : Time history of near ground temperature at three stations in the modeling region

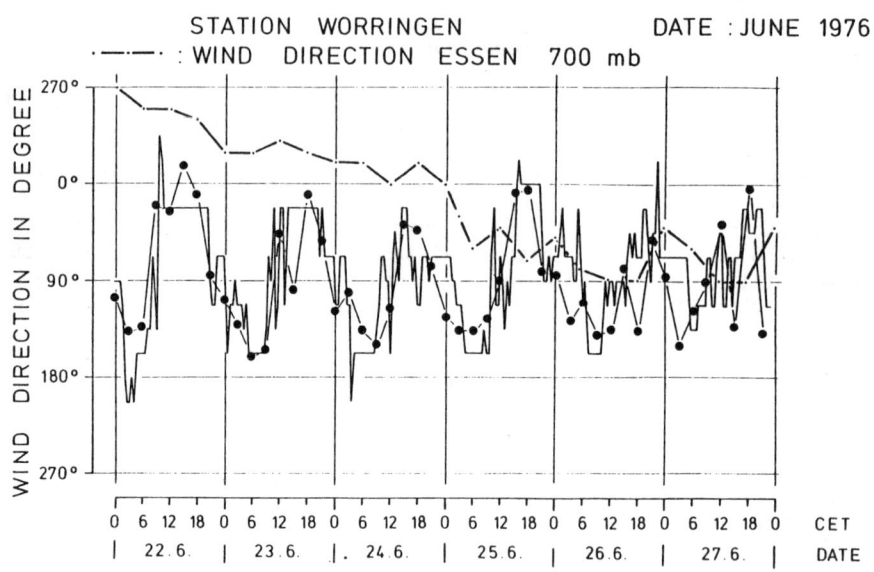

Fig.4 : Time history of measured and calculated wind direction at Worringen (curve derived from half hourly mean values). Calculated values are marked by full circles every third hour, further explanations in the text. 90° = wind from east .

the area (Fig.5) demonstrate the different behaviour of rural and urban sites: At all urban stations the ozone levels are low at night because NO emissions act to destroy ozone. The secondary peaks in the late afternoon indicate the influence of locally induced horizontal and vertical transport processes. At the two rural stations Ölberg and Michelsberg the nocturnal destruction of ozone is strongly reduced, therefore oxidant concentrations stay on a high level at night also.

While the solar radiation intensity is lowest on June 22 due to cloud coverage (1), the ozone levels observed at all stations except Michelsberg are the highest of the whole episode. Especially the Ölberg site shows a very pronounced oxidant peak in the late afternoon. A backward trajectory analysis using the wind data from Essen indicates that this peak can be explained by transport of pollutants from Düsseldorf and Cologne to the Ölberg site. A similiar constructed trajectory is shown in Fig.6 reaching the Eifel (Michelsberg site) at 19 CET, the time of the maximum ozone level on that day at that site. This trajectory also illustrates how the Cologne/Bonn area can be influenced by transport of precursors and oxidants from the Ruhr area in the north.

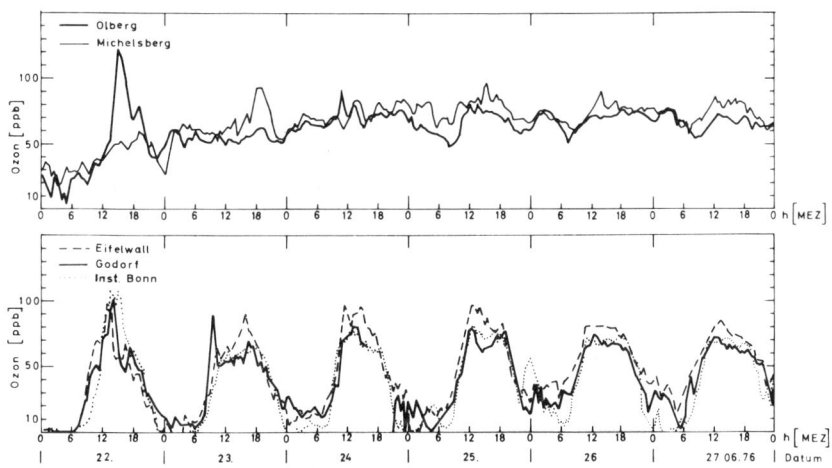

Fig.5 : Time history of ozone concentration at five monitoring stations (half-hourly mean values), see text for further explanations

On June 23 and 24 the air flow is mainly from northeast. Thus, the urban plume of Cologne is embedded in airmasses having morning origins in the Ruhr area. Fig.7 shows the measurement results of an early evening flight on June 23 downwind of Cologne/Bonn. Maximum ozone values up to 131 ppb are found approximately 30 km downwind of Cologne. In the NO_x concentration field the plume of a power plant southwest of Cologne can be recognized.

From June 25 to June 27 a steady easterly flow prevails so that the Cologne/Bonn area is no longer directly influenced by pollutant transport from the Ruhr area. However, the ozone levels at the rural stations (Fig.5) do not change significantly but a diurnal pattern develops with a midday peak and a minimum in the early morning.

Afternoon flight measurements downwind of the Ruhr area, Düsseldorf and Cologne (not shown) show distinct urban plumes with ozone levels up to 160 ppb. Downwind of Bonn maximum ozone levels are lowered by approximately 30 ppb, indicating a different emission situation (less industry). Upwind

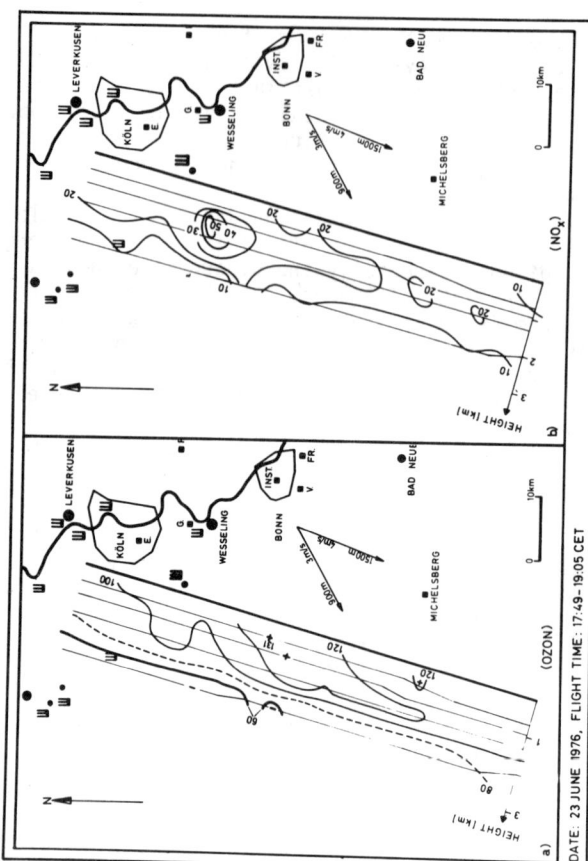

DATE: 23 JUNE 1976, FLIGHT TIME: 17:49-19:05 CET

Fig.7 : Vertical ozone (a) and NO$_x$ (b) distribution downwind of Cologne on late afternoon. Arrows indicate wind as measured at 19 CET over Essen.

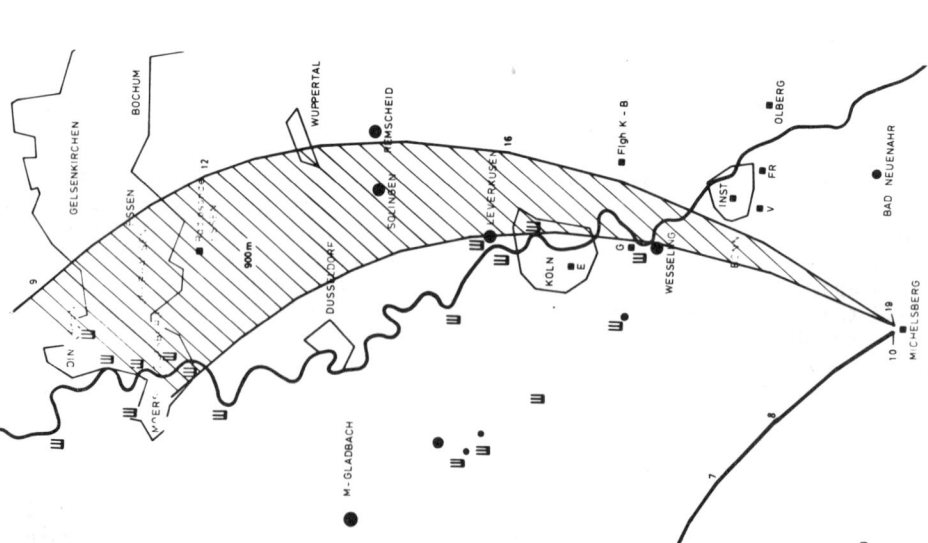

Fig.6 : Trajectory arriving at 19 CET over the Eifel. The shaded area indicates a one hour shift of the trajectory.

of the urban-industrial areas background concentrations of approx. 80 ppb ozone are observed. This level is also measured in the morning above the inversion. The vertical soundings show little excess of 50 ppb ozone above the daytime mixing layer.

In the paper presented here only the main results of the analysis are given. An extensive discussion can be found in (6).

5. PREPARATION OF METEOROLOGICAL INPUT DATA FOR A DISPERSION MODEL

The basis of the SAI airshed model (2) is the continuity equation, which expresses the conservation of mass of each pollutant in a turbulent fluid in which chemical reactions occur.

To calculate the advection terms in this equation, the model requires the horizontal wind components as a function of space and time. The vertical wind component is derived from the horizontal windfield.

Additionally, the meteorological input consists of the depth of the mixing layer, vertical temperature gradients, ground level temperature, a stability class for the lower atmosphere, the NO_2 photolysis rate and orographical parameters. However, here only some aspects of the preparation of the windfield are discussed.

The original plan was to limit the modeling region to the Cologne/Bonn area. But the data analysis showed that in this case the total amount of pollutant influx especially across the northern boundary from the Ruhr area can be as large as or even larger than the emissions in the modeling area itself. This would lead to significant computational errors since it is not possible to determine the boundary conditions exactly because of the limited data. Therefore the modeling area was expanded to the region shown in Fig.1 to make sure that all major emissions are within the modeling area. Based on practical considerations like availability of emission data and computer capacity this region is divided into 19 x 22 horizontal grids of 8 x 8 km.

Due to the complexity of the terrain within the modeling region and the little information on ground and upper level winds it is difficult to provide gridded windvalues as needed by the dispersion model on the base of measurements. Therefore it seems appropiate to apply meteorological mesoscale models to derive these data.Here, some preliminary results of the application of such models are presented.

Two models are used, first a modified version of the three dimensional mesoscale model of the University of Virginia (UVMM) (4) based on the solution of the mass, momentum and energy equations, and secondly REWISIM (REgionales WIndfeld SImulations Modell) (3), a one layer model, especially designed to simulate the regional flow pattern near ground due to the effects of orography, roughness, stability and differential heating.

To ensure that the calculated windfields are representative for the scale chosen for the dispersion calculations, the three dimensional model is also used with a 8 x 8 km grid . This gridsize produces a strong smoothing of the topography, especially in the southern part of the modeling area, where the Rhine valley is rather narrow.This can create problems, if measured data which always reflect subgrid irregularities are used for model verification. To demonstrate the influence of gridsize on the calculated windfields, REWISIM is applied to the region enclosed by dashed lines in Fig.1 with a gridsize of 2 x 2 km, which incorporates all important features of the topography.

REWISIM was run for all days of the episode. As an example the time history of the calculated wind direction is shown in Fig.4 (full circles) in comparison with the measurements at the station Worringen. The calculat-

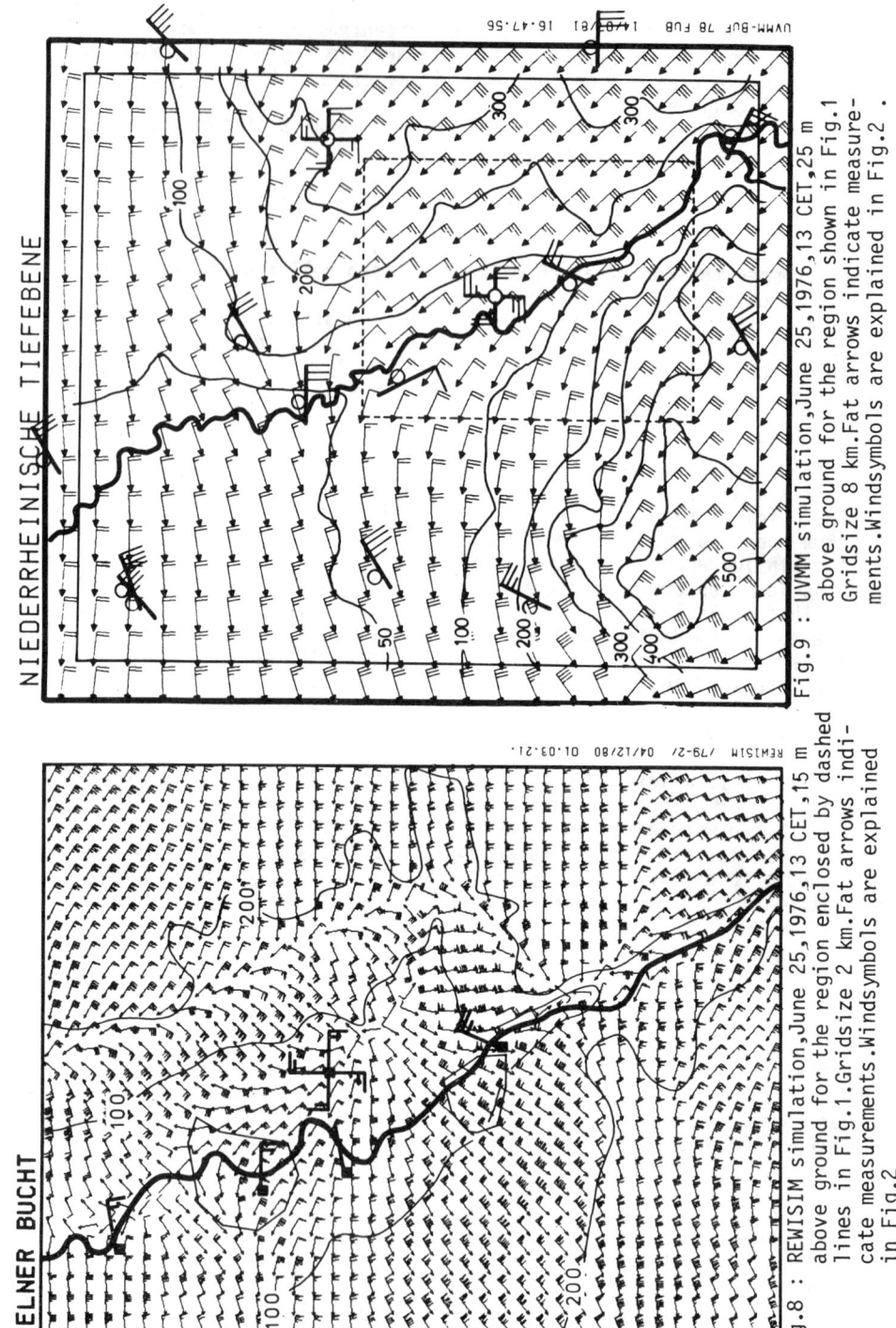

KOELNER BUCHT

NIEDERRHEINISCHE TIEFEBENE

REWISIM /79-2/ 04/12/80 01.03.21.

UVMM-BUF 78 FUB 14/03/81 16.47.56

Fig.8 : REWISIM simulation,June 25,1976,13 CET,15 m above ground for the region enclosed by dashed lines in Fig.1.Gridsize 2 km.Fat arrows indicate measurements.Windsymbols are explained in Fig.2

Fig.9 : UVMM simulation,June 25,1976,13 CET,25 m above ground for the region shown in Fig.1 Gridsize 8 km.Fat arrows indicate measurements.Windsymbols are explained in Fig.2 .

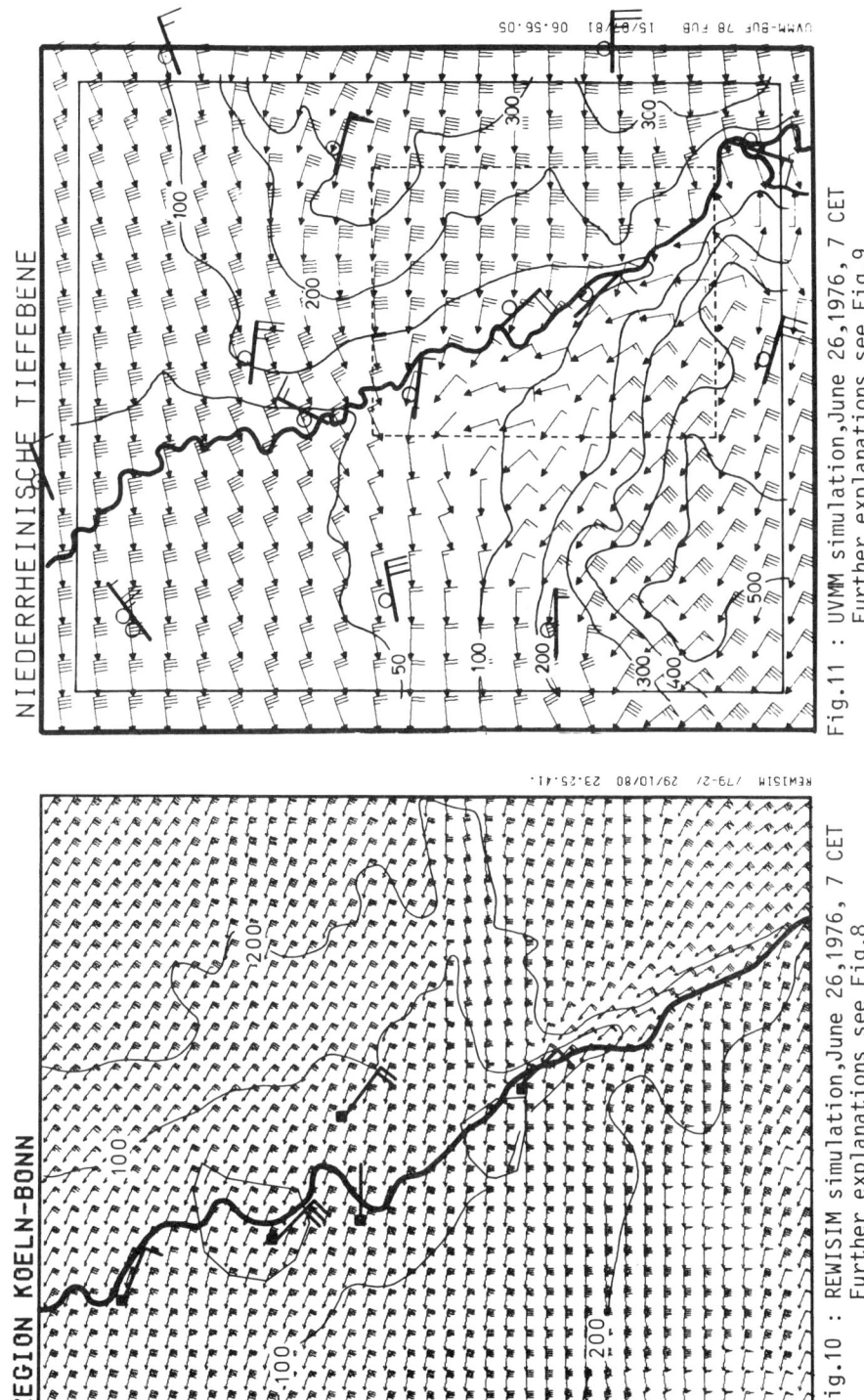

NIEDERRHEINISCHE TIEFEBENE

UVMM-BUF 78 FUB   15/03/81   06.56.05

Fig.11 : UVMM simulation,June 26,1976, 7 CET
Further explanations see Fig.9

REGION KOELN-BONN

REWISIM /79-2/   29/10/80   23.25.41.

Fig.10 : REWISIM simulation,June 26,1976, 7 CET
Further explanations see Fig.8

ed wind directions correspond quite well with the observations, whereas wind speeds are slightly overestimated, especially during weak wind conditions (no figure).

Calculated near ground wind fields using two different gridsizes are presented for two cases. Fig.8 illustrates a typical afternoon simulation of REWISIM. At this time of the day the surface temperatures have reached their peak values and the air motion is mainly driven by thermal forces. An irregular flow pattern has developed between Cologne and Bonn with weak and divergent winds. The correspondence with the few available measurements (fat arrows) is quite well. The same situation simulated with the UVMM on a 8 km grid is shown in Fig.9, where the sub-area of Fig.8 is enclosed by dashed lines. It is obvious that all the small scale features of Fig.8 do not appear in this calculation. Especially in the Bonn area the calculated wind direction does not correspond with the measurement as it does on the small gridsize. However, the main features of the near ground windfield are quite satisfactorily predicted by the UVMM also.

Small scale features as they appear in Fig.8 vanish in the late afternoon due to the enforced vertical mixing at this time as both measurements and REWISIM calculations show. Therefore calculated windfields on both grid scales correspond in a much higher degree (no figures).

Fig.10 and 11 demonstrate the airflow approx. two hours after sunrise. Insolation has not yet been strong enough to break up the night time inversion. Thus,the thermal forces are of minor importance at this time. Both the REWISIM (Fig.10) and the UVMM (Fig.11) simulations show the channeling effect of the Rhine valley. While in Fig.10 this topographic effect seems to be limited to the steep Rhine valley, in Fig.11 the influence of the change in terrain and roughness results in a rather wide zone of weak northerly winds. Although correspondence with measured wind data is quite satisfactory for both models, there are some remarkable differences between the calculated windfields in areas where no measurements are available.

6. SUMMARY AND CONCLUSIONS

The analysis of the photochemical smog episode has shown that it is necessary to include the Ruhr area as part of the modeling region since transport of pollutants from the north can play an important role for the air quality in the Cologne/Bonn region. Expansion of the modeling area implies a restriction on the model resolution dictated by the computer capacity available, since sophisticated dispersion models as the SAI airshed model have very high computational requirements.

The use of a coarse grid can create problems in those regions, where the physical processes are dominated by not resolvable subgrid effects.This is shown for the windfield by running two mesoscale models with different gridsizes. Under specific meteorological conditions some small scale features cannot be resolved on the large gridsize. This fact must be kept in mind when scale-consistent calculated windfields are used as input for dispersion models and the model results are compared with air quality data which reflect all the subgrid scale irregularities. However, it could be shown that the application of meteorological models to prepare input data for dispersion models is useful, especially in regions with limited data available.

7. ACKNOWLEDGEMENT

This study was supported by the Umweltbundesamt. The authors would like to thank Dipl.-Met. D. Heimann and Dr. L. Blumhagen for their help in running the computer models.

8. REFERENCES

1. K.H. Becker et.al., Untersuchungen über Smogbildungen, insbesondere über die Ausbildung von Oxidantien als Folge der Luftverunreinigungen in der Bundesrepublik Deutschland.
Forschungsbericht 79/104 02 502/03/04 Umweltbundesamt 1979
2. S. Reynolds et.al.,An Introduction to the SAI Airshed Model and Its Usage. Systems Applications Inc. March 1979
3. D. Heimann, Ein einfaches Modell zur Simulation regionaler Windsysteme Institut für Geophysikalische Wissenschaften FU Berlin,1978
4. Y. Mahrer and R.A. Pielke, Numerical Simulation of Air Flow Over Irregular Terrain, Beitr.z.Phys.d.Atm. 50, 1977
5. Emissionskataster Köln,Minister f.Arbeit,Gesundheit und Soziales Nordrhein-Westfalen 1976
6. B. Scherer and R. Stern, Untersuchung einer photochemischen Smogepisode im Raum Köln/Bonn. Institut f. Geophysikalische Wissenschaften der FU Berlin, 1980

URBAN OZONE - AN INDICATOR OF PHOTOCHEMICAL SMOG?

G. BROSE[1] and A. GHAZI[2]
Institut für Geophysik und Meteorologie
Universität Köln, 5 Köln 41, FRG

Present affiliation of authors:

1 - Rheinisch-Westfälischer Technischer Überwachungs-Verein
    e.V., 4300 Essen, FRG
2 - Commission of the European Communities,
    DG XII/E-MP, 1049 Brussels, Belgium

An analysis is presented of ground-level ozone
measurements made in Cologne (51°N, 6.5°E).
Special attention is paid to the general weather
situation prevailing during the time of measurements.
The observational results show that high ozone concentrations frequently occur when high pressure
systems cross Cologne. This suggests a possible source
of non-anthropogenic ozone. Anthropogenic surface
ozone is mainly observed in late spring and summer
on days with temperatures above 20°C associated with
high intensity of solar radiation. Preliminary calculations performed with a 1-D radiative photochemical
model are compared with the measurements. In order to
take effective steps to control photochemical smog,
it is of paramount importance to differentiate between
anthropogenic and natural ozone.

1. INTRODUCTION

In urban areas ground-level ozone is frequently used
as an indicator of photochemical smog. Boundary limits for
maximum ground-level ozone concentrations exist in several
countries; the US air quality standard (1971) specifies
80 ppb (volume parts per 10^9) maximum 1-hour mean, the
World Health Organisation (WHO, 1972) 60 ppb and the German
guideline, VDI 2310 (Verein Deutscher Ingenieure) 70 ppb
half-hour mean. The problem is however, that we have as
yet no means of differentiating between natural ozone and
that produced by anthropogenic processes. In an attempt to
clarify this problem, ground-level ozone measurements made
at Cologne were analysed. To trace the origin of the ozone
both anthropogenic and natural sources were considered.
Specially the dependence of ground-level ozone on the
movement of synoptic systems was studied. The results indicate that in general ozone concentrations increase when
a ridge of a surface anticyclone passes over the region,
the highest values occurring on the rear sides of high

pressure centres (Brose, 1981). These findings suggest a possible source of tropospheric ozone within these high pressure systems. During the passage of cold fronts, ozone concentrations were generally low (~30ppb), though occasionally - specially during the night - ground-level ozone concentrations were observed to increase.

2. GROUND-LEVEL OZONE IN COLOGNE AND ITS VARIATION WITH THE METEOROLOGICAL SITUATION

Cologne is centred in the industrial area of Nordrhein-Westfalen. To the north and south of the city are large industrial petrochemical concerns and it is therefore likely that the observed ground-level ozone is mainly of anthropogenic origin. Nevertheless, half-hour ozone concentrations above the German guideline, VDI 2310, of 70 ppb occur on average on less than 4% of the time per year. This is presumably partly due to a lack photochemically active solar radiation but also due to the wind which blows predominantly along the Rhine-valley in a SSE direction.

Ground-level ozone concentrations above the 70 ppb mark are mainly observed in the spring and summer. Figure 1 shows the variation of ground-level ozone concentrations, temperature, surface pressure and wind-direction observed in Cologne, in July 1980. As can be seen from this figure, when the surface air pressure begins to rise, the ozone concentration usually decreases initially but then increases and reaches a maximum as the pressure levels off (e.g. 17th July, 1980). Immediately following the passage of cold fronts (e.g. 27th July, 1980) ground-level ozone concentrations are usually observed to fall significantly. This is possibly caused by advection of cold air in the new air mass and the ground-level ozone concentrations observed then are probably of stratospheric origin. Behind cold fronts ground-level ozone concentrations are generally observed to increase (e.g. July 28th, 1980) (Paetzold, 1954).

In Cologne, ground-level ozone concentrations above the 70 ppb mark are usually observed when air temperatures are above $20^{\circ}C$ and when solar radiation is relatively high. High radiation alone, however, does not lead to high ozone values. It appears that air temperature is also very important for ozone production in the urban atmosphere.

Our investigations have shown that, on average, stratospheric ozone is responsible for about 30 ppb of the half-hourly mean ground-level ozone concentration. However, under certain meteorological conditions, such as turbulence at the tropopause level and during cyclogenesis, stratospheric ozone can produce ground-level ozone concentrations very much higher than this (Singh et al.,1975).Stratospheric air intrusions are particularly observed from March to June, the period corresponding to the time during which the total ozone concentration of the northern hemisphere goes through a maximum (Ghazi, 1980).

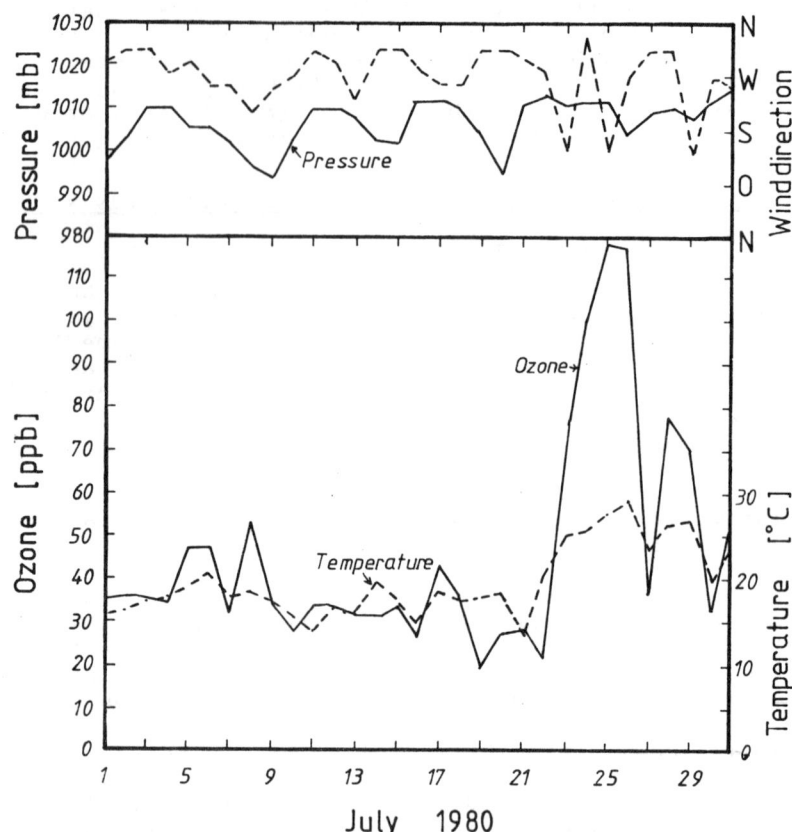

Fig.1 Variation of the daily maxima of the ground-
level ozone concentration, temperature, pressure
and wind-direction in Cologne in July 1980.

3. ESTIMATE OF THE PHOTOCHEMICAL PRODUCTION OF OZONE IN
   THE URBAN ATMOSPHERE

In the urban atmosphere, when there is a sufficient
concentration of nitrogen oxides and when the sun is shining,
the ozone concentration is approximately (within 10% ,
Calvert, 1976) given by:

$$[O_3] = \frac{[NO_2]}{[NO]} \frac{j}{k}$$

where j is the photolysis rate of NO_2 and k the reaction
rate of NO and O_3 . In table I we compare the calculated
ozone concentrations using the above equation with the
values measured in Cologne. The NO and NO_2 concentrations

used here were measured by the Amt für Umweltschutz in
Cologne. The photolysis rate, j, which depends on light
intensity (varying diurnally, seasonally and spatially)
was taken from Dickerson (1980) and the value for k from
the Upper Atmospheric Programs Bulletin (1979).

Table I

Date	NO ppb	NO$_2$	O$_3$calc.	O$_3$meas.
28.07.1980	21	57	55	58
2.08.1980	12	30	51	86

On days with stable meteorological conditions, i.e. low
wind speeds and high solar radiation, the calculated and
the observed ozone concentrations were in very good agreement (e.g. 28.7.1980). On other days however, calculated
ozone values were lower than the measured ones thus suggesting another source for the ground-level ozone.
    Fricke (1977) measured ozone concentrations in the
early morning in excess of 80 ppb at altitudes between
700m and 1500m above an inversion in the Cologne/Bonn
region. During the day, vertical mixing of the lower atmosphere increases due to solar heating of the ground, and
the ozone, which can be both of anthropogenic as well as
of stratospheric origin, can get to the ground where it
contributes to the surface-level ozone.

4. 1-DIMENSIONAL PHOTOCHEMICAL MODEL

    In order to simulate the vertical distribution of the
atmospheric ozone concentration we used a one-dimensional
radiative-photochemical model. The model which is based
on that of Isaacs et al. (1978) incorporates odd oxygen,
odd nitrogen, odd hydrogen and odd chlorine chemistry.
Radiative transfer calculations are performed including
multiple scattering and surface reflection effects.
The photochemistry used has essentially been dealt with
in some detail by Isaacs et al. (1978) and will therefore
not be further considered here. The governing equations
of the model are the continuity and diffusion equations.
    Using this model we calculated the variation of the
ozone concentration from ground-level to a height of
12 km. In figure 2 the calculated ozone profile is shown
using the temperature profile measured at Uccle, Belgium,
on the 27th July 1979. For a comparison we show for the
same day the measured ozone profile (published in the
Bulletin trimestrial of the Institute Royal Meteorologique
de Belgique, 1979). As can be seen the agreement between
both profiles is fairly good, though the model generally
predicts to low ozone concentrations below about 6 km.
The cause for this has still to be investigated.

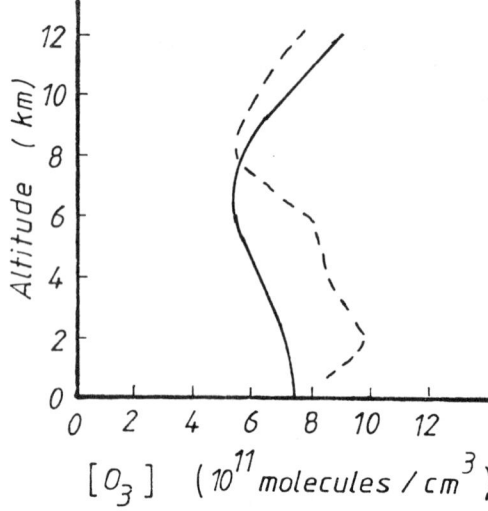

Figure 2
Calculated ozone profile
(solid line) and
measured ozone profile
(dashed line)

5. CONCLUSION

The investigations have shown that in Cologne in
situ photochemical ozone production does not usually lead
to half-hour ozone concentrations in excess of 70 ppb.
Only in spring and summer is this limit occasionally
surpassed. The results also suggest that in Cologne
ground-level ozone may not always be solely attributable
to in situ photochemical production. When studying
photochemical smog episodes ground-level ozone concentrations should therefore always be considered in
conjunction with other pollutants.

REFERENCES

Brose, G. Das bodennahe Ozon im Kölner Raum, Ph. D. Thesis,
University of Cologne (1981)

Dickerson, R.R.
Direct measurements of ozone and nitrogen
dioxide photolysis rates in the atmosphere,
Nat. Center for Atmos. Res., Boulder, Colo. (1980)

Fricke, W. Ergebnisse der Flugzeugmessungen
Paper presented at the "Kolloquium über Modell-
hafte Untersuchungen zur Bildung photochemischer
Luftverunreinigungen im Raum Köln/Bonn
(Oktober 1977)

Ghazi, A. Atlas der Globalverteilung des Gesamtozonbetrages
nach Satellitenmessungen (April 1970 - Mai 1972),
University of Cologne (1980)

Isaacs, R.G., N.D. Sze, H.K. Burke and N. Tripp
Report No. AFGL-TR-77-0293
Air Force Geophysics Laboratory Hanscom
Air Force Base, Bedford, Mass. (1978)

National Primary and Secondary Ambient Air
Quality Standards. Federal Register 36, 84.
Part II (US Government Printing Office,
Washington DC, 1971)

Technical Report No. 506 (World Health Organisation, Geneva, 1972)

Paetzold, H.K.
Über die Photochemie der Erdatmosphäre unter
besonderer Berücksichtigung der Ozonschicht
Habilitationsschrift, München 1954

Singh, H.B., W.B. Johnson and E.R. Reiter
The Relation of Oxidant to Meteorochemical
Processes: A Review of Available Research Results
and Monitoring Data. Interim Report, Project
No. 4432, Stanford Res. Institute, Menlo Park,
California (1975)

MODELE NUMERIQUE DE DISPERSION POUR L'ETUDE DE L'INTERACTION
ENTRE POLLUANTS GAZEUX RADIOACTIFS ET GOUTTELETTES
ISSUES DE REFRIGERANTS ATMOSPHERIQUES

J.P. GRANIER et A.E. SAAB
ELECTRICITE DE FRANCE - Direction des Etudes et Recherches (FRANCE)

Summary

We present an Eulerian model based on a solution of the diffusion
equation for three (or more) pollutants interacting together (for
instance, in its first application : water droplets issued from a
cooling tower, airborne pollutants issued from a stack and airborne
pollutants captured by droplets). It is used in steady state conditions and describe the dispersion of pollutants on a scale up to 20 km.
The lateral diffusion is assumed gaussian and the vertical diffusion
is described by a two dimensional finite difference model, using eddy
diffusivity coefficients.
A value of dry deposition velocity is prescribed for each pollutant.
Chemical reactions between pollutants can be taken into account.

1 - INTRODUCTION

    La réalisation par ELECTRICITE DE FRANCE de centrales nucléaires
équipées de tours de réfrigération atmosphérique du type "humide" pose le
problème de l'interaction entre les effluents gazeux radioactifs émis par
la centrale et le panache de vapeur d'eau issu du réfrigérant.
    Sous certaines conditions, les gouttelettes d'eau du panache humide
captent les effluents radioactifs émis par la centrale et peuvent ainsi
conduire à une contamination au niveau du sol lorsqu'elles atteignent
celui-ci.
    Nous nous proposons dans cette présente étude d'évaluer la contami-
nation globale au sol entraînée par les effluents radioactifs d'une
centrale nucléaire lorsqu'il y a interaction de ces effluents avec un
panache de réfrigérant humide. Pour ce faire, on utilisera l'équation
fondamentale de la diffusion appliquée :
    1/ aux effluents radioactifs,
    2/ aux gouttelettes d'eau du panache du réfrigérant,
    3/ aux gouttelettes d'eau ayant capté des particules radioactives.

2 - MODELE DE CALCUL

    Les variations de concentration d'un constituant minoritaire ou agent
de contamination dans un fluide porteur, en un point de l'espace et à un
instant donné sont régis par quatre catégories de processus.
    Les deux premiers processus sont les phénomènes d'advection
-transport des polluants par l'écoulement moyen- et de diffusion turbulente
due aux fluctuations du vecteur vent.
    Les deux dernières catégories correspondent à des phénomènes de
sources et de puits. Elles vont donc porter sur tous les processus d'appa-

rition ou de disparition locales ou diffuses qui comprennent aussi bien les phénomènes de rejets que les phénomènes de capture, dissociation ou recombinaison chimique et radioactivité.

Si χ représente la concentration du contaminant, l'équation de la diffusion peut s'écrire :

$$\frac{\partial \chi}{\partial t} = - \vec{V} . \overrightarrow{\nabla \chi} - \vec{\nabla} . \vec{F} + S - P \qquad [1]$$

avec :

$$\vec{F} = - |K| \; \rho \; \overrightarrow{\nabla \frac{\chi}{\rho}} \qquad [2]$$

terme d'advection   : $- \vec{V} . \vec{\nabla}$

terme de diffusion turbulente : $- \vec{\nabla} . \vec{F}$

terme de sources   : S

terme de puits   : P

Afin de simplifier la formulation mathématique du problème, nous ferons les hypothèses suivantes :
- La cheminée d'évacuation des effluents radioactifs et la tour de réfrigération sont assimilées à des points-sources situés respectivement à des hauteurs au-dessus du sol H et h et séparées d'une distance D le long de l'axe des x.
- Nous supposerons que l'atmosphère reçoit du réfrigérant des gouttes d'eau déjà formées qui ne changent pas de dimensions et ne s'évaporent pas au cours de leur transport dans l'atmosphère.
- Nous avons supposé en outre que le nuage de gouttelettes était monodispersé.

Ceci étant posé, nous sommes conduits à résoudre un système de trois équations aux dérivées partielles donnant la concentration Q (x, y, z) en polluants radioactifs, le nombre total de gouttes N (x, y, z) par unité de volume et la radioactivité R (x, y, z) contenue dans ces gouttes.

Les équations sont formulées en coordonnées cartésiennes, l'axe des z étant perpendiculaire à la surface du sol et l'axe des x dirigé dans la direction du vent moyen.

On supposera que la distribution des contaminants ou des gouttes dans la direction transversale par rapport à la direction du vent moyen est gaussienne, c'est-à-dire que chaque inconnue peut être mise sous la forme :

$$N \; (x, \; y, \; z) = \frac{1}{\sqrt{2\pi}\sigma y} \; e^{\frac{y^2}{2\sigma^2 y}} \; \overline{N} \; (x, \; z) \qquad [3]$$

et de façon similaire pour Q (x, y, z) et R (x, y, z).

σy représente l'écart-type de diffusion transversale, supposé identique pour les effluents gazeux et les gouttelettes.

Les valeurs de σy s'expriment en fonction de l'état de stratification thermique de l'atmosphère, d'après PASQUILL (1).

3 - RESOLUTION DU MODELE

Le cas envisagé ici correspond à l'influence maximale du panache de réfrigérant, lorsque le vent souffle le long d'un axe reliant la cheminée de la centrale à la tour de réfrigération.

Les équations qui régissent la diffusion turbulente des effluents radioactifs et des gouttelettes peuvent être mises sous la forme suivante, dans le cas d'un régime stationnaire :

$$u \frac{\partial \overline{N}}{\partial x} - \frac{\partial}{\partial z} Kz \frac{\partial \overline{N}}{\partial z} - w \frac{\partial \overline{N}}{\partial z} = M \, \delta(x) \, \delta(z-h) \qquad [4]$$

$$u \frac{\partial \overline{Q}}{\partial x} - \frac{\partial}{\partial z} Kz \frac{\partial \overline{Q}}{\partial z} = q \, \delta(x-D) \, \delta(z-D) - \frac{\beta \, \overline{N} \, \overline{Q}}{2\sqrt{\pi}\sigma y} \qquad [5]$$

$$u \frac{\partial \overline{R}}{\partial x} - \frac{\partial}{\partial z} Kz \frac{\partial \overline{R}}{\partial z} - w \frac{\partial \overline{R}}{\partial z} = \frac{\beta \, \overline{N} \, \overline{Q}}{2\sqrt{\pi}\sigma y} \qquad [6]$$

Dans le système d'équations [4] à [6] le premier terme représente le processus d'advection (transfert par le vent), le second terme représente le processus de diffusion turbulente. La diffusion dans le sens de l'écoulement est supposée négligeable par rapport à la diffusion verticale, perpendiculaire à l'écoulement.

Dans l'équation [4] le troisième terme représente l'effet dû à la chute des gouttes, w étant la vitesse de chute.

Les parties droites des équations [4] et [5] représentent les termes sources et puits ; les termes comprenant les fonctions de DIRAC reflètent la présence des points-sources gouttes et particules, q et M étant les débits de substances radioactives et de gouttelettes pénétrant dans l'atmosphère par unité de temps.

Les derniers termes des équations [5] et [6] représentent les puits (équation [5]) et les sources secondaires (équation [6]) sous la forme d'un terme de recombinaison (2) :

$$\frac{\beta \, \overline{N} \, \overline{Q}}{2\sqrt{\pi}\sigma y}$$

Ce terme caractérise la capture des particules radioactives par les gouttelettes d'eau du réfrigérant, β étant le coefficient de capture, fonction à la fois des dimensions des particules et des gouttelettes.

On remarquera que les éléments radioactifs rejetés à l'atmosphère par une centrale nucléaire ont une demi-période de vie beaucoup plus élevée que l'échelle de temps caractéristique du phénomène de diffusion, ce qui conduit à négliger l'effet de décroissance radioactive dans les termes puits.

Les conditions aux limites du système d'équations [4] à [6] sont les suivantes :

$$\left. \overline{Q} \right|_{x=0} = \left. \overline{N} \right|_{x=D} = \left. \overline{R} \right|_{x=D} = 0$$

$$\left. Kz \frac{d\overline{Q}}{dz} \right|_{z=z_0} = \left. \frac{\partial^2 \overline{N}}{\partial z^2} \right|_{z=z_0} = \left. \frac{\partial^2 \overline{R}}{\partial z^2} \right|_{z=z_0} = 0 \qquad [7]$$

$$\overline{Q}_{z \to \infty} = \overline{N}_{z \to \infty} = \overline{R}_{z \to \infty} = 0$$

z_0 désigne la hauteur de rugosité, hauteur au-dessus du sol où la vitesse du vent est nulle (hypothèse où la distribution verticale des vitesse obéit à une loi logarithmique).

Dans l'expression des conditions aux limites au niveau z_0, on suppose que les contaminants radioactifs sont complètement réfléchis et que les gouttelettes sont absorbées totalement.

Nous avons admis que les profils verticaux de vitesse et de diffusivité étaient donnés par les relations suivantes, dans le cas d'une atmosphère neutre (3) :

$$
\left\{
\begin{array}{l}
u = \dfrac{u_*}{K} \log \dfrac{z}{z_0} \quad \text{pour } z \leqslant 100 \text{ m} \\[3mm]
u(z) = u(100) \quad \text{pour } z > 100 \text{ m}
\end{array}
\right.
$$

$$
\left\{
\begin{array}{l}
Kz(z) = Ku_* z \quad \text{pour } z \leqslant 100 \text{ m} \\[3mm]
Kz(z) = Kz(100) \quad \text{pour } z > 100 \text{ m}
\end{array}
\right.
$$

. K désigne la constante de Von KARMAN (0,407),
. u_* désigne la vitesse de frottement.

La vitesse de frottement peut être reliée grossièrement à la vitesse du vent au niveau 10 m par la relation suivante :

$$u_* \# 0.15 \, u_{10}$$

Le système [4] à [6] est résolu numériquement avec les conditions aux limites [/] en faisant varier les paramètres suivants : la hauteur de la cheminée H, le coefficient de capture β et le flux de contaminants q. La hauteur du réfrigérant h et la distance D entre la cheminée et le réfrigérant sont fixées.

Le modèle de calcul permet d'obtenir les concentrations Q(x, y, 0) en polluants radioactifs au niveau du sol, ainsi que le flux de gouttelettes radioactives au niveau de la hauteur de rugosité, flux qui s'écrit :

$$
P = - Kz \left. \dfrac{\partial R}{\partial z} \right|_{z = z_0} + \left. WR \right|_{z = z_0}
$$

4 - ANALYSE DES RESULTATS

4.1 - Présentation des résultats

On a défini les valeurs "standard" suivantes pour les différents paramètres :

- hauteur du réfrigérant : h = 150 m
- hauteur de la cheminée : H = 50 m
- distance entre les deux sources : D = 200 m
- diamètre des gouttes : 50 μm
- débit de polluants radioactifs : q = 1 C_i/s
- vitesse de chute des gouttelettes : W = .26 m/s
- coefficient de capture : $\beta = 10^{-13}$ m³/s
- vitesse du vent : $u = \dfrac{.25}{.407} \log(Z/.025)$

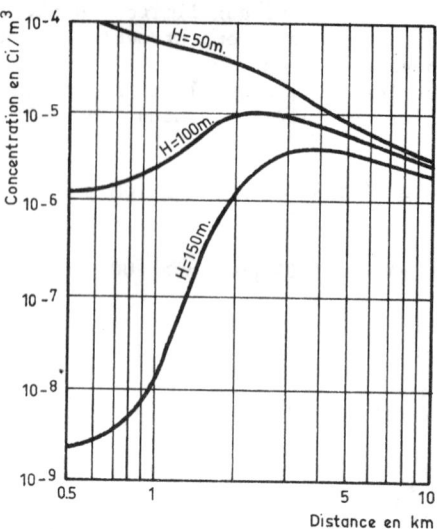

Figure 1 : Concentrations au sol en polluants gazeux radioactifs
en fonction de la distance à la source pour 3 hauteurs
H de cheminée.

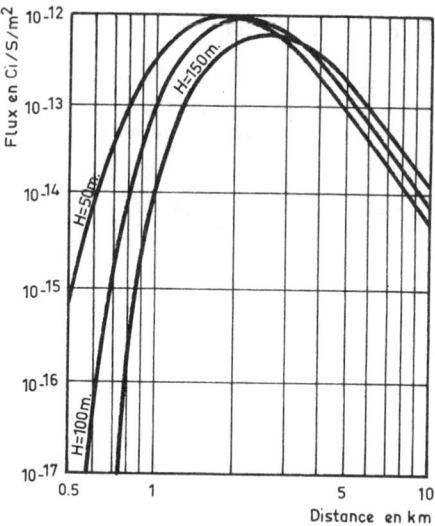

Figure 2 : Flux à travers le sol des gouttelettes radioactives
en fonction de la distance au réfrigérant (h = 150 m)
pour 3 hauteurs H de cheminée.

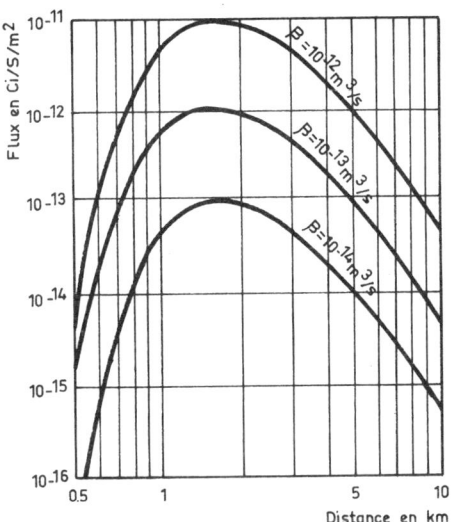

Figure 3 : Flux à travers le sol des gouttelettes radioactives
en fonction de la distance au réfrigérant (h = 150 m)
pour 3 valeurs du coefficient de captage β.

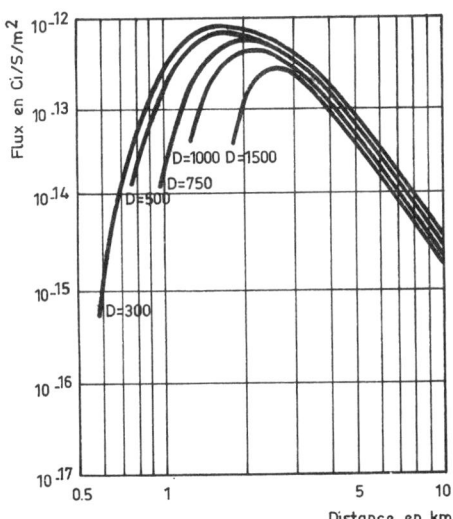

Figure 4 : Flux à travers le sol des gouttelettes radioactives
en fonction de la distance au réfrigérant (h = 150 m)
pour 2 valeurs de la distance entre la cheminée et
le réfrigérant.

Nous avons fait successivement varier H, β et D, les autres paramètres gardant leur valeur "standard".

- La figure 1 donne les concentrations au sol en polluants gazeux radio-actifs pour trois hauteurs de cheminée : 50, 100 et 150 m.
- La figure 2 donne le flux de gouttes radioactives au niveau du sol pour chacune des trois hauteurs de cheminée.
- Les figures 3 et 4 donnent le flux de gouttes radioactives en faisant varier successivement :

. Le coefficient de capture qui prend les trois valeurs suivantes :

$$10^{-12}, 10^{-13} \text{ et } 10^{-14} \text{ m}^3/\text{s (figure 3)}.$$

. La distance D entre les deux sources variant de 300 à 1500 m (figure 4).

4.2 - Commentaires

- Sur la figure 1 on retrouve des résultats connus : la concentration maximale varie approximativement comme l'inverse de la hauteur de la cheminée.
- Au vu des figures 2 à 4, nous pouvons tirer les conclusions suivantes en ce qui concerne le flux à travers le sol de gouttelettes radioactives.
- Ce flux dépend très faiblement -dans les conditions du calcul- de la hauteur de la cheminée. Sa valeur maximale est de l'ordre de 10^{-12} Ci/m^2/s entre 1.5 et 2 km. Cette valeur maximale est atteinte à une distance plus faible de la source que la concentration maximale en polluants radioactifs gazeux : ceci est dû à l'effet de gravité agissant sur les gouttelettes.
- La valeur maximale de ce flux est proportionnelle à la valeur du coefficient de capture β.
- Ce flux semble en outre peu dépendant de la distance entre la cheminée et le réfrigérant, tout au moins tant que celle-ci est de l'ordre de quelques centaines de mètres.

5 - CONCLUSIONS

Ce type de modèle basé sur la résolution bidimensionnelle de l'équation d'advection-diffusion présente deux avantages principaux :
- d'une part, grâce à la représentation de la dispersion au moyen d'un formalisme à gradient dans le plan vertical, ce modèle est bien adapté pour la prise en compte des diverses paramétrisations des phénomènes de dépôts et d'interactions physico-chimiques entre polluants gazeux et particu-laires,
- d'autre part, grâce à la structure bidimensionnelle du modèle utilisant une répartition gaussienne dans le plan transversal, il offre une simplicité et un faible coût d'exploitation.

Nous pouvons donc envisager d'adapter ce modèle à la modélisation des transformations physico-chimiques qui prennent place dans un panache issu d'une centrale thermique fonctionnant au fuel ou au charbon. Une revue des réactions à prendre en compte a été faite par COCKS et FLETCHER (4) et par ELGROTH et HOBBS (5).

REFERENCES

(1) F. Pasquill
   Atmospheric Diffusion.
   Ellis Horwood Publisher - 1974.

(2) E.K. Byutner - F.A. Gisina
   Effective coefficient of Aerosol particles capture by rain drops and
   cloud borne drops.
   Trudy L.G.M.I. n° 15 - 1963.

(3) A. Saab
   Etude des profils de diffusivité verticale dans la couche limite
   atmosphérique. Application à un modèle physionumérique de diffusion.
   Rapport E.D.F. F42-74/n° 11 - Avril 1974.

(4) A.T. Cocks - I.S. Fletcher
   A model of the gaz phase chemical reactions of power station plume
   constituents.
   Central Electricity Generating board.
   Rapport C.E.R.L. RD/L/R1999.

(5) M.W. Eltgroth - P.V. Hobbs
   Evolution of particles in the plumes of coal fired power plant.
   Atm. Env. - vol. 13 - p. 953-975.

NUMERICAL MODEL OF DISPERSION IN ATMOSPHERE OF CHEMICALLY REACTIVE
POLLUTANTS

C. BOVINI (TEMA, BOLOGNA,ITALY).
T. NANNI, M. TAGLIAZUCCA (INST. FISBAT-CNR, BOLOGNA,ITALY).

Abstract

A diffusion model of chemically reactive pollutants concerning the photochemical smog formation in atmosphere has been carried out. The pa-
rametrization of meteorological factors such as wind components and
diffusion coefficients, correctly simulates the behaviour of an actual
atmosphere. As far as the kinetic reaction mechanism is concerned, the
model describes the temporal evolution of NO, NO_2, HC, O_3. A quasi
stationary-state has been hypothesized for all the other atmospheric
constituents. The validity of the kinetic mechanism has been verified
by comparing the experimental data obtained in smog chamber. The limits of this simplified model have been checked with Heicklen's kinetic reaction model (1976). The model is valid for reaction times
of 5 or 6 hours and for concentration ratios HC/NC$>$ 1.

1. INTRODUCTION

The numerical model which simulates the atmospheric dispersion of chemically reactive minor constituents must describe the phenomenon in general
without going into details of the reaction mechanism. In particular, the
model which describes the formation of photochemical smog has been studied
based on the kinetic reaction mechanism proposed by Hecht-Seinfeld (1972)
(Table I).

2. STUDY OF SENSITIVITY OF THE SMOG PHOTOCHEMICAL MODEL

In the model proposed by Hecht-Seinfeld three fundamental hypotheses
are made. These are:
a) the chemical system is described by the evolution of NO, NO_2, O_3, HC
while the radicals HNO_2 and NO_3 are in a stationary-state;
b) the termination reaction can be neglected;
c) entire chain reaction cycles are sufficiently represented by stoichiometric reaction coefficients $\alpha, \beta, \gamma, \delta$.
These hypotheses need verification in order to evaluate the degree of
reliability of the kinetic reaction mechanism proposed in relation to the
various ratios of the chemical components concentration and of the evolution of the system in time. The verification of the reliability of the numerical model relative to the kinetic reaction mechanism only, takes place
through the simulation of experimental tests in smog chambers found in literature (fig. 1,2).

1) $NO_2 \xrightarrow{h\nu} NO + O$ $K_1 = *$

2) $O_3 \longrightarrow O_2 + O$ $K_2 = 2.3 \times 10^2$ min⁻¹

3) $O + C_2H_4 \xrightarrow{\;} HCO + CH_3 \xrightarrow{3O_2+NO} CH_2O+2HO_2+CO+NO_2$ $K_3 = 1.2 \times 10^3$ ppm⁻¹min⁻¹

4) $O + CH_2O \longrightarrow HO + HCO \xrightarrow{O_2} HO+HO_2+CO$ $K_4 = 2.4 \times 10^2$

5) $O + O_2 + M \longrightarrow O_3 + M$ $K_5 = 2.0 \times 10^6$ min⁻¹

6) $NO + O_3 \longrightarrow NO_2 + O_2$ $K_6 = 2.5 \times 10^1$ ppm⁻¹min⁻¹

7) $NO_2 + O_3 \longrightarrow NO_3 + O_2$ $K_7 = 4.6 \times 10^{-2}$

8) $NO + NO_3 \longrightarrow 2\,NO_2$ $K_r = 1.3 \times 10^4$

9) $NO_2 + NO_3 \longrightarrow N_2O_5$ $K_9 = 5.0 \times 10^3$

10) $O_3 + C_2H_4 \longrightarrow CH_2O + CH_2O_2 \longrightarrow CH_2O + CO + H_2O$ $K_{10} = 4.0 \times 10^{-3}$

11) $CH_2O \xrightarrow{h\nu} H_2 + CO$ $K_{11} = *$

12) $CH_2O \longrightarrow H + HCO \xrightarrow{2O_2} 2HO_2 + CO$ $K_{12} = 4.7 \times 10^{-3}$ min⁻¹

13) $HO + CH_2O \longrightarrow H_2O + HCO \xrightarrow{O_2} H_2O + HO_2 + CO$ $K_{13} = 2.2 \times 10^4$ ppm⁻¹min⁻¹

14) $HO + C_2H_4 \longrightarrow H_2O + C_2H_3 \xrightarrow{O_2+NO} CH_2O+CO+NO_2+HO+H_2O$ $K_{14} = 1.9 \times 10^3$

15) $HO + C_2H_4 \xrightarrow{\;} HOC_2H_4 \xrightarrow{2O_2+2NO} 2\,CH_2O + 2NO_2 + HO$ $K_{15} = 2.5 \times 10^3$

16) $HO + NO_2 \longrightarrow HONO_2$ $K_{16} = 1.2 \times 10^4$

17) $HO_2 + NO \longrightarrow HO + NO_2$ $K_{17} = 1.4 \times 10^3$

18) $HO_2 + NO_2 \longrightarrow HONO + O_2$ $K_{18} = 2.7 \times 10^2$

19) $HO + NO \longrightarrow HONO$ $K_{19} = 1.2 \times 10^4$

20) $HONO \xrightarrow{h\nu} HO + NO$ $K_{20} = *$

21) $HO + HONO \longrightarrow H_2O + NO_2$ $K_{21} = 1.0 \times 10^4$

22) $HO_2 + O_3 \longrightarrow HO + 2O_3$ $K_{22} = 2.2 \times 10^0$

23) $HO + O_3 \longrightarrow HO_2 + O_2$ $K_{23} = 8.3 \times 10^1$

24) $2\,HO \longrightarrow H_2O + O$ $K_{24} = 2.4 \times 10^3$

25) $HO_2 + HO \longrightarrow H_2O + O_2$ $K_{25} = 8.8 \times 10^4$

26) $2\,HO_2 \longrightarrow H_2O_2 + O_2$ $K_{26} = 4.9 \times 10^3$

Table I. The kinetic reaction mechanism. (*depends on the intensity of solar radiation.- **depends on the type of hydrocarbon mixture considered).

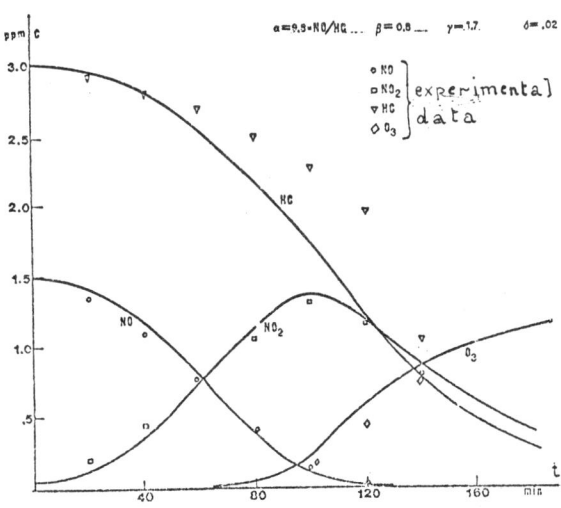

Fig.1–"Smog chamber" simulation.- Data for isobutylene from Westberg (1971).

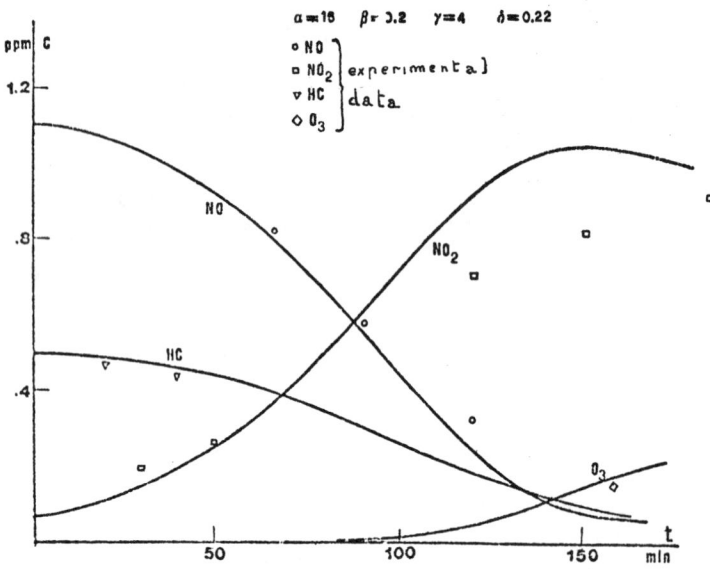

Fig. 2- Simulation in "smog chamber".- Data for propylene from Altshuller (1970).

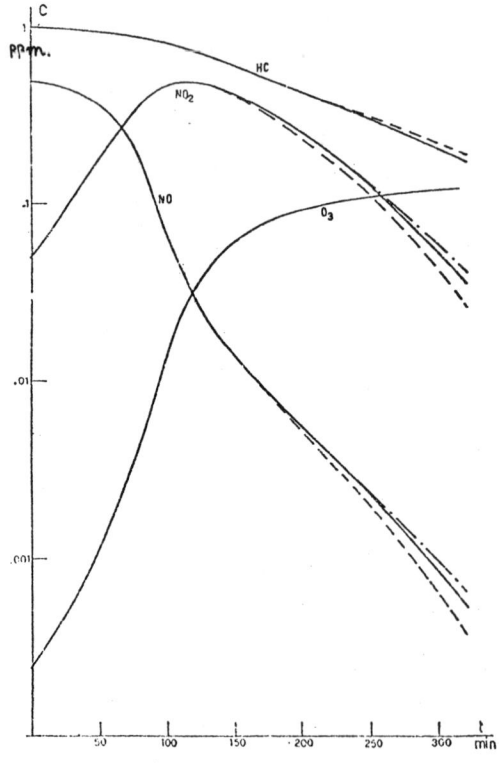

Fig. 3- Evolution of the system with steady-state (dashed lines); evolution of system without steady-state: base model (solid lines), with termination reactions (dotted lines).

3. QUASI-STATIONARITY HYPOTHESIS

In cases in which the simulation of the simplified numerical model was reliable, the integration of the equation representing the reagent system were carried out, without making hypotheses on the steady state of the radicals HNO_2 and NO_3. Fig.3 shows that the simplified model simulates the formation of photochemical smog in the same way as the model in which the steady state hypothesis was not made for integration time in order magnitude of about twice the time in which NO_2 reaches maximum concentration. For longer simulation times, the curves widen progressively. The percentual errors due to the quasi-stationarity hypothesis of the system become progressively greater after a certain time, but these are errors on quantities which are usually very small.

4. VALIDITY OF STOICHIOMETRIC COEFFICIENTS

The stoichiometric reaction coefficients α, β, γ, δ are defined in an empirical way in order to fit the experimental data into a considerably simplified reagent system but they introduce parameters which are not measurable into the system. The stoichiometric reaction coefficients depend not only on the type or mixture of hydrocarbons but also on the concentration relative of HC and NO. This is particularly important for α, which is the most sensitive to concentration ratio variation of all the stoichiometric coefficients.

Fig.4 shows three numerical simulations and an experimental curve relative to NO_2 evolution in the presence of propylene. Two of the curves with a solid line were obtained by choosing values of the parameters α, proposed in literature (α=9.8 HC/NO, Altshuller et al., 1967; Hecht-Seinfeld, 1972; α=16, Hecht et al., 1974) and with these, several exparimental tests were simulated satisfactorily. In this specific case these values of α do not give reliable simulations, on the other hand, the experimental curve (Alt

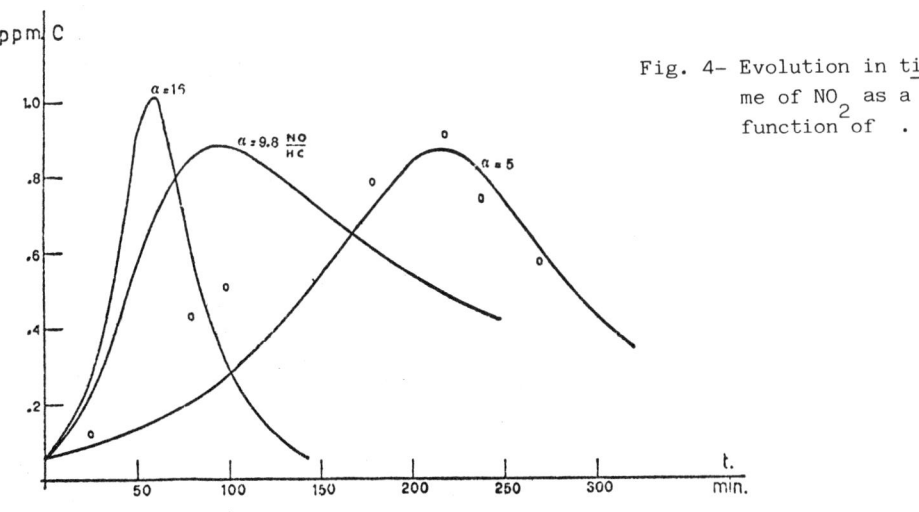

Fig. 4- Evolution in time of NO_2 as a function of .

shuller et al., 1971) is approximated by a valueα=5. Figure 5 shows the
evolution of the entire reagent system in the last case. It is therefore
possible to deduce that α is a function both of the kind of hydrocarbon and
of the concentration ratio of HC with NO. Thus in a free atmosphere it
varies both from point to point and time.

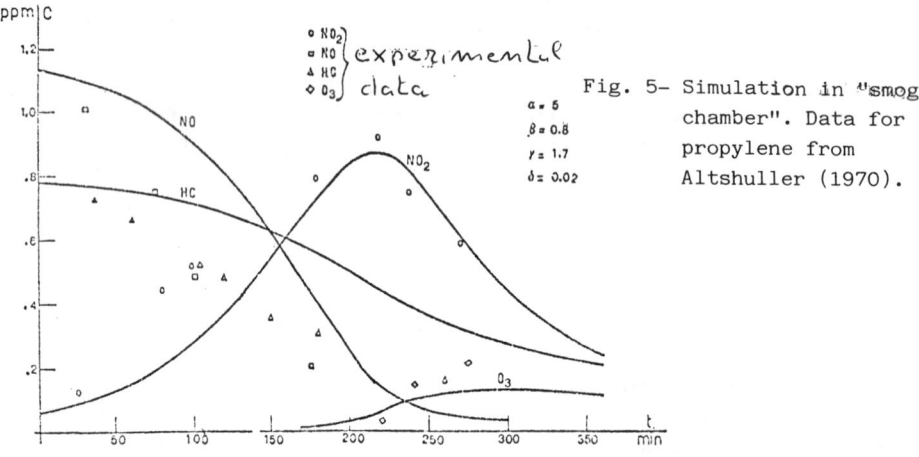

Fig. 5- Simulation in "smog
chamber". Data for
propylene from
Altshuller (1970).

Usually in the case in which the initial concentration ratio is HC_o/NO_o
$\geqslant 1$, good simulations for values of α found in literature can be obtained.
That is α does not change rapidly. In the opposite case, when $HC_o/NO_o \lneq 1$,
the numerical model is very sensitive to a correct choise of α . In fact
the case in which the numerical model does not work is very frequent in so
far as the hydrocarbons undergo a rapid decrease and tend to produce an ex
cess of HO_2 and OH. The reaction speed of HO_2 and OH becomes of the same or
der of magnitude as the speed of fundamental reactions, eq.1,2,3 in table I.
    Therefore it is necessary to define the function which ties α to the
concentration HC and NO before being able to apply the chemical reaction
model in the atmosphere where a subsequent verification of the numerical
simulation validity is difficult.

5. WEIGHT OF TERMINATIONS REACTIONS

    Termination reactions are taken into consideration in a kinetic model
proposed by Heicklen (1976) and are represented by reactions 21-26 in table
II.
    The numerical model in which hypothesis on the steady state are not
made and in which termination reactions are introduced simulates the evolu
tion of the reagent system which do not vary in the slightest from the sim
plified base system for a reaction time equal to twice the time of reaching
the maximum of NO_2 and varies from the system without steady state and wi
thout termination reactions only after an even greater time (figure 3).
    Therefore the introduction of termination reaction into the reagent
system take on a significant meaning only after relatively long simulation
times when their reaction velocity tend to be of the same order of magnitu

de as those of fundamental reactions. In fig.6, the evolution of photoche mical smog is shown with Heicklen's model in the same experimental condi tions shown in fig.5. From the comparison of evolution of the photochemical phenomenon in figures 5 and 6 it can be seen that in this specific case the different reaction kinetics proposed by Hecht-Seinfeld and Heicklen only imply a different trend of the NO_2 concentration after this has reached its maximum.

1) $NO_2 \xrightarrow{h\nu} NO + O$ $\qquad$ $K_1 = *$

2) $O_3 \longrightarrow O_2 + O$ $\qquad$ $K_2 = 2.3 * 10^{-2}$ $\quad$ min^{-1}

3) $O + C_2H_4 \longrightarrow HCO + CH_3 \xrightarrow{3O_2 + NO} CH_2O + 2HO_2 + CO + NO_2$ $\quad$ $K_3 = 1.2 * 10^3$ $\quad$ $ppm^{-1} min^{-1}$

4) $O + CH_2O \longrightarrow HO + HCO \xrightarrow{O_2} HO + HO_2 + CO$ $\quad$ $K_4 = 2.4 * 10^2$ $\quad$ "

5) $O + O_2 + M \longrightarrow O_3 + M$ $\qquad$ $K_5 = 2.0 * 10^6$ $\quad$ min^{-1}

6) $NO + O_3 \longrightarrow NO_2 + O_2$ $\qquad$ $K_6 = 2.5 * 10^1$ $\quad$ $ppm^{-1} min^{-1}$

7) $NO_2 + O_3 \longrightarrow NO_3 + O_2$ $\qquad$ $K_7 = 4.6 * 10^{-2}$ $\quad$ "

8) $NO + NO_3 \longrightarrow 2 NO_2$ $\qquad$ $K_8 = 1.3 * 10^4$ $\quad$ "

9) $NO_2 + NO_3 \longrightarrow N_2O_5$ $\qquad$ $K_9 = 5.6 * 10^3$ $\quad$ "

10) $O_3 + C_2H_4 \longrightarrow CH_2O + CH_2O_2 \longrightarrow CH_2O + CO + H_2O$ $\quad$ $K_{10} = 4.0 * 10^{-3}$ $\quad$ "

11) $CH_2O \xrightarrow{h\nu} H_2 + CO$ $\qquad$ $K_{11} = *$?

12) $CH_2O \longrightarrow H + HCO \xrightarrow{2O_2} 2HO_2 + CO$ $\quad$ $K_{12} = 4.7 * 10^{-3}$ $\quad$ min^{-1}

13) $HO + CH_2O \longrightarrow H_2O + HCO \xrightarrow{O_2} H_2O + HO_2 + CO$ $\quad$ $K_{13} = 2.2 * 10^4$ $\quad$ $ppm^{-1} min^{-1}$

14) $HO + C_2H_4 \longrightarrow H_2O + C_2H_3 \xrightarrow{2O_2 + NO} CH_2O + CO + NO_2 + HO + H_2O$ $\quad$ $K_{14} = 1.9 * 10^3$ $\quad$ "

15) $HO + C_2H_4 \longrightarrow HOC_2H_4 \xrightarrow{2O_2 + 2NO} 2 CH_2O + 2NO_2 + HO$ $\quad$ $K_{15} = 2.5 * 10^3$ $\quad$ "

16) $HO + NO_2 \longrightarrow HONO_2$ $\qquad$ $K_{16} = 1.2 * 10^4$ $\quad$ "

17) $HO_2 + NO \longrightarrow H_2O + NO_2$ $\qquad$ $K_{17} = 1.4 * 10^3$ $\quad$ "

18) $HO_2 + NO_2 \longrightarrow HONO + O_2$ $\qquad$ $K_{18} = 2.7 * 10^2$ $\quad$ "

19) $HO + NO \longrightarrow HONO$ $\qquad$ $K_{19} = 1.2 * 10^4$ $\quad$ "

20) $HONO \xrightarrow{h\nu} HO + NO$ $\qquad$ $K_{20} = *$ $\quad$ "

21) $HO + HONO \longrightarrow H_2O + NO_2$ $\qquad$ $K_{21} = 1.0 * 10^4$ $\quad$ "

22) $HO_2 + O_3 \longrightarrow HO + 2O_2$ $\qquad$ $K_{22} = 2.2 * 10^0$ $\quad$ "

23) $HO + O_3 \longrightarrow HO_2 + O_2$ $\qquad$ $K_{23} = 8.3 * 10^1$ $\quad$ "

24) $2 HO \longrightarrow H_2O + O$ $\qquad$ $K_{24} = 2.4 * 10^3$ $\quad$ "

25) $HO_2 + HO \longrightarrow H_2O + O_2$ $\qquad$ $K_{25} = 8.8 * 10^4$ $\quad$ "

26) $2 HO_2 \longrightarrow H_2O_2 + O_2$ $\qquad$ $K_{26} = 4.9 * 10^3$ $\quad$ "

Table II. Kinetic mechanism of reaction(Heicklen, 1976) . Ethylene.
(*depends on the solar radiation intensity).

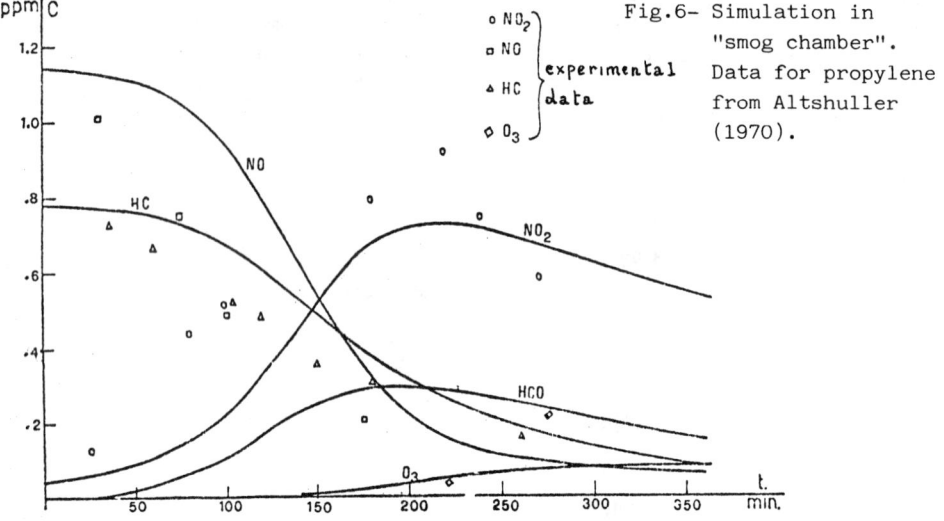

Fig.6- Simulation in "smog chamber". Data for propylene from Altshuller (1970).

Experimental data:
o NO_2
□ NO
▲ HC
◇ O_3

6. PHOTOCHEMICAL SMOG NUMERICAL MODEL

The simulation of photochemical atmospheric pollution essentially consists in the description of the behaviour of N species chemically reagent in the turbolent planetary boundary layer through N continuity equations

$$(1) \quad \frac{\partial \langle c_i \rangle}{\partial t} + \bar{u}_j \frac{\partial \langle c_i \rangle}{\partial x_j} = \frac{\partial}{\partial x_j} \left(K_{jj} \frac{\partial \langle c_i \rangle}{\partial x_j} \right) + R_j (c_1 , \dots c_n) + S_i (x_j, t)$$

where: $\bar{u}_j$ are the components along the coordinate axes of wind velocity;
K_{jj} are the coefficients of turbolent diffusion along the main axes;
R_j is the tranformation rate of the j-th chemical species;
S_i^j is the emission of the i-th species into the atmosphere.
In the case presented here we assume $\bar{u}_2 = \bar{u}_3 = 0$, $\bar{u}_1 (z)$ constant in time, and

$$\frac{\partial}{\partial x} (K_{jj} \frac{\partial c}{\partial x}) \ll \bar{u} \frac{\partial \langle c \rangle}{\partial x} \qquad \text{(Walters, 1969)}$$

The set of equations is solved with the following initial and boundary conditions:

$$\langle c \rangle (x, z) = 0 \qquad \text{for } t = 0$$
$$K_{jj} \frac{\partial \langle c \rangle}{\partial z} = 0 \qquad \text{for } z = 0, \text{ HML} \qquad \text{(HML= mixing layer height)}$$

The equation (1) is valid for reactions of second order only in the case where chemical reaction processes are slow compared with the turbolent transport and the length (L) and the time (T) of the characteristic scale of the medium concentration field are large with respect to the corresponding scales of turbolent transport.

If we define l_e and τ_e as lengths and time of the turbolence scale and $\tau_c = (K_{max} \cdot \langle c \rangle_{max})^{-1}$, the time of the reaction scale of second order, the result well be:

$$(2) \quad \tau_c \gg \tau_e ; \qquad T \gg \tau_e ; \qquad L \gg l_e .$$

In consequence equation (1) does not sufficiently describe the reagent system near to isolated sources of great intensity. The kind of kinetic mechanism which simulates the formation of photochemical smog has been applied to a model of gas dispersion in the atmosphere.

We chose to simulate the formation of photochemical smog produced by the contemporary emission of NO, NO_2, HC from a line at ground level. This choice was made due to the fact the kinetic mechanism proposed by Hecht et al. (1972) valid for a ratio HC/NO$>$1, typical of emissions from motor vehicles. The pollutants are dispersed around the source with a gaussian distribution to satisfy conditions (2). The values of the kinetic constants and stoichiometric coefficients which depend on solar radiation and on the kind of hydrocarbon mixture used in the simulation are:

$$K_1 = 0.4 \ min^{-1} \qquad K_7 = 5x \ 10^{-3} \ min^{-1} \qquad K_{11} = 7.3 \times 10^3 \ ppm^{-1} \ min^{-1}$$
$$K_{12} = 9.5x10^3 \ ppm^{-1} \ min^{-1} \quad K_{13} = 1.9x10^{-3} \ ppm^{-1} \ min^{-1} \quad K_{15} = 1.38x10^1 \ ppm^{-1} \ min^{-1}$$
$$\alpha = 12 \qquad \beta = 0.95 \qquad \gamma = 4 \qquad \varepsilon = 0.51$$

These values can be considered valid for hydrocarbon mixtures which are, on average, reactive (Reynolds et al., 1974).

The results of a simulation in neutral atmospheric conditions, with an emission $Q_{(NO)}$=100 ppm/m^3 . m, u(10)=1.5 m/sec, roughness equal to 10 cm and HML=600 m, and with an emission ratio HC/NO=2 and NO/NO_2=10, are illustrated in figures 7,8.

The meteorological parameters which characterize dispersion in the planetary boundary layer influence the rate of transformation through a more or less rapid dilution of the components.

From figures 7 and 8 can be seen that:
-the concentration of hydrocarbons at ground level strongly decreases with distance in the first 15 km. Therefore the rate of transformation undergoes, due to the progressive dilution of the components, a slowing down and the concentration is more uniform for greater distances.
-NO reacts with a medium rate on a transport time of several hours of about 10% h^{-1}; the instantaneous transformation rate near the source (1-2 km), where the concentrations are high, is of about 50% h^{-1}, as is deduced from the comparison of NO reactive concentration prifiles (solid line) and NO supposed inert (dashed line) in figures 7 and 8. This comparison shows how the reaction rate diminishes with height in relation to concentration reduction.
-NO_2 concentration at ground level does not vary much with distance. This trend is due to the action of two contrasting effects: the dispersion which tends to dilute, NO_2 and the chemical reaction which tends to produce it. The vertical profile of NO_2 tends to flatten as it moves away from the source.
-The ozone which at the beginning of the simulation had a zero concentration, always has a maximum at ground level (fig.8). Its concentration increases with distance and at the limit of the field considered reaches values of about 60 ppb. at ground level, which however does not yet represent the maximum which it presumably reaches at greater distances ($\sim$50 km).

The photochemical transformation rates obtained in this experiment do

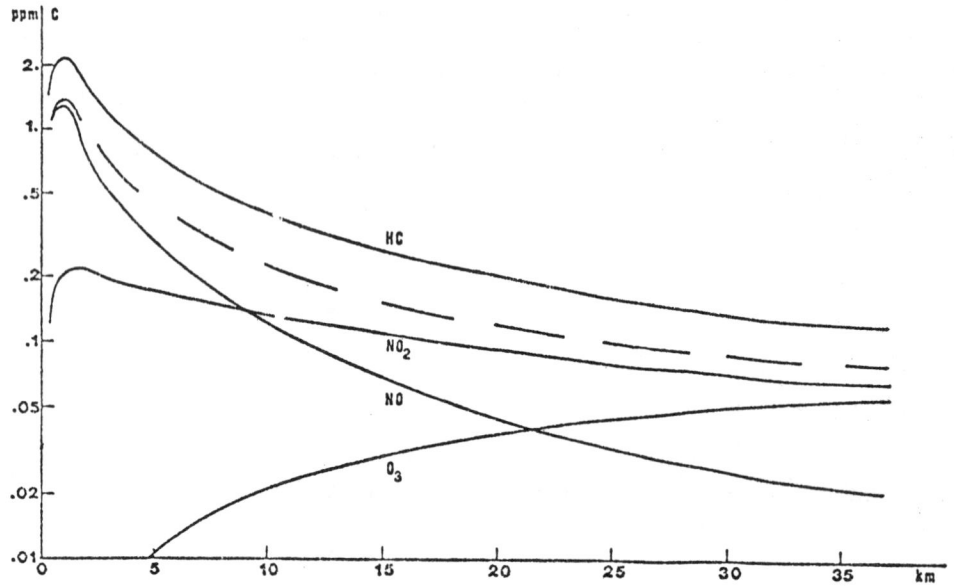

Fig. 7- Concentration distribution at ground level.

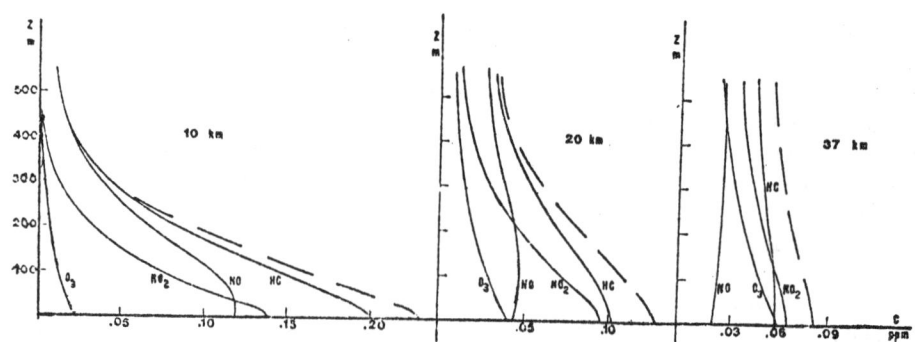

Fig. 8- Concentration vertical profiles. The scale of HC concentration is double.

not have general characteristics in so far as they are function of solar
radiation, the content of water vapour in the atmosphere, of CO_2 concentra
tions, of the reactivity of the hydrocarbons emitted, of the concentration
ratio between the various species and of the absolute values of the same.
In fact the effects due to chemical reactions of the second order are not
linear and in consequence the superimposition principle is not appllicable.

7. REFERENCES

(1) Altshuller, A.P., Bufalini J.J. (1971) Environ.Sci.Technol.,5, 39.

(2) Altshuller, A.P., Kopozynski S.L., Lonneman V.A., Becker T.L., Slater
R. (1967) Environ.Sci.Technol.,1, 899.

(3) Hecht, T.A., Seinfeld J.H. (1972) Environ.Sci.Technol., 6, 47.

(4) Hecht, T.A., Seinfeld J.H., Dodge M.C. (1974) Environ.Sci.Technol., 4,
327.

(5) Heicklen, J. (1976)Atmospheric Chemistry, 406, Academic Press, New York.

(6) Lamb R.G. (1973) Atmospheric Environment, 7, 235.

(7) Reynolds S.D., Liu M.K., Hecht T.A., Roth P.M., Seinfeld J.H. (1974),
Atmospheric Environment, 8, 563.

(8) Walters, T.S. (1969) Atmospheric Environment, 3, 461.

(9) Westberg, K., Cohen N., Wilson K.W. (1971) Science, 171, 1013.

ON THE RESULTS OF THE AEROSOL MEASURING NETWORK FOR PARTICULATE SULPHATES
AND NITRATES IN THE PERIOD 1979-1980 IN THE NETHERLANDS
PART 1

H. VAN DUUREN and F.G. RÖMER
N.V. KEMA, Environmental Research Department
The Netherlands

Summary

Total sulphate (sulphuric acid and sulphates) and nitrate aerosols
were measured during the period April 1979-September 1980 on 9 sites.
For the winter period, in particular for nitrate, distinctly higher
mean concentrations were found than for the summer periods. Over the
year maximum values for nitrate varied to a lesser extent than for
total sulphate on the measuring locations. The spatial distribution
for total sulphate aerosol was rather uniform, the opposite could
be found for nitrate, which concentration could be influenced by local
emissions of NO_x. A few examples are given of the contribution per
wind sector to the concentration of secondary components which show
the relatively strong influence of foreign sources. Some calculations
of the correlation between aerosol and precursor concentrations were
carried out. For a rural site the relation was somewhat better during
summer than for industrial areas. In the winter period the correlation
between the concentration of total sulphate and SO_2 was better for
industrial areas than for the rural site. Over short distances ($\sim$
100 km) the degree of oxidation of SO_2 in the transport direction
can strongly increase.

The variation of the concentration over the day (3-hour mean values)
measured with an automatic dichotomous sampler, was found to be very
strong under conditions of a rather polluted atmosphere. Peak values
of nitrate (NO_x peak/NO_3 low $\approx$ 60) were measured during the night
or early in the morning. The nitrate and 12h back O_3 concentrations
correlated rather well.

1. INTRODUCTION

The aerosol measuring network of the electricity utilities in The
Netherlands has been in operation since April 1979. It was mainly set up
to study:
-    the concentration variations of total sulphate (sulphuric acid and
     sulphates) and particulate nitrates on different locations;
-    the contribution from different - mostly foreign - source areas;
-    the relationship between the concentration of primary gases and ozone
     with the concentrations of secondary aerosols under different meteoro-
     logical conditions.
     A few selected data from the measuring period April 1979 - September
1981 are presented.
     Because of the relatively low number of aerosol data gathered in the
network (6-hour mean values on weekdays; 10h00-16h00 local time) attention
has also been paid to the concentration variation of these components during

the day. The importance of short sampling times - in particular during episodes of air pollution - will be explained in more detail.

2. THE AEROSOL MEASURING NETWORK

Details about measuring times and frequencies, as well as the siting were discussed earlier (1). Fig. 1 indicates the relatively limited distances between the sites and the nearby position of source areas in The Netherlands and the neighbouring countries from which important contributions to the pollutant levels can be expected.

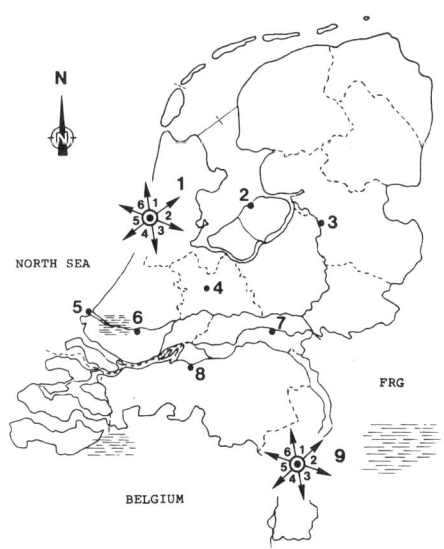

Figure 1 The aerosol measuring network of the electricity companies (sites 1...9) and the position of the main nearby source areas.

Concentrations of total sulphate (tSO_4) and nitrate (NO_3) measured in summer and in winter differ to a large extent as can be derived from table 1.
For the winter period we found distinctly higher concentrations for tSO_4 (range 1-80 µg/m³) than for the summer periods (1-57 µg/m³). Maximum values for the concentrations of NO_3 differ to a lesser extent (winter 55 µg/m³, summer 45 µg/m³).

Table 1 Frequency distribution for tSO₄ and NO₃ concentrations (µg/m³) for nine measuring stations during summer (April-September) and winter (October-March) periods. Number of measurements: 1050 per period.

period	tSO_4-percentiles					NO_3-percentiles				
	50	90	95	98	mean	50	90	95	98	mean
summer 1979	10	23	29	39	11	4	11	15	20	5
summer 1980	9	22	29	36	11	3	9	14	20	4
winter 1979/1980	11	30	37	46	14	7	19	23	30	8

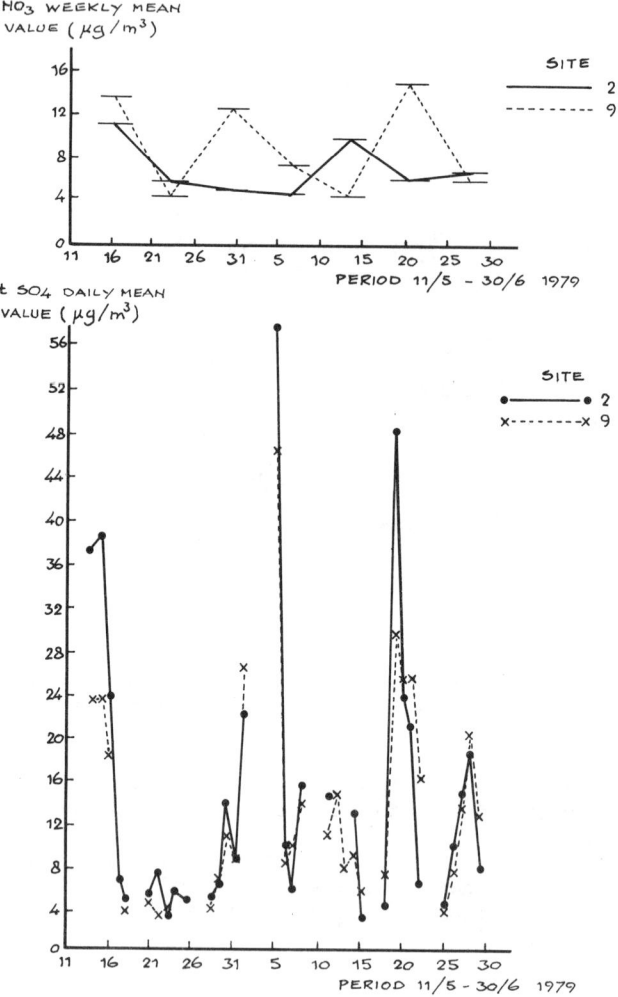

Figure 2 Trends of aerosol concentrations at two sites (tSO₄ daily mean and NO₃ weekly mean).

- 598 -

For most measuring sites the trend of the weekly mean concentrations for tSO_4, as well as the daily trend in the concentration, are quite similar. For the concentration of nitrate the opposite - also for weekly mean values - can be found. In Fig. 2 examples are shown for a period of eight weeks in 1979.

As is the case for site 9, which is situated near industrial areas, the deviation in the trend of the nitrate concentration has also been found quite often at the other sites.

An indication for the contribution from specific source areas to the measured pollutant levels at a separate site during a period can be found by determining the contribution for each of the 60 degrees wind sectors. Thus it follows from the example given in table 2 for site 1, that contribution to the mean concentration is mainly due to three transport zones. In this case for zone 2 with the highest mean concentration and the relatively high contribution to tSO_4, the sampled air masses will probably not originate from the most nearby source areas. In this case transport over longer distances from eastern Europe has been supposed.

Table 2   Contributions per 60 degree wind sector for sites 1 and 9 in the winter of 1979/1980 (see also Fig. 1). tSO_4 values

sector number	site 1: PC/100	main source area	site 9: PC/100	main source area
1	0.4		2.2	
2	3.8	NL, GDR, FRG	3.5	FRG, GDR
3	2.9	NL, FRG	0.5	
4	4.4	NL, B, F	4.2	B, F
5	1.0		1.8	
6	1.2		0.2	

C = mean concentration ($\mu g/m^3$); P = percentage of measurements per sector; PC/100 = contribution ($\mu g/m^3$) per sector; sector number 1 = 346°-45°, 2 = 46°-105°, etc.

The relationship between the concentration of aerosols and primary gases: in general literature shows the ambiguous correlation between these components. Nevertheless, we found some differences (table 3) for several measuring stations which can be characterized as follows:
  site 2: rural area
  site 6: highly industrialized Rijnmond area
  site 9: situated near several source areas

Table 3   Linear correlations for tSO_4 vs SO_2 and NO_3 vs NO_2 (6-hour mean values).

site	NO_3-NO_2: summer 1979		summer 1980		winter 1979/1980		tSO_4-SO_2: winter 1979/1980	
	n=:	r=:	n=:	r=:	n=:	r=:	n=:	r=:
2	113	0.56	112	0.63	107	0.55	105	0.43
6	94	0.46	111	0.39	112	0.67	119	0.69
9	107	0.41	117	0.45	118	0.66	118	0.74

It can be stated that for rural site 2 during summer the correlation NO_3 vs NO_2 is somewhat better than for sites 6 and 9. In the winter period the opposite has been derived, but it was found to be less clear. The correlation tSO_4 vs SO_2 gives a different picture. In this case the rather high correlation at sites 6 and 9 during winter is rather conspicuous. The values of the correlation coefficient given in table 3 are significant with a confidence of at least 99%.

In order to determine the oxidation rate of SO_2 from available data, an inventarisation was carried out of the degree of oxidation ($R = tSO_4$/($tSO_4 + 1,5\ SO_2$) under a few limiting conditions, as to wind direction, decrease (increase) in SO_n (tSO_4) concentrations. In table 4 some data are given of the rather strong increase of R.

Table 4 The degree of oxidation of SO_2

date	transport direction (site 6 → 3, 130 km)					
	tSO_4* site 6:	SO_2*	R	tSO_4 site 3:	SO_2	R
79-05-07	11	45	0.14	17	30	0.28
79-05-08	12	51	0.14	14	25	0.27
	transport direction (site 9 → 7, 95 km)					
	site 9:			site 7:		
79-09-12	17	39	0.23	35	20	0.53
79-10-30	11	55	0.12	31	33	0.47
79-11-22	25	83	0.16	29	46	0.30

* concentration μg/m³

3. THE DAILY TREND OF THE NITRATE AND SULPHATE AEROSOL CONCENTRATION

As mentioned before (1) the concentration of secondary aerosols can vary strongly during the day. Especially under episodic conditions trend measurements preferably with chemical speciation seem to be useful. In that case in stead of 6-hour or 24-hour mean values, which data can be used to study long term effects on ecosystems and on corrosion, there is a need for measurements of peak values to support studies to elucidate the possible impact of secondary aerosols (e.g. H_2SO_4) and HNO_3 on health and visibility.

During a period of two months (August-September 1980) trend measurements (3-hour mean values) of tSO_4, NO_3 and NH_4 were carried out at site 8 using an automatic dichotomous sampler (ADS)*. Within a period of 24 hours rather extreme ratios of 3-hour mean peak value and lowest 3-hour values can be found (NO_3 peak/NO_3 low $\approx$ 60; SO_4 peak/SO_4 low $\approx$ 20). In table 5 an example is given of the relatively strong change in concentration within a rather short period. This holds in particular for the smaller particles (< 2,5 μm). Concentrations found for both tSO_4 and NO_3 in the particle fraction > 2,5 μm mostly are low. For the concentration of nitrate sometimes the highest values can be found for the larger particles.

* At the same time a comparison was made between measurements of 6-hour mean values of high volume sampling (HVS) and ADS. The results for tSO_4 were satisfactory, for NO_3 the ADS values were higher: ADS = 1.58 HVS.

Table 5 Typical data for the concentration of tSO_4 and NO_3 aerosol ($\mu g/m^3$) sampled with a dichotomous sampler.

date	period*:								particle fraction	
	1	2	3	4	5	6	7	8	<2,5 µm	>2,5 µm
80-08-27 tSO_4	7.4	17.4	9.8	15.3	21.6	17.1	22.0	23.8	+	
	0.6	0.0	0.9	1.7	1.4	0.0	0.7	1.0		+
80-08-28 tSO_4	23.6	37.4	24.5	33.9	32.5	6.9	4.8	5.2	+	
	1.7	1.5	5.0	5.9	4.1	0.4	0.7	0.8		+
80-08-27 NO_3	6.6	9.5	9.4	9.5	6.3	1.3	11.9	31.9	+	
	1.6	1.1	1.0	3.7	3.1	1.8	2.5	1.2		+
80-08-28 NO_3	49.0	64.0	27.8	8.9	3.2	0.5	0.5	0.4	+	
	2.1	1.6	3.2	1.5	1.2	0.7	0.7	0.3		+

* 1 = 01h00 - 04h00 ... 8 = 22h00 - 01h00, local time

For NO_3 the peak values were found mostly at night or in the early morning, for tSO_4 the picture was less clear though the correlation between the two components for three periods (5 days) with higher concentrations was good (n = 40, r = 0.67).

In general it can be stated that during these periods measured values for tSO_4 and NO_3 can significantly correlate with SO_2 and NO_2 respectively. The relationship between NO_3 and O_3 was best for 12h back O_3 concentrations. Some results are given in table 6.

Table 6 The relationship between NO_3 and 12h back O_3 concentrations at site 8.

period	number of measurements	correlation coefficient
26-28 August 1980	16	0.84
3- 5 September 1980	16	0.53
29-30 September 1980	8	0.64

REFERENCES

1   Römer, F.G., Duuren, H. van, Elshout, A.J. Nitrate and sulphate levels in The Netherlands: Discussion of some recent measuring data, Proceedings of the first European Symposium on Physico-chemical Behaviour of Atmospheric Pollutants, Ispra 1979, B. Versino and H. Ott, ed., p. 458-464

ASSESSMENT AND CHEMICAL ANALYSIS OF WATER SOLUBLE ATMOSPHERIC FALL-OUT ON SOME SITES IN FRANCE WHERE BRACKISH-WATER COOLING TOWERS MAY BE CONSTRUCTED; CONTRIBUTION OF SALT DROPLETS EMISSION FROM THE TOWERS.

P. MASNIERE   J. DUBOIS

ELECTRICITE DE FRANCE
DIRECTION DES ETUDES ET RECHERCHES
6 quai Watier 78400 CHATOU

Summary

Recently, several studies were undertaken dealing with the environmental impact of large brackish-water cooling towers with drift eliminators (power plants of Chalk Point, Forked River). In France, such
towers were considered in 3 sites by " ELECTRICITE DE FRANCE ".
To quantify the salt emission escaping from the tower (especially Na,
Cl, Ca) with respect to the atmospheric ground level, previous quantitative information on the amount of natural fall-out deposited on the
soil within the area influenced by the proposed towers, is required.
Chemical analysis of water soluble atmospheric fall-out (collected in
a special rain gauge) were performed during periods of 1 to 3 years
according to the considered site. The deposition rate values may be
compared with the considered ones estimated in a predicting salt transport model established for these towers.
On the other hand it has been showed that exists inside a tower, above
the drift eliminators, a large population of small particles, below
the range seen by PILLS for instance (a few μm or less) , which contains
primarily mist, but also include drift components . An isokinetic sampler was operated inside several fresh water cooling towers to trap
both salt droplets and condensed water. Enrichment values (10 to 10^5)
versus concentrations in circulating water were pointed out for a few
elements (Mn, Cu, Zn ...).

1. INTRODUCTION

L'implantation en France de nouvelles Centrales électro-nucléaires
comportant plusieurs tranches de puissance unitaire de l'ordre de 1000 MWe
implique souvent un refroidissement en circuit fermé. Les réfrigérants, de
type humide, rejettent dans l'atmosphère des quantités importantes de vapeur d'eau porteuse de sels dissous. Des gouttes d'eau chargées en sels
sont alors susceptibles de retomber sous le trajet du panache. Ces dépôts
peuvent être importants comparés à ceux d'origine atmosphérique préexis-
tants, surtout dans le cas de réfrigérants alimentés en eau saumâtre.
C'est ainsi qu' " ELECTRICITE DE FRANCE " a réalisé diverses études relatives à l'impact de tours alimentées en eau d'estuaire (LOIRE, GIRONDE) ou
de rivières à forte teneur saline (MOSELLE). Les aspects abordés ici concernent l'obtention de données d'entrée dans un modèle numérique de simulations de retombées de gouttes utilisé par " ELECTRICITE DE FRANCE " :

a) évaluation des retombées atmosphériques solubles dans les 3 régions où des tours étaient envisagées (particulièrement en Na,Cl,Ca), afin de comparer les valeurs moyennes mesurées avec celles entrant à priori dans le modèle de calcul.

b) Evaluation des flux salins émis par des tours en fonctionnement, alimentées en eaux douces, afin de déterminer le taux de primage par rapport au débit d'eau de circulation (qui est important). Les mesures chimiques effectuées dans ce but à l'intérieur des tours mettent en évidence des processus plus ou moins marqués d'enrichissement en certains éléments (particulièrement en métaux) dans les gouttes émises au dessus des séparateurs.

1. INCIDENCE DES TOURS DE REFROIDISSEMENT ALIMENTEES EN EAU SAUMATRE SUR LES RETOMBEES ATMOSPHERIQUES GLOBALES

Peu de résultats sont disponibles dans la littérature permettant de valider les valeurs moyennes entrant dans le modèle de simulation utilisé pour des réfrigérants éventuellement installés sur des sites tels que BLAYAIS, PELLERIN et CATTENOM (figure 1) . Ces valeurs moyennes, calculées à partir de diverses courbes et en fonction de la distance à la côte, ne permettaient pas d'évaluer les retombées en Ca dont la détermination est nécessaire pour CATTENOM. De plus, ces valeurs moyennes masquent des disparités élevées suivant les périodes de prélèvements, particulièrement importantes à cerner durant des périodes de forte intrusion saline en estuaire.

1.1. Méthode de mesures des retombées atmosphériques solubles

En fonction du but recherché et des conditions de maintenance d'appareillage, seules des mesures de dépôts atmosphériques au niveau du sol ont été effectuées (à l'exclusion des mesures de concentrations dans l'air). Elles prennent en compte la totalité des retombées solubles, durant des périodes répétées de 2 à 3 semaines. Un réseau de jauges de dépôt du type préconisé par le C.F.R. (CNRS/CEA), et d'une section de capture de 0,1m 2, ont été installés dans les régions du BLAYAIS et du PELLERIN. La multiplication des collecteurs permet, à partir de moyennes locales, de s'affranchir de certains résultats aberrants parfois relevés.

Région	Nombre de récepteurs	Début des / Fin des mesures / mesures	Distance à la côte (axe Est-Ouest)	Eléments particulièrement recherchés
BLAYAIS	11 (part et d'autre de l'estuaire)	Juin 1977-Juin 1978 Sept.1978-Fév. 1979 Nov. 1979-Avr. 1980	9 à 60 km	Na, Cl
PELLERIN	5 (rive gauche de la Loire)	Sept.1976-Déc. 1979	2 à 30 km	Na, Cl
CATTENOM	1 (site de la Centrale)	Déc. 1977-Sept.1978	-	Na, Cl, Ca

FIGURE I

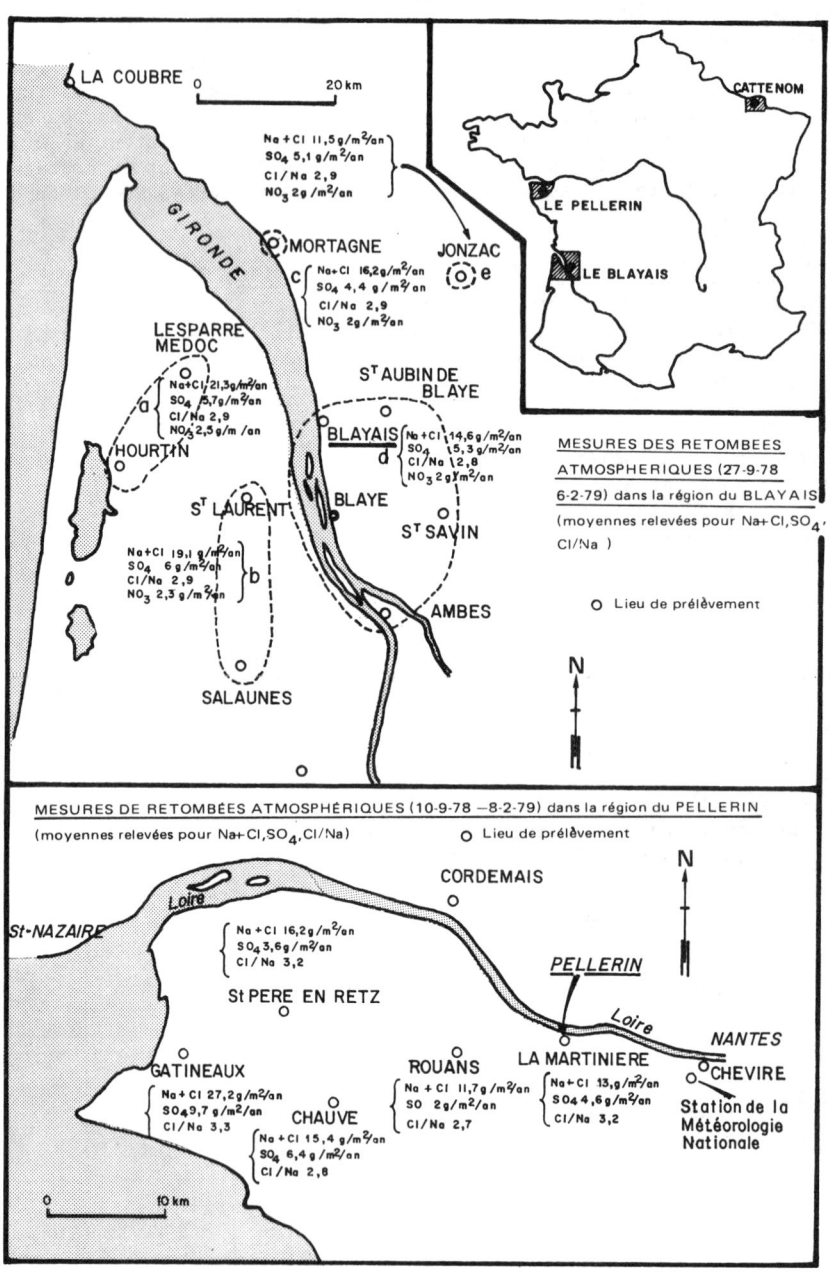

LA COUBRE ⊙

0 ⊢———⊣ 20 km

Na+Cl 11,5g/m²/an
SO₄ 5,1 g/m²/an
Cl/Na 2,9
NO₃ 2g/m²/an

GIRONDE

⊙ MORTAGNE   → JONZAC ⊙ e

c { Na+Cl 16,2g/m²/an
SO₄ 4,4 g/m²/an
Cl/Na 2,9
NO₃ 2g/m²/an

LESPARRE MEDOC ⊙

a { Na+Cl 21,3g/m²/an
SO₄ 5,7g/m²/an
Cl/Na 2,9
NO₃ 2,5g/m²/an

Sᵀ AUBIN DE BLAYE

BLAYAIS ⊙
d { Na+Cl 14,6g/m²/an
SO₄ 5,3g/m²/an
Cl/Na 2,8
NO₃ 2g/m²/an

HOURTIN ⊙

Sᵀ LAURENT

BLAYE

Sᵀ SAVIN

Na+Cl 19,1 g/m²/an
SO₄ 6 g/m²/an
Cl/Na 2,9
NO₃ 2,3 g/m²/an
} b

AMBES

SALAUNES

N

⊙

CATTENOM

LE PELLERIN

LE BLAYAIS

MESURES DES RETOMBEES
ATMOSPHERIQUES (27-9-78
6-2-79) dans la région du BLAYAIS
(moyennes relevées pour Na+Cl,SO₄,
Cl/Na)

⊙ Lieu de prélèvement

MESURES DE RETOMBÉES ATMOSPHÉRIQUES (10-9-78 — 8-2-79) dans la région du PELLERIN

(moyennes relevées pour Na+Cl,SO₄,Cl/Na)   ⊙ Lieu de prélèvement

N

Loire

St-NAZAIRE

CORDEMAIS ⊙

Na+Cl 16,2g/m²/an
SO₄ 3,6g/m²/an
Cl/Na 3,2

St PERE EN RETZ ⊙

PELLERIN

Loire

NANTES

GATINEAUX ⊙

Na+Cl 27,2g/m²/an
SO₄9,7g/m²/an
Cl/Na 3,3

ROUANS ⊙

Na+Cl 11,7g/m²/an
SO 2g/m²/an
Cl/Na 2,7

LA MARTINIERE ⊙

Na+Cl 13,g/m²/an
SO₄ 4,6g/m²/an
Cl/Na 3,2

⊙ CHEVIRE

Station de la
Météorologie
Nationale

CHAUVE ⊙

Na+Cl 15,4 g/m²/an
SO₄ 6,4 g/m²/an
Cl/Na 2,8

0 ⊢———⊣ 10 km

FIGURE II

RETOMBÉES EN Na et Cl EN FONCTION DE LA DISTANCE A LA COTE

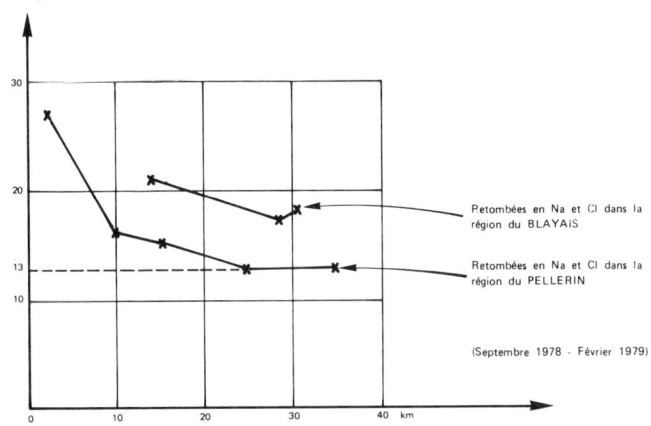

Retombées en Na et Cl dans la région du BLAYAIS

Retombées en Na et Cl dans la région du PELLERIN

(Septembre 1978 - Février 1979)

TABLEAU I : RETOMBÉES EN Na et Cl

g/m²/an

N°	SITE	PÉRIODE DE PRÉLEVEMENT	Na+Cl moyenne	Na+Cl maximum	Na+Cl minimum	SO$_4$ moyenne	NO$_3$ moyenne	Ca moyenne	$\frac{Na+Cl}{\Sigma^*}$ (%)	$\frac{SO_4}{\Sigma^*}$ (%)	$\frac{NO_3}{\Sigma^*}$ (%)	Cl/Na
1	Blayais	16/6/77 - 6/1/78	2,1									
2	Blayais	16/6/77 - 9/6/78	10,1	56,6	2,5	5,5	6,4	1,1	37 % moyenne	20 % moy.	24 % moy.	2,1
3	Blayais	27/9/78 - 6/2/79	14,6	24,4	1,1	5,9	2,0	1,9				2,8
4	Blayais	21/11/79-29/4/80	14,4	38,0	3,7	5,0 (3,8 à 9,2)	3,9	2,0	20 à 60 %	13 à 36 %		1,5
5	Pellerin	16/9/76 - 12/12/77	12,3	32,0	2,5				30 à 80 %			1,9
6	Pellerin	20/9/78 - 8/2/79	13,0	40,3	1,8	4,6	1,5	0,9	46 % moyenne	16 % moy.	5 % moy.	3,2
7	Pellerin	16/9/76 - 18/12/79	10,0	36,5	0,6	3,0 (0,3 à 10,6)		1,0	5 à 80 %			2,2
8	Pellerin	17/5/78 - 20/9/78	2,6	4,7	0,6	1,8	2,1	0,4	37 % moyenne	26 % moy.	30 % moy.	
9	Pellerin	30/5/79 - 19/9/79	2,8	8,9	0,6	2,7	2,7	0,8	26 % moyenne			
10	Cattenom	déc. 77 - sept. 78	2,3	8,0	0,4	6,6 (1,5 à 14,9)	3,0 (0,5 à 6)	3,8** (1,2 à 11)	13 % moyenne	38 % moy.	17 % moy.	3,8

* Σ représente la somme des teneurs en divers ions mesurées dans les dépôts.

** Ce chiffre est à minorer : du fait de la proximité d'un chantier de démolition, on a constaté une augmentation de 20 % du niveau moyen de particules en suspension entre 1977 et 1978 [appareil COLLECTRON aspirant 2 m³/h].

TABLEAU II : PRIMAGE ET ENRICHISSEMENT CONSTATÉS
POUR 4 RÉFRIGÉRANTS

SITES ÉTUDIÉS	BOUCHAIN (tour provisoire en toile)	COURRIERES	ANSEREUILLES (Tr. 2)	BUGEY (Tr. 4 - Ouest)
Type du réfrigérant	SCAM courant croisé	SCAM contre courant	SCAM mixte	HAMON contre courant
Hauteur de la tour Section au sommet	62 m 1780 m^2	95 m 855 m^2	93 m 1590 m^2	127 m 3670 m^2
Section de la tour au niveau où sont effectués les prélèvements	2640 m^2	2700 m^2	2640 m^2	6800 m^2
Vitesse de l'air au niveau où sont effectués les prélèvements	1,30 m/s (moyenne)	–	1,60 m/s (moyenne)	2 m/s (moyenne)
Débit de l'eau de circulation	30 880 m^3 / h	16 500 m^3 / h	18 000 m^3 / h	62 100 m^3 / h
Teneur en sels de l'eau de circulation Σ $(Na^+ + Mg^{++} + Ca^{++} + K^+ + SO_4^= NO_3^= + NH_4^+ + Cl^-)$	700 - 1000 mg/l	1800 - 2000 mg/l	1200 - 1400 mg/l	120 - 270 mg/l
Nombre d'heures de marche pour l'année	–	7200	6450	6000
Quantité de sels (kg/h) rejetés au niveau du prélèvement, pour diverses périodes	0,8 à 24,7	0,8 à 7,5	330 (moy.)	0,6 à 3,4
Taux de primage pour diverses périodes de prélèvement	3×10^{-5} à $8,5 \times 10^{-4}$	$2,6 \times 10^{-5}$ à $2,4 \times 10^{-4}$	$2,7 \times 10^{-4}$ (moyenne)	2×10^{-5} à 7×10^{-4}
Enrichissement constaté pour Zn (à partir de Zn/Na dans l'eau de circulation et dans les gouttes)	10 (moyenne du 5 au 19/11/80) 100 (du 13 au 14/1/81)	10^3	30 à 590 (moy. 190)	130 (moyenne)
Enrichissement constaté pour Mn	5 (du 5 au 19/11/80) 10 (du 13 au 14/1/81)	10 à 10^2	> 2500	170 (moyenne)
Enrichissement constaté pour Cu	10 (du 5 au 19/11/80) 2 (du 13 au 14/1/81)	–	3 à 44 (moyenne 19)	80 (moyenne)

Les teneurs en Na^+, Ca^{2+}, Mg^{2+}, K^+ sont mesurés, après filtration des échantillons recueillis dans des récipients en polyéthylène, par spectrométrie d'absorption atomique (PYE - UNICAM SP 1900) . Les teneurs en Cl^- et SO_4^{2-} sont déterminés par titrimétrie et turbidimétrie. Les teneurs en NH_4^+ et NO_3^-, quoique moins significatives, ont été également déterminées par colorimétrie afin d'obtenir un ordre de grandeur raisonnable des retombées globales solubles. Les diverses teneurs sont exprimées en $g/m^2/an$.

1.2. Teneurs en chlorures et sodium des dépôts atmosphériques solubles

Impact des gouttes rejetées par les tours

Les résultats sont regroupés dans le tableau II, ainsi que les contributions en calcium, sulfate et nitrate. Des résultats plus détaillés ont fait l'objet de divers rapports pour les 3 régions étudiées (1). Le modèle de calcul utilisé par E.D.F. (2) effectue une simulation pour une période de 10 années à partir de données trihoraires recueillies par des stations météorologiques proches des sites. Nous avons évalué l'impact de gouttes rejetées par des tours à partir des valeurs moyennes de retombées relevées pour les 3 régions concernées.

. CATTENOM : suivant les types de réfrigérants envisagés, les retombées artificielles maxima attendues varient de 75 à 223 mg/m^2/an en chlorures. Une valeur de retombées " naturelles " de 720 mg/m^2/an introduite a priori dans le modèle conduisait à une augmentation de 10 à 30% des retombées. La campagne de mesures donne une valeur moyenne de 190 mg/m^2/an qui implique donc une diminution d'un facteur de 2 à 3 de l'impact prévu.

. PELLERIN : durant la campagne de mesure, une valeur moyenne de 10g/m^2/ an est obtenue pour les retombées en Na et Cl et qui n'est plus que de 2,6 à 2,8 g/m^2/an durant les trois mois (Juin à Août) où l'intrusion saline dans la LOIRE atteint le niveau du site de la Centrale projetée. Suivant les types de réfrigérants envisagés, la contribution aux retombées préexistantes serait alors de 10% au maximum.

. BLAYAIS : suivant les campagnes de mesure, les valeurs moyennes de retombées en Na et Cl sont de 10 à 15 $g/m2$/an. Les tours ne contribueraient qu'à une augmentation de quelques % des retombées atmosphériques préexistantes.

1.3. Remarques

. Les valeurs moyennes obtenues masquent de grandes disparités : au BLAYAIS, par exemple sur une année les retombées varient de 2,5 à 56,6 g/m^2/an et constituent de 5 à 80% des retombées globales. Dans ces conditions, l'incidence de dépôt des gouttes de primage renvoie avant tout à une étude de la tolérance de la végétation à certains dépôts salins.
. Le rapport Cl/Na est supérieur dans la région de CATTENOM (3,8 contre environ 2 en région côtière), reflétant sans doute l'importance de la composante anthropogénique des retombées.
. L'influence des paramètres météorologiques est abordée par ailleurs (3)(4) : des retombées importantes en Cl et Na sont en général associées à certains secteurs de vent et à une forte pluviosité en région côtière.
. Par ailleurs un effet d'estuaire non négligeable a été constaté pour le BLAYAIS perturbant certaines données recueillies plus au sud à la station météorologique de BORDEAUX et entrant dans le modèle (2).

. Les résultats indiquent une rapide décroissance des teneurs en Na et Cl entre le rivage et les stations situées à une dizaine de km de celui-ci (ou plus). La figure II reprend les résultats acquis pour une même période dans les régions du BLAYAIS et du PELLERIN, distantes d'environ 250 km.
. De Février à Septembre 1977 on relève 2,5 g/m²/an de retombées en Na et Cl à CATTENOM contre 6,5 g/m²/an au PELLERIN.

1.4. Sulfates et Nitrates dans les retombées atmosphériques solubles

Certains résultats sont reportés dans le tableau I.

Sulfates : des valeurs moyennes de 3 à 5 g/m²/an sont relevées dans les régions du BLAYAIS et du PELLERIN, masquant d'importantes variations. Ils constituent 10 à 30% des retombées au BLAYAIS. Une certaine uniformité des retombées est constatée pour cette région sauf à proximité de la Centrale électrique d'AMBES. De plus fortes teneurs sont également relevés en bordure de côte dans la région du PELLERIN. A CATTENOM on relève des teneurs approchant celles relevées à AMBES entre Février et Juin 1978.

Nitrates : une certaine uniformité est également relevée dans la région du BLAYAIS. L'importance relative des teneurs en nitrates et sulfates dans les retombées sur les 2 régions d'estuaire est fluctuante : d'Avril à Octobre 1978, on relève 1,5 g/m²/an de sulfates contre 3 g/m²/an de nitrates au PELLERIN alors que de Novembre 1978 à Avril 1979 on relève 5,2 g/m²/an en sulfates contre 2g/m²/an en nitrates . A CATTENOM on relève en moyenne 3g/m²/an en nitrates et 6,6 g/m²/an en sulfates (Décembre 1977-Septembre 1978).

2. FLUX SALINS EMIS DANS L'ATMOSPHERE PAR DIVERSES TOURS DE REFROIDISSEMENT - PHENOMENES D'ENRICHISSEMENT.

Les constructeurs de tours garantissent un taux maximum pour le " primage " , défini comme le débit d'eau rejeté à l'extérieur sous forme de gouttes d'eau de circulation non arrêtées par les séparateurs, rapporté au débit d'eau de circulation.
La mesure des teneurs des précipitations recueillies dans les tours à l'aide de récepteurs du type préconisé par le C.F.R. et correspondant aux retombées des plus grosses gouttes qui ne peuvent être entraînées par le flux d'air à l'extérieur, montrent d'une part une grande hétérogénéité suivant l'endroit et le moment du prélèvement, et d'autre part un effet systématique de dilution par rapport aux teneurs de l'eau de circulation. Il est donc plus approprié de calculer les flux salins que les volumes d'eau s'échappant des tours pour évaluer le primage (5).

2.1. Mesures des flux salins

Ces flux ont été mesurés dans 4 tours où l'eau de circulation est moins chargée en sels qu'une eau de circulation saumâtre éventuellement utilisée dans les 3 régions étudiées plus haut. Le débit de sels est calculé à partir d'analyses chimiques des échantillons recueillis généralement à une quinzaine de mètres au dessus des séparateurs et de façon isocinétique à l'aide d'un appareil type cyclone (CALIDYNE).

Les caractéristiques des tours et les taux de primage calculés sont rassemblés dans le tableau II. Des prélèvements ont également été effectués par impaction sur filtres chauffés.
D'autre part, les fluctuations des rapports des divers ions entre eux, Na étant pris comme référence, ont été étudiées à partir des analyses

chimiques (spectrométrie d'absorption atomique, titrimétrie, colorimétrie, etc). En effet, des essais effectués dans d'autres réfrigérants (Pittsburg, Homer City) indiquent des anomalies concernant les teneurs relatives en Mg^{++}, Na^+, Ca^{++} (Mg/Na est par exemple différent dans les prélèvements d'air et dans l'eau de circulation). Mis à part le cas de Bugey où des difficultés spécifiques ont été rencontrées, les rapports X/Na (X = Mg,Ca,K, Cl, SO_4), s'ils sont légèrement différents dans les gouttes et l'eau de circulation, ne semblent pas remettre en cause la théorie des gouttes d'eau de primage, fraction d'eau de circulation n'ayant pas été piégée par les séparateurs et entraînée par le flux d'air.

2.2. Les légères anomalies relevées dans les rapport X/Na pour les éléments majeurs deviennent cependant notables pour NH_4, Mn, Cu, Zn. Les enrichissements constatés dans les gouttes sont évidents quoique très variables suivant les prélèvements et les tours et sont sans doute liés aux tailles des gouttes (affectant plus particulièrement celles de faibles diamètres).

Les teneurs en métaux dans l'air aspiré par la tour ne permettent d'expliquer que marginalement les enrichissements constatés (6).

Les enrichissements relevés sont rassemblés dans le tableau II. Les facteurs d'enrichissement affectent des éléments présents en très faibles quantités dans l'eau de circulation et l'impact sur l'environnement demeure très faible; à Courrières, par exemple, les résultats conduisent à un rejet de quelques grammes par heure de zinc dans l'atmosphère (mais ce métal ne contribue que pour 8×10^{-4} % à la salinité de l'eau de circulation.

REFERENCES

1) J.C. LOEWENSTEIN, P. MASNIERE, F. TRAVADE
   " Les conséquences sur l'environnement du fonctionnement de réfrigérants atmosphériques alimentés en eau saumâtre ".

   Société Hydrotechnique de France, XVIèmes journées de l'hydraulique, Paris, 16-17-18 Septembre 1980.

2) A. HODIN
   " Estimation par modèles numériques des retombées de sel dues aux réfrigérants - Rapport EDF-DER HE/32/79/06.

3) P. MASNIERE, C. DUTRANNOY
   " Mesures de retombées atmosphériques globales, sèches et humides, dans larégion de BRAUD-SAINT-LOUIS (Gironde) entre Juin 1977 et Juin 1978. - Rapport EDF-DER E 33/78/032.

4) P. MASNIE \. , C. DUTRANNOY
   " Mesures de retombées atmosphériques naturelles solubles, sèches et humides, dans la région du PELLERIN (Loire Atlantique) entre Septembre 1976 et Décembre 1979. - Rapport EDF-DER E 33/80/04.

5) G. MAFFIOLO, P. MASNIERE, D. GAUTIER, M.F. SOUCHET
   " Physicochimie des gouttes et aérosols émis par des réfrigérants humides à tirage naturel. Résultats de la campagne d'essais à la Centrale des Ansereuilles du 7 au 18 Mai 1979.
   Rapport EDF-DER HT/31.80.09; HE 33.80.03.

6) M.F. SOUCHET   Communication personnelle - Thèse de 3ème cycle 1981.

QUELQUES ASPECTS DE LA SIMULATION HYDRAULIQUE DE LA COUCHE LIMITE
ATMOSPHERIQUE ET DE PHENOMENES CONNEXES

A FEW INSIGHTS IN HYDRAULIC MODELLING OF THE ATMOSPHERIC BOUNDARY
LAYER AND RELATED PHENOMENA

P. BESSEMOULIN
Etablissement d'ETudes et de Recherches Météorologiques
Direction de la Météorologie - FRANCE

Summary

Many problems - in particular atmospheric diffusion of pollutants - involve
air flow and turbulence patterns over complex terrain. In general, mathematical models are not able to take into account such complicated boundary
conditions. This is the reason why so many efforts are devoted to physical
modelling. Very often, the order of magnitude of typical length scales of
practical problems is "some kilometers". So it is necessary to use relatively large length scale ratios, unless working with very large facilities.
     Using length scale ratios as low as 1/5000 results in small Reynolds
numbers. The main aim of this paper is to examine how such a drastic reduction of the Reynolds number influence the mean and turbulent properties of
the flow. Some conclusions are raised about modelling of diffusion.

1. INTRODUCTION

     Il existe de nombreux articles consacrés à la comparaison des caracté-
ristiques moyennes et turbulentes d'une couche limite obtenue en laboratoire et de celles observées dans les basses couches de l'atmosphère (voir par
exemple (10)).Généralement, ces études en similitude sont conduites à des
échelles de réduction de l'ordre de 1/500.
     Connaissant les possibilités et l'intérêt de telles études, il est ten-
tant d'augmenter le facteur de réduction afin de pouvoir reproduire un certain nombre de phénomènes à une échelle tombant dans le domaine méso-météo-
rologique - en particulier des effets de reliefs qui peuvent se faire sentir à plusieurs kilomètres en aval.
     D'autres auteurs (6,7) ont entrepris également récemment des simula-
tions à des échelles de réduction de cet ordre.
     Le présent article est destiné à examiner les possibilités et les li-
mites d'une similitude à des échelles aussi réduites.

2. LES BASES DE LA SIMILITUDE HYDRAULIQUE

     L'écoulement d'eau isotherme utilisé est l'homologue d'une atmosphère
adiabatique. S'agissant d'atmosphère neutre, le seul paramètre de similitude pertinent est le nombre de Reynolds. Compte tenu des échelles de réduction des maquettes utilisées (1/5000), les nombres de Reynolds du laboratoire peuvent être 10^3 à 10^5 fois plus faibles que dans l'atmosphère. Il
importe donc d'étudier en quoi cette différence d'ordre de grandeur intervient sur les caractéristiques turbulentes simulant celles de l'atmosphère.
Actuellement la similitude des écoulements est essentiellement basée sur :

a) L'utilisation de nombres de Reynolds suffisamment grands pour satisfaire à la relation $U_* \cdot Z_0/\nu > 3$, où U_* est la vitesse de frottement, Z_0 la rugosité et ν la viscosité cinématique. Ce critère assure en effet un écoulement aérodynamiquement rugueux.

Les nombres de Reynolds ainsi réalisés peuvent cependant être localement insuffisants si figurent sur la maquette des éléments profilés (sphères, cylindres...) dans le cas où l'échelle de réduction est trop grande.

b) La forme des profils de vitesse moyenne U (z) et d'intensité de turbulence $I = \sigma_u/U$ (z) en des zones relativement plates et uniformes de la maquette (σ_u étant l'écart-type des fluctuations de vitesse autour de la valeur moyenne de celle-ci), caractérisée soit par un exposant α p pour un profil en loi puissance, soit par une vitesse de frottement U_* et une rugosité Z_0 dans le cas d'un profil logarithmique. Pour une stabilité donnée (ici la neutralité), ces profils normalisés par rapport à une vitesse de référence dépendent essentiellement de la rugosité Z_0 . Les profils désirés - compte tenu du type de terrain parcouru par le vent avant d'atteindre ces zones - sont obtenus grâce à l'emploi de maquettes et d'avant-maquettes rugueuses et de générateurs de turbulence.

Grâce à ces systèmes, il est possible d'obtenir les gammes de paramètres suivants dans notre canal :

$$\delta \leqslant 0,15 \text{ m}$$
$$0,12 \leqslant \alpha_p \leqslant 0,55$$
$$0,15 \leqslant I_{sol} \leqslant 0,30$$
$$0,03 \leqslant I_\delta \leqslant 0,05$$

c) L'épaisseur de la couche limite est basée sur des formules du type de celle que proposent Blackadar et Tennekes (1) :

$$\delta = \frac{U_*}{4f}$$

où $f = 10^{-4}$ s^{-1} est le paramètre de Coriolis.

Avec une valeur typique de $U_* = 0.2$ m/s (vent modéré), on obtient $\delta = 500$ m, valeur proche de celle que recommande Counihan (2) ($\delta \simeq 600$ m quelles que soient la rugosité et la vitesse du vent).

Ces deux derniers paragraphes conduisent à considérer le rapport δ/Z_0 comme paramètre de similitude.

La veine hydraulique de Magny-les-Hameaux possède des dimensions relativement modestes en comparaison d'autres installations :
- longueur : 7 m hors tout, dont 3,50 m de chambre d'expérience
- largeur : 1,20 m
- profondeur : 0,50 m

L'exploration des écoulements s'effectue à l'aide :
- de sondes à film chaud
- d'anémomètres à laser
- d'une méthode de visualisation dite du "plan lumineux" (8) permettant d'effectuer des coupes horizontales (Fig.1) ou verticales (Figure 2) de l'écoulement.

3. CONDITIONS EXPERIMENTALES ET RESULTATS

Les mesures présentées ici ont été effectuées sur une maquette au 1/5000 d'un site de la Vallée du Rhône (FRANCE).

Les résultats concernent :
- un profil de vitesse moyenne obtenu avec un vélocimètre à laser, en aval d'un relief (mais sur une portion plane de la maquette)
- le traitement numérique d'un enregistrement de la vitesse (sonde à film chaud) en un point P situé légèrement en aval du précédent,

Fig. 1 - Visualisation de l'écoulement : coupe horizontale

Fig. 2 - Visualisation de l'écoulement : coupe verticale

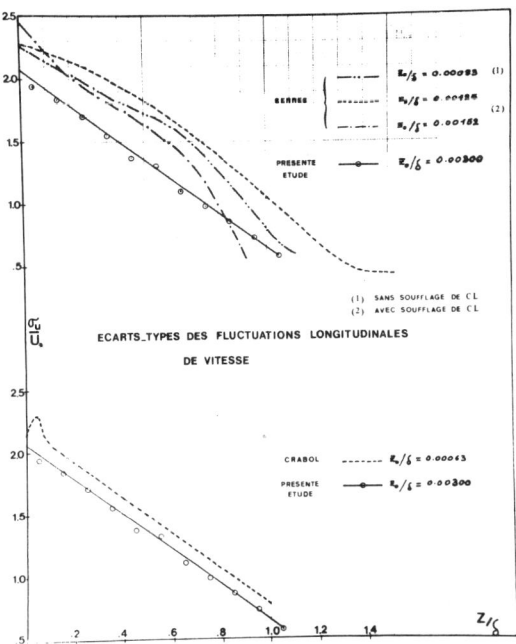

Fig. 3-4 — Variation de l'intensité de turbulence au
sein de la couche limite

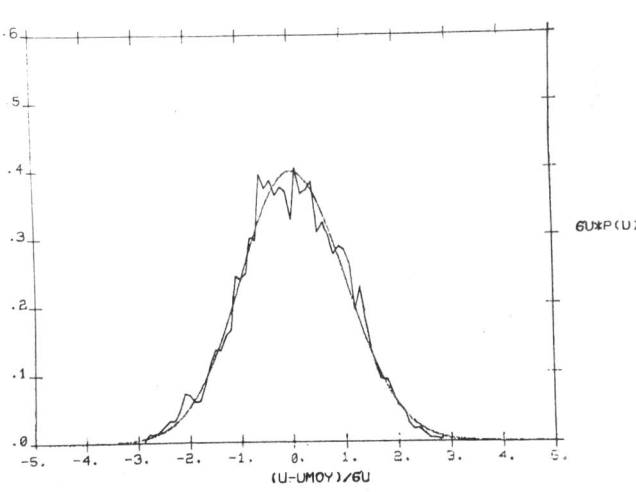

Fig. 5 — Densité de probabilité normalisée des
fluctuations de vitesse longitudinale

à une hauteur de 8 mm au-dessus de la maquette, correspondant à une hauteur de 40 m au-dessus du sol.

Les conditions expérimentales étaient les suivantes :
. U_∞ (vitesse hors couche limite) = 18.5 cm/s
. δ (épaisseur de la couche limite) = 10 cm, soit 500 m en réalité.

Les profils de vitesse moyenne et d'intensité de turbulence présentent les caractéristiques suivantes : mis sous forme de loi puissance, U (Z) le profil de vitesse fait apparaître un exposant voisin de 0,19. La vitesse de frottement U_* et la rugosité Z_0 sont estimées à l'aide d'un lissage du profil de vitesse au moyen de l'expression :
$$U (z) = 2.5 \; U_* . \text{Log} \frac{x}{Z_0}$$

On obtient ici :
$U_* = 1,273$ cm/s
$Z_0 = 0,03$ cm $\qquad (U_* . Z_0 / \nu = 3,82)$

Compte tenu de l'échelle de simulation, cette rugosité correspondrait à une rugosité-nature voisine de 1,5 m, cette valeur importante traduisant l'influence du relief amont (13).

L'intensité de turbulence, rapportée à la vitesse de frottement au lieu de la vitesse locale est représentée Fig.3 et Fig.4 ; elle est comparée respectivement à des résultats en soufflerie (11) et en hydraulique (3) obtenus à des échelles moins réduites.

On constate que les courbes trouvées lors de la présente étude sont systématiquement sous celles correspondant à ces études.

Ceci est dû au fait que la rugosité de notre maquette est plus forte, ainsi qu'en témoignent les valeurs du rapport Z_0 / δ :

SERRES (11)   : $1,25.10^{-3}$
CRABOL (3)   : $6,3 .10^{-4}$
Présente étude : $3. \quad 10^{-3}$

La densité de probabilité normalisée de la composante longitudinale de la vitesse en P ($\sigma_u . P$ (u) en fonction de $(u-\bar{u})/\sigma_u$ est représentée Figure 5, ainsi que la courbe de Gauss ayant le même écart-type. La dissymétrie ($\mu_3^2 \mu_2^{-3}$) est égale à 0,004 et l'aplatissement ($\mu_4 \mu_2^{-2}$) à 2,783 (contre respectivement 0 et 3 pour une courbe de Gauss).

Ce résultat est conforme aux observations de Serres : à la base d'une couche limite turbulente, la dissymétrie de la densité de probabilité de la vitesse est voisine de zéro. Elle prend des valeurs négatives croissantes en valeur absolue dès que l'intermittence de la turbulence commence à se faire sentir (Z > 0,5 δ). Quant au facteur d'aplatissement, il reste légèrement plus petit que trois dans la moitié inférieure de la couche limite.

Le spectre de turbulence normalisé (divisé par la variance de façon que $\int F_{(n)}/\sigma_u^2 \; dn = 1$ est présenté sur la Figure 6 en fonction de la fréquence réduite

On constate que pour $\frac{nz}{\bar{U}}$ variant de 10^{-1} à 3.10^{-1} , le spectre présente une pente voisine de $n^{-2/3}$ ainsi que le prévoit la théorie de Kolmogorov dans la zone inertielle. Au-delà de 7.10^{-1}, le spectre s'éloigne nettement de la courbe en puissance $-2/3$. On peut donc considérer que les structures turbulentes de dimensions longitudinales inférieures à 5000 $\times \frac{1}{10} (s^-) \times 11.5 (cm/s) \simeq$ 60m seront mal reproduites.

Rappelons cependant que la valeur $-2/3$ de cette pente est une condition nécessaire, mais non suffisante de l'existence d'un domaine inertiel.

La micro-échelle de Taylor λ_x calculée par la formule de Dryden :
$$\frac{1}{\lambda_x^2} = \frac{4\pi^2}{\bar{U}^2} \int_0^\infty n^2 \; F_{(n)} \, dn$$

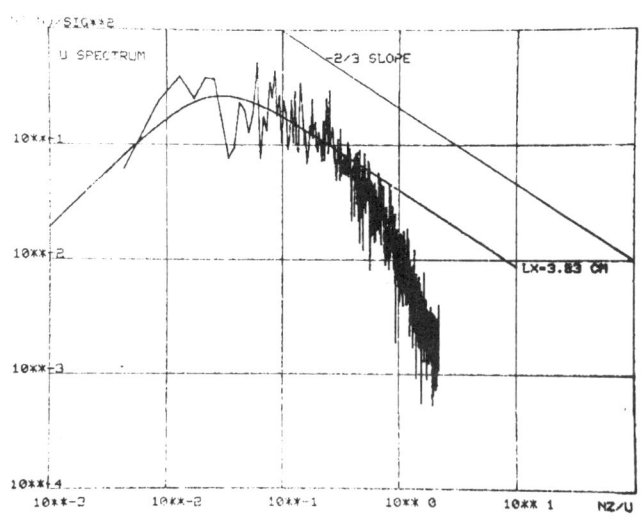

Fig. 6 – Spectre de turbulence de la composante longitudinale

s'élève à 3,3 mm, en bon accord avec les résultats de Raichlen (9).

Le nombre de Reynolds turbulent $R_\lambda = \frac{\lambda_x . \sigma_u}{\nu}$ est donc voisin de 70 (alors qu'il est de l'ordre de 5000 dans l'atmosphère), ce qui explique la faible étendue spectrale en $n^{-2/3}$. Sur la Figure 6, on a porté également le tracé du spectre de Karman couramment utilisé pour représenter sous forme analytique simple tant les spectres de laboratoire que les spectres atmosphériques (4) :

$$\frac{n \, F(n)}{\sigma_u^2} = \frac{4n \, \frac{L_x}{U}}{(1 + 70,8 \, \frac{n^2 L_x^2}{U^2})^{5/6}}$$

où L_x est l'échelle longitudinale de turbulence. Le début du spectre ($nZ/U > 0,3$), lissé par cette expression fournit une valeur égale à 3,83 cm correspondant à une échelle L_x = 191,5 m en réalité.

La formule préconisée par Counihan sur terrain plat donne (avec Z_0 = 1,5 m) :

Dans l'atmosphère, le calcul de l'échelle de turbulence est la plupart du temps effectué par intégration de la fonction d'autocorrélation R (t), mais en utilisant uniquement la partie positive de celle-ci avant le premier passage à zéro (4,5).

Cette méthode, appliquée aux résultats précédents, fournit une échelle de temps égale à 0,2021 s, et donc, si l'on applique l'hypothèse de Taylor (champ de turbulence figé), une échelle longitudinale égale à 0,2021 x U = 2,32 cm soit 116 m en vraie grandeur.

Cette valeur est alors tout à fait compatible avec la formule de Counihan (2) (et on aurait alors λ_x/L_x = 0,14).

Raichlen (9) et tout récemment Teunissen (12) ont déjà fait remarquer que l'estimation des échelles à partir de l'autocorrélation pouvait conduire à des différences significatives par rapport à celles dérivant des spectres.

4. POSSIBILITES D'ETUDES DE DIFFUSION

Outre la résolution de problèmes d'aérodynamique tels que l'influence de reliefs ou de structures diverses sur un écoulement, on est souvent amené à aborder en similitude la question de l'évolution à petite et moyenne échelle de polluants gazeux émis dans les basses couches de l'atmosphère. Le théorème de Taylor :

$$\overline{\chi^2(t)} = \overline{v^2} \, t^2 \int_0^\infty F_L(n) \left(\frac{\sin \pi n t}{\pi n t}\right)^2 dn$$

traduit le résultat fondamental suivant lequel la variance de la position de particules émises à partir d'une source ponctuelle continue (c'est-à-dire les dimensions du panache) dépend :
- de la variance des fluctuations turbulentes des composantes de la vitesse, variance reliée à l'intensité de turbulence I = $\sigma_u/\overline{u}$.

Les études passées ont confirmé que les profils verticaux de cette quantité sont bien reproduits dans nos couches limites.
- du spectre d'énergie lagrangien de chaque composante de la vitesse, ce dernier étant toutefois pondéré par un filtre "passe-bas" dont la fréquence de coupure diminue avec le temps de transfert. Autrement dit, pour de petits temps de transfert, toutes les échelles de turbulence contribuent à la diffusion.

Au fur et à mesure que le temps de transfert augmente, les plus grandes échelles de turbulence (gros tourbillons - basses fréquences) dominent progressivement la dispersion. Les tourbillons dont les dimensions sont inférieures à la largeur du panache auront donc peu d'effet sur la dispersion de celui-ci (sauf très près de la source).

Compte tenu des plus petits tourbillons correctement reproduits au 1/5000 dans nos installations, de l'ordre de 60 m, on ne peut donc envisager de simuler que des émissions ayant au moins cet ordre de grandeur - les réfrigérants atmosphériques des centrales nucléaires par exemple. Il convient alors de s'assurer que les paramètres de similitude de l'émission peuvent être respectés.

5. CONCLUSIONS

A des échelles de réduction voisines de 1/5000, les couches limites de laboratoire simulant la couche limite atmosphérique présentent des profils de vitesse et d'intensité de turbulence en similitude moyennant quelques précautions sur la valeur du nombre de Reynolds et de la rugosité du modèle. Les moments d'ordre supérieur (dissymétrie, aplatissement) semblent également correctement reproduits. Par contre dans le domaine spectral, seules les basses fréquences (gros tourbillons, ici > 60 m) sont simulées avec une puissance spectrale comparable à la réalité. Il en résulte donc que l'on ne pourra étudier des phénomènes de diffusion que dans le cas où les dimensions géométriques de l'émission sont au moins de l'ordre de grandeur des plus petits tourbillons reproduits correctement sur le modèle.

BIBLIOGRAPHIE

1 - BLACKADAR A.K. et TENNEKES H (1968) : Asymptotic similarity in neutral barotropic atmospheric boundary layer". Journal of Atmospheric Sciences, Vol 25, pp 1015-1020

2 - COUIHAN J. (1975) : "Adiabatic atmospheric boundary layer : a review and analysis of data from the period 1880-1972". Atmospheric Environment, Vol 9, N°10, pp 871-905.

3 - CRABOL B. (1978) : "Contribution à l'étude de la simulation en laboratoire des transferts de masse en atmosphère neutre". Thèse de doctorat de 3ème cycle, Université Pierre et Marie Curie - PARIS VIème.

4 - DUCHENE-MARULLAZ P. (1978) : "Extraits du compte rendu d'activité du CSTB en 1977"- Climatologie - Cahiers du CSTB N°1496, Livraison 188, Avril 1978 pp 335-337.

5 - KAIMAL J.C. (1973) : "Turbulence spectra, length scales and structure parameters in the stable surface layer". Boundary-layer Meteorology, Vol 4, pp 289-309.

6 - MERONEY R.N. (1980) : "Wind-tunnel simulation of the flow over hills and complex terrain". Journal of Industrial Aerodynamics, Vol 5 pp 297-321.

7 - MERONEY R.N. (1980) : "Physical simulation of dispersion in complex terrain and valley drainage flow situations". 11th ITM on Air Pollution Modeling and its applications. NATO/CCMS, Amsterdam.

8 - NICLOT C. (1980) : "Méthode d'analyse des caractéristiques dynamiques du vent en similitude hydraulique". Note EERM N°54.

9 - RAICHLEN F. (1967) : "Some turbulence measurements in water". Journal of Engineering,Mechanics Division, pp 73-97.

10 - SCHON J.P., REY C., MERY P. (1976) : "La simulation des basses couches de l'atmosphère ; possibilités et limites". XIVème Journées de l'Hydraulique. SHF.

11 - SERRES E. (1978) : "Etude de la simulation en soufflerie des basses couches de l'atmosphère. Application à la prévision de l'impact d'un site industriel sur l'environnement". Thèse de Docteur-Ingénieur, Université Claude Bernard de Lyon.

12 - TEUNISSEN H.W. (1980) : "Structure of mean winds and turbulence in the planetary boundary layer over rural terrain". Boundary-layer Meteorology,Vol. 19 pp 187-221.

13 - THOMPSON R.S. (1978) : "Note on the Aerodynamic Roughness length for complex terrain". Journal Applied Meteo, Vol 17, pp 1402-1403.

THE PRESENCE OF PAN IN LONG-RANGE TRANSPORTED POLLUTED AIR MASSES

P. GRENNFELT and U. SAMUELSSON
Swedish Water and Air Pollution Research Institute,
Göteborg, Sweden
T. NIELSEN
Chemistry Department, Risø National Laboratory,
Roskilde, Denmark
E.L. THOMSEN
Air Pollution Laboratory, National Agency of Environmental
Protection, Roskilde, Denmark

Summary

PAN was monitored in Göteborg and at Risø 20 km west of Copenhagen during the summer of 1980. At both sites the PAN concentrations were mostly below than 1 ppb. At some occasions it rose to 3-4 ppb. The PAN concentration at the two sites showed a very good covariation. Evaluation of the data with respect to meteorological conditions and the concentration of other pollutants showed that most of the PAN episodes were associated with long-range transport of polluted air masses.

1. INTRODUCTION

Besides ozone, peroxyacetyl nitrate (PAN) has been considered the most essential photochemical oxidant, and it is often considered to be a better photochemical pollution indicator than ozone. The reason for this is that high ozone concentrations may be produced by natural sources as well.

Increased ozone concentrations due to photochemical air pollution have been observed at many places in Scandinavia (1,2,3). The reason for these episodes has mainly been large scale photochemical oxidant formation during high pressure episodes. On these occasions ozone levels up to 150-200 ppb have been monitored. Very few episodes of substantial local oxidant formation have been observed.

In order to further evaluate the origin and consequences of the oxidant episodes in Scandinavia, PAN was monitored at one site in Denmark and one in Sweden during the summer of 1980. These measurements appear to be the first PAN monitoring in Scandinavia and the northernmost measurements undertaken so far.

2. EXPERIMENTAL

PAN was monitored at Risø, a rural area, about 20 km
west of Copenhagen and in Göteborg (Figure 1). The distance
between the two sites is 230 km. PAN was analyzed by electron
capture gas chromatography. The analyzing conditions were as
follows: Column: 120 cm x 4.6 mm o.d. glass tubing packed
with 5% Carbowax 400 on Chromosorb W-AW-DCMS (100-120 mesh).
Column temperature: ambient. Carrier gas: purified nitrogen.
Flow rate: 35 ml/min. Detection limit for PAN was 0.02 ppb
at Risø (5 ml air sample) and 0.1 ppb in Göteborg (2 ml air
sample). At Risø PAN was monitored throughout working hours
on 44 days in the period June 11 - September 23. The air
masses were manually injected. In Göteborg, the PAN concentrations were determined in the period July 17 - September 9,
using automated valve injection at a height of 3 m above
ground level. Analyses were performed every hour except for
a total of 97 hours. The experimental details concerning the
preparation, characterization and calibration of PAN standards are described elsewhere (Nielsen, T., Hansen, A.M. and
Thomsen, E.L., submitted for publication). For the evaluation
of the PAN data, meteorological and air pollution data from
places around Göteborg and at Risø were used. More details
are presented in Figure 1.

3. RESULTS AND DISCUSSION

The daily maximum PAN concentrations at both places
were in the range of 0.1-4 ppb. The daily maximum concentrations were 1 ppb or more on 12 days at Risø and on 10
days in Göteborg. The highest observed PAN concentration
during the total monitoring period was 4.2 ppb at Risø and
3.5 ppb in Göteborg. Despite the distance between the two
monitoring sites (230 km) the levels of PAN were very similar
and a significant correlation between daily maximum levels
was obtained (r = 0.72; p <0.001).
In order to explore the origin for the monitored PAN
concentrations the PAN data were analyzed with respect to
the meteorological situation and the occurrence of other
pollutants.
Since local PAN formation is favoured by high solar
radiation and low wind velocities, the PAN concentrations in
Göteborg (daily maximum and daily mean values)were compared
with cloud cover, wind velocity and temperature by means of
correlation coefficient calculations. These calculations
showed no correlation (/r/ = 0.05-0.19; p >0.05). Similar
results were obtained at Risø, where the PAN concentration
(daily maximum value) was compared to the number of sun
hours, wind speed and temperature (/r/ = 0.05-0.22; p >0.05).
The results indicate that the observed PAN concentration
might have been affected by other sources than local
formation.
The covariation between the PAN concentrations was
studied by drawing a wind concentration diagram for the PAN
concentration and the ground wind direction in Göteborg

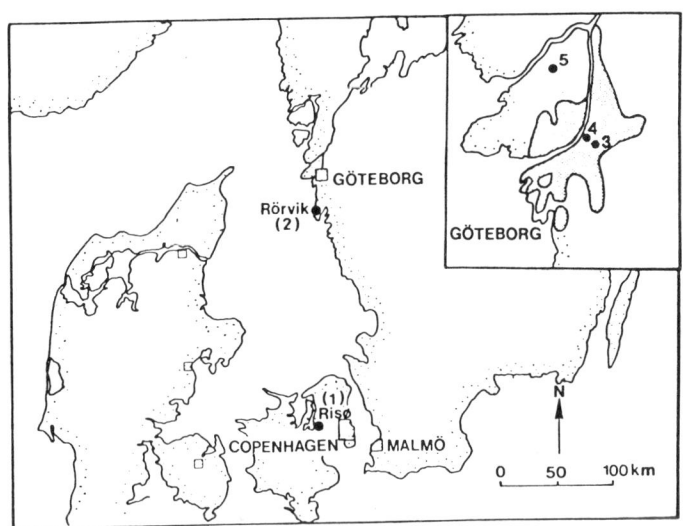

Figure 1. Map of Denmark and west Sweden showing the positions
of monitoring and meteorological stations. In the
corner a detailed map over the Göteborg area. PAN
was monitored at station 1 and 3. At site 2 (Rörvik),
"background" ozone and nitrogen dioxide concentrations and size-fractioned particle number were monitored. At site 4 local air pollution data for Göteborg was monitored as well as local wind situation
(20 m above street level). Remaining meteorological
parameters were collected at site 5.

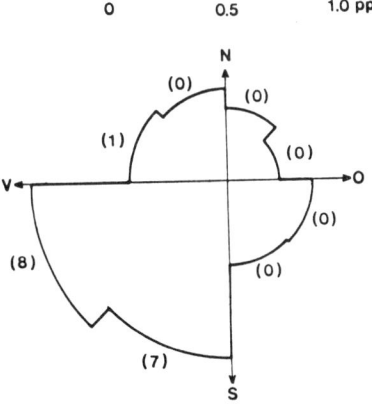

Figure 2. Wind-concentration diagram for average PAN concentrations in Göteborg for the period August 13 -
September 9. The numbers in brackets denote the
percentage of cases with the concentration of
PAN $\geq$ 2 ppb.

(Figure 2). This shows that the PAN concentration was significantly higher when the wind was blowing from the sector
180° - 270° than from other wind directions. Similar results
were obtained when a corresponding diagram was drawn for the
PAN concentrations and the 96 hours trajectories ending in
Göteborg at 07 and 19 MEST (Middle European Summer Time).
These calculations excluded the few cases, when trajectories
showed a successive curve or when they terminated only a small
distance from Göteborg.

The PAN concentrations in Göteborg showed a remarkably
small variation throughout the day. For the period August 13-
September 9, when all the occasions with PAN concentrations
above 1 ppb occurred, the mean afternoon level in Göteborg
was 0.62 ppb (12-16 MEST) and the average night-time level
was 0.50 ppb (23-05 MEST). It is reasonable to assume that
the formation rate of PAN is much larger at day-time than at
night, even if radicals persisting in photochemically
polluted air during the night, e.g. nitrogen trioxide, may
promote the formation of PAN (4,5). Using the rate constants
for chemical decomposition of PAN in the atmosphere (6) and
the data on observed diurnal variations of nitrogen monoxide,
nitrogen dioxide, temperature and the residence time of air
masses in Göteborg area (estimated from wind velocity observations), the depletion of PAN within Göteborg was estimated to be less than 10% at day-time as well as at night-
time. Thus the small diurnal variation was not caused by
high day-time depletion. Examination of the wind distribution
 at night and at day-time excludes differences in the wind
distribution as an explanation. Also the possibility of
stagnant air masses in Göteborg could be excluded. The small
diurnal variations, therefore, indicates that the transport
of PAN into Göteborg has been far more important than local
formation. In fact, most of the PAN recorded during the
night has probably been formed before sunset. This suggests
that the dominant sources for PAN formation are located more
than 200 km from Göteborg.

During the monitoring period three marked episodes with
increased PAN levels were observed which could be ascribed
to long range transport of pollutants. One of these will be
described further. This occurred during the period September
3-4, 1980. Figure 3 shows the relationships between the concentrations of PAN and ozone, the flux of photochemically
polluted air in over the Swedish west coast and its effect
of the visibility. At Risø, contributions from the main local
source, Copenhagen, are excluded by the south westerly wind
directions prevailing during the episode. The 850 mbar trajectories for air masses arriving at Risø and at Rörvik at
07 and 19 MEST show that the air masses had passed England
or the north-western part of the Continent (Figure 4).

The possibility of local PAN formation was thoroughly
evaluated by analyzing the local meteorological situation at
the two sites during the monitoring period. In Göteborg local
formation during typical land-sea breeze situations seems to
have occurred during two consecutive days. During these days
the maximum PAN concentrations were 1.0 and 1.7 ppb, respec-

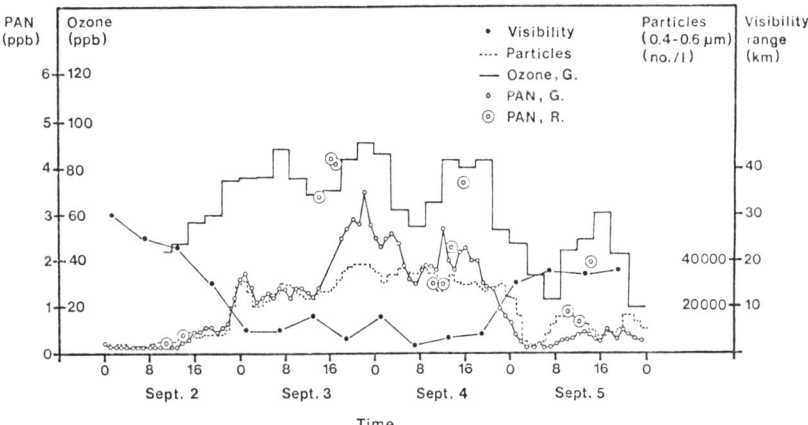

Figure 3. The variation of PAN and other parameters in
Göteborg (G) and at Risø (R) for the period
September 2-5, 1980.

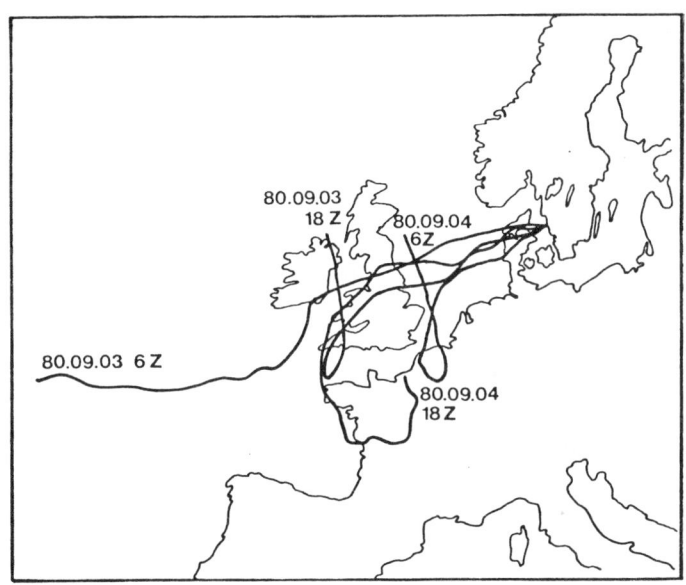

Figure 4. Trajectories (850 mb 96 h) for Rörvik
September 3-4 (0700 and 1900 MEST). The
trajectories indicate only roughly the transport
situation for the actual period.The trajectories
were calculated at the Norwegian Meteorological
Institute.

tively. The maximum ozone concentrations on these same days were 51 and 50 ppb in Göteborg and 56 and 73 ppb at Rörvik.

4. FINAL REMARKS

The PAN concentrations observed in Scandinavia seem to be more related to large scale formation and transport of pollutants than to local sources. These results are in good agreement with those observed for ozone, i.e. high concentrations seem to be formed mostly by large scale photochemical processes.

The monitoring of PAN occurred during a period with only a few situations favourable to the local formation of oxidants as well as to the formation during long-range transport. Moreover, the maximum ozone concentration was below what has been monitored in earlier periods. It seems possible, therefore, that more frequent PAN episodes and higher PAN concentrations than those monitored in the present investigation might occur in Scandinavia.

5. REFERENCES

(1) Grennfelt, P. (1976). Ozone episodes on the Swedish west coast. Proc.Int.Conf. on Photochemical Pollution and its Control. EPA 600/3-77-001a, 329-337 Research Triangle Park, North Carolina, USA.

(2) Schjoldager, J. (1980). Ambient ozone measurements in Norway 1975-1979. APCA paper, presented at the 73rd meeting of the Air Pollution Control Ass., Montr. Canada.

(3) Schjoldager, J., Dovland, H., Grennfelt,P. and Saltbones, J. (1981). Photochemical oxidants in north-western Europe 1976-79. A Pilot Project. Norwegian Institute for Air Research. Report 19/81, Lillestrøm.

(4) Platt, U., Perner, D., Winer, A.M., Harris, G.W. and Pitts, J.N.Jr.,(1980). Geophys.Res.Lett. 7, 89-92.

(5) Demerjian, K.L., Kerr, J.A. and Calvert, J,G.(1974). Adv. Environ. Sci. Technol. 4, 1-262.

(6) Cox, R.A. and Roffey, M.J.(1977). Environ. Sci. Technol. 11, 900-906.

THE TURNOVER OF SULPHUR DIOXIDE AND NITROGEN OXIDES IN THE ATMOSPHERIC BOUNDARY LAYER

Ø. Hov
Norwegian Institute for Air Research
P.O. Box 130, N-2001 Lillestrøm, Norway

Summary

An atmospheric boundary layer model is developed to study the vertical distribution and turnover of hydrocarbons, nitrogen oxides and sulphur dioxide in a situation with good convective mixing throughout the day. A log-linear grid is introduced with good resolution ($\Delta z \approx 1$ m) close to the ground, coarser towards the free troposphere. The degradation of 5 precursor non methane hydrocarbons is described, involving 40 chemical species and 80 chemical reactions. Hydroxyl has maximum concentration close to the ground, with some fall off with height. The maximum conversion rates of NO_x (sum of NO and NO_2) and SO_2 are approx. 32 and 4% hr^{-1} near the ground, corresponding to characteristic life times of approximately 6 and 50 hrs, respectively, in the surface layer. The production of nitric acid and sulphate aerosol is highly height dependent with a peak close to the ground. It is demonstrated that these vertical gradients are caused by the concentration of precursor sources near the ground.

1. INTRODUCTION

Laboratory and field evidence demonstrate that the photochemical gas phase oxidation of sulphur dioxide (SO_2) is an important, if not the most important, oxidation route for SO_2 in the atmosphere (1). The reaction with hydroxyl radical (OH) converts SO_2 at a rate of typically a few per cent per hour during the sunlit part of the day. Nitrogen dioxide (NO_2) reacts about an order of magnitude faster with OH than does SO_2, making photochemical gas phase oxidation the dominant atmospheric sink for nitrogen oxides (NO_x = NO + NO_2).

The vertical structure in the atmospheric boundary layer (ABL) of SO_2 and NO_x and the rate of formation of sulphate aerosol and nitric acid (HNO_3), will be studied in this paper. This will be done for a good weather situation when the vertical transport in the ABL is determined by the heating and cooling at the earth's surface throughout the day. A moderately polluted situation will be considered.

2. ABL MODEL

Vertical structure is of primary interest when studying the composition and chemical turnover in the ABL, because of the location of sources and sinks of turbulent energy and

chemical species at the lower boundary. Pronounced vertical
structure is observed for a number of important chemical
species, like ozone (2, 3, 4), SO_2 and sulphate aerosol (5, 6)
and nitrogen oxides (7, 8). Previous photochemical model
studies of the ABL have assumed uniform mixing vertically (9)
or applied some rough parameterisation (10) or introduced a
very coarse grid (11) to resolve the vertical concentration
gradients.

The model applied in the work reported in this paper, is
based on the requirement that each chemical specie satisfies
the continuity equation written on the form

$$\frac{\partial c}{\partial t} = \frac{\partial}{\partial z} MK_z \frac{\partial}{\partial z} \frac{c}{M} + P - Lc = \left(\frac{\partial c}{\partial t}\right)_d + \left(\frac{\partial c}{\partial t}\right)_{ch} \qquad (1)$$

where c is the concentration, K_z vertical eddy diffusion coefficient, P and L·c chemical production and sink terms (both
homogeneous and heterogeneous), emission and ground deposition,
and M is the air density at the given height. Subscripts d and
ch denote diffusion and chemistry, respectively.

Various assumptions are invoked to derive eq. 1, most
important horizontal homogeneity and the approximation that
turbulent transport is proportional with the mean concentration
field gradient. The assumptions leading to eq. 1 are summarized
in (12).

To obtain high resolution close to the ground, a coordinate transformation is introduced:

$$\zeta(z) = \frac{1}{k} \ln \left(\frac{z+z_o}{z_o}\right) + \frac{z}{\lambda_t} \qquad (2)$$

where k is von Karman's constant, z_o the roughness length and
λ_t a length scale which equals 40 m (13). With constant spacing
in ζ, eq. 2 gives a log-linear mesh in z. In the model there is
20 grid points between the lower boundary at 1 m and the top at
2 km, (see Fig. 1). The lowest level at 1 m is chosen because
this is usually the reference height when ground deposition
fluxes of airborne species (14) are estimated.

As upper boundary condition for eq. 1 is taken zero flux
through the 2 km height level. At the lower boundary the flux
equals the rate of deposition (v_g·c where v_g is the deposition
velocity of a specie with concentration c at 1 m height). As
deposition velocities are taken: 0.6 cms^{-1} during daytime and
0.3 cms^{-1} during the night for ozone (15), 0.8 cms^{-1} for SO_2
(16), 0.5 cms^{-1} for NO_2 (60% of SO_2, Grennfelt private communication), 0.2 cms^{-1} for PAN (17), 0.6 cms^{-1} for HNO_3 (18) and
0.1 cms^{-1} for sulphate aerosol (19). Values as high as 1.0 and
0.5 cms^{-1} for HNO_3 and sulphate, respectively, and as low as
0.1 cms^{-1} for NO_2 (20) are also tested in the model calculations.

The computation of the ABL meteorology is done along the
same lines as reported in (21). The diurnally varying height
profiles of K_z, temperature and air density are calculated,
assuming that the transport is driven by the diurnal variation
of surface heating (summer time high pressure situation, see
(12) for details).

The K_z profiles used when solving eq. 1, are shown in
Fig. 1.

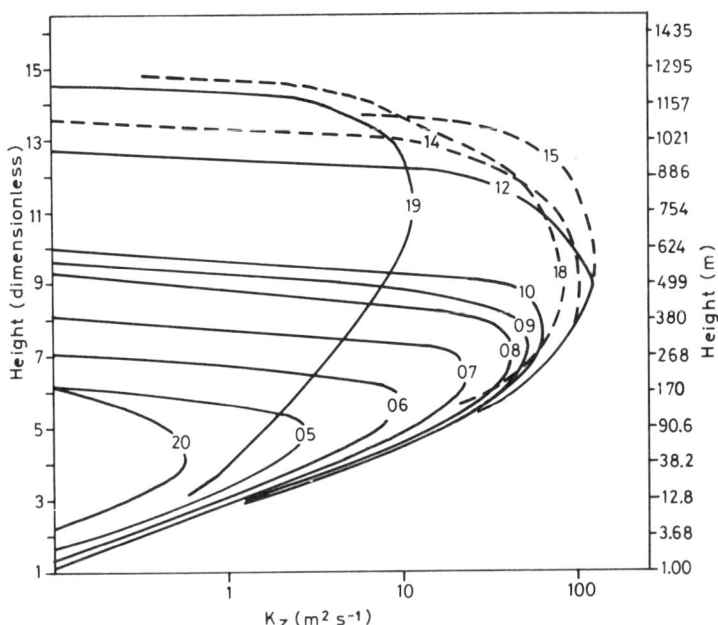

Figure 1: Calculated K_z profiles at various times of the day.

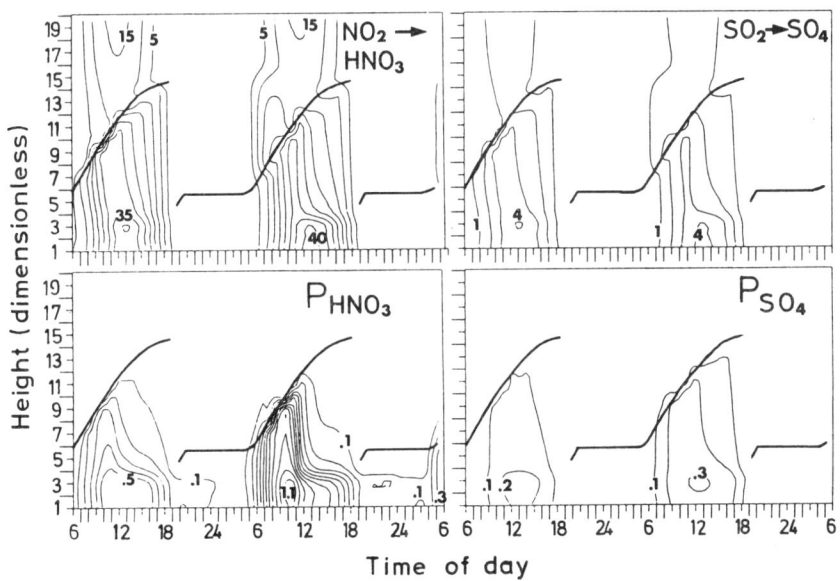

Figure 3: Rate of conversion of NO_2 and SO_2 through reaction with OH, in % hr^{-1}, together with rate of production of nitric acid and sulphate aerosol through the same reactions, in ppb hr^{-1}.

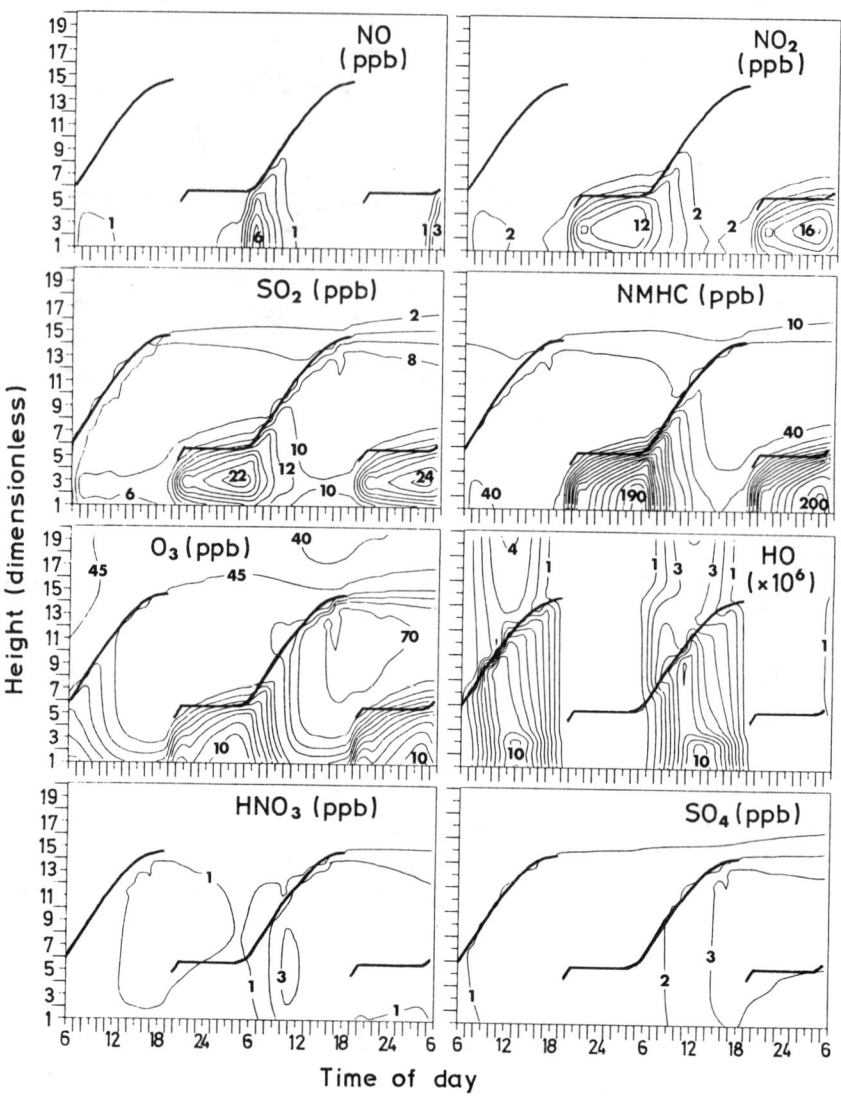

Figure 2: Contour plots of NO, NO$_2$, SO$_2$, non-methane hydro-carbons (NMHC), O$_3$, HNO$_3$ and sulphate aerosol in ppb, OH in 10^6 molecules cm^{-3}.

The chemical system which describes the chemical turnover, consists of approx. 40 species and 80 chemical reactions. Ten species are precursors: NO, NO_2, SO_2, CO, CH_4 and 5 non-methane hydrocarbons (NMHC)- C_2H_6, nC_4H_{10}, C_2H_4, C_3H_6 and m-xylene. The emissions of NMHC are distributed as 30% C_2H_6, 20% of each of nC_4H_{10}, C_2H_4 and m-xylene and 10% as C_3H_6 (by volume) (22). A detailed degradation of the various hydrocarbons is described in the model, together with a representative picture of the inorganic atmospheric chemistry based on recent data evaluations (23, 24, 25) and a comprehensive model of photochemical smog formation (26). Twelve species are photodissociated and diurnally varying height profiles of photodissociation rate coefficients for 50° latitude, mid summer are calculated using the method described in (27).

The numerical scheme used to solve eq. 1 is based on the Crank Nicholson implicit finite difference scheme combined with a modified version of a quasi steady state approximation method (28). The numerical methods used here, together with an assessment of numerical accuracy, is discussed at length in (29).

3. RESULTS, PRINCIPAL COMPONENTS

The intensity of precursor emissions is an influential factor when the accumulation of secondary pollutants is to be assessed. A NO_x flux of 3×10^{11} molecules $(cm^2 s)^{-1}$ (all as NO), NMHC and SO_2 fluxes 1.5 times the NO_x flux (by volume), and CO emissions 10 times the NO_x flux will be considered. These numbers are fairly close to a Eastern U.S. average (30). The precursor emissions are distributed at the 3 lowest grid points ($z \leq 12.8$ m).

The initial concentration field is established through a 2 days' model calculation with very low emissions of precursors (close to northern hemispheric average anthropogenic emissions). The results to be presented here start at 6 am and are carried on through two full diurnal cycles (48 hrs). In Fig. 2 is shown concentration contours as a function of height and time of day for 8 principal components: NO, NO_2, SO_2, NMHC, O_3, OH, HNO_3 and sulphate aerosol. Also shown is the calculated diurnal variation in mixing height. Primary species accumulate in the shallow nocturnal mixed layer. Species which are not deposited at the ground peak at the ground as the erosion of the low nocturnal inversion starts (NO, NHMC) while NO_2 and SO_2 which are removed at the surface, peak in the middle of the mixed layer. Ozone demonstrates the behaviour which is typical in a good weather situation in an ABL influenced by pollution (4): buildup during the day with a uniform level in the mixed layer and a fall off towards the ground due to deposition, trapping of ozone rich air above the inversion and erosion of ozone below because of the combined effect of deposition and removal through reaction with NO (12). As the convective elements gain intensity in the morning, ozone rich air from aloft is mixed to the ground at the same time as chemical generation takes places. Hydroxyl peaks around noon near the ground. Nitric acid and sulphate aerosol peak at approx. 10 and 13 µg/m³ (approx. 3 ppb of each), respectively, in the middle of the mixed layer.

Direct comparison with measurements is not possible, but the computed concentration ranges are fairly representative for what is found in a moderately polluted atmosphere. See (12) for a further discussion.

4. CONVERSION OF SO_2 AND NO_x

In Fig. 3 is shown the rate of conversion of SO_2 and NO_2 through reactions

$$NO_2 + OH \rightarrow HNO_3 \qquad k_i \quad = 1.0 \times 10^{-11} cm^3 \text{ (molecule s)}^{-1} \quad (25)$$

$$SO_2 + OH \rightarrow HSO_3 \qquad k_{ii} = 1.1 \times 10^{-12} cm^3 \text{ (molecule s)}^{-1} \quad (25)$$

in % hr^{-1}, together with the rate of formation of HNO_3 and sulphate aerosol in ppb hr^{-1}, as contour plots. Peak NO_2 to HNO_3 conversion is approximately 40% hr^{-1} (corresponding to approx. 32% hr^{-1} conversion of NO_x to HNO_3) and SO_2 to sulphate aerosol 4% hr^{-1}. These peak rates correspond to characteristic atmospheric loss times of about 3 hrs for NO_x and 25 hrs for SO_2, or 6 and 50 hrs if the diurnal average conversion is assumed to be 50% of the daily peak. These numbers apply only to the lowest deka meters. Further up in the ABL the conversion is slower due to lower OH-concentrations. This is even more evident in the lower part of Fig. 3 where HNO_3 and sulphate aerosol production in ppb hr^{-1} is plotted. The products $k_i [NO_2]$ [OH] and $k_{ii} [SO_2]$ [OH] then enter the expression, and all these concentrations peak near the ground. The absolute value of the nitric acid and sulphate aerosol production is therefore highly height dependent, even within an ABL efficiently mixed by convective motion. The vertical gradients are set up by the confinement of precursor sources to the lowest three grid points. If the assumption of instantaneous mixing of precursor emissions within the ABL is invoked, the vertical gradients apparent from Fig. 3 in particular through the afternoon, would vanish (Fig. 4 e, f). In Fig. 4 the effect on the SO_2 to sulphate conversion rate of a perturbation of a couple of model parameters is demonstrated. A lowering of the NO_2 deposition from .5 cms^{-1} to .1 cms^{-1} is seen to lower the conversion rate, because the NO_2 level is then higher and the loss of hydroxyl consequently more efficient. A doubling of the initial CO level (from 125 ppb to 250 uniform with height) has a similar effect. Calculations done for 60° latitude instead of 50°N cause a reduction in OH and hence also a slightly slower SO_2 conversion rate. If the precursor emissions are mixed instantaneously over the ABL, the vertical gradients in SO_2 conversion (Fig. 4 d), HNO_3 production (Fig. 4 e) and hydroxyl (Fig. 4 f) are removed. It is evident that the vertical distribution of precursor sources is decisive for the shape of the vertical profiles of secondary species.

In Fig. 5 is shown vertical profiles of sulphate aerosol and nitric acid at 6 hours' intervals on the second day of integration. Sulphate is shown for two choices of deposition velocities: .5 and .1 cms^{-1}. In the latter case the characte-

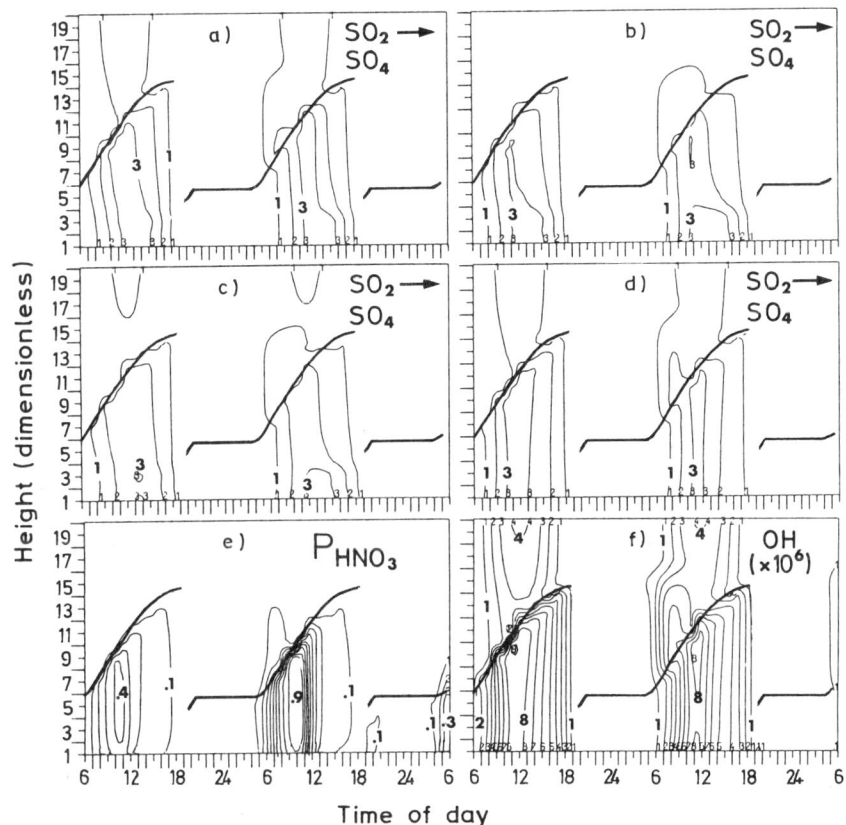

Figure 4: SO_2 to sulphate conversion rate in % hr^{-1} through reaction with OH in a situation where (a) ground removal of NO_2 is reduced to .1 cms^{-1}, (b) CO is doubled from 125 to 250 ppb initially, (c) at 60°N latitude, (d) emissions evenly distributed up to the mixing height, (e) rate of production of HNO_3 in ppb hr^{-1} in the same case as (d), (f) contours of the concentration of OH in 10^6 molecules cm^{-3}, in the same case as (d).

Figure 5: Height profiles of nitric acid and sulphate at selected times of the day, and with two choices of ground removal for sulphate (.5 and .1 cms^{-1}). The location of the mixing height is indicated by horizontal bars.

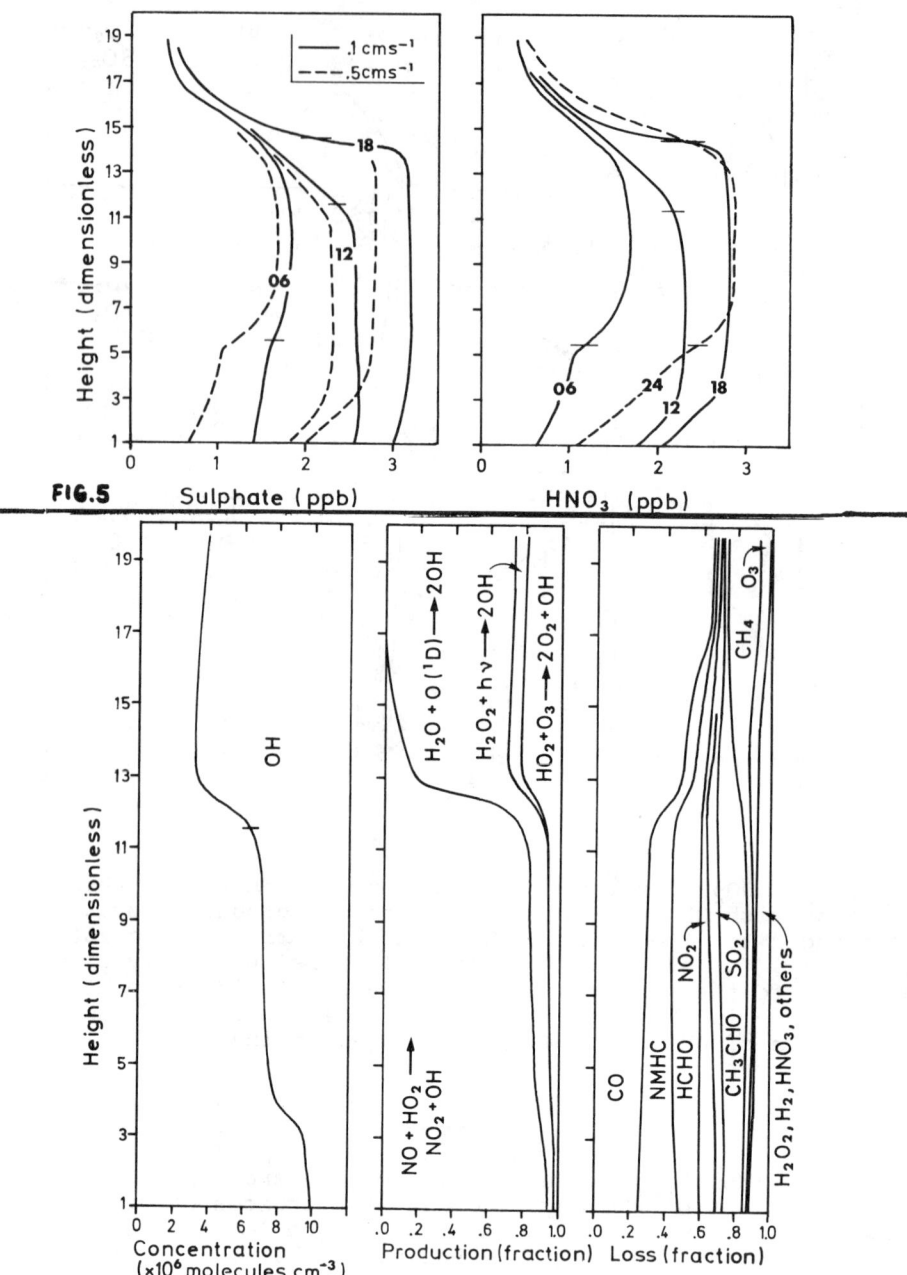

FIG.5

Sulphate (ppb)

HNO₃ (ppb)

Figure 6: Hydroxyl balance at noon on the second day of integration. To the left is shown the concentration profile with height, in the middle the relative significance of the various production reactions, to the right the relative influence of the sink processes.

ristic removal time through deposition is so long that the
concentration profile is hardly affected. In the first case,
however, the characteristic removal time is approx. 7 hrs at
night (mixing height about 125 m), in which case the sulphate
aerosol present at the surface is efficiently depleted. The
same effect is seen for HNO_3 where the ground deposition velocity is .6 cms^{-1} (18). During daytime the concentration profiles of sulphate and HNO_3 are quite uniform with height.

5. HYDROXYL CHEMISTRY

The abundance of hydroxyl is crucial for the efficiency
of the turnover of NO_x and SO_2. In Fig. 6 is given a schematic
picture of the processes affecting the OH-balance at noon on
the second day of integration, as a function of height. The
concentration profile is shown to the left, with three different height regimes with respect to concentration: Peak concentration in the lowest part of the ABL where the precursor
emissions are located, fall off and fairly uniform concentration
up to a level limited by the mixing height, and then again
quite uniform, but lower, concentration above the mixed layer.

To the right in Fig. 6 the relative importance of the
various production and loss processes of OH are shown as a
function of height. The reaction

$$NO + HO_2 \rightarrow OH + NO_2 \qquad (iii)$$

is the dominating production mechanism within the mixed layer.
Above the mixed layer

$$H_2O + O(^1D) \rightarrow 2OH \qquad (iv)$$

dominates.

The loss is fairly evenly distributed among OH reactions
with CO(v), NMHC(vi), HCHO(vii), NO_2(i), SO_2(ii) and CH_3CHO
(viii) within the mixed layer, further up reaction (v) and
reaction with CH_4 dominates.

It is worth nothing that in the model, the concentration
of NO_x and anthropogenic hydrocarbons is very low above the
mixed layer. Consequently, the role of reactions (iv) and (v)
is overestimated here compared to a more realistic atmosphere.

There is a principal difference between e.g. reaction
(iii) and reactions (i) and (iv). (iv) is a production mechanism for odd hydrogen (OH, HO_2 or H) while (i) is a sink.
Reaction (iii) is only a redistribution process of odd hydrogen.
The concentration of hydroperoxy radical usually dominates
that of hydroxyl by a factor 10-1000, however, making it
essential to keep track of all redistribution processes among
odd hydrogen species to assess the OH-balance, and not only
the processes affecting the overall odd hydrogen balance.

ACKNOWLEDGEMENT

Discussions with professor Ivar S.A. Isaksen, Institute
of Geophysics, University of Oslo, are gratefully acknowledged.

REFERENCES

(1) ISSA (1978) *Atmospheric Environment* 12, 10-12.
(2) Galbally, I. (1968) *Nature* 218, 456-457.
(3) Galbally, I. (1971) *Atmospheric Environment* 5, 15-25.
(4) EPA (1975) EPA-450/3-75-036.
(5) Garland, J.A. and Branson, J.R. (1976) *Atmospheric Environment* 10, 353-362.
(6) Georgii, H.W. (1978) *Atmospheric Environment* 12, 681-690.
(7) Husar, R.B., Patterson, D.E., Blumenthal, D.L., White, W.H. and Smith, T.B. (1977) *J. Appl. Meteor* 16, 1089-1096.
(8) White, W.H., Blumenthal, D.L., Anderson, J.A., Husar, R.B. and Wilson, W.E. Jr. (1977) EPA-600/3-77-001a.
(9) Graedel, T.E., Farrow, L.A. and Weber, T.A. (1976) *Atmospheric Environment* 10, 1095-1116.
(10) Mac Cracken, M.C., Wuebbles, D.J., Walton, J.J., Duewer, W.H. and Grant, K.E. (1978) *J. Appl. Meteor* 17, 254-273.
(11) Schiavone, J.A. and Graedel, T.E. (1981) *Atmospheric Environment* 15, 163-176.
(12) Hov, Ø. (1981) One dimensional vertical model for ozone and other gases in the atmospheric boundary layer. Submitted for publication.
(13) Taylor, P.A. (1969) *J. Atm. Sci.* 26, 427-431.
(14) Sehmel, G.A. (1980) *Atmospheric Environment* 14, 983-1011.
(15) Garland, J.A. and Derwent, R.G. (1979) *Quart. J. Roy. Met. Soc.* 105, 169-183.
(16) Garland, J.A. (1977) *Proc.R.Soc.Lond.A.* 354, 245-268.
(17) Garland, J.A. and Penkett, S.A. (1976) *Atmospheric Environment* 10, 1127-1131.
(18) Huebert, B.J. and Lazrus, A.L. (1978) *Geophys. Res. Lett.* 5, 577-580.
(19) Garland, J.A. (1978) *Atmospheric Environment* 12, 349-362.
(20) Bøttger, A., Ehhalt, D.H. and Gravenhorst, G. (1978) Berichte der Kernforschungsanlage Jülich, Nr. 1558, KFA Jülich, West Germany.
(21) Busch, N.E., Chang, S.W. and Anthes, R.A. (1976) *J.Appl.Meteor.* 15, 909-919.
(22) Eliassen, A., Hov, Ø., Isaksen, I.S.A., Saltbones, J. and Stordal, F. (1981) A Lagrangian long range transport model with atmospheric boundary layer chemistry. Submitted to J. Appl. Meteor.
(23) Hampson, R.F. and Garvin, D. (1978) NBS. Special Publication 513, Washington, D.C.
(24) Atkinson, R., Darnall, K.R., Lloyd, A.C., Winer, A.M. and Pitts, J.N., Jr. (1979) *Advances in Photochemistry* 11, 375-488. John Wiley & Sons.
(25) NASA (1981) JPL, California Institute of Technology, Pasadena. JPL-Publication 81-3.
(26) Derwent, R.G. and Hov, Ø. (1980) *Environ. Sci. Technol.* 14, 1360-1366.
(27) Isaksen, I.S.A., Midtbø, K.H., Sunde, J. and Crutzen, P.J. (1977) *Geophysica Norvegica* 31, 11-26.
(28) Hesstvedt, E., Hov, Ø. and Isaksen, I.S.A. (1978) *Int.J.Chem.Kinet.* 10, 971-994.
(29) Hov, Ø. (1981) Numerical solution of the diffusion equation for chemically reactive atmospheric species. Submitted for publication.
(30) Clark, T.L. (1980) *Atmospheric Environment* 14, 197-226.

CHAIRMEN'S SUMMARIES

IDENTIFICATION AND ANALYSIS OF POLLUTANTS

A. Liberti

The increasing awareness of the complexity of the atmosphe
ric environment renders the analytical work a challenging field.
There is therefore a variety of contributions to exploit new
techniques and new instruments to obtain basic information
which may clarify the meaning of pollution indices conventional
ly measured to evaluate air quality.
The contributions related to this topic may be classified
according to the following lines:
- development of new instruments and apparatus
- improvement of methods for the determination of pollutants
- study of analytical procedures for specific problems.
Most of them are related to two research areas, which are
the evaluation of particulated matter and the identification
and determination of organic compounds in the environment.
The main achievements are shortly summarized.
A noticeable improvement to the measurament of total suspended particulated matter, according to the gravimetric method, which is the standard procedure of some countries of the
european community, has been described. It makes use of a beta
rays gauge for mass measurement, which is already applied in
commercial devices, but has a number of interesting features
which overcome the limitations of available instruments. They
are a sensitive and accurate record of air flow and volume, cor
rected for variations of temperatura and flow. It is provided
with a feedback control to keep constant the flow rate at a
given value, according to various international regulations,
and a full automatization, which permits either to transmit or
store the data.
An airborne particulate can be characterized according to
its shape and its composition by laser microraman spectroscopy.
A microscope is used to select a particle and upon it a laser
beam is focussed; the backscattered radiation is analyzed in a
conventianl Laser Raman spectrometer through its Raman spectrum. This experimental device can be used not only to microprobe any material but also to follow chemical reactions of ga
ses on particles, which yield structural modification of the so
lid phase.
A new device has been developed to sample and analyse metals
present in atmosferic particulated matter. Air is pumped through an electrostatic precipitator; the dust ia directly precipi-
tated upon on electrothermal furnace (a simple graphite tube),
which in further set into an atomic absorption spectrometer and
analysed. The small volume of air which has to be sampled to
obtain a detectable signal and the high sensitivity obtained by

direct atomic absorption are the most interesting features of this device. It seems this sampling procedure might permit to carry on the determination of atmospheric metals in a fast way and with a high degree of sensitivity comparable to the sampling of organic compound by means of absorption traps thermally desorbed and analyzed by G.C.

The determination of OH in ambient air for the specific activity of this radical to react in the atmosphere with a large number of trace gases is of a great environmental importance. It has been shown this measurament can be carried out by long path differential absorption technique by using an argon ion laser optoacoustically modulated as light source.

It has been investigated the sampling of particulated matter carried out in order to remove the effect due to SO_2, which can react either with the filter material or the dust collected.

It has been shown this aim can be reached by the use of a denuder, costructed by internally covering an absorption tube with a mixture of tetrachoromercurate and maleic buffer.

An analytical procedure has been developed to determine nitrate in atmospheric aerosols by high volume air sampling on a glass fiber filter and further water elution. The sample is analyzed after Cd/Hg reduction; this procedure which is recommended in respect to the brucine method permirs to measure the aerosol nitrate in background regions.

The determination of organic compounds in the atmoshpere is by for the more complicate problems on account of the large number of species which may exhist at a trace level and on the contributions of natural as well of anthropogenic emissions. In most cases the cycle of these compounds is not known and cannot be desribed on account of the side reactions which might occur and the different sink mechanism they undergo either in gas phase or in absorbed state. Almost any area has a different organic compounds spectrum strongly modified by local conditions and emissions.

Typical examples of the variations of this spectrum are the determinations of specific classes of organic polluntants, as it has been shown in the evaluation of chlorine and sulfur containing molecules in atmospheric samples of various areas and of hydrocarbons in different urban sites. The experimental evidence for the presence of acrylicnitrile and acetonitrile at the ppt level has been reported.

It has been found very useful on a metodological stand point the use of the relative retention data from linear temperature programmed GC as a mean to identify chromatgraphic peaks and the use of combined detectors.

Toxic species arising from combustion of urban wastes as well from any incineration process have been evaluated, this problem being strictly related to the impact these processes might heve upon the environment. Investigations carried out in this line indicate that trace amounts of species such as polychlorodibenzodioxins (PCDD) and polychlorodibenzofurans are always found in the incinerator emissions (fumes, vapours, fly ash and sludge from scrubbers) in various amounts. Their origin is attributed to a pyrolitc synthesis of species indicated as precursors, which can be phenolic compounds and chorine donor species. Chlorophenols are produced which by heating yield the

above compounds. Though the process as a whole is well understood the figures reported on the content of PCDD and PCDF are
mainly indicative on account of the many variables affecting
the sampling, the extraction step and the analytical determina
tion which are not yet fully clarified. The determination of
the above compounds in atmospheric aerosols has not yet achieved due to technical difficulties in the sample collection.

A critical examination of the various projects outlines
the need for standardization of the sampling procedures and to
optimize analytical procedure for various polluntants. This
need is strongly felt for the determination of any species at
traces level and mainly when organic compounds have to be analysed.

CHEMICAL AND PHOTOCHEMICAL REACTIONS

R.A. Cox

1. Introduction

Working Group 2 of COST 61a-bis is concerned with the study of chemical and photochemical processes involving atmospheric pollutants. These relate primarily to the transformation of trace pollutants in the free atmosphere. Over the past two years two discussion meetings involving participants working in this field have been held, at AERE Harwell and at the University of Leuven, Belgium. The following summary is based on papers presented at these discussion meetings and at the 2nd European Symposium.

2. Nature of Pollutants Investigated

Chemical reactions involving a wide range of pollutants, as well as certain naturally present atmospheric species, particularly free radicals, have been studied. The species studied fall broadly into the following categories which are listed together with the relevant areas of atmospheric interest:

1. Organics	– Degradation/Persistence Natural Hydrocarbons Photochemical Oxidant formation
2. Nitrogen Oxides	– NO_2 pollution Background tropospheric Chemistry Photochemical Oxidant formation Acid Precipitation
3. Sulphur Compounds	– SO_2 dispersion/oxidation Acid Precipitation Global Sulphur Cycle The Atmospheric Aerosol
4. Miscellaneous Compounds	– e.g. PAN, Ozone, HO_2 radicals important in atmospheric cycles and as pollutants

3. Progress in Topics covered by Working Group 2

3.1 OH Radical Reactions

A considerable amount of work has been performed in projects in Working Group 2, on the development and implementation of methods for the determination of rate coefficients for OH radical reactions with organic compounds. Since OH radicals are the primary attacking species initiating the atmospheric transformation of a wide variety of organic and inorganic compounds, th OH rate coefficients are important physico-chemical parameters for any serious quantitative discussion of the atmospheric behaviour of these compounds. In particular the OH rate coefficients are used to assess (a) reactivity of organic compounds with respect to photo-oxidant formation and (b) degradation rates and lifetimes for volatile pollutants in the troposphere.

Two basic techniques have been used to determine OH rate coefficients i.e. direct measurement techniques and relative rate measurements. Over 50 new rate constant determinations have been performed in the COST 61a-bis programme, using these techniques.

3.1.1 Direct measurement techniques for OH rate coefficients

In these techniques time resolved measurements of the concentration of the reacting species (usually OH) are made. It is usual to arrange for the more stable reactant to be present in excess over OH so that OH decay exhibits pseudo first order kinetics. This has required the development of sensitive techniques which allow measurements of very low concentration of OH radicals, so that radical + radical reactions and secondary reactions of OH with primary reaction products do not complicate the kinetics. OH detection by resonance absorption has been used in Project D 40 to determine OH rate constants with a number of inorganic and organic species and the resonance fluorescence technique has been used in Project D1. The latter technique gives higher sensitivity whilst the absorption method is more suitable for high pressure work. Project D1 has investigated OH reactivity with a number of less volatile organic species including epoxides and organic acids, which is a new area presenting additional experimental problems which have been successfully overcome.

3.1.2 Relative rate measurements for OH reactions

These measurements rely on measurements of the rates of reaction of OH with an unknown compound relative to that with one for which the OH rate coefficient is known.

Systems in which the OH radical is in steady state are normally used, with rate determination from concentration-time behaviour of reactants of products. The most widely used system for OH reactions of atmospheric interest is the 'smog chamber' method with photochemical generation of a steady state OH concentration from photolysis of NO_2 and olefin + air mixtures (Projects D10, I3) or of HONO + air/N_2 mixtures. (Projects UK5, UK17). Both these sources have given satisfactory results but the HONO method offers some advantage inasmuch as the OH concentration generated is higher and there is consequently less problem from removal of substrates by other active species e.g. O (^{3}P) and O_3. An alternative source of OH radicals in this type of experiment is the thermal decomposition of peroxynitric acid in the presence of NO, which has been developed in Project D10. This is particularly useful for compounds which are photolabile themselves and cannot be easily investigated in photochemical systems. Gas chromatography and chemiluminescence were the main measurement techniques used in these studies but the new Fourier Transform Infra-Red facility installed in Project D10 is showing excellent potential for these studies.

3.? 3 Errors and reliability of OH rate measurements

At the Leuven Meeting there was some discussion of the errors and reliability of rate coefficients for OH and other free radical reactions. It was concluded that the techniques discussed above, when applied in the appropriate manner were capable of yielding data of good precision (repeatability to $\pm$ 5%) and with an overall experimental error of $\pm$ 30 %. The problem of systematic error was however always present and there are still a number of unresolved differences between measurements made in different laboratories. The sources of systematic error can only be discovered and avoided by continued investigation.

3.2 Organic Reaction Mechanisms

Information on the detailed decay mechanisms for organic compounds in the atmosphere is required for modelling the behaviour of these compounds with respect to formation of photo-oxidant and other secondary pollutants, and also for the elucidation of the oxidation or degradation products. Products may have important consequences from the toxicological or ecotoxicological point of view, other than those of the pollutants originally released.

Working Party 2 projects have concentrated mainly on the decay mechanisms of relatively simple hydrocarbons species e.g. Butadiene (Project I3), Ethene (Project B8), Aldehydes (UK 17), Ketones (UK 5) and Aromatics (NL7). The general picture that has emerged for the oxidation of simpler hydrocarbon species is that, following attack by an active species (e.g. OH, O (^{3}P)), the organic compound is converted to a radical, R$^{\cdot}$, which reacts with O_2 to form a peroxy radical i.e.

$$OH + RH \quad \rightarrow R^{\cdot} + H_2O \quad \text{abstraction}$$
$$\rightarrow R^{\cdot}HOH \quad \text{addition}$$
$$\overset{\bullet}{R}\text{ (or } R^{\cdot}HOH) + O_2 \quad \rightarrow RO_2$$

- 641 -

Under atmospheric conditions the peroxy radicals react rapidly with NO to form NO_2 together with an alkoxy radical, $RO^{\cdot}$

$$RO_2 + NO \rightarrow RO^{\cdot} + NO_2$$

The fate of the alkoxy radicals has been the subject of work in projects UK12 and UK5. The radicals may isomerise, decompose or react with O_2, the pathways depending on the structure and reactivity or the organic structural chain. Knowledge of this important aspect of the mechanism of degradation of organics is improving but far from complete.

The decay mechanism of aromatic hydrocarbons following radical attack has been the subject of study in Project NL7. The important finding is that rupture of the aromatic ring to form, among other products, α di-ketones, is a major pathway in the oxidation of aromatics. A simplified model has been developed to describe formation of O_3 and PAN in toluene-NO_x photochemical systems. More work is necessary in this field to determine the oxidation products of a number of different aromatics.

There is now some information relating to the mechanism of oxidation of more complex organic compounds containing other functional groups such as halogen, amines, organic acids, epoxides, alcohols and ethers. All these are commonly used organic chemicals and a full description of their atmospheric behaviour requires more information of this type.

The elucidation of reaction mechanism relies heavily on identification and analysis of products. This has been mainly achieved using gas chromatography but mass spectrometry has been used to identify both stable and free radical products in the reaction of OH with C_2H_4 (Proj. B8). There is scope here for the application of three new techniques which could assist in analysis of the rather complex mixtures of reactants and products in these systems. These are FTIR, Gas Chromatography coupled with Mass Spectrometry, and High Pressure Liquid Chromatography. The first of these has already been used in project D8.

3.3 Inorganic Reaction Mechanisms and Chemistry

3.3.1 Sulphur Species

The main activity on homogeneous reactions of Sulphur species is centred round the role of CS_2, OCS and H_2S in the atmospheric Sulphur cycle. Work carried out under project UK5 (not yet reported) has provided information on the rate of OH attack on CS_2 and other sulphur species.

Heterogeneous oxidation of SO_2 in cloud water droplets and in aqueous aerosols has been subjected to continued investigation in projects NL6, UK24 and UK32. The latest data from UK24 presented at the 2nd European Symposium suggest a rapid Chloride-ion-catalysed oxidation of SO_2 in solutions of high ionic strength. This is not consistent with previous observations and conclusions concerning SO_2 oxidation

on marine aerosols containing NaCl and this apparent anomaly requires
clarification.

3.3.2 Nitrogen Species

The chemistry of the nitrogen oxides in the lower atmosphere presents
a complex picture because of the variety of chemical species involved
and the numerous chemical and photochemical transformations they can
undergo. Although much work has been done over the years and there is
a considerable body of information on the atmospheric chemistry of
NO_x, recent observations in the atmosphere, in particular those reported at the 2nd Discussion meeting of Working Group 2 by investigators from KFA Jülich, suggest that knowledge of NO_x chemistry is incomplete. A similar conclusion was reached by Linquist et. al.
(Ispra Proceedings EUR 6621 en (1980)), and is evident from measurements of NO, NO_2 and O_3 in rural areas (D. Steadman, U. of Michigan),
P. Warneck (Project D25) (private communication). The uncertainties
arise in the following areas:

(a) Night-time Chemistry of NO_2 – NO_3

From measurements of the NO_3 radical behaviour in the atmosphere
it was found that the N_2O_5 levels estimated are about 10 times
lower than calculated from available literature data. Above a
threshold level of about 50-60 % RH the lifetime of NO_3 is very
short ($\sim$1 min) and even under dry conditions the lifetime of NO_3
is $\leq$ 30 min. There is clearly unknown reactions playing an important role in night-time NO_x chemistry. Heterogeneous reactions such as $N_2O_5 + H_2O \xrightarrow{surface} 2\ HNO_3$ could partially explain
the phenomenon but this would provide a large sink for NO_2, which
is not observed under many conditions. In view of the need to
describe the atmospheric behaviour of NO_2 for modelling dispersion of this toxic pollutant as well as for an understanding of
photo-oxidant formation, an understanding of the night-time
sinks for NO_x near the ground is of great importance.

(b) Nitrous Acid Formation

Formation of nitrous acid, HONO, in the atmosphere is of importance since it provides a potentially strong source of OH radicals through its U.V. photolysis

$$HONO + h\nu \rightarrow HO + NO$$

Results from Project D11 presented at 2nd Discussion meeting
indicated that HONO was formed heterogeneously in stored air
samples and could accelerate photo-oxidation when the air was
subjected to photolysis. At the same meeting measurements of
HONO in the atmosphere were reported. HONO was undetectable at
unpolluted locations but formation rates up to 0.25 ppb hr^{-1}
were observed in moderately NO_x polluted areas. The available

data for the reaction

$$H_2O + NO + NO_2 - 2 HONO$$

were insufficient to account for observed HONO formation. A
significant correlation between the product of relative humidity and SO_2 concentration was found which led to the suggestion
of involvement of sulphur acids in HONO formation. An alternate
route is the reaction of NO with HNO_3.

It is clear that more laboratory work is needed to characterise
the unknown reactions involved in NO_3 and HONO chemistry, so
that night-time and daytime HO_x chemistry can be elucidated.

3.4 Photochemistry and Photodissociation processes

3.4.1 Aldehydes and Ketones

Photodissociation of aldehydes and ketones is an important source of
free radicals in the sunlight atmosphere as well as being a route for
removal of these partially oxidised organic species. The simplest
species, formaldehyde, has been the subject of intensive study in
view of its importance in the tropospheric oxidation of CH_4. The recent work from Project D27 has provided definitive values for the
absorption cross-sections and primary quantum yields for photodissociation through the two channels:

$$HCHO + h\nu \rightarrow H + HCO$$

$$HCHO + h\nu \rightarrow H_2 + CO$$

The photolysis of higher aldehydes and ketones under atmospheric conditions has been subjected to preliminary study as reported in the
previous discussion meeting. New information on the behaviour of the
triplet state of acetone obtained in Project IRL4 demonstrates clearly the complexity of the processes involved in the photolysis of carbonyl compounds. The complex dependence of photodissociation quantum
yield on pressure, temperature and wavelength of excitation, dictates
great caution in extrapolating laboratory results to atmospheric conditions. Carefully designed atmospheric simulation experiments with
particular emphasis on correct reproduction of the solar spectrum
seems to be the only practical way of assessing atmospheric photolysis rates for higher aldehydes and ketones at the present time.

3.4.2 Other photodissociation processes

At present the emphasis is on photolysis of organic species for which
data are required for realistic urban oxidant production models and
for assessing decay rates for these compounds in air. Among the inorganic species, photolysis of NO_3, HONO, HO_2NO_2 and CS_2 are currently
of interest for atmospheric chemistry. More work in this area would
be uselful, particularly of the type conducted in project D12, where
photolysis rates in sunlight are measured.

3.4.3 Photochemical Simulation Experiments

A valuable contribution to our understanding of the nature of atmospheric photochemical processes has come from 'smog chamber' type simulation experiments. These have been conducted to investigate ozone production from polluted air samples (projects D8 and D13) and to compare them with model calculations (which are used as a basis for oxidant control strategy). The main conclusion reached is that extrapolation of models to real atmospheric conditions has to be treated with caution, because many of the variables in the Smog Chambers differ from the free atmosphere. Useful qualitative information concerning the build-up of oxidants and their relationship to precursors can be gained however. Similar types of experiment have been developed to investigate particulate formation in ambient air collected in Teflon Bags and subjected to natural sunlight (Project CCR3). These experiments show clearly that naturally emitted terpenes can give rise to particulate in sunlight, but their role in ozone production has not been fully clarified.

3.5 Ozone Reactions

Reactions with ozone are important for a number of atmospheric pollutants notably NO, NO_2 and unsaturated hydrocarbons. The rates and mechanisms of ozone-alkene reactions have been the subject of study for many years. New insight into the mechanism has been obtained using ESR and Infrared analysis of products from the cis-2-butane + O_3 reaction (H.H. Gunter, Zurich - project submitted for approval in COST 61a bis). The observed products were consistent with a mechanism in which the dominant path is a Criegee type split of the C=C bond into methyldioxirane and acetaldehyde, but with a minor route involving a transient Oxy-peroxy biradical (reciprocal lifetime 6 s^{-1}), as originally suggested by O'Neill and Blumstein.

Little work has been conducted recently on the rate constants for ozone reactions. This can constitute an importante sink mechanisms for degradation of unsaturated organic species both natural and man-made, in the atmosphere and estimation of atmospheric life-time with respect to removal by O_3 requires rate information. It would be desirable to obtain and compile a data base for O_3 reaction rate coefficients, similar to that assembled for OH reactions.

3.6 Heterogeneous Reactions

The study of heterogeneous reactions of atmospheric interest has not featured heavily in the Working Party 2 programme. However a report from Dr. ten Brink (ECN Petten, Project NL1, working party 3) describes interesting experiments investigating the removal of radicals on aerosol particles. Twin Smog Chambers were used for this study and the rates of oxidation of NO to NO_2 in a typical smog mixture (NO_x + toluene) were compared when one of the chambers was charged with 20 mg/m^3

of NaCl aerosol. The lower rate in the presence of aerosol indicated removal of radical species. A removal efficiency at the surface of $\gamma \sim 0.1$ was deduced from the results. The techniques used require validation and refinement but this type of experiment could give useful data.

The extent of aerosol scavenging in atmospheric free radical chemistry has been an area of controversy and discussion for some time. There is very little experimental information in this acknowledgedly difficult area. Most models of boundary layer photochemistry omit aerosol scavenging and this may in fact be justified. However further work is definitely required in this area, so that a more quantitative treatment of removal of radicals by aerosol particles can be made.

4. Conclusions

Although progress has not been uniform in all the areas studied, the amount of effort has to a large extent followed current requirements and priorities in the field of tropospheric chemistry. Thus the role of the OH radical in determining the fate of organics and their 'reactivity' in photo-oxidant formation has been widely recognised as an area of high priority for data acquisition. Similarly the understanding of photochemical processes and rates, which is a fundamental part of atmospheric chemistry, has improved and the work initiated in the Working Group should lead to more progress in the remainder of the programme. Areas which require more input are (a) Organic Product identification, which will benefit from the application of newly available analytical techniques, and (b) Nitrogen Oxide Chemistry, which needs input from studies both of homogeneous and heterogeneous (aerosol) reactions. The recent work on SO_2 oxidation on aerosols and in related systems, does not appear to be entirely consistent with earlier work and a continued effort is required here. An area of chemistry not represented in the current programme is tropospheric halogen chemistry. In view of growing concern about the significance of Hydrogen Chloride and Methyl Bromide emission this area might justify a small level of effort.

AEROSOLS

G. Madelaine

Au cours de la session concernant les aérosols, quinze communications provenant de quinze laboratoires différents correspondant à huit pays ont été retenues. Cette situation est à rapprocher de celle rencontrée lors du 1er Symposium du COST 61A bis qui s'est tenu en 1979, où l'on a dénombré un nombre moitié de laboratoires participants.

Ainsi aux sept laboratoires initiaux ayant présentés le résultat de leur travaux se sont ajoutés huit autres démontrant ainsi l'intérêt que suscite une telle action.

Si on examine le contenu des papiers présentés, on peut voir que les principaux domaines préoccupant les spécialistes de la physico-chimie des aérosols atmosphériques sont représentés :

- la caractérisation physique et dynamique de l'aérosol présent dans différentes atmosphères ;

- la caractérisation chimique en fonction de la dimension des particules.

En effet, si on veut obtenir des renseignements et comprendre la physico-chimie atmosphérique, il est impératif de posséder une connaissance complète de la formation (origine mécanique, réaction en phase gazeuse, etc...) de l'évolution et de la disparition des particules présentes dans l'atmosphère. De plus, la caractérisation de la concentration et la granulométrie des particules peuvent donner une indication sur l'origine naturelle ou anthropogène de l'aérosol.

C'est ainsi que plusieurs papiers présentés ont contribués à caractériser l'aérosol continental (D.39, F.28 et CCR 3) et l'aérosol marin (NL 1, F.6). Les particules présentes dans une atmosphère urbaine ou polluée ont été étudiées par plusieurs laboratoires et les résultats présentés apportent ainsi une importante et originale contribution à la connaissance de l'aérosol dit "de pollution" (D. UK25. Youg.).

Il semble que l'effort commencé il y a quelques années dans ce domaine se poursuive.

Le second point important est l'identification de la nature chimique des particules notamment en fonction de leur dimension. Cet aspect n'a pas été négligé non plus lors de ce symposium et des papiers Belges (B 24), Yougoslaves, Français (F.28, F.7) Britaniques (UK 18) ont couvert une partie de ce champ d'étude très important.

Enfin on signalera une étude originale de l'Irlande (IRL 2 et 3) qui a traité de la détection des fumées à l'aide de techniques intéressantes.

En conclusion, il semble que les travaux menés sur les aérosols au sein du COST 61Abis correspondent bien aux sujets traités par la Communauté Scientifique Internationale. Toutefois, un effort devrait encore être poursuivi notamment en ce qui concerne la détermination de la nature chimique des particules et de son évolution par des mesures in situ et en laboratoire.

La transformation gaz-particule a été bien étudiée de façon relativement détaillée pour le SO_2 (action COST 61Abis) il reste cependant de grandes lacunes à combler dans le domaine de la transformation des composés soufrés organiques; des hydrocarbures et des oxydes de l'azote conduisant à la production de nitrate.

Enfin, un effort devra être fait dans le domaine de la métrologie physique des aérosols et notamment dans la région des particules inférieures à 0,1 μm. Une standardisation conduisant à l'utilisation d'un language identique par les différents laboratoires serait utile.

POLLUTANT CYCLES

S. Beilke

Residence times of tropospheric gases range from less than
one second for some radicals to millions of years for persistent gases such as N_2 or He.If we confine ourselves to
gases which are of interest to project COST 61a bis(i.e.non
persistent gases which have or are thought to have measurable
environmental effects),the range of residence times is still
between a few hours and several years.Of the large number of
gases meeting the criteria for inclusion in project COST 61a
bis,only relatively few are being investigated by the 38
European research projects assigned to Working Party 4
(Pollutant Cycles) or were dealt with during this session.
As a complete study of the tropospheric cycles will be a
lengthy process lasting many decades,only some aspects can
be highlighted here under special consideration of the contributions within project COST 61a bis.
Very little is known about cycles of organic gases such as
hydrocarbons although they are known to play an important
role in tropospheric chemistry.The bulk of hydrocarbons inter-
acts in a complicated manner with other pollutant gases.
For example,there is evidence from many investigations that
a series of pollutant gases may be produced by oxidation of
hydrocarbons such as isoprene,ethane ect. contributing most
likely to an appreciable extent to the global production of
these gases(CO,H_2,CH_2O,ect.).
Very much as a first attempt to combine the available information about concentration profiles and effective source-and
sink strenghts published in the literature,an extensive review paper was presented by Bruckmann(1981) during this sym-
posium.The main objective was to define "the islands of
knowledge in the ocean of unknown quantities" providing a
sound basis for other authors to find their activities
properly placed in this fascinating research field of organic
pollutants.
Of the extremely wide field of hydrocarbon cycles some aspects
regarding natural sources are under investigation within
COST 61a bis.Considerable progress was achieved regarding
natural hydrocarbon emissions by trees and other vegetation
(Derwent and Hov,1979;Termonia and Istas,1980),soils
(Van Cleemput et al.,1981) and by the ocean(Rudolph,1981).
Another example for the complicated interaction of hydrocarbons with other pollutant gases are the groups of aldehydes
and ozone.Oxidation pathways produce aldehydes(Schmidt and
Lowe,1981) and also ozone(Lopez et al.,1981).The presence of
such compounds enhances the chemical reactivity of the
troposphere and hence the fate of other pollutant gases.The
chemistry of formaldehyde(CH_2O) is a good example to show
that pollutant gas cycles are coupled and a complete reaction
scheme is necessary.In order to better understand the compli-

cated interaction of the various pollutant gas cycles,investigations into the physico-chemical behaviour of such compounds as aldehydes and ozone are needed on a local,regional and global scale.

Another important pollutant gas from the point of view of environmental effects is CO_2.As the residence time of CO_2 in the troposphere is long(ca. 10 years) compared to times for mixing within this reservoir,increase of CO_2-concentrations due to anthropogeneous activities is a global problem. There are reasons to believe that global CO_2-concentrations in the atmosphere were ca. at the present level for long periods of time.For example,analysis of deep Antarctic ice cores did not show large variations of CO_2 indicating that atmospheric CO_2 was rather constant during the last milleniums except for the last Ice Age(ca.15000-20000 years ago) in which case atmospheric CO_2 was ca. 50% of the present level (Delmas et al.,1981). During the last 100 years the CO_2-level has increased by 10% from ca. 290 ppm to 330 ppm due to burning of fossil fuels and other CO_2 released to the atmosphere from anthropogeneous activities.Since 1957 CO_2-increase in the atmosphere has been monitored globally at sites of low local interferences by vegetation or anthropogeneous activities(for exemple Mauna Loa Observatory).Since 1980 a new CO_2-research station has been operating in the oceanic environment of the small island of Amsterdam($37^0S,77^0E$, 55 km^2,Indian Ocean).The chosen site is not influenced by anthropogeneous pollution(industry,air and boat traffic) or by local natural processes(very reduced vegetation,no active volcanism) providing therefore an ideal place to study the CO_2-cycle in a pure marine environment (Gaudry and Lambert,1980). It is hoped that first results will be available by the end of COST 61a bis which improve our knowledge on the marine CO_2-cycle. The investigation of long-term CO_2-trends provides an useful element for studying the atmospheric CO_2-cycle.The trends reported by baseline stations give a present rate of CO_2-increase of ca. 1 ppm/year.In spite of large biogenic and anthropogeneous interferences it was shown by Grosch et al., (1981) that a development of such a trend is also possible for an area like Germany confirming the above increase rate. Due to some important knowledge gaps in the global CO_2-cycle, an extrapolation to predict future CO_2-levels is difficult to make.On the basis of fossil fuel consumption trends some authors have predicted an CO_2-increase in the troposphere by a factor of 2 to 3 by the middle of the next century.Such an increase will most likely result in a change of the global climate.As a consequence,the search for other energy sources than fossil fuels should be intensified in order to meet the increasing energy demands including an increased use of nuclear energy.

Another pollutant gas investigated within COST 61a bis is carbon monoxide(CO).As far as its global cycle is concerned, considerable progress was achieved by contributions of

Marenco(1980,1981),Volz et al.(1981) and Derwent(1981).
For example,measurements of CO-emissions by African(Nigerian)
soils have shown that these soils act as CO-sources regardless
of temperature which is in contrast to most European soils
which are CO-sinks in the temperature range below 30°C
(Marenco,1981).If this result were representative for most
tropical and subtropical regions,previous global assessments
of dry CO-deposition would have to be taken with causion since
they are based on dry deposition velocities determined for
different European soils.

Another long-lived pollutant gas with potential environmental effects is mercury(Hg).Attention was first drawn to mercury after a contamination of fishing areas in some countries
by sewages containing organic mercury comounds.As a result,
intensified investigations were started to quantitatively
determine the atmospheric mercury cycle which was essentially
unknown before.
Our knowledge on the tropospheric mercury cycle has considerably increased by the important contributions of Brosset(1981a;
1981b) within project COST 61a bis.It seems now to be evident
that on a regional and global scale mercury is mainly produced
as Hg^0 at the earth surface by soils(degassing of Hg^0) and to
a lesser extent by industrial processes(chloralkali industry
and electro industry).In the atmosphere Hg^0 undergoes a series
of chemical reactions forming such compounds as $HgCl_2$,CH_3HgCl,
$(CH_3)_2Hg$ ect.which are much more soluble in droplets than Hg^0
suggesting that mercury removal from the atmosphere proceeds
to a large extent by wet deposition of soluble mercury com-
pounds.Part of it is being reduced in the soils to Hg^0 and
reemitted to the atmosphere.
It is hoped that the mercury cycle can be closed to some extent
by the end of project COST 61a bis.

In contrast to most long-lived gases,the investigation of
short-lived pollutant gases is from the point of view of environmental effects more a local or regional problem rather
than a global one.

In the past investigations of short-lived pollutant gases
were concentrated on atmospheric cycles of sulfur dioxide(SO_2)
because SO2 was emitted in much larger quantities than other
gases in most countries of Western Europe.
As an example,the physico-chemical behaviour of SO_2 was most
extensively studied within project COST 61a(1972-1976)of the
Commission of the European Communities.Concerning its tropospheric cycles a series of contributions were presented providing useful elements for studying the SO_2-budget on a local
and rgional scale.In the same way its environmental effects
such as acidification of precipitation,visibility reduction
due to sulfate,health effects,damage to materials including
vegetation have been investiagted within other projects.
In spite of a large number of publications including the considerable progress achieved within project COST 61a,there are
still important knowledge gaps in the field of atmospheric SO_2
some of which can most likely be filled within project COST

61a bis.The important question,to what extent do natural
sulfur emissions contribute to global atmospheric sulfur
emissions,can most likely be answered by the end of COST 61a
bis.On the basis of some important contributions,the conclusion can be drawn that volcanic emissions of SO_2 were underestimated in the past by a factor of ca. 10 in the global
sulfur budget(Sabroux et al.,1980;Jaeschke,1980;Carbonelle,
1981).
One the other hand,biogenic sulfur emissions were slightly
overemphasized.It had previously been widely assumed that
H_2S emitted by the ocean was the dominant biogenic sulfur
emission.Recent investigations within COST 61a bis seem to
show that the most important biogenic sulfur gas emissions
are $(CH_3)_2S$ from the ocean(Nguyen et al.,1980) and reduced
sulfur compounds such as H_2S and CS_2 from tropical soils
(Delmas et al.,1980;Gravenhorst and Varhelyi,1980) and to a
lesser extent from soils in a moderate climate(Servant,1981).
Useful results regarding the atmospheric sulfur cycle are
also expected by the proposal of Granat(1981) which was
added to COST 61a bis later than all others.

Since a few years attention has been foccussed more on the
chemistry of nitrogen oxides.Interest in atmospheric chemistry
of $NO_x(NO + NO_2)$ and other nitrogen compounds was originally
restricted to only a few areas in connection with photochemical smog formation.Today investigation of the physico-chemical
behaviour of NO_x is concentrated on a regional scale not only
from the point of view of environmental effects which are most
likely similiar to those of $SO_2$and sulfate.
There are some other points which make the NO_x-problem important:
In contrast to the emissions of SO_2,NO_x emissions have steadily been increased during the last two decades in most countries
of Western Europe.For some areas of the European Community
even a downward movement of SO_2-emissions can be observed due
to emission reducing measures.

Figure 1 shows yearly mean concentrations of SO_2(upper figure)
and NO_x(lower figure) for a representative site in the city
of Frankfurt which should reflect at least to some extent the
emission strenght in this area.
Although no updated emission inventories for SO_2 and NO_x are
available for the Federal Republic of Germany,measurements
by correlation spectroscopy seem to indicate that NO_x-emissions
are ca. at the same level today as SO_2-emissions(Beilke,1982).

Another aspect for the preferential investigation of NO_x should
be considered:In contrast to SO_2 and other sulfur compounds,
the cycle of NO_x exerts a strong influence on the cycles of
many other pollutant gases in the troposphere.
As examples,the complicated interaction of NO_x with ozone and
the formation of OH radicals due to photolytic dissociation
of HNO_2 accumulated during the night should be mentioned
(Kessler et al.,1981).
In spite of the fundamental importance of NO_x,our knowledge
of its cycle is only fragmentary and some of the aspects of
the NO_x-budget are still not understood.

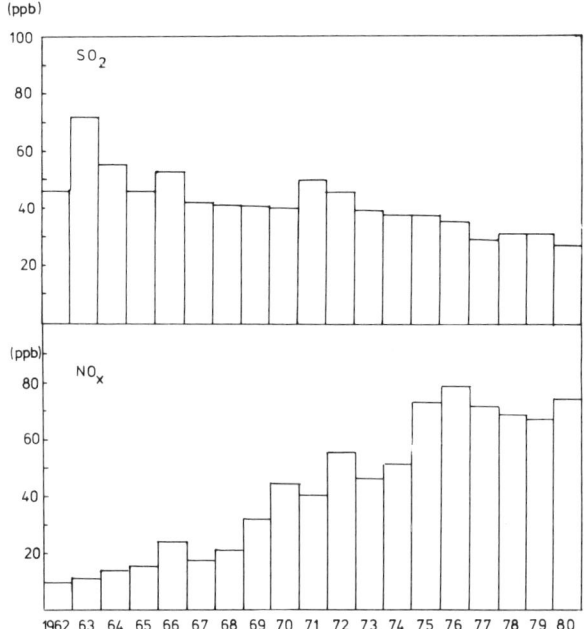

<u>FIGURE 1</u> : Yearly mean concentrations of SO_2(upper figure) and
NO_x(lower figure) for the period 1962-1980 measured
at Central station(city of Frankfurt).

Source:Umweltbundesamt,Pilotstation Frankfurt

For example,the means by which nitrogen oxides are transformed into nitrates are controversial and the rates of such
processes are not well known.
Another example is that up to now only two sinks for NO_x have
been recognized to play a role for removal of NO_x from the
troposphere:a.) daytime oxidation of NO_2 via OH to HNO_3
followed most likely by a heterogeneous removal of HNO_3 and
b.) dry deposition of NO_2.
These removal mechanisms are most likely too slow to account
for the observed tropospheric concentrations of nitrogen oxides in the light of their emissions suggesting that unsuspec-
ted removal mechanisms may be operating.Two of such possible
additional sinks for NO_x were proposed by Kessler et al.(1981)
which should be operating only during nighttime.This important
paper has emphasized the night-time chemistry of NO_x which is
important for the accumulation of such compounds as NO_3 and
HNO_2.
A potentially significant impact of nitrogen oxides is the
effect they may have on precipitation chemistry.The papers
of Perseke et al.(1981) and Müller et al.(1981) have shown
that substantial amounts of nitrate are found in rainwater
but we don't know how it gets there.

It is hoped that some of the knowledge gaps in the tropospheric NO_x-cycle can be filled by the end of project COST
61a bis.

REFERENCES:

Beilke,S.(1982)
  Measurements of emission strengths of area sources by
  correlation spectroscopy.To be published in 1982.

Brosset,C.(1981a)
  The atmospheric mercury cycle.Paper presented during 2nd
  meeting of Working Party 4 in Aussois(France) on 28th/29th
  April,1981.Project COST 61a bis.

Brosset,C.(1981b)
  Airborne mercury and its origin.Paper presented during
  2 nd European Symposium on Physico-chemical Behaviour of
  Atmospheric Pollutants.Varese(Italy),29 Sept.-1 Oct.1981.
  COST 61a bis.

Bruckmann,P.(1981)
  Cycles of organic gases in the atmosphere.Paper presented
  during 2 nd European Symposium on Physico-Chemical Behaviour
  of Atmospheric Pollutants.Varese(Italy),29 Sept.-1 Oct.1981.
  COST 61 a bis.

Carbonelle,J.(1981)
  Preliminary results on contribution of Mt.Etna and Mt.
  Stromboli to atmospheric CO_2 and SO_2.Paper presented during
  2 nd meeting of Working Party 4 in Aussois(France) on
  28th/29th April,1981.Project COST 61a bis.

Delmas,R.;Baudet,J. and Servant,J.(1980)
  Emissions and concentrations of H_2S in the air of tropical
  forest of the Ivory Coast and of temperate regions(France).
  Paper presented by A.Marenco during 1 st meeting of Working
  Party 4 in Berlin(Germany) on May 6th/7th,1980.
  COST 61a bis.

Delmas,R.;Aristarain,A.and Legrand,M.(1981)
  Acidity of polar precipitation.Paper presented during 2 nd
  meeting of Working Party 4(Pollutant Cycles) in Aussois
  (France) on 28th/29th April,1981.COST 61a bis.

Derwent,R.G. and Hov,Ö.(1979)
  The contribution from natural hydrocarbons to photochemical
  air pollution formation in the United Kingdom.Proceedings
  of the 1 st European Symposium on Physico-Chemical Behaviour
  of Atmospheric Pollutants.Ispra(Italy),16-18 Oct,1979.
  pp.367-382.COST 61a bis.

Derwent,R.G.(1981)
  Tropospheric OH radicals and the budgets of hydrogene,
  methane,carbon monoxide and halocarbons.Paper presented
  during 2 nd meeting of Working Party 4(Pollutant Cycles)
  in Aussois(France) on 28th/29th April,1981.COST 61a bis.

Gaudry,A.and Lambert,G.(1980)
  Installation d'une station de mesure de CO_2 a l'ile d'

Amsterdam(Ocean Indien).Premier resultats.Paper presented
during 1 st meeting of Working Party 4(Pollutant Cycles)
in Berlin(Germany) on 6th/7th May,1980.COST 61a bis.

Gravenhorst,G.and Varhelyi,G.(1980)
Natural sulfur sources and the global atmospheric sulfur
cycle.Paper presented during 1 st meeting of Working
Party 4(Pollutant Cycles) in Berlin(Germany) on 6th/7th
May,1980.COST 61a bis.

Granat,L.(1981)
Summer and winter budget for sulfur over Europe-an
indication of large seasonal variations of its residence time.Paper presented during 2 nd meeting of
Working Party 4(Pollutant Cycles) in Aussois(France)
on 28th/29th April,1981.COST 61a bis.

Grosch,W.;Fleck,W.and Jost,D.(1981)
The increase of carbon dioxide at rural sites of Ger-
many.Paper presented during 2 nd European Symposium on
Physico-Chemical Behaviour of Atmospheric Pollutants.
Varese(Italy),29 Sept.-1 Oct.1981.COST 61a bis.

Jaeschke,W.(1980)
Sulfur emissions from Mt.Etna.Paper presented during
1 st meeting of Working Party 4(Pollutant Cycles) on
May 6th/7th,1980 in Berlin(Germany).COST 61a bis.

Kessler,C.;Perner,D.and Platt,U.(1981)
Spectroscopic measurements of nitrous acid and formalde-
hyde-implications for urban photochemistry.Paper presented during 2 nd European Symposium on Physico-Chemical Behaviour of Atmospheric Pollutants.Varese(Italy),
29 Sept.-1 Oct.1981.COST 61a bis.

Lopez,A.;Prieur,S.and Fontan,J.(1981)
Study of the ozone source in the planetary boundary
layer.Paper presented during 2 nd European Symposium
on Physico-Chemical Behaviour of Atmospheric Pollutants.
Varese(Italy), 29 Sept.-1 Oct.1981.COST 61a bis.

Marenco,A. and Delauny,J.C.(1980)
Study of natural sources of atmospheric CO.
Paper presented during 1 st meeting of Working Party 4
(Pollutant Cycles) on May 6th/7th in Berlin(Germany).
COST 61a bis.

Marenco,A.(1981)
Experimental evidence of natural production of CO by
soils from measurements in Africa.Paper presented
during 2 nd meeting of Working Party 4(Pollutant Cycles)
in Aussois(France) on 28th/29th April,1981.COST 61a bis.

Müller,J.;Reuver,H.and D.Jost(1981)
Measurements of F^-,Cl^-,NO_3^- and $SO_4^=$ - ions in rainwater
and particulate matter by aid of ionic-chromatography.
Paper published in the proceedings of 2 nd European
Symposium on the Physico-Chemical Behaviour of Atmos-
pheric Pollutants.Varese(Italy),29 Sept.-1 Oct.1981.
COST 61 a bis.

Nguyen,B.C.;Bonsang,B.;Gaudry,A. and G.Lambert(1980)
Gaseous marine sulphur compounds in the atmospheric sulphur
cycle.Paper presented during 1 st meeting of Working Party
4(Pollutant Cycles) on May 6th/7th,1980 in Berlin(Germany).
COST 61a bis.

Perseke,C.;Georgii,H.W.and E.Rohbock(1981)
Investigation of the regional distribution of wet depo-
sition of pollutants.Paper presented during 2 nd European
Symposium on Physico-Chemical Behaviour of Atmospheric
Pollutants.Varese(Italy), 29 Sept.-1 Oct.1981.COST 61a bis.

Rudolph,J.(1981)
Measurements of the large scale distribution of C_2-C_5
hydrocarbons in unpolluted air.Paper presented during
2 nd meeting of Working Party 4(Pollutant Cycles) in
Aussois(France) on 28th/29th April,1981.COST 61a bis.

Sabroux,J.C.;Carbonelle,J.and Zettwoog,P.(1980)
Active volcanism and the geochemical cycles of carbon,
sulfur,chlorine and fluorines.Paper presented during 1 st
meeting of Working Party 4(Pollutant Cycles),Berlin,
May 6th/7th,1980.COST 61a bis.

Schmidt,U.and Lowe,D.C.(1981)
Vertical profiles of formaldehyde in the troposphere.
Paper presented during 2 nd European Symposium on Physico-
Chemical Behaviour of Atmospheric Pollutants.Varese(Italy),
29 Sept.-1 Oct.1981.COST 61a bis.

Servant,J.(1981)
Daily variations of the H_2S content in atmospheric air at
ground level in France.Paper presented during 2 nd meeting
of Working Party 4(Pollutant Cycles) in Aussois(France)
on 28th/29th April,1981.COST 61a bis.

Termonia,M.and Istas,J.R.(1980)
Emissions of organic gases from vegetation.Paper presented
during 1 st meeting of Working Party 4(Pollutant Cycles),
Berlin,May 6th/7th,1980.COST 61a bis.

Van Cleemput,O.;El-Sebaay,A.S.and L.Baert(1981)
Production of gaseous hydrocarbons in soil.Paper presented
during 2 nd European Symposium on Physico-Chemical Behaviour
of Atmospheric Pollutants.Varese(Italy),29 Sept.-1 Oct.1981.
COST 61a bis.

Volz,A.;Ehhalt,D.and R.G.Derwent(1981)
OH-radicals via natural ^{14}CO.Paper presented by D.Perner
during 2 nd meeting of Working Party 4(Pollutant Cycles)
in Aussois(France) on April 28th/29th,1981.COST 61a bis.

TRANSPORT AND MODELLING - FIELD EXPERIMENTS

A.J. ELSHOUT

It is the aim of COST 61a bis to bring together on a cooperative way in a co-ordinated action relevant information on the physico-chemical behaviour of pollutants in the atmosphere, for a better understanding of the impact of emitted air pollutants on the environment (health, ecosystems, weather and climate), which can be used in environmental policy and to lay down the necessary measurements.

The progress in this COST action should in the next years be judged in the light of this relevance and I think that one of the objectives for the last symposium of this action, in 1983, has to be reviewing the "state of art" of our joint knowledge of the different components and systems, which are of importance for further environmental policy decisions.

The projects in Working Group 5 of this action enclose work on dispersion and transport from modelling and field experiments, with the purpose to describe the processes as they occur in the real atmosphere. With respect to the contributions in this section it can be concluded that there has been done a tremendous amount of work in the different projects to come to more adequate sampling and measuring methods. There is a clear progress in the description of the spatial distributions of pollutants and changes in time, from both the results of airborne and remote sensing measurements and modelling activities. Nevertheless, there is still an urgent need for a better integration of these different measurement and modelling activities. The modelling studies to describe the physico-chemical behaviour of pollutants in the real atmosphere, including plumes and clouds, need input parameters which can be derived from measurements. Validation of the models needs also measurement results from well defined experiments in the atmosphere.

From the modelling side the information has to come to select and describe the parameters that are of special importance to measure, in order to have the possibility to analyse and to generalize the measurement results for description of the atmospheric physico-chemical processes of interest.

If we make a short review of the knowledge of the behaviour of the two main air pollutants of interest: SO_2, NO_x, especially in relation to dispersion and transport of plumes, the following can be established.

The conversion of SO_2 in plumes has been studied for many years in various countries in plume tracing studies with instrumented aircrafts. The earliest studies resulted in conversion rates between 1 and 50% h^{-1}. However, it could be shown later that most of these measurements had significant short-comings in sampling and analysis.

During the last five years, with the availability of more adequate sampling and measuring methods, the reported range for the conversion rate of SO_2 to sulfates (including sulfuric acid) is substantially narrowed. Most of the recent studies on oxidation of SO_2 in plumes from coal-fired power plants for example, have resulted in oxidation rates in summer time of 0.4% h^{-1} to a maximum of 5.0% h^{-1}. Taking into account the diurnal cycle, the overall rate in summer is about 1% h^{-1} and in winter about 0.2% h^{-1}. The oxidation rates for isolated power plant plumes can be well-explained by gas phase oxidation of SO_2 with OH radicals. Modelling calculations based on the reaction rates as measured in laboratory experiments give comparable results with the above-mentioned measurements in this case.

In situations with very high cloud coverage and high relative humidity, there are indications that rates appear that are higher than 1% h^{-1}, which suggest an increase in the conversion rates due to heterogeneous chemistry. An example is given by Kallend in his contribution to this symposium.

Not much is known about the in-cloud oxidation chemistry of SO_2 after interaction between plumes and clouds in the real atmosphere, but there is evidence, primarily from laboratory experiments, that the major route for the production of atmospheric sulfates in through oxidation of SO_2 in cloud water, with oxidation rates in summer time to the order of 10-20% h^{-1}.

Conversion rates of NO_x in isolated power plant plumes, as deduced from airborne measurements of nitrate concentrations and first order removal curves for NO_x, have only recently been reported.

Results from studies at the Widows Creek and Colbert coal-fired power stations in the USA, show NO_x conversion rates of 13 and 11% h^{-1} respectively. Nitrate formation rates were found from 2.5% h^{-1} in the morning to 7.5% h^{-1} in the afternoon, with the remark that volatile nitrates like nitric acid have been collected here with less than unit efficiency.

Just like for the oxidation of SO_2, the homogeneous gas phase reaction of NO_2 with OH radicals is rapid enough to account for the oxidation rate measured.

To determine NO_x conversion rates in urban-industrial areas, we have, in The Netherlands, carried out airborne measurements in the plume of the Rijnmond area. These measurements were focussed on that part of the industrial area occupied with the major oil refineries and chemical plants. Because the SO_2 removal (transformation) rate is relatively low with regard to that of NO_x - as already known from laboratory experiments - this component has been used as a tracer in the transported air mass for the determination of the NO_x conversion rate.

The NO_x conversion rate has been determined for three measuring days with O_3 ambient concentrations at the time of measurement of 25-30 ppb. Considering any errors in wind velocity estimates (deduced from surface wind velocity) and in relatively low NO_x and SO_2 concentration, the NO_x conversion rate will have been at least 15% h^{-1}. Under conditions with maximum O_3 concentrations of 80-100 ppb, a more accurate conversion rate of 19% h^{-1} was estimated through direct measurement at the place of the average transport velocity in the mixing layer with a weather balloon. These results are comparable with the results from airborne measurements in the Boston plume, where the expected NO_x concentration was calculated from the initial urban-industrial tracer/NO_x ratio and the downwind concentrations. Based on these experiments applicable to daylight hours under photochemical conditions with ambient O_3 concentrations of 90-140 ppb, the rate of NO_x conversion varied between 14 to 24% h^{-1}, with an average rate and life time of 18% h^{-1} and 5.8 h respectively.

The here mentioned conversion rates of SO_2 and NO_x are comparable with the results given by Hov in his contribution to this symposium, concerning the turnover of sulfur dioxide and nitrogen oxides in the atmospheric boundary layer.

At the moment there is very little information about the in-cloud oxidation of nitrogen oxides in real atmospheric situations.

In relation to this it is of importance to refer to earlier statements on our first symposium in 1979, forwarded by Eggleton: "Also the whole subject of atmospheric chemical reactions in the aqueous phase (cloud and fog droplets) is lacking. Despite work on the aqueous phase oxidation of sulfur dioxide in COST 61a, there are still uncertainties and no clear view has been obtained of the importance of this process in the atmosphere. The subject of organic compounds in rain has received very little attention so

far and this is likely to be of considerable importance", and by Beilke:
"An important point for elucidating precipitation scavenging as a sink of
tropospheric pollutants is to measure composition and pH of cloud water.
Such measurements should be carried out with priority otherwise we do not
advance in understanding wet deposition as a sink of pollutants".

It has to be concluded that at the moment there are uncertainties about
the oxidation rates of SO_2 and NO_x in cloud water, the part of O_2, O_3 and
H_2O_2 in the oxidation, the H_2O_2 concentrations in the atmosphere and the
interactions between different pollutants in the cloud water, also in relation to the pH.

These point are of importance to look into the evaluation of influences
of different types of emission of local and long distance sources on the
formation of acid rain and needs further study. This type of information
has its relevance to alternative control strategies for acid rain.

I have a hope that a number of projects of this COST action can give
some relevant information in this field in the next two years, which can
be reported at the next symposium.

LIST OF PARTICIPANTS

ALLEGRINI, I.   C.N.R.
                                       Area di Ricerca di Roma
                                       I - ROMA

ANGELETTI, G.   Commission of the European Communities
                                       Directorate General Research, Science
                                       and Development
                                       200, rue de la Loi
                                       B - 1049 BRUXELLES

AUGUSTIN, H.   EERM
                                       73-77 rue de Lèvres
                                       F - 92100 BOULOGNE

BARNES, I.   Gesamthochschule
                                       Gauss-Strasse 20
                                       D - 56 WUPPERTAL

BATT, L.   Chemistry Dept. of Aberdeen
                                       Meston Walk
                                       GB - ABERDEEN

BAUDER, A.   Lab. Phys. Chemie
                                       Eth-Zentrum
                                       CH - 8092 ZUERICH

BECKER, K.   Universität Wuppertal
                                       Gauss-Strasse 20
                                       D - 56 WUPPERTAL 1

BEILKE, S.   Umweltbundesamt
                                       Feldbergstrasse 45
                                       D - 6 FRANKFURT

BESSEMOULIN, P.   French Met. Office
                                       Centre de Recherche de Magny
                                       F - 78470 ST. REMY LES CHEVREUSE

BIEHL, M.   KFA
                                       PTU, Postfach 1913
                                       D - 5170 JUELICH

BIGNOZZI, C.   Centro Fotochimica CNR
                                       Via Borsari 46
                                       I - FERRARA

BISHOP, G.   Commission of the European Communities
                                    Joint Research Centre
                                    I - 21020 ISPRA (VA)

BOULAUD, D.   CEA (LPA/SPT)
                                    B.P. no 6
                                    F - 92260 FONTENAY-AUX-ROSES

BOURDEAU, P.   Commission of the European Communities
                                    Directorate General Research, Science
                                    and Development
                                    200, rue de la Loi
                                    B - 1049 BRUSSELS

BROLL, A.   Max-Planck-Institut für Chemie
                                    Saarstrasse 23
                                    D - 6500 MAINZ

BROSE, G.   RWTUV
                                    Steubenstrasse 53
                                    D - 43 ESSEN 1

BROSSET, C.   IVL
                                    P.O.Box 5207
                                    S - 402 24 GOTHENBURG

BRUCKMANN, P.   Landesamt für Immissionsschutz
                                    Wallneyer Str. 6
                                    D - 43 ESSEN 1

CAMAGNI, P.   Commission of the European Communities
                                    Joint Research Centre
                                    I - 21020 ISPRA (VA)

CHIORBOLI, C.   Centro Fotochimica
                                    CNR Università Ferrara
                                    Via Borsari 46
                                    I - FERRARA

CICCIOLI, P.   CNR
                                    Istituto Ing. Atm.
                                    Area delle Ricerca di Roma
                                    I - ROMA

CLARKE, A.   Leeds University
                                    Department of Fuel and Energy
                                    GB - LEEDS LS2

CLAYTON, P.   Department of Industry
                                    Warren Spring Laboratory Gunnels
                                    Wood Rd
                                    GB - STEVENAGE

COLLIN, J.   Université de Liège
                                    Institut de Chimie B6
                                    Sart-Tilman
                                    B - 4000 LIEGE

COX, R.	UKAEA - EMS Division AERE Harwell GB - DIDCOT, Oxfordshire
DE BORTOLI, M.	Commission of the European Communities Joint Research Centre I - 21020 ISPRA (VA)
DERWENT, R.	AERE Harwell GB - DIDCOT, Oxfordshire
DE WISPELAERE, C.	Wetenschapsbeleid Wetenschapsstraat 8 B - 1040 BRUSSEL
DUBOIS, J.	Electricité de France 6, Quai Watier F - 78400 CHATOU
ELICHEGARAY, C.	Université Paris VII F - 75251 PARIS
ELSHOUT, A.	NV KEMA NL - 6800 ET ARNHEM
FINK, E.	Gesamthochschule Wuppertal Gauss-Strasse 20 D - 56 WUPPERTAL
FLYGER, H.	Danish Air Pollution RISO DK - 4000 ROSKILDE
FONDERIE, V.	K.U. Leuven Celestijnenlaan 200 F B - 3030 HEVERLEE
FONTAN, J.	Université Paul Sabatier 118, rue de Narbonne F - 31063 TOULOUSE
FUGAS, M.	Institute for Medical Research and Occupational Health Mose Puade 158 YU - 41000 ZAGREB
FUMAROLA, G.	Università Genova Ist. Ing. Chimica Via Opera Pia I - GENOVA
GEISS, F.	Commission of the European Communities Joint Research Centre I - 21020 ISPRA (VA)

GHAZI, A.

Commission of the European Communities
Directorate General Research, Science
and Development
200, rue de la Loi
B - 1049 BRUSSELS

GIOVANELLI, G.

CNR
Lab. FISBAT
Via dei Castagnoli 1
I - BOLOGNA

GRANIER, J.

Electricité de France
6, Quai Watier
F - 78400 CHATOU

GRENNFELT, P.

Swedisch Water and Air Pollution
Research Institute (IVL)
P.O.Box 5207
S - 402 24 GOTHENBURG

GUICHERIT, R.

IMG-TNO
97, Schoennkerstraat
NL - DELFT

GUILLOT, P.

Commission of the European Communities
Directorate General Research, Science
and Development
200, rue de la Loi
B - 1049 BRUSSELS

HEINTZENBERG, J.

Dept. of Meteorology
University of Stockholm
S - STOCKHOLM

HOV, Ø.

Norvegian Institute for Air Research
Box 130
N - 2001 LILLESTROEM

HRSAK, J.

Institute for Medical Research
and Occupational Health
Mose Puade 158
YU - 41000 ZAGREB

HUEBLER, G.

KFA Jülich
ICH 3
Postfach 1913
D - 5170 JUELICH

ISRAEL, G.

Technische Universität Berlin
Sekr. KF 2
Str. des 17. Juni 135
D - 1000 BERLIN 12

JANSSENS, J.

Dept. Chemistry
Universitaire Instelling Antwerpen
B - 2610 WILRIJK

KALLEND, A.

CEGB
Kelvin Avenue
GB - LEATHERHEAD

KESSLER, C.

Kernforschungsanlage
Postfach 1913
D - 5170 JUELICH

KNOEPPEL, H.

Commission of the European Communities
Joint Research Centre
I - 21020 ISPRA (VA)

KOECHLER, C.

Commission of the European Communities
Joint Research Center
I - 21020 ISPRA (VA)

LEWIN, E.

Danish Air Pollution Lab.
RISØ
DK - 4000 ROSKILDE

LIBERTI, A.

Ist. Ing. Atm.
CNR
C.P. 10
I - 00016 MONTEROTONDO Stazione

LOEBEL, J.

VDI-Kommission Reinh. Luft
D - 4000 DUESSELDORF

LOHSE, C.

Commission of the European Communities
Joint Research Centre
I - 21020 ISPRA (VA)

LORENZ, K.

University of Goettingen
Tammannstr. 6
D - 3400 GOETTINGEN

MADELAINE, G.

LPA/SPT CEN
B.P. no 6
F - 92260 FONTENAY-AUX-ROSES

MAGDONELLE, F.

Commission of the European Communities
Directorate General Environment,
Consumer Protection and Nuclear Safety
200, rue de la Loi
B - 1049 BRUSSELS

MASNIERE, P.

Electricité de France
6, Quai Watier
F - 78400 CHATOU

METAYER, Y.

CEA (LPA/SPT)
B.P. no 6
F - 92260 FONTENAY-AUX-ROSES

MEURRENS, A.

Inst. Hygiène et Epidémiologie
14, rue Wytsmans
B - 1050 BRUXELLES

MORELLI, J.   Lab. Chimie Minérale des Milieux
Naturels - CNRS
Université Paris VII
F - 75251 PARIS CEDEX 05

MOUVIER, G.   Université Paris VII
F - 75251 PARIS CEDEX 05

MULLER, M.   Mission Etudes et Recherches
Ministère de l'Environnement
F - 92521 NEUILLY-SUR-SEINE

NEUBER, E.   Inst. of Meteorology and Geophysics
Feldbergstr. 47
D - 6000 FRANKFURT

NICOLAY, D.   Commission of the European Communities
Directorate General Information Market
and Innovation
Jean Monnet Building B4/072 - P.O.B. 1907
L - 2920 LUXEMBOURG

OTT, H.   Commission of the European Communities
Directorate General Research, Science
and Development
200, rue de la Loi
B - 1049 BRUSSELS

PAYRISSAT, M.   Commission of the European Communities
Joint Research Centre
I - 21020 ISPRA (VA)

PENKETT, S.   UKAEA
AERE Harwell
GB - DIDCOT, Oxfordshire

PEPERSTRAETE, H.   SCK/CEN Nuclear Study Centre
B - 2400 MOL

PERSEKE, C.   Inst. of Meteorology
Feldbergstr. 47
D - 6 FRANKFURT/M

PETERSEN, B.   Air Pollution Laboratory
RISØ
DK - 4000 ROSKILDE

POSSANZINI, M.   CNR
Via Salaria Km 29, 300
I - 00016 MONTEROTONDO Stazione

PUXBAUM, H.   Technische Universität
Getreidemarkt 9
A - 1060 WIEN

REITER, R.	Fraunhofer Institut für atmosphärische Umweltforschung D - 8100 GARMISCH-PARTENKIRCHEN
RENOUX, A.	Université Paris XII Lab. Physique des Aerosols Av. du Général de Gaulle F - 94000 CRETEIL
RESTELLI, G.	Commission of the European Communities Joint Research Centre I - 21020 ISPRA (VA)
ROEMER, F.	NV KEMA Environmental Research Department Utrechtseweg 310 NL - ARNHEM
ROETH, R.	Universität Essen Kosaheng. 2 D - 5170 JUELICH
SANDRONI, S.	Commission of the European Communities Joint Research Centre I - 21020 ISPRA (VA)
SCHERER, B.	Institut für Geophysikalische Wissen- schaften der Freien Universität Berlin Thielallee 50 D - 1000 BERLIN 33
SCHMIDT, U.	KFA JUELICH GMBH Inst. für Chemie 3 Postfach 1913 D - 5170 JUELICH
SCHURATH, U.	Inst- für Physikalische Chemie Universität Bonn Wegelerstrasse 12 D - 5300 BONN
SCOTT, J.	University College Dublin Belfield IRL - DUBLIN 4
SIDEBOTTOM, H.	University College Dublin Chemistry Dept. UCD Belfield IRL - DUBLIN 4
STANGL, H.	Commission of the European Communities Joint Research Centre I - 21020 ISPRA (VA)
STINGELE, A.	Commission of the European Communities Joint Research Centre I - 21020 ISPRA (VA)

STUHL, F.	Ruhr University Physikalische Chemie Postfach 102148 D - 4630 BOCHUM 1
TEN BRINK, H.	Neth. Energy Res. Centre P.O.Box 1 NL - 1755 ZG PETTEN
TERMONIA, M.	IRC/ISO Museumlaan 5 B - 1980 TERVUREN
TORSI, G.	Ist. Chim. Anal. Università Bari Via Amendola 173 I - BARI
TYMEN, G.	CEC Lab. Physique Aerosols Faculté des Sciences F - 29283 BREST CEDEX
VAN CAUWENBERGHE, K.	University of Antwerp B - 2610 WILRIJK
VAN CLEEMPUT, O.	Rijksuniversiteit Gent Coupure 533 B - 9000 GENT
VANDENDRIESSCHE, S.	K.U. Leuven Lab. Anal. Anorg. Scheikunde Celestijnenlaan 200 F B - 3030 HEVERLEE
VANDERBORGHT, B.	SCK/CEN B - 2400 MOL
VAN VAECK, L.	Universitaire Instelling Antwerpen Universiteitsplein 1 B - 2610 WILRIJK
VERSINO, B.	Commission of the European Communities Joint Research Centre I - 21020 ISPRA (VA)
WAHNER, A.	Ruhr Universität Postfach 102148 D - 4630 BOCHUM
WALSH, J.	University College Dublin Chem. Engineering Dept. Upper Merrion Str. IRL - DUBLIN 2
WANGE, D.	Max-Plank-Institut Saarstr. 23 D - 6500 MAINZ

WARNECK, P.

Max-Planck-Institut für Chemie
Saarstrasse 23
D - 6500 MAINZ

ZANDER, R.

Université de Liège
Inst. d'Astrophysique
5, av. de Cointe
B - 4200 LIEGE

ZELLNER, R.

University of Goettingen
Tammannstr. 6
D - 3400 GOETTINGEN

ZETZSCH, C.

Ruhr Universität
Postfach 102148
D - 4630 BOCHUM